Boron-Based Compounds

Boron-Based Compounds

Potential and Emerging
Applications in Medicine

Edited by

Evamarie Hey-Hawkins
Department of Inorganic Chemistry, Leipzig University
Leipzig, Germany

Clara Viñas Teixidor
Spanish Council for Scientific Research, Institut de Ciència de Materials de Barcelona
Barcelona, Spain

This edition first published 2018
© 2018 John Wiley & Sons Ltd

All rights reserved. No part of this publication may be reproduced, stored in a retrieval system, or transmitted, in any form or by any means, electronic, mechanical, photocopying, recording or otherwise, except as permitted by law. Advice on how to obtain permission to reuse material from this title is available at http://www.wiley.com/go/permissions.

The rights of Evamarie Hey-Hawkins and Clara Viñas Teixidor to be identified as the authors of the editorial material in this work has been asserted in accordance with law.

Registered Office(s)
John Wiley & Sons, Inc., 111 River Street, Hoboken, NJ 07030, USA
John Wiley & Sons Ltd, The Atrium, Southern Gate, Chichester, West Sussex, PO19 8SQ, UK

Editorial Office
9600 Garsington Road, Oxford, OX4 2DQ, UK

For details of our global editorial offices, customer services, and more information about Wiley products visit us at www.wiley.com.

Wiley also publishes its books in a variety of electronic formats and by print-on-demand. Some content that appears in standard print versions of this book may not be available in other formats.

Limit of Liability/Disclaimer of Warranty
In view of ongoing research, equipment modifications, changes in governmental regulations, and the constant flow of information relating to the use of experimental reagents, equipment, and devices, the reader is urged to review and evaluate the information provided in the package insert or instructions for each chemical, piece of equipment, reagent, or device for, among other things, any changes in the instructions or indication of usage and for added warnings and precautions. While the publisher and authors have used their best efforts in preparing this work, they make no representations or warranties with respect to the accuracy or completeness of the contents of this work and specifically disclaim all warranties, including without limitation any implied warranties of merchantability or fitness for a particular purpose. No warranty may be created or extended by sales representatives, written sales materials or promotional statements for this work. The fact that an organization, website, or product is referred to in this work as a citation and/or potential source of further information does not mean that the publisher and authors endorse the information or services the organization, website, or product may provide or recommendations it may make. This work is sold with the understanding that the publisher is not engaged in rendering professional services. The advice and strategies contained herein may not be suitable for your situation. You should consult with a specialist where appropriate. Further, readers should be aware that websites listed in this work may have changed or disappeared between when this work was written and when it is read. Neither the publisher nor authors shall be liable for any loss of profit or any other commercial damages, including but not limited to special, incidental, consequential, or other damages.

Library of Congress Cataloging-in-Publication Data

Names: Hey-Hawkins, Evamarie, 1957– editor.
Title: Boron-based compounds : potential and emerging applications in medicine / edited by Evamarie Hey-Hawkins, University of Leipzig, Leipzig, Germany, Clara Viñas Teixidor, Institut de Ciencia de Materials de Barcelona, Barcelona, Spain.
Description: First edition. | Hoboken, NJ, USA : Wiley, [2018] | Includes bibliographical references and index. |
Identifiers: LCCN 2017055295 (print) | LCCN 2018009581 (ebook) | ISBN 9781119275589 (pdf) | ISBN 9781119275596 (epub) | ISBN 9781119275558 (cloth)
Subjects: LCSH: Boron compounds. | Boron compounds–Diagnostic use. | Boron compounds–Therapeutic use.
Classification: LCC QD181.B1 (ebook) | LCC QD181.B1 B663 2018 (print) | DDC 615.2/671–dc23
LC record available at https://lccn.loc.gov/2017055295

Cover design: Wiley
Cover image: Designer, MSc Christoph Selg

Set in 10/12pt Warnock by SPi Global, Pondicherry, India

Printed in the UK by Bell & Bain Ltd, Glasgow

10 9 8 7 6 5 4 3 2 1

Contents

List of Contributors *xiii*
Preface *xvii*

Part 1 Design of New Boron-based Drugs *1*

1.1 Carboranes as Hydrophobic Pharmacophores: Applications for Design of Nuclear Receptor Ligands *3*
Yasuyuki Endo
1.1.1 Roles of Hydrophobic Pharmacophores in Medicinal Drug Design *3*
1.1.2 Carboranes as Hydrophobic Structures for Medicinal Drug Design *4*
1.1.3 Estrogen Receptor Ligands Bearing a Carborane Cage *5*
1.1.3.1 Estrogen Agonists *5*
1.1.3.2 Estrogen Antagonists and Selective Estrogen-Receptor Modulators (SERMs) *7*
1.1.4 Androgen Receptor Ligands Bearing a Carborane Cage *7*
1.1.4.1 Androgen Antagonists *7*
1.1.4.2 Improvement of Carborane-Containing Androgen Antagonists as Candidates for Anti–Prostate Cancer Therapy *9*
1.1.5 Retinoic Acid Receptor (RAR) and Retinoic Acid X Receptor (RXR) Ligands Bearing a Carborane Cage *11*
1.1.5.1 RAR Agonists and Antagonists *11*
1.1.5.2 RXR Agonists and Antagonists *12*
1.1.6 Vitamin D Receptor Ligands Bearing a Carborane Cage *12*
1.1.7 Determination of the Hydrophobicity Constant π for Carboranes and Quantitative Structure–Activity Relationships in ER Ligands *13*
1.1.7.1 Determination of the Hydrophobicity Constant π for Carboranes *13*
1.1.7.2 Quantitative Structure–Activity Relationships of Carboranylphenols with Estrogenic Activity *14*
1.1.8 Conclusion and Prospects *16*
References *16*

1.2 Boron Cluster Modifications with Antiviral, Anticancer, and Modulation of Purinergic Receptors' Activities Based on Nucleoside Structures 20

Anna Adamska-Bartłomiejczyk, Katarzyna Bednarska, Magdalena Białek-Pietras, Zofia M. Kiliańska, Adam Mieczkowski, Agnieszka B. Olejniczak, Edyta Paradowska, Mirosława Studzińska, Zofia Sułowska, Jolanta D. Żołnierczyk, and Zbigniew J. Lesnikowski

- 1.2.1 Introduction 20
- 1.2.2 Boron Clusters as Tools in Medicinal Chemistry 21
- 1.2.3 Modification of Selected Antiviral Drugs with Lipophilic Boron Cluster Modulators and New Antiviral Nucleosides Bearing Boron Clusters 23
- 1.2.4 *In Vitro* Antileukemic Activity of Adenosine Derivatives Bearing Boron Cluster Modification 26
- 1.2.5 Adenosine–Boron Cluster Conjugates as Prospective Modulators of Purinergic Receptor Activity 28
- 1.2.6 Summary 30
- Acknowledgments 30
- References 30

1.3 Design of Carborane-Based Hypoxia-Inducible Factor Inhibitors 35

Guangzhe Li, Hyun Seung Ban, and Hiroyuki Nakamura

- 1.3.1 Introduction 35
- 1.3.2 Boron-Containing Phenoxyacetanilides 36
- 1.3.2.1 Synthesis of Boron-Containing Phenoxyacetanilides 36
- 1.3.2.2 Biological Activity of Boron-Containing Phenoxyacetanilides 38
- 1.3.3 Target Identification of GN26361 39
- 1.3.3.1 Design of GN26361 Chemical Probes 39
- 1.3.3.2 Synthesis of GN26361 Chemical Probes 39
- 1.3.3.3 Target Identification of GN26361 40
- 1.3.4 Carborane-Containing HSP60 Inhibitors 40
- 1.3.4.1 Design of *Ortho-* and *Meta-*Carborane Analogs of GN26361 40
- 1.3.4.2 Synthesis of GN26361 Analogs 41
- 1.3.4.3 HIF Inhibitory Activity of Carborane Analogs of GN26361 46
- 1.3.4.4 HSP60 Inhibitory Activity of Carborane Analogs of GN26361 47
- 1.3.5 Carborane-Containing Manassantin Mimics 49
- 1.3.5.1 Synthesis of Carborane-Containing Manassantin Mimics 49
- 1.3.5.2 Biological Activity of Carborane-Containing Manassantin Mimics 51
- 1.3.6 Carborane-Containing Combretastatin A-4 Mimics 52
- 1.3.6.1 Design of *Ortho-*Carborane Analogs of Combretastatin A-4 52
- 1.3.6.2 Synthesis of Carborane Analogs of Combretastatin A4 53
- 1.3.6.3 HIF Inhibitory Activity of Carborane Analogs of Combretastatin A4 54
- 1.3.7 Conclusion 56
- References 57

1.4 Half- and Mixed-Sandwich Transition Metal Dicarbollides and *nido-*Carboranes(−1) for Medicinal Applications 60

Benedikt Schwarze, Marta Gozzi, and Evamarie Hey-Hawkins

- 1.4.1 Introduction 60
- 1.4.2 Synthetic Approaches to *nido-*Carborane $[C_2B_9H_{12}]^-$ Derivatives 66

1.4.3	Biologically Active Organometallic *nido*-Carborane Complexes and Organic *nido*-Carborane Derivatives *72*	
1.4.3.1	Biologically Active Half- and Mixed-Sandwich Metallacarborane Complexes *72*	
1.4.3.1.1	Half-Sandwich Complexes of Rhenium(I) and Technetium(I)-99 m as Radio-Imaging and Radiotherapeutic Agents *72*	
1.4.3.1.2	*nido*-Carborane(−1) Anions as Pharmacophores for Metal-Based Drugs *79*	
1.4.3.2	Biologically Active Compounds Containing a *nido*-Carborane(−1) Core *83*	
1.4.3.2.1	Radiotherapy and Radio-Imaging *83*	
1.4.3.2.2	Pharmacophores *91*	
1.4.4	Conclusions and Future Challenges *95*	
	Appendix: Abbreviations *96*	
	Acknowledgments *97*	
	References *98*	

1.5 Ionic Boron Clusters as Superchaotropic Anions: Implications for Drug Design *109*

Khaleel I. Assaf, Joanna Wilińska, and Detlef Gabel

1.5.1	Introduction *109*	
1.5.2	Water Structure and Coordinating Properties *110*	
1.5.3	Host–Guest Chemistry of Boron Clusters *112*	
1.5.4	Ionic Boron Clusters in Protein Interactions *116*	
1.5.4.1	Interactions of Boron Clusters with Lipid Bilayers *118*	
1.5.5	Implications for Drug Design *120*	
1.5.5.1	Binding to Proteins *120*	
1.5.5.2	Penetration through Membranes *120*	
1.5.5.3	Computational Methods *120*	
1.5.6	Conclusions *121*	
	References *121*	

1.6 Quantum Mechanical and Molecular Mechanical Calculations on Substituted Boron Clusters and Their Interactions with Proteins *126*

Jindřich Fanfrlík, Adam Pecina, Jan Řezáč, Pavel Hobza, and Martin Lepšík

1.6.1	Introduction *126*	
1.6.2	Plethora of Noncovalent Interactions of Boron Clusters *127*	
1.6.3	Computational Methods *129*	
1.6.3.1	Advanced Methods for Models Systems in Vacuum *129*	
1.6.3.2	Approximate Methods for Extended Systems Including the Environment *129*	
1.6.3.2.1	SQM Methods *130*	
1.6.3.2.2	MM Methods *130*	
1.6.3.2.3	Solvation and Ion Models *131*	
1.6.3.2.4	pKa Calculations *131*	
1.6.3.2.5	Docking and Scoring *131*	
1.6.4	Boron Cluster Interactions with Proteins *131*	
1.6.5	Conclusions *135*	
	References *135*	

Part 2 Boron Compounds in Drug Delivery and Imaging *139*

2.1 Closomers: An Icosahedral Platform for Drug Delivery *141*
Satish S. Jalisatgi
2.1.1 Introduction *141*
2.1.2 Synthesis and Chemistry of $[closo\text{-}B_{12}H_{12}]^{2-}$ *143*
2.1.3 Hydroxylation of $[closo\text{-}B_{12}H_{12}]^{2-}$ *143*
2.1.4 Ether and Ester Closomers *144*
2.1.5 Carbonate and Carbamate Closomers *145*
2.1.6 Azido Closomers *146*
2.1.7 Methods of Vertex Differentiation for Multifunctional Closomers *148*
2.1.7.1 Vertex Differentiation by Selective Derivatization of $[closo\text{-}B_{12}(OH)_{12}]^{2-}$, 1 *148*
2.1.7.2 Vertex Differentiation by Functionalizing the B_{12}^{2-} Core Prior to the Cage Hydroxylation *151*
2.1.8 Conclusions *155*
References *155*

2.2 Cobaltabisdicarbollide-based Synthetic Vesicles: From Biological Interaction to *in vivo* Imaging *159*
Clara Viñas Teixidor, Francesc Teixidor, and Adrian J. Harwood
2.2.1 Introduction *159*
2.2.2 A Synthetic Membrane System *160*
2.2.3 Crossing Lipid Bilayers *161*
2.2.4 Visualization of COSAN within Cells *162*
2.2.5 COSAN Interactions with Living Cells *163*
2.2.6 Enhancing Cellular Effects of COSAN *165*
2.2.7 Tracking the *in vivo* Distribution of I-COSAN *167*
2.2.8 Discussion and Potential Applications *168*
2.2.9 Summary *170*
Appendix: Abbreviations *170*
Acknowledgments *171*
References *171*

2.3 Boronic Acid–Based Sensors for Determination of Sugars *174*
Igor B. Sivaev and Vladimir I. Bregadze
2.3.1 Introduction *174*
2.3.2 Interactions of Boronic Acids with Carbohydrates *175*
2.3.3 Fluorescence Carbohydrate Sensors *179*
2.3.3.1 Intramolecular Charge Transfer Sensors *180*
2.3.3.2 Photoinduced Electron Transfer Sensors *186*
2.3.3.3 Fluorescence Resonance Energy Transfer Sensors *196*
2.3.4 Colorimetric Sensors *197*
2.3.5 Conclusions *199*
References *199*

2.4	**Boron Compounds in Molecular Imaging** *205*	
	Bhaskar C. Das, Devi Prasan Ojha, Sasmita Das, and Todd Evans	
2.4.1	Introduction *205*	
2.4.2	Molecular Imaging in Biomedical Research *207*	
2.4.3	Molecular Imaging Modalities *207*	
2.4.4	Boron Compounds in Molecular Imaging *210*	
2.4.5	Boron-Based Imaging Probes *213*	
2.4.5.1	Boron-Based Optical Probes *213*	
2.4.5.2	Boron-Based Nuclear Probes *216*	
2.4.5.3	Boron-Based MRI Probes *219*	
2.4.5.4	Boron-Based Molecular Probes for Disease *220*	
2.4.6	Future Perspectives *224*	
	Appendix: Companies Offering Imaging Instruments and Reagents *225*	
	References *225*	
2.5	**Radiolabeling Strategies for Boron Clusters: Toward Fast Development and Efficient Assessment of BNCT Drug Candidates** *232*	
	Kiran B. Gona, Vanessa Gómez-Vallejo, Irina Manea, Jonas Malmquist, Jacek Koziorowski, and Jordi Llop	
2.5.1	Boron Neutron Capture Therapy *232*	
2.5.1.1	Boron: The Element *232*	
2.5.1.2	The Principle behind Boron Neutron Capture Therapy (BNCT) *232*	
2.5.2	Boron Clusters *235*	
2.5.2.1	Boranes *236*	
2.5.2.2	Carboranes *236*	
2.5.2.3	Metallocarboranes *238*	
2.5.3	Nuclear Imaging: Definition and History *238*	
2.5.3.1	Radioactivity: The Basis of Nuclear Imaging *239*	
2.5.3.2	Single-Photon Emission Computerized Tomography *240*	
2.5.3.3	Positron Emission Tomography *241*	
2.5.3.4	Multimodal Imaging *243*	
2.5.3.5	Nuclear Imaging and the Need for Radiolabeling *243*	
2.5.4	Radiolabeling of Boron Clusters *244*	
2.5.4.1	Radiohalogenation *245*	
2.5.4.1.1	Radioastatination of Boron Clusters *245*	
2.5.4.1.2	Radioiodination of Boron Clusters *246*	
2.5.4.1.3	Radiofluorination of Boron Clusters *248*	
2.5.4.1.4	Radiobromination of Boron Clusters *250*	
2.5.4.2	Radiometallation *251*	
2.5.4.2.1	Radiolabeling using other Radionuclides *253*	
2.5.5	The use of Radiolabeling in BNCT Drug Development: Illustrative Examples *254*	
2.5.6	Conclusion and Future Perspectives *259*	
	References *259*	

Part 3 Boron Compounds for Boron Neutron Capture Therapy 269

3.1 Twenty Years of Research on 3-Carboranyl Thymidine Analogs (3CTAs): A Critical Perspective 271
Werner Tjarks
3.1.1   Introduction 271
3.1.2   Boron Neutron Capture Therapy 272
3.1.3   Carboranes 273
3.1.4   Rational Design of 3CTAs 274
3.1.5   Synthesis and Initial Screening of 3CTAs as TK1 Substrates 275
3.1.5.1 First-Generation 3CTAs 275
3.1.5.2 Second-Generation 3CTAs 277
3.1.6   Enzyme Kinetic and Inhibitory Studies 279
3.1.7   Cell Culture Studies 279
3.1.8   Metabolic Studies 280
3.1.9   Cellular Influx and Efflux Studies 282
3.1.10 *In vivo* Uptake and Preclinical BNCT Studies 282
3.1.11 Potential Non-BNCT Applications for 3CTAs 284
3.1.12 Conclusion 285
        Acknowledgments 286
        References 286

3.2 Recent Advances in Boron Delivery Agents for Boron Neutron Capture Therapy (BNCT) 298
Sunting Xuan and Maria da Graça H. Vicente
3.2.1   Introduction 298
3.2.1.1 Mechanisms of BNCT 298
3.2.1.2 General Criteria for BNCT Agents 299
3.2.1.3 Main Categories of BNCT Agents 299
3.2.2   Amino Acids and Peptides 300
3.2.3   Nucleosides 304
3.2.4   Antibodies 305
3.2.5   Porphyrin Derivatives 307
3.2.5.1 Porphyrin Macrocycles 307
3.2.5.2 Chlorin Macrocycles 314
3.2.5.3 Phthalocyanine Macrocycles 316
3.2.6   Boron Dipyrromethenes 318
3.2.7   Liposomes 321
3.2.8   Nanoparticles 324
3.2.9   Conclusions 330
        References 331

3.3 Carborane Derivatives of Porphyrins and Chlorins for Photodynamic and Boron Neutron Capture Therapies: Synthetic Strategies 343
Valentina A. Ol'shevskaya, Andrei V. Zaitsev, and Alexander A. Shtil
3.3.1   Introduction 343
3.3.2   Recent Synthetic Routes to Carboranyl-Substituted Derivatives of 5,10,15,20-Tetraphenylporphyrin 344

3.3.3	Synthesis of Carborane Containing Porphyrins and Chlorins from Pentafluorophenyl-Substituted Porphyrin *350*	
3.3.4	Carborane Containing Derivatives of Chlorins: New Properties for PDT and Beyond *355*	
3.3.4.1	Carborane Containing Derivatives of Pyropheophorbide *a* and Pheophorbide *a* *356*	
3.3.4.2	Carborane Containing Derivatives of Chlorin e_6 *358*	
3.3.4.3	Carborane Containing Derivatives of Purpurin-18 and Bacteriopurpurinimide *362*	
3.3.5	Conclusion *364*	
	Acknowledgments *364*	
	References *365*	
3.4	**Nanostructured Boron Compounds for Boron Neutron Capture Therapy (BNCT) in Cancer Treatment** *371*	
	Shanmin Gao, Yinghuai Zhu, and Narayan Hosmane	
3.4.1	Introduction *371*	
3.4.2	Boron Neutron Capture Therapy (BNCT) *373*	
3.4.2.1	Principles of BNCT *373*	
3.4.2.2	Liposome-Based BNCT Agents *375*	
3.4.2.3	Carbon Nanotubes *377*	
3.4.2.4	Boron and Boron Nitride Nanotubes *377*	
3.4.2.5	Magnetic Nanoparticles-Based BNCT Carriers *379*	
3.4.2.6	Other Boron-Enriched Nanoparticles *382*	
3.4.3	Summary and Outlook *383*	
	References *383*	
3.5	**New Boronated Compounds for an Imaging-Guided Personalized Neutron Capture Therapy** *389*	
	Nicoletta Protti, Annamaria Deagostino, Paolo Boggio, Diego Alberti, and Simonetta Geninatti Crich	
3.5.1	General Introduction on BNCT: Rationale and Application *389*	
3.5.2	Imaging-Guided NCT: Personalization of the Neutron Irradiation Protocol *392*	
3.5.2.1	Positron Emission Tomography *392*	
3.5.2.2	Single-Photon Emission Computed Tomography *395*	
3.5.2.3	Magnetic Resonance Imaging and Spectroscopy *396*	
3.5.2.3.1	^{1}H-MRI *396*	
3.5.2.3.2	MRS Spectroscopy *398*	
3.5.2.3.3	^{10}B and ^{11}B NMR *398*	
3.5.2.4	Optical Imaging *400*	
3.5.2.5	Boron Microdistribution *401*	
3.5.3	Targeted BNCT: Personalization of *in vivo* Boron-Selective Distribution *402*	
3.5.3.1	Small-Sized Boron Carriers *402*	
3.5.3.2	Nanosized Boron Carriers *406*	
3.5.4	Combination of BNCT with Other Conventional and Nonconventional Therapies *407*	

3.5.4.1	Chemotherapy	*408*
3.5.4.2	Photodynamic Therapy (PDT)	*408*
3.5.4.3	Standard Radiotherapy	*408*
3.5.5	Conclusions	*410*
	References	*410*

3.6 Optimizing the Therapeutic Efficacy of Boron Neutron Capture Therapy (BNCT) for Different Pathologies: Research in Animal Models Employing Different Boron Compounds and Administration Strategies *416*
Amanda E. Schwint, Andrea Monti Hughes, Marcela A. Garabalino, Emiliano C.C. Pozzi, Elisa M. Heber, and Veronica A. Trivillin

3.6.1	BNCT Radiobiology	*416*
3.6.2	An Ideal Boron Compound	*417*
3.6.3	Clinical Trials, Clinical Investigations, and Translational Research	*418*
3.6.4	Boron Carriers	*419*
3.6.5	Optimizing Boron Targeting of Tumors by Employing Boron Carriers Approved for Use in Humans	*422*
3.6.6	BNCT Studies in the Hamster Cheek Pouch Oral Cancer Model	*422*
3.6.6.1	The Hamster Cheek Pouch Oral Cancer Model	*422*
3.6.6.2	BNCT Mediated by BPA	*425*
3.6.6.3	BNCT Mediated by GB-10 or by GB-10 + BPA	*425*
3.6.6.4	Sequential BNCT	*428*
3.6.6.5	Tumor Blood Vessel Normalization to Improve Boron Targeting for BNCT	*430*
3.6.6.6	Tumor Blood Vessel Normalization + Seq-BNCT	*432*
3.6.6.7	Electroporation + BNCT	*432*
3.6.6.8	Assessing Novel Boron Compounds	*433*
3.6.7	BNCT Studies in a Model of Oral Precancer in the Hamster Cheek Pouch for Long-Term Follow-up	*435*
3.6.8	BNCT Studies in a Model of Liver Metastases in BDIX Rats	*438*
3.6.9	BNCT Studies in a Model of Diffuse Lung Metastases in BDIX Rats	*441*
3.6.10	BNCT Studies in a Model of Arthritis in Rabbits	*442*
3.6.11	Preclinical BNCT Studies in Cats and Dogs with Head and Neck Cancer with no Treatment Option	*445*
3.6.12	Future Perspectives	*446*
	References	*446*

Index *462*

List of Contributors

Anna Adamska-Bartłomiejczyk
Laboratory of Molecular Virology and Biological Chemistry, IMB PAS, Łódź, Poland

Diego Alberti
Department of Molecular Biotechnology and Health Sciences, University of Torino, Italy

Khaleel I. Assaf
Department of Life Sciences and Chemistry, Jacobs University Bremen, Germany

Hyun Seung Ban
Biomedical Genomics Research Center, Korea Research Institute of Bioscience and Biotechnology, Daejeon, Republic of Korea

Katarzyna Bednarska
Laboratory of Experimental Immunology, IMB PAS, Łódź, Poland

Magdalena Białek-Pietras
Laboratory of Molecular Virology and Biological Chemistry, IMB PAS, Łódź, Poland

Paolo Boggio
Department of Chemistry, University of Torino, Italy

Vladimir I. Bregadze
A.N. Nesmeyanov Institute of Organoelement Compounds, Russian Academy of Sciences, Moscow, Russia

Bhaskar C. Das
Departments of Medicine and Pharmacological Sciences, Icahn School of Medicine at Mount Sinai, New York, USA

Sasmita Das
Molecular Bio-nanotechnology, Imaging and Therapeutic Research Unit, Veteran Affairs Medical Center, Kansas City, USA

Annamaria Deagostino
Department of Chemistry, University of Torino, Italy

Yasuyuki Endo
Faculty of Pharmaceutical Sciences, Tohoku Medical and Pharmaceutical University, Sendai, Japan

Todd Evans
Department of Surgery, Weill Cornell Medical College of Cornell University, New York, USA

Jindřich Fanfrlík
Institute of Organic Chemistry and Biochemistry of the Czech Academy of Sciences, Prague, Czech Republic

Detlef Gabel
Department of Life Sciences and Chemistry, Jacobs University Bremen, Germany

Shanmin Gao
School of Chemistry and Materials Science, Ludong University, China

Marcela A. Garabalino
Department of Radiobiology, National Atomic Energy Commission, San Martín, Buenos Aires Province, Argentina

Simonetta Geninatti Crich
Department of Molecular Biotechnology and Health Sciences, University of Torino, Italy

Vanessa Gómez-Vallejo
Radiochemistry and Nuclear Imaging Group, CIC biomaGUNE, San Sebastian, Spain

Kiran B. Gona
Radiochemistry and Nuclear Imaging Group, CIC biomaGUNE, San Sebastian, Spain Cardiovascular Molecular Imaging Laboratory, Section of Cardiovascular Medicine and Yale Cardiovascular Research Center, Yale University School of Medicine, New Haven, USA
Veterans Affairs Connecticut Healthcare System, West Haven, USA

Marta Gozzi
Universität Leipzig, Institut für Anorganische Chemie, Germany

Adrian J. Harwood
School of Biosciences and Neuroscience and Mental Health Research Institute, Cardiff University, Cardiff, UK

Elisa M. Heber
Department of Radiobiology, National Atomic Energy Commission, San Martín, Buenos Aires Province, Argentina

Evamarie Hey-Hawkins
Universität Leipzig, Institut für Anorganische Chemie, Germany

Pavel Hobza
Institute of Organic Chemistry and Biochemistry of the Czech Academy of Sciences, Prague, Czech Republic
Regional Centre of Advanced Technologies and Materials, Palacký University, Olomouc, Czech Republic

Narayan Hosmane
Department of Chemistry and Biochemistry, Northern Illinois University, DeKalb, USA

Satish S. Jalisatgi
International Institute of Nano and Molecular Medicine, School of Medicine, University of Missouri, Columbia, USA

Zofia M. Kiliańska
Department of Cytobiochemistry, Faculty of Biology and Environmental Protection, University of Łódź, Poland

Jacek Koziorowski
Department of Radiology and Department of Medical and Health Sciences, Linköping University, Sweden

Martin Lepšík
Institute of Organic Chemistry and Biochemistry of the Czech Academy of Sciences, Prague, Czech Republic

Zbigniew J. Lesnikowski
Laboratory of Molecular Virology and Biological Chemistry, IMB PAS, Łódź, Poland

List of Contributors | **xv**

Guangzhe Li
School of Pharmaceutical Science and Technology, Dalian University of Technology, China

Jordi Llop
Radiochemistry and Nuclear Imaging Group, CIC biomaGUNE, San Sebastian, Spain

Jonas Malmquist
Department of Radiology and Department of Medical and Health Sciences, Linköping University, Sweden

Irina Manea
Colentina Clinical Hospital, Bucuresti, Romania

Adam Mieczkowski
Department of Biophysics, Institute of Biochemistry and Biophysics, Polish Academy of Sciences, Warsaw, Poland

Andrea Monti Hughes
Department of Radiobiology, National Atomic Energy Commission, Buenos Aires Province, Argentina

Hiroyuki Nakamura
Laboratory for Chemistry and Life Science, Institute of Innovative Research, Tokyo Institute of Technology, Yokohama, Japan

Devi Prasan Ojha
Departments of Medicine and Pharmacological Sciences, Icahn School of Medicine at Mount Sinai, New York, USA

Agnieszka B. Olejniczak
Screening Laboratory, IMB PAS, Łódź, Poland

Valentina A. Ol'shevskaya
A.N. Nesmeyanov Institute of Organoelement Compounds, Russian Academy of Sciences, Moscow, Russia

Edyta Paradowska
Laboratory of Molecular Virology and Biological Chemistry, IMB PAS, Łódź, Poland

Adam Pecina
Institute of Organic Chemistry and Biochemistry of the Czech Academy of Sciences, Prague, Czech Republic

Emiliano C.C. Pozzi
Department of Radiobiology, National Atomic Energy Commission, San Martín, Buenos Aires Province, Argentina

Nicoletta Protti
Nuclear Physics National Institute (INFN), University of Pavia, Italy

Jan Řezáč
Institute of Organic Chemistry and Biochemistry of the Czech Academy of Sciences, Prague, Czech Republic

Benedikt Schwarze
Universität Leipzig, Institut für Anorganische Chemie, Germany

Amanda E. Schwint
Department of Radiobiology, National Atomic Energy Commission, San Martín, Buenos Aires Province, Argentina

Alexander A. Shtil
Blokhin National Medical Research Center of Oncology, Moscow, Russia

Igor B. Sivaev
A.N. Nesmeyanov Institute of Organoelement Compounds, Russian Academy of Sciences, Moscow, Russia

Mirosława Studzińska
Laboratory of Molecular Virology and Biological Chemistry, IMB PAS, Łódź, Poland

List of Contributors

Zofia Sułowska
Laboratory of Experimental Immunology, IMB PAS, Łódź, Poland

Francesc Teixidor
Institut de Ciència de Materials de Barcelona (ICMAB-CSIC), Bellaterra, Spain

Werner Tjarks
Division of Medicinal Chemistry & Pharmacognosy, The Ohio State University, Columbus, USA

Veronica A. Trivillin
Department of Radiobiology, National Atomic Energy Commission, San Martín, Buenos Aires Province, Argentina

Maria da Graça H. Vicente
Department of Chemistry, Louisiana State University, Baton Rouge, USA

Clara Viñas Teixidor
Institut de Ciència de Materials de Barcelona (ICMAB-CSIC), Bellaterra, Spain

Joanna Wilińska
Department of Life Sciences and Chemistry, Jacobs University Bremen, Germany

Sunting Xuan
Department of Chemistry, Louisiana State University, Baton Rouge, USA

Andrei V. Zaitsev
A.N. Nesmeyanov Institute of Organoelement Compounds, Russian Academy of Sciences, Moscow, Russia

Yinghuai Zhu
School of Pharmacy, Macau University of Science and Technology, Taipa, Macau, China

Jolanta D. Żołnierczyk
Department of Cytobiochemistry, Faculty of Biology and Environmental Protection, University of Łódź, Poland

Preface

Today, medicinal chemistry is still clearly dominated by organic chemistry, and most commercial drugs are purely organic molecules, which, besides carbon and hydrogen, can incorporate nitrogen, oxygen, sulfur, phosphorus, and halogens, all of which are to the right of carbon in the periodic table, whereas boron is located to the left. Boron and carbon are elements that have the ability to build molecules of unlimited size by covalent self-bonding. However, commercial boron-based drugs are still rare. Bortezomib, tavaborole (AN2690), crisaborole (AN2728), epetraborole (AN3365), SCYX-7158 (AN5568), 4-(dihydroxyboryl)phenylalanine (BPA), and sodium mercapto-undecahydro-*closo*-dodecaborate (BSH) are used as drugs, the last two compounds in boron neutron capture therapy (BNCT). All of these boron-containing drugs are derivatives of boronic acids except BSH, which contains an anionic boron cluster. While the pharmacological uses of boron compounds have been known for several decades, recent progress is closely related to the discovery of further boron-containing compounds as prospective drugs. While first developments of the medicinal chemistry of boron were stipulated by applications in BNCT of cancers, knowledge accumulated during the past decades on the chemistry and biology of bioorganic and bioinorganic boron compounds laid the foundation for the emergence of a new area of study and application of boron compounds as skeletal structures and hydrophobic pharmacophores for biologically active molecules. These and other recent findings clearly show that there is still a great, unexplored potential in medicinal applications of boron-containing compounds.

This book summarizes the present status and further promotes the development of new boron-containing drugs and boron-based materials for diagnostics by bringing together renowned experts in the field of medicinal chemistry of boron compounds. It aims to provide a balanced overview of the vibrant and growing field of the emerging and potential applications of boron compounds in medicinal chemistry and chemical biology. The book is aimed at academics and professional researchers in this field, but also at scientists who want to get a better overview on the state of the art of this rapidly advancing area. It contains reviews of important topics, which are divided into three main sections: (1) "Design of New Boron-based Drugs", (2) "Boron Compounds in Drug Delivery and Imaging", and (3) "Boron Compounds for Boron Neutron Capture Therapy".

The first section, "Design of New Boron-Based Drugs", consists of six reviews dealing with the use of carborane derivatives for the development of novel drugs. In his review (Chapter 1.1), Yasuyuki Endo, one of the pioneers in the development of carboranes as hydrophobic pharmacophores almost 20 years ago, describes the development of a variety of potent nuclear receptor ligands with carborane structures as hydrophobic moieties. Nucleoside drugs have been in clinical use for several decades and have become cornerstones of treatment for patients with cancer or viral infections. One of the new developments in the medicinal chemistry of nucleosides is derivatives comprising a boron component such as a boron cluster, as described in the review by Zbigniew J. Lesnikowski and coworkers (Chapter 1.2), whose group has long-standing expertise in the introduction of boron clusters into molecules with diverse biological activity, where they serve as pharmacophores, building blocks, and modulators of the physicochemical and biological properties. An alternative approach to battling cancer is described by Hiroyuki Nakamura *et al.* in their chapter on the design of carborane-based hypoxia-inducible factor (HIF) inhibitors (Chapter 1.3). Overexpression of HIF1α has been observed in human cancers, including brain, breast, colon, lung, ovary, and prostate cancers; thus, HIF1α is a novel target of cancer therapy, and the Nakamura group has shown carborane-based HIF1 inhibitors to be very promising targets. Another emerging type of boron-based drugs are metallacarboranes. The group of Evamarie Hey-Hawkins has been involved in carborane chemistry for more than 20 years. In Chapter 1.4, they report recent examples of biologically active half- and mixed-sandwich metallacarborane complexes of the dicarbollide ligand, as well as hybrid organic-inorganic compounds containing a *nido*-carborane(−1) as appended moiety. Their potentially beneficial properties, such as stability in aqueous environments and new binding modes due to their lipophilicity, are described. Prospective applications in radio-imaging, radiotherapy, and drug design are envisaged. In Chapter 1.5, Detlef Gabel and coworkers focus on ionic boron clusters that are soluble in water as well as in nonpolar solvents. This highly interesting feature sets them apart from other ionic and nonionic pharmacophores and renders them interesting new entities for drug design. The final review (Chapter 1.6) by Pavel Hobza, Martin Lepšík, and coworkers on the current status of structure-based computer-aided drug design tools for boron-cluster-containing protein ligands concludes this first section.

In the second section, the focus is on "Boron Compounds in Drug Delivery and Imaging". Satish S. Jalisatgi, a collaborator of Frederick Hawthorne, who was the pioneer of boron cluster chemistry almost 60 years ago, gives an overview of closomer drug delivery platforms based on an icosahedral polyhedral borane scaffold (Chapter 2.1). The resulting monodisperse nanostructures are capable of performing a combination of therapeutic, diagnostic, and targeting functions, which is highly useful for emerging applications. A complementary approach is described in the review by Clara Viñas Teixidor (Chapter 2.2), one of the founders of EuroBoron conference, and her colleagues. The anionic boron-based cobaltabis(dicarbollide) can form atypical monolayer membranes with the shape of vesicles and micelles with similar dimensions to those seen in nature, but of a very different chemical composition. These vesicles interact with liposomes and biological membranes to accumulate inside living cells. Their particular properties offer new opportunities for the development of nanoscale platforms to directly introduce new functionality for use in cancer therapy, drug design, and molecular delivery systems.

Diabetes is a chronic disease that has devastating human, social, and economic consequences. A tight control of blood glucose is the most important goal in dealing with diabetes. The majority of blood glucose monitoring tools relies on the glucose oxidase enzyme (GOx), but they have some drawbacks. A powerful approach for detecting glucose in fluids is the development of boronic acid-based saccharide sensors. The main principles of their design and factors governing their selectivity are discussed by Igor B. Sivaev and Vladimir I. Bregadze in Chapter 2.3.

Drug development is a lengthy process requiring identification of a biological target, validation of the target, and development of pharmacological agents designed and subsequently confirmed by *in vivo* studies. Molecular and functional imaging applied in the initial stages of drug development can provide evidence of biological activity and confirm on-target drug effects. In their contributions, Bhaskar C. Das *et al.* focus on various boron-containing molecular probes used in molecular imaging (Chapter 2.4), and Jordi Llop *et al.* provide an overview of nuclear imaging techniques, as well as the different radiolabeling strategies reported so far for the incorporation of positron and gamma emitters into boron clusters (Chapter 2.5). Finally, some illustrative examples on how radiolabeling and *in vivo* imaging can aid in the process of drug development are described, focusing on BNCT drug candidates containing boron clusters, linking this chapter to the third section dedicated to "Boron Compounds for Boron Neutron Capture Therapy".

Cancer is the second leading cause of death globally, and was responsible for 8.8 million deaths in 2015. Treatment typically comprises surgery, radiotherapy, and chemotherapy. BNCT is a unique binary therapy that was developed during the last five to six decades. With the availability of accelerator-based neutron sources at clinics, selective boron compounds for use in BNCT will become very important. In this third section, several novel classes of potential BNCT agents are described. Werner Tjarks critically reviews aspects of the design, synthesis, and biological evaluation of 3-carboranyl thymidine analogs (3CTAs) as boron delivery agents for BNCT over a time span of approximately 20 years (Chapter 3.1). Potential future non-BNCT applications of 3CTAs are also discussed, linking this review to the first section on boron-based drug design. Maria da Graça H. Vicente and Sunting Xuan describe different classes of third-generation boron delivery agents with enhanced tumor-localizing properties, which are under investigation for use in BNCT (Chapter 3.2), and the contribution by Valentina A. Ol'shevskaya and colleagues deals with synthetic approaches leading to tumor-selective boronated porphyrins and chlorins with potential applications in diagnosis, drug delivery, and treatment. This study emphasizes the role of boron in rendering the photoactivatable tetrapyrrolic scaffolds more potent in photodynamic therapy (Chapter 3.3). A highly innovative approach is described in the review by Narayan Hosmane, one of the founders of Boron in the Americas (BORAM), and his coworkers covering the recent developments in the use of nanoparticles as adjuncts to boron-containing compounds in BNCT, involving boron nanotubes (BNTs) and boron nitride nanotubes (BNNTs) (Chapter 3.4). For further implementation of BNCT at the clinical level, new specifically targeted boron carriers for BNCT, conjugated with functional groups detectable by highly sensitive imaging tools, are required. This allows the determination of the local boron concentration, which is crucial to personalize the treatment for each patient. Simonetta Geninatti Crich and coworkers cover this important topic in Chapter 3.5. Furthermore, *in vivo* research in appropriate animal models is important to expand BNCT radiobiology and optimize its therapeutic efficacy for different pathologies. This highly interdisciplinary topic is covered by Amanda E. Schwint and coworkers in their comprehensive contribution in Chapter 3.6.

We are very grateful to all the authors for their contributions and their patience. Last but not least, we would like to thank the Wiley team, especially Sarah Higginbotham and Emma Strickland, for their continuous support in planning and compiling this book, which gives a timely overview of the evolving potential and emerging applications of boron-based compounds in medicine.

Evamarie Hey-Hawkins and
Clara Viñas Teixidor

Part 1

Design of New Boron-based Drugs

1.1

Carboranes as Hydrophobic Pharmacophores: Applications for Design of Nuclear Receptor Ligands

Yasuyuki Endo

Faculty of Pharmaceutical Sciences, Tohoku Medical and Pharmaceutical University, Sendai, Japan

1.1.1 Roles of Hydrophobic Pharmacophores in Medicinal Drug Design

A pharmacophore is a partial structure in which important functional groups and hydrophobic structure are arranged in suitable positions for binding to a receptor [1]. Typically, hydrophilic functional groups of the pharmacophore interact with the receptor by hydrogen bonding and/or ionic bonding, and the hydrophobic structure interacts with a hydrophobic surface of the receptor. While hydrogen bonding plays a key role in specific ligand–receptor recognition, the hydrophobic interaction between receptor and drug molecule is especially important in determining the binding affinity. The difference of binding constants between a ligand having a suitable hydrophobic group and a ligand without such a group can be as large as 1000-fold. In medicinal drug design, the hydrophobic structures are often composed of aromatic and heteroaromatic rings, which also play a role in fixing the arrangement of functional groups appropriately for binding to the receptor. On the other hand, three-dimensional hydrophobic structures are not yet widely used in drug design, even though they could be well suited for interaction with the three-dimensional hydrophobic binding pockets of receptors. It is noteworthy that various steroid hormones target distinct steroid hormone receptors owing to differences of functionalization of the hydrophobic steroidal skeleton. The binding of the natural ligand 17β-estradiol to human estrogen receptor-α (ERα) is illustrated in Figure 1.1.1 as an example. The large number of steroid hormones may be a consequence of evolutionary diversification of the functions of the steroidal skeleton. In this context, we aimed to establish a new three-dimensional hydrophobic skeletal structure for medicinal drug design.

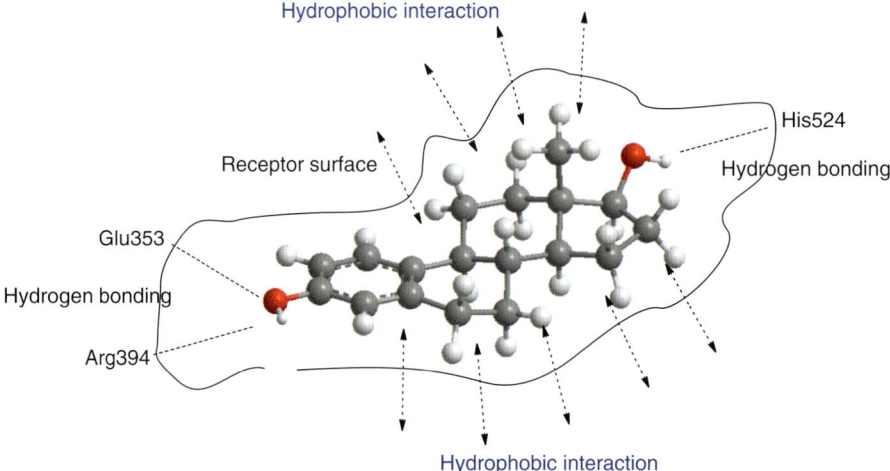

Figure 1.1.1 Interactions of ligand with receptor (example for 17β-estradiol with estrogen receptor-α).

1.1.2 Carboranes as Hydrophobic Structures for Medicinal Drug Design

In the past three decades, there has been increasing interest in globular molecules. In the 1980s, dodecahedrane, which consists of sp^3 carbons, was synthesized [2]; and in the 1990s, the chemistry of fullerene C$_{60}$, which also consists of sp^2 carbons, was explored [3]. However, the former is not easy to synthesize, while the latter molecule may have limited application because of its large molecular size. On the other hand, icosahedral carboranes [4] are topologically symmetrical, globular molecules, and have been known for more than half a century. The B–B and C–B bonds of 12-vertex carboranes are approximately 1.8 angstroms in length, and the molecular size of carboranes is somewhat larger than adamantane or the volume of a rotated benzene ring. Carboranes have a highly electron-delocalized hydrophobic surface, and are considered to be three-dimensional aromatic compounds [5] or inorganic benzenes. The structures of these compounds are illustrated in Figure 1.1.2. But, although the use of boron derivatives for boron neutron capture therapy (BNCT) of tumors has a long history [6], relatively little attention has yet been paid to the possible use of carboranes as components of biologically active molecules, despite their desirable hydrophobic character, spherical geometry, and convenient molecular size for use in the design and synthesis of medicinal drugs.

Carboranes have three isomers, *ortho-*, *meta-*, and *para-*carboranes (Figure 1.1.2), and their rigid and bulky cage structures hold substituents in well-defined spatial relationships. The two carbon atoms of carboranes have relatively acidic protons, which can readily be substituted with other organic groups [7]. Substituents can also be introduced selectively at certain boron atoms, to construct structures having three or more substituents, as illustrated in Figure 1.1.3 [8]. Carbocyclic skeletons often rearrange under acidic conditions, whereas carborane cage skeletons do not rearrange even in the presence of strong Lewis acids. Adamantane and bicyclo[2,2,2]octane are also available as hydrophobic skeletons, and substituents can readily be introduced at bridgehead carbons of adamantane, but selective introduction at other carbons is difficult, and chirality is also an issue.

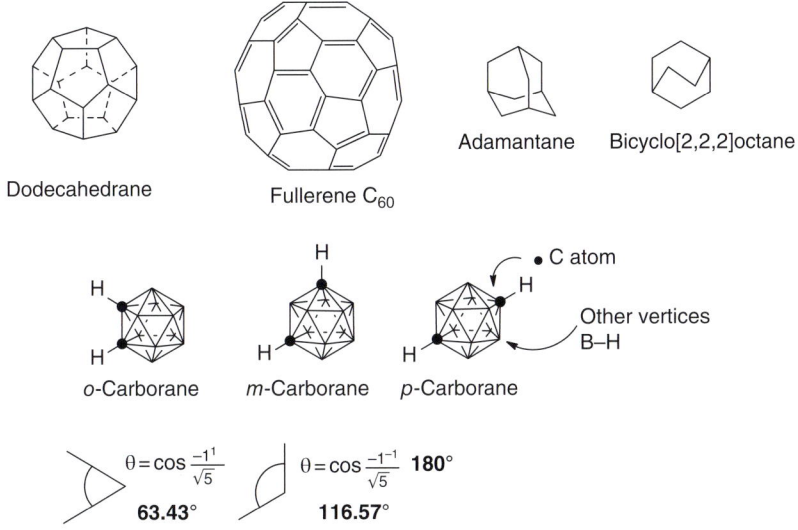

Figure 1.1.2 Structures of globular molecules and characteristics of carboranes.

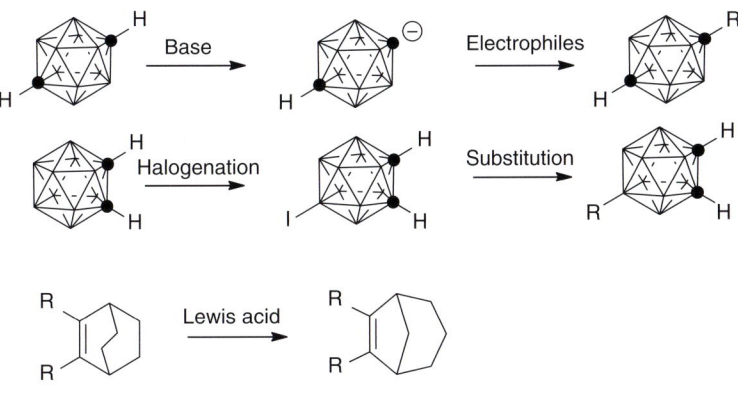

Rearrangement of hydrocarbon skeleton (for example)

Figure 1.1.3 Advantages of carborane skeleton for synthesis.

1.1.3 Estrogen Receptor Ligands Bearing a Carborane Cage

1.1.3.1 Estrogen Agonists

Estrogen mediates a wide variety of cellular responses through its binding to a specific estrogen receptor (ER). The hormone-bound ER forms an active dimer, which binds to the ER-responsive element of DNA and regulates gene transcription. Endogenous estrogen, such as 17β-estradiol, plays an important role in the female reproductive system, and also in bone maintenance, the central nervous system, and the cardiovascular system. Recent studies on the three-dimensional structure of the complex formed by estradiol and the human ERα ligand-binding domain have identified the structural requirements for estrogenic activity [9]. 17β-Estradiol is oriented in the ligand-binding

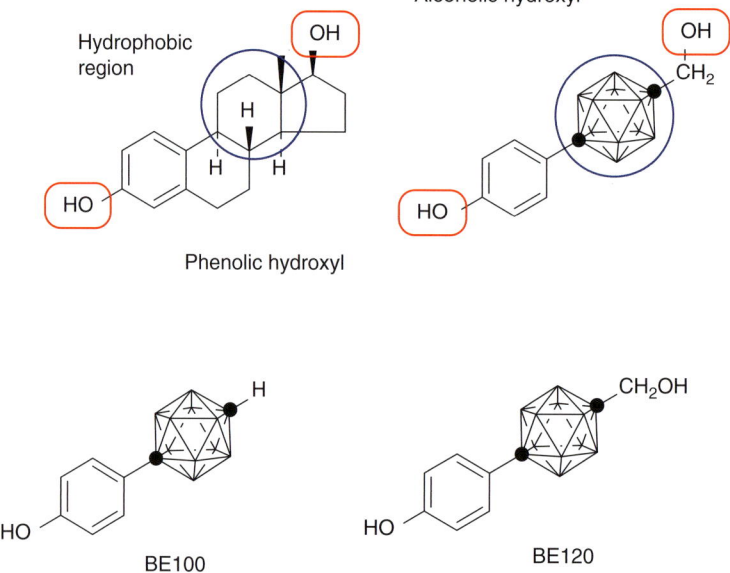

Figure 1.1.4 Structures of β-estradiol and designed molecule bearing *p*-carborane.

pocket by two types of contacts: hydrogen bonding from the phenolic hydroxyl group to Glu353 and Arg394, and from the 17β-hydroxyl group to the nitrogen of His524, and hydrophobic interaction along the body of the skeleton (see Figure 1.1.1). Therefore, we designed a simple compound with a 4-phenolic residue and a hydroxymethylated *p*-carborane, together with some derivatives (Figure 1.1.4) [10].

The estrogenic activities of the synthesized compounds were examined by means of receptor binding assays. Surprisingly, the simple 4-(*p*-carboranyl)phenol BE100 exhibited potent ERα-binding affinity, comparable with that of estradiol, and the most active compound, BE120, was several times more potent than estradiol. In transcriptional assay, the simple 4-(p-carboranyl)phenol BE100 exhibited potent agonistic activity, comparable with that of estradiol. The activity was increased by the introduction of a hydroxylmethyl group onto carbon of the carborane cage, and the resulting compound, BE120, was at least 10-fold more potent than estradiol. In a docking simulation of BE120 with the receptor based on the crystal structure of the estradiol–ERα complex, the phenolic hydroxyl group and hydroxymethyl group of BE120 appeared to play similar roles to those in the case of estradiol. The higher activity of BE120 suggests that the carborane cage binds to the hydrophobic cavity of the receptor more tightly than does the equivalent structure of estradiol [11].

BE120 also showed potent *in vivo* effects. Uterine atrophy due to estrogen deficiency or ovariectomy is blocked by estrogen administration, and this forms the basis of a typical *in vivo* assay for estrogenic activity. Estradiol and BE120 at 100 ng per day both restored the uterine weight, indicating that BE120 reproduces the biological activity of estradiol. Similarly, decrease of the bone mineral density of ovariectomized mice was blocked by administration of either estradiol or BE120, with similar potency [12].

1.1.3.2 Estrogen Antagonists and Selective Estrogen-Receptor Modulators (SERMs)

Since estrogen agonists increase the risk of carcinogenesis in breast and uterus [13], estrogen antagonists can be used as anticancer agents. On the other hand, estrogen agonists may be useful for the control of osteoporosis, if the risk of carcinogenesis can be avoided. Therefore, there is great interest in SERMs that selectively affect different organs, especially agents with agonistic activity in bone, but no effect or antagonistic activity in the reproductive organs. Among SERMs so far developed, tamoxifen is used to treat breast cancer [14], and raloxifene to treat osteoporosis [15].

The balance of activities depends on the precise ligand–receptor complex structure, which influences subsequent binding with co-factors and other proteins, leading to different physiological actions. In the case of tamoxifen [14], the bulky dimethylaminoethoxyphenyl group plays a key role in the antagonistic activity. Taking this into account, we designed compounds containing o- and m-carborane skeletons, as shown in Figure 1.1.5.

The o-carborane derivative BE362 inhibited the activity of estradiol in the concentration range of 10^{-7} M in a transcriptional activity assay, being equipotent with tamoxifen. The m-carborane derivative BE262 was somewhat less potent than BE362. In this assay, synthetic intermediate BE360 also exhibited antagonistic activity, although its potency was somewhat weaker than that of BE362 [16]. In spite of its very simple structure, BE360 exhibited strong binding affinity for ER [17]. Therefore, we focused on BE360 as a candidate SERM. Loss of bone mineral density of ovariectomized mice was blocked by administration of BE360 at 1–30 mg/day [18]. BE360 was 1000-fold less potent than estradiol, but was almost equipotent with the osteoporosis drug raloxifene. On the other hand, BE360 did not affect uterine weight at this concentration. Thus, BE360 is a promising lead compound for development of therapeutically useful SERMs.

We next investigated structural development of BE360. Insertion of a methylene group (BE380) changed the partial agonist–antagonist character of BE360 to weak agonist, and insertion of two methylene units generated a potent antagonist (BE381). Replacing the carborane cage with a bicyclo[2,2,2]octane skeleton caused a drastic change of biological activity, affording a potent full agonist (BE1060). It seems clear that altering the three-dimensional hydrophobic core structure is a promising strategy for control of the agonist–antagonist activity balance toward ER [19].

In addition, we have recently reported that BE360 has antidepressant and antidementia effects through enhancement of hippocampal cell proliferation in olfactory bulbectomized mice [20]. Thus, BE360 may have potential for treatment of depression and neurodegenerative diseases, such as Alzheimer's disease.

1.1.4 Androgen Receptor Ligands Bearing a Carborane Cage

1.1.4.1 Androgen Antagonists

Like estrogen, androgen mediates cellular responses through binding to a specific androgen receptor (AR). The hormone-bound AR forms a dimer, which binds to the AR-responsive element of DNA and regulates gene transcription. Endogenous androgen, such as testosterone and dihydrotestosterone, plays an important role in the

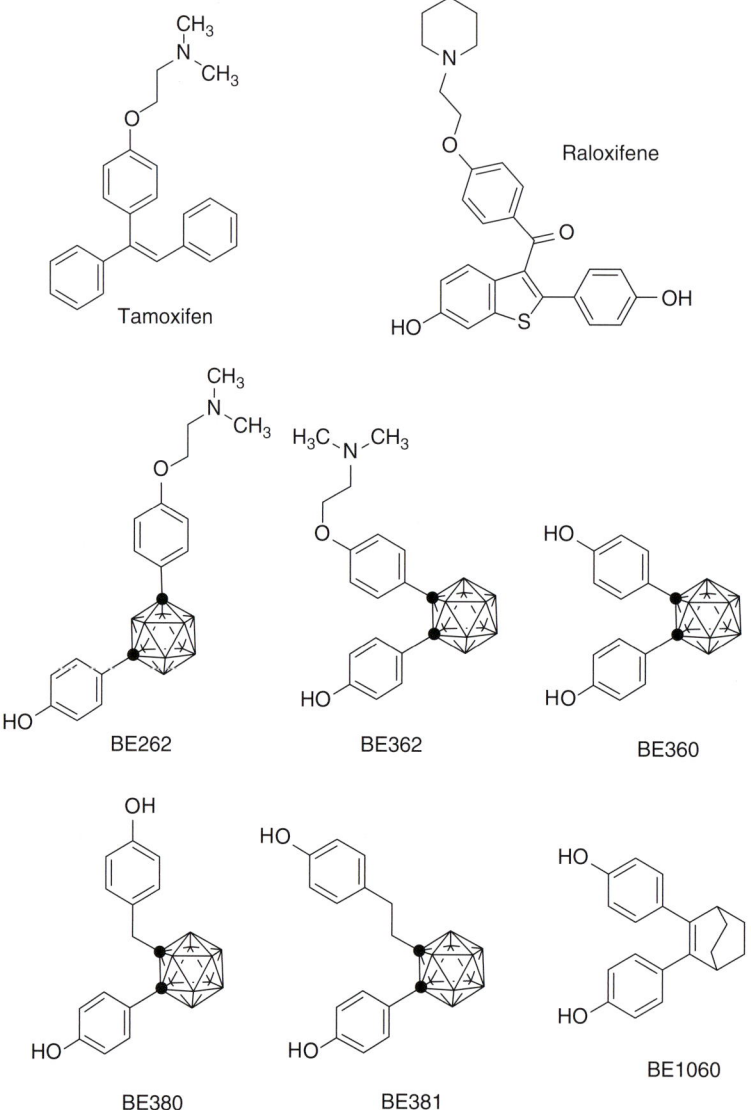

Figure 1.1.5 Structures of selective estrogen receptor modulators: tamoxifen and raloxifene, and designed molecules bearing carborane.

male reproductive system, and also in prostate enlargement, body hair growth, and muscle development. The X-ray structure of the complex of AR ligand-binding domain with an androgen agonist has been reported [21]. The overall structure of the ligand-binding domain is very similar to that of ER, but there are differences in the structures surrounding the ligand-binding pocket. One of the differences between AR and ER ligands is that the aliphatic cyclohexene A-ring of the steroid skeleton bears an 18-methyl group, so that the structure is bulky compared with the flat aromatic A-ring of estrogen. In addition, a ketone is present instead of the phenolic hydroxyl group in

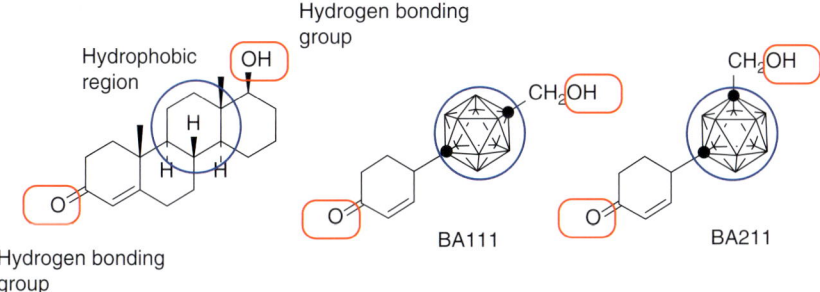

Figure 1.1.6 Structures of testosterone and designed molecules bearing carborane.

the case of estrogen ligand. Therefore, we designed new AR ligand candidates bearing a carborane skeleton, cyclohexenone, and a hydroxymethyl group (Figure 1.1.6). Among the synthesized compounds, BA111 and BA211, bearing a carboranyl group at the 4-position of the cyclohexenone ring, exhibited binding activity to human AR (hAR). These compounds did not show agonistic activity in growth promotion assay using androgen-dependent SC3 cells (Shionogi carcinoma-3, a human prostate cancer cell line), but they inhibited testosterone-promoted cell growth of SC3. The potency of the antagonistic activity was comparable to that of 4-hydroxyflutamide [22].

Next, we designed a new type of carborane-containing compounds with the aim of discovering more potent AR antagonists [23]. Typical potent nonsteroidal androgen antagonists, such as hydroxyflutamide and bicalutamide, contain a benzene ring bearing an electron-withdrawing nitro-, cyano-, or trifluoromethyl group. Therefore, we focused on aromatic derivatives incorporating a *p*-carborane moiety (Figure 1.1.7). Among these compounds, BA321 and BA341, bearing an electron-withdrawing group at the 3-position of the benzene ring, exhibited potent binding activity to AR. The affinities were 10-fold stronger than those of hydroxyflutamide. Structure–activity studies of BA321 and BA341 were guided by transcription assay. We found that BA321 and BA341 showed potent antagonistic activity in the concentration range of 1×10^{-7}–10^{-5} M, and they dose-dependently inhibited the activity of dihydrotestosterone. Their potencies were 10-fold stronger than those of hydroxyflutamide. These novel aromatic, carborane-containing AR modulators should be useful tools for analysis of AR–ligand interactions, and also as scaffolds for development of clinically useful nonsteroidal androgen antagonists.

1.1.4.2 Improvement of Carborane-Containing Androgen Antagonists as Candidates for Anti–Prostate Cancer Therapy

Our designed androgen antagonists still required improvement for potential clinical use. Androgen agonists, including male hormone, are promoters of prostate cancer, and therefore antagonists are used clinically as anticancer agents to treat prostate cancer. On the other hand, agonists of the female hormone, estrogen, might also be effective in opposing the action of androgen agonists. Therefore, we designed AR–ER dual ligands using a *meta*-carborane skeleton (Figure 1.1.7). The *meta*-carborane BA812 and the *para*-carborane BA818 both exhibited potent androgen antagonist and estrogen

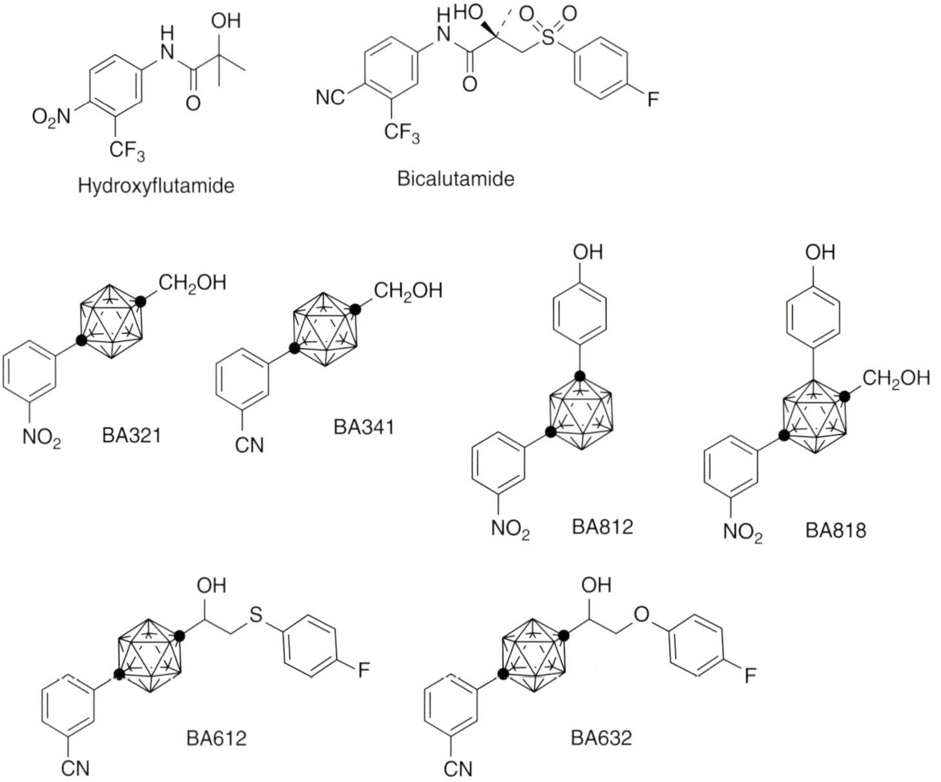

Figure 1.1.7 Structures of typical androgen receptor antagonists: hydroxyflutamide and bicalutamide, and designed molecules bearing carborane.

agonist activities (i.e., they are dual ligands). However, BA812 showed AR-antagonistic activity at the concentration of 1×10^{-6} M, and ER-agonistic activity at 1×10^{-8} M, whereas BA818 showed AR-antagonistic activity at 5×10^{-8} M, and ER agonistic activity at 5×10^{-7} M. Thus, BA812 is closer to an ER agonist, and BA818 is closer to an AR antagonist [24]. Recently, we reinvestigated the binding affinity of BA321 to ERα and found that this compound showed activity at 1×10^{-7} M. We also reported *in vivo* ER-agonistic activity of BA321 [25].

A common problem is that prostate cancer becomes resistant to anti-androgen therapy. One reason for this is mutation of threonine 877 to alanine (T877A) in the receptor [26]. Thus, although BA321 and BA341 show very strong inhibitory activity toward wild-type tumor cells, they work as agonists toward receptor-mutated cells. This is similar to the case of hydroxyflutamide. On the other hand, bicalutamide works as antagonist toward both wild-type and mutant cells. However, bicalutamide resistance has also been reported to occur. In the X-ray crystal structure of the mutant cells bound to bicalutamide [27], the cyano group binds to Gln711 and Arg752 through one molecule of water. Therefore, we designed carborane-containing molecules having another benzene ring as mimics of bicalutamide. Among the synthesized compounds, BA632 and BA612 strongly inhibited LNCap cells (a human prostate cancer cell line, including mutation T877A), which have a mutated receptor, at the concentration of 4×10^{-7} M [28]. Since BA632 and BA612 did not show any agonistic activity in functional assays, they seem to be pure AR full antagonists and are therefore candidates for treatment of anti-androgen withdrawal syndrome.

1.1.5 Retinoic Acid Receptor (RAR) and Retinoic Acid X Receptor (RXR) Ligands Bearing a Carborane Cage

1.1.5.1 RAR Agonists and Antagonists

Retinoids are used as therapeutic agents in the fields of dermatology and oncology. They modulate specific gene transcription through binding to the RARs. Retinoidal actions are also modulated by RXRs. RAR–RXR heterodimers bind to the RA-responsive element of DNA to regulate biological actions. Thus, the initial key step is the binding of retinoid to the ligand-binding domain of RAR. All-*trans*-retinoic acid is oriented in the ligand-binding pocket by two types of contact, hydrogen bonding at carboxylic acid and hydrophobic interaction. We investigated the use of a carborane cage as the hydrophobic region.

A synthetic aromatic retinoid, AM80, is 10 times more potent than the native RAR ligand, all-*trans*-retinoic acid (Figure 1.1.8) [29]. In both molecules, the distance between the important hydrogen-bonding carboxylic acid and the end of the bulky hydrophobic structure is the same: 14 angstroms. Therefore, we designed molecules with a diphenylamine skeleton, in which the geometry and distance from the hydrogen-bonding carboxylic acid to the hydrophobic region, carborane, resemble those of AM80 and retinoic acid. In biological activity assay, the synthetic compounds exhibited potent

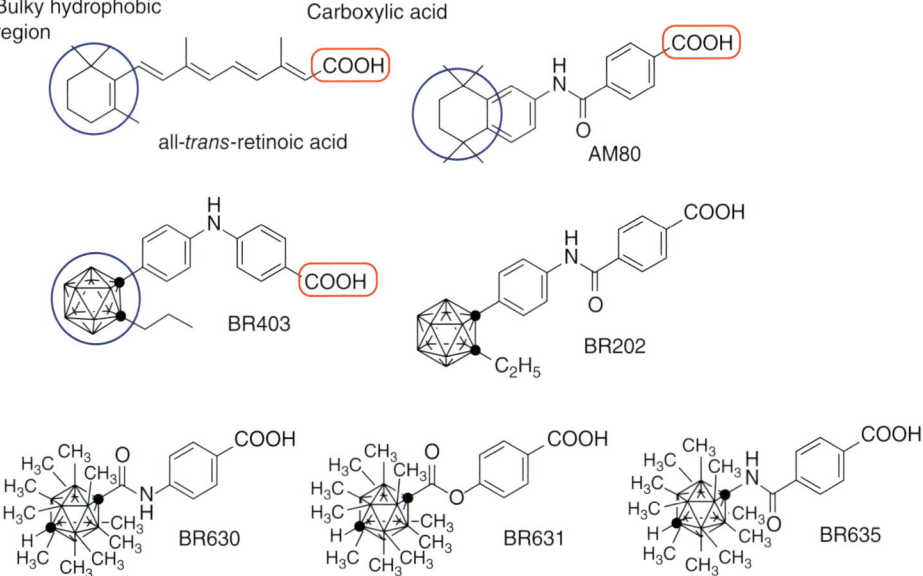

Figure 1.1.8 Structures of all-*trans*-retinoic acid and designed molecules bearing carborane as RAR ligands.

differentiation-inducing activity toward human leukemia (HL60) cells. The agonistic activity was increased by the introduction of a small alkyl group on the carborane cage. For example, the EC$_{50}$ value of the most active compound, BR403, is 10^{-9} M [30], which is comparable to that of all-*trans*-retinoic acid. On the other hand, the activity was completely abolished by the introduction of a methyl group on the nitrogen atom.

Introduction of a bulky substituent into an agonist molecule often transforms it into an antagonist, because a compound with a bulky substituent may bind to the receptor but induce a critical conformational change. The synthetic compound BR202 is an example of this. Polymethylcarboranes [31] may also have potential as a new bulky and hydrophobic unit for biologically active molecules, especially for antagonist design. We designed several polymethylcarborane-containing molecules based on the skeletal structure of AM80. Among them, BR630 showed strong retinoid antagonistic activity with an IC$_{50}$ value of 2×10^{-8} M. BR631 and BR635 were also antagonists. The polymethylcarboranyl group seems to serve effectively as a bulky and hydrophobic structure favoring the appearance of antagonistic activity [32].

1.1.5.2 RXR Agonists and Antagonists

The native ligand for RXRs, which modulate retinoidal actions, is 9-*cis*-retinoic acid. We designed and synthesized novel RXR-selective antagonists bearing a carborane moiety. The synthetic compound BR1211 (Figure 1.1.9) itself has no differentiation-inducing activity toward HL60 cells and does not inhibit the activity of RAR agonists. However, BR1211 inhibited the synergistic activity of an RXR agonist with AM80 in an HL60 cell differentiation-inducing assay [33]. Transactivation assay using RARs and RXRs suggested that the inhibitory activity of BR1211 resulted from selective antagonism at the RXR site of RXR–RAR heterodimers. BR1211 is a useful tool in the fields of embryology [34] and neurosciences [35,36].

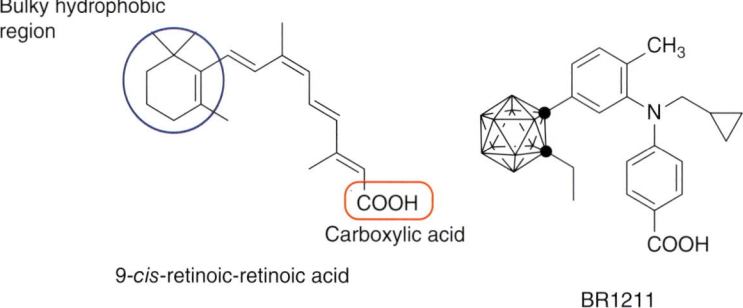

Figure 1.1.9 Structures of 9-*cis*-retinoic acid and designed molecule bearing carborane as RXR ligands.

1.1.6 Vitamin D Receptor Ligands Bearing a Carborane Cage

Vitamin D is involved in many physiological processes, including calcium homeostasis, bone metabolism, and cell proliferation and differentiation. Its actions are modulated by vitamin D receptor (VDR), which regulates the expression of specific target genes.

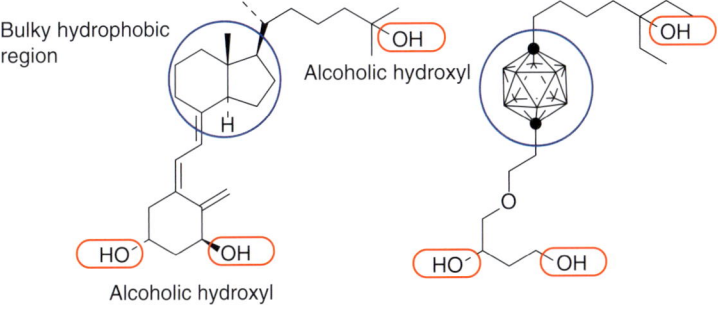

Figure 1.1.10 Structures of 1α,25-dihydroxyvitamin D₃ and designed molecule bearing carborane as VDR ligands.

VDR and its ligands have significant roles in the pathogenesis and therapy of osteoporosis, arthritis, psoriasis, and cancer.

The active form of vitamin D is 1α,25-dihydroxyvitamin D$_3$, which is a metabolically activated form of vitamin D$_3$. It binds to VDR by hydrogen bonding via hydroxyl groups at both ends of the molecule, and by hydrophobic interaction with the seco-steroidal skeleton. We designed a novel VDR agonist bearing *p*-carborane as a hydrophobic core structure (Figure 1.1.10). It exhibited moderate vitamin D activity, comparable to that of the native hormone, despite its simple and flexible structure. Effective hydrophobic interaction of the carborane cage with the hydrophobic region of VDR was confirmed by X-ray crystallography [37].

1.1.7 Determination of the Hydrophobicity Constant π for Carboranes and Quantitative Structure–Activity Relationships in ER Ligands

1.1.7.1 Determination of the Hydrophobicity Constant π for Carboranes

In connection with the application of carboranes in medicinal chemistry, we considered that all carboranes have essentially the same geometry and similar hydrophobic character. However, upon investigation of the ER-binding activity of various simple carboranylphenols, we found that *p*-carboran-1-ylphenol (BE100) exhibited very potent activity (as strong as that of estradiol). On the other hand, *o*-carboran-3-ylphenol showed 100 times weaker activity than *p*-carboran-1-ylphenol. In other words, the position of substitution on the carborane cage (*o*-, *m*-, and *p*-carboranes) affected the biological activities. Therefore, we quantitatively evaluated the hydrophobic character of various types of carboranes (Figure 1.1.11) [38] by determining the partition coefficients of their phenol derivatives, which exhibited strong binding affinity for ER. We measured the partition coefficients (log*P*) of carboranylphenols by means of a high-performance liquid chromatography (HPLC) method [39], because

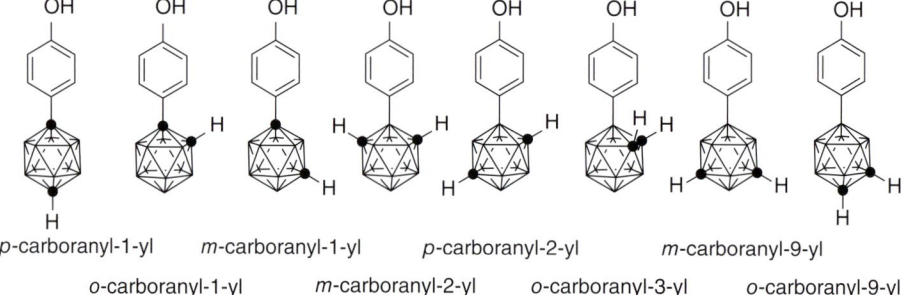

Figure 1.1.11 Structures of carboranylphenols used for determination of hydrophobic parameters.

Table 1.1.1 Partition constant logP, hydrophobic parameter π, pK_a, and binding affinity for ERα of carboranylphenols

Compound	logP	π	pK_a	IC_{50}
4-(o-carboran-9-yl)phenol	4.13	+2.71	10.25	3.91
4-(m-carboran-9-yl)phenol	4.62	+3.20	10.19	23.38
4-(o-carboran-3-yl)phenol	5.16	+3.74	9.61	71.61
4-(p-carboran-2-yl)phenol	5.46	+4.04	9.91	6.88
4-(m-carboran-2-yl)phenol	5.53	+4.11	9.67	9.57
4-(m-carboran-1-yl)phenol	5.71	+4.29	9.46	3.60
4-(o-carboran-1-yl)phenol	5.75	+4.33	9.16	2.85
4-(p-carboran-1-yl)phenol	5.89	+4.47	9.56	0.59

Note: Relative IC_{50} values based on competitive inhibition to 4 nM of ^{3}H-estradiol binding to ERα.

that is suitable for highly hydrophobic compounds in the range of logP above 4. We calculated the Hansch–Fujita hydrophobic parameters π of various carboranyl groups. The values varied from +2.71 to +4.47 (Table 1.1.1), which is within the same range as seen for hydrocarbons. p-Carboran-1-ylphenol was the most hydrophobic, and o-carboran-9-ylphenol was the most hydrophilic.

1.1.7.2 Quantitative Structure–Activity Relationships of Carboranylphenols with Estrogenic Activity

The carboranylphenols all have the same molecular geometry and do not show conformational variations, so they should be suitable for quantitative structure–activity study. We had anticipated that the principal factor influencing the binding affinity would be hydrophobicity, and this was broadly the case. However, two carboranylphenols with low hydrophobicity exhibited unexpectedly high binding affinity. The relationship

Figure 1.1.12 Correlation between log$P$ and ER-binding affinity of carboranylphenols.

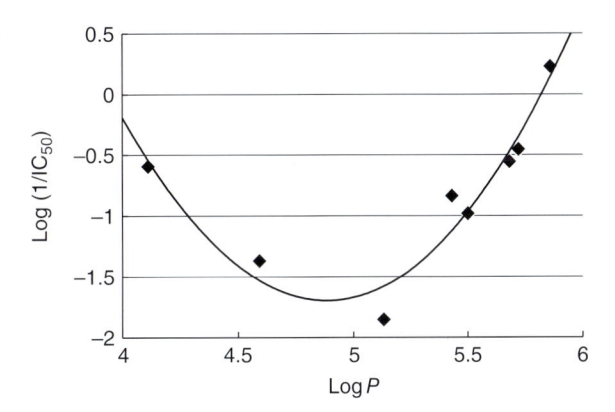

between log$P$ and the binding affinity was not first-order. Therefore, we fitted the data to the following parabolic equation (see Figure 1.1.12) by regression analysis.

$$\log(1/IC_{50}) = 1.93(\pm 0.276)(\log P)^2 - 18.9(\pm 2.76)(\log P) + 44.4(\pm 6.80)$$
$$(n = 8, R^2 = 0.923, s = 0.206)$$

This equation shows that the principal factor determining the binding affinity of carboranylphenols is hydrophobicity.

On the other hand, the acidity of the phenolic group may affect the binding affinity for ER through its influence on hydrogen bonding, in addition to the hydrophobicity of the carboranes. However, the acidity of the phenolic group pKa did not correlate with binding affinity. Thus, both log$P$ and pKa are key parameters for quantitative structure–activity relationship (QSAR) analysis in this system.

$$\log(1/IC_{50}) = 2.00(\pm 0.158)(\log P)^2 - 19.1(\pm 1.57)(\log P)$$
$$+ 0.779(\pm 2.30)(pK_a) + 36.3(\pm 4.54)$$
$$(n = 8, R^2 = 0.980, s = 0.117)$$

This equation has a very high regression coefficient ($R^2 = 0.980$) and reflects the major role of hydrophobicity, with some contribution of electronic factor [40]. The increase of the activity in the low log$P$ range can be explained by the difference in the natures of CH and BH in the carborane cage. The two carbon vertices bear relatively acidic protons and have hydrogen-bonding interactions with His524 of the receptor. In the past, extensive QSAR analysis of estrogens has been reported, but the relationships are not necessarily easy to interpret, because of the variety of skeletal structures and conformations. The present results may effectively represent the microscopic QSAR at the ligand-binding cavity of the receptor, free of the influences of membrane transportation and conformational factors.

1.1.8 Conclusion and Prospects

In conclusion, we have developed a variety of potent nuclear receptor ligands with carborane structures as hydrophobic moieties. These compounds have already applied as biological tools, and one of them, BE360, has antidepressant and antidementia effects in animals [19]. However, there are several issues that should be considered before clinical trials. One is the difference in the receptor interactions of hydrocarbons and carboranes. Another is to understand the metabolism and stability of carboranes in the living body. We have already conducted preliminary studies with a liver metabolic enzyme mixture (S-9), but extensive studies will be necessary. It is 17 years since we first applied carboranes for medicinal drug design, but the field is far from mature. Carboranes are novel structural fragments for drug design, and they appear to have great potential for use in the field in a wide variety of medicinal drug designs.

References

1 Textbook: See Patrick, G. L., An Introduction to Medicinal Chemistry 2nd ed., **2001**, Oxford University Press Inc., New York.
2 Ternansky, R. J.; Balogh, D. W.; Paquette, L. A. Dodecahedrane. *J. Am. Chem. Soc.* **1982**, *104*, 4503–4504. Paquette, L. A.; Ternansky, R. J.; Balogh, D. W.; Kentgen, G. Total synthesis of dodecahedrane. *J. Am. Chem. Soc.* **1983**, *105*, 5446–5450.
3 Kroto, H. W.; Heath, J. R.; O'Brien, S. C.; Curl, R. F.; Smalley, R. E. C_{60}: Buckminsterfullerene. *Nature* **1985**, *318*, 162–163.
4 Heying, T. L.; Ager, J. W.; Clark, S. L.; Mangold, D. J.; Goldstein, H. L.; Hillman, M.; Polak, R. J.; Szymanski, J. W. A New Series of Organoboranes. I. Carboranes from the Reaction of Decaborane with Acetylenic Compounds. *Inorg. Chem.* **1963**, *2*, 1089–1092.
5 Poater, J.; Sola, M.; Vinas, C.; Teixidor, F. π Aromaticity and Three-Dimensional Aromaticity: Two Sides of the Same Coin? *Angew. Chem. Int. Ed.* **2014**, *53*, 12191–12195.
6 For review, Soloway, A. H.; Tjarks, W.; Barnum, B. A.; Rong, F.-G.; Barth, R. F.; Codogni, I. M.; Wilson, J. G. The Chemistry of Neutron Capture Therapy. *Chem. Rev.* **1998**, *98*, 1515–1562.
7 For review, Scolz, M.; Hey-Hawkins, E. *Chem. Rev.* **2011**, *111*, 7035–7062., Olid, D.; Nunez, R.; Vinas, C.; Teixidor, F. Methods to Produce B–C, B–P, B–N and B–S Bonds in Boron Clusters. *Chem. Soc. Rev.* **2013**, *42*, 3318–3336.
8 Zakharkin, L. I.; Koveredou, A. I.; Ol'Shevskaya, V. A.; Shaugumbekova, S. Synthesis of *B*-organo-substituted 1,2-, 1,7-, and 1,12-Dicarba-closo- dodecarboranes(12). *J. Organomet. Chem.* **1982**, *226* 217–226., Zheng. Z.; Jiang. W.; Zinn, A. A.; Knobler, C. B.; Hawthorne, M. F. Facile Electrophilic Iodination of Icosahedral Carboranes. Synthesis of Carborane Derivatives with Boron-Carbon Bonds via the Palladium-Catalyzed Reaction of Diiodocarboranes with Grignard Reagents. *Inorg. Chem.* **1995**, *34*, 2095–2100., Jiang, W.; Knobler, C. B.; Curtis, C. E.; Mortimer, M. D.; Hawthorne, M. F. Iodination Reactions of Icosahedral *para*-Carborane and the Synthesis of Carborane Derivatives with Boron-Carbon Bonds. *Inorg. Chem.* **1995**, *34*, 3491–3498.

9 Brzozowski, A.M.; Pike, A.C.W.; Dauter, Z.; Hubbard, R.E.; Bonn, T.; Engstrom, O.; Ohman, L.; Greene, G.L.; Gustafsson, J.; Carlquist, M. Molecular Basis of Agonism and Antagonism in the Oestrogen Receptor. *Nature* **1997**, *389*, 753–758.

10 Endo, Y.; Iijima, T.; Yamakoshi, Y.; Yamaguchi, M.; Fukasawa, H.; Shudo, K. Potent Estrogenic Agonists Bearing Dicarba-closo-dodecaborane as a Hydrophobic Pharmacophore. *J. Med. Chem.* **1999**, *42*, 1501–1504.

11 Endo, Y.; Iijima, T.; Yamakoshi, Y.; Kubo, A.; Itai, A. Structure-activity Study of Estrogenic Agonist Bearing Dicarba-*closo*-dodecaborane. Effect of Geometry and Separation Distance of Hydroxyl Groups at the Ends of Molecules. *BioMed. Chem. Lett.*, **1999**, *9*, 3313–3318.

12 Endo, Y.; Iijima, T.; Yamakoshi, Y., Fukasawa, H.; Miyaura, C.; Inada, M.; Kubo, A.; Itai, A. Potent Estrogen Agonists Based on Carborane as a Hydrophobic Skeletal Structure. A New Medicinal Application of Boron Clusters. *Chem. Biol.* **2001**, *8*, 341–355.

13 Feigelson, H. S.; Henderson, B. E. Estrogens and Breast Cancer. *Carcinogenesis* **1996**, *17*, 2279–2284.

14 Shiau, A. K.; Barstad, D.; Loria, P. M.; Cheng, L.; Kushner, P. J.; Agard, D. A.; Greene, G. L. The Structural Basis of Estrogen Receptor/Coactivator Recognition and the Antagonism of This Interaction by Tamoxifen. *Cell* **1998**, *95*, 927–937.

15 Grese, T.A.; Sluka, J.P.; Bryant, H.U.; Cullinan, G.J.; Glasebrook, A.L.; Jones, C.D. Matsumoto, K.; Palkowitz, A.D.; Sato, M.; Termine, J.D.; Winter, M.A.; Yang, N.N..; Dodge, J.A. Molecular Determinants of Tissue Selectivity in Estrogen Receptor Modulators. *Proc. Natl. Acad. Sci. U.S.A.* **1997**, *94*, 14105–14110.

16 Endo, Y.; Yoshimi, T.; Iijima, T.; Yamakoshi, Y. Estrogenic Antagonists Bearing Dicarba-*closo*-dodecaborane as a Hydrophobic Pharmacophore. *BioMed. Chem. Lett.* **1999**, *9*, 3387–3392.

17 Endo, Y.; Yoshimi, T.; Miyaura, C. Boron Clusters for Medicinal Drug Design: Selective Estrogen Receptor Modulators Bearing Carborane. *Pure Appl. Chem.* **2003**, *75*, 1197–1205.

18 Hirata, M.; Inada, M.; Matsumoto, C.; Takita, M.; Ogawa, T.; Endo, Y.; Miyaura, C. A Novel Carborane Analog, BE360, with a Carbon-Containing Polyhedral Boron-cluster is a New Selective Estrogen Receptor Modulator for Bone. *Biochem. Biophys. Res. Commun.* **2009**, *380*, 218–222.

19 Endo, Y.; Yoshimi, T.; Ohta, K.; Suzuki, T.; Ohta, S. Potent Estrogen Receptor Ligands Based on Bisphenols with a Globular Hydrophobic Core. *J. Med. Chem.* **2005**, *48*, 3941–3944.

20 Nakagawasaia, O.; Nemoto, W.; Onogi, H.; Moriya, T.; Lin, J.-R. Odaira, T.; Yaoita, F.; Ogawa, T.; Ohta, K.; Endo, Y.; Tan-No, K. BE360, a New Selective Estrogen Receptor Modulator, Produces Antidepressant and Antidementia Effects through the Enhancement of Hippocampal Cell Proliferation in Olfactory Bulbectomized Mice. *Behav. Brain. Res.* **2016**, *297*, 315–322.

21 Sack, J. S.; Kish, K. F.; Wang, C.; Attar, R. M.; Kiefer, S. E.; An, Y.; Wu, G. Y.; Scheffler, J. E.; Salvati, M. E.; Krystek, S. R., Jr.; Weinmann, R.; Einspahr, H. M. Crystallographic Structures of the Ligand-binding Domains of the Androgen Receptor and Its T877A Mutant Complexed with the Natural Agonist Dihydrotestosterone. *Proc. Natl. Acad. Sci. U.S.A.* **2001**, *98*, 4904–4909.

22 Fujii, S.; Hashimoto, Y.; Suzuki, T.; Ohta, S.; Endo, Y. A New Class of Androgen Receptor Antagonists Bearing Carborane in Place of a Steroidal Skeleton. *Bioorg. Med. Chem. Lett.* **2005**, *15*, 227–230.

23 Fujii, S.; Goto, T.; Ohta, K.; Hashimoto, Y.; Suzuki, T.; Ohta, S.; Endo, Y. Potent Androgen Antagonists Based on Carborane as a Hydrophobic Core Structure. *J. Med. Chem.* **2005**, *48*, 4654–4662.

24 Unpublished results.

25 Watanabe, K.; Hirata, M.; Tominari, T.; Matsumoto, C.; Endo, Y.; Murphy, G.; Nagase, H.; Inada, M.; Miyaura, C. BA321, A Novel Carborane Analog That Binds to Androgen and Estrogen Receptors, Acts as a New Selective Androgen Receptor Modulator of Bone in Male Mice. *Biochem. Biophys. Res. Commun.* **2016**, *478*, 279–285.

26 Suzuki, H.; Akakura, K.; Komiya, A.; Aida, S.; Akimoto, S.;Shimazaki, J. Codon 877 Mutation in the Androgen Receptor Gene in Advanced Prostate Cancer: Relation to Antiandrogen Withdrawal Syndrome. *Prostate* **1996**, *29*, 153–158.

27 Bohl, C. E.; Gao, W.; Miller, D. D.; Bell, C. E.; Dalton, J. T. Structural Basis for Antagonism and Resistance of Bicalutamide in Prostate Cancer. *Proc. Natl. Acad. Sci. U.S.A.* **2005**, *102*, 6201–6206.

28 Goto, T.; Ohta, K.; Fujii, S.; Ohta, S.; Endo, Y. Design and Synthesis of Androgen Receptor Full Antagonists Bearing a *p*-Carborane Cage: Promising Ligands for Anti-Androgen Withdrawal Syndrome. *J. Med. Chem.* **2010**, *53*, 4917–4926.

29 Fukasawa, H.; Iijima, T.; Kagechika, H.; Hashimoto, Y.; Shudo, K. Expression of the Ligand-Binding Domain-Containing Region of Retinoic Acid Receptors Alpha, Beta and Gamma in *Escherichia coli* and Evaluation of Ligand-Binding Selectivity. *Biol. Pharm. Bull.*, **1993**, *16*, 343–348.

30 (a) Endo, Y.; Iijima, T.; Ohta, K.; Kagechika, H.; Kawachi, E.; Shudo, K. Dicarba-*closo*-dodecaboranes as a Pharmacophore. Novel Potent Retinoidal Agonists. *Chem. Pharm. Bull.* **1999**, *47*, 585–587. (b) Endo, Y; Iijima, T.; Yaguchi, K.; Kawachi, E.; Kagechika, H.; Kubo, A.; Itai, A. Structure–Activity Study of Retinoid Agonists Bearing Substituted Dicarba-*closo*-dodecaborane. Relation between Retinoidal Activity and Conformation of Two Aromatic Nuclei. *BioMed. Chem. Lett.* **2001**, *11*, 1307–1311.

31 Jiang, W.; Knobler, C. B.; Mortimer, M. D.; Hawthorne, M. F. A Camouflaged Icosahedral Carborane: Dodecamethyl-1,12-dicarba-*closo*-dodecaborane(12) and Related Compounds. *Angew. Chem., Int. Ed. Engl.* **1995**, *34*, 1332–1334.

32 Endo, Y.; Yaguchi, K.; Kawachi, E.; Kagechika, H. Polymethylcarborane as a Novel Bioactive Moiety: Derivatives with Potent Retinoid Antagonistic Activity. *BioMed. Chem. Lett.* **2000**, *10*, 1733–1736.

33 Ohta, K.; Iijima, T.; Kawachi, E.; Kagechika, H.; Endo, Y. Novel Retinoid X Receptor (RXR) Antagonists Having a Dicarba-*closo*-dodecaborane as a Hydrophobic Moiety. *BioMed. Chem. Lett.* **2004**, *14*, 5913–5918.

34 Calleja, C.; Messaddeq, N.; Chapellier, B.; Yang, H.; Krezel, W.; Li, M.; Metzger, D.; Marcrez, B.; Ohta, K.; Kagechika, H.; Endo, Y.; Mark, M.; Ghyselinck, N. B.; Chambon, P. Genetic and Pharmacological Evidence That a Retinoic Acid Cannot Be the RXR-Activating Ligand in Mouse Epidermis Keratinocytes. *Genes Dev.* **2006**, *20*, 1525–1538.

35 Wietrzych-Schindler, M.; Szyszka-Niagolov, M.; Ohta, K.; Endo, Y.; Pérez, E.; de Lera, A. R.; Chambon, P.; Krezel, W. Retinoid X Receptor Gamma Is Implicated in Docosahexaenoic Acid Modulation of Despair Behaviors and Working Memory in Mice. *Biol. Psychiatry* **2011**, *69*, 788–794.

36 Certoa, M.; Endo, Y.; Ohta, K.; Sakurada, S.; Bagetta, G.; Amantea, D. Activation of RXR/PPARγ Underlies Neuroprotection by Bexarotene Inischemic Stroke. *Pharmacol. Res.* **2015**, *102*, 298–307.

37 Fujii, S.; Masuno, H.; Taoda, Y.; Kano, A.; Wongmayura, A.; Nakabayashi, M.; Ito, N.; Shimizu, M.; Kawachi, E.; Hirano, T.; Endo, Y.; Tanatani, A,; Kagechika, H. Boron Cluster-based Development of Potent Nonsecosteroidal Vitamin D Receptor Ligands: Direct Observation of Hydrophobic Interaction between Protein Surface and Carborane. *J. Am. Che. Soc.* **2011**, *133*, 20933–20941.

38 Yamamoto, K.; Endo, Y. Utility of Boron Clusters for Drug Design. Hansch-Fujita Hydrophobic Parameters π of Dicarba-*closo*-dodecaboranyl Groups. *BioMed. Chem. Lett.* **2001**, *11*, 2389–2392.

39 OECD Guideline for Testing of Chemicals 117, Partition Coefficient (*n*-octanol/water), High Performance Liquid Chromatography (HPLC) Method, March **1989**.

40 Endo, Y.; Yamamoto, K.; Kagechika, H. Utility of Boron Clusters for Drug Design. Relation between Estrogen Receptor Binding Affinity and Hydrophobicity of Phenols Bearing Various Types of Carboranyl Groups. *BioMed. Chem. Lett.*, **2003**, *13*, 4089–4092.

1.2

Boron Cluster Modifications with Antiviral, Anticancer, and Modulation of Purinergic Receptors' Activities Based on Nucleoside Structures

Anna Adamska-Bartłomiejczyk,[1,§] Katarzyna Bednarska,[2] Magdalena Białek-Pietras,[1] Zofia M. Kiliańska,[3] Adam Mieczkowski,[4] Agnieszka B. Olejniczak,[5] Edyta Paradowska,[1] Mirosława Studzińska,[1] Zofia Sułowska,[2] Jolanta D. Żołnierczyk,[3] and Zbigniew J. Lesnikowski[1,]*

[1] *Laboratory of Molecular Virology and Biological Chemistry, IMB PAS, Łódź, Poland*
[2] *Laboratory of Experimental Immunology, IMB PAS, Łódź, Poland*
[3] *Department of Cytobiochemistry, Faculty of Biology and Environmental Protection, University of Łódź, Łódź, Poland*
[4] *Department of Biophysics, Institute of Biochemistry and Biophysics, PAS, Warsaw, Poland*
[5] *Screening Laboratory, IMB PAS, Łódź, Poland*
* Correspondence to: Laboratory of Molecular Virology and Biological Chemistry, IMB PAS, Lodowa 106, 93-232 Łódź, Poland. Email: zlesnikowski@cbm.pan.pl
§ *Present address: Department of Biomolecular Chemistry, Faculty of Medicine, Medical University of Lodz, Mazowiecka 6/8, 92-215 Lodz, Poland*

1.2.1 Introduction

Nucleoside analogs have been in clinical use for several decades and have become cornerstones of treatment for patients with cancer or viral infections [1,2]. This is complemented with nucleoside antibiotics, a large family of microbial natural products and synthetic derivatives derived from nucleosides and nucleotides [3].

The approval of several new nucleoside drugs over the past decade demonstrates that this class of compounds still possesses strong potential [1,2]. The potential of nucleosides in chemotherapy is enhanced by development of new chemistries for nucleoside modification, better understanding of molecular mechanisms of nucleoside drugs' actions [4], and pro-drug technology [5,6]. One of the new developments in the medicinal chemistry of nucleosides is nucleoside derivatives comprising a boron component [7]. The boron part can contain a single boron atom [8] or several boron atoms in the form of a boron cluster (Figure 1.2.1) [9–11].

Boron-containing nucleosides were originally designed as prospective boron carriers for boron neutron capture therapy (BNCT) of tumors [10]. As boron-rich donors in boron-carrying molecules, dicarba-*closo*-dodecaboranes ($C_2B_{10}H_{12}$) (**1–3**) are frequently used due to their chemical and biological stability and physicochemical versatility. More recently, dodecaborate [$(B_{12}H_{12})^{2-}$] (**4**) and metallacarboranes such as 3-cobalt-bis(1,2-dicarbollide)ate [$Co(C_2B_9H_{11})_2^{2-}$] (**4**) (Figure 1.2.1), complexes of carboranes and

Boron-Based Compounds: Potential and Emerging Applications in Medicine, First Edition.
Edited by Evamarie Hey-Hawkins and Clara Viñas Teixidor.
© 2018 John Wiley & Sons Ltd. Published 2018 by John Wiley & Sons Ltd.

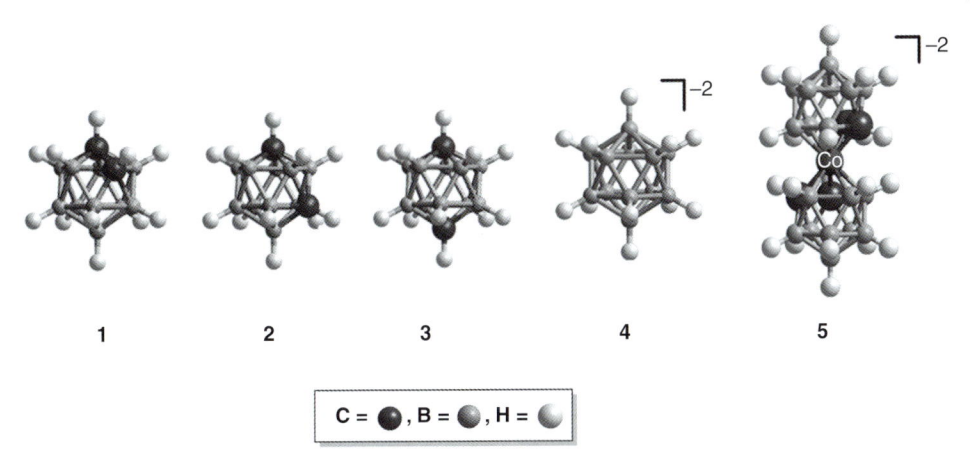

Figure 1.2.1 Examples of boron clusters used in medicinal chemistry: dicarba-*closo*-dodecaborane (carborane) isomers *ortho*- (**1**), *meta*- (**2**), *para*- (**3**), and ($C_2B_{10}H_{12}$) *closo*-dodecaborate ($B_{12}H_{12}^{2-}$) (**4**), and 3-cobalt-*bis*(1,2-dicarbollide)ate (**5**).

metal ions, are also attracting attention of medicinal chemists [11]. In this chapter, new applications and biological activities of nucleoside–boron cluster conjugates beyond BNCT will be highlighted.

1.2.2 Boron Clusters as Tools in Medicinal Chemistry

Boron is an element that is generally not observed in the human body (its content does not exceed 18 mg in an average individual [12]); however, it possesses considerable potential for the facilitation of new biological activity and for use in pharmaceutical drug design. There are two types of boron-containing bioactive molecules: the first contain a single boron atom, while in the second boron is present in the form of a boron cluster. The principle for biological activity of these molecules due to the presence of boron is different for each compound type. In the compounds containing a single boron atom, the ability of boron to readily convert from a neutral and trigonal planar sp2 form to an anionic tetrahedral sp3-hybridized form under physiological conditions is utilized. This feature provides the basis for the use of boron as a mimic of carbon-based transition states and in the design of inhibitors of various enzyme-catalyzed hydrolytic processes [13,14]. In boron cluster–containing compounds, the properties are used: not those of a single, separate boron atom, but rather the features of the boron atoms in the cluster as a whole, and the ability of the cage to elicit unique interactions with protein targets [11,15,16].

The properties of boron clusters that are useful in drug design include: (1) the ability to form unique noncovalent interactions, including dihydrogen bond formation due to the hydridic character of H atoms, σ-hole bonding, and ionic interactions – as a result, the types of interactions of boron cluster–containing compounds with biological targets may differ from those of purely organic molecules [17–19]; (2) spherical or ellipsoidal geometry and rigid three-dimensional (3D) arrangement: these offer versatile platforms for 3D molecular construction; (3) lipophilicity, amphiphilicity, or hydrophilicity: these qualities depend on the type of boron cluster used, which allows the tuning of pharmacokinetics and bioavailability; (4) chemical stability and simultaneous susceptibility to functionalization; (5) bioorthogonality, stability in biological environments, and a decreased susceptibility to metabolism; (6) high boron content, which is important for BNCT; and (7) resistance to ionizing radiation, a feature that is important for the design of radiopharmaceutical agents [11,14].

Finally, the fact that polyhedral boron hydrides are manmade molecules and are unfamiliar to life has additional potential advantages. This is because the active substances bearing boron cluster modification are less likely to be prone to the development of resistance and are expected to be more stable in biological systems compared to carbon-based molecules. While pathogens, such as bacteria and viruses, are eventually capable of evolving resistance against almost any molecule that attacks them, one could hypothesize that this process would take longer for boron-based compounds.

Technically, two major avenues in the search for new biologically active molecules containing boron clusters are exploited. The first is based on the modification of natural products, previously identified molecular tools, or clinically used drugs with the hope of finding compounds with improved biological or pharmacokinetic properties, for example pro-drugs aciclovir (ACV), cidofovir (CDV), ganciclovir (GCV), and Tamiflu® (discussed in this chapter) (Figure 1.2.2) and several other clinically used drugs modified with boron clusters [11]. The second approach focuses on screening available collections of compounds in search for novel structures with desired activities and properties.

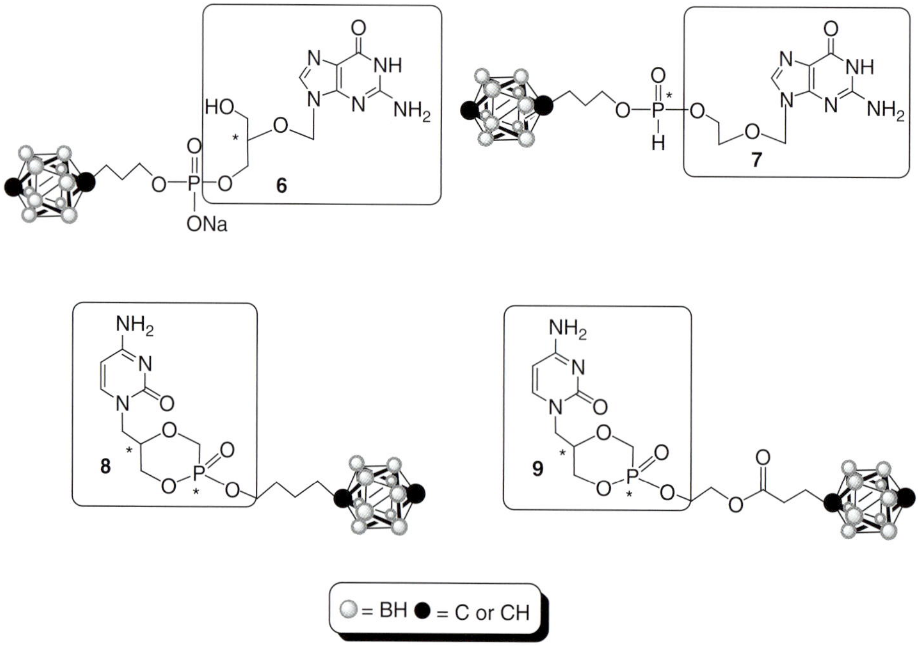

Figure 1.2.2 Anti-HCMV boron cluster prodrugs: ganciclovir (GCV) (**6**), acyclovir (ACV) (**7**), and cidofovir (CDV) (**8, 9**); the parent drug fragment is marked with a frame, and the center of chirality is marked with a star.

This approach is hampered currently by the lack of easily available high-throughput screening (HTS) libraries of boron cluster–containing compounds. A third approach, based on rational drug design supported with *in silico* methods, still needs to be developed for boron-based drugs to match the level of methods that are available for purely organic molecules. Progress in the computational chemistry of boron clusters is discussed in some other chapters of this book.

Boron is not a solution for every drug discovery problem, but there is a good chance that it will become a useful addition to the medicinal chemistry toolbox. The current status of boron in medicinal chemistry resembles that of fluorine three decades ago; now, fluorinated compounds are synthesized in pharmaceutical research on a routine basis, and comprise a substantial fraction of pharmaceuticals on the market. The great variety of biological activities of molecules and drug analogs bearing a single boron atom or boron clusters demonstrates that there is immense potential in boron medicinal chemistry that is awaiting further exploration.

1.2.3 Modification of Selected Antiviral Drugs with Lipophilic Boron Cluster Modulators and New Antiviral Nucleosides Bearing Boron Clusters

Anti–infectious disease drugs bearing an essential boron component form an area of medicinal chemistry still awaiting exploration. Potent antiviral [1] and anticancer activity [2,4], demonstrated by the number of nucleoside analogs with modified sugar and/or nucleobase residues, encouraged synthesis of the carborane derivatives of these molecules. In this direction, various modified sugar residues were introduced to the carborane-containing nucleosides, and the obtained derivatives were tested for antiviral and anticancer activity [20]. These modifications include, among others, oxathialane carboranyl uridines [21] β-D, α-D, β-L, and α-L of 5-carboranyl-2′,3′-didehydro-2′,3′-dideoxyuridine (D4CU) [22]; 5-*o*-carboranyl-1-(2-deoxy-2-fluoro-β-D-arabinosyl)uracil (CFAU) and its α-isomer [23]; 5-carboranyl-1-(β-D-xylofuranosyl)uracil [24]; and 5-*o*-carboranyl-2′,3′-dideoxy-2′-(phenylthio)uracil and its *nido*-form [25]. These early works did not, however, produce compounds with improved properties; in contrast, the boron cluster derivatives often expressed lower antiviral activity than the original compounds.

As of April 2016, antiviral drugs have been approved to treat nine human infectious diseases (human immunodeficiency virus [HIV], hepatitis B [HBV] and C virus [HCV], human cytomegalovirus [HCMV], herpes simplex virus [HSV], human papillomavirus [HPV], respiratory syncytial virus [RSV], varicella zoster virus [VZV], and influenza virus) [1,26]. Among them, five are targeted against DNA viruses (HBV, HCMV, HSV, HPV, and VZV), three against RNA viruses (HCV, RSV, and influenza virus), and one against a retrovirus (HIV). Herein, we focus on drugs against HCMV, belonging to the *Herpesviridae* family.

The seroprevalence of HCMV infection ranges from 45 to 95% in worldwide populations depending on country and socioeconomic status of the individual [27]. The infection is spread by contact with body fluids (blood, saliva, urine, or breast milk), by organ transplants, by placenta (transmission to the fetus during pregnancy), or by sexual contact. In hosts with impaired immune functions, such as transplant recipients, persons infected with HIV, or those undergoing anticancer chemo- and/or radiotherapy, the full pathogenic potential of the virus may be realized. Fever, pneumonia, diarrhea, and ulcers in the digestive tract (possibly causing bleeding) are the most common complications; others include hepatitis, inflammation of the brain (encephalitis), behavioral changes, coma, seizures, and visual impairment and blindness. Hence, HCMV is a recognized cause of morbidity and mortality in immunocompromised individuals.

Cytomegalovirus infection also may be acquired prenatally or perinatally. HCMV is the most common cause of viral intrauterine infection, affecting from 0.4% to 2.3% of live-born infants. Severe infections may occur among congenitally infected fetuses and infants due to immaturity of the immune system in neonates [28]. Although most infants with HCMV infection are asymptomatic, about 10% of infants with congenital HCMV infection are symptomatic at birth, including intrauterine growth restriction, microcephaly, jaundice, petechiae, hepatosplenomegaly, periventricular calcifications, chorioretinitis, pneumonitis, and hepatitis. Some babies without signs at birth may develop later long-term health problems, such as sensorineural hearing loss, visual impairment or blindness, intellectual disability, or dyspraxia. Antiviral treatment may decrease the risk of health problems in some neonates with symptomatic HCMV infection at birth.

More recently, effects of persistent HCMV infection on immunosenescence and association of HCMV with cardiovascular diseases such as hypertension and atherosclerosis focus attention on, and unveil another face of HCMV [29].

The current anti-HCMV treatments, in addition to foscarnet (known for many years), include ganciclovir, valganciclovir, cidofovir, the experimental drug maribavir, and the off-label use of leflunomide. Immune globulin against cytomegalovirus is also used, usually in combination with ganciclovir to treat HCMV pneumonia. High doses of the anti-HSV drug acyclovir have also been used as prophylactic against HCMV with some success. Fomivirsen, an antisense oligonucleotide, was approved for HCMV retinitis as an intravitreal treatment in 1996, but the latter was withdrawn from the market [30].

While currently available systemic anti-HCMV agents are effective against the virus, their use is limited by toxicities, most notably bone marrow suppression, renal impairment, hematologic effects, nephrotoxicity, and neutropenia. The limited number of anti-HCMV drugs, the side effects, and the development of resistance create a need for a new generation of more efficient and safer drugs against this opportunistic and widely spread pathogen.

To approach these limitations and to recognize the potential of boron clusters as modulators of known antiviral drugs, methods for the synthesis of ganciclovir, acyclovir, and cidofovir phosphates modified with boron clusters have been developed [31]. All of the anti-HCMV drugs belong to an acyclic family of nucleoside analogs.

The target ganciclovir phosphate modified with *para*-carborane cluster **6** was obtained in a four-step procedure involving: (1) protection of 6 N amino and hydroxyl functions of ganciclovir, (2) phosphonylation/phosphorylation, (3) boron cluster addition, and (4) removal of the protecting group. The acyclovir phosphonate modified with *para*-carborane cluster **7** was obtained in a simple two-step procedure based on transformation of acyclovir into the corresponding monoester of H-phosphonic acid, then esterification of the resultant intermediate with 1-(3-hydroxypropyl)-*para*-carborane, an alcohol-bearing boron cluster.

Cidofovir **8** modified with a *para*-carborane cluster attached through an alkyl linker was obtained in a convenient, simple, two-step procedure involving transformation of cidofovir in acyclic form into cyclic derivative, which was next converted into cyclic triester **8** in the reaction with 1-(3-bromopropyl)-*para*-carborane used as a boron cluster donor. Compound **9** containing a *para*-carborane cluster attached through a linker equipped with carboxylic acid ester function is another example of cidofovir modification with a boron cluster. It was obtained in an esterification reaction between cyclic cidofovir and the ethylene glycol-(3-*para*-carboranyl)propionic acid ester using a suitable condensing agent.

The cytotoxicity assays performed on confluent monolayers of MRC-5, Vero, L929, LLC-MK2, and A549 cells revealed no effect on growth of cells at concentration up to 1000 µM ($CC_{50} > 1000$ µM). No major differences between cytotoxicity for compounds **6–9** and unmodified GCV, ACV, and CDV were observed.

The derivatives **6–9** were then tested, first with an HCMV plaque reduction assay in comparison with GCV, ACV, and CDV, respectively. The GCV phosphate **6** exhibited a similar degree of anti-HCMV activity as unmodified GCV; the compounds **7** and **8** showed lower antiviral activity against HCMV compared with ACV and CDV, respectively. Interestingly, CDV pro-drug **9** was almost 200-fold more active against HCMV than derivative **8**. The compounds **6** and **7** were also effective in inhibiting the HSV1 replication. GCV, ACV, and CDV phosphates modified with *para*-carborane cluster **6–9** were not inhibitors of HPIV3 and VSV replication even at a concentration of 1000 µM.

In a search for new nucleoside entities with antiviral activities, several methods for the synthesis of boron cluster derivatives of pyrimidine nucleosides have been developed. The methods employed include: (1) attachment of the carborane moiety at the 2′ position of the nucleoside via a formacetal linkage (compound **10**) [32]; (2) tethering of the *nido*-carborane or metallacarborane group to the nucleobase via dioxane ring opening of the oxonium derivative of the boron cluster (compound **11**) [33,34]; (3) Sonogashira coupling between 5-iodo-uridine or 5-iodo-2′-deoxyuridine and terminal alkynes bearing boron clusters (compounds **12, 13**), followed by intramolecular cyclization (compound **14**) [35]; (4) a "click chemistry" approach based on a modified Huisgen 1,3-dipolar cycloaddition (compounds **15–16**) [34,36]; and (5) *de novo* formation of a metallacarborane complex via the reaction of a nucleoside bearing a boron cluster ligand with the metal (rhenium(I)) tricarbonyl complex (compound **17**) [37,38] (Figure 1.2.3).

The obtained libraries of compounds were tested for cytotoxicity in MRC-5, A549, LLC-MK2, L929, and Vero cell lines, and for antiviral activity against a panel of DNA and RNA viruses (HCMV, HSV1, human parainfluenza virus-3 [HPIV3], encephalomyocarditis virus [EMCV], and VSV). The most potent novel compound identified was 5-[(1,12-dicarba-*closo*-dodecaboran-2-yl)ethyn-1-yl]-2′-deoxyuridine (**10**), with an IC_{50} value of 5.5 µM and a selectivity index higher than 180 [35]. It is unusual that it exhibited antiviral activity against HCMV but was not active against HSV1, another virus belonging to the *Herpesviridae* family; it was also inactive toward HPIV3 and EMCV.

A series of 5-ethynyl-2′-deoxyuridines [39] and its cyclic derivatives [40], with *closo*-dodecaborate and cobalt-bis-dicarbollide boron clusters as well as *closo*-dodecaborate conjugates with non-natural adenosine analog 7-deaza-8-aza-2′-deoxyadenosine [41], have also been synthesized. The cytotoxicity of the obtained compounds in several cell lines and antiviral activity against HCMV, HSV1, HPIV3, EMC, and VSV were tested. Most of the designed compounds have shown low cytotoxicity and low or moderate antiviral activity.

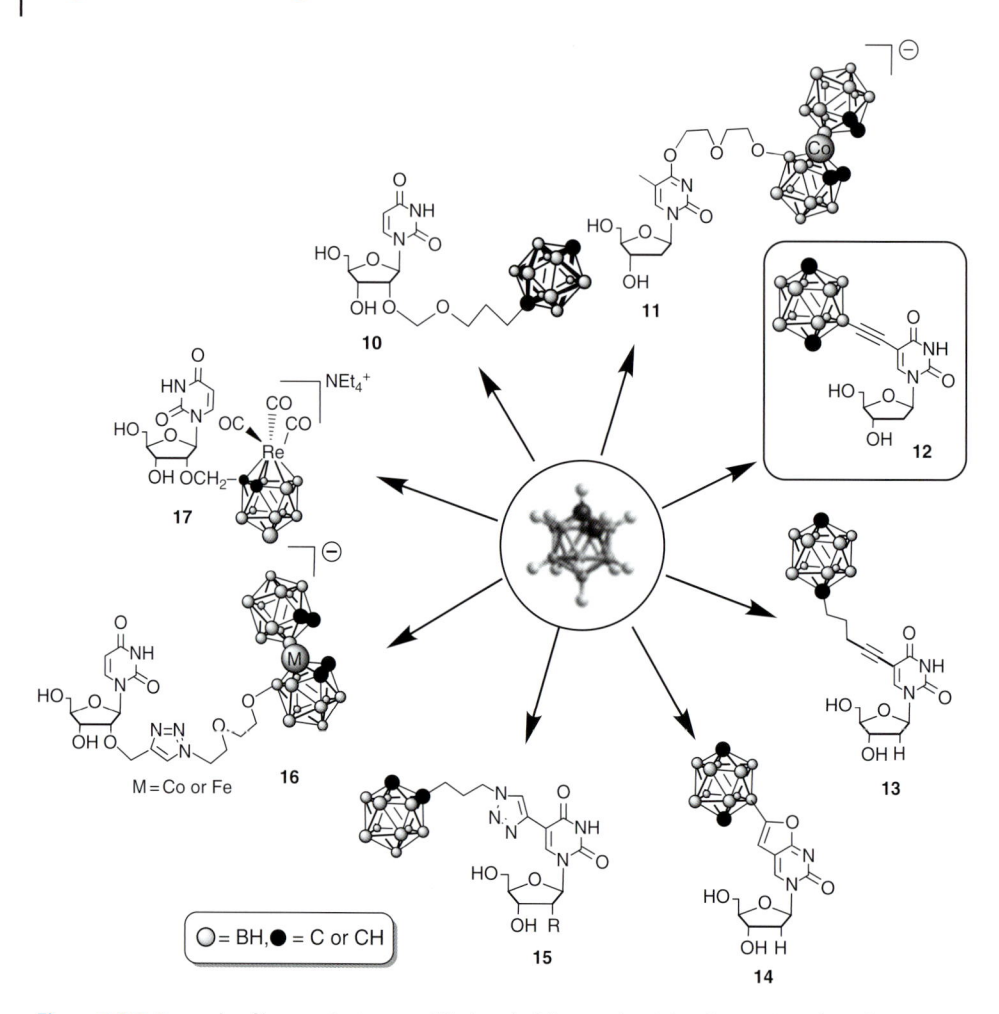

Figure 1.2.3 Example of boron cluster–modified pyrimidine nucleosides. *Source*: Lesnikowski et al. (2016, pp. 35–38) [61].

1.2.4 *In Vitro* Antileukemic Activity of Adenosine Derivatives Bearing Boron Cluster Modification

Chronic lymphocytic leukemia (CLL) is the most common type of adult leukemia in western countries. CLL cells are typically B cells in origin, with the T-cell variant of the disease occurring rarely [42,43]. It is characterized by a highly heterogeneous course and does not always require chemotherapy [43,44]. Nevertheless, it is thought to be incurable since relapse usually takes place after an initial remission. CLL in turn may transform into more aggressive lymphomas (Richter's syndrome) or leukemias (e.g., prolymphocytic leukemia [PLL]).

Although purine nucleoside analogs routinely used in CLL treatment (cladribine, fludarabine) have notably improved progression-free survival (PFS), overall survival

(OS) was not altered significantly. Among the anticancer drugs used to treat these leukemias, the majority are nucleoside derivatives applied alone or in combinations with other drugs and antibodies [45–47]. Examples of medicinal preparations based on nucleoside structure include cytarabine (cytosine arabinoside, ara-C, or Cytosar®), cladribine (2-chloro-2′-deoxyadenosine, or Litak®), clofarabine (2-chloro-2′-deoxy-2′-fluoroadenosine, or Clolar®), and fludarabine (2-fluoro-arabinoadenosine 5′-phosphate/F-ara-AMP, or Fludara®); the last three are adenosine derivatives [47,48].

At present, the gold standard for CLL therapy is the combination of purine nucleoside analog fludarabine, cyclophosphamide (an alkylating agent), and rituximab (FCR), or cladribine, cyclophosphamide, and rituximab (CCR) [45,46]. In patients not eligible for such intensive treatment because of comorbidities, less effective drugs might be applied. Thus, there is still a strong need to identify new targets for antileukemic chemotherapy and to develop novel anticancer compounds and treatment strategies. Among new approaches assigned for CLL, much attention is paid to agents with the ability to turn on the apoptosis process [49–51].

We proposed the use of carboranes for modification of selected nucleoside designs. In this direction, libraries of modified adenosine, 2′-deoxyadenosine, and arabinoadenosine derivatives were synthesized, and their cytotoxicity and proapoptotic potential against peripheral blood mononuclear cells (PBMCs) of CLL and PLL patients, as well as those of healthy volunteers, were investigated.

The boron cluster–modified purine nucleosides were prepared according to the following methods: (1) Sonogashira-type cross-coupling between 2- or 8-halogenoadenosine and a boron cluster donor containing a triple bond (compounds **19, 20**, and **26**) [34,52], (2) substitution of chlorine at carbon 6 of adenosine with alkylamine containing a boron cluster (compound **21**) [52], (3) amide bond formation between exo-amine group and boron cluster carboxylic acid (**22**), (4) tethering of the *nido*-carborane to the nucleobase via dioxane ring opening of the oxonium derivative of the boron cluster (compound **23**) [34], (5) a "click chemistry" approach based on a modified Huisgen 1,3-dipolar cycloaddition (compound **24**) [34,36], and (6) attachment of the carborane moiety at the 2′ position of the nucleoside via a formacetal linkage (compound **25**) [32].

The synthesized libraries of boron clusters containing adenosine derivatives were evaluated for their cytotoxicity against primary peripheral mononuclear cells from the blood of patients with leukemias (CLL or PLL). It was found that modification of adenosine with a boron cluster at C-2 (compound **19**) and C-8 (compound **26**) of the purine ring leads to products with high cytotoxicity against CLL and PLL cells at 15–20 µM, whereas normal cells of healthy donors are resistant to this concentration [53]. Interestingly, the adenosine derivatives bearing a boron cluster at C-2′ of the sugar residue and at the amino group at C-6 of the nucleobase are cytotoxic to both cell types or show some small activity in leukemic and normal cells, respectively. Complementary biological studies demonstrated that compounds **19** and **26** possess (similarly as clinically used antileukemic drugs cladribine and fludarabine) high proapoptotic potential. The obtained data revealed also that compounds **19** and **26** are potent activators of caspase 3, the enzyme that is involved in proteolysis of numerous proteins (e.g., PARP1 and Mcl1) and to activate a key endonuclease, DNA fragmentation factor/caspase-activated DNase (DFF/CAD), whose functions are relevant for the formation of the apoptotic ladder [53]. The observed characteristics of these compounds might be of clinical importance.

1.2.5 Adenosine–Boron Cluster Conjugates as Prospective Modulators of Purinergic Receptor Activity

Adenosine receptors (AR; also called purinergic P1 receptors) are members of the G-protein family of receptors, which also include many well-known receptors such as dopamine receptors, adrenergic receptors, histamine receptors, and serotonin receptors. They are responsible for the transduction of a diverse array of signals into the cells by activating one or more heterotrimeric G proteins located on the cytoplasmic face of the plasma membrane and subsequently interacting with the effector system, including ion channels, phospholipases, and adenylate cyclase [54]. Similar to other G protein-coupled receptors, ARs consist of seven transmembrane helices, which accommodate the binding site for ligands. Each helix is constituted by approximately 21 to 28 amino acids. The transmembrane helices are connected by three extracellular and three cytoplasmic loops of unequal size of amino acid chain. The N-terminal is on the extracellular side and the C-terminal on the cytoplasmic side of the membrane [55]. The development of potent and selective synthetic agonists and antagonists of ARs has been the subject of medicinal chemistry research for more than three decades. In addition, allosteric enhancers of agonist action could allow the effects of endogenous adenosine to be selectively magnified in an event-responsive and temporally specific manner, which might have therapeutic advantages compared with agonists. AR action might also be modulated not only by direct-acting ligands, but also by inhibition of the metabolism of extracellular adenosine or its cellular uptake [56].

Although selective AR modulators have promise for numerous therapeutic applications, in practice this goal, as in the case of many other targets, is not easy to achieve. One reason for this is the ubiquity of ARs and the possibility of side effects. In addition, species differences in the affinity of putatively selective ligands complicate preclinical testing in animal models. As a result of continuous approaches to solve these problems, studies of the modulation of cellular function by purines have increased exponentially over the past decades as a large family of discrete molecular targets has been identified using molecular biological and pharmacological tools [57]. In consequence of these efforts, several pharmacological compounds specifically targeting individual ARs – either directly or indirectly – have now entered the clinic. However, only one AR-specific agent – the adenosine A2A receptor agonist regadenoson (Lexiscan; Astellas Pharma) – has so far gained approval from the US Food and Drug Administration (FDA) [58]. Therefore, there is a long way ahead to explore fully the potential of ARs as targets for clinically useful molecules. We hope that works on the chemistry and biology of innovative adenosine modification bearing boron clusters will contribute to this endeavor.

In a search for new types of AR modulators, a series of adenosine derivatives bearing electroneutral 1,12-dicarba-*closo*-dodecaborane (**3**) or a metallacarborane complex, 3-cobalt-*bis*(1,2-dicarbollide)ate (**5**), were synthesized. Modifications of the nucleobase at the 6-N, 2-C, or 8-C positions, or the sugar residue at the 2′-C position, with suitable boron clusters tethered via different types of linkers were prepared (Figure 1.2.4). Taking the advantage that the A2A AR has already been recognized as a mediator of adenosine-dependent effects on platelet aggregation, the obtained library of adenosine derivatives was tested for the effects of these modifications on the functions of

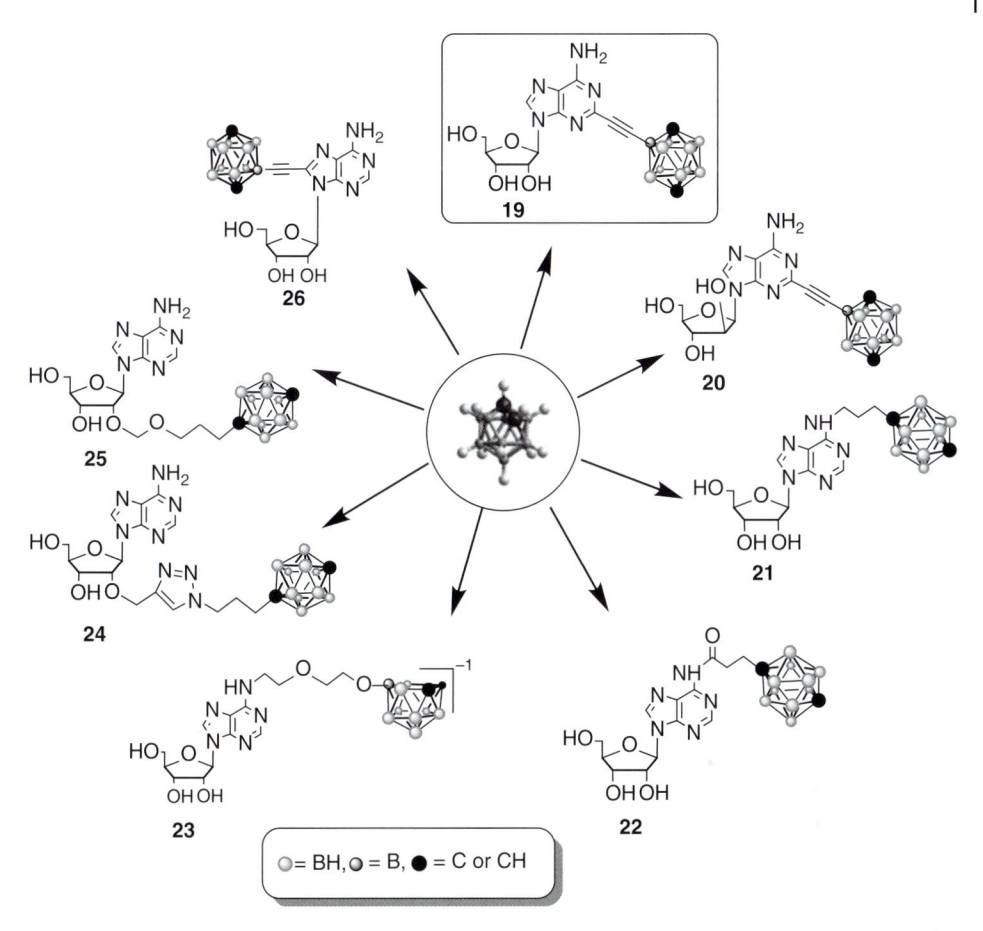

Figure 1.2.4 Examples of boron cluster–modified adenosine (purine nucleoside). *Source*: Lesnikowski et al. (2016, pp. 35–38) [61].

stimulated platelets [59] and, in addition, on the production of reactive oxygen species (ROS) by human neutrophils [52]. The modification of adenosine at the 2′-C position with 1,12-dicarba-*closo*-dodecaborane (**25**) was shown to result in efficient inhibition of platelet aggregation induced by thrombin or adenosine diphosphate (ADP). Remarkably, inhibition of ROS production in activated neutrophils by adenosine modified at the 6-N position (**21**) was also observed. The high affinities of the selected compounds for AR A_{2A} were established [52], which may suggest the possible involvement of the A_{2A} receptor in the observed biological activities of the adenosine–boron cluster conjugates. It was found that compounds **21** and **25** displaced a specific receptor A_{2A} agonist [3H]CGS 21680 in a radioligand binding assay with inhibitory constants (Ki) of 9.9 nM and 1.04 nM, respectively, and displayed 3- and 26-fold higher affinities for the A_{2A} receptor than CGS 21680 did, respectively [52].

These preliminary findings, and the new chemistry proposed, form the basis for the development of a new class of adenosine analogs that can modulate human blood platelet activities and extend the range of innovative molecules available for testing as agents that affect inflammatory processes and prospective modulators of AR activity. Described recently, another example of purinergic receptor targeting with boron cluster–containing ligands, such as of 1-[(5-methoxy-2-chlorobenzamido)methyl]-1,2-carborane or 10-[(5-methoxy-2-chlorobenzamido)methyl]-7,8-*nido*-carborane, further illustrates the potential of boron cluster–containing compounds as purinergic receptors' modulators [60].

1.2.6 Summary

Chemotherapeutic nucleosides are chemically modified analogs (antimetabolites) of natural nucleosides that are endogenous metabolites involved in many essential cellular processes, such as DNA and RNA synthesis, and in purinergic signaling. The modifications used in the synthesis of chemotherapeutic nucleosides can be divided generally into two classes: (1) structural changes of the canonical nucleoside structure (e.g., acyclic nucleosides, nucleobase-modified analogs such as azanucleosides, nucleosides with a truncated nucleobase, and sugar-modified analogs such as arabino-, L-, or carbonucleosides) and (2) chemically changed analogs (e.g., iodinated, fluorinated, or chlorinated derivatives; thianucleosides; and lipophilic analogs with long, alkyl chain substituent). Often, both types of modifications are applied in the same molecule. Modification of nucleosides with a boron cluster belongs to the second class of nucleoside derivatives. Exploration of this class of nucleosides as potentially chemotherapeutic is in the early stage; however, the great variety of biological activities of molecules and drug analogs bearing boron clusters discovered so far [7,11,14–16] demonstrates that there is immense potential in boron cluster medicinal chemistry that is awaiting further exploration.

Acknowledgments

This work was supported in part by the National Science Centre, Poland, grant 2014/13/B/NZ1/03989; contributions from the Statutory Fund of IMB PAS are also gratefully acknowledged. The equipment used was sponsored in part by the Centre for Preclinical Research and Technology (CePT), a project cosponsored by European Regional Development Fund and Innovative Economy, The National Cohesion Strategy of Poland.

References

1 E. De Clercq, G. Li. Approved antiviral drugs over the past 50 years. *Clin. Microbiol. Rev.*, **2016**, *29*, 695–747.
2 L.P. Jordheim, D. Durantel, F. Zoulim, C. Dumontet. Advances in the development of nucleoside and nucleotide analogues for cancer and viral diseases. *Nat. Rev. Drug Discov.*, **2013**, *12*, 447–464.
3 G. Niu, H. Tan. Nucleoside antibiotics: biosynthesis, regulation, and biotechnology. *Trends Microbiol.*, **2015**, *23*, 110–119.

4 J. Shelton, X. Lu, J.A. Hollenbaugh, J. Hyun Cho, F. Amblard, R.F. Schinazi. Metabolism, biochemical actions, and chemical synthesis of anticancer nucleosides, nucleotides, and base analogs. *Chem. Rev.*, **2016**, doi:10.1021/acs.chemrev.6b00209.
5 F. Li, H. Maag, T. Alfredson. Prodrugs of nucleoside analogues for improved oral absorption and tissue targeting. *J. Pharm. Sci.*, **2008**, *97*, 1109–1134.
6 U. Pradere, E.C. Garnier-Amblard, S.J. Coats, F. Amblard, R.F. Schinazi. Synthesis of nucleoside phosphate and phosphonate prodrugs. *Chem. Rev.*, **2014**, *114*, 9154–9218.
7 A.A. Druzina, V.I. Bregadze, A.F. Mironov, A.A. Semioshkin. Synthesis of conjugates of polyhedral boron hydrides with nucleosides. *Russ. Chem. Rev.*, **2016**, *85*, 1229–1254.
8 A.R. Martin, J-J. Vasseur, M. Smietana. Boron and nucleic acid chemistries: merging the best of both worlds. *Chem. Soc. Rev.*, **2013**, *42*, 5684–5713.
9 Z.J. Lesnikowski, J. Shi, R. Schinazi. Nucleic acids and nucleosides containing carboranes. *J. Organomet. Chem.*, **1999**, *581*, 156–169.
10 R.F. Barth, W. Yang, A.S. Al-Madhoun, J. Johnsamuel, Y. Byun, S. Chandra, D.R. Smith, W. Tjarks, S. Eriksson. Boron-containing nucleosides as potential delivery agents for neutron capture therapy of brain tumors. *Cancer. Res.*, **2004**, *64*, 6287–6295.
11 Z.J. Leśnikowski. Challenges and opportunities for the application of boron clusters in drug design. *J. Med. Chem.*, **2016**, *59*, 7738–7758.
12 J. Emsley. Nature's Building Blocks: An A-Z Guide to the Elements. Oxford University Press, Oxford, UK, **2011**.
13 B.C. Das, P. Thapa, R. Karki, C. Schinke, S. Das, S. Kambhampati, S.K. Banerjee, P. Van Veldhuizen, A. Verma, L.M. Weiss, T. Evans. Boron chemicals in diagnosis and therapeutics. *Future Med. Chem.*, **2013**, *5*, 653–676.
14 Z.J. Leśnikowski. Recent developments with boron as platform for novel drug design. *Expert Opin. Drug Discov.*, **2016**, *11*, 569–578.
15 M. Scholz, E. Hey-Hawkins. Carbaboranes as pharmacophores: properties, synthesis, and application strategies. *Chem. Rev.*, **2011**, *111*, 7035–7062.
16 F. Issa F, M. Kassiou, L.M. Rendina. Boron in drug discovery: carboranes as unique pharmacophores in biologically active compounds. *Chem. Rev.*, **2011**, *111*, 5701–5722.
17 R. Sedlák, J. Fanfrlik, A. Pecina, D. Hnyk, P. Hobza, M. Lepšík. Noncovalent Interactions of Heteroboranes. In: Boron: The Fifth Element. D. Hnyk, M.L. McKee (Eds.). Springer International Publishing, Berlin, **2016**, pp. 219–239.
18 J. Fanfrlík, M. Lepšík, D. Horinek, Z. Havlas, P. Hobza. Interaction of carboranes with biomolecules: formation of dihydrogen bonds. *ChemPhysChem.*, **2006**, *7*, 1100–1105.
19 R. Lo, J. Fanfrlik, M. Lepsik, P. Hobza. The properties of substituted 3D-aromatic neutral carboranes: the potential for sigma-hole bonding. *PhysChemChemPhys*, **2015**, *17*, 20814–20821.
20 Z.J. Lesnikowski, J. Shi, R.F. Schinazi. Nucleic acids and nucleosides containing carboranes. *J. Organomet. Chem.*, **1999**, *581*, 156–169.
21 R.F. Schinazi, N. Goudgaon, J. Soria, D. Liotta. Synthesis, antiviral activity, cytotoxicity, and cellular pharmacology of 5-carboranyl pyrimidine nucleosides. *Advances in Neutron Capture Therapy*, **1993**, 285–288.
22 J.-C.G. Graciet, J. Shi, R.F. Schinazi. Synthesis and biological properties of the four optical isomers of 5-o-carboranyl-2′,3′-didehydro-2′,3′-dideoxyuridine. *Nucleos. Nucleot.*, **1998**, *17*, 711–727.
23 G. Fulcrand-El Kattan, N.M. Goudgaon, N. Ilksoy, J.-T. Huang, K.A. Watanabe, J.-P. Sommadossi, R.F. Schinazi. Synthesis and biological properties of 5-o-carboranyl-1-(2-deoxy-2-fluoro-β-D-arabinofuranosyl)uracil. *J. Med. Chem.*, **1994**, *37*, 2583–2588.

24 N.M. Goudgaøn, Y.A. El-Kattan, X. Xia, J. McAtee, J. Soria, S.-J. Wey, D.C. Liotta, R.F. Schinazi. A general synthetic method of 5-carboranyluracil nucleosides with potential antiviral activity and use in neutron capture therapy. *Nucleos. Nucleot.*, **1997**, *16*, 2133–2150.

25 K. Imamura, Y. Yamamoto. Synthesis and in vitro evaluation of sugar-modified carboranyluridines. *Bioorg. Med. Chem.*, **1996**, *6*, 1855–1858.

26 M. Woolhouse, F. Scott, Z. Hudson, R. Howey, M. Chase-Topping. Human viruses: discovery and emergence. *Philos. T. R. Soc. B.*, **2012**, *367*, 2864–2871.

27 M.J. Cannon, D.S. Schmid, T.B. Hyde. Review of cytomegalovirus seroprevalence and demographic characteristics associated with infection. *Rev. Med. Virol.*, **2010**, *20*, 202–213.

28 A. Kenneson, M.J. Cannon. Review and meta-analysis of the epidemiology of congenital cytomegalovirus (CMV) infection. *Rev. Med. Virol.*, **2007**, *17*, 253–276.

29 P. Sansoni, R. Vescovini, F. Fagnoni, J. Nikolich-Zugich, and 36 others. New advances in CMV and immunosenescence. *Exp. Gerontol.*, **2014**, *55*, 54–62.

30 A. Ahmed. Antiviral treatment of cytomegalovirus infection. *Infect. Disord. Drug Targets*, **2011**, *5*, 475–503.

31 A.B. Olejniczak, A. Adamska, E. Paradowska, M. Studzinska, P. Suski, Z.J. Lesnikowski. Modification of selected anti-HCMV drugs with lipophilic boron cluster modulator. *Acta Pol. Pharm. Drug Res.*, **2013**, *70*, 489–504.

32 B.A. Wojtczak, A. Semenyuk, A.B. Olejniczak, M. Kwiatkowski, Z.J. Lesnikowski. General method for the synthesis of 2′-O-carboranyl nucleosides. *Tetrahedron Lett.*, **2005**, *46*, 3969–3972.

33 A.B. Olejniczak, J. Plesek, Z.J. Lesnikowski. Nucleoside-metallacarborane conjugates for base-specific metal-labeling of DNA. *Chem. Eur. J.*, **2007**, *13*, 311–318.

34 B.A. Wojtczak, A.B. Olejniczak, Z.J. Lesnikowski. Nucleoside Modification with Boron Clusters and Their Metal Complexes in Current Protocols in Nucleic Acid Chemistry. S.L. Beaucage. (Ed.). John Wiley & Sons, Hoboken, NJ, **2009**, Unit 4.37, pp. 1–26.

35 M. Białek-Pietras, A.B. Olejniczak, E. Paradowska, M. Studzińska, P. Suski, A. Jabłońska, Z.J. Leśnikowski. Synthesis and *in vitro* antiviral activity of lipophilic pyrimidine nucleoside/carborane conjugates. *J. Organomet. Chem.*, **2015**, *798*, 99–105.

36 B.A. Wojtczak, A. Andrysiak, B. Grűner, Z.J. Lesnikowski. Chemical ligation – a versatile method for nucleoside modification with boron clusters. *Chem. Eur. J.*, **2008**, *14*, 10675–10682.

37 J.F. Valliant, P. Morel, P. Schaffer, J. Kaldis. Carboranes as ligands for the preparation of organometallic Tc and Re radiopharmaceuticals: synthesis of [M(CO)(3)(eta-2,3-C(2)B(9)H(11))](-) and rac-[M(CO)(3)(eta-2-R-2,3-C(2)B(9)H(10))](-) (M = Re, (99)Tc; R = CH(2)CH(2)CO(2)H) from [M(CO)(3)Br(3)](2-). *Inorg. Chem.*, **2002**, *41*, 628–630.

38 A.B. Olejniczak. Nucleoside-metallacarborane conjugates: synthesis of a uridine-bearing 3,3,3-(CO)$_3$-*closo*-3,1,2-ReC$_2$B$_9$H$_{10}$ complex. *ARKIVOC*, **2012** (*viii*), 90–97.

39 Semioshkin, A. Ilinova, I. Lobanova, V. Bregadze, E. Paradowska, M. Studzińska, A. Jabłońska, Z.J. Lesnikowski. Synthesis of the first conjugates of 5-ethynyl-2′-deoxyuridine with *closo*-dodecaborate and cobalt-bis-dicarbollide boron clusters. *Tetrahedron*, **2013**, *69*, 8034–8041.

40 Ilinova, A. Semioshkin, I. Lobanova, V.I. Bregadze, A.F. Mironov, E. Paradowska, M. Studzinska, A. Jabłonska, M. Białek-Pietras, Z.J. Lesnikowski. Synthesis, cytotoxicity

and antiviral activity studies of the conjugates of cobalt bis(1,2-dicarbollide)(-I) with 5-ethynyl-2′-deoxyuridine and its cyclic derivatives. *Tetrahedron*, **2014**, *70*, 5704–5710.

41 J. Laskova, A. Kozlova, M. Białek-Pietras, M. Studzińska, E. Paradowska, V. Bregadze, Z.J. Leśnikowski, A. Semioshkin. Reactions of *closo*-dodecaborate amines. Towards novel bis-(*closo*-dodecaborates) and *closo*-dodecaborate conjugates with lipids and non-natural nucleosides. *J. Organomet. Chem.*, **2016**, *807*, 29–35.

42 J.D. Hoyer, C.W. Ross, C-Y. Li, T.E. Witzig, R.D. Gascoyne, G.W. Dewald, C.A. Hanson. True T-cell chronic lymphocytic leukemia: a morphologic and immunophenotypic study of 25 cases. *Blood*, **1995**, *86*, 1163–1169.

43 S.Y. Kristinsson, P.W. Dickman, W.H. Wilson, N. Caporaso, M. Björkholm, O. Landgren. Improved survival in chronic lymphocytic leukemia in the past decade: a population-based study including 11.179 patients diagnosed between 1997–2003 in Sweden. *Haematologica*, **2009**, *94*, 1259–1265.

44 B. Eichhorst, T. Robak, E. Montserrat, P. Ghia, P. Hillmen, M. Hallek, C. Buske. Chronic lymphocytic leukemia: ESMO Guidelines for diagnosis, treatment and follow-up. *Ann. Oncol.* (Suppl. 5), **2015**, 74–84.

45 M.J. Keating, S. O'Brien, M. Albitar, S. Lerner, W. Plunkett, F. Giles, M. Andreeff, J. Cortes, S. Faderl, D. Thomas, C. Koller, W. Wierda, M.A. Detry, A. Lynn, H. Kantarjian. Early results of a chemoimmunotherapy regimen of fludarabine, cyclophosphamide, and rituximab as initial therapy for chronic lymphocytic leukemia. *J. Clin. Oncol.*, **2005**, *18*, 4079–4088.

46 T. Robak, P. Smolewski, B. Cebula, A. Szmigielska-Kaplon, K. Chojnowski, J.Z. Błoński. Rituximab combined with cladribine or with cladribine and cyclophosphamide in heavily pretreated patients with Indolent lymphoproliferative disorders and mantle cell lymphoma. *Cancer*, **2006**, *107*, 1542–1550.

47 V. Gandhi, W. Plunket. Combination strategies for purine nucleoside analogs. In Chronic Lymphoid Leukemias. B.D. Cheson (Ed.). Marcel Dekker, New York, **2001**, pp. 195–208.

48 E. Van den Neste, S. Cardoen, F. Offner, F. Bontemps. Old and new insights into the mechanisms of action of two nucleoside analogs active in lymphoid malignancies: fludarabine and cladribine. *Int. J. Oncol.*, **2005**, *4*, 1113–1124.

49 R. Castejón, J.A. Vargas, M. Briz, E. Berrocal, Y. Romero, J.C. Gea-Banacloche, M.N. Fernández, A. Durantez. Induction of apoptosis by 2-chlorodeoxyadenosine in B cell chronic lymphocytic leukemia. *Leukemia*, **1997**, *11*, 1253–1257.

50 I. Marzo, P. Perez-Galan, P. Giraldo, D. Rubio-Felix, A. Anel, J. Naval. Cladribine induces apoptosis in human leukemia cells by caspase-dependent and -independent pathways acting on mitochondria. *Biochem. J.*, **2001**, *35*, 537–546.

51 J.D. Żołnierczyk, A. Borowiak, P. Hikisz, B. Cebula-Obrzut, J.Z. Błoński, P. Smolewski, T. Robak, Z.M. Kiliańska. Promising anti-leukemic activity of atorvastatin. *Oncol. Rep.*, **2013**, *29*, 2065–2071.

52 K. Bednarska, A.B. Olejniczak, A. Piskala, M. Klink, Z. Sulowska, Z.J. Lesnikowski. Effect of adenosine modified with a boron cluster pharmacophore on reactive oxygen species production by human neutrophils. *Bioorg. Med. Chem.*, **2012**, *20*, 6621–6629.

53 J.D. Żołnierczyk, A.B. Olejniczak, A. Mieczkowski, J.Z. Błoński, Z.M. Kiliańska, T. Robak, Z.J. Leśnikowski. *In vitro* antileukemic activity of novel adenosine derivatives bearing boron cluster modification. *Bioorg. Med. Chem.*, **2016**, *24*, 5076–5087.

54 H. Piirainen, Y. Ashok, R.T. Nanekar, V-P. Jaakola. Structural features of adenosine receptors: from crystal to function. *BBA – Biomembranes*, **2011**, *1808*, 1233–1244.
55 M.L. Trincavelli, S. Daniele, C. Martini. Adenosine receptors: what we know and what we are learning. *Curr. Top. Med. Chem.*, **2010**, *10*, 860–877.
56 K.A. Jacobson, M.F. Jarvis (Eds.). Purinergic Approaches in Experimental Therapeutics. Wiley-Liss, New York, **1997**.
57 K.A. Jacobson, Z.-G. Gao. Adenosine receptors as therapeutic targets. *Nat. Rev. Drug Discovery*, **2006**, *5*, 247–264.
58 J.F. Chenm, H.K. Eltzschig, B.B. Fredholm. Adenosine receptors as drug targets: what are the challenges? *Nat. Rev. Drug. Discov.*, **2013**, *12*, 265–286.
59 K. Bednarska, A.B. Olejniczak, B.A. Wojtczak, Z. Sułowska, Z.J. Leśnikowski. Adenosine and 2′-deoxyadenosine modified with boron cluster pharmacophores as new classes of human blood platelet function modulators. *ChemMedChem.*, **2010**, *5*, 749–756.
60 S.M. Wilkinson, H. Gunosewoyo, M.L. Barron, A. Boucher, M. McDonnell, P. Turner, D.E. Morrison, M.R. Bennett, I.S. McGregor, L.M. Rendina, M. Kassiou. The first CNS-active carborane: A novel P2X7 receptor antagonist with antidepressant activity. *ACS Chem. Neurosci.*, **2014**, *5*, 335–339.
61 Z. Lesnikowski *et al.* Book of Abstracts, XXII International Round Table on Nucleosides, Nucleotides and Nucleic Acids (IRT) 2016, July 18–22, **2016**, Paris, pp. 35–38.

1.3

Design of Carborane-Based Hypoxia-Inducible Factor Inhibitors

Guangzhe Li,[1] Hyun Seung Ban,[2] and Hiroyuki Nakamura[3]

[1] *School of Pharmaceutical Science and Technology, Dalian University of Technology, Dalian, China*
[2] *Biomedical Genomics Research Center, Korea Research Institute of Bioscience and Biotechnology, Daejeon, Republic of Korea*
[3] *Laboratory for Chemistry and Life Science, Institute of Innovative Research, Tokyo Institute of Technology, Yokohama, Japan*

1.3.1 Introduction

Cancer is a disease involving uncontrolled cell growth. In a solid tumor, cancer cells are deprived of oxygen due to their rapid growth. As a result, the oxygen concentration of the tumor regions distant from blood vessels is significantly lower than that of healthy tissues, causing hypoxia in tumor. Tumor hypoxia upregulates a number of genes involved in tumor angiogenesis, cellular energy metabolism, metastasis, cell proliferation, and resistance to apoptosis. Hypoxia-inducible factor (HIF) is a transcriptional factor that plays an important role as an oxygen sensor in cells.

HIF is a heterodimer composed of α subunits (HIF1α, -2α, and -3α) and a β subunit (HIF1β, which is also known as an aryl hydrocarbon receptor nuclear translocator [ARNT]) [1,2]. Under aerobic conditions, posttranslational hydroxylation of proline residues in HIF1α by prolyl hydroxylase (PHD) induces ubiquitination by the von Hippel–Lindau (VHL) tumor suppressor protein, a component of the E3 ubiquitin ligase complex, resulting in the oxygen-dependent degradation through a ubiquitin–proteasome pathway. Under hypoxic conditions, HIF1α does not undergo the oxygen-dependent degradation due to the inactivation of PHD; instead, it translocates into the nucleus, where it dimerizes with the constitutively expressed HIF1β to form a heterodimeric complex. HIF binds to the hypoxia response element (HRE) DNA sequence with co-activators to activate various genes, including glucose transporters, glycolytic enzymes, angiogenic growth factors, and several molecules involved in apoptosis and cell proliferation [3,4]. The activated HIF plays pivotal roles in various pathological conditions, including inflammation, cardiovascular disorder, and cancer. Indeed, overexpression of HIF1α has been observed in human cancers, including brain, breast, colon, lung, ovary, and prostate cancers [5]; thus, HIF1α is a novel target of cancer therapy.

We have studied a boron-based medicinal drug design. A boron atom has a vacant orbital and readily interconverts between the neutral sp2 and anionic sp3 hybridization

Boron-Based Compounds: Potential and Emerging Applications in Medicine, First Edition.
Edited by Evamarie Hey-Hawkins and Clara Viñas Teixidor.
© 2018 John Wiley & Sons Ltd. Published 2018 by John Wiley & Sons Ltd.

states, resulting in a new stable interaction between a boron atom and a donor molecule through a covalent bond. Furthermore, three-center two-electron bonds such as B–H–B and B–B–B generate various stable boron clusters, particularly carboranes (dicarba-*closo*-dodecaboranes: $C_2B_{10}H_{12}$), that exhibit remarkable thermal stability. The icosahedral geometry and exceptional hydrophobicity of carboranes may allow their use as hydrophobic pharmacophores in biologically active molecules [6]. This chapter examines our recent development of carborane-based design of HIF1 inhibitors.

1.3.2 Boron-Containing Phenoxyacetanilides

1.3.2.1 Synthesis of Boron-Containing Phenoxyacetanilides

As a novel HIF1 inhibitor, AC1-001 was reported by Lee and coworkers [7]. This compound involves an adamantyl moiety and a benzoic acid moiety in the molecule (Figure 1.3.1). The similarity of steric and electronic characteristics between adamantane and carboranes was focused on, and boron-containing phenoxyacetanilides **1** and **2** based on the structure of AC1-001 were designed as shown in Figure 1.3.1. Compound **1** has a boronic acid instead of a carboxylic acid in the molecule. In addition, an adamantyl group of compound **1** is replaced by an *ortho*-carborane moiety in compound **2**.

Synthesis of compound **1** is shown in Scheme 1.3.1. Adamantylphenol **3** was converted to carboxylic acid **4**, and the aniline moiety was introduced into **4** using ethyl chloroformate (ECF) and *N*-methylmorphine (NMM). The resulting phenoxyacetanilide **5** was treated with pinacolatodiboron under the Suzuki–Miyaura-type diboron coupling condition to introduce a boronic ester at the 5 position of the aniline ring, and the following hydrogenation gave the boron ester **6** in 63% yield in two steps. Deprotection of pinacol ester using BBr_3 gave the desired adamantyl boronic acid **1** in 72% yield.

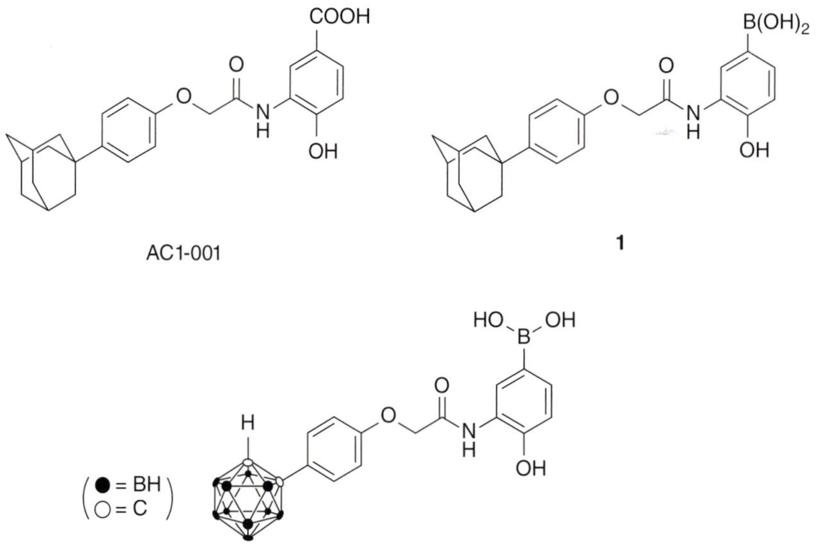

Figure 1.3.1 Design of HIF1 inhibitors based on the structure of AC1-001.

Synthesis of *ortho*-carboranyl boronic acid **2** is shown in Scheme 1.3.2. In order to introduce the *ortho*-carborane moiety instead of an adamantyl group, 4-iodophenol was chosen as a starting material and converted the ethyl ester **7**. Sonogashira coupling was carried out to introduce an ethynyl moiety at the 4 position of the benzene ring of **7**, and the ethynyl carboxylic acid **8** was obtained after being subjected to basic conditions. 5-Pinacolatoboryl-2-hydroxyaniline was reacted with **8** using isobutyl chloroformate (IBCF), and the resulting borate was treated with KHF_2 in methanol to give the boronic acid **2** (GN25361) in 83% yield.

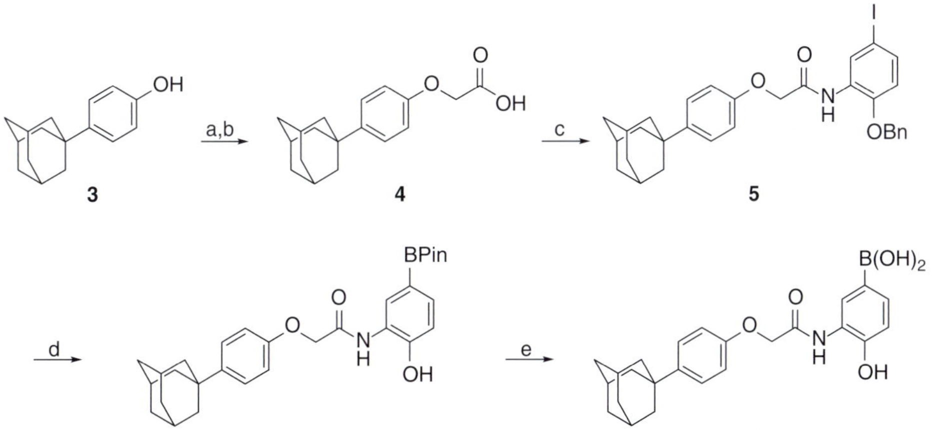

Scheme 1.3.1 Reagent and conditions: (a) $ClCH_2CO_2Et$, K_2CO_3, DMF, r.t.; (b) LiOH, THF/H_2O; (c) 2-benzyloxy-5-iodoaniline, ECF, NMM, TEA, THF, 20 °C, 72%; (d) i. pinacolatodiboron, $PdCl_2$(dppf), AcOK, DMF, 80 °C, 63%; ii. Pd/C, H_2, EtOH, 99%; (e) BBr_3, CH_2Cl_2, 0 °C, 72%.

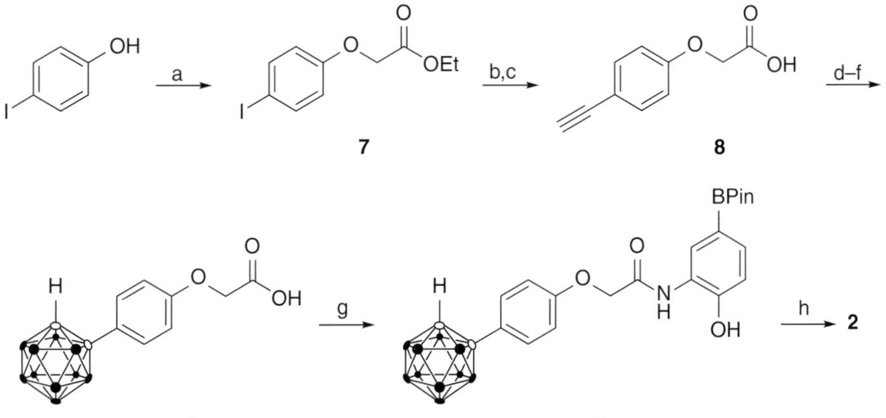

Scheme 1.3.2 Reagent and conditions: (a) $ClCH_2CO_2Et$, K_2CO_3, DMF, r.t.; (b) ethynyltrimethylsilane, $PdCl_2(PPh_3)_2$, CuI, *N,N*-diethylamine, DMF, microwave, 120 °C, 93%; (c) LiOH, THF–H_2O, r.t., 82%; (d) BnBr, Na_2CO_3, DMF, r.t., 93%; (e) $B_{10}H_{14}$, CH_3CN, toluene, reflux, 31%; (f) Pd/C, H_2, EtOH, 96%; (g) 5-pinacolatoboryl-2-hydroxyaniline, IBCF, 30%; (h) KHF_2, MeOH, 83%.

1.3.2.2 Biological Activity of Boron-Containing Phenoxyacetanilides

Synthesized compounds **1**, **2**, and **6** were evaluated for their ability to inhibit hypoxia-induced HIF1 transcriptional activity using a cell-based reporter assay in HeLa cells expressing HRE-dependent firefly luciferase reporter construct (HRE-Luc) and constitutively expressing cytomegalovirus (CMV)-driven Renilla luciferase reporter. In addition, the cell growth inhibition (GI_{50}) of compounds **1**, **2**, and **6** was also determined by MTT assay to investigate the effects of compounds on cell viability. AC1-001 was used as a positive control for comparison. The results are shown in Table 1.3.1. In the case of adamantyl derivatives, both the boronic acid **1** and its pinacol ester **6** displayed significant inhibition of HIF1 transcriptional activity (4.6 and 3.1 µM, respectively) similar to AC1-001 (3.1 µM). Interestingly, the carboranyl derivatives are more potent: the IC_{50} value of the boronic acid **2** is 0.74 µM. In all cases, significant cell growth inhibition was not observed ($GI_{50} > 10$ µM).

The effect of compound **2** against hypoxia-induced HIF1α accumulation was examined by western blot analysis, and the messenger RNA (mRNA) expression level of HIF1α and vascular endothelial growth factor (VEGF) by reverse transcription polymerase chain reaction (RT-PCR) analysis in HeLa cells. As shown in Figure 1.3.2,

Table 1.3.1 Inhibition of HIF1 transcriptional activity in cell-based HRE reporter gene assay and cell growth inhibition

Compound	IC_{50} (µM)[a]	GI_{50} (µM)[b]
1	4.6 ± 0.86	15.2 ± 1.13
2 (GN26361)	0.74 ± 0.24	16.2 ± 0.70
6	3.1 ± 0.15	16.0 ± 0.47
AC1-001	3.1 ± 0.07	51.6 ± 0.89

[a] HeLa cells stably transfected with HRE-Luc were used. The drug concentration required to inhibit the relative light unit by 50% (IC_{50}) was determined from semi-logarithmic dose–response plots, and results represent the mean ± SD of triplicate samples.
[b] HeLa cells were incubated for 48 h with various concentrations of compounds under normoxic conditions, and viable cells were determined by MTT assay.

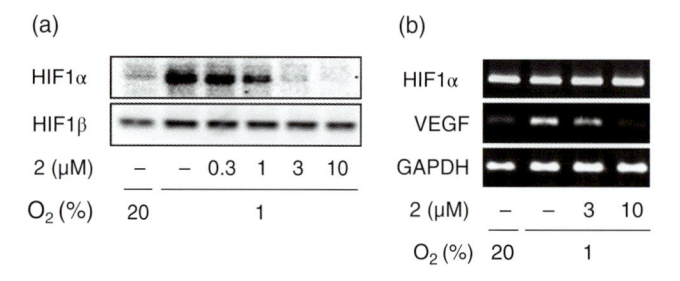

Figure 1.3.2 Effect of compound **2** on hypoxia-induced accumulation of HIF1α protein (a) and expression of VEGF and HIF1α mRNAs (b) in HeLa cells.

compound **2** suppressed hypoxia-induced HIF1α accumulation and expression of VEGF mRNA in a concentration-dependent manner without affecting the expression levels of HIF1α mRNA, indicating that compound **2** inhibits the hypoxia-induced expression of VEGF via suppression of HIF1α accumulation.

1.3.3 Target Identification of GN26361

1.3.3.1 Design of GN26361 Chemical Probes

In the course of phenotype-based drug discovery, chemical biology techniques are useful methods for identification of target molecules and clarification of mechanisms of action. Until now, various methods including affinity chromatography, the genetic interactions method, drug affinity responsive target stability, and activity-based probes have been reported [8,9]. Based on the chemical biologic approach, the mechanism of action of **2** in HIF inhibition was clarified. As shown in Figure 1.3.3, multifunctional chemical probes of GN26361 substituted with two functional moieties were designed. One is a benzophenone for covalent binding with a target protein through photoaffinity labeling [10]. The other is an acetylene moiety for conjugation with azide-linked fluorophore through the click reaction (Figure 1.3.3) [11].

1.3.3.2 Synthesis of GN26361 Chemical Probes

Synthesis of the chemical probes is shown in Scheme 1.3.3. The carboxylic acid **11** was reacted with the aminobenzophenones **12a** (4-isomer) or **12b** (3-isomer) using 1-(3-dimethylaminopropyl)-3-ethylcarbodiimide hydrochloride (EDCI) and l-hydroxybenzotriazole (HOBt) to give the corresponding aryloxyacetanilides **13a** and **13b** in 81 and 71% yields, respectively [12]. Diisopropylethylamine (DIPEA) was an effective base for these amide formations. Decarborane coupling reaction proceeded at the ethynyl moiety of **13** in the presence of acetonitrile as a Lewis base under toluene reflux to give the corresponding carborane derivatives **14a** and **14b** in 65 and 45% yields, respectively. Debenzylation of compound **14** was carried out in the presence of Pd/C under hydrogen atmosphere, and the resulting phenols **15a** and **15b** were treated with propargyl bromide under a basic condition to afford the **16a** and **16b** in 37 and 61% yields, respectively.

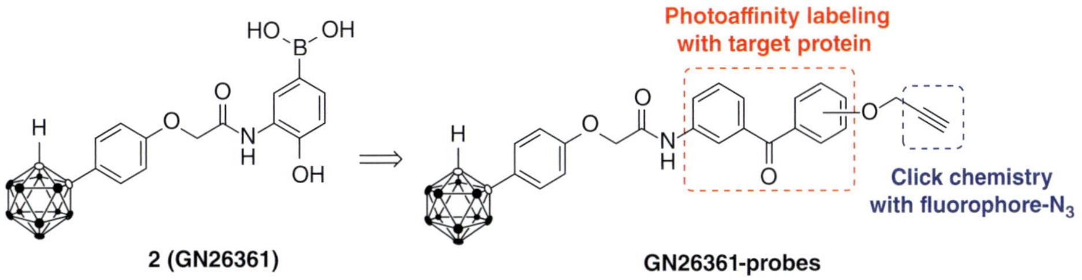

Figure 1.3.3 Design of GN26361 chemical probes for clarification of mechanism of action in HIF inhibition.

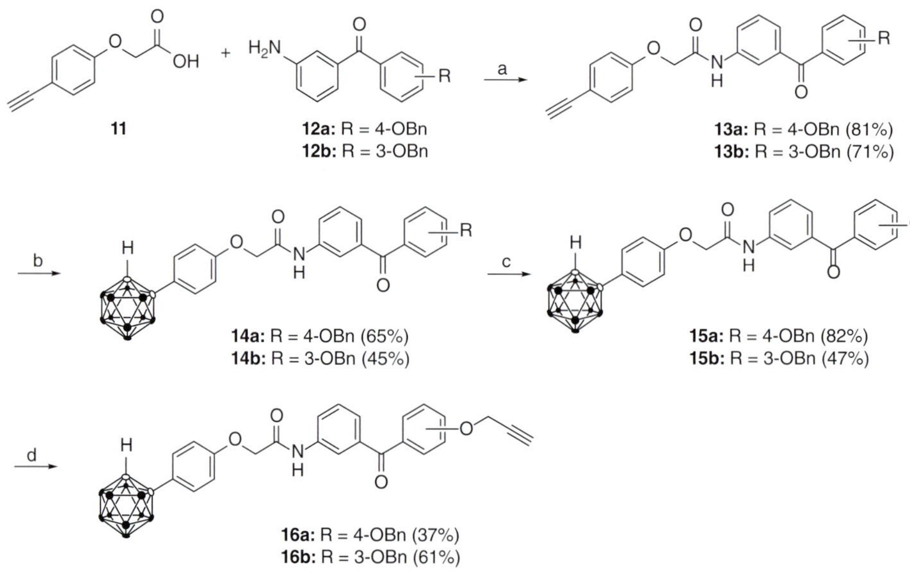

Scheme 1.3.3 Reagent and conditions: (a) EDCl, HOBt, DIPEA, DMF, r.t., overnight; (b) $B_{10}H_{14}$, CH_3CN, toluene, reflux; (c) Pd/C, H_2, EtOH/THF, r.t.; (d) propalgyl bromide, K_2CO_3, acetone, reflux.

1.3.3.3 Target Identification of GN26361

To test whether the synthesized probes are suitable for target identification of GN26361, the HIF inhibitory activity of the probes was determined. Western blot analysis using anti-HIF1α antibody revealed that both **16a** and **16b** showed HIF inhibitory activity similar to that of **2** (Figure 1.3.4a) [13]. Using these probes, photoaffinity labeling in HeLa cell lysate and click conjugation with Alexa Fluor 488 azide were performed, and proteins were separated by sodium dodecyl sulfate–polyacrylamide gel electrophoresis (SDS-PAGE) and visualized by direct in-gel fluorescent detection. As shown in Figure 1.3.4b, a specific protein bound to the probes was detected and identified as heat shock protein-60 (HSP60) by mass spectrometry. Because HSP60 was identified as one of the target proteins of **2**, the effect on human HSP60 chaperone activity was examined by using the porcine heart malate dehydrogenase (MDH) refolding assay. An *in vitro* HSP60 enzyme assay revealed that **2** significantly suppressed HSP60 activity at the same inhibitory concentrations for transcriptional activity of HIF1 and accumulation of HIF1α protein (Figure 1.3.4c). These results suggest that HSP60 is a direct target of **2** and is associated with HIF1α.

1.3.4 Carborane-Containing HSP60 Inhibitors

1.3.4.1 Design of *Ortho*- and *Meta*-Carborane Analogs of GN26361

Since HSP60 was identified as a primary target of **2** (GN26361) and inhibition of the HSP60 chaperone activity was clarified to result in the degradation of HIF1α protein through an oxygen-independent pathway, the further structure–activity relationship

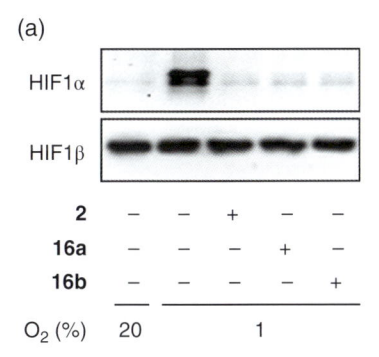

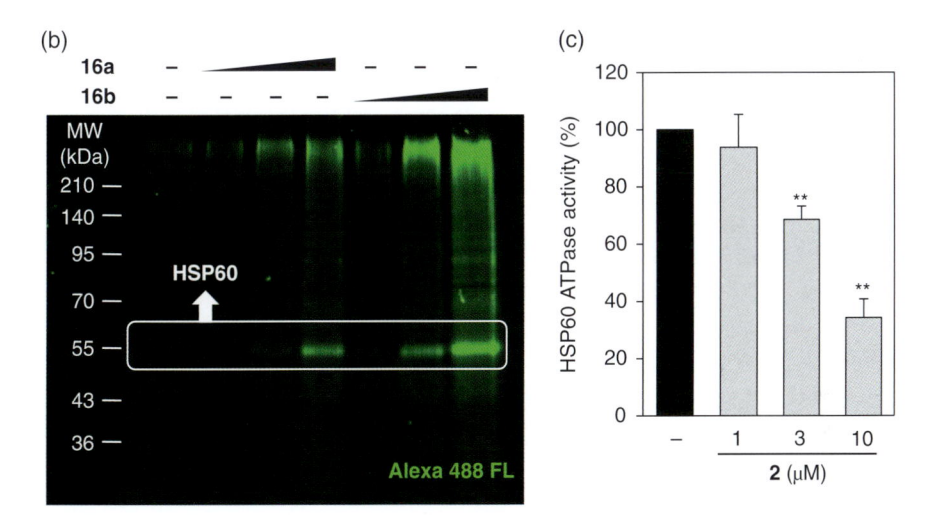

Figure 1.3.4 Identification of target protein of **2** (GN26361) using chemical probes. (a) Effect of chemical probes **16a** and **16b** on the hypoxia-induced HIF1α accumulation. The levels of each protein were detected by immunoblot analysis. (b) Fluorescent imaging of protein bound to the chemical probes. HeLa cell lysate was irradiated for 30 min at 360 nm in the presence of each probe (30, 100, and 300 μM), and then the probe was conjugated with Alexa Fluor 488 azide by click reaction. (c) Inhibition of HSP60 by **2**. Recombinant HSP60 (1 μM) was reacted with ATP (1 μM) for 30 min, and then ATP content was determined by luciferase reaction.

based on the structure was studied to develop potent HIF1α and HSP60 inhibitors. As shown in Figure 1.3.5, two types of analogs, *ortho-* and *meta-*carboranylphenoxyacetanilides, were designed [14–16].

1.3.4.2 Synthesis of GN26361 Analogs

The key intermediate bromoacetanilide **21** was synthesized from 4-bromo-2-nitrophenol **17**, as shown in Scheme 1.3.4. Benzyl protection of **3** was carried out using benzyl bromide to give the benzyl ethers **18**. The Suzuki–Miyaura-type diboron coupling reaction of **18** proceeded in the presence of dichloro(diphenylphosphino-ferrocene) palladium(II) ($PdCl_2dppf$) catalyst to afford **19**, and reduction of the nitro group of **19** with Fe powder gave **20**. Aniline **20** was treated with bromoacetylbromide to give **21**.

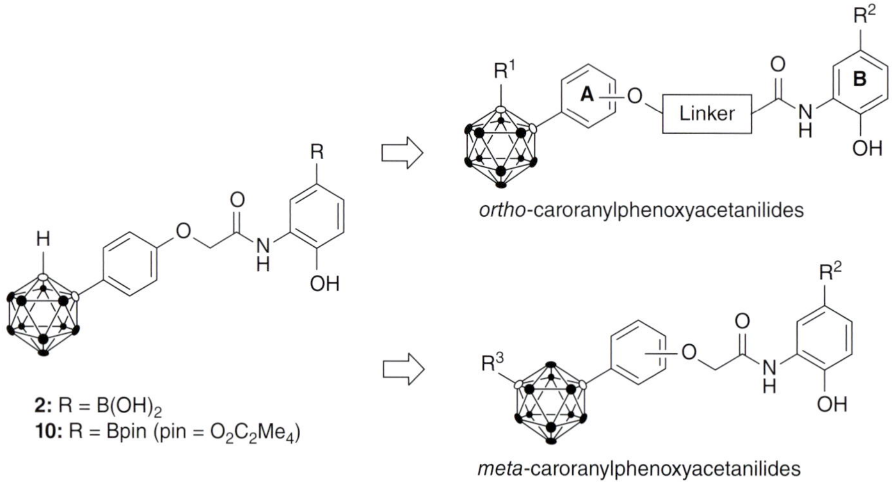

2: R = B(OH)$_2$
10: R = Bpin (pin = O$_2$C$_2$Me$_4$)

Figure 1.3.5 Design *ortho*- and *meta*-carborane analogs of GN26361.

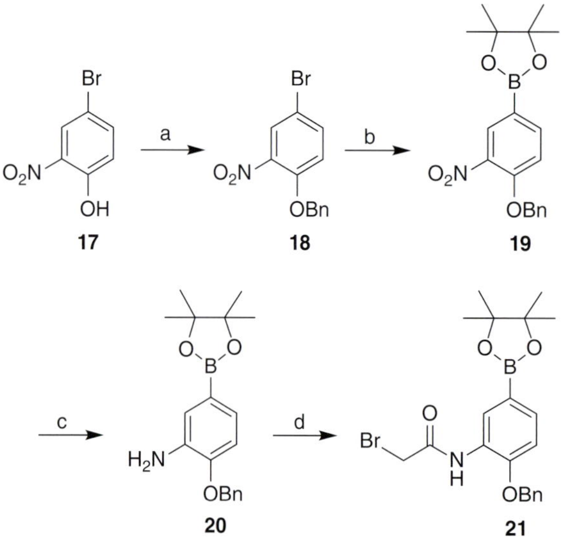

Scheme 1.3.4 Reagents and conditions: (a) BnBr, K$_2$CO$_3$, acetone, reflux, overnight, 85%; (b) 2,2′-Bi-1,3,2-dioxaborolane, PdCl$_2$, dppf, AcOK, dioxane, reflux, overnight, 93%; (c) Fe, NH$_4$Cl, EtOH–H$_2$O, reflux, 3 h, 74%; (d) bromoacetylbromide, pyridine, DMAP, r.t., 1.5 h, 80%.

ortho-Carboranylphenoxyacetanilides **27a–f** were synthesized from iodoanisoles **22** shown in Scheme 1.3.5. Sonogashira coupling of **22** with ethynyltrimethylsilane proceeded in the presence of PdCl$_2$(PPh$_3$)$_2$ catalysts, and the resulting alkynes were treated with tetrabutylammonium fluoride (TBAF) to remove the trymethylsilyl (TMS) group, giving ethynylanisoles **23**. Decaborane coupling of **23** was carried out in the presence of *N,N*-dimethylaniline in chlorobenzene under microwave conditions to give the *ortho*-carboranylanisoles **24**. The treatment of **24** with *n*-butyllithium generated lithiated

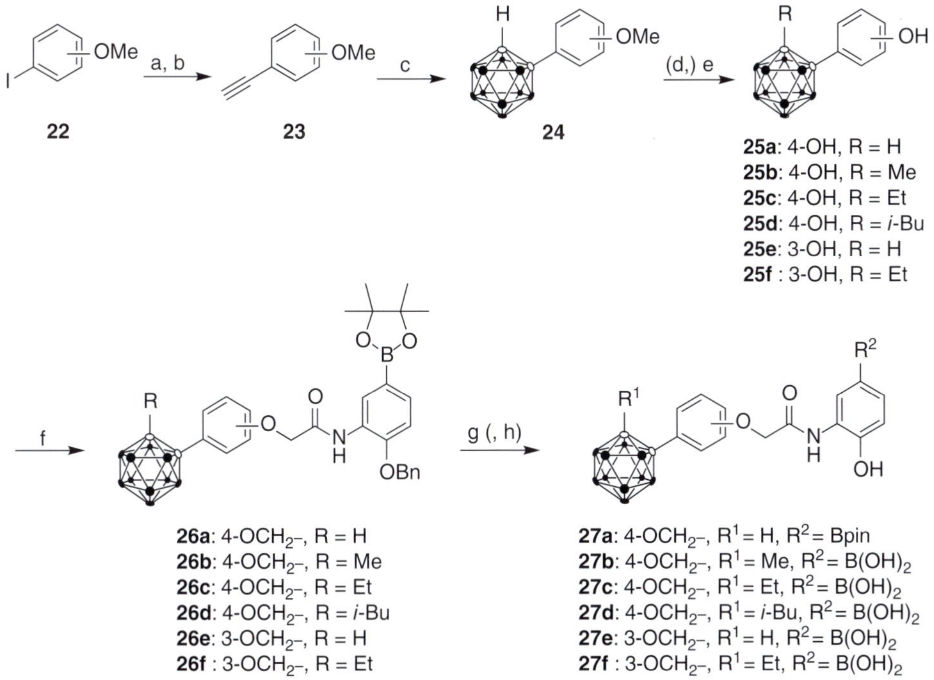

Scheme 1.3.5 Reagents and conditions: (a) ethynyltrimethylsilane, $PdCl_2(PPh_3)_2$, PPh_3, CuI, diethylamine, THF, reflux, 5 h, 95%; (b) TBAF, THF, r.t., 30 min, 86%; (c) $B_{10}H_{14}$, N,N-dimethylaniline, chlorobenzene, microwave, 130 °C, 10 min, 76%; (d) (i) n-BuLi, THF, −10 °C, 30 min, (ii) RI, 1.5 h; (e) BBr_3, CH_2Cl_2, r.t., overnight, 90–98%; (f) **21**, NaH, THF, r.t., 2 h; (g) H_2, Pd/C, MeOH–THF, r.t., overnight, 60–85%. (h) (i) KHF_2 (4 M), MeOH, r.t., 2 h, (ii) HCl (1 N), r.t., overnight, 32–65%.

ortho-carboranyl intermediate, which reacted with alkyl iodides to give di-substituted *ortho*-carboranes. Deprotection of a methoxy group was carried out using boron tribromide to give **25a–f**. Compound **25a–f** reacted with boromoacetylanilide **21** under the basic condition, and the resulting *ortho*-carboranylphenoxyacetanilides **26a–f** were hydrogenated using Pd/C under hydrogen atmosphere. Finally, the pinacol esters were hydrolyzed to give **27b–f** as *ortho*-derivatives of **2**.

ortho-Carboranylphenoxy derivatives with an ethylene linker was synthesized from 4-(*ortho*-carboranyl)phenol **25a** (Scheme 1.3.6). 4-(*ortho*-carboranyl)phenol **25a** was treated with methyl propiolate using 1,4-diazabicyclo[2.2.2]octane (DABCO) as a base to give the conjugated methyl ester **28**,(17) which was hydrolyzed with LiOH in aqueous THF solution. The resulting carboxylic acid **29** was converted to the acid chloride, which reacted with **30** to give the corresponding pinacolatoboron ester **27g**. Deprotection of pinacolatoboronester followed by hydrogenesis gave 4-(*ortho*-carboranyl)phenoxypropananilide **27h**. Interestingly, deboronated product **27i** was obtained when pinacolatoboron ester **27g** treated with KHF_2 was reacted with HCl for overnight.

ortho-Carboranylphenoxy derivative with a propylene linker was also synthesized, as shown in Scheme 1.3.7. 4-(*ortho*-carboranyl)phenol **25a** was treated with ethyl 4-bromobutyrate using K_2CO_3 as a base, and the resulting ethyl ester **31** was hydrolyzed to

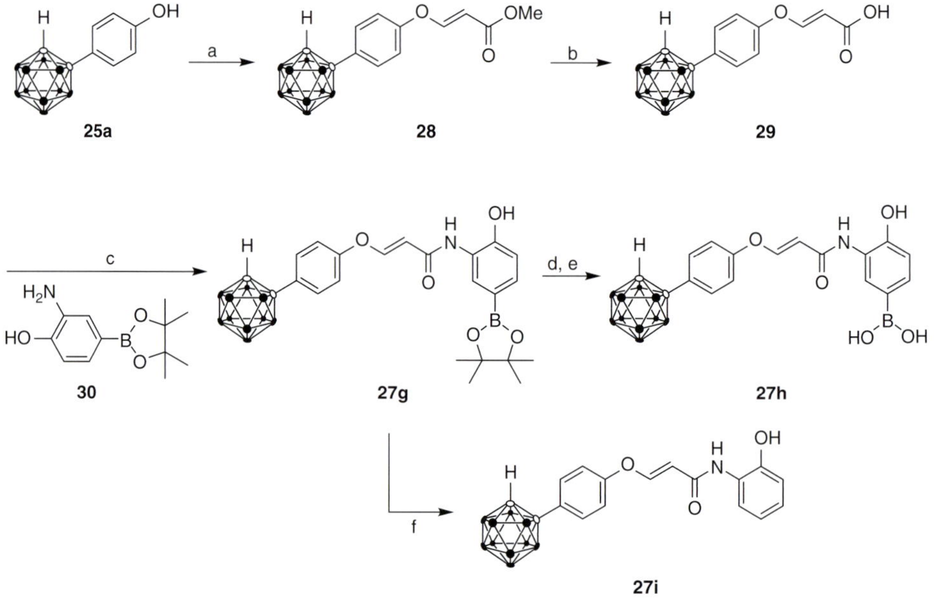

Scheme 1.3.6 Reagents and conditions: (a) methyl propiolate, DABCO, CH$_2$Cl$_2$, r.t., overnight, 80%; (b) LiOH, THF–H$_2$O, r.t., overnight, 60%; (c) (i) oxalyl chloride, CH$_2$Cl$_2$, 0 °C, 2 h, (ii) **30**, triisobutylamine, THF, r.t., 2 h, 28%; (d) (i) KHF$_2$ (4 M), MeOH, r.t., 30 min, (ii) HCl (1 N), 1 h; (e) H$_2$, Pd/C, MeOH–THF, r.t., overnight; (f) (i) KHF$_2$ (4 M), MeOH, r.t., 2 h, (ii) HCl (1 N), r.t., overnight, 64%.

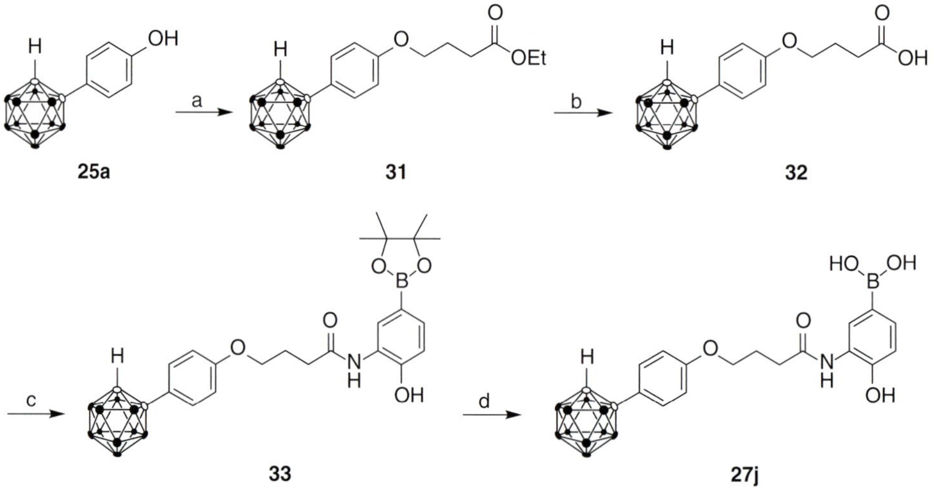

Scheme 1.3.7 Reagents and conditions: (a) Ethyl 4-bromobutyrate, K$_2$CO$_3$, DMF, r.t., overnight, 57%; (b) LiOH, THF–H$_2$O, r.t., overnight, 93%; (c) **30**, HATU, DIPEA, DMF, r.t., overnight, 48%; (d) (i) KHF$_2$ (4 M), MeOH, r.t., 2 h, (ii) HCl (1 N), r.t., overnight, 58%.

afford the corresponding carboxylic acid **32**. The amide bond formation of **32** with aniline **30** was carried out using 1-[Bis(dimethylamino)methylene]-1H-1,2,3-triazolo[4,5-b] pyridinium-3-oxidhexafluorophosphate (HATU) as a condensation reagent, and the resulting pinacolatoboron ester **33** was treated with KHF$_2$ followed by HCl to give 4-(*ortho*-carboranyl)phenoxybutananilide **27j**.

ortho-Carboranylphenoxy derivative with two carborane frameworks was also synthesized as shown in Scheme 1.3.8. Sonogashira coupling of 3,5-dibromoanisole **34** with ethynyltrimethylsilane and deprotection of the TMS group gave 3,5-diethynylanisole **35**. Decaborane coupling of **35** proceeded under the microwave conditions, and the resulting anisole **36** was converted to the corresponding phenol **37** by treatment with BBr$_3$. Alkylation of **37** with **21** was carried out using NaH as a base, and hydrogenesis of the resulting **38** afforded 3,5-(di-*ortho*-carboranyl)phenoxyacetanilide derivative **27k**.

meta-Carboranylphenoxyacetanilides **43a–h** were synthesized from *meta*-carborane **39** shown in Scheme 1.3.9. The C-arylation of meta-carborane through its copper derivative was developed by Wade and coworkers [18]. *meta*-Carborane **39** was treated with *n*-BuLi in dimethoxyethane (DME) at 0 °C to give the corresponding lithiated carborane, which reacted with CuCl to generate the C-copper derivative of *meta*-carborane. Coupling reaction of the copper derivative with iodoanisoles proceeded in the presence of pyridine under refluxed conditions to give (methoxyphenyl)-*meta*-carboranes **40** [19]. Lithiation of **40** with *n*-BuLi proceeded in tetrahydrofuran (THF) at −10 °C, and the resulting lithiated carborane reacted with alkyl iodides to give the corresponding di-substituted *meta*-carboranes. The methoxy group was deprotected by boron tribromide to give (hydroxyphenyl)-*meta*-carboranes **41a–h** [19]. Alkylation of (hydroxyphenyl)-*meta*-carboranes **41a–h** with

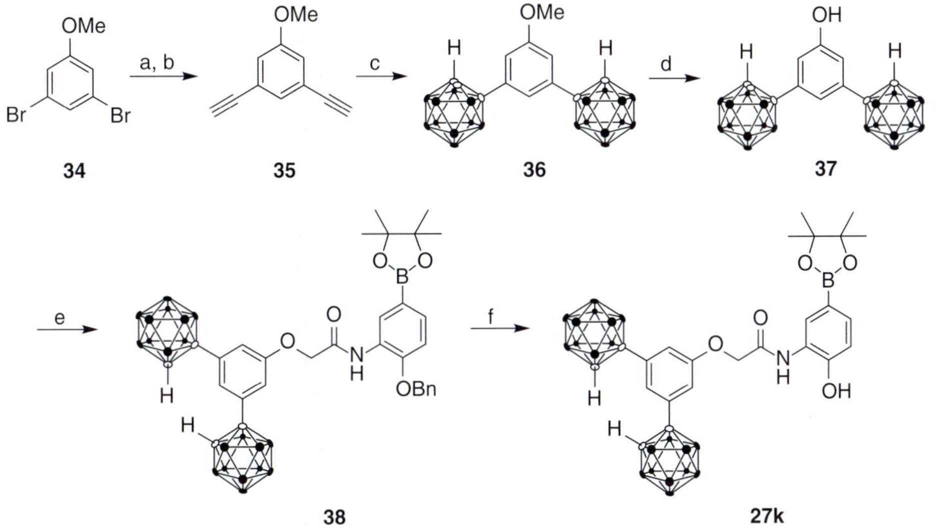

Scheme 1.3.8 Reagents and conditions: (a) ethynyltrimethylsilane, PdCl$_2$(PPh$_3$)$_2$, PPh$_3$, CuI, diethylamine, THF, reflux, 5 h, quant.; (b) KOH, MeOH, r.t., 6 h, 78%; (c) B$_{10}$H$_{14}$, N,N-dimethylaniline, chlorobenzene, microwave, 130 °C, 37%; (d) BBr$_3$, CH$_2$Cl$_2$, r.t., overnight, 100%; (e) **21**, NaH, THF, r.t., 2 h, 72%; (f) H$_2$, Pd/C, MeOH–THF, r.t., overnight, 54%.

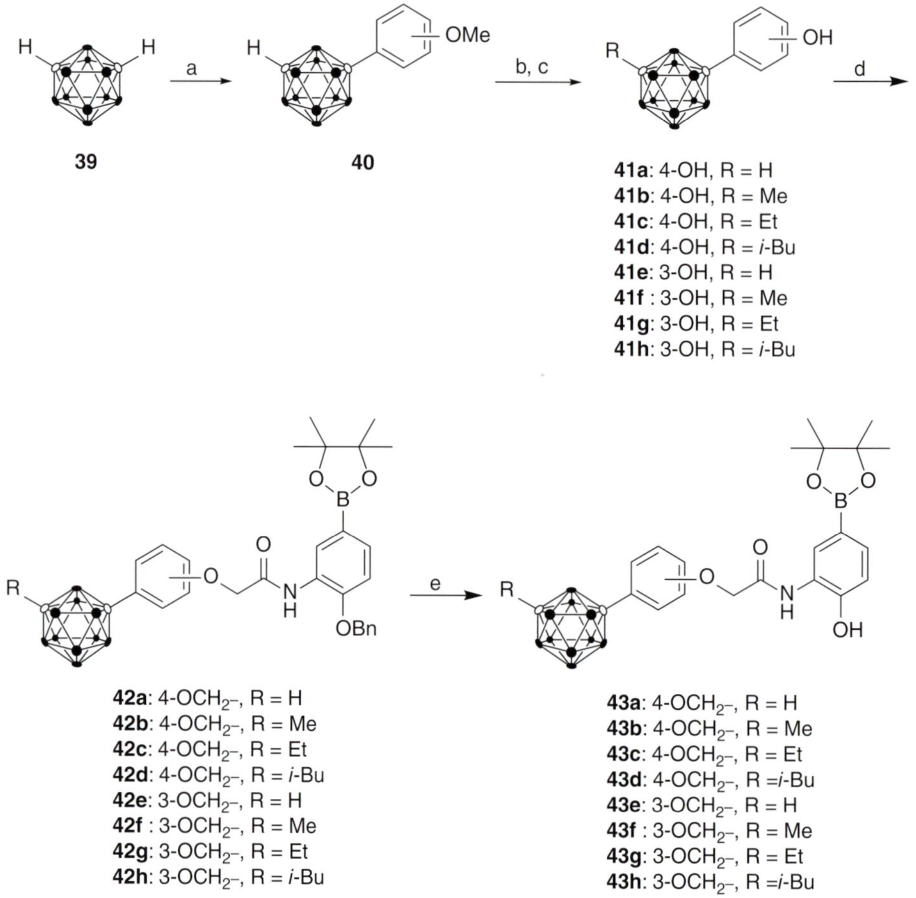

Scheme 1.3.9 Synthesis of *meta*-carborane analogs. Reagents and conditions: (a) (i) n-BuLi, DME, 0 °C, 30 min, (ii) CuCl, r.t., 1.5 h, (iii) 4-iodoanisole, pyridine, reflux, 48 h, 26%; (b) (i) n-BuLi, THF, −10 °C, 30 min, (ii) RI, −10 °C, 1.5 h; (c) BBr$_3$, CH$_2$Cl$_2$, r.t., overnight, 40-91%; (d) **21**, NaH, THF, r.t., 2 h; (e) H$_2$, Pd/C, MeOH–THF, r.t., overnight.

bromacetylanilide **21** proceeded in THF at room temperature under basic conditions, and the benzyl group of the resulting phenoxy ethers **42** was deprotected by hydrogenesis to give the *meta*-carboranylphenoxyacetanilides **43a–h**.

1.3.4.3 HIF Inhibitory Activity of Carborane Analogs of GN26361

Synthesized substituted *ortho*-carboranylphenoxy analogs **27a–k** and *meta*-carboranylphenoxyacetanilides **43a–h** were evaluated for their ability to inhibit hypoxia-induced HIF1 transcriptional activity using a cell-based reporter assay in HeLa cells expressing HRE-Luc and constitutively expressing CMV-driven Renilla luciferase reporter. GN26361 was used as a positive control for comparison. The results are summarized in Table 1.3.2. The IC$_{50}$ of **2** toward HIF1 transcriptional activity was 2.2 ± 0.2 µM, whereas the pinacol ester **27a** was more potent than **2** with an IC$_{50}$ of 1.3 ± 0.1 µM. The higher

Table 1.3.2 Inhibition of HIF1 transcriptional activity in HeLa cell-based HRE and CMV dual luciferase assay and cell growth inhibition

Compound	HRE-Luc[a] IC$_{50}$ (µM)	Compound	HRE-Luc IC$_{50}$ (µM)
2 (GN26361)	2.2 ± 0.2	27j	14.6 ± 2.7
27a	1.3 ± 0.1	27k	16.4 ± 0.6
27b	1.9 ± 0.2	43a	2.28 ± 0.52
27c	1.4 ± 0.2	43b	1.52 ± 0.32
27d	1.8 ± 0.3	43c	0.83 ± 0.12
27e	4.2 ± 0.8	43d	0.73 ± 0.01
27f	1.1 ± 0.3	43e	4.80 ± 0.80
27g	0.53 ± 0.1	43f	1.34 ± 0.29
27h	2.4 ± 0.6	43g	1.29 ± 0.44
27i	1.3 ± 0.2	43h	0.55 ± 0.03

[a] HeLa cells expressing HRE-dependent firefly luciferase reporter construct (HRE-Luc) and constitutively expressing CMV-driven Renilla luciferase reporter with SureFECT Transfection Reagent were established with Cignal™ Lenti Reporter (SABiosciences, Frederick, MD), according to the manufacturer's instructions.

inhibitory activity of **27a** toward HIF1 transcription is probably due to the lipophilicity of the pinacol ester functional group, which affects the transmembrane property. The alkyl-substituted *ortho*-carboranylphenoxyacetanilide analogs **27b–d** and **27f** showed slight enhancement of the inhibitory activity compared to **2** and **27e**, respectively. The IC$_{50}$ of compound **27h** having an ethylene linker was slightly decreased to 2.4 ± 0.6 µM, whereas that of compound **27j** having a propylene linker dropped to 14.6 ± 2.7 µM, indicating that methylene and ethylene linkers are appropriate lengths for the potency of HIF1 inhibitors. The highest inhibition was achieved by compound **27g**: the IC$_{50}$ was 0.53 ± 0.07 µM. These results indicate that the creation of rigid conformation by introducing an unsaturated ethylene linker in the molecule is attributed to the potency of HIF1 inhibitors. Interestingly, deboronated derivative **27i** also inhibited HIF1 transcriptional activity with an IC$_{50}$ of 1.3 ± 0.2 µM. However, 3,5-(di-*ortho*-carboranyl) substituent was not effective; the IC$_{50}$ of compounds **27k** dropped to 16.4 ± 0.6 µM. Inhibitory potency of *meta*-carboranylphenoxyacetanilides **43a–h** was evaluated. The sterically larger substituents of the R group, such as the methyl (**43b**, **43f**), ethyl (**43c**, **43g**), and *i*-butyl (**43d**, **43h**) groups, increased their inhibitory activity more than compounds **43a** and **43e**, respectively. Especially, *i*-butyl-substituted *meta*-carboranylphenoxyacetanilide **43h** exhibited significant inhibitory activity against the hypoxia-induced HIF1 transcription, and the IC$_{50}$ was 0.55 ± 0.03 µM.

1.3.4.4 HSP60 Inhibitory Activity of Carborane Analogs of GN26361

The effects of compounds **27f**, **27g**, **27i**, and di-carboranyl compound **27k** on human HSP60 chaperone activity were examined by using the porcine heart MDH refolding assay (Figure 1.3.6) [20,21]. Inhibition ratio of each compound at 2 µM concentration

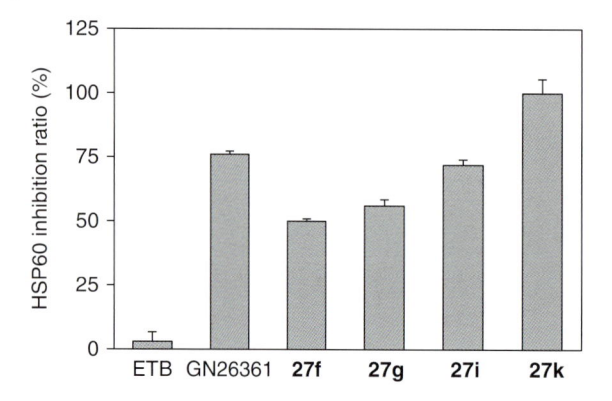

Figure 1.3.6 HSP60 chaperone activity analyzed by using MDH as substrate.

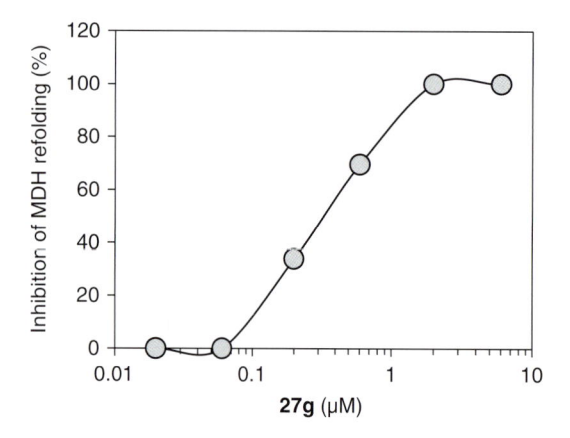

Figure 1.3.7 Dose-dependent inhibition of human HSP60 chaperone activity by compound **27g**. IC_{50} value was calculated to be 0.35 ± 0.08 µM.

against HSP60 chaperone activity is indicated by the vertical axis. Although epolactaene *tert*-butyl ester (ETB) exhibited significant inhibition (~50%) at 6 µM, a weak inhibition was observed at 2 µM. Moderate inhibition was observed in the cases of compounds **27f** and **27g**, whereas an inhibition ratio of compound **27i** was similar to that of GN26361. The best result was observed in the case of compound **27k**; 100% inhibition was observed at 2 µM, indicating that the sterically bulky bis-*ortho*-carborane moiety is more effective on inhibition of the HSP60 chaperone activity than the mono-*ortho*-carborane moiety. The dose-dependent inhibition of HSP60 chaperone activity revealed that compound **27k** inhibited ~70% of HSP60 chaperone activity at 0.6 µM concentration, and the IC_{50} was calculated to be 0.35 ± 0.08 µM (Figure 1.3.7). So far, compound **27k** is the highest inhibitor of HSP60 chaperone activity among the reported four compounds, ETB [21], mizoribine [22], EC3016 [23], and GN26361 [24].

Inhibitors of HIF transcriptional activity and HSP60 chaperone activity were developed based on **2** as a lead compound. Among the compounds synthesized, 4-(*ortho*-carboranyl)phenoxypropenanilide **27g** suppressed HIF1 transcriptional activity with an IC_{50} of 0.53 ± 0.1 µM (Table 1.3.2), although the inhibition of HSP60 chaperone activity was not as high as that of **2** (Figure 1.3.6). In contrast is the inhibition of HIF1 transcriptional activity of 3,5-(di-ortho-carboranyl)phenoxyacetanilide **27k** with an IC_{50} of $16.4 \pm 0.6\,\mu M$; however, that inhibited HSP60 chaperone activity with an IC_{50} of $0.35\,\mu M$. It is known that the stability and activity of HIF1α are regulated not only by HSP60 but also by HSP90 [25], which is the main chaperone of HIF1α. Therefore, HIF1 inhibition may be involved to some degree in HSP60 inhibition, and each inhibitory effect of the compounds might be attributed to different mechanisms, respectively.

1.3.5 Carborane-Containing Manassantin Mimics

1.3.5.1 Synthesis of Carborane-Containing Manassantin Mimics

Manassantins were isolated from *Saururus cernuus* L. (Saururaceae), a native aquatic/wetland plant. They have been shown to have many medically relevant activities, including inhibition of hypoxia-induced HIF1 transcriptional activity [26]. Manassantin A consists of the core tetrasubstituted *syn-anti-syn*-THF unit with C_2 symmetric side chains (Figure 1.3.8), and it possesses the highest inhibitory activity of hypoxia-induced HIF1 transcription (the reported IC_{50} is 3 nM) among the natural product-based inhibitors of HIF1 [27]. However, the rather long steps necessary for its chemical synthesis are mainly due to the construction of the THF skeleton and in manassantin A [28,29]. Thus, compounds **44** and **45**, which contain icosahedral 1,2-dicarba-*closo*-dodecaborane (*ortho*-carborane) and 1,7-dicarba-*closo*-dodecaborane (*meta*-carborane) frameworks as an alternative skeleton into manassantin A, were designed.

Synthesis of *ortho*-carborane derivative **44** is shown in Scheme 1.3.10. The *ortho*-carborane skeleton can be derived from the coupling reaction of decaborane ($B_{10}H_{14}$) and alkynes. Therefore, the diaryl substituted alkynes were first synthesized by Sonogashira coupling of terminal alkyne **46** and aromatic iodide **47**. The reaction

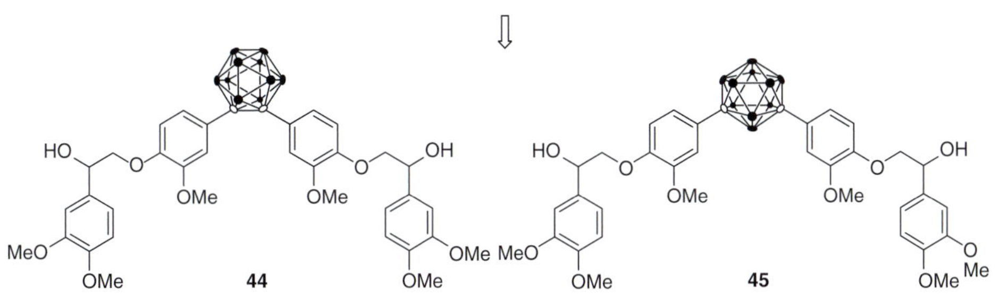

Figure 1.3.8 Design of carborane-containing manassantin mimics **44** and **45**.

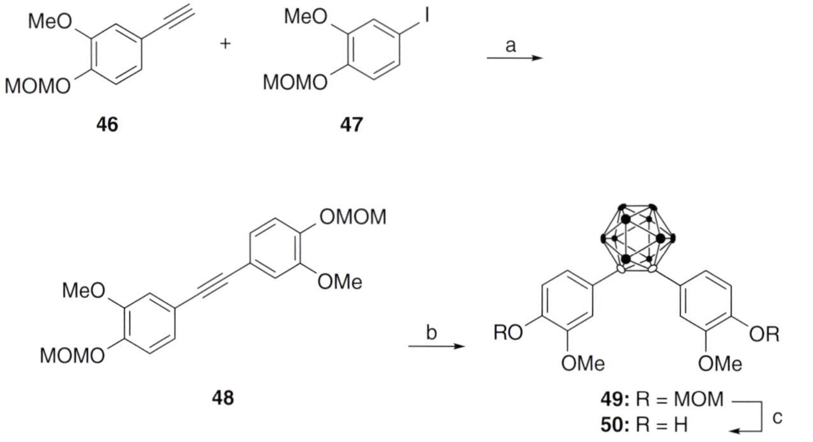

Scheme 1.3.10 Reagents and conditions: (a) Pd(PPh$_3$)$_4$ (10 mol%), CuI, TEA, THF, reflux, overnight, 84%; (b) B$_{10}$H$_{14}$, N,N-dimethylaniline, chlorobenzene, MW, 120 °C, 15 min, 52%; (c) conc. HCl/MeOH, CH$_2$Cl$_2$, 12 h, 87%.

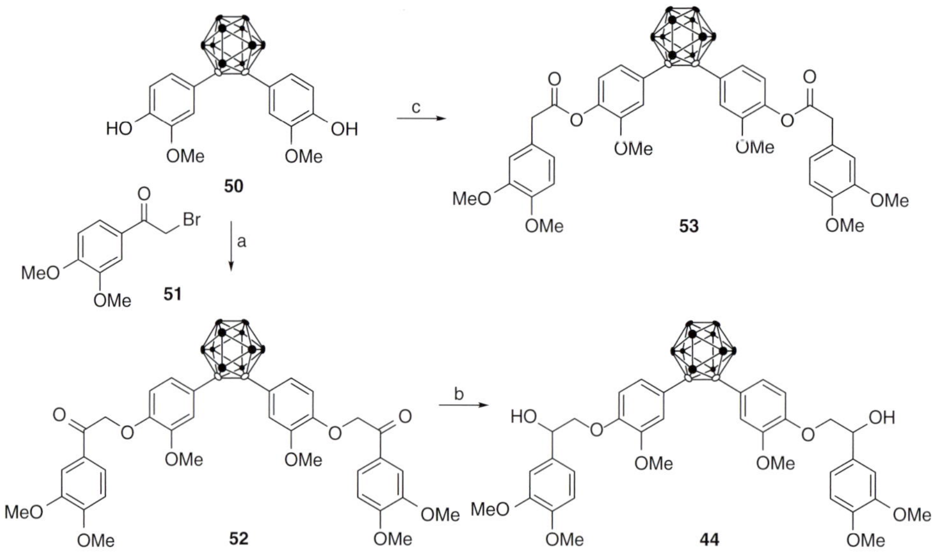

Scheme 1.3.11 Reagents and conditions: (a) **51**, K$_2$CO$_3$, DMF, r.t., 2 h, 51%; (b) NaBH$_4$, MeOH, 0 °C, 1.5 h, 92%; (c) homoveratric acid, DCC, DMAP, THF, 50 °C, 3 h, 52%.

proceeded in the presence of 10 mol% of Pd(PPh$_3$)$_4$ catalyst, and the corresponding disubstituted alkyne **48** was obtained in 84% yield. The coupling reaction of alkyne **48** and decaborane proceeded under microwave (MW) irradiation condition to give **49** in 52% yield. Deprotection of **49** was carried out under acidic condition to afford **50** in 87% yield.

Compound **50** was reacted with 2-bromo-1-(3,4-dimethoxyphenyl)ethanone **51** in DMF under the basic condition to give compound **52** in 51% yield. Reduction of a carbonyl group in **52** with NaBH$_4$ afforded compound **44** in 92% yield. In addition, compound **50** was also reacted with homoveratric acid using N,N'-dicyclohexylcarbodiimide (DCC) and N,N-dimethyl-4-aminopyridine (DMAP) to give compound **53** in 52% yield (Scheme 1.3.11).

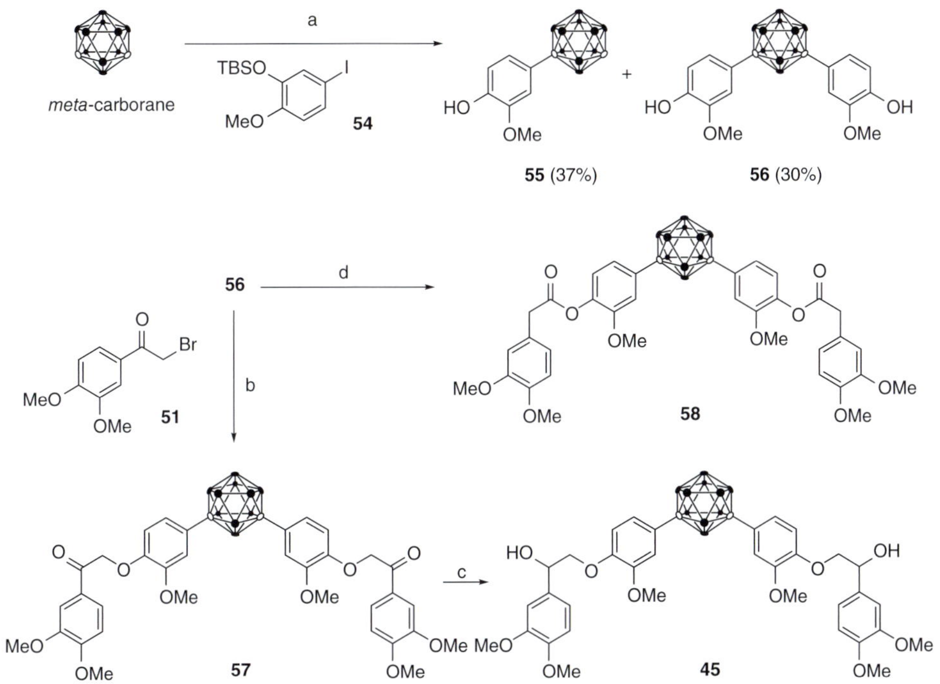

Scheme 1.3.12 Reagents and conditions: (a) (i) n-BuLi, CuCl, pyridine/DME, **54**, (ii) TBAF, THF; (b) **51**, K$_2$CO$_3$, DMF, r.t., 4 h, 90%; (c) NaBH$_4$, MeOH, 0 °C, 1.5 h, 64%; (d) homoveratric acid, EDCI, HOBt, DIPEA, DMF, r.t., overnight, 99%.

Synthesis of *meta*-carborane derivative **45** is shown in Scheme 1.3.12. Introduction of aromatic functional groups into *meta*-carborane was achieved by the Ullmann-type reaction [30]. *meta*-Carborane was treated with 2.1 equivalents of *n*-butyl lithium, and the generating dilithiated *meta*-carborane was transmetallated with CuCl (2.1 equiv.). The resulting cuprous *meta*-carborane was reacted with aromatic iodide **54** under the reflux conditions to give the monoarylated and the diarylated *meta*-carboranes, **55** and **56**, in 37 and 30% yields, respectively. Alkylation of diarylated *meta*-carborane **56** with bromoacetylphenone **51** gave the corresponding *meta*-carborane **57** in 90% yield. Compound **57** was converted to **45** by reduction with NaBH$_4$ in 64% yield. Compound **58** was also synthesized by the condensation reaction of **56** and homoveratric acid using EDCI, HOBt, and DIPEA in 99% yield.

1.3.5.2 Biological Activity of Carborane-Containing Manassantin Mimics

The ability of synthesized carboranes to inhibit the hypoxia-induced HIF1 transcriptional activity was examined by using a cell-based HRE reporter gene assay. HeLa cells expressing HRE-Luc and constitutively expressing CMV-driven Renilla luciferase reporter were used for this experiment. The results are summarized in Table 1.3.3. Among the compounds synthesized, compounds **44**, **45**, and **56** showed significant inhibition of the hypoxia-induced HIF1 transcriptional activity with IC$_{50}$ values of 3.2, 2.2, and 5.1 µM, respectively. The *meta*-carborane framework was more effective than the *ortho*-carborane framework. Although both compounds **44** and **45** possessed relatively

Table 1.3.3 Inhibition of HIF1 transcriptional activity in cell-based HRE reporter gene assay and cell growth inhibition

Compound	IC$_{50}$ (μM)[a]	GI$_{50}$ (μM)[b]
44	3.2 ± 1.1	3.8 ± 0.03
45	2.2 ± 1.6	7.5 ± 0.3
50	17 ± 1.5	14.3 ± 0.4
52	>100	>100
53	84 ± 8.4	41.7 ± 1.5
56	5.1 ± 0.6	13.8 ± 0.7
57	50 ± 9.1	>100
58	76 ± 12.0	>100
YC1	1.5 ± 0.7	n.d.

[a] HeLa cells, expressing HRE-dependent firefly luciferase reporter construct (HRE-Luc) and constitutively expressing CMV-driven Renilla luciferase reporter, were incubated for 12 h with or without drugs under normoxic or hypoxic conditions. The fluorescence intensity of HRE-Luc was normalized to that of constitutively expressing CMV-driven Renilla luciferase control. YC1 was used as the positive control.
[b] HeLa cells were incubated for 48 h with various concentrations of compounds under normoxic condition, and viable cells were determined by MTT assay.
n.d., No data.

high cell growth inhibition with GI$_{50}$ values of 3.8 and 7.5 μM, respectively, compound **56** showed moderate cell growth inhibition toward HeLa cells (13.8 μM). It should be noted that inhibitory activity of compounds **44** and **45** toward hypoxia-induced HIF1 transcriptional activity is weaker than that of manassantin A. Indeed, the difference in relative configuration of at least one of the two side chains in manassantin (erythro and threo isomers) has only a minor influence on the activity of HIF1 inhibition, suggesting that the chirality of compounds **44** and **45** would not significantly affect the activity, but that hydroxy groups at the benzyl position are required for HIF1 inhibitory potency.

Effects of compounds **44** and **45** against hypoxia-induced HIF1α protein accumulation and the expression levels of HIF1α mRNA were examined by western blot analysis and RT-PCR analysis, respectively. Both compounds **44** and **45** similarly suppressed hypoxia-induced HIF1α accumulation in a concentration-dependent manner without affecting the expression levels of HIF1α mRNA under hypoxia in HeLa cells (Figure 1.3.9). These results indicate that the inhibition of hypoxia-induced HIF1 transcriptional activity is induced by compounds **44** and **45** through a degradation pathway of HIF1α protein under hypoxia.

1.3.6 Carborane-Containing Combretastatin A-4 Mimics

1.3.6.1 Design of *Ortho*-Carborane Analogs of Combretastatin A-4

Microtubules play a role in organization of cytoplasm, intracellular transport, and chromosome segregation [31]. Microtubules have been recognized as a target for cancer

(a)

44 (μM)	–	–	3	10	30	100	–	–	–	–
45 (μM)	–	–	–	–	–	–	3	10	30	100
O$_2$ (%)	20						1			

(b)

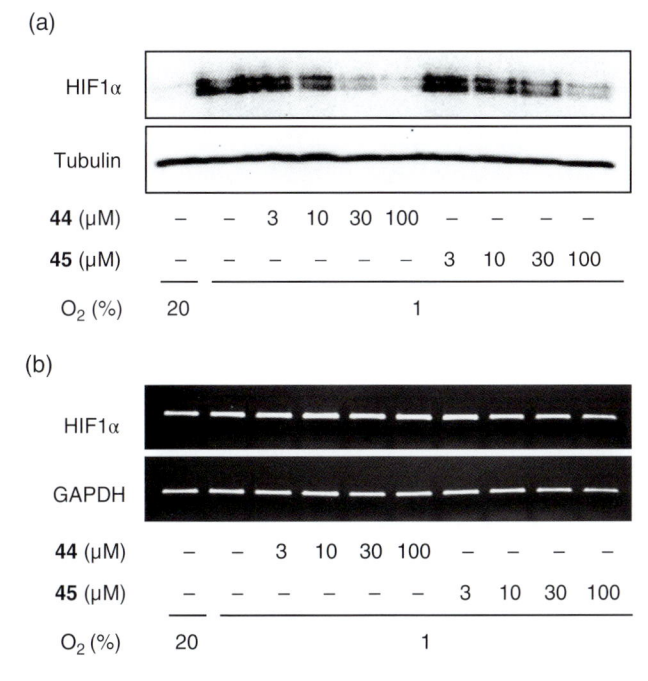

44 (μM)	–	–	3	10	30	100	–	–	–	–
45 (μM)	–	–	–	–	–	–	3	10	30	100
O$_2$ (%)	20						1			

Figure 1.3.9 Effects of compounds **44** and **45** on hypoxia-induced accumulation of HIF1α protein and expression of HIF1α mRNA in HeLa cells. (a) The levels of each protein were detected by western blot analysis with HIF1α- or tubulin-specific antibodies. Tubulin was used as the loading control, and "–" was DMSO used as the control. (b) HIF1α and GAPDH mRNA expression was detected by RT-PCR.

therapy due to their important functions in mitosis and cell division, and many microtubule-targeted compounds such as colchicine, combretastatin, vinblastine, and taxol have been reported [32]. Combretastatin A-4 (CA-4) is natural *cis* stilbenes isolated from the South African tree *Combretum caffrum* (Figure 1.3.4), and it binds to tubulin at the binding site of colchicine leading to tubulin depolymerization [33]. CA-4 induces cell-cycle arrest and apoptosis of cancer cells by disrupting microtubule dynamics [34]. To date, various derivatives of CA-4 have been reported, and combretastatin A-4 phosphate (CA-4P) is a water-soluble CA-4 prodrug under Phase II clinical trials for cancer therapy [35,36]. The *cis* stilbene scaffold in CA-4 was focused, and carborane analogs through the replacement of the *cis* geometry of stilbene framework by *ortho*-carborane were designed as shown in Figure 1.3.10.

1.3.6.2 Synthesis of Carborane Analogs of Combretastatin A4

The synthesis of *ortho*-carborane analogs **61** is shown in Scheme 1.3.13. The *ortho*-carborane scaffold can be generated from the coupling of decaborane ($B_{10}H_{14}$) with alkynes. Therefore, diaryl-substituted alkynes **60** via the Sonogashira coupling of terminal alkynes **59** with aromatic iodides (ArI) were initially synthesized. The reaction of 5-ethynyl-1,2,3-trimethoxybenzene **59a** and ArI proceeded in the presence of the Pd(PPh$_3$)$_4$ catalyst (10 mol%) in THF to give the corresponding diaryl-substituted alkynes **60a–d** in 42–90% yields. In a similar manner, alkynes **60e–g** were obtained from 4-ethynyl-1,2-dimethoxybenzene **6b** in 46–87% yields [37]. The substituted

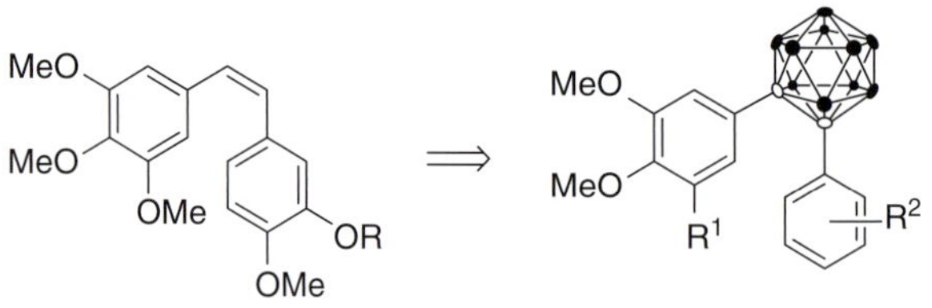

Combretastatin A-4 (R = OH)
CA-4P (R = PO₃Na)

ortho-Carborane analogs

Figure 1.3.10 Design of ortho-carborane analogs of combretastatin A4.

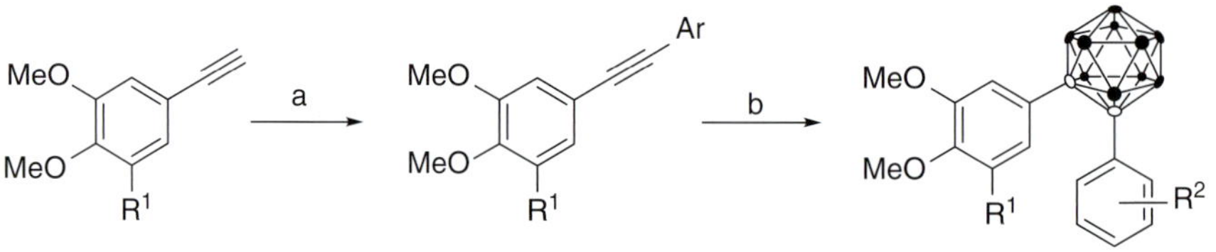

59a: R^1 = OMe
b: R^1 = H

60a–g

61a: R^1 = OMe, R^2 = 4-OMe, 3-OH
b: R^1 = OMe, R^2 = 4-OMe
c: R^1 = OMe, R^2 = 3-OMe
d: R^1 = H, R^2 = 2-OMe
e: R^1 = H, R^2 = 4-OMe
f: R^1 = H, R^2 = 3-OMe
g: R^1 = H, R^2 = 2-OMe

Scheme 1.3.13 Synthesis of ortho-carborane analogs. Reagents and conditions: (a) ArI, Pd(PPh₃)₄, CuI, TEA/THF, reflux, 4–6 h; (b) (i) $B_{10}H_{14}$, CH_3CN/toluene, 110 °C, 24 h or (ii) $B_{10}H_{14}$, N,N-dimethylaniline/chlorobenzene, microwave, 150 °C, 15 min.

decaborane coupling of alkyne **60a** with decaborane was carried out in the presence of acetonitrile as the Lewis base under toluene reflux conditions [38]. However, **61a** was obtained only in 20% yield. Furthermore, in the cases of **60b–d**, the corresponding diaryl-substituted ortho-carboranes **61b–d** were obtained in even lower yields (5–12%) under similar conditions.

In order to increase the yields of diaryl-substituted ortho-carboranes, the reaction conditions of the decaborane coupling were optimized by using 1,2-diphenylethyne **62** as the starting material. As shown in Table 1.3.4, the combination of N,N-dimethylaniline and chlorobenzene worked effectively. Indeed, the coupling of decaborane with 1,2-diphenylethyne proceeded smoothly in the presence of three equivalents of N,N-dimethylaniline as the Lewis base in chlorobenzene by microwave irradiation for 15 min, giving 1,2-diphenyl-ortho-carborane **63** in 75% yield (entry 7 in Table 1.3.4). With the optimum conditions, diaryl-substituted alkynes **60a–g** underwent coupling with decaborane by microwave irradiation in the presence of N,N-dimethylaniline in chlorobenzene to give the corresponding ortho-carborane analogs **61a–g** in 50–78% yields.

1.3.6.3 HIF Inhibitory Activity of Carborane Analogs of Combretastatin A4

The effects of the synthesized carborane analogs on the hypoxia-induced HIF1 activation were determined in HeLa cells stably transfected with the HRE-reporter gene, and

Table 1.3.4 Optimization of the reaction conditions for decaborane coupling

$$\text{Ph}\equiv\text{Ph} \xrightarrow[\text{Conditions}]{B_{10}H_{14}} \text{63}$$

62

Entry	Base	Solvent	Temp/°C	Time	Yield (%)
1	CH$_3$CN	Toluene	110[a]	5 h	30
2	CH$_3$CN	Toluene	110[a]	22 h	30
3	CH$_3$CN	Toluene	110[b]	30 min	7
4	CH$_3$CN	Toluene	110[b]	1 h	35
5	CH$_3$CN	Toluene	150[b]	1 h	31
6	CH$_3$CH$_2$CN	Toluene	110[b]	1 h	30
7	N,N-dimethylamiline	Cholorobenzene	150[b]	15 min	75

[a] Reactions were carried out under toluene reflux conditions.
[b] Reactions were carried out under microwave-irradiated conditions in a sealed tube.

Table 1.3.5 Inhibition of HIF1 transcriptional activity by *ortho*-carborane analogs of combretastatin A4 in hypoxic cancer cells

Compound	HRE-Luc[a] IC$_{50}$ (µM)	Compound	HRE-Luc IC$_{50}$ (µM)
61a	0.6	61e	0.4
61b	1.2	61f	5.5
61c	5.1	61g	4.6
61d	0.8	YC1	1.5

[a] HeLa cells stably expressing HRE reporter gene were incubated for 12 h under hypoxic condition. The luciferase activity was determined with luciferin and adenosine triphosphate (ATP), and the drug concentration required to inhibit luciferase activity by 50% (IC$_{50}$) was determined.

the result is shown in Table 1.3.5. All synthesized carborane analogs significantly inhibited the hypoxia-induced HIF1 activation. Particularly, carborane analogs **61a**, **61b**, **61d**, and **61e** showed more potent HIF1 inhibitory activity than a known HIF inhibitor, YC1, and their IC$_{50}$ values were 0.4–1.2 µM.

The RT-PCR analysis revealed that carborane analogs **61a** and **61d** suppressed expression of VEGF mRNA without affecting the level of HIF1α mRNA (Figure 1.3.11a). In addition, carborane analogs **61a** and **61d** inhibited the hypoxia-induced accumulation of HIF1α in a concentration-dependent manner (Figure 1.3.11b). To further determine whether carborane analog binds to tubulin, biotinylated probes were prepared (Figure 1.3.11c). In HeLa cells, photoaffinity labeling and pull-down assay

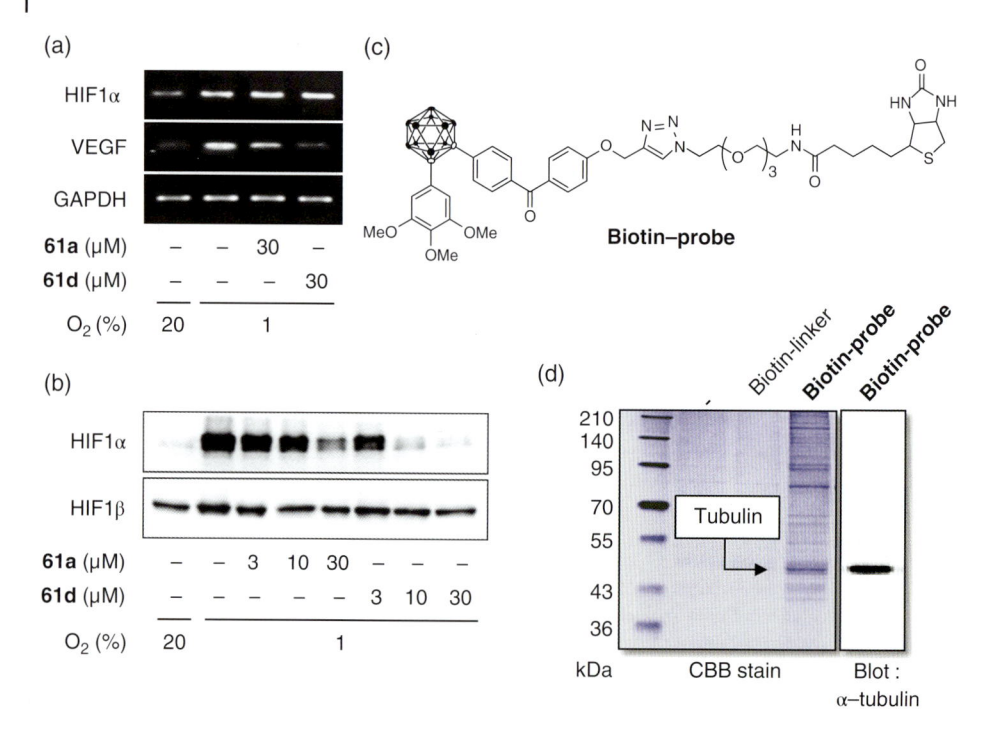

Figure 1.3.11 Effect of carborane analogs **61a** and **61d** on the hypoxia-induced HIF1 activation and tubulin binding of biotin probe. (a) HeLa cells were incubated with carborane analogs under hypoxic condition for 4 h. The levels of HIF1α and VEGF mRNA were detected by RT-PCR analysis. (b) The levels of each protein were detected by immunoblot analysis. (c) The structure of a biotinylated probe of a carborane analog. (d) Pull-down assay was performed to detect protein bound to a biotinylated carborane analog in HeLa cell lysate. The proteins were separated by SDS-PAGE, and tubulin was detected by immunoblot analysis.

resulted in binding of protein with the biotinylated probe as shown in Figure 1.3.11d. Further electrospray ionization–liquid chromatography–mass spectrometry (ESI-LC-MS) analysis and immunoblot analysis revealed that the major band around 50 kDa was tubulin. These results indicate that tubulin is a primary target protein of carborane analogs, and the HIF1 inhibitory activity of carborane analogs is mediated by regulation of the tubulin pathway [39].

1.3.7 Conclusion

To data, carborane has been investigated as a unique pharmacophore in drug discovery. For example, a conjugation of carborane enhanced binding affinity of various ligands to their receptors, such as estrogen receptor [40], androgen receptor [41], retinoic acid receptor [42], and vitamin D receptor [43]. In this chapter, various carborane-based HIF inhibitors are described. GN26361 and its derivatives suppressed the hypoxia-induced accumulation of HIF1α and expression of target genes via inhibition of HSP60 activity. Furthermore, carborane analogs of manassantin and combretastatin A4 showed a potent inhibitory activity against accumulation of HIF1α in hypoxic cancer cells. It is suggested that these carborane-based HIF inhibitors provide further insight into the application of carborane frameworks in drug discovery.

References

1. Iwai K, Yamanaka K, Kamura T, Minato N, Conaway RC, Conaway JW, et al. Identification of the von Hippel-Lindau tumor-suppressor protein as part of an active E3 ubiquitin ligase complex. *Proc Natl Acad Sci USA* 1999;96(22):12436–12441.
2. Maxwell PH, Wiesener MS, Chang GW, Clifford SC, Vaux EC, Cockman ME, et al. The tumour suppressor protein VHL targets hypoxia-inducible factors for oxygen-dependent proteolysis. *Nature* 1999;399(6733):271–275.
3. Wenger RH. Cellular adaptation to hypoxia: O2-sensing protein hydroxylases, hypoxia-inducible transcription factors, and O2-regulated gene expression. *FASEB J* 2002;16(10):1151–1162.
4. Zhong H, De Marzo AM, Laughner E, Lim M, Hilton DA, Zagzag D, et al. Overexpression of hypoxia-inducible factor 1alpha in common human cancers and their metastases. *Cancer Res* 1999;59(22):5830–5835.
5. Semenza GL. Targeting HIF-1 for cancer therapy. *Nat Rev Cancer* 2003;3(10):721–732.
6. Scholz M, Hey-Hawkins E. Carbaboranes as pharmacophores: properties, synthesis, and application strategies. *Chem Rev* 2011;111(11):7035–7062.
7. Lee K, Lee JH, Boovanahalli SK, Jin Y, Lee M, Jin X, et al. (Aryloxyacetylamino)benzoic acid analogues: a new class of hypoxia-inducible factor-1 inhibitors. *Journal of Medicinal Chemistry* 2007;50(7):1675–1684.
8. Schenone M, Dancik V, Wagner BK, Clemons PA. Target identification and mechanism of action in chemical biology and drug discovery. *Nat Chem Biol* 2013;9(4):232–240.
9. Lomenick B, Olsen RW, Huang J. Identification of direct protein targets of small molecules. *ACS Chem Biol* 2011;6(1):34–46.
10. Kotzyba-Hibert F, Kapfer I, Goeldner M. Recent trends in photoaffinity labeling. *Angewandte Chemie International Edition in English* 1995;34(12):1296–1312.
11. Kolb HC, Finn MG, Sharpless KB. Click chemistry: diverse chemical function from a few good reactions. *Angew Chem Int Ed Engl* 2001;40(11):2004–2021.
12. Shimizu K, Maruyama M, Yasui Y, Minegishi H, Ban HS, Nakamura H. Boron-containing phenoxyacetanilide derivatives as hypoxia-inducible factor (HIF)-1alpha inhibitors. *Bioorg Med Chem Lett* 2010;20(4):1453–1456.
13. Ban HS, Shimizu K, Minegishi H, Nakamura H. Identification of HSP60 as a primary target of o-carboranylphenoxyacetanilide, an HIF-1alpha inhibitor. *J Am Chem Soc* 2010;132(34):11870–11871.
14. Nakamura H, Yasui Y, Ban HS. Synthesis and biological evaluation of ortho-carborane containing benzoxazole as an inhibitor of hypoxia inducible factor (HIF)-1 transcriptional activity. *Journal of Organometallic Chemistry* 2013;747:189–194.
15. Li G, Azuma S, Sato S, Minegishi H, Nakamura H. ortho-Carboranylphenoxyacetanilides as inhibitors of hypoxia-inducible factor (HIF)-1 transcriptional activity and heat shock protein (HSP) 60 chaperon activity. *Bioorganic & Medicinal Chemistry Letters* 2015;25(13):2624–2628.

16 Li G, Azuma S, Minegishi H, Nakamura H. Synthesis and biological evaluation of meta-carborane-containing phenoxyacetanilides as inhibitors of hypoxia-inducible factor (HIF)-1 transcriptional activity. *Journal of Organometallic Chemistry* 2015;798, Part 1:189–195.

17 Cong Z-q, Nishino H. Manganese(III)-mediated direct introduction of 3-oxobutanamides into methoxynaphthalenes. *Synthesis* 2008;2008(17):2686–2694.

18 Coult R, Fox MA, Gill WR, Herbertson PL, MacBride JH, Wade K. C-Arylation and C-heteroarylation of icosahedral carboranes via their copper (I) derivatives. *Journal of Organometallic Chemistry* 1993;462(1):19–29.

19 Endo Y, Iijima T, Yamakoshi Y, Yamaguchi M, Fukasawa H, Shudo K. Potent estrogenic agonists bearing dicarba-closo-dodecaborane as a hydrophobic pharmacophore. *Journal of Medicinal Chemistry* 1999;42(9):1501–1504.

20 Nielsen KL, Cowan NJ. a single ring is sufficient for productive chaperonin-mediated folding in vivo. *Molecular Cell* 1998;2(1):93–99.

21 Nagumo Y, Kakeya H, Shoji M, Hayashi Y, Dohmae N, Osada H. Epolactaene binds human Hsp60 Cys442 resulting in the inhibition of chaperone activity. *Biochem J* 2005;387(Pt 3):835–840.

22 Itoh H, Komatsuda A, Wakui H, Miura AB, Tashima Y. Mammalian HSP60 is a major target for an immunosuppressant mizoribine. *Journal of Biological Chemistry* 1999;274(49):35147–35151.

23 Chapman E, Farr GW, Furtak K, Horwich AL. A small molecule inhibitor selective for a variant ATP-binding site of the chaperonin GroEL. *Bioorg Med Chem Lett* 2009;19(3):811–813.

24 Ban HS, Shimizu K, Minegishi H, Nakamura H. Identification of HSP60 as a primary target of o-carboranylphenoxyacetanilide, an HIF-1α inhibitor. *Journal of the American Chemical Society* 2010;132(34):11870–11871.

25 Liu YV, Baek JH, Zhang H, Diez R, Cole RN, Semenza GL. RACK1 competes with HSP90 for binding to HIF-1alpha and is required for O(2)-independent and HSP90 inhibitor-induced degradation of HIF-1alpha. *Mol Cell* 2007;25(2):207–217.

26 Hodges TW, Hossain CF, Kim Y-P, Zhou Y-D, Nagle DG. Molecular-targeted antitumor agents: the Saururus cernuus dineolignans manassantin B and 4-O-demethylmanassantin B are potent inhibitors of hypoxia-activated HIF-1. *Journal of Natural Products* 2004;67(5):767–771.

27 Dale GN, Yu-Dong Z. Natural product-based inhibitors of hypoxia-inducible factor-1 (HIF-1). *Current Drug Targets* 2006;7(3):355–369.

28 Hanessian S, Reddy GJ, Chahal N. Total synthesis and stereochemical confirmation of manassantin A, B, and B1. *Organic Letters* 2006;8(24):5477–5480.

29 Kim H, Kasper AC, Moon EJ, Park Y, Wooten CM, Dewhirst MW, et al. Nucleophilic addition of organozinc reagents to 2-sulfonyl cyclic ethers: stereoselective synthesis of manassantins A and B. *Organic Letters* 2009;11(1):89–92.

30 Ogawa T, Ohta K, Iijima T, Suzuki T, Ohta S, Endo Y. Synthesis and biological evaluation of p-carborane bisphenols and their derivatives: structure–activity relationship for estrogenic activity. *Bioorganic & Medicinal Chemistry* 2009;17(3):1109–1117.

31 Jordan MA, Wilson L. Microtubules as a target for anticancer drugs. *Nat Rev Cancer* 2004;4(4):253–265.

32 Perez EA. Microtubule inhibitors: differentiating tubulin-inhibiting agents based on mechanisms of action, clinical activity, and resistance. *Molecular Cancer Therapeutics* 2009;8(8):2086–2095.

33 Pettit GR, Singh SB, Hamel E, Lin CM, Alberts DS, Garcia-Kendall D. Isolation and structure of the strong cell growth and tubulin inhibitor combretastatin A-4. *Experientia* 1989;45(2):209–211.

34 Simoni D, Romagnoli R, Baruchello R, Rondanin R, Rizzi M, Pavani MG, et al. Novel combretastatin analogues endowed with antitumor activity. *J Med Chem* 2006;49(11):3143–3152.

35 Abma E, Daminet S, Smets P, Ni Y, de Rooster H. Combretastatin A4-phosphate and its potential in veterinary oncology: a review. *Vet Comp Oncol* 2015.

36 Mooney CJ, Nagaiah G, Fu P, Wasman JK, Cooney MM, Savvides PS, et alb A phase II trial of fosbretabulin in advanced anaplastic thyroid carcinoma and correlation of baseline serum-soluble intracellular adhesion molecule-1 with outcome. *Thyroid* 2009;19(3):233–240.

37 Lawrence NJ, Ghani FA, Hepworth LA, Hadfield JA, McGown AT, Pritchard RG. The synthesis of (E) and (Z)-combretastatins A-4 and a phenanthrene from Combretum caffrum. *Synthesis* 1999:1656–1660.

38 Hill WE, Johnson FA, Novak RW. Kinetics and mechanism of carborane formation. *Inorganic Chemistry* 1975;14(6):1244–1249.

39 Nakamura H, Tasaki L, Kanoh D, Sato S, Ban HS. Diaryl-substituted ortho-carboranes as a new class of hypoxia inducible factor-1alpha inhibitors. *Dalton Trans* 2014;43(13):4941–4944.

40 Fujii, S., Goto, T., Ohta, K., Hashimoto, Y., Suzuki, T., Ohta, S., Endo, Y. 2005. Potent androgen antagonists based on carborane as a hydrophobic core structure. *J Med Chem* 48, 4654–4662.

41 Endo, Y., Iijima, T., Yamakoshi, Y., Fukasawa, H., Miyaura, C., Inada, M., Kubo, A., Itai, A. 2001. Potent estrogen agonists based on carborane as a hydrophobic skeletal structure. *A new medicinal application of boron clusters. Chem Biol* 8, 341–255.

42 Iijima, T., Endo, Y., Tsuji, M., Kawachi, E., Kagechika, H., Shudo, K. 1999. Dicarba-closo-dodecaboranes as a pharmacophore: retinoidal antagonists and potential agonists. *Chem Pharm Bull* 47, 398–404.

43 Fujii, S., Masuno, H., Taoda, Y., Kano, A., Wongmayura, A., Nakabayashi, M., Ito, N., Shimizu, M., Kawachi, E., Hirano, T., Endo, Y., Tanatani, A., Kagechika, H. 2011. Boron cluster-based development of potent nonsecosteroidal vitamin D receptor ligands: direct observation of hydrophobic interaction between protein surface and carborane. *J Am Chem Soc* 133, 20933–20941.

1.4

Half- and Mixed-Sandwich Transition Metal Dicarbollides and *nido*-Carboranes(–1) for Medicinal Applications

Benedikt Schwarze,[1] Marta Gozzi,[1] and Evamarie Hey-Hawkins*

Universität Leipzig, Institut für Anorganische Chemie, Leipzig, Germany
[1] Contributed equally
* Corresponding author. Universität Leipzig, Institut für Anorganische Chemie, Johannisallee 29, 04103 Leipzig, Germany.
Email: hey@uni-leipzig.de

1.4.1 Introduction

Today, medicinal chemistry is still clearly dominated by organic chemistry, and most of the marketed drugs are purely organic molecules that can incorporate nitrogen, oxygen, and halogens besides carbon and hydrogen. On the other hand, commercial boron-based drugs are still rare [1]. Besides bortezomib, tavaborole (AN2690), crisaborole (AN2728), epetraborole (AN3365), and SCYX-7158 (AN5568) [2], L-4-(dihydroxyboryl)phenylalanine (BPA) and sodium mercapto-undecahydro-*closo*-dodecaborate (BSH) are used as drugs in boron neutron capture therapy (BNCT) [3–5].

Like carbon, boron readily forms compounds with covalent boron–hydrogen bonds and also boron–boron interactions. However, in contrast to hydrocarbons, boranes prefer the formation of polyhedral clusters with fascinating globular architectures [6]. Most boranes are unstable in aqueous environment; an exception is *closo*-$B_{12}H_{12}^{2-}$. In contrast, polyhedral carboranes, in which two BH^- units of *closo*-$B_{12}H_{12}^{2-}$ are replaced by two CH vertices (*closo*-$C_2B_{10}H_{12}$,[1] dicarba-*closo*-dodecaborane(12), carborane or carbaborane), have remarkable biological stability. Furthermore, the two carbon atoms are versatile starting points for various organic modifications. Carboranes are of special interest due to their unique properties that cannot be found in organic counterparts. These unique properties are based on the element boron, due to its inherent electron deficiency, lower electronegativity, and smaller orbital size compared to carbon. Of the borane clusters and heteroboranes, the three dicarba-*closo*-dodecaborane(12) isomers (*ortho* (1,2-), *meta* (1,7-), and *para* (1,12-dicarba-*closo*-dodecaborane(12)), each of which has specific electronic properties, have attracted much interest. Besides the use

---

1 Nomenclature adopted for carborane clusters (according to IUPAC convention): *closo*- = 12-vertex icosahedral cluster, with (*n*–1) skeletal electron pairs (*n* = total number of vertices); *nido*- = 11-vertex open-face cluster, with (*n*–2) skeletal electron pairs (*n* = total number of vertices); and *ortho*-, *meta*-, *para*- = 1,2-, 1,7-, 1,12-dicarba-*closo*-dodecaborane(12), respectively.

Boron-Based Compounds: Potential and Emerging Applications in Medicine, First Edition.
Edited by Evamarie Hey-Hawkins and Clara Viñas Teixidor.
© 2018 John Wiley & Sons Ltd. Published 2018 by John Wiley & Sons Ltd.

of icosahedral carborane clusters (here, *ortho*-carborane), anionic *nido* clusters (7,8-$C_2B_9H_{11}^{2-}$), derived from the neutral icosahedral *closo* clusters 1,2-dicarba-*closo*-dodecaborane(12) by deboronation followed by deprotonation, offer the possibility to form metallacarboranes via coordination of the C_2B_3 pentagonal face to a metal atom or cation [7].

For medicinal applications, carboranes are mainly used to design BNCT agents [8–10]. Combining the cluster with a tumor-targeting vector [11] is a common principle and a prerequisite for therapeutic application [12,13]. Furthermore, carboranes were found to be very good scaffolds for diagnostic and therapeutic labeling [14]. In addition, the use of carboranes as pharmacophores is increasingly being studied [15–20]. For applications in medicine, their favorable properties are (1) hydrophobicity, which (due to the presence of hydridic B–H units) is a beneficial property for transport across cell membranes and the blood–brain barrier (BBB) as well as for hydrophobic interactions with binding pockets in enzymes [21]; (2) the inorganic nature of carborane clusters, which prevents enzymatic degradation; (3) their non-toxicity, which lowers the toxic side effects of the potential drugs; (4) the relatively high neutron capture cross-section for application as BNCT agents in cancer treatment; and (5) the diversity of three isomers (*ortho*-, *meta*-, and *para*-carborane or 1,2-, 1,7-, and 1,12-dicarba-*closo*-dodecaborane(12), respectively) with distinct properties. The carborane cluster can be regarded as a three-dimensional aromatic counterpart of a phenyl ring; the volume of a carborane cluster is, however, approximately 50% larger than the volume of a benzene molecule rotating around one of the C_2 axes [22]. The strongest electron-withdrawing effect is observed at the C-vertex in the *ortho* isomer, whilst this effect is weakest in the *para* isomer. Depending on the position of substitution, the carborane moiety may behave as an electron acceptor, as in the case of the *ortho*-carboran-1-yl substituent ($\sigma_i = +0.38$), or as an electron donor, such as the *ortho*-carboran-9-yl substituent ($\sigma_i = -0.23$)[2] [23]. The different charge distribution throughout the cluster vertices can influence the reactivity of the carborane cluster in substitution reactions allowing asymmetric selective substitution of either the carbon or specific boron positions. The independent substitution of the C- and the B-vertices of the carborane cluster is easily possible since the (electronic) character of the hydrogen bound to either the boron or carbon atom differs. However, substitution reactions of *closo*- and *nido*-carboranes differ. For mono-substitution at the C-atoms, the *closo* cluster is treated with 1.00 eq. of *n*-BuLi, followed by addition of a suitable precursor of the desired substituent (Scheme 1.4.1). This chemistry was extensively reviewed by Teixidor and Viñas in 2005 [24] and Scholz et al. in 2011 [25]. When the corresponding 7,8-dicarba-*nido*-dodecahydroundecaborate(−1) (*nido*-carborane) is reacted with *n*-BuLi, the proton of the five-membered open face (C_2B_3) is removed, leading to the lithium salt of the dicarbollide anion ($Li_2[C_2B_9H_{11}]$) [26]. Therefore, *closo* clusters are usually first functionalized and then deboronated to obtain substituted *nido*-carboranes.

Another approach for preparing *C*-substituted *o*-carboranes is the reaction of *arachno*-borane ($[B_{10}H_{12}]^{2-}$) structures with functionalized alkynes [27].

On the other hand, the derivatization of the boron atoms follows different synthetic approaches [28]. In general, three types of substitution reactions are applied for

---

2 σ_i indicates, in all cases, the induction constant at the defined position of the carborane cluster.

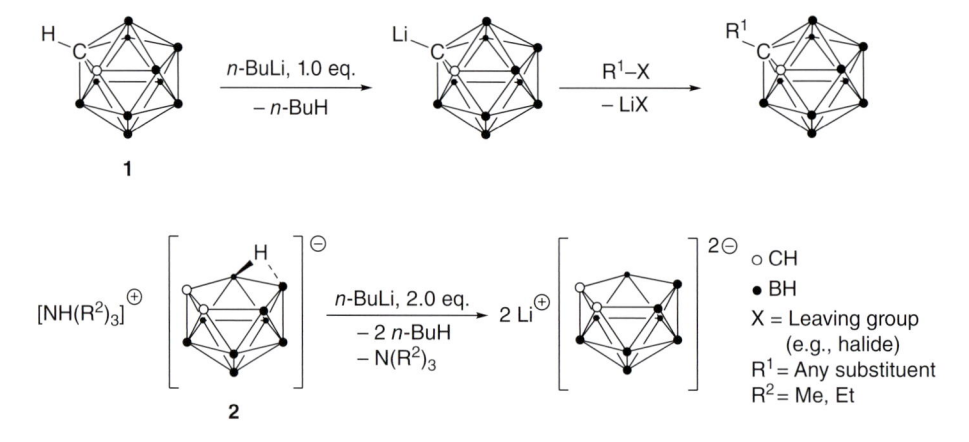

Scheme 1.4.1 Treatment of 1,2-dicarba-*closo*-dodecaborane(12) (**1**) and 7,8-dicarba-*nido*-dodecahydro undecaborate(–1) (**2**) with *n*-BuLi [25,26].

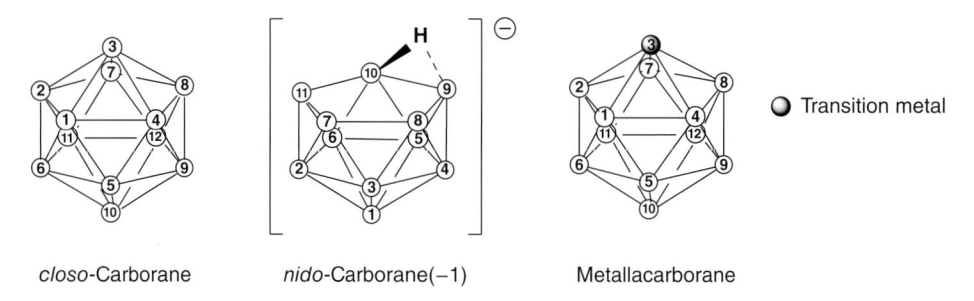

Figure 1.4.1 Numbering scheme for *closo*- (left) and *nido*-carborane(–1) (middle) clusters. Metallacarboranes, where the metal center is "π-like-coordinated" to the upper belt, are classified as *closo* structures (right). For a detailed description of the numbering scheme in metallacarborane complexes refer to [6c]. Here, only one isomer is shown.

nido-carborane species [29]. First is the electrophilic substitution, which is characterized by the reaction of electrophiles (such as *N*-halosuccinimides, organophosphorus or organosulfur halides) with the atoms of highest electronegativity (B(9) or B(11)), leading to asymmetric reaction products, 9-R-*nido*-7,8-$C_2B_9H_{11}^3$. The substitution mechanism is thought to be very similar to that in arenes [28a].

The second approach is the oxidative substitution employing mild oxidants (e.g., $FeCl_3$ [30], $CuSO_4$ [31], or $HgCl_2$ [32]) in order to oxidize the hydride in B–H followed by the substitution with nucleophiles (such as MeCN, THF, Me_2S [33], etc.). Mechanistically, the reaction is thought to proceed via oxidation of the *nido*-carborane(–1), producing a transient 11-vertex *closo*-cluster intermediate that is highly fluxional and interconvertible according to the *dsd* mechanistic concept [34]. Subsequently, the intermediate structure is attacked by nucleophiles (e.g., Me_2S) at a boron vertex next to the carbon

---

3 For the numbering scheme of *closo*-carboranes and *nido*-carboranes, see Figure 1.4.1.

atoms, resulting in the formation of the corresponding *nido* derivative (under these conditions, four possible isomers are formed). In addition, the C–C unit can also "move" within the cluster stabilized through the vertex flip mechanism [33,35].

The third type of reaction is the so-called precise hydride abstraction followed up by the attack of a Lewis base. Typically, Lewis acids like $BF_3 \cdot OEt_2$, $AlCl_3$, and H^+ tend to abstract a hydride from a B–H unit. The cooperation of electrophilic Lewis acids and nucleophilic donors is called *electrophile-induced nucleophilic substitution* (EINS) [28a]. Until recently, the use of two distinct molecules was necessary for this reaction, one nucleophile and one electrophile; in 2014, Frank et al. successfully combined both functionalities into one molecule by using heterodienes, resulting in either symmetric or asymmetric reaction products depending on the substrate [36].

While in the beginning of carborane chemistry a library of compounds was synthesized to examine their reactivity (as summarized by Grimes in 2011) [37], today, the intended structures are more and more often designed according to their potential application, whereby analogies to "classical" organic and coordination chemistry are used.

In Scheme 1.4.2, only a few selected examples are shown for applications of 12- and 11-vertex carborane clusters in medicine (**i–iv**), materials science (**v** and **vi**), and catalysis (**vii–ix**). The carborane moiety in compound **i** has a very simple structure; however, it is coupled to a modified neuropeptide Y ($[F^7,P^{34}]$-NPY), which facilitates the introduction of boron-containing molecules into breast cancer cells via the Trojan horse strategy [11]. Since $[F^7,P^{34}]$-NPY is selective for only one receptor subtype (the hY_1-receptor, which is overexpressed in breast cancer cells), this combined approach incorporates high affinity, high selectivity, and high internalization, which are the prerequisites for application as BNCT agents. The synthesis of asborin (**ii**), the carborane analog of aspirin, utilized the phenyl ring imitation strategy. In comparison to its analog (aspirin), asborin can be used as a new lead structure to fine-tune drug activity and incorporate new properties in order to overcome the problem of specificity of the classical AKR inhibitors, since AKR families have a highly conserved binding domain (discussed in Chapter 1.4.3.2). In this case, the carborane moiety in asborin not only allows to address the hydrophobic and big AKR TIM barrel, but the highly reactive acetyl group also improves lysine acetylation, necessary for a biological impact [16]. A very prominent moiety that is used in all applications is the cobaltabis(dicarbollide) (COSAN, $[3,3'\text{-Co}(1,2\text{-}C_2B_9H_{11})_2]^-$) structure, due to its robustness and the easiness of functionalization. Shown here is an example from the medicinal area (compound **iii**), which is functionalized with radioactive iodine-125 and a short polyethylene glycol (PEG) chain and is used for imaging [38]. As this chemistry is already extensively reviewed elsewhere [39], COSAN derivatives will not be included here. The Ru(*p*-cymene) (*p*-cymene = 1-*i*-Pr-4-Me-C_6H_4) fragment is a prominent moiety used in complexes for treatment of cancer. In 2009, Wu et al. combined Ru(*p*-cymene) with a dithiolated, electron-deficient *o*-carborane moiety, functionalized with a ferrocenyl unit (**iv**), which should also increase the biological activity [40].

Compounds **v** and **vi** exhibit luminescent properties and could, therefore, be employed in materials science as well as in medicine (as fluorescence markers and boron-rich compounds). The presence and nature of appended star-shaped carborane moieties within a dendritic framework determine the extent of luminescence, also depending on the second substituent at the second cage carbon atom [41,42].

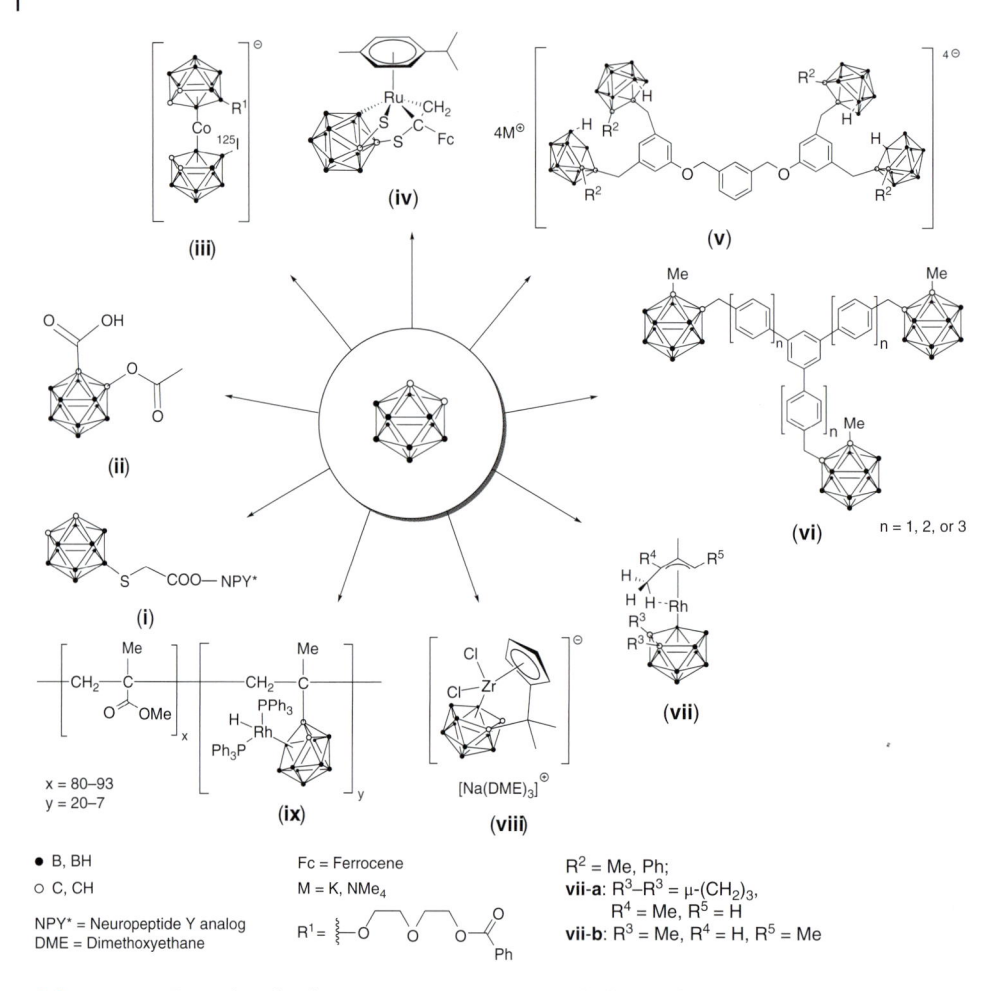

Scheme 1.4.2 Examples of carborane-containing compounds derived from 12-vertex 1,2-dicarba-*closo*-dodecaborane(12) and 7,8-dicarba-*nido*-dodecahydroundecaborate(−1) with applications in catalysis, medicine, and materials science [11,16,38,40–45].

In catalysis, the substituted carborane clusters are used either as an electron-withdrawing backbone in suitable ligands (*exo*-coordination) or as a ligand in metallocene-like structures (**vii–ix**) [43–45]. Very recently, we published a review on catalytically active metallacarborane complexes, like **vii**, which catalyzes hydroformylation reactions [46]. Other rhodium complexes (e.g., **ix**, which is embedded in a polymer matrix) catalyze the hydrogenation and isomerization of unsaturated hydrocarbons [45]. Compound **viii** is a carborane analog of zirconocene dichloride and, similarly to many group 4 metallocenes, is active in ethylene polymerization processes in the presence of methylalumoxane (MAO) as co-catalyst [44].

One main motivation for the use of $[C_2B_9H_{11}]^{2-}$ clusters (dicarbollide anion, Cb^{2-}), as ligands for coordination to a metal center to design compounds for either catalytic or biological applications, is the isolobal analogy with the cyclopentadienyl(−1) ligand (Cp^-) [47]. Both Cb^{2-} and Cp^- ligands can donate six electrons to the metal center, and

Figure 1.4.2 General structure of η^5-coordinating Cb^{2-} and Cp^- ligands to a metal center.

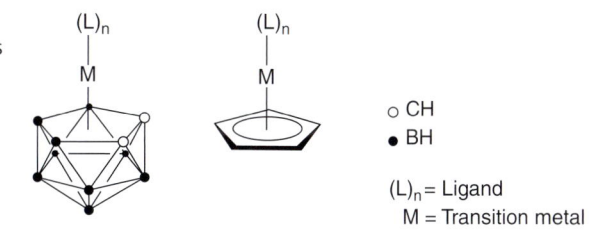

○ CH
● BH

$(L)_n$ = Ligand
M = Transition metal

both can coordinate via the pentagonal face (Figure 1.4.2), although the different charges, sizes, symmetries, and hybridization of the orbitals lead to mostly distinct activities and reactivities, as described recently (see [46] and references therein). However, in terms of geometries, Cb^{2-} and Cp^- complexes show many similarities.

Half-, mixed-, and full-sandwich as well as bent-metallocene structures are found throughout the chemistry of metallacarboranes [37,48]. Recently, the use of energy decomposition analysis (EDA) has given further insights into the strength and nature of the metal–ligand bonding interaction, for a small series of mixed-sandwich d^9 complexes of type $[3\text{-}(\eta^5\text{-}C_5R_5)\text{-}3,1,2\text{-}MC_2B_9H_{11}]$, where M = Co^{III}, Rh^{III}, Ir^{III} and R = H, Me [49]. In brief, the EDA method is based on the bonding interaction between fragments (e.g., A and B in the molecule A–B), which is described as the sum of three energy components,

$$\Delta E_{int} = \Delta E_{elstat} + \Delta E_{Pauli} + \Delta E_{orb}$$

where ΔE_{int} is the interaction energy between A and B, ΔE_{elstat} the quasi-classical electrostatic interaction energy, ΔE_{Pauli} the Pauli's repulsion energy between the fragments, and ΔE_{orb} the energy gain due to orbital mixing [50]. The three energy decomposition terms express much useful chemical information on the fragments and on their interaction, such as the electrostatic/covalent bonding character, the extent of conjugation and aromaticity, and the extent of σ-donation and π-backbonding in transition metal complexes.

Kudinov et al. studied the metal–ligand bonding interaction, considering as fragments either $[MCp]^{2+}$ and $[Cb]^{2-}$, or $[MCb]^+$ and $[Cp]^-$ (Figure 1.4.3) [49]. Fragmentation pattern 1 allowed to study the M–Cb bonding interaction, and fragmentation pattern 2 the M–Cp interaction. The results indicate a stronger bonding interaction between the metal center and the Cb^{2-} ligand than for the corresponding M–Cp interaction, the latter being on average 320 kcal/mol weaker than the relative M–Cb interaction (ΔE_{int}, Table 1.4.1). Moreover, the M–Cb interaction exhibits a higher electrostatic character than the corresponding M–Cp interaction (ΔE_{elstat}, Table 1.4.1).

Previous reviews of metallacarboranes of 11-vertex clusters have covered full-, half-, and mixed-sandwich complexes, often together with *exo*-coordinated metal complexes [51]. In this review, we focus on half- and mixed-sandwich metallacarborane complexes of the ligands $[C_2B_9H_{12/11}]^{-/2-}$ (and in some cases, also $[C_3B_8H_{12}]^-$), as well as on hybrid organic-inorganic molecules containing the $[C_2B_9H_{12}]^-$ cluster, with potential biological relevance. Their synthesis, chemical reactivity, and mode of action in biological media can differ significantly from, for example, the COSAN-type structures (i.e., full-sandwich complexes) [52].

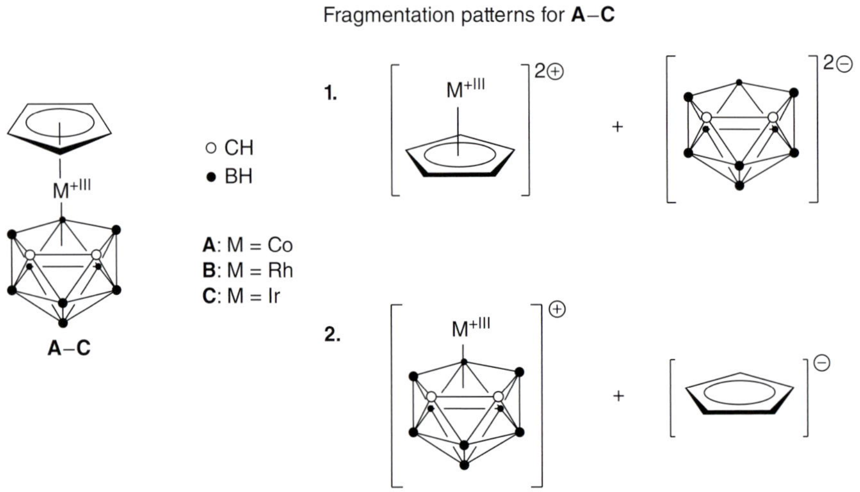

Fragmentation patterns for A–C

○ CH
● BH

A: M = Co
B: M = Rh
C: M = Ir

Figure 1.4.3 Mixed-sandwich group 9 metallacarborane complexes incorporating a Cp⁻ ligand subject to EDA analysis by Kudinov et al. [49]. The two fragmentation patterns studied are shown.

Table 1.4.1 Selected results of EDA analysis on structures A, B, and C from Figure 1.4.3 [49]

Complex[a]	Fragmentation pattern 1			Fragmentation pattern 2		
	A	B	C	A	B	C
ΔE_{int}	−572.68	−552.36	−579.00	−252.74	−235.64	−255.78
ΔE_{elstat}[b]	−510.75 (62.6%)	−537.34 (62.8%)	−598.64 (62.9%)	−258.45 (56.2%)	−265.50 (58.0%)	−316.94 (57.9%)
D_e[c]	−563.08	−538.30	−561.75	−242.55	−227.96	−238.85

[a] Geometry optimization and frequency calculation were done using PBE/L2; Morokuma Ziegler energy decomposition analysis, as implemented in ADF 2006, was calculated at the BP86/TZ2P level of theory. Computational details of the study can be found in Ref. [49]. Energy values are given in kcal/mol.
[b] Values in brackets indicate the percentage of electrostatic interaction with respect to the total attractive interactions.
[c] D_e is the bond dissociation energy, according to the formula $D_e = -(\Delta E_{int} + \Delta E_{prep})$, where ΔE_{prep} is the fragment preparation energy.

1.4.2 Synthetic Approaches to *nido*-Carborane [C₂B₉H₁₂]⁻ Derivatives

When *nido*-$[C_2B_9H_9R_2]^{2-}$ (dicarbollide anion) derivatives (with R = any substituent) are used as ligands for π-type complexation to a metal center or as a prosthetic group or pharmacophore (in their protonated form, *nido*-$[C_2B_9H_{10}R_2]^-$), the synthesis starts from the commercially available parent *closo*-carboranes, 1,2- or 1,7-dicarba-*closo*-dodecaborane(12) [53].

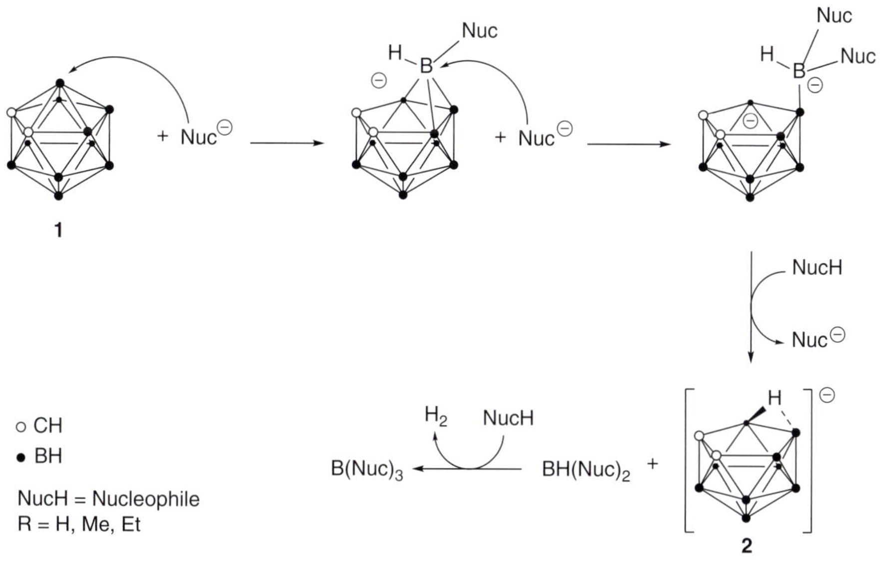

Scheme 1.4.3 Mechanism of deboronation of commercially available *ortho*-carborane [25].

○ CH
● BH

NucH = Nucleophile
R = H, Me, Et

One of the most electropositive BH groups is removed from the 12-vertex cluster via regioselective nucleophile-promoted abstraction (decapping or deboronation)[4] [25]. Two equivalents of a nucleophile (or base) are required to remove the BH vertex (Scheme 1.4.3), which is finally eliminated as borane B(Nuc)$_3$ (Nuc = EtO$^-$, MeO$^-$, etc.). The borane is observed as a very characteristic sharp signal (FWHH ca. 10 Hz) in the ^{11}B NMR spectrum (in the region between circa 30 ppm and circa 20 ppm, depending on the type of nucleophile), which is well separated from the broader signals of the *nido* cluster (**2**), which are found at a much higher field (ca. −38 to −5 ppm; Figure 1.4.4). Sources of line broadening in polyhedral boranes and carboranes, with respect to the natural line width, were already studied by Grimes *et al.* [54,55]. An ensemble of phenomena, such as partially or totally collapsed multiplets, scalar contributions to the spin–spin relaxation time, and unresolved ^{11}B–^{11}B spin–spin coupling, all of which are temperature-dependent factors, contributes to the observed line shape and line width [54]. As a result, it is often not possible to resolve $^1J_{BB}$ couplings from one-dimensional ^{11}B NMR spectra, making ^{11}B,^{11}B 2D NMR experiments a very powerful tool for the study of bonding interactions and, therefore, electronic structures of such boron-based polyhedra [55].

Deboronation of *ortho*-carborane produces 7,8-dicarba-*nido*-dodecahydroundecaborate(−1), while *meta*-carborane yields 7,9-dicarba-*nido*-dodecahydroundecaborate(−1); both isomers have the two carbon atoms in the open upper C$_2$B$_3$ face (Figure 1.4.1) [53].

Isomerization of such *nido*-carboranes to cage systems, where the second carbon atom occupies a position on the lower boron belt, can take place during complexation

---

4 For the *ortho* isomer, B(3) or B(6) are the most electropositive boron vertices, being adjacent to both carbon atoms. For the *meta* isomer, these are B(2) and B(3).

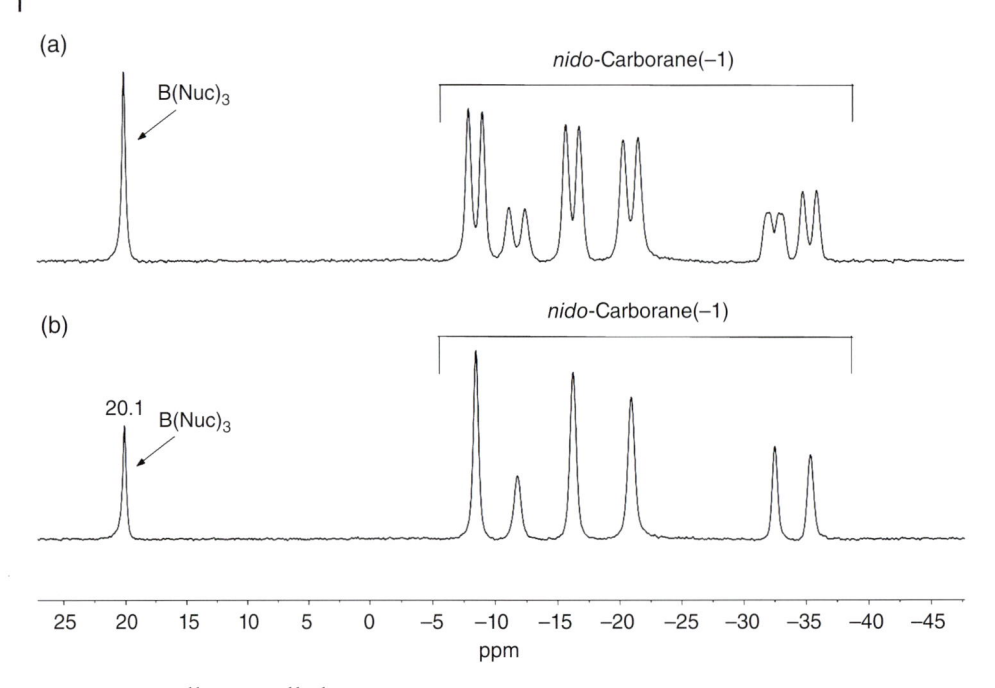

Figure 1.4.4 (a) ^{11}B and (b) ^{11}B{^{1}H} NMR spectra of a C-mono-substituted *nido*-carborane(−1) anion (in CD$_3$CN at 128.4 MHz). The signal corresponding to the borane derivative is observed at 20.1 ppm, whereas the signals of the *nido*-carborane(−1) are found in the region between −35.4 and −8.5 ppm with the pattern 2:1:2:2:1:1.

to a metal center, due to steric effects and the conditions of the complexation reaction [56,57]. Microwave heating or simply heating at reflux has led in many cases to formation of 2,1,8- and 1,7,9-isomers of the metallacarborane complex, instead of the expected isomers, with all carbon atoms located on the upper face [14,58].

One of the most interesting examples in this category is surely the microwave-assisted synthesis of dicarbollide-based radiopharmaceuticals of Re^I and ^{99m}Tc^I, pioneered by the Valliant group [57], in which a dependence of the cage rearrangement process on the type of substituent(s) at the cage carbon vertex or vertices was observed. When reacted under microwave heating with the metal precursor [M(CO)$_3$(H$_2$O)$_3$]Br (with M = Re, ^{99m}Tc), the 1-(2-methoxyphenyl)-piperazine-substituted *nido*-carborane gave the rearranged metallacarborane complexes *rac*-8-[*N*-[7-[4-(2-methoxyphenyl)-piperazin-1-yl]butyl]]-2,2,2-tricarbonyl-2-M-2,1,8-dicarba-*closo*-dodecaborate(−1)-8-carboxamide (**7, 8**) and *rac*-8-[(4-(2-methoxyphenyl)-piperazin-1-yl)methyl]-2,2,2-tricarbonyl-2-M-2,1,8-dicarba-*closo*-dodecaborate(−1) (**5, 6**), where the linker between the cage carbon atom and the 1-(2-methoxyphenyl)piperazine moiety is either an amide (**7, 8**) or a methylene group (**5, 6**), respectively (Figure 1.4.5) [59,60].

The migration of the carbon vertex bearing the substituent with a methylene group is attributed to a release of steric strain, although also electronic contributions are not ruled out, as stated by the authors themselves [59]; when the spacer contains an amide group, it is suggested that the electron-withdrawing nature of the amide group might lower the energy barrier for isomerization, as it had previously been observed for

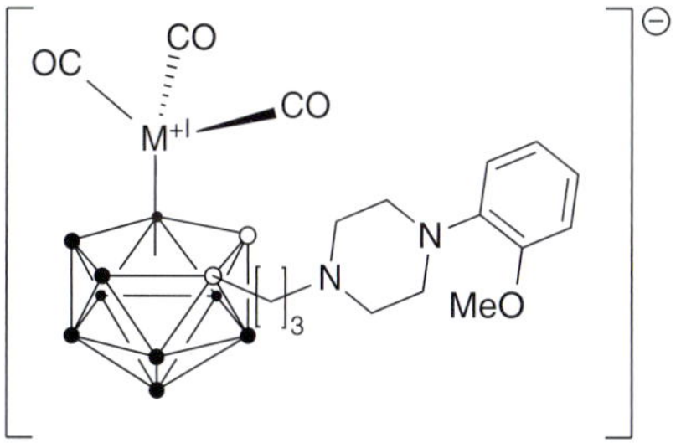

3, 4

3: M = Re
4: M = ^{99m}Tc
5: M = Re
6: M = ^{99m}Tc
7: M = Re
8: M = ^{99m}Tc

○ C or CH
● BH

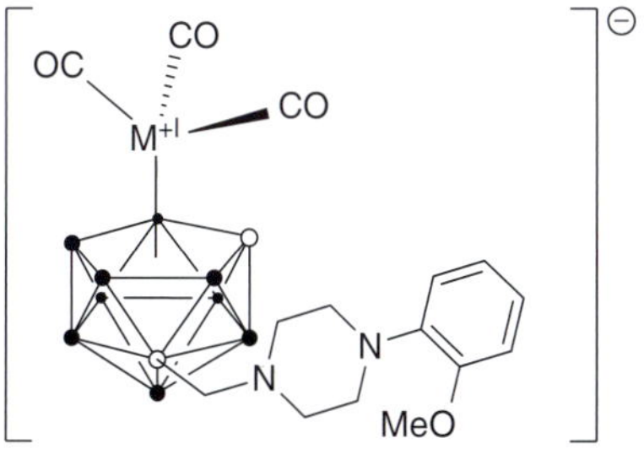

5, 6

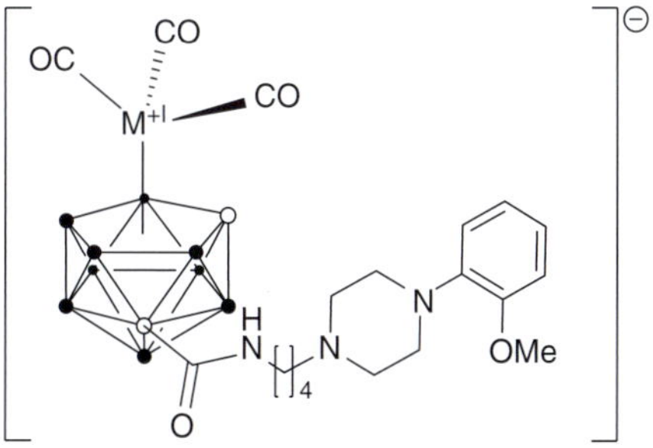

7, 8

Figure 1.4.5 Anionic Re^I– and $^{99m}Tc^I$–dicarbollide complexes, where the cluster bears a 1-(2-methoxyphenyl)-piperazine substituent at one carbon vertex. The counterion (Na^+) is omitted [59,60].

other rhenacarborane complexes, such as K[2,2,2-$(CO)_3$-8-C_5H_4N-2,1,8-*closo*-$ReC_2B_9H_{10}$] [57]. Isomerization to the 2,1,8-configuration may also contribute to a release of steric strain between the $Re(CO)_3$ fragment and the *nido* cluster.

No isomerization was observed with a longer alkyl linker, namely a propyl group, under the same reaction conditions, and *rac*-1-[(4-(2-methoxyphenyl)piperazin-1-yl)-propyl]-3,3,3-tricarbonyl-3-M-3,1,2-dicarba-*closo*-dodecaborate(−1) (**3**, **4**) was isolated as a racemate.

Preliminary studies on the isomerization mechanism according to the specific metal center for the cluster bearing the 1-(2-methoxyphenyl)piperazine group and the methylene spacer (**5**, **6**) have been performed [59]. For Re^I, analysis of the products of the complexation reaction (performed in the temperature range 100–200 °C) by ^{1}H NMR spectroscopy, at a fixed reaction time (10 min), gave no evidence of cluster isomerization prior to complexation, nor of formation of the 3,1,2-isomer. Only formation of the 2,1,8-isomer was detected by ^{1}H NMR spectroscopy. On the other hand, analysis of the reaction mixture starting from [$^{99m}Tc(CO)_3(H_2O)_3$]Br by high-performance liquid chromatography (HPLC) gave evidence of formation of the metallacarborane complex with a 3,1,2-configuration prior to isomerization to 2,1,8, which occurred at temperatures higher than 190 °C. Reasons for this different behavior are postulated to be due to the different redox potentials of the two metals, but this hypothesis is still being tested.

Since the pioneering work of Hawthorne and coworkers [53], several different methods have been developed for the decapping of *ortho-* and *meta*-carborane clusters. Today, nucleophiles such as amines, fluorides, alkoxides, and combinations thereof have become standard reagents used for deboronation of the carborane clusters [25,61]. In several cases, the choice of the deboronating agent is strongly influenced by the presence and type of substituent(s) on the carbon vertices. Therefore, tailoring of the deboronation method (i.e., reagents and conditions) is often required to gain access in high yields to the desired decapped clusters, while at the same time avoiding undesired side reactions, such as cleavage of the C-bound substituents, especially for electron-withdrawing groups.

The presence of an electron-withdrawing group in the α (or β) position to one or two carbon vertices often promotes deboronation, even with weak bases and nucleophiles, which do not attack the unsubstituted cluster. This is, for example, the case in asborin [62], for which aqueous hydrolysis competes with the deboronation reaction, and water itself acts as the deboronating agent. Also, α-amide [19,63] and α-carboxyl groups [64] have been shown to promote BH removal in wet methanol, acetonitrile, and dimethyl sulfoxide (DMSO). These observations must be considered when applying such *closo* clusters in aqueous biological systems, because deboronation not only results in a negative charge but also changes the whole electronic structure and polarizability of the *nido*-carborane moiety.

Genady et al. recently studied the deboronation of carborane-functionalized tetrazines in wet DMSO-d_6 by ^{1}H NMR spectroscopy [63]. The reaction was completed at room temperature after 30 min, which led the authors to scale up the procedure to laboratory scale, using a MeCN–H_2O mixture at 35 °C for 24 h (62% yield).

The use of fluorides (NaF, CsF, and TBAF) in polar protic solvents (EtOH or EtOH–H_2O) was shown to promote selective deboronation without cleavage of the C-bound groups for a multitude of substituents [65], such as alkylene spacers [59,66] and amide [60], carboxy, and hydroxy groups [67].

Standard deboronation conditions (NaOH in MeOH) have been used by Valliant and coworkers to remove a BH vertex from carborane–carbohydrate conjugates bearing a C-bound benzamide substituent, allowing isolation of the *nido* product in 76% yield [68].

In the search for milder reaction conditions that would give access to radiolabeled metallacarboranes without cleaving the vector, Valliant et al. (2012) introduced guanidine-based substituents at the C-vertex, with a methylene linker (Figure 1.4.6) [69].

The presence of an internal base (guanidine) is regarded to be beneficial for the subsequent complexation step, as it should facilitate removal of the bridging proton from the upper C_2B_3 face of the *nido*-carborane cluster. Three different N-substituted guanidine derivatives were introduced selectively at one C-vertex to give [1-($CH_2NHC(NH_2)$-NHR)-1,2-$C_2B_{10}H_{10}$][TFA] (with R = H, Et) and [1-($CH_2NHC(NH)NHNO_2$)-1,2-$C_2B_{10}H_{10}$].

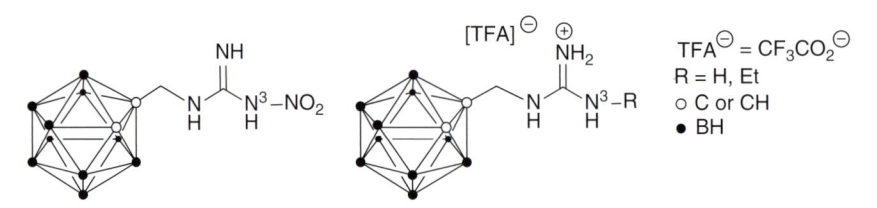

Figure 1.4.6 *closo*-Carboranes bearing a guanidine substituent at one C-vertex. The nitrogen atom N^3 of the guanidine group carries different substituents to modify basicity properties [69].

The nature of the substituent at N^3 of the guanidine group modulates its basicity: the N-nitro guanidine derivative is the least basic one, and the N-ethyl-substituted guanidine derivative the most basic. In all three cases, decapping was achieved by heating a water–ethanol solution of the compound in the presence of NaF to 85 °C for 3–5 h. Almost-quantitative yield was obtained in the case of the N-ethyl-substituted guanidine species, which was attributed to the higher basicity of the N-alkyl-guanidine within the series and to the electron-withdrawing properties of the guanidine group. Reaction of the C-guanidine-substituted *nido*-carboranes with Re^I and $^{99m}Tc^I$ complexes under mild reaction conditions (room temperature for 3 h, or 35 °C for 1 h) also supported the hypothesis that an internal basic group promotes abstraction of the bridging proton and facilitates complexation with $[Re(CO)_3]^+$ and $[^{99m}Tc(CO)_3]^+$ fragments under much milder conditions than the well-established microwave heating approaches [14].

The third 12-vertex isomer, 1,12-dicarba-*closo*-dodecaborane(12), with the two carbon vertices in *para* position, is the most symmetric and least polarized of the three isomers. The common synthetic procedures for decapping of *ortho* and *meta* isomers in high yields (i.e., KOH in ethanol solution) fail to give access to the corresponding *nido* isomer (2,9-$[C_2B_9H_{12}]^-$) of *para*-carborane. However, yields higher than 95% of the 2,9-*nido*-carborane were obtained by Hawthorne *et al.* by heating a benzene solution of *para*-carborane for 42 hours in the presence of KOH and 18-crown-6 ether [70]. Protic solvents like ethanol were found to tremendously decrease the yield of the reaction.

In 2007, Welch *et al.* reported the reduction of the diphenyl-substituted *para*-carborane at room temperature with sodium in tetrahydrofuran (THF), and subsequent complexation with Ru(*p*-cymene) [71]. Density functional theory (DFT) calculations of the reaction pathways showed that the cluster undergoes cage rearrangement to two $[C_2B_{10}]^{2-}$ type anions, namely the 1,7- and 4,7-isomers, which in turn produce three more isomers and are finally isolated as five different supra-icosahedral metallacarborane isomers [72], among which the 4,1,11-isomer was postulated but not isolated.

This last example by Welch and coworkers shows that even though nowadays much is known about the chemistry and the reaction mechanisms of the 12-vertex *closo*-carboranes, there are still new structures to be discovered. This is especially true for metallacarborane complexes, where the possibilities of combining a metal–ligand fragment with a suitable dianionic ligand can extend well beyond the chemistry of the 11-vertex dicarbollide ligand. Here, the 11-vertex tricarbollide $[C_3B_8H_{11}]^-$ should be mentioned [58], which is not obtained by cage degradation (removal of a BH vertex), but by insertion of the third carbon vertex into neutral *arachno*-5,6-$[C_2B_8H_{12}]$ or anionic *arachno*-5,6-$[C_2B_8H_{11}]^-$. Nucleophilic attack at the most electropositive carbon vertex using NaCN or *t*-BuNC together with hydrogenation of the resulting C=N bond gave C-amino-substituted 7,8,9-$[C_3B_8H_{10}L]$ clusters (L = amine) [73]. Complexation can be achieved in two ways: the *nido*-tricarborane cluster can be deprotonated using $NaOH–TlNO_3$ and then reacted with $[CpFe(CO)_2]I$, or it can be directly reacted at reflux temperature with the dimer, $[CpFe(CO)_2]_2$, resulting in a mixed-sandwich iron(II)–cyclopentadienyl complex bearing an 11-vertex tricarbollide moiety [74]. Cage isomerization to the 1-L-1,7,9-$[C_3B_8]$ type (L = amine) occurred upon complexation, yielding the "*para*-substituted" ferratricarborane complexes.

1.4.3 Biologically Active Organometallic *nido*-Carborane Complexes and Organic *nido*-Carborane Derivatives

In this section, most recent examples of organometallic complexes incorporating a transition metal–*nido*-carborane unit (metallacarboranes), which have been studied for their biological activities, as well as examples of biologically active hybrid organic-inorganic molecules containing the *nido*-carborane cluster alone are discussed. A specific focus is on half- and mixed-sandwich transition metal complexes with a pentahapto-coordinating $[C_2B_9H_9R_2]^{2-}$ or $[C_3B_8H_8R_3]^-$ ligand (with R = any substituent) (Chapter 1.4.3.1) [58,75,76]. On the other hand, studies of the biological activity of compounds containing the *nido*-carborane alone cover a wider spectrum of applications and approaches (Chapter 1.4.3.2), ranging from BNCT to pharmacophores in drug design, radio-imaging applications, and combinations thereof [11,19,77].

Recently, Leśnikowski published a detailed review on the most recent developments in drug design based on the use of boron clusters [15d]. Many examples of biologically active compounds are described, ranging from substituted *closo-ortho*-carboranes and appended *nido*-carboranes to a few examples of metallacarboranes of the COSAN type.

nido-Carborane conjugates of biomolecules were already described by Grimes [15c]. These include non-classical antifolates [78] and derivatives of 5,10,15,20-tetraphenyl-porphyrins [79] as antibacterial and anticancer agents, or thymidine analogs [80], adenosine conjugates [81], and carbon nanotubes with appended *nido*-carborane [82] as potential BNCT agents.

An overlap between the present chapter and the reviews by Leśnikowski [15d] and Grimes [15c] is as far as possible avoided. However, a few examples and some other aspects will be recalled here, as otherwise the present review would be incomplete.

1.4.3.1 Biologically Active Half- and Mixed-Sandwich Metallacarborane Complexes

The metallacarborane complexes described here have been investigated predominantly as potential novel radio-imaging agents, for diagnostics and prospective radiotherapies.

In addition, two examples of mixed-sandwich metallacarborane complexes are described, where the *nido*-carborane ligand is used as a pharmacophore to impart potency and high selectivity for the desired biological target(s), and to increase cellular and BBB uptake.

1.4.3.1.1 Half-Sandwich Complexes of Rhenium(I) and Technetium(I)-99 m as Radio-Imaging and Radiotherapeutic Agents

The development of radio-imaging agents is a very broad and varied ensemble of interdisciplinary fields, ranging from biochemistry to chemistry, materials science, and imaging technologies. A comprehensive description of the progress of radio-imaging techniques exceeds the scope of this chapter, but can be found elsewhere from a medical-oncological perspective [83] and from a biochemical perspective [84].

One of the approaches that have been explored most extensively in the last decade is the development of so-called molecular imaging probes (MIPs), agents used to visualize, characterize, and quantify biological processes *in vivo*, at the molecular and cellular level [84]. The potential of such an approach is straightforward. It allows detection and quantification of biological pathways and/or status in a non-invasive way; in other words, the system is left intact after detection. A necessary condition for the development of efficient MIPs is in fact that the biological system is not perturbed by such agents, so that the biochemical information is preserved. A well-known example is the use of positron emission tomography (PET), single-photon emission computed tomography (SPECT), and magnetic resonance imaging (MRI) technologies as routine primary diagnostic methods.

In brief, an MIP is composed of three fragments [84]. The signal agent produces the signal for image detection; it can be a radionuclide (e.g., ^{125}I or ^{99m}Tc), a fluorescent molecule, or a magnetic core, depending on the detection technique. The targeting vector is the one responsible for interaction with the biological target and, therefore, for the biological response; it is frequently a small molecule, such as a sugar moiety or a peptide. Finally, the prosthetic group (or linker) has a duplex role: on one hand, it provides the linking element between the other two fragments (signal agent and targeting vector) mentioned before, which means that it has to provide the necessary biological stability to the MIP molecule as a whole, and it has to be chemically compatible with both the signal agent and the targeting vector; and, on the other hand, it has a great influence on the biodistribution of the probe.

MIPs can be used to characterize healthy as well as structurally mutated cells and tissues (e.g., as a result of a pathological status, such as tumors or neuronal diseases) (*anatomic imaging*), or they can be used to detect physiological abnormalities (e.g., overexpression of a specific receptor or altered biochemical pathways) that are associated with the appearance of a tumorigenic status (*oncologic imaging*) [83].

This specification serves as a classification for the anionic rhenium(I)- and technetium(I)-99 m–*nido*-carborane complexes described in this chapter. Since most of these complexes were designed as potential highly selective MIPs for receptors that are primarily found within the central nervous system (CNS), with the ultimate scope of achieving further insights into several neuronal diseases (Parkinson's, Alzheimer's, etc.), they are classified as *oncologic-imaging agents*.

Regardless of the specific signal agent and prosthetic group, design strategies for clinically relevant MIPs, in general, must take into account: (1) high *specificity* for the target(s), in terms of specific molecular recognition that should favor high, fast, and specific uptake of the agent; (2) high *affinity* for the target, in terms of thermodynamic stability and kinetic inertness of the bonding interaction; (3) high *sensitivity*, so that only very low concentrations of the MIP are necessary for efficient detection; (4) rapid blood clearance; (5) high chemical stability *in vivo*, to allow the probe to reach its biological target intact; and (6) low toxicity [84].

In particular, bioorganometallic target-specific *radio-imaging agents* are being extensively studied as potential novel MIPs [85]. About ten years ago, Alberto and Gmeiner independently developed promising MIPs, in which the prosthetic group was based on a cyclopentadienyl ligand [86,87]. Here, the focus was on the development of an efficient synthetic approach to the desired target Re^I– and $^{99m}Tc^I$–Cp complexes [87] or on the rational design of ferrocenyl- or ruthenocenyl-containing molecules [86]; these compounds are among the first organometallic complexes designed to target specific receptors within the CNS. Selective probing of target receptors within the CNS could in fact lead to improved understanding and characterization of many neuronal disorders (Huntington's disease, schizophrenia, Alzheimer's disease, etc.) [88] and is, therefore, a commonly sought strategy [85]. The cyclopentadienyl ligand is generally regarded to possess potentially beneficial properties for the design of metal-containing radiopharmaceuticals for two main reasons. First, the Cp$^-$ ligand stabilizes rhenium and technetium-99m in the oxidation state +I, corresponding to a low-spin d^6 electron configuration; such complexes are usually kinetically inert, and thus potentially highly chemically stable *in vivo* [89]. Second, the Cp$^-$ ligand is a small molecule with a low molecular weight, which could minimize the problem of poor binding affinity for biological receptors that has been encountered for bulky metal complexes (with, e.g., bifunctional chelating ligands) [89].

The Valliant group has pioneered the research of Re^I- and ^{99m}Tc^I-based target-specific radio-imaging agents, in which the prosthetic group is a dicarbollide ligand [14]. A small library of anionic rhenium(I)- and technetium(I)-99m–dicarbollide complexes has been created over the past decade. Typically, the complexation to the $[M(CO)_3]^+$ (M = Re,^{99m}Tc) fragment is performed in aqueous solution under microwave heating (150–200 °C), starting from the respective *nido*-carborane precursor and $[M(CO)_3(H_2O)_3]Br$ [57]. Many of these complexes have been specifically designed and tested for interactions with chosen biological targets [75].

Most of the early work of the Valliant group has already been thoroughly reviewed by Grimes [90] and Armstrong [14], including for instance the remarkable examples of rhena- and technetacarborane complexes designed as probes for estrogen receptors [91] and those by Hawkins *et al.* incorporating an iodine-labeled bipyridyl ligand and a nitrosyl group at the metal center, designed to enhance transport across the BBB [75]. Therefore, only the most recent examples of anionic rhenium(I)– and technetium(I)-99m–dicarbollide complexes as potential MIPs are described here, namely a group of complexes where the dicarbollide ligand bears a 1-(2-methoxyphenyl)piperazine substituent [59,60], and complexes in which the cluster is tethered to a guanidine group [69].

Valliant and coworkers have applied the so-called *bio-isosteric replacement strategy* in their research [59,60], which basically means that known promising radiopharmaceuticals were modified using a dicarbollide cluster in place of a Cp ring, while preserving and/or modulating the biological activity. Bio-isosteric replacement is nowadays a very common approach in drug design, and numerous examples can be found in the literature on medicinal organic chemistry [92] as well as for polyhedral boron clusters [15c,d,25].

The 1-(2-methoxyphenyl)piperazine group is already well known for being a rather potent and selective inhibitor of the 5-HT1A subtype of the serotonin receptor, primarily located in brain tissues [93]. Therefore, this piperazine core has often been incorporated as the targeting vector within molecules designed as potential inhibitors or MIPs of serotonin receptors (Figure 1.4.7) [86,89]. The rhenium(I) complexes **3**, **5**, and **7**, as well as the corresponding technetium(I)-99m complexes **4**, **6**, and **8** (see Figure 1.4.5 and Figure 1.4.7), were also designed in an attempt to obtain highly selective MIPs for serotonin receptors (5-HTx class), instead of α-adrenergic receptors – or, in other words, to develop radiopharmaceuticals that could well differentiate between receptors *within* the CNS (5-HTx) and receptors *within and outside* the CNS (α-receptors) [40]. In fact, despite the extensive research for selective MIPs for target proteins within the CNS [83,87], the understanding of the exact function of serotonin

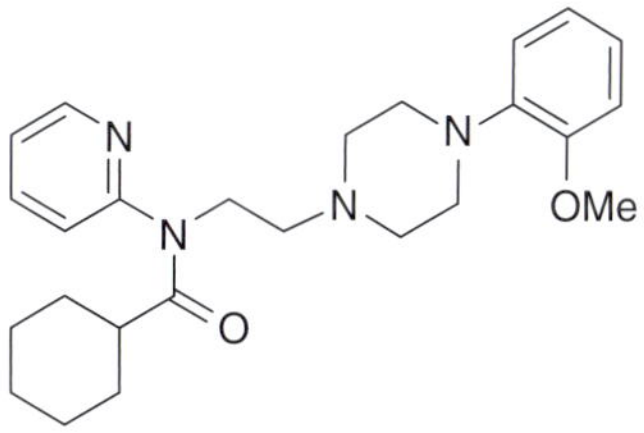

WAY100635

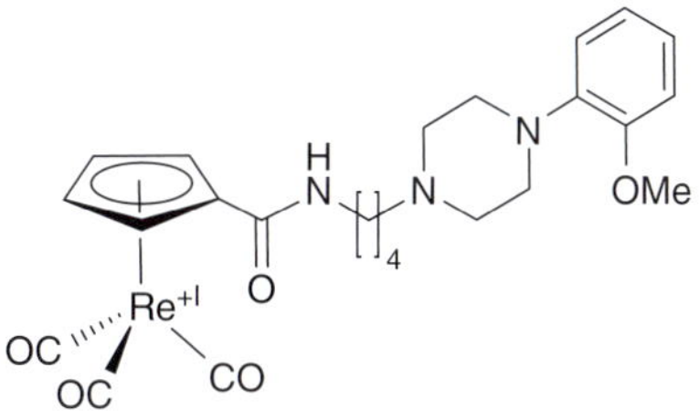

WAY–[CpRe(CO)3] derivative

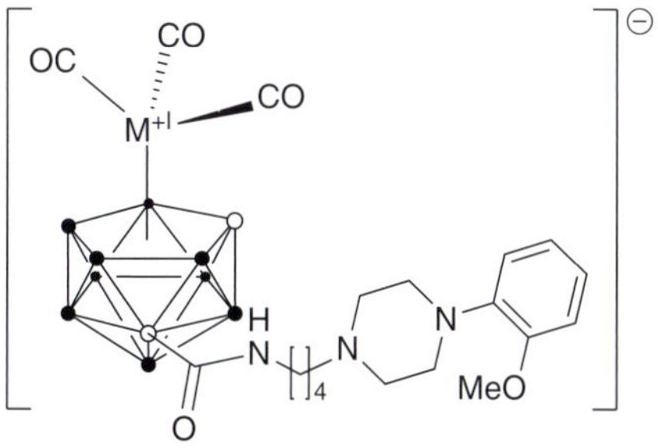

7, **8**

7: M = Re
8: M = ^{99m}Tc

○ C or CH
● BH

Figure 1.4.7 WAY100635 [N-[2-[4-(2-methoxyphenyl)piperazin-1-yl]ethyl]-N-pyridin-2-ylcyclohexanecarboxamide] (top left), a potent and selective 5-HT1A receptor antagonist [93]; Alberto's WAY–[CpRe(CO)$_3$] derivative (top right) [87]; and Valliant's Re^{I}– and $^{99m}Tc^{I}$–dicarbollide complexes **7** and **8** [59,60]. The counterions (Na^+) of **7** and **8** are omitted.

and α-adrenergic receptors in neuronal and pain-associated diseases has been so far impaired, also, by the lack of suitable MIPs [94].

An important feature of the complexes is their capacity to cross the BBB and access the CNS, which means that they should be rather lipophilic. The function of the dicarbollide ligand is therefore twofold. First, the lipophilicity of the target complexes [59], compared to analogous Cp-based MIPs [93], is increased, as the hydridic B–H bonds render these clusters extremely hydrophobic [25]. Second, the cluster itself can form stable bonds with the target vector (piperazine unit) and the metal center (Re^{I} or $^{99m}Tc^{I}$) in aqueous environment, which offers advantages for the synthesis of the complex itself [95], and for conferring the necessary stability to the MIP agent in biological media to reach the target unaltered [59,60].

Stability studies in ethanol–saline solution (0.5 mL EtOH, 0.9% NaCl in 4.5 mL H_2O) were performed on **4**, **6**, and **8** (see Figure 1.4.5 and Figure 1.4.7) [60]. The complexes showed no sign of decomposition (studied by HPLC analysis) for up to 6 h; this is ample proof that the MIP is stable in aqueous environment, under experimental conditions that resemble those of a biological medium.

Partition coefficients (log*D*) between *n*-octanol and a phosphate buffer (0.02 M, pH = 7.4) were also determined for **4**, **6**, and **8**, to assess specifically the ability of the complexes to cross the BBB and to predict their rate of clearance from nontarget tissues [59]. Typically, for a radiotracer to be considered as a potential efficient MIP,

Table 1.4.2 Binding affinities (K_i) of rhenacarborane complexes **3**, **5**, and **7** and reference compound WAY100635 for 5-HT1A, 5-HT2B, and 5-HT7 serotonin receptors and α_{1A}, α_{1B}, α_{1D}, and α_{2C} adrenergic receptors

	K_i (nM)[a]						
	Serotonin receptors			Adrenergic receptors			
Compound	5-HT1A	5-HT2B	5-HT7	α_{1A}	α_{1B}	α_{1D}	α_{2C}
3	834 ± 54	>1000	–[b]	>1000	–[b]	270 ± 15	575 ± 25
5	118 ± 8	211 ± 16	–[b]	115 ± 6	122 ± 7	174 ± 9	244 ± 10
7	–[b]	128 ± 7	40 ± 5	17 ± 1	435 ± 39	21 ± 1	39 ± 2
WAY100635	0.5 ± 0.04	302 ± 11	248 ± 23	62 ± 4	162 ± 9	36 ± 1	562 ± 28

[a] K_i = Inhibition constant (*in vitro*). K_i values were measured in triplicate.
[b] Less than 50% inhibition in the primary binding assay.
Source: Valliant et al. (2011) [60].

log$D$ values should range from 2.0 to 3.5 [96]. Values of log$D$ between 2.26 and 2.6 were found for the $^{99m}Tc^I$–dicarbollide complexes **4**, **6**, and **8**, suggesting potential effective transport across the BBB.

Compounds **3**, **5**, and **7** were screened for binding affinity toward several key receptors found in the CNS, including serotonin (5-HTx series) and α-adrenergic receptors. WAY100635 was also tested in the same experiment as positive control, as this compound is a known potent and selective 5-HT1A inhibitor [97], which was already used for analogous binding affinity tests for potential serotonin receptor inhibitors [93]. Selected binding affinity values (K_i in nM) are presented in Table 1.4.2.[5]

In comparison with the positive control WAY100635, none of the tested rhenacarborane complexes showed competitive binding affinities with either of the serotonin receptors. Complex **7**, which bears an amide group between the cluster carbon atom and the arylpiperazine unit, showed high affinities for α_{1A}, α_{1D}, and α_{2C} adrenergic receptors, in the low nanomolar range. These results were quite unexpected, since serotonin and adrenergic receptors show a high degree of homology [98] and, therefore, pose a serious problem for the design of target-specific MIPs. Here, the role of the dicarbollide ligand was even more beneficial, since it could confer the necessary lipophilicity to cross the BBB and preferentially target adrenergic receptors within the CNS over serotonin receptors. However, when the metal center is in oxidation state + I, the resulting complexes are anionic, which of course lowers their lipophilicity. Therefore, studies to increase the lipophilicity of such WAY–dicarbollide conjugates are now being conducted [60]. Introduction of a charge-compensating group at a selected B-vertex, such as a sulfonium, ammonium, or phosphonium group [99], could allow the synthesis of neutral complexes for target-selective MIPs within the CNS.

With the purpose of finding milder reaction conditions to give access to radiolabeled metallacarborane complexes with a wider spectrum of vectors (see Chapter 1.4.1.5), the

---

5 The complete series of CNS receptors tested, and relative binding affinity values (K_i), can be found in the supporting information of Ref. [60].

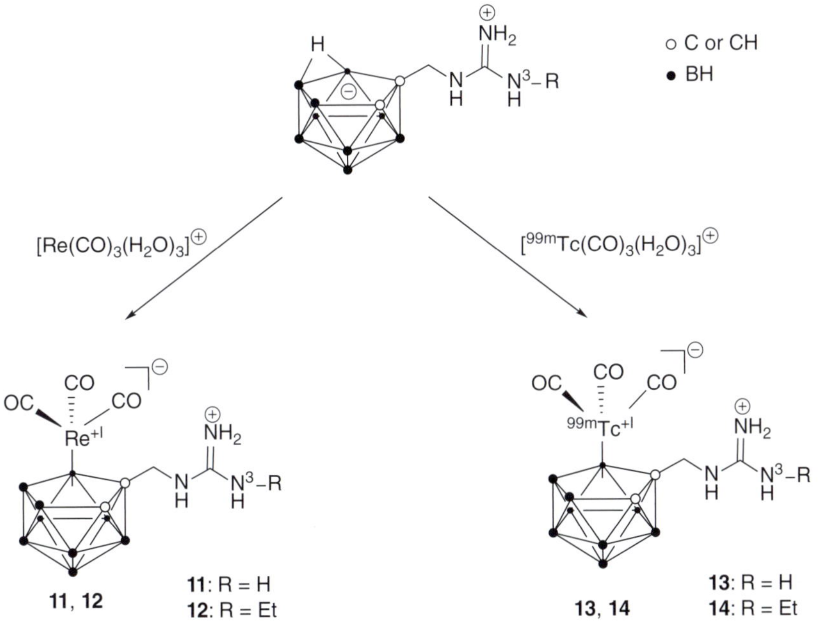

Scheme 1.4.4 Synthesis of complexes **11–14** (reaction conditions: aq. EtOH, room temperature [3 h] or 35 °C [1 h]), according to Ref. [69]. The nitrogen atom N³ of the guanidine group carries different substituents to modify basicity properties.

synthesis of rhenium- and technetium-99 m–dicarbollide complexes bearing a guanidine substituent at the C-vertex was pursued [69]. This study was prompted by the observation that the presence of an ancillary base at the cluster, such as pyridine, can lower the reaction temperature of the complexation reaction [57]. The choice of guanidine as base was justified on the basis of its basicity (pK_a = 13–14), stability *in vivo*, easiness of access, and the possibility to perform derivatizations [69].

Synthesis of the target rhenium(I) and technetium(I)-99 m complexes was carried out in aqueous ethanolic solution at room temperature and at 35 °C at pH 10 (Scheme 1.4.4). The presence of an electron-withdrawing group R on the guanidine nitrogen atom N³ prevents complexation of [Re(CO₃)]⁺ at ambient conditions and also at 35 °C, most likely because the basicity is too low to assist in the removal of the bridging proton from the *nido* cluster [69]. On the other hand, unsubstituted and N-alkyl-substituted guanidine derivatives afforded the target Re^I– and ^{99m}Tc^I–dicarbollide complexes in moderate yields (30 to 69%). No isomerization to the 2,1,8-isomer was observed under these reaction conditions.

Like the WAY–technetium–dicarbollide conjugates **4, 6,** and **8** [60], complexes **13** and **14** were found to be stable for up to 8 h in EtOH–saline solution [69], and, therefore, biodistribution studies were carried out on healthy female CD1 mice. Both **13** and **14** were rapidly cleared from the bloodstream, but showed different tissue accumulation profiles, with **13** showing highest accumulation in the gall bladder (42.11 ± 16.99 %ID/g at 4 h), whereas **14** accumulated preferentially in the bladder [69]. This suggests that the two complexes might have different clearance pathways, which may be due to the different guanidine substituents at the clusters, as all other factors were equal.

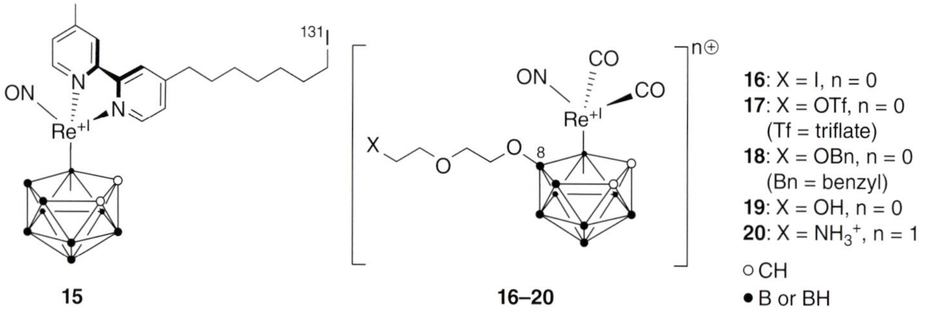

Figure 1.4.8 Rhenacarborane complexes incorporating a nitrosyl ligand at the rhenium(I) center (**15–20**) and an oxygen-based tether at the β–B vertex (**16–20**) [75,76]. Position of the β–B vertex (B(8)) is shown. The counterion ([BF$_4$]$^-$) of **20** is omitted.

16: X = I, n = 0
17: X = OTf, n = 0
  (Tf = triflate)
18: X = OBn, n = 0
  (Bn = benzyl)
19: X = OH, n = 0
20: X = NH$_3^+$, n = 1

○ CH
● B or BH

Another approach, which is currently being investigated in the Jelliss group, is the introduction of the targeting vector at the β-B vertex[6] on the upper belt of the corresponding rhenacarborane complex [76]. First, a tether is regioselectively attached at B(8) of the [Re(CO)$_3$(Cb)]$^-$ core, followed by targeted modification of the terminal group of the tether according to the desired biological function of the complex. It seems that modifying the B(8) vertex via introduction of a tailored tether represents a smart and efficient synthetic strategy to gain access to rhenacarborane complexes that incorporate biologically relevant vectors, and at the same time avoid the problem of vector disruption under the harsh conditions of the complexation reaction. The main motivation for this investigation was to simplify the synthesis of nitrosyl rhenacarborane complexes, designed previously by the same group as potential delivery vehicles of high boron content to the CNS [75].

So far, a series of β-B-tethered rhenacarborane complexes has been obtained by ring-opening reaction of the cyclic oxonium species [3,3,3-(CO)$_3$-*closo*-Re(8-O(CH$_2$)$_3$CH$_2$-3,1, 2-C$_2$B$_9$H$_{10}$)] with an appropriate nucleophile (LiOBn, NMe$_4$OH·5H$_2$O, AgOTf, etc.), as is routinely performed with COSAN-type complexes [100]. Subsequent nitrosylation gave access to a small library of new complexes (some relevant examples are given in Figure 1.4.8) [76].

In particular, compounds **19** and **20** have been envisaged as very useful synthons for the design of amino acid and peptide conjugates [76], and **16** is a very important complex, because it could be labeled with iodine-131 (instead of non-radioactive iodine) and, therefore, be used as a marker for biodistribution studies of the rhenacarborane complex.

These examples of rhenacarboranes and technetacarboranes as potential selective radioimaging agents show that nowadays, drug design strategies that are routine in medicinal organic and organometallic chemistry, such as the bio-isosteric replacement approach, can be successfully extended also to metallacarborane-based medicinal chemistry. Furthermore, the inherent properties of the C$_2$B$_9$ clusters (e.g., regioselective functionalization, robustness, and biocompatibility) can lead to biologically active hybrid organic-inorganic compounds with unexpected binding modes with the biological target and unprecedented modulation of the biological activity of the parent organic compound [19,60].

---

6 β-B vertex = unique boron atom (symmetry independent, B(8); see Figure 1.4.1 Metallacarborane) on the upper C$_2$B$_3$ belt.

Last but not least, the rhenacarborane complexes described here could potentially be used as *radiotherapeutic agents*, since ^{188}Re is a radioactive isotope that can be used for radiotherapies [14]. Therefore, the synthesis of rhenium(I) and technetium(I)-99 m complexes is often carried out in parallel, with the idea to achieve a synergistic combination of target-selective radio-imaging and radiotherapeutic agents.

1.4.3.1.2 *nido*-Carborane(−1) Anions as Pharmacophores for Metal-Based Drugs

Studies of the cytotoxic activity or enzyme inhibition properties of metallacarboranes, which incorporate *one* dicarbollide anion coordinated in a pentahapto fashion to a metal center, are extremely sparse in the literature. Only a few examples are found in Russell Grimes' monograph on carboranes [101]. A small library of bent-metallocene-type tantalum(V) and niobium(V) complexes, incorporating a mixed-ligand system of cyclopentadienyl(−1) and a six-vertex sub-icosahedral dicarbollide, $[C(R)C(R')B_4H_4]^{2-}$ (R = R' = Me, Et, SiMe$_3$), was tested against several suspended tumor cell lines (P388, HI-60, murine L1210, etc.) and showed moderate to potent activity *in vitro* [102]. Other examples mentioned are ten-vertex tricarbollide complexes of vanadium(IV) and iron(III) [103].

In other recent reviews on the use of carboranes in medicine [14,15d,51a], no examples of half- or mixed-sandwich metallacarboranes that were screened for cytotoxic activity are found. This is in contrast with the vast literature available on COSAN-type icosahedral metallacarboranes, which have been extensively studied over the last 15 years, thanks to the encouraging results of HIV-1 protease inhibition studies of the cobalt(III) bis(dicarbollide) sandwich complexes by Cigler *et al.* [104].

Recently, Kaplánek *et al.* have studied a small library of bis(dicarbollide) complexes, together with boranes and mixed-sandwich ferracarborane complexes, as potential novel isoform-selective nitric oxide synthases (NOS) inhibitors [58]. Three isoforms of human NOS are known; each is located in different tissues and has specific physiological functions [105]. Neuronal NOS (nNOS) is found predominantly in neurons in brain tissues and modulates the learning processes, the memory, and neurogenesis [106]. Inducible NOS (iNOS) can potentially be expressed in every type of tissue, upon appropriate stimulation (e.g., bacterial lipopolysaccharide, cytokines, etc.); when iNOS is induced in macrophages, it is responsible for cytotoxic effects in the physiological immune response against bacteria, viruses, and certain types of tumor [107]. Endothelial NOS (eNOS) is expressed in vascular endothelial tissues and regulates, among others, the blood pressure [106]. All three isoforms catalyze the conversion of L-arginine to L-citrulline and nitric oxide (NO), following the reaction steps shown in Scheme 1.4.5.

The development of inhibitors of NOS has received increasing interest since the 1990s [105]. The importance of isoform-specific inhibitors became evident already at the early stages, due to the fact that NO is implicated in many different and independent physiological functions, according to the specific cells where it is produced, or, in other words, to the specific synthase enzyme (as discussed in this chapter) [107]. Overproduction of NO by nNOS is associated with a variety of neuronal pathologies, such as Alzheimer's, Parkinson's, and Huntington's diseases, whereas NO overproduction by iNOS is associated with an inflammatory status of the immune system, such as arthritis [107].

Therefore, inhibition of nNOS and/or iNOS enzymes is often the required function for a novel drug, which, at the same time, should show minimal affinity for the third

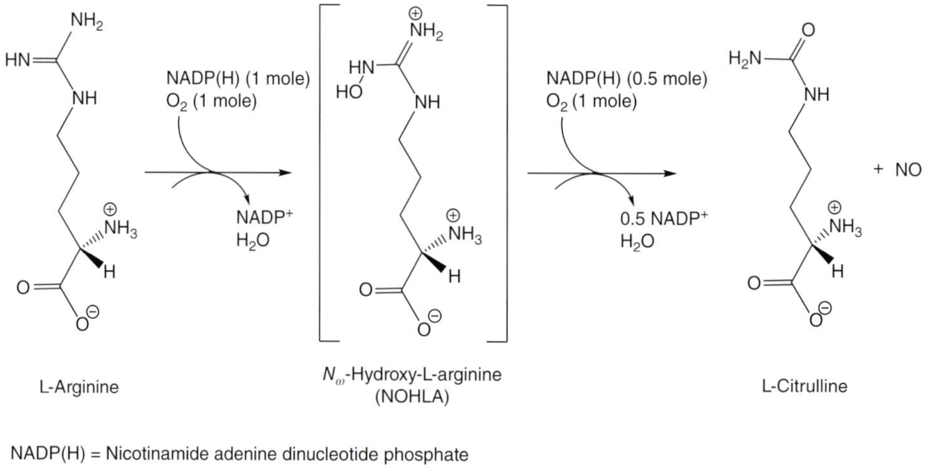

Scheme 1.4.5 Conversion of L-arginine to L-citrulline and nitric oxide (NO) as catalyzed by NOS. Flavin, (6R)-5,6,7,8-tetrahydrobiopterin, and heme act as co-factors in both steps (not shown). *Source:* Adapted from Ref. [107].

isoform (eNOS), so that it would not interfere with the regulation of blood homeostasis [106,107]. The major challenge here is that unfortunately all three isoforms contain highly preserved residues and overall cavity shape in both co-factors and L-arginine binding sites. However, certain local differences became apparent on studying the crystal structures of the oxygenase N-terminal domains of the three NOS [105], such as the Asp382 site at iNOS, which is replaced by an Asn residue in eNOS. Due to these minor local topological differences, the development of isoform-selective NOS inhibitors, as well as the definitions of *NOS-selective inhibitor* itself and of enzyme inhibition protocols, is still an open quest [105,108]. Up to date, to the best of our knowledge, there is no drug marketed as a selective NOS inhibitor, even though many compounds with known NOS inhibition activities (Figure 1.4.9) can be purchased and are also used for research purposes [109]. For recent reviews on the newest developments in structure–activity relationship studies on nitric oxide synthases and NOS inhibitors, refer to References [106] and [110,111].

The approach used by Kaplánek et al. was to design potentially selective NOS inhibitors that target not the catalytic cavities of the enzymes, but the region *just outside* the binding pocket. Although this approach is not exactly new [112], the originality of their work was the use of an icosahedral metallacarborane fragment for the synthesis of the inhibitor, which therefore acts as a *pharmacophore*. The high hydrophobicity of these clusters could potentially provide the necessary lipophilicity to the drug for efficient transport across the BBB, and the possibility of regioselective three-dimensional functionalization could allow modulation of the biological activities based on different specific binding interactions with the substrate (NOS isoforms). Several types of basic binding motifs, guanidines, urea, alkyl amines, and so on were selected, and several substitution patterns at the clusters were explored. The presence of a basic amine has proven to be beneficial for increasing inhibitory potency and selectivity toward nNOS, over both iNOS and eNOS [107]. Arginine was not chosen as a substituent, to try to minimize non-isoform-selective binding interactions [58].

N_ω-Nitro-L-arginine methyl ester
(**L-NAME**)

N_ω-Methyl-L-arginine
(**L-NMMA**)

N-([3-(Aminomethyl)phenyl]methyl)ethanimidamide
(**1400W**)

S-{2-[(1-iminoethyl)amino]ethyl}-L-homocysteine
(**GW274150**)

Figure 1.4.9 Selected examples of organic compounds with known nitric oxide synthase (NOS) inhibition activities, available on the market for laboratory use [105,107,111].

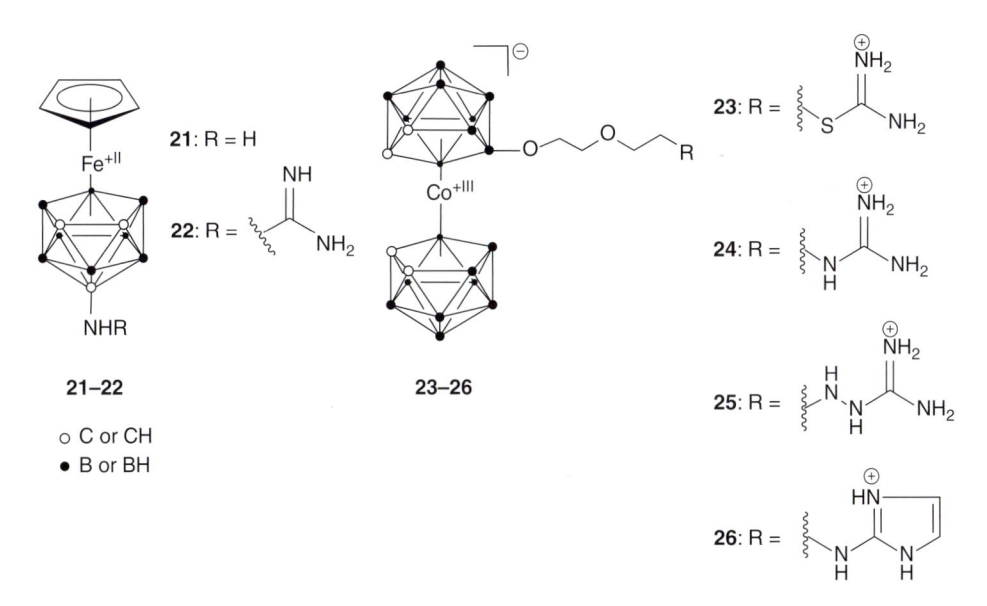

21–22

○ C or CH
● B or BH

23–26

Figure 1.4.10 Selected examples of mixed-sandwich ferracarborane (**21–22**) and full-sandwich cobaltacarborane (**23–26**) complexes tested as NOS inhibitors [58].

The two mixed-sandwich ferracarborane complexes tested (**21** and **22**; Figure 1.4.10) showed very different biological activity. Particularly, complex **21**, bearing an –NH$_2$ group in *para* position with respect to the iron center, showed no inhibition of NOS activity, regardless of the specific isoform, under either competitive or noncompetitive binding assay conditions. On the contrary, complex **22**, with a guanidine substituent, showed complete inhibition of all NOS isoforms, even though in a nonspecific manner, at 10 μM concentration. Compounds **23–26** all showed binding affinities for the NOS

Table 1.4.3 Selectivity ratios for nitric oxide synthases (NOS) inhibition of compounds **23–26** [58]

Complex	Selectivity ratios*	
	eNOS/iNOS	eNOS/nNOS
23	8.5	1.1
24	5.1	0.46
25	8.7	1.6
26	8.1	2.8

* Selectivity ratios are given in Ref. [58]. Values are calculated as $IC_{50(eNOS)}/IC_{50(iNOS)}$ or $IC_{50(eNOS)}/IC_{50(nNOS)}$, respectively.

enzymes, and a slight preferential binding affinity for the eNOS isoform over the other two (Table 1.4.3).

The fact that these specific complexes (**21–26**) show nonspecific NOS inhibition or slightly higher affinities for the eNOS isoform over iNOS and nNOS, whereas normally the opposite is required for efficient therapies [107], does not necessary imply that metallacarboranes might not be useful in the synthesis of promising NOS inhibitors. The importance of this investigation resides, in our opinion, in the fact that for the first time metallacarborane complexes of the mixed-sandwich type were designed and tested for selective enzyme inhibition. As stated by the same authors [58], the study is a *proof-of-principle*, because it shows that one tricarbollide anion in combination with a face-coordinated metal center can act as an effective pharmacophore in drug design. This strongly supports the idea that the whole ensemble of metallacarborane complexes (i.e., full-, mixed-, and half-sandwich) could definitely show attractive properties for application as enzyme inhibitors.

We have recently prepared a small series of mixed-sandwich ruthenacarborane complexes, incorporating different arene ligands [113], to combine the known, well-studied cytotoxic activity of ruthenium(II)–arene fragments (arene = benzene, p-cymene, biphenyl, tetrahydro-anthracene, etc.) [114] with a dicarbollide ligand, which could have a profound influence on the biological activity of the complexes *per se* and/or upon changing the substitution pattern at the cluster vertices [25].

Complexes **27–29** (Figure 1.4.11) were screened against three types of tumorigenic cell lines (i.e., B16: murine skin melanoma; HCT116: human colon carcinoma; and MCF-7: human hormone-dependent breast adenocarcinoma) and three

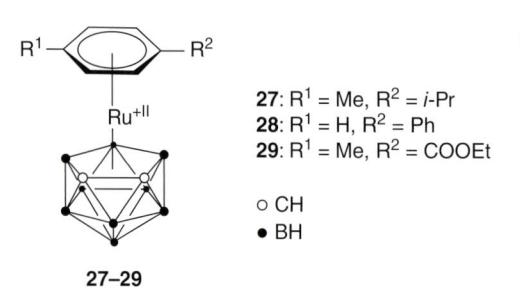

Figure 1.4.11 Ruthenacarborane complexes (**27–29**) incorporating different arene ligands, tested as potential novel cytotoxic agents [113].

27: R^1 = Me, R^2 = i-Pr
28: R^1 = H, R^2 = Ph
29: R^1 = Me, R^2 = COOEt

○ CH
● BH

non-tumorigenic cell lines (i.e., human MRC-5: human fetal fibroblasts; murine MLEC: murine lung endothelial cells; and peritoneal Mf: murine peritoneal macrophages). All three complexes were found to be active against HCT116 and MCF-7 cell lines, with IC_{50} values in the low micromolar range (6–32 µM), whereas viability of the non-tumorigenic cell lines was not significantly impaired upon treatment with compounds **27–29**.

No specific modulation of the cytotoxic activity could be found for different arene ligands incorporating the unsubstituted dicarbollide ligand. Further investigations on the mode of action as well as on the possible delivery routes to the cells (i.e., on the biological targets) are currently being carried out. Preliminary results suggest a caspase-dependent apoptotic cell-death pathway [113].

As for the work by Kaplánek *et al.* discussed in this section [58], this investigation of the biological activity of ruthenacarborane complexes represents a *proof-of-principle* since it shows for the first time that the combination of a dicarbollide ligand with a ruthenium(II)–arene fragment imparts promising cytotoxic properties to the complex. These and related ruthenacarborane complexes will now be studied, to fully assess and possibly explain their cytotoxic activity, especially in comparison with analogous ruthenocene complexes, such as the ones designed and studied by Micallef *et al.* [115].

Metallacarborane complexes are, therefore, emerging as very interesting compounds for medical applications, thanks to their many attractive properties, including (1) the general chemical and biological stability of bis(dicarbollide) sandwich complexes of cobalt(III); (2) their amphiphilic nature, due to the highly hydrophobic carborane cages and the ionic character of the complex; and (3) the wide plethora of possible exoskeletal functionalizations. Moreover, the promising results obtained upon studying the biological activity of such complexes encourage further analogous studies on the whole group of "Cp-like" complexes (i.e., half- and mixed-sandwich complexes).

1.4.3.2 Biologically Active Compounds Containing a *nido*-Carborane(−1) Core

Exact classification of the *nido*-carborane-containing compounds described in this section is not an easy task. Most recently, after Grimes published his monograph on carboranes [37], several studies have appeared in the literature where these hybrid organic–inorganic compounds were used for a target-specific molecular or enzymatic recognition, in other words for selective/specific binding affinities, in the search for optimal drugs (inhibitors), or imaging, radio-imaging, and radiotherapeutic agents. Biodistribution studies are often carried out, together with the development of optimal methods for the detection and quantification of the boronated compounds *in vivo*.

Therefore, the following examples are organized into two subgroups: (1) radiotherapy and radio-imaging, and (2) pharmacophores. Nonetheless, cross-over between these two categories can be envisaged and will, therefore, also be discussed in this section.

1.4.3.2.1 Radiotherapy and Radio-Imaging

A major challenge for the development of efficient BNCT agents is to selectively deliver a sufficiently high amount of boron to tumor tissues without high accumulation in healthy tissues or blood. One possible approach is to synthesize *nido*-carborane(−1) containing conjugates of different classes of biomolecules.

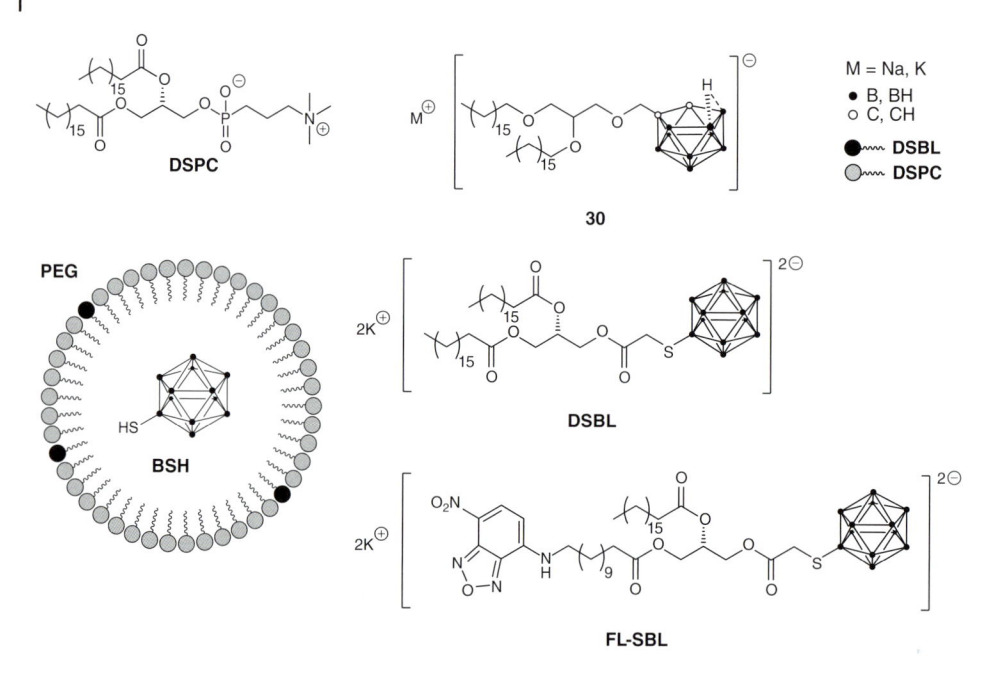

Figure 1.4.12 BSH-encapsulation with liposomes made of 10% distearoyl boron lipid (DSBL) or fluorescence-labeled boron lipid (FL-SBL) and 90% distearoylphosphatidylcholine (DSPC) [118].

In 2013, Białek-Pietras et al. reported the syntheses, cytotoxicity, and biodistribution studies of a series of cholesterol–COSAN and cholesterol–nido-carborane(−1) conjugates [116]. The two strategies pursued in the topic of cholesterol–carborane alteration are the conjugation to cholesterol and the mimesis of rigid parts of this natural product [117]. The delivery of these conjugates to malignant cells is achieved by incorporation into liposome membranes. In the same year, Nakamura and coworkers showed that liposomes made of 10% distearoyl boron lipid (DSBL) (i.e., nido-carborane(−1) or icosahedral borane(−2)-modified distearoyl lipids) and 90% distearoylphosphatidylcholine (DSPC), inspired by the zwitterionic DSPC, form high-boron-containing liposomes filled with BSH as a boron delivery vehicle (Figure 1.4.12) [118]. The nido-carborane(−1) and COSAN-type moieties ($[M(\eta^5\text{-Cb})_2]^-$, M = Co, Fe, Cr) were modified with dioxane and combined through an oxonium ring-opening reaction with 3-thiocholesterol. This approach led to a boron concentration (43.0 ppm) in tumor cells, which is much higher than the necessary concentration for an efficient BNCT. Furthermore, it is beneficial that these conjugates are only weakly cytotoxic (IC_{50} = 0.25–0.86 mM), even though an unusually high accumulation in the spleen was observed.

The Valliant group in contrast has been exploring two other main approaches, namely *target-vector recognition* and the *pre-targeting strategy*. In both approaches, the nido-carborane(−1) cluster acts as the prosthetic group, carrying the radionuclide at one B-vertex of the open upper face, and the vector at one of the C-atoms. The vector is either a biomolecule (e.g., glucose or a urea-containing small peptide) for the target–vector recognition approach, or an organic group (e.g., tetrazine) that is able to react *in vivo* with specific cell tags (pre-targeting strategy). An important advantage of using *nido*-carborane(−1) instead of aryl-based systems for radiohalogenation is the higher stability of the B–I bond over the C–I bond (B–I is ca. 100 kJ/mol stronger), which is reflected by the higher stability of the radio-iodinated carboranes/*nido*-carboranes(−1) in solution (up to 24 h).

In 2009, bifunctional radiolabeled carbohydrate derivatives of carborane clusters were reported by the Valliant group aimed at designing robust, selective radiopharmaceuticals, using a vector to enhance *target-to-not-target* ratios (benzamide substituent at a *C*-vertex) and a strong ligand (*nido*-carborane(−1)), able to form a stable covalent bond with the radionuclide (iodine-125) [68]. In this case, the nido-carborane(−1) acts as the prosthetic group for the target–vector recognition approach. Glucose was chosen to increase hydrophilicity of the *nido*-carborane(−1)-containing molecule, because the high hydrophobicity of the cluster can drastically reduce the selectivity. Within the synthetic route, the desired alkyne is prepared first carrying the glucose and the benzoic acid benzyl ester groups, which was then reacted with *arachno*-borane as described in Chapter 1.4.1.1. The *closo*-carborane derivative was subsequently deboronated and the sugar moiety deprotected with NaOH in MeOH, followed by halogenation of the *nido*-carborane(−1) species. However, when lipophilicity studies on the benzamide *nido*-carborane(−1) derivatives with and without the glucose conjugate were undertaken, the desired hydrophilicity was indeed increased by the sugar moiety, but the uptake into the melanoma cells was considerably lowered compared to the non-glycosylated species [68].

Recently, small peptide conjugates of the *nido*-carborane(−1) cluster, where the peptide incorporates a urea-based group, have been reported by Valliant and coworkers (Figure 1.4.13) [77]. The target was a receptor called prostate-specific membrane antigen (PSMA), a transmembrane protein that is predominantly expressed in the normal human prostate epithelium, but overexpressed in the corresponding primarily damaged tissues and also in metastases [119]. However, inhibition of PSMA was not the only purpose; a specific binding to tumor cells was intended in order to use these conjugates as BNCT agents, in which the carborane cluster is used as the boron-rich moiety and the peptide (Lys–Urea–Glu) as a vector linked by an amide bond either directly or through a pentyl spacer [77].

Starting from 1-hydroxy-2-carboxy-1,2-dicarba-*closo*-dodecaborane(12), the short, protected urea-linked dipeptide (*S*)-2-(3-((*S*)-5-amino-1-carboxypentyl)ureido)-pentanedioic acid was first introduced, and subsequently the carborane cluster deboronated via reflux in water for 15 min followed by cation exchange with NaOH. This decapping approach works well when electron-withdrawing substituents are directly attached at the cluster carbon atoms [77].

Compounds **31–34** were tested with *in vitro* competitive binding assays using LNCaP (androgen-sensitive human prostate adenocarcinoma) cells. None of the new inhibitors reached the IC$_{50}$ value for PMPA (Table 1.4.4 and Figure 1.4.13); however, **32a** (73.2 nM) is almost as active as PMPA (63.9 nM). The substitution of a boron vertex impairs the performance, but not significantly, and therefore **32b** can be used to trace the molecule in cells with ^{123}I as a quantitative probe. From the IC$_{50}$ values in Table 1.4.4, it can be concluded that the *nido*-carborane(−1) species **32a** and **34a** are more active than the corresponding *closo*-carboranes **31** and **33**, and the pentyl spacer (compounds **33** and **34a**) also decreases the activity of the potent inhibitors.

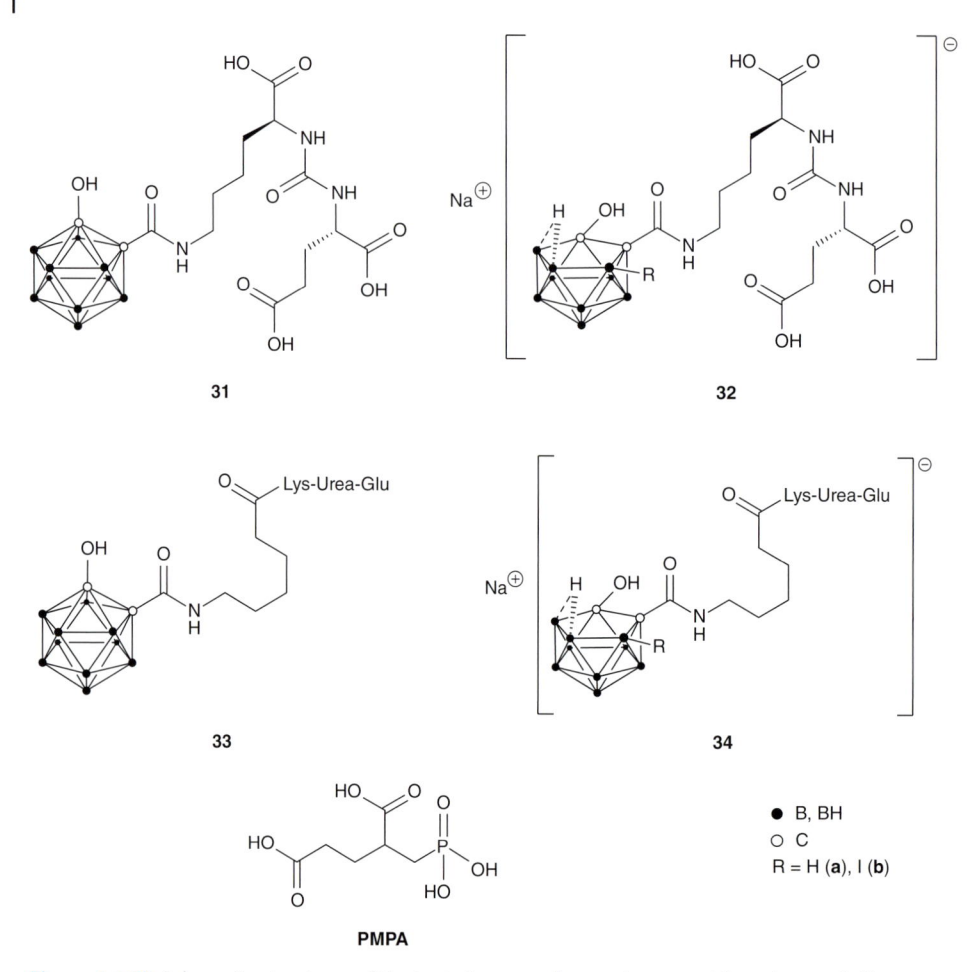

Figure 1.4.13 Schematic structures of the tested potent *closo*-carborane, *nido*-carborane(−1), and iodo-*nido*-carborane(−1) conjugates and 2-(phosphonomethyl)pentane-1,5-dioic acid (PMPA) [77].

Table 1.4.4 IC_{50} values for PMPA (a known PSMA inhibitor) and compounds **31–34** [77]

Compound	IC_{50} (nM)
PMPA (control)	63.9
31	161.0
32a	73.2
32b	80.7
33	206.6
34a	109.7
34b	240.8

Even though all these molecules are active in the nanomolar range, their selectivity needs to be improved, because besides the accumulation in tumor cells, they also show very high concentrations in the gall bladder and thyroid.

Another approach, which is also pursued by the Valliant group, is the pretargeting *bio-orthogonal chemistry strategy*. This route uses certain types of reactions that would not occur in biological systems, because the reacting functional groups are not naturally present. The main criteria for chemical ligation reactions within the concept of bio-orthogonality are the absence of side reactions with naturally occurring functional groups (selectivity), the biological function of the targeted molecule should not be affected (biological inertness), and the formed connection or bond should be stable under biological conditions (chemical inertness). Additionally, the ligation reaction should reach completion faster than the time scale of cellular processes (minutes) prior to metabolism and clearance (kinetic aspects); and, finally, the reacting molecules need to be designed so that they can easily access the reactive site [120]. Famous examples are Staudinger ligation, click chemistry (copper-free), and cycloaddition reactions (nitrone dipole, norbornene/oxanorbornadiene, and [4 + 1]), tetrazine ligation, tetrazole photoclick, quadricyclane ligation, and so on [120]. Also, transition metals such as copper, rhodium, ruthenium, and palladium can be used for ligation reactions in living systems [121].

In 2015, Valliant and coworkers published a series of radiohalogenated *nido*-carborane(−1), unsubstituted *nido*-carborane(−1), as well as *ortho*- and *para*-dicarba-*closo*-dodecaborane(12) derivatives, which contain one tetrazine-based substituent, aiming to conduct click chemistry ([4 + 2] inverse-electron demand Diels–Alder type reactions) *in vivo* between carborane- or *nido*-carborane(−1)-containing tetrazines and *trans*-cyclooctene (Tz-TCO) (Scheme 1.4.6), which proceeds rapidly and in quantitative yield [63]. To obtain the isotopically labeled species, the *ortho*-dicarba-*closo*-dodecaborane(12)-substituted tetrazine derivative was stirred in a H_2O–MeCN mixture (1:1) at 35 °C for 24 h for deboronation, followed by a labeling reaction with $Na^{125}I$ or $Na^{123}I$ and Iodogen®. The resulting compounds showed binding to the human H520 lung-cancer cell line *in vitro*, which were labeled with a *trans*-cyclooctene (TCO)-modified antibody. The ^{123}I-labeled species could be used to trace the distribution in healthy mice showing minimal loss of iodine and high, unspecific uptake in liver and gall bladder. Compounds with improved performance are now being pursued.

A rather interesting and quite isolated example are the studies of Calabrese *et al.* on the unsubstituted *nido*-carborane(−1) used as counterion in combination with a series of fluorescent delocalized lipophilic cations (DLCs), for use as stand-alone anticancer drugs or in BNCT applications [122]. Molecules such as Nile blue, dequalinium, rhodamine-123, or the tetraphenyl phosphonium cation are positively charged molecules with an extended conjugated system and lipophilic character, which are known to enter mitochondria in cells and can thus be used for visualization and activity assessment [123]. The working principle of this approach is based on the fact that mitochondria in tumor cells exhibit a 60 mV higher mitochondrial membrane potential (180–200 mV *in vitro* and 130–150 mV *in vivo*) compared to healthy epithelial cells [124], which is necessary for adenosine triphosphate (ATP) synthesis by oxidative phosphorylation [125]. Therefore, the energy (free Gibbs energy) for membrane penetration is reduced for DLCs. This process can be described physically by the Nernst equation, which predicts that an increase of 61.5 mV of membrane potential results in a tenfold enrichment of

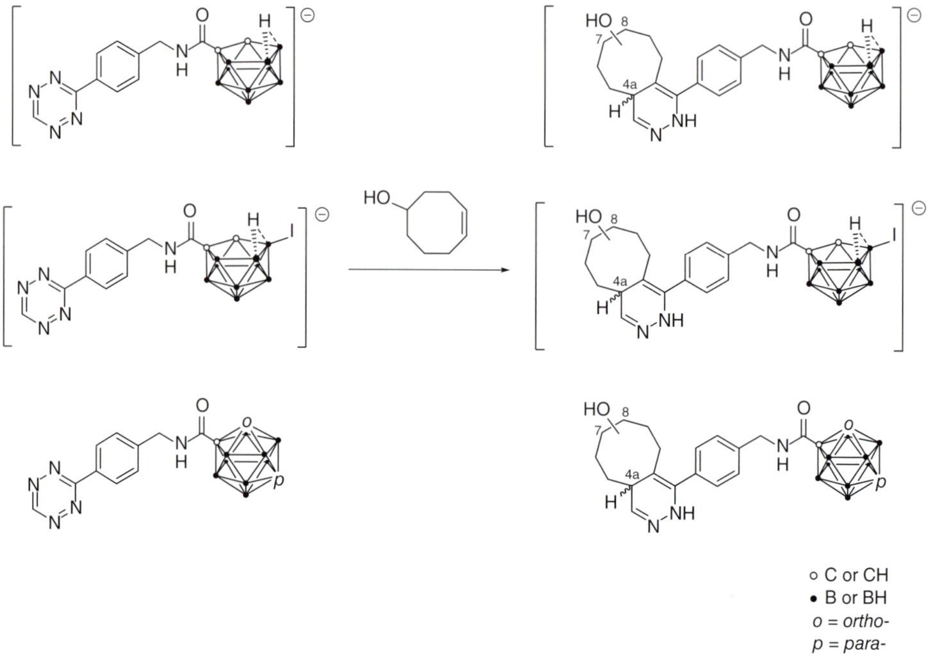

Scheme 1.4.6 Schematic structure of the *closo*-carborane- and *nido*-carborane(−1)-containing tetrazines and *trans*-cyclooctene (Tz-TCO) in a click reaction. Four possible isomers are formed: 4a-(R)-7-ol, 4a-(R)-8-ol, 4a-(S)-7-ol, and 4a-(S)-8-ol [63].

lipophilic cations within the mitochondrial matrix of malignant cells [126]. Additionally, the accumulation is further promoted by the plasma membrane potential (about 30–60 mV).

Basic examples for this type of compounds are rhodamine-123 and Nile blue *nido*-carborane(−1) salts (Scheme 1.4.7), which stain tumor cells and show growth-restricting behavior [127]. Furthermore, dequalinium chloride, commonly used as mouthwash and ointment, shows anticarcinoma activity and prolonged survival of mice with implanted mouse bladder carcinoma MB49 [128]. Adams *et al.* first prepared salts combining a *nido*-carborane(−1) moiety and dequalinium, and studied their uptake into cancer cells and their activity [129].

Based on these preliminary studies, Calabrese *et al.* prepared a variety of DLC salts (Scheme 1.4.7), and they investigated in their first *in vitro* study in 2008 the ^{10}B uptake into human prostate epithelial carcinoma (PC3) cells compared to healthy cells (PNT2) by inductively coupled plasma mass spectrometry (ICP-MS) (Table 1.4.5) [122].

In all cases, the ^{10}B uptake into malignant cells (PC3) was much higher than for healthy cells, with a sufficient amount of ^{10}B for effective BNCT (10^9 ^{10}B atoms per cell) [132], supporting the validity of this underlying principle. The bis-*nido*-carborane(−1) dequalinium salt (**47**) results in the highest amounts of ^{10}B, presumably due to two *nido*-carborane(−1) counterions [122]. However, biodistribution studies need to be performed for all tested compounds.

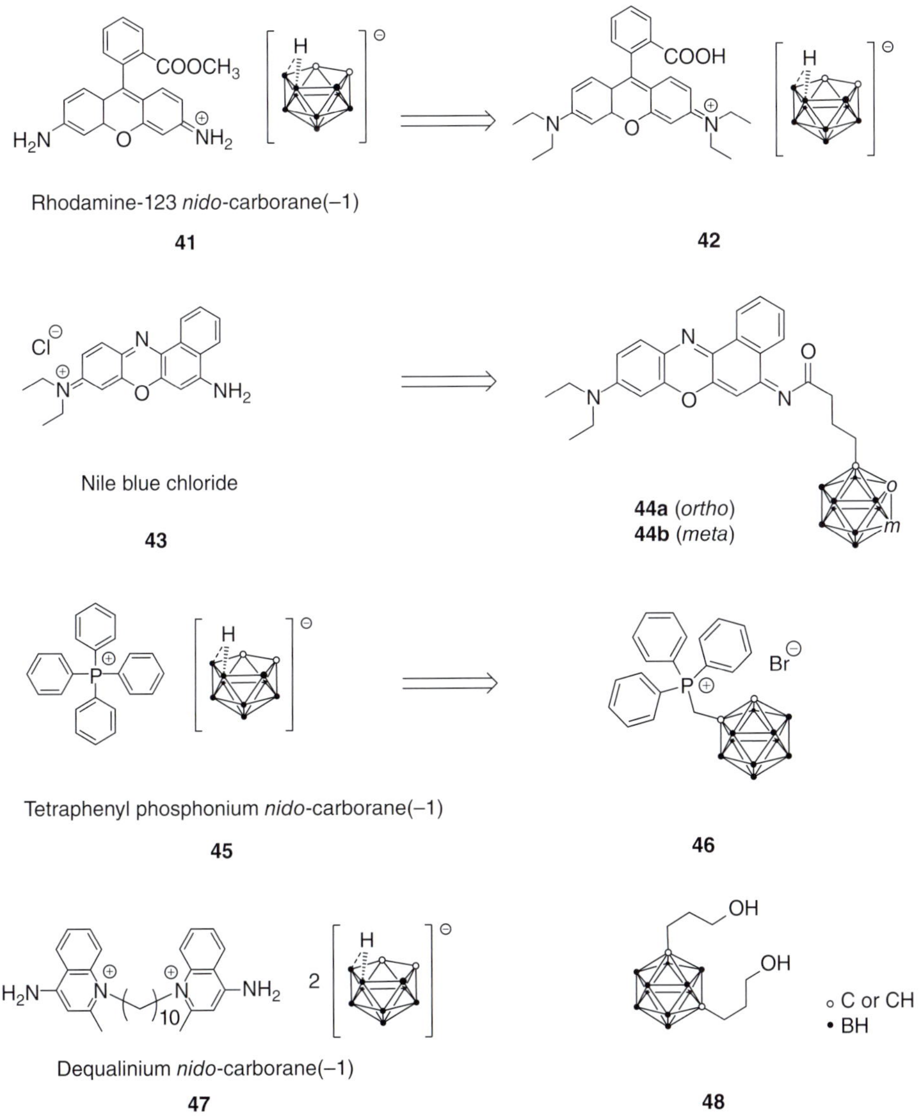

Scheme 1.4.7 Schematic structures of the DLC salts studied by Calabrese et al. [122,130,131].

The most promising compound, the bis(*nido*-carborane(−1)) dequalinium salt (**47**), was further investigated. The usage of liposomes as a carrying vehicle for **47** was reported in 2013 [130]. The principle is the same as discussed above for BSH (see Figure 1.4.12); however, in this case, the liposomes were made of phosphatidylcholine (PC) or dimyristoylphosphatidylcholine (DMPC). Extensive studies were performed to give a preferably complete picture of the biological behavior of the loaded liposomes *in vitro* on human glioblastoma U-87 MG cells (human Uppsala 87 malignant glioma), such as atomic force microscopy (AFM), dynamic light scattering, and ζ-potential

Table 1.4.5 Uptake efficiency (%) by PC3 and PNT2 cells, amounts of ^{10}B taken up by PC3 cells (atoms/cell), and accumulation ratios (PC3:PNT2) [122]

Compound	^{10}B uptake efficiency (%) by PC3 cells	^{10}B uptake efficiency (%) by PNT2 cells	Amount of ^{10}B taken up by PC3 cells (atoms/cell)	Accumulation ratios (PC3:PNT2)
41	68	17	$4.5 \cdot 10^{11}$	3.0
44a	87	28	$2.6 \cdot 10^{11}$	3.1
44b	75	45	$2.7 \cdot 10^{11}$	2.2
45	75	43	$4.5 \cdot 10^{11}$	3.1
47	69	37	$6.3 \cdot 10^{11}$	4.2

Note: The uptake efficacy was evaluated through the percentage of ^{10}B that was taken up by cells with respect to the total amount of administered boron, calculated based on the analysis of the supernatants via ICP-MS.

(zeta-potential) measurements, to determine morphology, size, and electrokinetics. Other techniques such as spectrofluorimetry gave information about miscibility of the *nido*-carborane(−1) species with liposomes; monitoring the release of calcein in serum and phosphate-buffered saline (pH 7.4) gave details on membrane integrity, whereas turbidity measurements reflected the physical stability. In addition, the ^{10}B load of the targeted cells was determined by ICP-MS [130]. These *in vitro* results obviously need to be followed by *in vivo* tests to give a proper understanding of the potency of the compounds for future application.

The most recent investigation by Calabrese *et al.* on this topic is the evaluation of the pharmacological anticancer profile of DLC-functionalized *nido*-carboranes(−1) for the normal MRC-5 and Vero cell types, cancer (U-87 MG, HSC-3), and primary glioblastoma cancer stem cultures (EGFR^{pos} and EGFR^{neg}) *in vitro* and their mode of action [131]. It was shown that the applied compounds have a cell growth-arrest effect on the two cancer cell lines compared to the two normal ones and selective growth inhibition on EGFR^{pos} and EGFR^{neg} cells without causing apoptosis. Normal cells kept their proliferation potential, in contrast to cancer cells. The mechanism of action of DLC-*nido*-carborane(−1) species proceeds through the activation of the p53–p21 axis,[7] which are both involved in the stop signal generation for the division of cells after exposure to damaging agents. After removal of the *nido*-carborane(−1) species from the treated cells, the agent-triggered cell-cycle arrest by p53–p21 was detected. Table 1.4.6 summarizes the IC$_{50}$ data obtained within the described study [131]. Compound **48** (Scheme 1.4.7) was used as a control reference for all other substances due to its low cytotoxicity.

Although the IC$_{50}$ values seem to be relatively low, an acceptable proportion of cell deaths of malignant cells was only observed at higher concentrations ($5 \cdot 10^{-5}$ and 10^{-4} M), especially for cultures treated with **46** and **47**. However, a highly selective cytotoxic pattern was observed for **45** and EGFR^{neg} compared to Vero.

---

[7] p53: tumor suppressor gene; p21: cyclin-dependent kinase inhibitor (CKI).

Table 1.4.6 Results of the cytotoxicity studies (IC$_{50}$ values) for the tested DLC salts and reference compound **48** [131]

	IC$_{50}$ [M]				
Cell type	42	45	46	47	48
U-87 MG[a]	$4.5 \cdot 10^{-5}$	$4.5 \cdot 10^{-6}$	$8.35 \cdot 10^{-6}$	$3.2 \cdot 10^{-6}$	$>10^{-4}$
HSC3[b]	$8.13 \cdot 10^{-5}$	$2.8 \cdot 10^{-6}$	$8.18 \cdot 10^{-6}$	$3.3 \cdot 10^{-6}$	$>10^{-4}$
MRC-5[c]	$2.4 \cdot 10^{-5}$	$8.1 \cdot 10^{-5}$	$2.1 \cdot 10^{-5}$	$5 \cdot 10^{-6}$	$>10^{-4}$
Vero[d]	$7.15 \cdot 10^{-5}$	$>10^{-4}$	$1.45 \cdot 10^{-5}$	$5.6 \cdot 10^{-6}$	$>10^{-4}$
EGFR^{neg} [e]	n.a.[f]	$7 \cdot 10^{-7}$	$2 \cdot 10^{-6}$	$6 \cdot 10^{-7}$	$>10^{-4}$
EGFR^{pos} [g]	n.a.[f]	$4.5 \cdot 10^{-6}$	$2.5 \cdot 10^{-6}$	$1.9 \cdot 10^{-6}$	$>10^{-4}$

[a] Human epithelial glioblastoma grade IV; astrocytoma cells.
[b] Human oral squamous carcinoma cells.
[c] Human normal lung fibroblast cells.
[d] Monkey normal kidney cells.
[e] Human brain cancer stem cells.
[f] n.a. = not analyzed.
[g] Human brain cancer stem cells.

1.4.3.2.2 Pharmacophores

Although the use of the parent *closo*-carboranes as pharmacophores in drug design is still dominant in the field, increasing examples are appearing in the literature, where the *nido*-carborane(−1) cluster is instead used as a pharmacophore.

Asborin, the carborane analog of aspirin, was reported by Scholz et al. in 2011 [62]. The main question was whether asborin would display the same mechanism of action with the biological targets (COX-1 and COX-2) as aspirin (i.e., acetylation at specific serine residues). Therefore, the acetylation pattern of both COX isoforms was investigated separately via LC-ESI (electrospray ionization) MS, after incubation with asborin. Preferential acetylation sites were found to be different than those of aspirin, namely six lysine and one serine residues located outside the active site were modified in COX-1, and no serine but five lysine residues in COX-2 [62]. Furthermore, the hydrolysis mechanism of asborin was studied via ^{1}H and ^{11}B{^{1}H} NMR spectroscopic experiments, revealing competitive deacetylation and deboronation reactions resulting in the *nido*-carborane(−1) analog of aspirin (*nido*-asborin) and salicylic acid (*nido*-salborin) as the final product (Scheme 1.4.8) [62].

Since asborin did not act as a COX inhibitor, further cytotoxicity investigations inspired by the potential of the aspirin derivative Co-ASS, which is active as an anticancer agent [133], were conducted. Sulforhodamine-B microculture colorimetric assays were performed to assess the IC$_{50}$ values for different types of cancer cell lines compared to aspirin, salicylic acid, and the established cisplatin (Table 1.4.7).

Aspirin, salicylic acid, and the product of the deboronation of asborin, boric acid, were not active (Table 1.4.7). The carborane-containing compounds by contrast were more active, with the parent *closo* isomers acting at even lower concentration than the *nido* species, but all are much less active than cisplatin. This behavior might be related

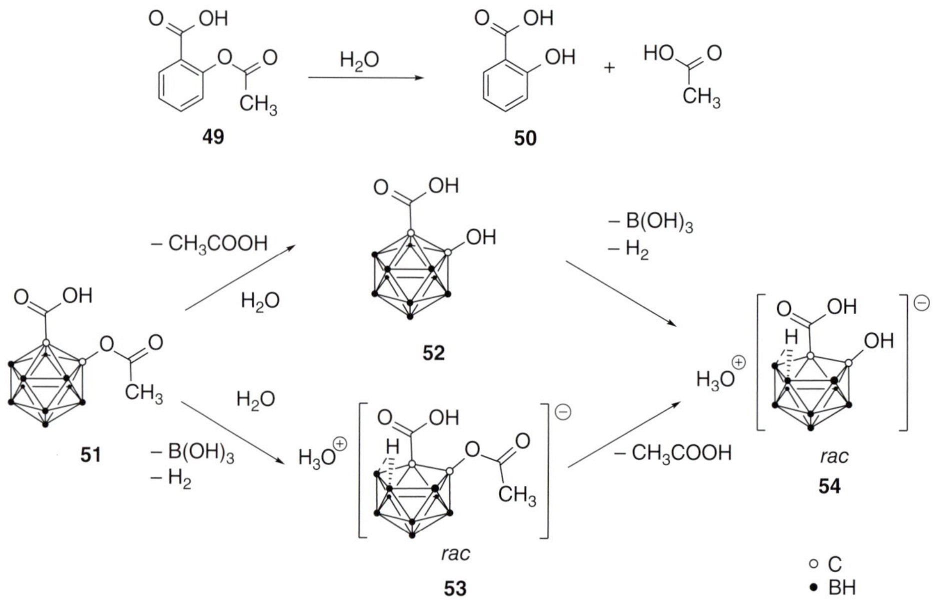

Scheme 1.4.8 Hydrolysis mechanism of aspirin and asborin [62].

Table 1.4.7 Results of the cytotoxicity studies (IC$_{50}$ values) of aspirin and asborin as well as their hydrolysis products [62]

Compound	IC$_{50}$ [µM]				
	518A2[a]	FaDu[b]	HT-29[c]	MCF-7[d]	SW1736[e]
49	>1000	>1000	>1000	>1000	>1000
50	>1000	>1000	>1000	>1000	>1000
51	139.1 ± 13.6	123.7 ± 17.2	132.5 ± 16.9	96.8 ± 8.0	162.3 ± 14.4
52	100.1 ± 15.6	100.3 ± 12.7	121.7 ± 13.9	91.9 ± 3.4	142.4 ± 14.0
53	247.2 ± 16.8	270.3 ± 19.7	247.8 ± 8.6	269.0 ± 4.7	451.0 ± 11.1
54	169.0 ± 2.8	218.1 ± 3.0	232.7 ± 4.3	182.1 ± 6.9	156.8 ± 10.2
B(OH)$_3$	>1000	>1000	>1000	>1000	>1000
Cisplatin	1.52 ± 0.19	1.21 ± 0.14	0.63 ± 0.03	2.03 ± 0.11	3.20 ± 0.24

[a] Human skin malignant melanoma.
[b] Human epithelial cell line from a squamous cell carcinoma of the hypopharynx.
[c] Human colorectal adenocarcinoma cell line with epithelial morphology.
[d] Human breast adenocarcinoma cells.
[e] Human thyroid carcinoma cells.

to the acetylation pattern [62]. Additionally, cell-death mechanistic studies revealed ultimately that the apoptotic pathway was preferred.

The Valliant group investigated the influence of polar groups on the target–vector specific recognition for a series of C-hydroxy carboranes, including 1-carboranol,

55a (ortho)
55b (para)

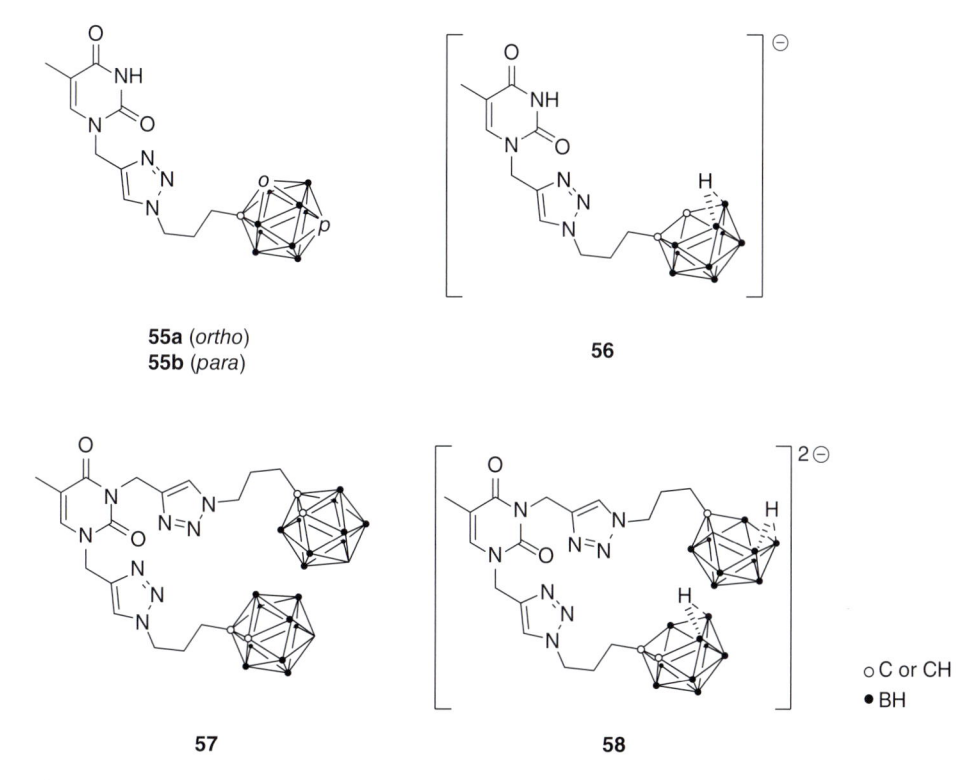

56

57

58

○ C or CH
● BH

Figure 1.4.14 Schematic structures of the *closo*-carborane– and *nido*-carborane(–1)–thymine conjugates by Leśnikowski and coworkers (the counterion Cs$^+$ was omitted) [66].

salborin, and asborin [67]. To obtain the target compounds, *closo*-carborane is substituted at the *C*-vertices, appropriately followed by decapping with methods developed for *C*-hydroxy- and *C*-carboxyl-substituted clusters (either EtOH/H$_2$O (3:2, v/v) with NaF at 160 °C in the microwave, or stirring in H$_2$O at 100 °C for 10 min). Finally, the resulting *nido* clusters were iodinated. The conducted biodistribution studies suggested a consistent influence of the hydroxy group in determining the selectivity of the compound toward the desired target, which in turn indicates that polar groups enhance target-specific interactions, allow enhanced renal clearance, and prevent accumulation in liver and other often unspecifically addressed organs.

Another very interesting example of the use of *nido*-carborane(–1) as pharmacophore is a series of thymine conjugates reported by the Leśnikowski group in 2016 [66]. The biological activity of a series of thymine–*closo*-carborane and thymine–*nido*-carborane(–1) conjugates was studied to exploit, on one hand, the anti-tubercular/antimycobacterial properties of thymine derivatives found by Pochet *et al.* [134], and, on the other hand, the hydrophobicity and inorganic nature of the carborane cluster.

The thymine derivatives **55a/55b** and **57** (Figure 1.4.14) were synthesized through click chemistry via copper(I)-catalyzed Huisgen–Meldal–Sharpless 1,3-dipolar cycloaddition. Deboronation of **55a** and **57** with CsF in EtOH under reflux conditions overnight led to **56** and **58**, respectively.

Table 1.4.8 Results of the cytotoxicity test (CC_{50} value) of compounds **55–58** with different mycobacterial cell lines [66]

| Comp. | CC$_{50}$ (µM) | | | | | ^{10}B/^{11}B uptake (mg/kg) |
	MRC-5[a]	Vero[b]	LLC-MK2[c]	A549[d]	L929[e]	M. smegmatis
55a	25.8 ± 4.2	1.2 ± 0.8	28.0 ± 0.5	57.7 ± 2.4	0.76 ± 0.2	6892
55b	61.3 ± 0.8	23.0 ± 0.3	37.7 ± 5.1	84.7 ± 3.1	57.8 ± 1.1	3927
56	77.7 ± 31.7	392.0 ± 12.7	747.3 ± 20.9	327.0 ± 7.1	162.5 ± 9.2	41.66
57	66.7 ± 0.5	14.5 ± 0.3	6.8 ± 1.1	19.1 ± 0.9	29.1 ± 8.9	594.8
58	189.3 ± 21.9	108.2 ± 6.8	398.8 ± 10.7	271.0 ± 8.5	55.8 ± 1.1	45.67

[a] Normal human lung fibroblast cells.
[b] Normal monkey kidney cells.
[c] Rhesus monkey kidney epithelial cells.
[d] Adenocarcinomic human alveolar basal epithelial cells.
[e] Adipose mouse fibroblast cell line.

Compounds **55–58** were tested *in vitro* against *Mycobacterium tuberculosis* thymidylate kinase (TMPKmt) and against mycobacterial growth in saprophytic *Mycobacterium smegmatis* and pathogenic *M. tuberculosis* strains. Compound **58** was the most active one against TMPKmt, but at the same time it was the least active one in bacterial cells, while the best anti-mycobacterial effect was shown by compounds **55b** and **57**, which were, however, not active against TMPKmt. This indicated that anti-mycobacterial activity does not go along with TMPKmt inhibition. Table 1.4.8 summarizes the results of the *in vitro* cytotoxicity assays. The *nido*-carborane(−1)-containing species have a high CC_{50} value[8] compared to their *closo* analogs, probably due to their much lower cellular uptake, determined via ICP-MS after 24 h exposure of the compound to *M. smegmatis* cells, which also correlates with their lower hydrophobicity. Thymine on its own does not have any cytotoxic effect.

Compounds **56** and **58** showed low toxicity on all used cell lines; **55a** and **57** led to high death rates for Vero, LLC-MK2, and L929, but low death rates for human MRC-5 and A549 tissues (Table 1.4.8). Compound **55b** is only moderately toxic against all tested mycobacterial cell lines.

Table 1.4.8 also contains information on ^{10}B/^{11}B uptake, indicating that the hydrophobicity of a *closo*-carborane cluster in particular could potentially increase the drug uptake in *M. tuberculosis*, whose cell walls frequently exhibit low permeability.

The last example discussed here for *nido*-carboranes(−1) as pharmacophores is indoborin, the carborane analog of indomethacin, which was published by Hey-Hawkins and coworkers [19] and recently also reviewed by Leśnikowski [15d]. The importance of the carbonyl group in promoting high binding affinity between a *nido*-carborane(−1) bioactive molecule and the biological target (enzyme) also was suggested recently by Neumann *et al.* (2016) [19], based on the crystal structure of recombinant murine

---

8 The 50% cytotoxic concentration (CC_{50}) is defined as the concentration required to reduce the cell growth by 50% compared to untreated controls.

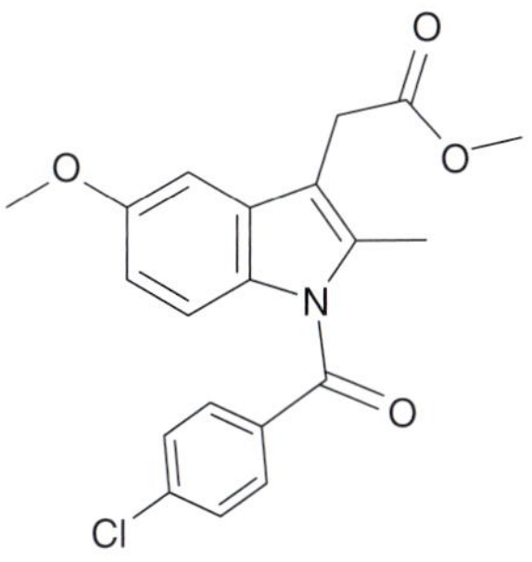

Figure 1.4.15 Indomethacin methyl ester (top), a potent COX-1 inhibitor, and two *nido*-carborane(−1)-containing analogs (**59** and **60**; bottom [only one enantiomer is shown]). The counterions (Na$^+$) for **59** and **60** are omitted [19].

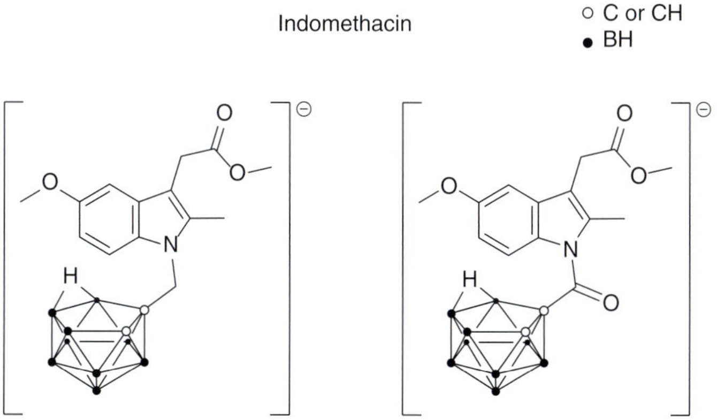

COX-2 enzyme complexed with a *nido*-carborane(−1) analog of the COX inhibitor indomethacin, namely 7-{[5-methoxy-2-methyl-3-(methoxycarbonylmethyl)-1*H*-indolyl]carbonyl}-7,8-dicarba-*nido*-dodecahydroundecarborate(−1) (Figure 1.4.15).

Neumann *et al.* suggested the primary importance of the polar interaction between the carbonyl group in **60** and Arg120 of COX-2 for the inhibition activity of **60** [19]. In support of this hypothesis, compound **59**, which contains a methylene bridge instead of an amide group and, therefore, cannot exhibit polar interactions with Arg120, showed much lower IC$_{50}$ values for COX-2 in comparison with **60** (>4 µM vs. 0.051 µM, respectively) [19]. It would, therefore, be interesting to study the binding mode of complex 7 (see Chapter 1.4.1.5) with the respective biological target(s) to be able to suggest a structure–activity relationship, which could explain the role of the carbonyl group in the binding affinity.

1.4.4 Conclusions and Future Challenges

The literature presented in this contribution has shown that half- and mixed-sandwich metallacarboranes have several possible applications in medicine, ranging from BNCT to pharmacophores in drug design, radio-imaging applications, and combinations thereof. As dicarbollide ligands offer a wide range of structural diversity through different substitution patterns, the presented applications are also wide and varied. While the reported studies on applications in medicine are very promising, only a small number of known half- and mixed-sandwich metallacarboranes has been tested for potential biological or medicinal applications; this leaves room for further research and improvement.

Carboranes have always been quite expensive, costing almost half of the price of gold. However, use of these compounds could be very useful in specific applications, such as medical applications, which utilize only minute amounts. Nevertheless, research into half- and mixed-sandwich metallacarboranes should not be restricted to medicinal applications. There are numerous possible applications for these compounds besides those presented here, for example in catalysis [46], and these promising areas of research certainly deserve further attention.

Appendix: Abbreviations

AFM	atomic force microscopy
AKR	aldo-keto reductase
Arg	arginine
ASS	Acetylsalicylsäure, acetylsalicylic acid
ATP	adenosine triphosphate
BBB	blood–brain barrier
Bn	benzyl
BNCT	boron neutron capture therapy
BPA	L-4-(dihydroxyboryl)phenylalanine
BSH	sodium mercapto-undecahydro-*closo*-dodecaborate
n-BuLi	n-butyllithium
Cb or Cb^{2-}	$[C,C\text{-R},\text{R}'\text{-C}_2\text{B}_9\text{H}_9]^{2-}$, with R = H, any substituent, R' = H, any substituent
CC$_{50}$	half-maximal cytotoxic concentration
CNS	central nervous system
cod	1,5-cyclooctadiene
COSAN	cobaltabis(dicarbollide)
COX	cyclooxygenase
Cp or Cp$^-$	$C_5H_5^-$, cyclopentadienyl
p-cymene	1-isopropyl-4-methylbenzene, paracymene
DFT	density functional theory
DLC	delocalized lipophilic cation
DMPC	dimyristoylphosphatidylcholine
DMSO	dimethyl sulfoxide
DSBL	distearoyl boron lipid
DSPC	distearoylphosphatidylcholine
EDA	energy decomposition analysis
EGFR	epidermal growth factor receptor
Et	ethyl
EtOH	ethanol
FL-SBL	fluorescence-labeled boron lipid
FWHH	full width at half height

Glu	glutamic acid
HIV	human immunodeficiency virus
HPLC	high-performance liquid chromatography
IC_{50}	half-maximal inhibitory concentration
ICP-MS	inductively coupled plasma mass spectrometry
K_i	inhibition constant
Lys	lysine
MAO	methylalumoxane $[Al(CH_3)_xO_y]_n$
Me	methyl
MeCN	acetonitrile
MeOH	methanol
MIP	molecular imaging probe
MRI	magnetic resonance imaging
NADP	nicotinamide adenine dinucleotide phosphate
NMR	nuclear magnetic resonance
NOS	nitric oxide synthase
eNOS	endothelial nitric oxide synthase
iNOS	inducible nitric oxide synthase
nNOS	neuronal nitric oxide synthase
NPY	neuropeptide Y
Nuc	nucleophile
OAc	acetate
PC	phosphatidylcholine
PE	polyethylene
PEG	polyethylene glycol
PET	positron emission spectroscopy
Ph	phenyl
pH	logarithmic proton concentration
pK_a	logarithmic acid dissociation constant
PMPA	2-(phosphonomethyl)pentane-1,5-dioic acid
i-Pr	isopropyl
PSMA	prostate-specific membrane antigen
SPECT	single-photon emission computed tomography
TBAF	tetrabutylammonium fluoride
TCO	*trans*-cyclooctene
TFA	trifluoroacetate
THF	tetrahydrofuran
TIM	triosephosphate isomerase
TMPKmt	*Mycobacterium tuberculosis* thymidylate kinase

Acknowledgments

Financial support from the Fonds der Chemischen Industrie (VCI, doctoral grant B.S.), the Saxon Ministry of Science and the Fine Arts (SMWK, Landesgraduiertenförderung, doctoral grant M.G.), and the Graduate School Building with Molecules and Nano-objects (BuildMoNa) is gratefully acknowledged.

References

1 Jones B, Adams S, Miller GT, Jesson MI, Watanabe T, Wallner BP. Hematopoietic stimulation by a dipeptidyl peptidase inhibitor reveals a novel regulatory mechanism and therapeutic treatment for blood cell deficiencies. Blood 2003;102(5):1641–1648.

2 *Bortezomib (PS-341)*: (a) Teicher BA, Ara G, Herbst R, Palombella VJ, Adams J. The proteasome inhibitor PS-341 in cancer therapy. Clin. Cancer Res. 1999;5:2638–2645; (b) Lightcap ES, McCormack TA, Pien CS, Chau V, Adams J, Elliott PJ. Proteasome inhibition measurements: clinical application. Clin. Chem. 2000;46(5):673–683; (c) Hideshima T, Richardson P, Chauhan D, Palombella VJ, Elliott PJ, Adams J, Anderson KC. The proteasome inhibitor PS-341 inhibits growth, induces apoptosis, and overcomes drug resistance in human multiple myeloma cells. Cancer Res. 2001;61: 3071–3076; (d) Adams J, Behnke M, Chen SW, Cruickshank AA, Dick LR, Grenier L, Klunder JM, Ma YT, Plamondon L, Stein RL. Potent and selective inhibitors of the proteasome: dipeptidyl boronic acids. Bioorg. Med. Chem. Lett. 2016;8(4):333–338; *Tavaborole (AN2690)*: (e) Baker SJ, Zhang Y-K, Akama T, Lau A, Zhou H, Hernandez V, Mao W, Alley MRK, Sanders V, Plattner JJ. Discovery of a new boron-containing antifungal agent, 5-fluoro-1,3-dihydro-1-hydroxy-2,1-benzoxaborole (AN2690), for the potential treatment of onychomycosis. J. Med. Chem. 2006;49(15):4447–4450; (f) Rock FL, Mao W, Yaremchuk A, Tukalo M, Crepin T, Zhou H, Zhang YK, Hernandez V, Akama T, Baker SJ, Plattner JJ, Shapiro L, Martinis SA, Benkovic SJ, Cusack S, Alley MR. An antifungal agent inhibits an aminoacyl-tRNA synthetase by trapping tRNA in the editing site. Science 2007;316(5832):1759–1761; *Crisaborole (AN2728)*: (g) Pfizer Inc. Available from: https://www.pfizer.com/products/product-detail/eucrisa (17/02/2018); (h) Bolger GB. RACK1 and β-arrestin2 attenuate dimerization of PDE4 cAMP phosphodiesterase PDE4D5. Cell. Signal. 2016;28(7):706–712; (i) Draelos ZD, Stein Gold LF, Murrell DF, Hughes MH, Zane LT. Post hoc analyses of the effect of crisaborole topical ointment, 2% on atopic dermatitis: associated pruritus from phase 1 and 2 clinical studies. J. Drugs Dermatol. 2016;15(2):172–176; *Epetraborole (AN3365)*: (j) Baker SJ, Hernandez VS, Sharma R, Nieman JA, Akama T, Zhang Y-K, Plattner JJ, Alley MRK, Singh R, Rock F. Boron-containing small molecules. PCT Int. Appl. 2008, WO A1 20081224;*SCYX-7158 (AN5568)*: (k) Jacobs RT, Nare B, Wring SA, Orr MD, Chen D, Sligar JM, Jenks MX, Noe RA, Bowling TS, Mercer LT, Rewerts C, Gaukel E, Owens J, Parham R, Randolph R, Beaudet B, Bacchi CJ, Yarlett N, Plattner JJ, Freund Y, Ding C, Akama T, Zhang Y-K, Brun R, Kaiser M, Scandale I, Don R. SCYX-7158, an orally-active benzoxaborole for the treatment of stage 2 human African trypanosomiasis. PLoS Negl. Trop. Dis. 2011;5(6):e1151; (l) Wring S, Gaukel E, Nare B, Jacobs R, Beaudet B, Bowling T, Mercer L, Bacchi C, Yarlett N, Randolph R, Parham R, Rewerts C, Plattner JJ, Don R. Pharmacokinetics and pharmacodynamics utilizing unbound target tissue exposure as part of a disposition-based rationale for lead optimization of benzoxaboroles in the treatment of stage 2 human African trypanosomiasis. Parasitology 2014;141(1):104–118.

3 Barth RF, Coderre JA, Vicente MG, Blue TE. Boron neutron capture therapy of cancer: current status and future prospects. Clin. Cancer Res. 2005;11(11):3987–4002.

4 Yokoyama K, Miyatake S-I, Kajimoto Y, Kawabata S, Doi A, Yoshida T, Asano T, Kirihata M, Ono K, Kuroiwa T. Pharmacokinetic study of BSH and BPA in simultaneous use for BNCT. J. Neurooncol. 2006;78:227–232.

5 Barth RF. Boron neutron capture therapy at the crossroads: challenges and opportunities. Appl. Radiat. Isot. 2009;67:S3–S6.

6 (a) Lipscomb WN. Die Borane und ihre Derivate (Nobel-Vortrag). Angew. Chem. 1977;89:685–696; (b) Williams RE. The polyborane, carborane, carbocation continuum: architectural patterns. Chem. Rev. 1992;92(2):177–207; (c) Grimes RN. Structure and Bonding. In: *Carboranes*, 2nd ed. Heidelberg: Elsevier, 2011, 7–20.
7 Hawthorne MF, Young DC, Andrews TD, Howe DV, Pilling RL, Pitts AD, Reintjes M, Warren LF, Jr., Wegner PA. π-Dicarbollyl derivatives of the transition metals metallocene analogs. J. Am. Chem. Soc. 1968;90(4):879–896.
8 Sauerwein WAG, Wittig A, Moss R, Nakagawa Y. Neutron Capture Therapy: Principles and Applications. Berlin: Springer, 2012.
9 Hosmane NS, Boron Science: New Technologies and Applications. Boca Raton, FL: CRC Press, 2012.
10 Moss RL. Critical review, with an optimistic outlook, on boron neutron capture therapy (BNCT). Appl. Radiat. Isot. 2014;88:2–11.
11 (a) Ahrens VM, Frank R, Boehnke S, Schütz CL, Hampel G, Iffland DS, Bings NH, Hey-Hawkins E, Beck-Sickinger AG. Receptor-mediated uptake of boron-rich neuropeptide Y analogues for boron neutron capture therapy. ChemMedChem 2015;10(1):164–172; (b) Frank R, Ahrens VM, Boehnke S, Beck-Sickinger AG, Hey-Hawkins E. Charge-compensated metallacarborane building blocks for conjugation with peptides. ChemBioChem 2016;17:308–317; (c) Ahrens VM, Frank R, Stadlbauer S, Beck-Sickinger AG, Hey-Hawkins E. Incorporation of ortho-carbaboranyl-N_ε-modified L-lysine into neuropeptide Y receptor Y_1- and Y_2-selective analogues. J. Med. Chem. 2011;54:2368–2377; (d) Frank R, Hey-Hawkins E. A convenient route towards deoxygalactosyl-functionalised *ortho*-carbaborane: synthesis of a building block for peptide conjugation. J. Organomet. Chem. 2015;798:46–50; (e) Frank R, Ahrens VM, Boehnke S, Hofmann S, Kellert M, Saretz S, Pandey S, Sárosi MB, Bartók Á, Beck-Sickinger AG, Hey-Hawkins E. Carbaboranes – more than just phenyl mimetics. Pure Appl. Chem. 2015;87(2):163–171.
12 Valliant JF, Guenther KJ, King AS, Morel P, Schaffer P, Sogbein OO, Stephenson KA. The medicinal chemistry of carboranes. Coord. Chem. Rev. 2002;232:173–230.
13 Hawthorne MF, Lee MW. A critical assessment of boron target compounds for boron neutron capture therapy. J. Neurooncol. 2003;62(1–2):33–45.
14 Armstrong AF, Valliant JF. The bioinorganic and medicinal chemistry of carboranes: from new drug discovery to molecular imaging and therapy. Dalton Trans. 2007;4240–4251.
15 (a) Wermuth CG, Ganellin CR, Lindberg P, Mitscher LA. Glossary of terms used in medicinal chemistry. Pure Appl. Chem. 1998;70(5):1129–1143; (b) Issa F, Kassiou M, Rendina LM. Boron in drug discovery: carboranes as unique pharmacophores in biologically active compounds. Chem. Rev. 2011;111:5701–5722; (c) Grimes RN. Carboranes as pharmacophores. In: *Carboranes*, 2nd ed. Heidelberg: Elsevier; 2011, 1053–1057; (d) Leśnikowski ZJ. Challenges and opportunities for the application of boron clusters in drug design. J. Med. Chem. 2016;59(17):7738–7758.
16 Scholz M, Steinhagen M, Heiker JT, Beck-Sickinger AG, Hey-Hawkins E. Asborin inhibits aldo/keto reductase 1A1. ChemMedChem 2011;6:89–93.
17 Laube M, Neumann W, Scholz M, Lönnecke P, Crews B, Marnett LJ, Pietzsch J, Kniess T, Hey-Hawkins E. 2-carbaborane-3-phenyl-1H-indoles – synthesis via McMurry reaction and cyclooxygenase (COX) inhibition activity. ChemMedChem 2013;8:329–335.
18 Neumann W, Frank R, Hey-Hawkins E. One-pot synthesis of an indole-substituted 7,8-dicarba-*nido*-dodecahydroundecaborate(−1). Dalton Trans. 2015;44:1748–1753.

19 Neumann W, Xu S, Sárosi MB, Scholz MS, Crews BC, Ghebreselasie K, Banerjee S, Marnett LJ, Hey-Hawkins E. *nido*-Dicarbaborate induces potent and selective inhibition of cyclooxygenase-2. ChemMedChem 2016;11:175–178.

20 (a) Fujii S, Yamada A, Nakano E, Takeuchi Y, Mori S, Masuno H, Kagechika H. Design and synthesis of nonsteroidal progesterone receptor antagonists based on *C*, *C'*-diphenylcarborane scaffold as a hydrophobic pharmacophore. Eur. J. Med. Chem. 2014;84:264–277; (b) Gómez-Vallejo V, Vázquez N, BabuGona K, Puigivila M, González M, SanSebastián E, Martin A, Llop J. Synthesis and in vivo evaluation of ^{11}C-labeled (1,7-dicarba-*closo*-dodecaboran-1-yl)-*N*-{[(2*S*)-1-ethylpyrrolidin-2-yl]methyl}ami-de. J. Label Compd. Radiopharm. 2014;57:209–214.

21 (a) Chen W, Mehta SC, Lu DR. Selective boron drug delivery to brain tumors for boron neutron capture therapy. Adv. Drug Deliv. Rev. 1997;26:231–247; (b) Tjarks W. The use of boron clusters in the rational design of boronated nucleosides for neutron capture therapy of cancer. J. Organomet. Chem. 2000;614–615:37–47; (c) Wilkinson SM, Gunosewoyo H, Barron ML, Boucher A, McDonnell M, Turner P, Morrison DE, Bennett MR, McGregor IS, Rendina LM, Kassiou M. The first CNS-active carborane: a novel P2X7 receptor antagonist with antidepressant activity. ACS Chem. Neurosci. 2014;5:335–339; (d) Xuan S, Zhao N, Zhou Z, Fronczek FR, Vicente MGH. J. Med. Chem. 2016;59:2109–2117.

22 (a) Lunato AJ, Wang J, Woollard JE, Anisuzzaman AKM, Ji W, Rong F-G, Ikeda S, Soloway AH, Erikson S, Ives DH, Blue TE, Tjarks W. Synthesis of 5-(carboranylalkylmercapto)-2'-deoxyuridines and 3-(carboranylalkyl)thymidines and their evaluation as substrates for human thymidine kinases 1 and 2. J. Med. Chem. 1999;42(17):3378–3389 (therein ref. 16: Hofmann M, University of Georgia, Athens, GA, unpublished results: according to *ab initio* computations at the MP2(fc)/6-31G* level, the diameter of benzene is 4.968 Å [H(C1)–H(C4)] and those of the *o*-carborane clusters are 5.48 Å [H(C1)–H(B12)], 5.75 Å [H(B3)–H(B12)], and 5.279 Å [H(B4)–H(B11)]), respectively); (b) Poater J, Solà M, Viñas, Teixidor F. π Aromaticity and three-dimensional aromaticity: two sides of the same coin? Angew. Chem. Int. Ed. 2014;53:12191–12195.

23 (a) Zakharkin LI, Kalinin VN, Snyakin AP, Kvasov BA. Effect of solvents on the electronic properties of 1-*o*-, 3-*o*- and 1-*m*-carboranyl groups. J. Organomet. Chem. 1969;18(1):19–26; (b) Teixidor F, Barberà G, Vaca A, Kivekäs R, Sillanpää R, Oliva J, Viñas C. Are methyl groups electron-donating or electron-withdrawing in boron clusters? Permethylation of *o*-carborane. J. Am. Chem. Soc. 2005;127(29):10158–10159.

24 Teixidor F, Viñas C. Product subclass 40: carboranes and metallacarboranes. In: Kauffmann DE, Matteson DS (Editors). Science of Synthesis: Hoube–Weyl Methods of Molecular Transformations. Organometallics: Boron Compounds (Category 1, Vol. 6). Stuttgart: Georg Thieme Verlag, 2005, 1235–1275.

25 Scholz M, Hey-Hawkins E. Carbaboranes as pharmacophores: properties, synthesis, and application strategies. Chem. Rev. 2011;111:7035–7062.

26 Fox MA, Hughes AK, Johnson AL, Paterson MAJ. Do the discrete dianions $C_2B_9H_{11}^{2-}$ exist? Characterisation of alkali metal salts of the 11-vertex *nido*-dicarboranes, $C_2B_9H_{11}^{2-}$, in solution. J. Chem. Soc., Dalton Trans. 2002;2009–2019.

27 Bould J, Laromaine A, Bullen NJ, Viñas C, Thornton-Pett M, Sillanpää R, Kivekäs R, Kennedy JD, Teixidor F. Borane reaction chemistry: alkyne insertion reactions into

boron-containing clusters. Products from the thermolysis of [6,9-(2-HC≡C–C$_5$H$_4$N)$_2$-*arachno*-B$_{10}$H$_{12}$]. Dalton Trans. 2008;1552–1563.

28 (a) Semioshkin AA, Sivaev IB, Bregadze VI. Cyclic oxonium derivatives of polyhedral boron hydrides and their synthetic applications. Dalton Trans. 2008:977–992; (b) Olid D, Núñes R, Viñas C, Teixidor F. Methods to produce B–C, B–P, B–N and B–S bonds in boron clusters. Chem. Soc. Rev. 2013;42(8):3318–3336.

29 (a) Jasper SA Jr., Mattern J, Huffman JC, Todd LJ. Palladium-mediated substitution of the *closo*-B$_{12}$H$_{12}$(−2) and *nido*-7,8-C$_2$B$_9$H$_{12}$(−1) ions by PMe$_2$Ph: the single-crystal structure studies of 1,7-(PMe$_2$Ph)$_2$-*closo*-B$_{12}$H$_{10}$ and 9-PMe$_2$Ph-*nido*-7,8-C$_2$B$_9$H$_{11}$. Polyhedron 2007;26:3793–3798; (b) Semioshkin AA, Sivaev IB, Bregadze VI. Cyclic oxonium derivatives of polyhedral boron hydrides and their synthetic applications. Dalton Trans. 2008;977–992.

30 (a) Young DC, Howe DV, Hawthorne MF. Ligand derivatives of (3)-1,2-dicarbadodecahydroundecaborate(−1). J. Am. Chem. Soc. 1969;91(4):859–862; (b) Kang HC, Lee SS, Knobler CB, Hawthorne MF. Syntheses of charge-compensated dicarbollide ligand precursors and their use in the preparation of novel metallacarboranes. Inorg. Chem. 1991;30(9):2024–2031.

31 Meshcheryakov VI, Kitaev PS, Lyssenko KA, Starikova ZA, Petrovskii PV, Janoušek Z, Corsini M, Laschi F, Zanello P, Kudinov AR. (Tetramethylcyclobutadiene)cobalt complexes with monoanionic carborane ligands [9-L-7,8-C$_2$B$_9$H$_{10}$] (L = SMe$_2$, NMe$_3$ and py). J. Organomet. Chem. 2005;690:4745–4754.

32 Zakharkin LI, Kalinin VN, Zhigareva GG. Izv. Akad. Nauk SSSR Ser. Khim. 1979; 2376–2377 [Oxidation of dicarbadodecahydro-*nido*-undecaborate anions by mercuric chloride in tetrahydrofurane and pyridine. Russ. Chem. Bull. 1979;28:2198–2199 (English trans.)].

33 Grüner B, Holub J, Plešek J, Štíbr B, Thornton-Pett M, Kennedy JD. Dimethylsulfide-dicarbaborane chemistry. Isolation and characterisation of isomers [9-(SMe$_2$)-*nido*-7,8-C$_2$B$_9$H$_{10}$-X-Me] (where X = 1, 2, 3 and 4) and some related compounds. An unusual skeletal rearrangement. Dalton Trans. 2007;4859–4865.

34 (a) Tolpin EI, Lipscomb WN. Fluxional behavior of undecahydroundecaborate(2-) (B$_{11}$H$_{11}^{2−}$). J. Am. Chem. Soc. 1973;95(7):2384–2386; (b) King RB. Chemical applications of topology and group theory. 11. Degenerate edges as a source of inherent fluxionality in deltahedra. Inorg. Chim. Acta 1981;49:237–240; (c) Volkov O, Dirk W, Englert U, Paetzold P. Reactions of the undecaborate anion [B$_{11}$H$_{11}$]$^{2−}$. In: Davidson MG, Hughes AK, Marder TB, Wade K (Editors). Contemporary Boron Chemistry. Cambridge: Royal Society of Chemistry Books, 2000, 159–162.

35 (a) Wille AE, Sneddon LG. Proton-sponge-initiated reactions of 6,8-dicarba-*arachno*-nonaborane(13) and 6,7-dicarba-*arachno*-nonaborane(13) with methyl propynoate: one-step syntheses of the 6-(MeOOCCH$_2$)-5,6,7-tricarba-*arachno*-decaborane(12) and 6-(MeOOCCH$_2$)-5,6,10-tricarba-*nido*-decaborane(10) tricarbaboranes. Collect. Czech. Chem. Commun. 1997;62:1214–1228; (b) Bratsev VA. Skeletal rearrangements following electrophilic alkylation of 7,8- and 7,9-dicarbollide anions (a review). In: Davidson MG, Hughes AK, Marder TB, Wade K (Editors). Contemporary Boron Chemistry. Cambridge: Royal Society of Chemistry Books, 2000, 205–211; (c) Bakardjiev M, Holub J, Štíbr B, Hnyk D, Wrackmeyer B. Diphosphacarbollide analogues of the C$_5$H$_5^−$ anion: isolation of the *nido*-di- and triphosphacarboranes 7,8,9-P$_2$CB$_8$H$_{10}$,

[7,8,9-P$_2$CB$_8$H$_9$]$^-$, [7,8,10-P$_2$CB$_8$H$_9$]$^-$, and 7,8,9,10-P$_3$CB$_7$H$_8$. Inorg. Chem. 2005;44(16):5826–5832.

36 Frank R, Adhikari AK, Auer H, Hey-Hawkins E. Electrophile-induced nucleophilic substitution of the *nido*-dicarbaundecaborate anion *nido*-7,8-C$_2$B$_9$H$_{12}$ by conjugated heterodienes. Chem. Eur. J. 2014;20:1440–1446.

37 Grimes RN. *Carboranes*, 2nd ed. Heidelberg: Elsevier, 2011

38 Gona KB, Zaulet A, Gomez-Vallejo V, Teixidor F, Llop J, Viñas C. COSAN as a molecular imaging platform: synthesis and "*in vivo*" imaging. Chem. Commun. 2014;50:11415–11417.

39 (a) Rezacova P, Cígler P, Matejicek P, Lepsik M, Pokorna J, Grüner B, Konvalinka J. Part I, Chapter 3: Medicinal applications of carboranes: inhibition of HIV protease (pp. 40–71); Grüner B, Rais J, Selucky P, Lucanikova M. Part VI, Chapter 19: Recent progress in extraction agents based on cobalt bis(dicarbollides) for partitioning of radionuclides from high-level nuclear waste. In: Hosmane NS (Editor). Boron Science: New Technologies and Applications. Boca Raton, FL: CRC Press, 2012; (b) Sivaev IB, Bregadze VI. Chemistry of cobalt bis(dicarbollides): a review. Collect. Czech. Chem. Commun. 1999;64:783–805.

40 Wu D-H, Wu C-H, Li Y-Z, Guo D-D, Wang X-M, Yan H. Addition of ethynylferrocene to transition-metal complexes containing a chelating 1,2-dicarba-*closo*-dodecaborane-1,2-dichalcogenolate ligand: *in vitro* cooperativity of a ruthenium compound on cellular uptake of an anticancer drug. Dalton Trans. 2009;285–290.

41 Dash BP, Satapathy R, Gaillard ER, Maguire JA, Hosmane NS. Synthesis and properties of carborane-appended *C*3-symmetrical extended πsystems J. Am. Chem. Soc. 2010;132(18):6578–6587.

42 (a) Lerouge F, Viñas C, Teixidor F, Núñes R, Abreu A, Xochitiotzi E, Santillan R, Farfán N. High boron content carboranyl-functionalized aryl ether derivatives displaying photoluminescent properties. Dalton Trans. 2007;19:1898–1903; (b) Lerouge F, Ferrer-Ugalde A, Viñas C, Teixidor F, Sillanpää R, Abreu A, Xochitiotzi E, Farfán N, Santillan R, Núñez R. Synthesis and fluorescence emission of neutral and anionic di- and tetra-carboranyl compounds. Dalton Trans. 2011;40:7541–7550.

43 Galkin KI, Lubimov SE, Godovikov IA, Dolgushin FM, Smol'yakov AF, Sergeeva E, Davankov VA, Chizhevsky IT. New acyclic (π-allyl)-*closo*-rhodacarboranes with an agostic CH$_3$···Rh bonding interaction that operate as unmodified rhodium-based catalysts for alkene hydroformylation. Organometallics 2012;31:6080–6084.

44 Wang Y, Liu D, Chan H-S, Xie Z. Synthesis, structural characterization, and reactivity of group 4 metallacarboranes containing the ligand [Me$_2$C(C$_5$H$_4$)(C$_2$B$_9$H$_{10}$)]$^{3-}$. Organometallics 2008;27:2825–2832.

45 Kalinin VN, Mel'nik OA, Sakharova AA, Frunze TM, Zakkharkin LI, Borunova NV, Sharf VZ. Izv. Akad. Nauk. SSSR Ser. Khim. 1984;1966–1967 [Polymeric phosphinocarborane Rh(I) complexes as catalysts for the hydrogenation and isomerization of unsaturated hydrocarbons. Russ. Chem. Bull. 1984;33:1966–1967 (English trans.)].

46 Gozzi M, Schwarze B, Hey-Hawkins E. Half- and mixed-sandwich metallacarboranes in catalysis. In: Hosmane NS, Eagling R. (Editors). Handbook of Boron Chemistry in Organometallics, Catalysis, Materials and Medicine, Volume 2: Boron in Catalysis, World Scientific, 2018, ISBN: 978-1-78634-441-0.

47 (a) Brown DA, Fanning MO, Fitzpatrick NJ. Molecular orbital theory of organometallic compounds. 15. A comparative study of ferrocene and π-cyclopentadienyl-(3)-1,2-dicarbollyliron. Inorg. Chem. 1978;17(6):1620–1623; (b) Hoffmann R. Building bridges between inorganic and organic chemistry. Angew. Chem. Int. Ed. Engl. 1982;21(10):711–724.

48 Corsini M, Fabrizi de Biani F, Zanello P. Mononuclear metallacarboranes of groups 6–10 metals: analogues of metallocenes electrochemical and X-ray structural aspects. Coord. Chem. Rev. 2006;250:1351–1372.

49 Loginov DA, Starikova ZA, Corsini M, Zanello P, Kudinov AR. (Cyclopentadienyl) metalla-dicarbollides 3-(η-C_5R_5)-3,1,2-$MC_2B_9H_{11}$(M = Co, Rh, Ir): synthesis, electrochemistry, and bonding. J. Organomet. Chem. 2013;747:69–75.

50 von Hopffgarten M, Frenking G. Energy decomposition analysis. WIREsComput. Mol. Sci. 2012;2:43–62.

51 (a) Grimes RN. Carboranes in the chemist's toolbox. Dalton Trans. 2015;44:5939–5956; (b) Gao SM, Hosmane NS. Dendrimer and nanostructure-supported carboranes and metallacarboranes: an account. Russ. Chem. Bull. Int. Ed. 2014;63(4):788–810.

52 (a) Farràs P, Juárez-Pérez EJ, Lepšík M, Luque R, Núñez R, Teixidor F. Metallacarboranes and their interactions: theoretical insights and their applicability. Chem. Soc. Rev. 2012;41:3445–3463; (b) Bühl M, Holub J, Hnyk D, Macháček J. Computational studies of structures and properties of metallaboranes. 2. Transition-metal dicarbollide complexes. Organometallics 2006;25(9):2173–2181.

53 Hawthorne MF, Young DC, Garrett PM, Owen DA, Schwerin SG, Tebbe FN, Wegner PA. The preparation and characterization of the (3)-1,2- and (3)-1,7-dicarbadodecahydroundecaborate(−1) ions. J. Am. Chem. Soc. 1968;90(4):862–868.

54 Weiss R, Grimes RN. Sources of line width in boron-11 nuclear magnetic resonance spectra: scalar relaxation and boron-boron coupling in B_4H_{10} and B_5H_9. J. Am. Chem. Soc. 1978;100(5):1401–1405.

55 Venable TL, Hutton WC, Grimes RN. Atom connectivities in polyhedral boranes elucidated via two-dimensional J-correlated ^{11}B-^{11}B FT NMR: a general method. J. Am. Chem. Soc. 1982;104:4716–4717.

56 Safronov AV, Dolgushin FM, Petrovskii PV, Chizhevsky IT. Low-temperature "1,2 → 1,7" isomerization of sterically crowded icosahedral closo-((2,3,8-$η^3$):(5,6-$η^2$)-norbornadien-2-yl)rhodacarborane via the formation of a pseudocloso intermediate: molecular structures of [3,3-((2,3,8-$η^3$):(5,6-$η^2$)-$C_7H_7CH_2$)-1,2-(4'-MeC_6H_4)$_2$-3,1,2-pseudocloso-Rh$C_2B_9H_9$] and 1,2 → 1,7 isomerized products. Organometallics 2005;24(12):2964–2970.

57 Armstrong AF, Valliant JF. Microwave-assisted synthesis of tricarbonyl rhenacarboranes: steric and electronic effects on the 1,2 → 1,7 carborane cage isomerization. Inorg. Chem. 2007;46(6):2148–2158.

58 Kaplánek R, Martásek P, Grüner B, Panda S, Rak J, Siler Masters BS, Král V, Roman LJ. Nitric oxide synthases activation and inhibition by metallacarborane-cluster-based isoform-specific affectors. J. Med. Chem. 2012;55:9541–9548.

59 Louie AS, Harrington LE, Valliant JF. The preparation and characterization of functionalized carboranes and Re/Tc-metallocarboranes as platforms for developing molecular imaging probes: structural and cage isomerism studies. Inorg. Chim. Acta 2012;389:159–167.

60 Louie AS, Vasdev N, Valliant JF. Preparation, characterization, and screening of a high affinity organometallic probe for α-adrenergic receptors. J. Med. Chem. 2011;54:3360–3367.

61 Grimes RN. 11-Vertex open clusters. In: Carboranes, 2nd ed. Heidelberg: Elsevier, 2011, 197–226.

62 Scholz MS, Kaluđerović GN, Kommera H, Paschke R, Will J, Sheldrick WS, Hey-Hawkins E. Carbaboranes as pharmacophores: similarities and differences between aspirin and asborin. Eur. J. Med. Chem. 2011;46:1131–1139.

63 Genady AR, Tan J, El-Zaria ME, Zlitni A, Janzen N, Valliant JF. Synthesis, characterization and radiolabeling of carborane functionalized tetrazines for use in inverse electron demand Diels–Alder ligation reactions. J. Organomet. Chem. 2015;791:204–213.

64 Schaeck JJ, Kahl SB. Rapid cage degradation of 1-formyl and 1-alkyloxycarbonyl substituted 1,2-dicarba-*closo*-dodecaboranes by water or methanol in polar organic solvents. Inorg. Chem. 1999;38(12):204–206.

65 Fox MA, Wade K. Deboronation of 9-substituted-ortho- and -meta-carboranes. J. Organomet. Chem. 1999;573(1–2):279–291.

66 Adamska A, Rumijowska-Galewicz A, Ruszczynska A, Studzinska M, Jablonska A, Paradowska E, Bulska E, Munier-Lehmann H, Dziadek J, Leśnikowski ZJ, Olejniczak AB. Anti-mycobacterial activity of thymine derivatives bearing boron clusters. Eur. J. Med. Chem. 2016;121:71–81.

67 El-Zaria ME, Janzen N, Blacker M, Valliant JF. Synthesis, characterisation, and biodistribution of radioiodinated C-hydroxy-carboranes. Chem. Eur. J. 2012;18:11071–11078.

68 Green AEC, Parker SK, Valliant JF. Synthesis and screening of bifunctional radiolabelled carborane-carbohydrate derivatives. J. Organomet. Chem. 2009;694:1736–1746.

69 El-Zaria ME, Janzen N, Valliant JF. Room-temperature synthesis of Re(I) and Tc(I) metallocarboranes. Organometallics 2012;31:5940–5949.

70 Busby DC, Hawthorne MF. The crown ether promoted base degradation of *p*-carborane. Inorg. Chem. 1982;21(11):4101–4103.

71 Zlatogorsky S, Ellis D, Rosair GM, Welch AJ. Unexpectedly facile isomerisation of [7,10-Ph$_2$-7,10-*nido*-C$_2$B$_{10}$H$_{10}$]$^{2-}$ to [7,9-Ph$_2$-7,9-*nido*-C$_2$B$_{10}$H$_{10}$]$^{2-}$. Chem. Commun. 2007;2178–2180.

72 Zlatogorsky S, Edie MJ, Ellis D, Erhardt S, Lopez ME, Macgregor SA, Rosair GM, Welch AJ. The mechanism of reduction and metalation of *para* carboranes: the missing 13-vertex MC$_2$B$_{10}$ isomer. Angew. Chem. Int. Ed. 2007;46:6706–6709.

73 (a) Štibr B, Holub J, Teixidor F, Viñas C. Tricarbollides: compounds of the eleven-vertex series of tricarbaboranes. J. Chem. Soc. Chem. Commun. 1995;7:795–796; (b) Štibr B, Holub J, Císařová I, Teixidor F, Viñas C, Fusek J, Plzák Z. The derivatives of the 7,8,9-9- series of tricarbollides. Preparation and structural characterization of the 11-vertex tricarbaboranes 7-L-nido-7,8,9-C$_3$B$_8$H$_{10}$ (L = amines). Inorg. Chem. 1996;35(12):3635–3642.

74 Holub J, Grüner B, Císařová I, Fusek J, Plzák Z, Teixidor F, Viñas C, Štibr BA. Series of the twelve-vertex ferratricarbollides [2-(η^5-C$_5$H$_5$)-9-X-closo-2,1,7,9-FeC$_3$B$_8$H$_{10}$] (where X = H$_2$N, MeHN, Me$_2$N, Bu^{t}HN, Bu^t(Me)N): a highly stable metallatricarbaborane system with amine functions in the para position to the metal center. Inorg. Chem. 1999;38(12):2775–2780.

75 Hawkins PM, Jelliss PA, Nonaka N, Shi X, Banks WA. Permeability of the blood-brain barrier to a rhenacarborane. JPET 2009;329(2):608–614.
76 Pruitt DG, Baumann SM, Place GJ, Oyeamalu AN, Sinn E, Jelliss PA. Synthesis and functionalization of nitrosylrhenacarboranes towards their use as drug delivery vehicles. J. Organomet. Chem. 2015;798:60–69.
77 El-Zaria ME, Genady AR, Janzen N, Petlura CI, Beckford Vera DR, Valliant JF. Preparation and evaluation of carborane-derived inhibitors of prostate specific membrane antigen (PSMA). Dalton Trans. 2014;43:4950–4961.
78 Reynolds RC, Campbell SR, Fairchild RG, Kisliuk RL, Micca PL, Queener SF, Riordan JM, Sedwick WD, Waud WR, Leung AKW, Dixon RW, Suling WJ, Borhani DW. Novel boron-containing, non-classical antifolates: synthesis and preliminary biological and structural evaluation. J. Med. Chem. 2007;50(14):3283–3289.
79 Ol'shevskaya VA, Zaitsev AV, Luzgina VN, Kondratieva TT, Ivanov OG, Kononova EG, Petrovskii PV, Mironov AF, Kalinin VN, Hofmann J, Shtil AA. Novel boronated derivatives of 5,10,15,20-tetraphenylporphyrin: synthesis and toxicity for drug-resistant tumor cells. Bioorg. Med. Chem. 2006;14:109–120.
80 Tjarks W, Tiwari R, Byun Y, Narayanasamy S, Barth RF. Carboranyl thymidine analogues for neutron capture therapy. Chem. Commun. 2007;4978–4991.
81 Wojtczak BA, Olejniczak AB, Wang L, Eriksson S, Leśnikowski ZJ. Phosphorylation of nucleoside metallacarborane and carborane conjugates by nucleoside kinases. Nucleosides Nucleotides Nucleic Acids 2013;32:571–588.
82 Yinghuai Z, Peng AT, Carpenter K, Maguire JA, Hosmane NS, Takagaki M. Substituted carborane-appended water-soluble single-wall carbon nanotubes: new approach to boron neutron capture therapy drug delivery. J. Am. Chem. Soc. 2005;127:9875–9880.
83 Kelloff GJ, Krohn KA, Larson SM, Weissleder R, Mankoff DA, Hoffman JM, Link JM, Guyton KZ, Eckelman WC, Scher HI, O'Shaughnessy J, Cheson BD, Sigman CC, Tatum JL, Mills GQ, Sullivan DC, Woodcock J. The progress and promise of molecular imaging probes in oncologic drug development. J. Clin. Cancer Res. 2005;11(22):7967–7985.
84 Chen K, Chen X. Design and development of molecular imaging probes. Curr. Top. Med. Chem. 2010;10(12):1227–1236.
85 G. Jaouen, M. Salmain. Bioorganometallic Chemistry: Applications in Drug Discovery, Biocatalysis, and Imaging. Weinheim: Wiley-VCH, 2015.
86 Schlotter K, Boeckler F, Hübner H, Gmeiner P. Fancy Bioisosteres: Metallocene-derived G-protein-coupled receptor ligands with subnanomolar binding affinity and novel selectivity profiles. J. Med. Chem. 2005;48(11):3696–3699.
87 Bernard J, Ortner K, Spingler B, Pietzsch H-J, Alberto R. Aqueous synthesis of derivatized cyclopentadienyl complexes of technetium and rhenium directed toward radiopharmaceutical application. Inorg. Chem. 2003;42(4):1014–1022.
88 Allardyce CS, Dorcier A, Scolaro C, Dyson PJ. Development of organometallic (organo-transition metal) pharmaceuticals. Appl. Organomet. Chem. 2005;19(1):1–10.
89 Jurisson SS, Lydon JD. Potential technetium small molecule radiopharmaceuticals. Chem. Rev. 1999;99(9):2205–2218.
90 Grimes RN. Carboranes in molecular imaging and radiotherapy. In: *Carboranes*, 2nd ed. Heidelberg: Elsevier, 2011, 1071–1074.
91 Causey PW, Besanger TR, Valliant JF. Synthesis and screening of mono- and di-aryl technetium and rhenium metallocarboranes: a new class of probes for the estrogen receptor. J. Med. Chem. 2008;51(9):2833–2844.

92 Meanwell NA. Synopsis of some recent tactical application of bioisosteres in drug design. J. Med. Chem. 2011;54:2529–2591.

93 Heimbold I, Drews A, Syhre R, Kretzschmar M, Pietzsch H-J, Johannsen B. A novel technetium-99 m radioligand for the 5-HT$_{1A}$ receptor derived from desmethyl-WAY-100635 (DWAY). Eur. J. Nucl. Med. 2002;29(1):82–87.

94 Michelotti GA, Price DT, Schwinn DA. α_1-Adrenergic receptor regulation: basic science and clinical implications. Pharmacol. Ther. 2000;88:281–309.

95 Valliant JF, Morel P, Schaffer P, Kaldis JH. Carboranes as ligands for the preparation of organometallic Tc and Re radiopharmaceuticals. synthesis of [M(CO)$_3$(η^5-2,3-C$_2$B$_9$H$_{10}$)]$^-$ and rac-[M(CO)$_3$(η^5-2-R-2,3-C$_2$B$_9$H$_{10}$)]$^-$ (M) Re, ^{99}Tc; R = CH$_2$CH$_2$CO$_2$H) from [M(CO)$_3$Br$_3$]$^{2-}$. Inorg. Chem. 2002;41(4):628–630.

96 Pike VW. PET Radiotracers: crossing the blood-brain barrier and surviving metabolism. Trends Pharmacol. Sci. 2009;30(8):431–440.

97 Forster EA, Cliffe IA, Bill DJ, Dover GM, Jones D, Reilly Y, Fletcher A. A pharmacological profile of the selective silent 5-HT1A receptor antagonist, WAY-100635. Eur. J. Pharmacol. 1995;281:81–88.

98 Trumpp-Kallmeyer S, Hoflack J, Bruinvels A, Hibert M. Modeling of G-protein-coupled receptors: application to dopamine, adrenaline, serotonin, acetylcholine, and mammalian opsin receptors. J. Med. Chem. 1992;35(19):3448–3462.

99 Timofeev SV, Sivaev IB, Prikaznova EA, Bregadze VI. Transition metal complexes with charge-compensated dicarbollide ligands. J. Organomet. Chem. 2014;751:221–250.

100 Franken A, Plešek J, Fusek J, Semrau M. Cobaltacarboranes with intramolecular monophosphorus bridges 8,8′-μ-Me$_2$P(1,2-C$_2$B$_9$H$_{10}$)$_2$-3-Co, 6,6′-μ-Me$_2$P(1,7-C$_2$B$_9$H$_{10}$)$_2$-2-Co and the respective non-bridged trimethylphosphine derivatives (8-Me$_3$P-1,2-C$_2$B$_9$H$_{10}$)-3-Co-(1,2-C$_2$B$_9$H$_{11}$) and (6-Me$_3$P-1,7-C$_2$B$_9$H$_{10}$)-2-Co-(1,7-C$_2$B$_9$H$_{11}$). Collect. Czech. Chem. Commun. 1997;62:1070–1079.

101 Grimes RN. Antitumor agents. In: Carboranes, 2nd ed. Heidelberg: Elsevier, 2011, 1058–1062.

102 Hall IH, Tolmie CE, Barnes BJ, Curtis MA, Russell JM, Finn MG, Grimes RN. Cytotoxicity of tantalum(V) and niobium(V) small carborane complexes and mode of action in P388 lymphocytic leukemia cells. Appl. Organomet. Chem. 2000;14:108–118.

103 (a) Hall IH, Warren AE, Lee CC, Wasczcak MD, Sneddon LG. Cytotoxicity of ferratricarbadecaboranyl complexes in murine and human tissue cultured cell lines. Anticancer Res. 1998;18(2A):951–962; (b) Hall IH, Durham RW Jr., Tram M, Mueller S, Ramachandran BM, Sneddon LG. Cytotoxicity and mode of action of vanada- and niobatricarbadecaboranyl monohalide complexes in human HL-60 promyelocytic leukemia cells. J. Inorg. Biochem. 2003;93:125–131.

104 Cígler P, Kožíšek M, Řezáčová P, Brynda J, Otwinowski Z, Pokorná J, Plešek J, Grüner B, Dolečková-Marešová L, Máša M, Sedláček J, Bodem J, Kräusslich H-G, Král V, Konvalinka J. From nonpeptide toward noncarbon proteaseinhibitors: metallacarboranes as specific and potent inhibitors of HIV protease. PNAS 2005;102:15394–15399.

105 Alderton WK, Cooper CE, Knowles RG. Nitric oxide synthases: structure, function and inhibition. Biochem. J. 2001;357:593–615.

106 Förstermann U, Sessa WC. Nitric oxide synthases: regulation and function. Eur. Heart J. 2012;33:829–837.

107 Erdal EP, Litzinger EA, Seo J, Zhu Y, Ji H, Silverman RB. Selective neuronal nitric oxide synthase inhibitors. Curr. Top. Med. Chem. 2005;5(7):603–624.

108 Salerno L, Sorrenti V, Di Giacomo C, Romeo G, Siracusa MA. Progress in the development of selective nitric oxide synthase (NOS) inhibitors. Curr. Pharm. Des. 2002;8(3):177–200.

109 (a) Sigma Aldrich. Available from: https://www.sigmaaldrich.com/life-science/biochemicals/biochemical-products.html?TablePage=14573075 (17/02/2018).

110 Barbanti P, Egeo G, Aurilia C, Fofi L, Della-Morte D. Drugs targeting nitric oxide synthase for migraine treatment. Expert Opin. Investig. Drugs 2014;23(8):1141–1148.

111 Wegener G, Volke V. Nitric oxide synthase inhibitors as antidepressants. Pharmaceuticals 2010;3:273–299.

112 Flinspach M, Li H, Jamal J, Yang W, Huang H, Hah J-M, Gomez-Vidal JA, Litzinger EA, Silverman RB, Poulos TL. Structural basis for dipeptide amide isoform-selective inhibition of neuronal nitric oxide synthase. Nature Struct. Mol. Biol. 2004;11(1):54–59.

113 Gozzi M, Schwarze B, Sárosi M-B, Lönnecke P, Drača D, Maksimović-Ivanić D, Mijatović S, Hey-Hawkins E. Antiproliferative activity of (η^6-arene)-ruthenacarborane sandwich complexes against HCT116 and MCF-7 cell lines. Dalton Trans. 2017;46: 12067–12080.

114 (a) Süss-Fink G. Arene ruthenium complexes as anticancer agents. Dalton Trans. 2010;39:1673–1688; (b) Van Rijt SH, Sadler PJ. Current applications and future potential for bioinorganic chemistry in the development of anticancer drugs. Drug. Discov. Today 2009;14:1089–1097.

115 Micallef LS, Loughrey BT, Parsons PC, Williams ML. Synthesis, spectroscopic characterization, and cytotoxic evaluation of pentasubstituted ruthenocenyl esters. Organometallics 2010;29(23):6237–6244.

116 Białek-Pietras M, Olejniczak AB, Tachikawa S, Nakamura H, Leśnikowski ZJ. Towards new boron carriers for boron neutron capture therapy: metallacarboranes bearing cobalt, iron and chromium and their cholesterol conjugates. Bioorg. Med. Chem. 2013;21:1136–1142.

117 Thirumamagal BTS, Zhao XB, Bandyopadhyaya AK, Narayanasamy S, Johnsamuel J, Tiwari R, Golightly DW, Patel V, Jehning BT, Backer MV, Barth RF, Lee RJ, Backer JM, Tjarks W. Receptor-targeted liposomal delivery of boron-containing cholesterol mimics for boron neutron capture therapy (BNCT). Bioconjugate Chem. 2006;17(5):1141–1150.

118 Koganei H, Ueno M, Tachikawa S, Tasaki L, Seung Ban H, Suzuki M, Shiraishi K, Kawano K, Yokoyama M, Maitani Y, Ono K, Nakamura H. Development of high boron content liposomes and their promising antitumor effect for neutron capture therapy of cancers. Bioconjugate Chem. 2013;24:124–132.

119 Silver DA, Pellicer I, Fair WR, Heston WDW, Cordon-Cardo C. Prostate-specific membrane antigen expression in normal and malignant human tissues. Clin. Cancer Res. 1997;3:81–85.

120 (a) Sletten EM, Bertozzi CR. From mechanism to mouse: a tale of two bioorthogonal reactions. Acc. Chem. Res. 2011;44(9):666–676; (b) Sletten EM, Bertozzi CR.

Bioorthogonal chemistry: fishing for selectivity in a sea of functionality. Angew. Chem. Int. Ed. 2009;48:6974–6998.

121 Yang M, Li J, Chen PR. Transition metal-mediated bioorthogonal protein chemistry in living cells. Chem. Soc. Rev. 2014;43:6511–6526.

122 Calabrese G, Gomes A, Barbu E, Nevell T, Tsibouklis J. Carborane-based derivatives of delocalised lipophilic cations for boron neutron capture therapy: synthesis and preliminary in vitro evaluation. J. Mater. Chem. 2008;18(40):4864–4871.

123 Liberman E, Topaly V, Tsofina L, Jasaitis A, Skulachev V. Mechanism of coupling of oxidative phosphorylation and the membrane potential of mitochondria. Nature 1969;222(5198):1076–1078.

124 (a) Modica-Napolitano JS, Singh K. Mitochondrial dysfunction in cancer. Mitochondrion 2004;4(5–6):755–762; (b) Chang LO, Schnaitman CA, Morris HP. Comparison of the mitochondrial membrane proteins in rat liver and hepatomas. Cancer Res. 1971;31(2):108–113.

125 Modica-Napolitano JS, Aprille JR. Basis for the selective cytotoxicity of rhodamine 123. Cancer Res. 1987;47(16):4361–4365.

126 Modica-Napolitano JS, Aprille JR. Delocalized lipophilic cations selectively target the mitochondria of carcinoma cells. Adv. Drug Deliv. Rev. 2001;49(1–2):63–70.

127 Crossley ML, Dreisbach PF, Hofmann CM, Parker RP. Chemotherapeutic dyes. III. 5-Heterocyclicamino-9-dialkylaminobenzo [a]phenoxazines 1. J. Am. Chem. Soc. 1952;74(3):573–578.

128 Weiss MJ, Wong JR, Ha S, Bleday R, Salem RR, Steele GD Jr., Chen LB. Dequalinium, a topical antimicrobial agent, displays anticarcinoma activity based on selective mitochondrial accumulation. Proc. Natl. Acad. Sci. USA 1987;84:5444–5448.

129 Adams DM, Ji W, Barth RF, Tjarks W. Comparative in vitro evaluation of dequalinium B, a new boron carrier for neutron capture therapy (NCT). Anticancer Res. 2000;20(5B):3395–3402.

130 Theodoropoulos D, Rova A, Smith JR, Barbu E, Calabrese G, Vizirianakis IS, Tsibouklis J, Fatouros DG. Towards boron neutron capture therapy: the formulation and preliminary in vitro evaluation of liposomal vehicles for the therapeutic delivery of the dequalinium salt of bis-*nido*-carborane. Bioorg. Med. Chem. Lett. 2013;23:6161–6166.

131 Tseligka ED, Rova A, Amanatiadou EP, Calabrese G, Tsibouklis J, Fatouros DG, Vizirianakis IS. Pharmacological development of target-specific delocalized lipophilic cation functionalized carboranes for cancer therapy. Pharm. Res. 2016;33:1945–1958.

132 Hawthorne MF. New horizons for therapy based on the boron neutron capture reaction. Mol. Med. Today 1998;4:174–181.

133 Ott I, Kircher B, Bagowski CP, Vlecken DH, Ott EB, Will J, Bensdorf K, Sheldrick WS, Gust R. Modulation of the biological properties of aspirin by formation of a bioorganometallic derivative. Angew. Chem. Int. Ed. Engl. 2009;48:1160–1163.

134 Pochet S, Douge L, Labesse G, Delepierre M, Munier-Lehmann H. Comparative study of purine and pyrimidine nucleoside analogues acting on the thymidylate kinases of mycobacterium tuberculosis and of humans. ChemBioChem 2003;4:742–747.

1.5

Ionic Boron Clusters as Superchaotropic Anions: Implications for Drug Design

Khaleel I. Assaf, Joanna Wilińska, and Detlef Gabel

Department of Life Sciences and Chemistry, Jacobs University Bremen, Bremen, Germany

1.5.1 Introduction

Boron clusters have become key units in various research fields, ranging from material science to medicine [1,2]. Cage-like boron clusters (Figure 1.5.1) can be classified as *closo-*, *nido-*, *arachno-*, *hypho-*, and so on based on the completeness of the polyhedron, with the *closo-* cluster being closed. With one or two missing vertices, boron clusters are named *nido-* or *arachno-*, respectively. They have characteristic three-dimensional polyhedral geometries, with delocalized electron-deficient structures [3–6]. Boron clusters, in their neutral and ionic forms, have shown a unique stability and low toxicity [3,4,6,7]. The ability of the ^{10}B isotope to emit α particles after absorbing neutrons makes them ideal for pharmaceutical and medical applications, in particular for boron neutron capture therapy (BNCT) [8–10].

Boron clusters interact with biomolecules, including biomembranes and proteins. This offers more possibilities in medicinal use than just BNCT. In this chapter, we summarize the current research on the noncovalent interactions of common boron clusters, in particular ionic ones, with supramolecular macrocycles, lipid bilayers, and proteins. We draw conclusions for drug design, and point out areas of future research.

Polyhedral boron clusters can be divided into neutral and anionic ones. Neutral clusters include *o-*, *p-*, and *m-*carboranes ($C_2B_{10}H_{12}$). These isomers are highly hydrophobic. *Closo-*dodecaborates ($B_{12}X_{12}^{2-}$) are water-soluble dianionic clusters with icosahedral structure, and they are nontoxic anions [11]. Shortly after their discovery, they were introduced as potential BNCT agents [12,13]. The decahydro-*closo-*decaborate anion ($B_{10}X_{10}^{2-}$) is another member of the $B_nH_n^{2-}$ clusters; it is water-soluble as sodium salt. Metalla bisdicarbollides, a different class of anionic boron clusters, are sandwiches of two $[C_2B_9H_{11}]^{2-}$ (biscarbollide) clusters with a metal ion in the center (e.g., cobalta bisdicarbollide anions); these clusters have recently emerged in medicinal chemistry as HIV protease inhibitors [14–16].

Boron-Based Compounds: Potential and Emerging Applications in Medicine, First Edition.
Edited by Evamarie Hey-Hawkins and Clara Viñas Teixidor.
© 2018 John Wiley & Sons Ltd. Published 2018 by John Wiley & Sons Ltd.

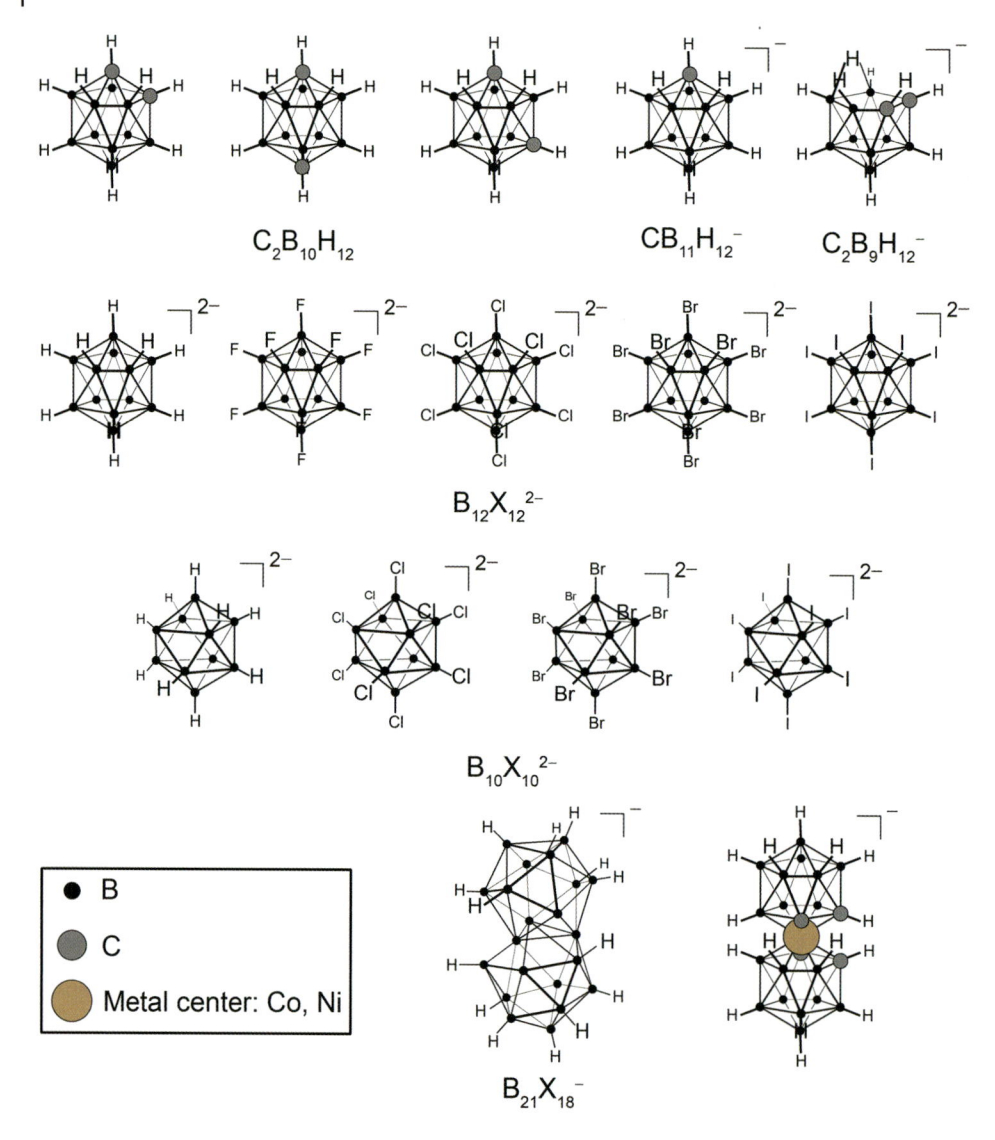

Figure 1.5.1 Chemical structures of selected neutral and ionic boron clusters: *closo* $C_2B_{12}H_{12}$, *closo* $[CB_{11}H_{12}]^-$, *nido* $[C_2B_9H_{12}]^-$, *closo* $[B_{12}H_{12}]^{2-}$, *closo* $[B_{10}H_{10}]^{2-}$, and *closo* $[B_{21}H_{18}]^-$, and metalla-bisdicarbollides $[3,3'-M(1,2-C_2B_9H_{11})_2]^-$.

1.5.2 Water Structure and Coordinating Properties

Boron cluster anions have recently been recognized as a new class of weakly coordinating anions [7,17–23]. The classification of anions as weakly coordinating means that they have low charge densities and a high degree of electron delocalization; therefore, large boron-based anions are prototypes of weakly coordinating anions (Figure 1.5.2). Among weakly coordinating anions (oxyanions, polyoxometalates, and iodide anions), boron-based anions display a range of novel applications, including their use for stabilization of very reactive and labile cations [7,18].

Most boron clusters are water-soluble, except for the uncharged carboranes. Their behavior in water has remained obscure to a large degree. The three-dimensional aromaticity of the boron clusters is explained by the delocalized three-center, two-electron bonding. The B–H bonds in these types of clusters have a different charge distribution compared to regular aromatic compounds such as benzene, as the hydrogen atoms hold a partial negative charge and thus have a hydridic character [25]. In water, this allows the cluster to make unconventional hydrogen bonding with water molecules, namely a dihydrogen bond (Figure 1.5.3) [26].

Recently, dodecaborate anions have been classified as superchaotropic anions based on their ability to increase the solubility of simple organic molecules in water (*salt-in effect*). Chaotropic agents are thought to break the water structure and decrease its order. The chaotropic nature of the dodecaborate anions was confirmed by modeling their ionic properties as devised by Marcus (Table 1.5.1) [24]. The water-structural entropies for ionic solvation (ΔS_{struct}) for the chaotropic anions are positive, indicating a decrease in the order of the water structure, which reflects an effective loss of hydrogen bonds around the anion (negative values of ΔHB). Conventional chaotropic anions, such as SCN^-, ClO_4^-, and PF_6^-, break one hydrogen bond. The dodecaborate anions are effectively able to break two or more hydrogen bonds in their surroundings.

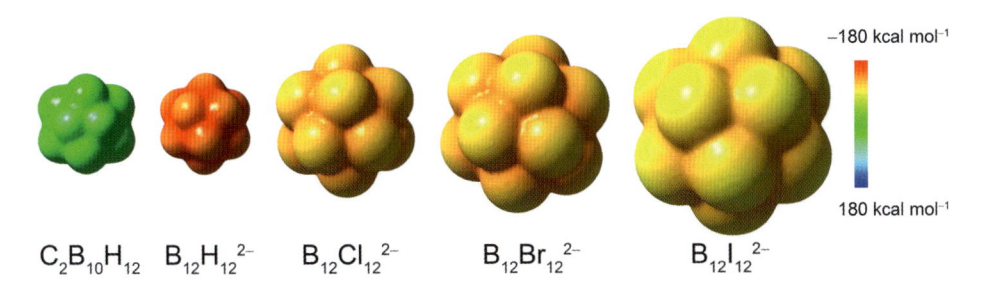

Figure 1.5.2 Electrostatic potential map of carborane and dodecaborate anions [24]. $C_2B_{10}H_{12}$ $B_{12}H_{12}^{2-}$ $B_{12}Cl_{12}^{2-}$ $B_{12}Br_{12}^{2-}$ $B_{12}I_{12}^{2-}$

Figure 1.5.3 Water structure around dodecaborate anion $B_{12}H_{12}^{2-}$. "A" indicates dihydrogen bonds; "B" indicates a bifurcated dihydrogen bond [26]. *Source*: Reprinted with permission from Ref. [26]. Copyright © 2012 American Chemical Society.

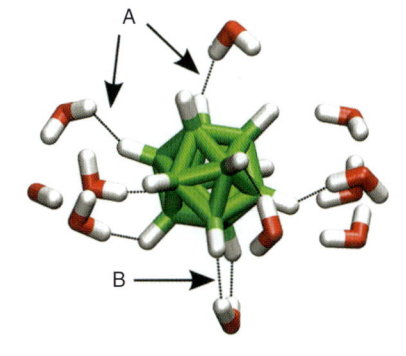

Table 1.5.1 Water-structural entropies of different anions (ΔS_{struct}) and net effects on the number of surrounding hydrogen bonds (ΔHB) [24]

Anion	$T\Delta S_{struct}°/$ (kcal/mol)	ΔHB	Anion	$T\Delta S_{struct}°/$ (kcal/mol)	ΔHB
F^-	−2.1	0.11	SO_4^{2-}	−6.7	0.78
Cl^-	4.1	−0.76	CO_2^{2-}	−4.0	0.40
Br^-	5.8	−1.01	HPO_4^{2-}	−4.4	0.46
I^-	8.3	−1.37	$B_{12}H_{12}^{2-}$	14.8	−2.31
SCN^-	5.9	−1.03	$B_{12}Cl_{12}^{2-}$	17.1	−2.63
BF_4^-	6.6	−1.12	$B_{12}Br_{12}^{2-}$	18.3	−2.81
ClO_4^-	7.6	−1.27	$B_{12}I_{12}^{2-}$	19.3	−2.95
PF_6^-	9.3	−1.51	PO_3^{3-}	−9.6	1.20

The interaction of different dodecaborate clusters, of the type $B_{12}X_{12}^{2-}$ and $B_{12}X_{11}Y^{2-}$ (X = H, Cl, Br, I; and Y = SH, OH, NR_3^+), with chromatographic column matrices, including unmodified and modified column materials, has been investigated [27,28]. Most of the dodecaborate clusters are retained strongly on size exclusion and anion exchange columns, indicating strong interactions, whereas they interact very weakly with silica gel materials [28]. Moreover, modified silica matrices containing phosphatidylcholine groups have been used as a model to study the interaction of boron clusters with membranes, mimicking the surface of lipid membranes [27]. In comparison with organic molecules, dodecaborate clusters eluted slower on these columns. Brominated and iodinated dodecaborate could not be eluted from immobilized phosphatidylcholine liposomes [27].

1.5.3 Host–Guest Chemistry of Boron Clusters

The encapsulation of carboranes, as highly hydrophobic residues, by macrocyclic host molecules with hydrophobic cavities, such as cyclodextrins [29–35], calixarenes [36,37], cucurbiturils [38], and cyclotriveratrylene [39], has been explored to a large extent. The shallow cavities of cyclotriveratrylene and calixarenes offer room for o-carborane to be complexed. These host–guest complexes are driven by the hydrophobic effect and stabilized through nonclassical C–H···π hydrogen bonding [36,37,39,40], in which the C–H vectors of the carborane point toward the aromatic rings, as shown in their X-ray diffraction (XRD) structures (Figure 1.5.4) [36,39].

Cyclodextrins (CDs), which are cyclic oligosaccharide-based molecules with a cone shape (Figure 1.5.5), have shown the capability to bind carboranes and their derivatives with relatively high affinities [29–33,41]. Harada and Takahashi were the first to study the complexation of o-carborane with CDs; they reported that carborane forms 1:1 inclusion complexes with β-CD and γ-CD, and a 2:1 complex with the smallest homolog, α-CD [35]. C-Hydroxycarboranes (mono- and di-substituted) have also been investigated with β-CD in aqueous solution [33], in which high affinities ($> 10^6 M^{-1}$)

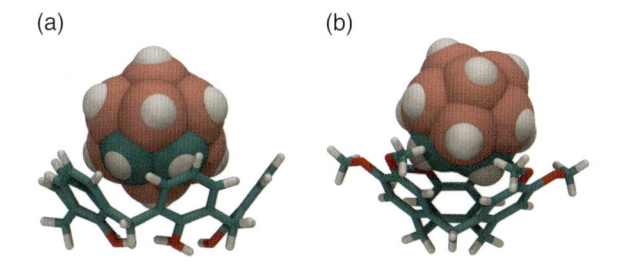

Figure 1.5.4 XRD structures of the *o*-carborane complex with (a) calix[5]arenes [36] and (b) cyclotriveratrylene [39]. The C–H vectors of the carborane point toward the aromatic rings of the host cavity.

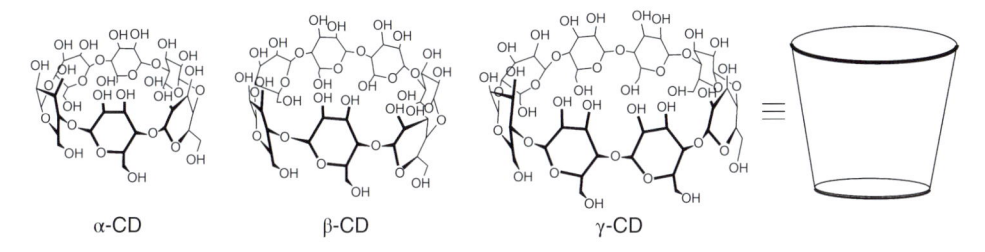

α-CD   β-CD   γ-CD

Figure 1.5.5 Chemical structure of common cyclodextrins (CDs).

were reported, which have been attributed to the hydrophobic effect and hydrogen bonding. 1-Phenyl-*o*-carborane was also found to form a 2:1 complex with β-CD as indicated by Nuclear Overhauser Enhancement Spectroscopy (NOESY) nuclear magnetic resonance (NMR) experiments. Recently, Sadrerafi et al. have used a dye displacement technique to study the complexation of different substituted carboranes (*o*-, *m*-, *p*-carboranes) with β-CD, and compared their binding affinities with the well-known CD binder adamantane and its derivatives [29]. The measured binding constants for the unsubstituted carboranes were higher than that of adamantane, and followed the order of: *o*- > *m*- > *p*-carborane [29]. This trend was found to parallel the dipole moment of these carboranes [42].

Most recently, Assaf et al. have shown that dodecaborate clusters of the type $B_{12}X_{12}^{2-}$ and $B_{12}X_{11}Y^{2-}$ (X = H, Cl, Br, I; and Y = OH, SH, NR^{3+}) act as strong binders to CDs in aqueous solution, an unexpected and striking observation considering their di-anionic nature and their high water solubility [43,44]. The binding constants with γ-CD (see Table 1.5.2) even exceed those of highly hydrophobic guests, such as carborane and adamantane, which hints to a different driving force for complexation [29]. The high affinities could be traced back to the chaotropic effect, as well as the high polarizability of the dodecaborate anions [29]. In fact, dodecaborate anions are water-soluble, preferring water over the *n*-octanol phase [26]; hence, they cannot be classified as hydrophobic anions. The chaotropic nature of the dodecaborate anions was supported by salting-in experiments, revealing them as "superchaotropes," and in salting-in experiments they act even more strongly than the known conventional chaotropic anions (SCN^-, PF_6^-, or ClO_4^-). The complexation showed a different thermodynamic signature, as measured

Table 1.5.2 Association constants (K_a) of dodecaborate cluster anions with γ-CDs and associated thermodynamic parameters (in kcal/mol)

Borate cluster[a]	$K_a/(10^3 \, M^{-1})$	$\Delta H°$	$T\Delta S°$	$\Delta G°$
$B_{12}H_{11}OH^{2-}$	0.62^b			
$B_{12}H_{11}N(nPr)_3^-$ [d]	1.1^b			
$B_{12}H_{11}NH_3^-$	1.7^b			
$B_{12}H_{12}^{2-}$	2.0^b			
$B_{12}H_{11}SH^{2-}$	$7.8,^b \, 9.2^c$	−5.7	−0.30	−5.4
$B_{12}Cl_{12}^{2-}$	17^c	−14.4	−8.6	−5.8
$B_{12}Br_{12}^{2-}$	960^c	−21.4	−13.3	−8.1
$B_{12}I_{12}^{2-}$	67^c	−25.0	−18.4	−6.6
$B_{12}I_{11}NH_3^-$	25^c			

[a] Measured as sodium salts at 25 °C for a 1:1 complexation model.
[b] ^{1}H NMR titration in D_2O.
[c] Measured by isothermal titration calorimetry (ITC) in neat water.
[d] Potassium salt. From Ref. [24].

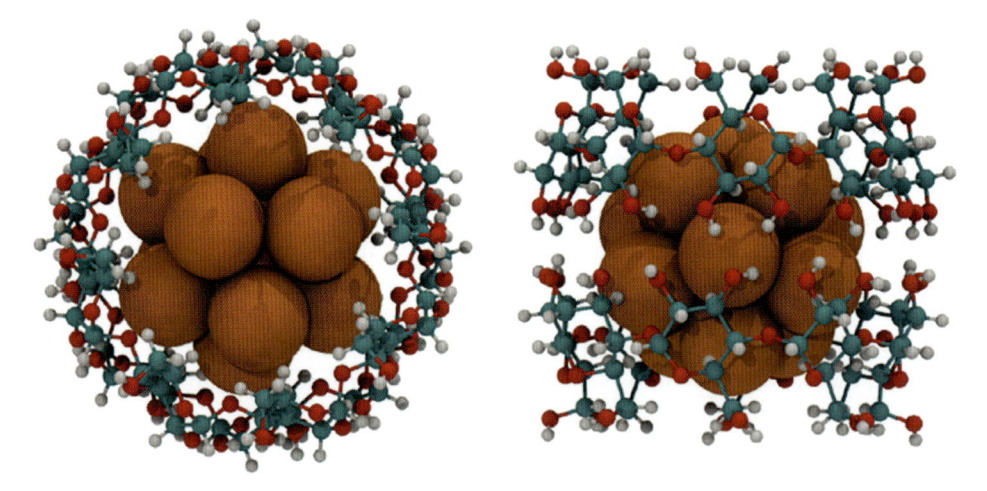

Figure 1.5.6 XRD structures of the inclusion entrapment of the $B_{12}Br_{12}^{2-}$ cluster into the γ-CD dimer [24].

by isothermal titration calorimetry, compared to the complexation of hydrophobic guests: high negative enthalpies and entropies. Chaotropic anions decrease the water structure, and the relocation of the anions into the CD cavity leads to more pronounced water structure recovery effects (chaotropic effect). In the solid state, $B_{12}Br_{12}^{2-}$ was found to be entrapped by two γ-CD rings (Figure 1.5.6). Halogenated dodecaborate clusters have also been found to form stable complexes with large-ring cyclodextrins (LRCDs: δ-, ε-, and ζ-CD), with affinities exceeding that for γ-CD (44). Warneke et al. have followed on the binding of dodecaborate anions to CDs and other macrocycles with hydrophobic cavities, and provided evidence for the stability of these anionic complexes in the gas phase [45].

Based on the high affinity of the dodecaborate clusters to CDs, a new class of water-soluble dye molecules, in which the cluster ($B_{12}H_{12}$) is tethered to a chromophore (7-nitrobenzofurazan [NBD]), has been synthesized (Figure 1.5.7a) [46]. The binding of the dodecaborate–dye molecules to CDs was studied by ultraviolet (UV)-visible, fluorescence, and NMR titrations. Both dyes form stable complexes with CDs ($>10^5 M^{-1}$), in which the dodecaborate is encapsulated inside the cavity as indicated by NMR and semiempirical calculations. The noticeable changes in the course of UV-visible and fluorescence spectral titrations upon complexation of the dyes with CDs allow them to be used as host–dye reporter pairs for indicator displacement applications (Figure 1.5.7b). With this, the CD–dye pair could be used to determine the affinity of organic and inorganic guest molecules to CDs in a fast and easy manner [46].

The host–guest chemistry of boron clusters with CDs has recently emerged as a tool for efficient noncovalent immobilization of biomolecules on solid surfaces, in analogy to other host–guest systems, such as cucurbit[7]uril/ferrocene [47,48]. For example, o-carborane was linked to a peptide to be immobilized, while β-CD was anchored to a glass or gold surface [30]. The carborane–peptide bioconjugate was immobilized on the β-CD functionalized surface through the complexation of the carborane inside the CD cavity. The binding of the carborane to β-CD was supported by infrared reflection absorption spectroscopy (IR-RAS) and quartz crystal microbalance with dissipation monitoring (QCM-D) experiments [30]. Rendina and coworkers synthesized Pt(II) complexes containing carborane residue and studied their complexation with CDs (Figure 1.5.8) [49–52]. Furthermore, a ternary supramolecular assembly consists of a hexanucleotide (d(GTCGAC)$_2$), a boronated 2,2′:6′,2″-terpyridineplatinum(II) complex containing p-carborane, and β-CD was designed (Figure 1.5.8a) in which the carborane moiety is encapsulated by β-CD [51]. ^{1}H NMR suggested that the Pt(II)–terpyridine moiety intercalates between the nucleotide bases, while the β-CD carborane complexes stay intact.

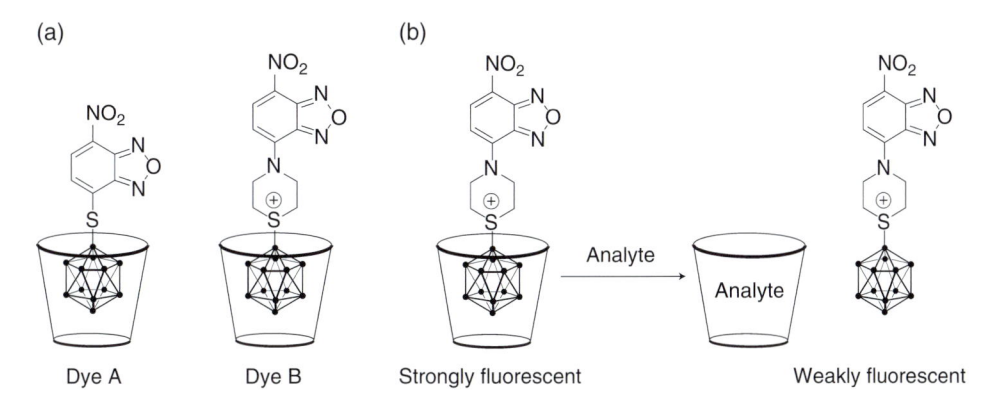

Figure 1.5.7 (a) Dodecaborate-anchor 7-nitrobenzofurazan (NBD) dyes A and B for CD binding (46). (b) Illustration represents the principle of the indicator displacement assay: the encapsulation of the dye inside the host cavity changes its optical properties (e.g., fluorescence); the addition of an analyte is expected to revert the optical changes.

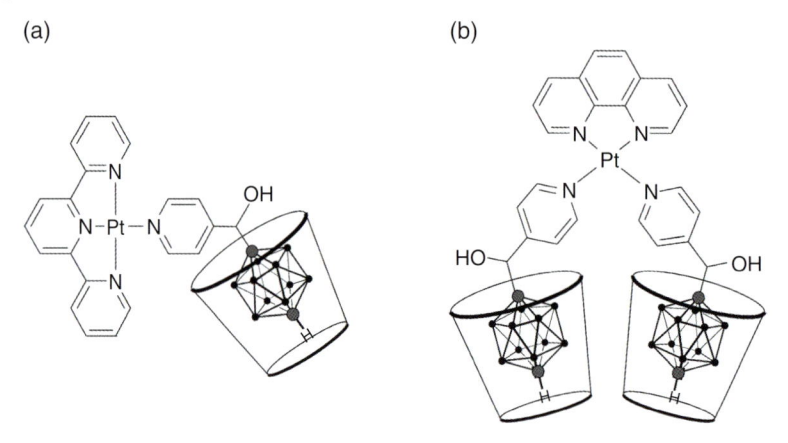

Figure 1.5.8 Pt(II) complexes containing carborane residue. *Source*: Redrawn after Ref. [52].

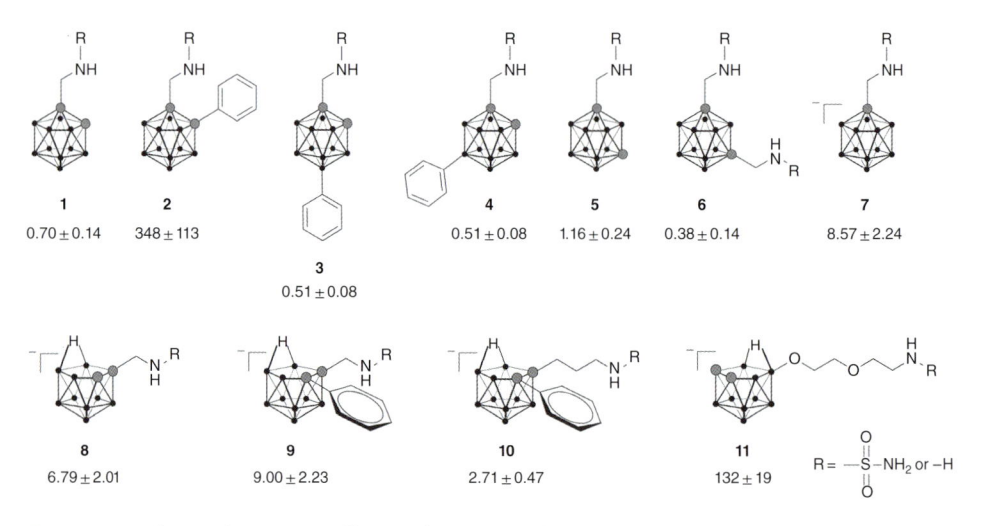

1	**2**			**4**	**5**	**6**	**7**
0.70 ± 0.14	348 ± 113			0.51 ± 0.08	1.16 ± 0.24	0.38 ± 0.14	8.57 ± 2.24

3
0.51 ± 0.08

8	**9**	**10**	**11**	
6.79 ± 2.01	9.00 ± 2.23	2.71 ± 0.47	132 ± 19	R = —S(=O)(=O)—NH₂ or —H

Figure 1.5.9 Chemical structures of boron clusters tested as inhibitors for carbonic anhydrase [54]. The *in vitro* inhibition constants (K_i in μM) of the sulfonamide-based derivatives with the human CA isozyme II (CAII) are given below the structures.

1.5.4 Ionic Boron Clusters in Protein Interactions

Boron clusters have been used as three-dimensional scaffolds, replacing bulky aromatic systems. For example, carboranes have been utilized as stable hydrophobic pharmacophores and were found to have potential use as precursors for designing new enzyme inhibitors [53–57]. Boron clusters functionalized with sulfamide (Figure 1.5.9), synthesized to mimic isoquinoline sulfonamide inhibitors, showed inhibitory activity toward carbonic anhydrases (CAs), with inhibition constants (K_i) in the micromolar range [54]. The binding of the cluster to the enzyme's active site was evidenced by crystal structure analysis (Figure 1.5.10).

Quantum mechanics and molecular mechanics (QM/MM) calculations have been performed to understand the interactions involved in the complexation of **1** and **8** with CAs (hCAII). The complexation was found to be stabilized by dispersion interactions and dihydrogen bonds within the active site [58]. Interestingly, the *nido* cage binds stronger than the *closo* cage by 59 kcal/mol, where the binding of the anionic cage is stabilized via electrostatic interactions [58]. Nevertheless, as the clusters bind in a region of the protein that is dominated by hydrophobic residues, conventionally one would have assumed that the neutral cluster would show better binding.

Recently, Hey-Hawkins and coworkers have introduced the *nido*-carborane cage into the cyclooxygenase (COX) inhibitors [59]. The newly synthesized inhibitors (Figure 1.5.11) showed a higher inhibitory potency relative to the parent inhibitor, indomethacin.

Also for this enzyme, the pocket where the negatively charged cluster binds is lined mostly with hydrophobic side chains (see Figure 1.5.12). Even more pronounced is the hydrophobicity of the pocket in the HIV protease where the cobaltabisdicarbollide cluster binds (Figure 1.5.13) [53].

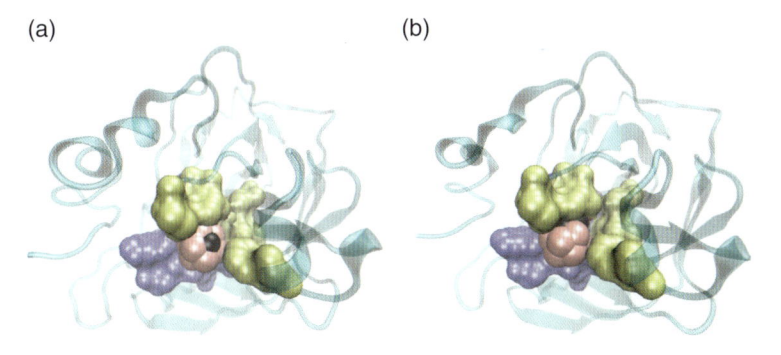

Figure 1.5.10 Crystal structure of CAII in complex with **5** (a) and **8** (b); PDB codes: 4MDL and 4MDM. Yellow: hydrophobic side chains; blue: hydrophilic side chains; pink: boron; black: carbon. *Source:* From Ref. [21].

Indomethacin   12   13

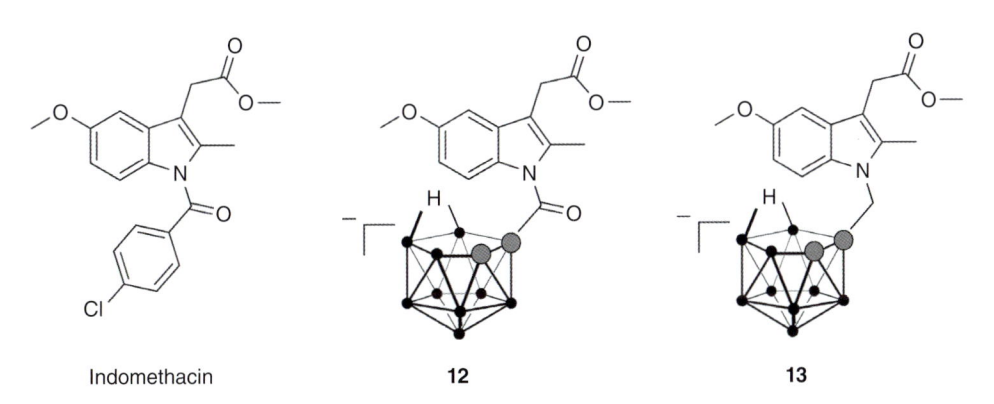

Figure 1.5.11 COX inhibitors: indomethacin and its *nido*-carborane derivatives [59].

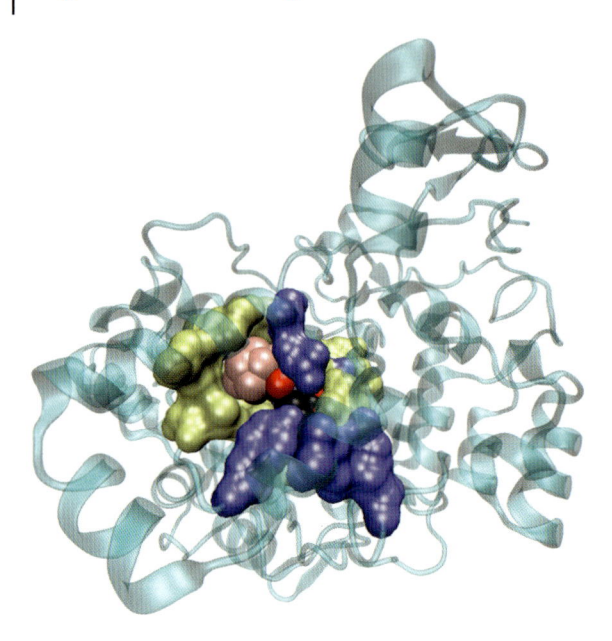

Figure 1.5.12 Binding of **12** to COX. PDB code: 4Z01. Yellow: hydrophobic side chains; blue: hydrophilic side chains; pink: boron; black: carbon. *Source*: From Ref. [59].

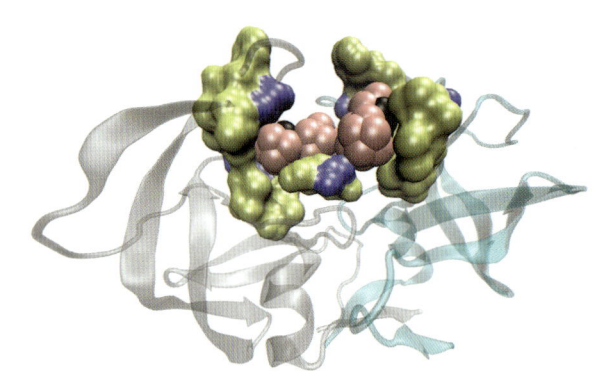

Figure 1.5.13 Cobaltabisdicarbollide binding to HIV protease. PDB code: 1ZTZ. Yellow: hydrophobic side chains; blue: hydrophilic side chains; pink: boron; black: carbon. *Source*: From Ref. [14].

1.5.4.1 Interactions of Boron Clusters with Lipid Bilayers

With their unique chemical and physical properties, dodecaborate anions have been found to interact with lipid bilayers. The interaction with different liposomes has been intensively investigated by Gabel and coworkers [11,60–62]. $Na_2B_{12}H_{11}SH$ (BSH), which is clinically used for neutron capture therapy (BNCT), interacts with liposomes causing a slow release for the liposome content [11,60,61]. Millimolar concentrations of the BSH were required to accelerate the release of carboxyfluorescein from dipalmitoylphosphatidylcholine (DPPC) and dimyristoylphosphatidylcholine (DMPC) liposomes [60,61]. Using cryo-transmission electron microscopy, BSH was found to induce structural changes [61]. Trialkylammonioundecahydrododecaborates with short and long alkyl chains interact with DPPC liposomes. For example, trialkylammonioundecahydrododecaborates with short alkyl chains result in the formation of large bilayer sheets, while clusters with longer alkyl chains tend to induce the formation of open or multilayered liposomes [62]. Recently, halogenated dodecaborate anions have also shown the capability to induce liposome leakage; this allows them to be used as agents for triggering the release of liposomal contents [11]. Their large sizes compared to the parent cluster, $Na_2B_{12}H_{12}$, and the low charge densities make them interact strongly with the liposome; the interaction increases with the size of the anion, parallel to their chaotropic character. The mechanism has not yet, however, been fully understood; it has been suggested that the halogenated clusters induce pores in the lipid membranes (Figure 1.5.14).

Metallacarboranes of the type cobaltabisdicarbollide, $[3,3'\text{-}Co(1,2\text{-}C_2B_9H_{11})_2]^-$, known as COSANs, are able to passively cross synthetic and living cells' membranes without causing breakdown of membrane barrier properties [63,64]. COSANs have amphiphilic properties; thus, they self-assemble and form monolayer vesicles [65,66]. These COSAN vesicles interact with liposomes, linking liposomes together. This interaction causes morphological changes of the COSAN assembly from vesicles to a planar lamellar structure, and then diffusion to the liposome interior [66].

Boron-based lipids have been designed and synthesized, based on *nido*-carboranes and *closo*-dodecaborate anions as head groups, allowing for their utilization for

(a)

(b)

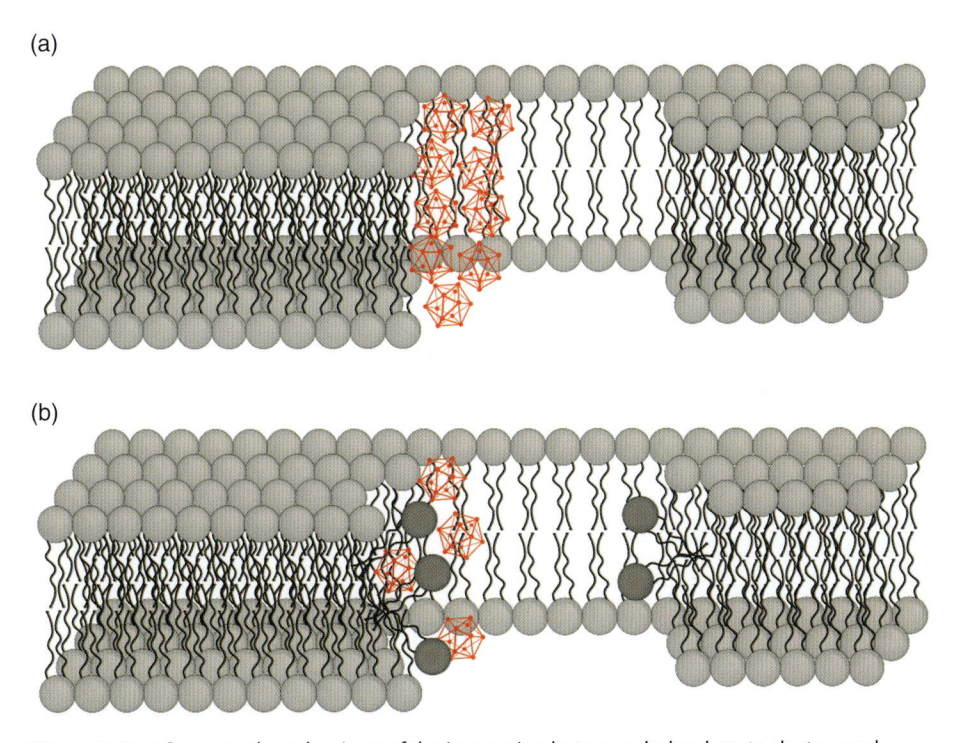

Figure 1.5.14 Suggested mechanisms of the interaction between dodecaborate clusters and liposome. (a) A barrel-stave pore; (b) a toroidal pore. *Source*: From Ref. [11].

liposomal boron delivery and more boron content to be used for BNCT [67–69]. At higher concentrations, several of these lipids are able to induce selectively hemorrhage in experimental murine tumors, leaving the other organs intact [70].

1.5.5 Implications for Drug Design

1.5.5.1 Binding to Proteins

Besides the examples of specific boron cluster compounds as protein ligands given in this chapter, unspecific binding has been observed [71]. These interactions should be mapped more carefully, preferably with photoreactive groups [72], in order to understand better the interaction modes that are represented by boron cluster compounds. Promiscuous binding is not desirable, yet it will be able to highlight structural motifs and molecular interactions capable of binding boron clusters. In fragment-based drug design, promiscuous binding is frequently observed.

1.5.5.2 Penetration through Membranes

The strong binding of ionic boron clusters to the hydrophobic interior of CDs, and the ability to influence the structure and integrity of lipid membranes, might be seen as an indication that ionic boron clusters are able to penetrate the hydrophobic environment of cell membranes. Such properties might be highly desirable, as it must be expected that several of the boron clusters will not only convey water solubility to the compounds in which they are present, but also allow these compounds to penetrate the different membranes that must be overcome for a drug to reach its target. The octanol–water distribution coefficient K_{OW} for several of these compounds has been measured, and it showed that solubility in a hydrophobic environment can be achieved with such compounds [62,73]. Recently, Genady *et al.* have shown that a fluorescent dye can penetrate cellular membranes and accumulate in mammalian cells [74].

1.5.5.3 Computational Methods

A few attempts have been described in the literature of how to handle boron clusters in computational methods routinely used in medicinal chemistry [75,76]. For docking programs, force fields for each functional group in organic structures have been developed and are routinely applied.

For boron, and especially for boron clusters, no such force fields exist. Thus, the literature data use carbon force fields [75,76]. With this, a boron cluster is treated almost as an adamantane. On the basis of the vastly different properties of boron clusters in comparison to, for example, adamantanes or carboxylic acids, computational results with carbon force fields imitating boron clusters must be considered quite unreliable. This is illustrated by the big difference in nicotinamide phosphoryltransferase inhibitors containing either an adamantane or a (neutral) carborane [77] or the lack of activity of adamantyl, *m*-, and *p*-carboranyl derivatives of indomethacin as compared to the *o*-derivative [78]. More appropriate force fields might have to be developed for neutral boron clusters, and are required for ionic boron clusters, before they can be incorporated into docking programs.

1.5.6 Conclusions

Boron cluster have properties not found in any organic compounds or pharmacophores derived from them. This is especially true for ionic boron clusters. They show good water solubility (depending on their counter-ions), but at the same time interact very weakly with water. Thus, they cannot be categorized into the conventional categories of "hydrophilic" and "hydrophobic." Rather, they represent a new class of compounds, thriving in both the water world and the non–water world. They can thus be seen as "hydroneutral."

It might be expected that considerable time passes before drug designers can appreciate these properties, as these properties run against the conventional wisdom of drug design. Once this is realized, however, we expect boron clusters to offer new, so far unimaginable, opportunities to design drugs.

References

1 Hosmane NS. Boron science: new technologies and applications. Boca Raton, FL: CRC Press; 2012. 1–825 p.
2 Dash BP, Satapathy R, Maguire JA, Hosmane NS. Polyhedral boron clusters in materials science. New J Chem. 2011;35(10):1955–1972.
3 Douvris C, Michl J. Update 1 of: chemistry of the carba-closo-dodecaborate(–) anion, $CB_{11}H_{12}^-$. Chem Rev. 2013;113(10):179–233.
4 Korbe S, Schreiber PJ, Michl J. Chemistry of the carba-closo-dodecaborate(–) anion, $CB_{11}H_{12}^-$. Chem Rev. 2006;106(12):5208–5249.
5 Núñez R, Romero I, Teixidor F, Viñas C. Icosahedral boron clusters: a perfect tool for the enhancement of polymer features. Chem Soc Rev. 2016. doi:10.1039/C6CS00159A
6 King RB. Three-dimensional aromaticity in polyhedral boranes and related molecules. Chem Rev. 2001;101(5):1119–1152.
7 Reed CA. Carboranes: a new class of weakly coordinating anions for strong electrophiles, oxidants, and superacids. Acc Chem Res. 1998;31:133–139.
8 Hawthorne MF. The role of chemistry in the development of boron neutron-capture therapy of cancer. Angew Chem Int Ed. 1993;32(7):950–984.
9 Barth RF, Soloway AH, Fairchild RG, Brugger RM. Boron neutron-capture therapy for cancer – realities and prospects. Cancer. 1992;70(12):2995–3007.
10 Barth RF, Soloway AH, Fairchild RG. Boron neutron-capture therapy of cancer. Cancer Res. 1990;50(4):1061–1070.
11 Awad D, Bartok M, Mostaghimi F, Schrader I, Sudumbrekar N, Schaffran T, et al. Halogenated dodecaborate clusters as agents to trigger release of liposomal contents. ChemPlusChem. 2015;80(4):656–664.
12 Sivaev IB, Bregadze VI, Kuznetsov NT. Derivatives of the closo-dodecaborate anion and their application in medicine. Russ Chem Bull. 2002;51(8):1362–1374.
13 Lechtenberg B, Gabel D. Synthesis of a $B_{12}H_{11}S^{2-}$ containing glucuronoside as potential prodrug for BNCT. J Organomet Chem. 2005;690(11):2780–2782.
14 Kozísek MCP, Lepsík M, Fanfrlík J, Rezácová P, Brynda J, Pokorná J, Plesek J, Grüner B, Grantz Sasková K, Václavíková J, Král V, Konvalinka J. Inorganic polyhedral

metallacarborane inhibitors of HIV protease: a new approach to overcoming antiviral resistance. J Med Chem. 2008;51(15):4839–4843. Epub 2008/07/05.

15 Rezácová P PJ, Brynda J, Kozísek M, Cígler P, Lepsík M, Fanfrlík J, Rezác J, Grantz Sasková K, Sieglová I, Plesek J, Sícha V, Grüner B, Oberwinkler H, Sedlácek' J, Kräusslich HG, Hobza P, Král V, Konvalinka J. Design of HIV protease inhibitors based on inorganic polyhedral metallacarboranes. J Med Chem. 2009;52(22):7132–7141.

16 Cígler P KM, Rezácová P, Brynda J, Otwinowski Z, Pokorná J, Plesek J, Grüner B, Dolecková-Maresová L, Mása M, Sedlácek J, Bodem J, Kräusslich HG, Král V, Konvalinka J. From nonpeptide toward noncarbon protease inhibitors: metallacarboranes as specific and potent inhibitors of HIV protease. Proc Natl Acad Sci USA. 2005;102(43):15394–15399.

17 Jelínek R, Baldwin P, Scheidt WR, Reed CA. New weakly coordinating anions. 2. Derivatization of the carborane anion $CB_{11}H_{12}^-$. Inorg Chem. 1993;32:1982–1990.

18 Bolli C, Derendorf J, Jenne C, Scherer H, Sindlinger CP, Wegener B. Synthesis and properties of the weakly coordinating anion $[Me_3NB_{12}H_{11}]-$. Chemistry Eur J. 2014;20(42):13783–13792.

19 Stasko D, Reed CA. Optimizing the least nucleophilic anion. A new, strong methyl$^+$ reagent. J Am Chem Soc. 2002;124(7):1148–1149.

20 Stasko D, Hoffmann SP, Kim KC, Fackler NL, Larsen AS, Drovetskaya T, et al. Molecular structure of the solvated proton in isolated salts. Short, strong, low barrier (SSLB) H-bonds. J Am Chem Soc. 2002;124(46):13869–13876.

21 Gentil S, Crespo E, Rojo I, Friang A, Vinas C, Teixidor F, et al. Polypyrrole materials doped with weakly coordinating anions: influence of substituents and the fate of the doping anion during the overoxidation process. Polymer. 2005;46(26):12218–12225.

22 Stoyanov ES, Hoffmann SP, Juhasz M, Reed CA. The structure of the strongest Broensted acid: the carborane acid $H(CHB_{11}Cl_{11})$. J Am Chem Soc. 2006;128:3160–3161.

23 Larsen AS, Holbrey JD, Tham FS, Reed CA. Designing ionic liquids: imidazolium melts with inert carborane anions. J Am Chem Soc. 2000;122:7264–7272.

24 Assaf KI, Ural MS, Pan F, Georgiev T, Simova S, Rissanen K, et al. Water structure recovery in chaotropic anion recognition: high-affinity binding of dodecaborate clusters to gamma-cyclodextrin. Angew Chem Int Ed Engl. 2015;54(23):6852–6856. Epub 2015/05/08.

25 Farras P, Vankova N, Zeonjuk LL, Warneke J, Dülcks T, Heine T, et al. From an icosahedron to a plane: flattening dodecaiodo-dodecaborate by successive stripping of iodine. Chemistry Eur J. 2012;18(41):13208–13212. Epub 2012/09/11.

26 Karki K, Gabel D, Roccatano D. Structure and dynamics of dodecaborate clusters in water. Inorg Chem. 2012;51(9):4894–4896. Epub 2012/04/11.

27 Fan P, Stolte S, Gabel D. Interaction of organic compounds and boron clusters with new silica matrices containing the phosphatidylcholine headgroup. Anal Methods. 2014;6(9):3045–3055.

28 Fan P, Neumann J, Stolte S, Arning J, Ferreira D, Edwards K, et al. Interaction of dodecaborate cluster compounds on hydrophilic column materials in water. J Chromatogr A. 2012;1256(0):98–104.

29 Sadrerafi K, Moore EE, Lee MW. Association constant of b-cyclodextrin with carboranes, adamantane, and their derivatives using displacement binding technique. J Inclusion Phenom Macrocyclic Chem. 2015;83(1–2):159–166.

30 Neirynck P, Schimer J, Jonkheijm P, Milroy LG, Cigler P, Brunsveld L. Carborane-β-cyclodextrin complexes as a supramolecular connector for bioactive surfaces. J Mater Chem B. 2015;3(4):539–545.

31 Vaitkus R, Sjoberg S. Large inclusion constants of b-cyclodextrin with carborane derivatives. J Inclusion Phenom Macrocyclic Chem. 2011;69(3–4):393–395.

32 Ohta K, Konno S, Endo Y. Complexation of a-cyclodextrin with carborane derivatives in aqueous solution. Chem Pharm Bull. 2009;57(3):307–310.

33 Ohta K, Konno S, Endo Y. Complexation of b-cyclodextrin with carborane derivatives in aqueous solution. Tetrahedron Lett. 2008;49(46):6525–6528.

34 Kusukawa T, Fujita M. Encapsulation of large, neutral molecules in a self-assembled nanocage incorporating six palladium(II) ions. Angew Chem Int Ed. 1998;37(22):3142–3144.

35 Harada A, Takahashi S. Preparation and properties of inclusion complexes of 1,2-dicarbadodecaborane(12) with cyclodextrins. J Chem Soc, Chem Commun. 1988(20):1352–1353.

36 Hardie MJ, Raston CL. Supramolecular assemblies of 1,2-dicarbadodecaborane(12) with bowl-shaped calix[5]arene. Eur J Inorg Chem. 1999(1):195–200.

37 Clark TE, Makha M, Raston CL, Sobolev AN. Solution and solid state studies on the binding of isomeric carboranes $C_2B_{10}H_{12}$ by *p*-Bu-t-calix[5] arene. Dalton Trans. 2006(46):5449–5453.

38 Blanch RJ, Sleeman AJ, White TJ, Arnold AP, Day AI. Cucurbit[7]uril and o-carborane self-assemble to form a molecular ball bearing. Nano Lett. 2002;2(2):147–149.

39 Blanch RJ, Williams M, Fallon GD, Gardiner MG, Kaddour R, Raston CL. Supramolecular complexation of 1,2-dicarbadodecaborane(12). Angew Chem Int Ed. 1997;36(5):504–506.

40 Raston CL, Cave GW. Nanocage encapsulation of two ortho-carborane molecules. Chemistry. 2004;10(1):279–282. Epub 2003/12/26.

41 Frixa C, Scobie M, Black SJ, Thompson AS, Threadgill MD. Formation of a remarkably robust 2 : 1 complex between β-cyclodextrin and a phenyl-substituted icosahedral carborane. Chem Commun. 2002(23):2876–2877.

42 Ohta K, Goto T, Yamazaki H, Pichierri F, Endo Y. Facile and efficient synthesis of C-hydroxycarboranes and C,C'-dihydroxycarboranes. Inorg Chem. 2007;46(10):3966–3970.

43 Assaf KI, Ural MS, Pan F, Georgiev T, Simova S, Rissanen K, *et al*. Water structure recovery in chaotropic anion recognition: high-affinity binding of dodecaborate clusters to g-cyclodextrin. Angew Chem Int Ed. 2015;54(23):6852–6856. Epub 2015/05/08.

44 Assaf KI, Gabel D, Zimmermann W, Nau WM. High-affinity host–guest chemistry of large-ring cyclodextrins. Organic and Biomolecular Chemistry. 2016. doi:10.1039/C6OB01161F.

45 Warneke J, Jenne C, Bernarding J, Azov VA, Plaumann M. Evidence for an intrinsic binding force between dodecaborate dianions and receptors with hydrophobic binding pockets. Chem Commun. 2016;52(37):6300–6303. Epub 2016/04/19.

46 Assaf KI, Suckova O, Al Danaf N, von Glasenapp V, Gabel D, Nau WM. Dodecaborate-functionalized anchor dyes for cyclodextrin-based indicator displacement applications. Org Lett. 2016;18(5):932–935. Epub 2016/02/24.

47 Hwang I, Baek K, Jung M, Kim Y, Park KM, Lee DW, *et al*. Noncovalent immobilization of proteins on a solid surface by cucurbit[7]uril-ferrocenemethylammonium pair, a potential replacement of biotin-avidin pair. J Am Chem Soc. 2007;129(14):4170–4171.

48 Young JF, Nguyen HD, Yang LT, Huskens J, Jonkheijm P, Brunsveld L. Strong and reversible monovalent supramolecular protein immobilization. ChemBioChem. 2010;11(2):180–183.

49 Ching HYV, Clarke RJ, Rendina LM. Supramolecular b-cyclodextrin adducts of boron-rich DNA metallointercalators containing dicarba-closo-dodecaborane(12). Inorg Chem. 2013;52(18):10356–10367.

50 Ching HYV, Clifford S, Bhadbhade M, Clarke RJ, Rendina LM. Synthesis and supramolecular studies of chiral boronated platinum(ii) complexes: insights into the molecular recognition of carboranes by β-cyclodextrin. Chemistry Eur J. 2012;18(45):14413–14425.

51 Ching HYV, Buck DP, Bhadbhade M, Collins JG, Rendina LM. A ternary supramolecular system containing a boronated DNA-metallointercalator, b-cyclodextrin and the hexanucleotide d(GTCGAC)(2). Chem Commun. 2012;48(6):880–882.

52 Ching HY, Buck DP, Bhadbhade M, Collins JG, Rendina LM. A ternary supramolecular system containing a boronated DNA-metallointercalator, beta-cyclodextrin and the hexanucleotide d(GTCGAC)2. Chem Commun. 2012;48(6):880–882. Epub 2011/12/06.

53 Mader P, Pecina A, Cígler P, Lepšík M, Šícha V, Hobza P, et al. Carborane-based carbonic anhydrase inhibitors: insight into CAII/CAIX specificity from a high-resolution crystal structure, modeling, and quantum chemical calculations. BioMed Res Int. 2014;2014:389869. Epub 2014/10/14.

54 Brynda J, Mader P, Šícha V, Fábry M, Poncová K, Bakardiev M, et al. Carborane-based carbonic anhydrase inhibitors. Angew Chem Int Ed. 2013;52(51):13760–13763.

55 Ban HS, Nakamura H. Boron-based drug design. Chem Rec. 2015;15(3):616–635.

56 Gabel D. Boron clusters in medicinal chemistry: perspectives and problems. Pure Appl Chem. 2015;87(2):173–179.

57 Lesnikowski ZJ. Recent developments with boron as a platform for novel drug design. Expert Opin Drug Discovery. 2016;11(6):569–578.

58 Pecina A, Lepsik M, Řezáč J, Brynda J, Mader P, Rezacova P, et al. QM/MM Calculations reveal the different nature of the interaction of two carborane-based sulfamide inhibitors of human carbonic anhydrase II. J Phys Chem B. 2013;117:16096–16104.

59 Neumann W, Xu S, Sárosi MB, Scholz MS, Crews BC, Ghebreselasie K, et al. nido-Dicarbaborate induces potent and selective inhibition of cyclooxygenase-2. ChemMedChem. 2016;11(2):175–178.

60 Gabel D, Awad D, Schaffran T, Radovan D, Daraban D, Damian L, et al. The anionic boron cluster $B_{12}H_{11}SH^{2-}$ as a means to trigger release of liposome contents. ChemMedChem. 2007;2(1):51–53.

61 Awad D, Damian L, Winterhalter M, Karlsson G, Edwards K, Gabel D. Interaction of Na2B12H11SH with dimyristoyl phosphatidylcholine liposomes. Chem Phys Lipids. 2009;157(2):78–85.

62 Schaffran T, Li J, Karlsson G, Edwards K, Winterhalter M, Gabel D. Interaction of N,N,N-trialkylammonioundecahydro-closo-dodecaborates with dipalmitoyl phosphatidylcholine liposomes. Chem Phys Lipids. 2010;163(1):64–73.

63 Verdia-Baguena C, Alcaraz A, Aguilella VM, Cioran AM, Tachikawa S, Nakamura H, et al. Amphiphilic COSAN and I2-COSAN crossing synthetic lipid membranes: planar bilayers and liposomes. Chem Commun. 2014;50(51):6700–6703.

64 Tarres M, Canetta E, Paul E, Forbes J, Azzouni K, Vinas C, et al. Biological interaction of living cells with COSAN-based synthetic vesicles. Sci Rep. 2015;5.

65 Uchman M, Ďorďovič V, Tošner Z, Matějíček P. Classical amphiphilic behavior of nonclassical amphiphiles: a comparison of metallacarborane self-assembly with SDS micellization. Angew Chem Int Ed. 2015;54(47):14113–14117.

66 Bauduin P, Prevost S, Farràs P, Teixidor F, Diat O, Zemb T. A theta-shaped amphiphilic cobaltabisdicarbollide anion: transition from monolayer vesicles to micelles. Angew Chem Int Ed. 2011;50(23):5298–5300.

67 Lee JD, Ueno M, Miyajima Y, Nakamura H. Synthesis of boron cluster lipids: closo-dodecaborate as an alternative hydrophilic function of boronated liposomes for neutron capture therapy. Org Lett. 2007;9(2):323–326.

68 Justus E, Awad D, Hohnholt M, Schaffran T, Edwards K, Karlsson G, et al. Synthesis, liposomal preparation, and in vitro toxicity of two novel dodecaborate cluster lipids for boron neutron capture therapy. Bioconjugate Chem. 2007;18(4):1287–1293.

69 Schaffran T, Lissel F, Samatanga B, Karlsson G, Burghardt A, Edwards K, et al. Dodecaborate cluster lipids with variable headgroups for boron neutron capture therapy: synthesis, physical-chemical properties and toxicity. J Organomet Chem. 2009;694(11):1708–1712.

70 Schaffran T, Jiang N, Bergmann M, Küstermann E, Süss R, Schubert R, et al. Hemorrhage in mouse tumors induced by dodecaborate cluster lipids intended for boron neutron capture therapy. Int J Nanomed. 2014;9:3583–3590.

71 Goszczynsik TM, Kowalski K, Lesnikowski ZJ, Boratyriski J. Solid state, thermal synthesis of site-specific protein-boron cluster conjugates and their physicochemical and biochemical properties. Biochim Biophys Acta, Gen Subj. 2015;1850(2):411–418.

72 Ban HS, Shimizu K, Minegishi H, Nakamura H. Identification of HSP60 as a primary target of o-carboranylphenoxyacetanilide, an HIF-1α inhibitor. J Am Chem Soc. 2010;132(34):11870–11871.

73 Schaffran T, Justus E, Elfert M, Chen T, Gabel D. Toxicity of N,N,N-trialkylammoniododecaborates as new anions of ionic liquids in cellular, liposomal and enzymatic test systems. Green Chem. 2009;11:1458–1464.

74 Genady AR, Ioppolo JA, Azaam MM, El-Zaria ME. New functionalized mercaptoundecahydrododecaborate derivatives for potential application in boron neutron capture therapy: synthesis, characterization and dynamic visualization in cells. Eur J Med Chem. 2015;93(0):574–583.

75 Tiwari R, Mahasenan K, Pavlovicz R, Li C, Tjarks W. Carborane clusters in computational drug design: a comparative docking evaluation using AutoDock, FlexX, Glide, and Surflex. J Chem Inf Model. 2009;49(6):1581–1589. Epub 2009/05/20.

76 Calvaresi M, Zerbetto F. In silico carborane docking to proteins and potential drug targets. J Chem Inf Model. 2011;51(8):1882–1896. Epub 2011/07/22.

77 Lee MW, Sevryugina YV, Khan A, Ye SQ. Carboranes increase the potency of small molecule inhibitors of nicotinamide phosphoribosyltranferase. J Med Chem. 2012;55(16):7290–7294.

78 Scholz M, Blobaum AL, Marnett LJ, Hey-Hawkins E. Synthesis and evaluation of carbaborane derivatives of indomethacin as cyclooxygenase inhibitors. Bioorg Med Chem. 2011;19(10):3242–3248.

1.6

Quantum Mechanical and Molecular Mechanical Calculations on Substituted Boron Clusters and Their Interactions with Proteins

Jindřich Fanfrlík,[1] Adam Pecina,[1] Jan Řezáč,[1] Pavel Hobza,[1,2] and Martin Lepšík[1],*

[1] *Institute of Organic Chemistry and Biochemistry of the Czech Academy of Sciences, 16610, Prague 6, Czech Republic*
[2] *Regional Centre of Advanced Technologies and Materials, Palacký University, 77146 Olomouc, Czech Republic*
* *Corresponding author: RNDr. Martin Lepšík, PhD, Institute of Organic Chemistry and Biochemistry of the Czech Academy of Sciences, Flemingovo nám. 2, 16610 Prague 6, Czech Republic. Email: lepsik@uochb.cas.cz*

1.6.1 Introduction

Medicinal chemistry entails not only synthesis but also compound design. The role of rational drug design has been boosted in recent decades through the use of computer-aided drug design, either ligand- or structure-based [1]. The former area makes heavy use of statistics and chemi-informatics to set up dependencies between the physico-chemical properties of the compounds and their biological activities. Although some properties, such as the lipophilicity (expressed as the n-octanol/water partition coefficient, log$P$), can also be evaluated computationally by quantum mechanical (QM) or molecular mechanical (MM) methods, ligand-based drug design is not the focus of this chapter. On the contrary, we discuss here structure-based drug design, which uses three-dimensional (3D) structures of protein–ligand complexes to estimate affinities. The geometries are most often determined experimentally by X-ray crystallography or nuclear magnetic resonance (NMR). In computer-aided structure-based drug design, the ligand's binding pose within the protein is predicted by docking, a task that has practically been mastered for the broad organic chemistry space [2]. Scoring is thereafter used to assess which of the poses represent the native complex and to rank the compounds by affinity. In contrast, this task has, until recently, been deemed unsolved [3]. We have approached the low reliability of scoring by developing a QM-based scoring function [4]. Its fundamental principle is QM treatment of protein–ligand noncovalent interactions and solvation [5]. We showed that such an approach can be advantageously used to unequivocally identify the ligand native pose [6], reproduce binding affinity in a series of ligands to various protein targets [5], and describe nonclassical noncovalent interactions, such as halogen bonds [7] or even covalently binding inhibitors [8]. Based on this extensive experience of ours with organic ligands and a decade-long experience with calculations of boron clusters bound to proteins [9], we affirm that QM scoring is a general solution to the affinity prediction of boron cluster/protein binding.

Boron-Based Compounds: Potential and Emerging Applications in Medicine, First Edition.
Edited by Evamarie Hey-Hawkins and Clara Viñas Teixidor.
© 2018 John Wiley & Sons Ltd. Published 2018 by John Wiley & Sons Ltd.

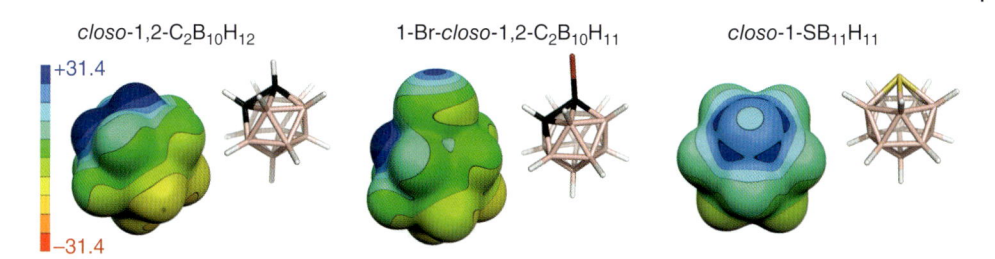

closo-1,2-C$_2$B$_{10}$H$_{12}$ 1-Br-closo-1,2-C$_2$B$_{10}$H$_{11}$ closo-1-SB$_{11}$H$_{11}$

+31.4

−31.4

Figure 1.6.1 Molecular structures of (substituted) heteroboranes and their computed electrostatic potentials (ESPs) on 0.001 a.u. molecular surface at the HF/cc-pVDZ level. Color range of ESPs in kcal/mol.

The major tools of structure-based drug design, docking and scoring, have been optimized over decades of heavy use on the broad organic chemistry space, which dwells on ten elements from the Mendeleev periodic table. However, parameters are missing or inaccurate when other elements, such as boron, come into play. This may be of marginal importance for drugs containing single boron atoms [10], but it becomes catastrophic for boron hydride clusters (boranes), which are slowly finding their way into drug design as nontraditional pharmacophores [11–14]. The simple solutions, such as replacing all boron atoms with carbons [15] and zeroing all their charges [16], are very crude and do not respect the boron cluster chemistry. We rather advocate a systematic QM approach based on detailed understanding of borane structures and their noncovalent interactions [17–19].

Boranes come in a variety of sizes and shapes. The most common are closed cages (*closo*-B$_n$H$_n$$^{2-}$). The exoskeletal hydrogens can be replaced by, for example, halogens, via reactions distinct from those known in organic chemistry [20–23]. Heteroboranes are formed when notionally neutral {BH} vertices are substituted by various heterovertices, such as {CH}$^+$, {S}$^{2+}$, or metals, giving rise to carboranes, thiaboranes, or metallacarboranes, respectively (Figure 1.6.1). The iconic heteroboranes are neutral dicarbaboranes *closo*-C$_2$B$_{10}$H$_{12}$ (Figure 1.6.1a) occurring as three isomers, *ortho*- (1,2), *meta*- (1,7), and *para*- (1,12) [24,25]; and {Co[C$_2$B$_9$H$_{11}$]$_2$}$^-$ metallacarborane, a so-called COSAN with three rotational isomers, *cisoid*, *gauche*, and *transoid* [26,27]. In analogy to boranes, heteroboranes can be further substituted on their exoskeletal hydrogens (Figure 1.6.1). Boranes have unique properties stemming from their electron structure and charge distribution (Figure 1.6.1). They are held together by three-center two-electron bonding, resulting in boron atom hypervalency, electron delocalization, electron deficiency, and 3D aromaticity [28–30].

1.6.2 Plethora of Noncovalent Interactions of Boron Clusters

Due to their peculiar characteristics, boron clusters exhibit unique types of noncovalent interactions (dihydrogen bonding, σ-hole bonding) [19] besides the classical ones such as hydrogen bonding. The classical R–H···Y hydrogen bonding can in principle occur for some rare heteroboranes (e.g., *closo*-1-NB$_{11}$H$_{12}$), in which the N–H group of the azaborane acts as the H-bond donor. Heteroboranes engage more often in weak hydrogen bonds [31] of C–H···O or C–H···π type (Figure 1.6.2A),

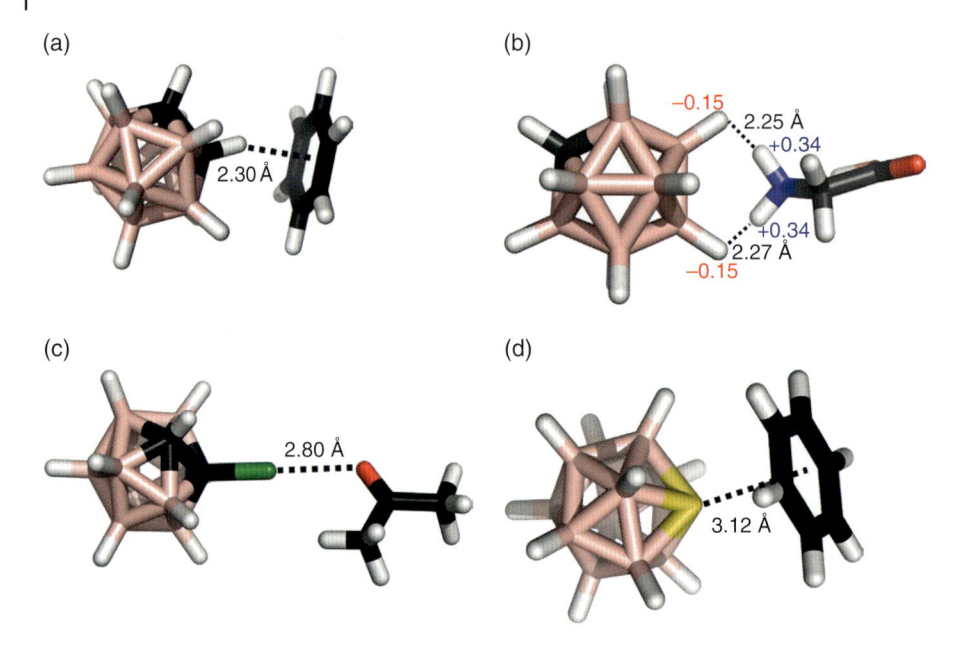

Figure 1.6.2 Model systems of protein–heteroborane interactions. (a) C–H···π. (b) Dihydrogen bonding. (c) Halogen bonding (Br···O). (d) Chalcogen bonding (S···π). Black: carbon; white: hydrogen; yellow: sulfur; pink: boron; green: bromine. Distances in Å. Partial charges in e.

where the C–H group of carboranes acts as the H-bond donor. Such carborane interactions have been evidenced by crystal structures and explored in detail computationally [32]. They can be expected in proteins where the H-bond acceptors would be backbone carbonyls, side-chain hydroxyls, or aromatic side chains of Phe, Tyr, His, and Trp.

The R-H···H-Z dihydrogen bonding is a special type of hydrogen bonding (nonclassical) where R are mostly electronegative (e.g., O, S, N) and Z electropositive (e.g., B) elements. It is brought about by the low electronegativity of boron (lower than that of hydrogen) which results in the boron-bound hydrogens having hydridic character (Figure 1.6.2b) [33,34]. Dihydrogen bonds are of high importance for boron clusters since all aliphatic hydrogens in biomolecules carry partial positive charge [18]. Even the nonpolar C–H groups of benzene engage in C–H···H–B type of dihydrogen bonds with heteroboranes [32].

Furthermore, substituted heteroboranes can form σ-hole bondings such as halogen and chalcogen bonding (Figure 1.6.2c and 1.6.2d) [35–39]. These are interactions of quantum origin where the region of positive electrostatic potential (σ-hole) on partially negatively charged halogen atoms interacts with the electron donor [40]. QM calculations suggested that neutral carboranes halogenated on carbon vertices had highly positive σ-holes and were predicted to form strong halogen bonds [36,37]. Highly positive σ-holes were computed for heteroboranes with chalcogen and pnictogen atoms incorporated into boron clusters [37]. We have recently demonstrated the dominant role of chalcogen and halogen bonding in crystal packing of phenyl-substituted thiaborane and brominated carboranes, respectively [38,39]. σ-Hole interactions of heteroboranes might also be employed in drug design.

1.6.3 Computational Methods

To describe reliably noncovalent interactions and understand their geometrical and energetical characteristics, we must use advanced QM methods, which are, however, prohibitively costly for more than a few hundreds of atoms. Therefore, the interactions of interest are usually studied in small model systems in vacuum. To capture the behavior in protein binding sites, including the physiological environment, we must resort to more approximate methods to increase the computational efficiency.

1.6.3.1 Advanced Methods for Models Systems in Vacuum

Second-order Møller–Plesset (MP2) and density functional theory (DFT) are the most widely used QM methods for various types of calculations that involve boron clusters [17]. MP2 covers large portions of correlation energy (both intra- and intermolecular) and was very successful for H-bonded small model complexes, while it had problems with large stacked aromatic complexes or when transition metals were present. Thus, in the case of larger metallaboranes, the choice of basis set for MP2 is crucial even for geometry optimization since timing and basis set superposition error start to play important roles. In contrast, DFT methods are much better suited for the calculations of metallaboranes, although they cover only part of intramolecular correlation energy. DFT methods reliably reproduce the geometries of studied molecules but do not describe reliably noncovalent interactions of boron clusters.

A quantitative description of σ-hole bonding, dihydrogen bonding, and stacking interactions is very demanding, as London dispersion energy plays an important role. Dispersion is a nonlocal correlation effect that is not described at the DFT level of theory. One of the most successful approaches to overcome this drawback of DFT is the addition of an empirical dispersion correction (D) [41,42] to the plain DFT results. It was shown for several DFT-D methods that they provided stabilization energies with respect to the "golden standard" CCSD(T)/CBS reference values [43] with an error of about 10% [44,45]. The performance of the dispersion parameters for boron was checked by computing benchmark CCSD(T)/CBS interaction energies for σ-hole interactions of heteroboranes [37]. The DFT-D3 values at the TPSS/TZVPP level had a root mean square error (RMSE) of 0.43 kcal/mol, which is 9.5%. This suggests that D3 parameters for boron are reliable, which makes DFT-D3 methods very useful tools for various types of applications. When the size of the studied systems requires smaller basis sets, we recommend using the DFT-D3 method at the BLYP/DZVP level [39], which has an acceptable RMSE of 1.08 kcal/mol for σ-hole interactions of heteroboranes. For comparison, the BLYP/SVP functional/basis set combination has comparable time requirements but a considerably bigger RMSE of 1.70 kcal/mol.

1.6.3.2 Approximate Methods for Extended Systems Including the Environment

To be able to treat noncovalent interactions in larger parts of the protein binding sites, several alternative approaches have been developed. One is the use of fragmentation [46,47], another comprises hybrid QM/MM methods [48] and yet another uses semiempirical QM (SQM) methods [5]. Besides that, the effect of the surrounding solvent and ions needs to be taken into account using explicit (all-atom) or implicit (continuum) models. The ligand's pKa values sometimes need to be considered, especially when they change upon binding to the protein.

1.6.3.2.1 SQM Methods

The increase in efficiency of SQM methods is obtained by reducing the number of two-electron integrals, thus limiting calculations to valence electrons only and considering only a subminimal basis set. Therefore, SQM methods augmented with linear-scaling algorithms are able to treat up to 10,000 atoms in a reasonable time [49]. The accuracy of the description of noncovalent interactions is, however, compromised. We have overcome this limitation by a careful and systematic parametrization of empirical corrections for SQM methods using comparison to benchmark data in databases of noncovalent interactions of model systems [43]. The resulting PM6 method with the D3H4X corrections quantitatively describes dispersion, hydrogen, and halogen bonding [50–52], and can thus be used for many difficult cases in computational drug design [5–7]. The SQM treatment provides a natural description of QM effects that are difficult to describe with MM force fields. At the same time, SQM methods are efficient enough for fast calculations of systems with thousands of atoms. We have successfully used it for the ranking and scoring of protein–ligand complexes [4,5,53]. In these cases, however, the ligands were organic molecules. When applied to interactions of boron clusters, large errors were observed. One of the reasons is that the PM6 method predicts inversed partial charges on B and H atoms. We are currently working on reparametrization of boron in the PM6 method to correct this problem. Preliminary results show that the partial charges can be corrected, and accuracy of description of noncovalent interactions significantly improves.

Another efficient SQM method applicable to this type of treatment is the self-consistent-charge tight binding (SCC-DF-TB) method [54]. The method is less empirical than other SQM methods, including PM6, and thus provides more reliable results even for difficult systems (e.g., metalloproteins). The parameter set needed for working with boron compounds is, however, still in development, but the preliminary results are encouraging.

To conclude, although SQM methods with empirical corrections are a useful tool in computational drug design, at present, no SQM methods are readily applicable to boron cluster interactions. Our preliminary data, however, show that both PM6 and SCC-DF-TB can be extended to make these calculations fast and reliable.

1.6.3.2.2 MM Methods

MM methods require parameters for electrostatics (atomic partial charges) and van der Waals interactions (σ and ε Lennard–Jones parameters). These are stored in a force field and have been developed and extensively tested for biomolecules [55] and organic ligands [56]. However, for other molecules, only generic parameters are available in the universal force field (UFF) [57], where the accuracy, by definition, cannot be as high. Although the methods for force-field parameter derivation are known, caution has to be exercised for boron clusters. Partial atomic charges are derived from QM calculations. Restrained fit to the electrostatic potential (RESP) method [58] correctly gives a slightly negative charge on boron-bound hydrogens [18,59,60]. However, when different QM setups are used, such as natural population analysis (NPA), inverse charges are obtained, which is inconsistent with their dihydrogen bonding [18]. Lennard–Jones parameters can either be taken from the UFF, to generate structures to be recalculated by QM [18], or optimized to fit experimental log$P$ data [60].

1.6.3.2.3 Solvation and Ion Models

Besides the protein–ligand noncovalent interactions, the effect of the solvent (including the ions surrounding the protein and the ligand) needs to be taken into account. Molecular dynamics (MD)-based free-energy simulations with explicit solvent molecules and ions can be used to obtain accurate values of solvation energies of organic ligands [61]. MD was also used to study logP_{ow} [60] and surfactant behavior of boron clusters [59]. However, apart from issues with boron cluster parametrization discussed above [60], the limitation of such approaches may be incomplete sampling. This drawback is overcome in implicit solvent models that are about one order of magnitude faster [62–64]. The accuracy of implicit solvent models depends on the method: QM-based models, such as COSMO [65] or SMD [62], are more reliable than MM-based models, such as generalized Born (GB) [62] or Poisson–Boltzmann (PB), as we and others found for sets of organic molecules [66,67]. The implicit solvent models were parameterized and validated on experimental solvation free energies, partition coefficients, or other macroscopic properties of simple organic compounds and ions [62,65]. Such experimental data are, however, very rare for boron clusters [60]. This situation hinders a reliable validation of the accuracy and performance of implicit solvent models for substituted boron clusters. Another way of comparing the calculated solvation free energies with experimental data is by using dissolution free energies. However, that situation is complicated by the problem of accurately evaluating lattice energies [68].

1.6.3.2.4 pKa Calculations

Protonation/deprotonation phenomena upon ligand binding can influence the free energies substantially [69]. To determine the prevalent ligand protonation state and its change upon binding is thus important. The effects can be direct, such as a functional group protonation, or indirect, such as influence of one functional group on the pKa of a neighboring group. The approach to evaluating pKa's consists of (1) calculating the gas-phase acidity [70] and (2) assessing the solvation effect by use of explicit or implicit solvent models. The lack of reference for the solvation free energies of boron clusters mentioned in this chapter can lead to quite poor pKa predictions [71].

1.6.3.2.5 Docking and Scoring

Docking based on the geometric matching between the ligand and the protein can give some insights into boron cluster binding possibilities [16]. However, for scoring, all the problems with boron clusters, MM, and implicit solvation are present, and the results need to be validated by experiments or QM calculations.

1.6.4 Boron Cluster Interactions with Proteins

The major source of geometries of protein–heteroborane complexes is X-ray crystallography. However, the structural data in this area are scarce with only a handful of examples. Although the experimental insight into the binding modes is invaluable, most of these data require further computational refinement to understand the structural and energetical determinants of binding.

The very first crystal structure of a protein–heteroborane complex was that of HIV-1 protease (HIV PR) with anionic COSAN metallacarborane [72]. The binding mode was surprising as two COSAN molecules bound asymmetrically to the C_{2v} symmetrical HIV PR dimer. Moreover, two HIV PR dimers were arranged in a tetramer in which the pair of COSAN molecules was in close contact with the other COSAN pair from the other PR dimer (Figure 1.6.3a). It thus became evident why previous docking could not have succeeded in binding geometry predictions (M. Lepšík, unpublished results). But even having the X-ray structure was not sufficient to understand the binding mode in atomistic detail, as carbon and boron atoms were indistinguishable. Our task was thus to evaluate the energies of circa 400,000 possible rotamer combinations of bound COSAN molecules. Although this task was beyond the capabilities of a brute-force QM approach, we succeeded in overcoming this by fragmentation and DFT-D QM/MM optimizations and identified 81 rotamer combinations with favorable energies. They were partially defined by the closeness of the Na^+ counterion, which was found to be important in stabilizing the inhibitors (Figure 1.6.3a) [73]. This computational setup paved the way for a further study in which we rationalized the favorable profile of carboranes against resistance [74]. The HIV PR–COSAN crystal structure pointed to the design of more potent compounds. Linking the two COSAN cages by linker led to a more potent inhibitor (GB80). The crystal structure of the HIV PR–GB80 complex showed that, unlike in the previous crystal structure (Figure 1.6.3a), the two COSAN cages bound *symmetrically* within the symmetrical HIV PR dimer (Figure 1.6.3b) [75]. The flexible organic linker connecting the two COSAN cages could not be observed in the crystallographic data due to its flexibility and had to be modeled by MD, allowing the COSAN cages to rotate. The stabilization energies were subsequently calculated with the QM/MM approach and yielded five nearly isoenergetic conformers of the linker (Figure 1.6.3b) [75].

The second protein–heteroborane crystal structure was that of dihydrofolate reductase (DHFR) with a carborane-based trimethoprim analog [76]. The crystallographic data was of such a high quality that they allowed the authors to identify the single rotamer out of the possible five that were present. QM calculations somewhat vaguely concluded that electrostatic interactions played an important role, without specifying any details [76]. We assume that QM-based scoring could shed more light on the nature of the carborane–DHFR interactions.

Human carbonic anhydrase II (hCAII) was another enzyme whose crystal structures in complex with carborane-based sulfamide inhibitors were obtained (Figure 1.6.4) [77,78]. Even though the binding of these inhibitors is mainly driven via the sulfamide interaction with the Zn^{2+} ion of the enzyme, we found by QM calculations that the carborane parts of the inhibitors had different natures of binding [79]. One contained neutral 1,2-dicarba-*closo*-dodecaborane, and the second a negatively charged 7,8-*nido*-dicarbaundecaborante cage (i.e., one BH vertex was missing from the closed cage). The binding of the *nido* cage was driven by electrostatics. It also formed strong dihydrogen bonds. The most significant was the interaction with the NH_2 group of Asn67 (Figure 1.6.4b). The energy of this N–H⋯H–B dihydrogen bond was computed to be about −4.2 kcal/mol, and the H⋯H distance was as short as 1.7 Å. On the other side, the binding of the neutral *closo* cage was mainly driven by

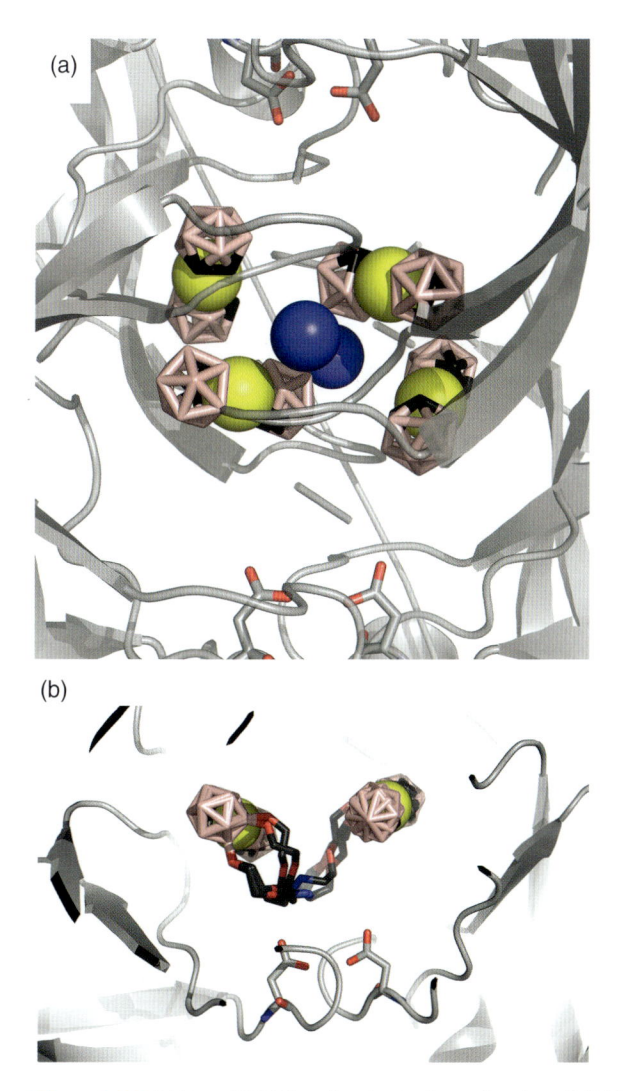

Figure 1.6.3 (a) Four COSAN molecules in the active site of HIV PR (gray ribbon) [73]. (b) Five low-energy conformers of a dual-COSAN inhibitor (GB80) in HIV PR obtained by MD and QM/MM calculations [75]. Yellow spheres: Co^{3+}; blue spheres: Na^+; gray/black: carbon; pink: boron; blue: nitrogen. Hydrogen atoms are omitted for clarity.

dispersion, and it only formed weak dihydrogen bonds (H···H distance over 2.2 Å) with nonpolar C–H groups of the protein. Even though the *closo* cage had weaker individual noncovalent binding interactions, it had bigger overall binding affinity due to the lower desolvation penalty [79]. Furthermore, the indirect effect of the carborane cage on the pKa of the sulfamide group was assessed by a QM-based implicit solvent COSMO-RS model. The calculated values were in the range of 8–9, comparable with electron-withdrawing phenyl substituents. It thus suggested that the electron-deficient cages have a similar effect, thus contributing to the potency of the carborane-based inhibitors [79].

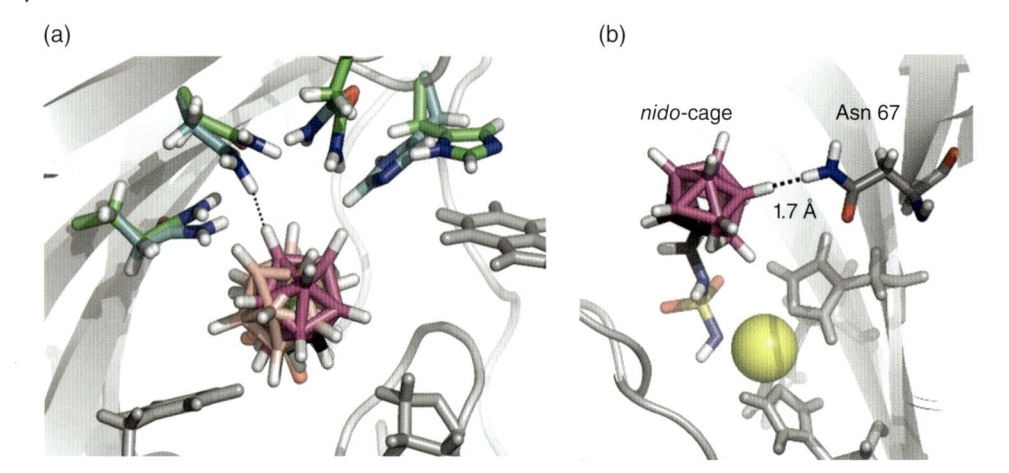

Figure 1.6.4 (a) An overlay of hCAII–inhibitor QM/MM optimized structures. The *nido* and *closo* carborane cages of the inhibitor are in magenta and pink, respectively. (b) Dihydrogen bonding in the hCAII–inhibitor complex.

Such knowledge about the different nature of binding of heteroborane cages was further utilized in rational design of inhibitors selective toward the cancer-specific human carbonic anhydrase IX (hCAIX) isoenzyme [78]. We modeled the binding of a neutral 1,2-dicarba-*closo*-dodecaborane compound into a CAIX isoenzyme that differs from CAII in the shape of the active-site cavity caused by variations of six amino acids. We showed the different positions of the *closo*-cage in hCAII and hCAIX. The interacting contributions of all important amino acids in the opposite site of the hCAIX active site were determined by virtual glycine scan and compared with those of hCAII.

Decade-long research on carboranes modifying steroid hormones yielded active ligands for a dozen targets (reviewed in Ref. [14]). The only experimental binding mode determination was for a vitamin D receptor with a nonsecosteroidal ligand with a *p*-carborane core. The authors claimed to "directly observe hydrophobic interaction" with nonpolar amino acids without supplying further details [80]. We consider this term unfortunate because it can confuse dispersion interactions (the authors seem to have meant this) and entropy-driven hydrophobic effects [81]. While the former can be addressed by QM calculations of interaction energies and by evaluating the dispersion contributions for individual amino acids, the latter needs to include the solvation, at least by implicit methods or, better yet, by explicit ones.

Recently, structural insights have been obtained for carborane-substituted nonsteroidal anti-inflammatory drug (NSAID) indomethacin in complex with COX-2 enzyme [82]. The *nido*-carborane binds in a hydrophobic region of the enzyme. The crystallographic data did not allow the authors to distinguish boron and carbon atoms, and thus they modeled the *nido*-carborane compound as two enantiomers. Using docking and DFT calculations, they attempted to evaluate their binding free energy, but the results were rather inconclusive.

1.6.5 Conclusions

In this chapter, we have reviewed the current status of structure-based computer-aided drug design of boron cluster–containing protein ligands. The main problem concerns a reliable description of the unusual electronic structure of electron-deficient boron clusters, including the hydridic character of boron-bound hydrogens, caused by the low electronegativity of boron. Today's software codes for docking/scoring also approximate semi-empirical QM methods but do not describe these inherently quantum phenomena. The solution that we advocate starts with the use of high-level QM methods on model systems. These offer a quantitative description of the wide array of the unique heteroborane noncovalent interactions, such as dihydrogen bonding and σ-hole bonding. Besides enabling us to understand the nature of these nonclassical noncovalent interactions of heteroboranes, these results can serve for parametrization of approximate semi-empirical QM methods as well as MM and docking/scoring approaches. Such fast methods are needed for the inclusion of larger portions of protein active sites and their dynamics and for predicting protein–heteroborane binding modes.

The few available crystallographic structures of protein–heteroborane complexes lent themselves to QM calculations that, in some instances, shed light on the nature of heteroborane binding and refined the details of their binding modes. With the rapid development of QM methods, we bear a strong hope that in the near future, the field of computer-aided boron cluster drug design will come of age and offer reliable descriptions of geometries and affinities of protein–heteroborane complexes in the prospective mode.

In the end, it is worth noting that boron is a "drug-compatible element." It is a constituent of several medicaments, both organic (bortezomib against cancer, or tavaborole as an antimycotic) and inorganic (mercaptoundecahydrododecaborate and BSH for boron neutron capture therapy). The extremely low abundance of boron in human bodies (0.00001% lower than that of N, C, O, and H) means that no human enzymes have evolved to process boron-containing compounds in the human body – they are catabolically stable. Similarly, no appreciable binding of boron clusters to human proteins renders them nontoxic to humans. In conclusion, the favorable marriage of properties of boron and the development of QM methods gives promise that new potent and specific heteroborane-based protein ligands will in the near future turn into boron-based drugs.

References

1 Young, D. *Computational Chemistry: A Practical Guide for Applying Techniques to Real World Problems*; John Wiley & Sons Inc (Verlag), **2001**.
2 Yuriev, E.; Agostino, M.; Ramsland, P. A. *J Mol Recognit* **2011**, *24*, 149.
3 Wang, Z.; Sun, H. Y.; Yao, X. J.; Li, D.; Xu, L.; Li, Y. Y.; Tian, S.; Hou, T. *J. Phys Chem Chem Phys* **2016**, *18*, 12964.
4 Fanfrlik, J.; Bronowska, A. K.; Rezac, J.; Prenosil, O.; Konvalinka, J.; Hobza, P. *J Phys Chem B* **2010**, *114*, 12666.
5 Lepsik, M.; Rezac, J.; Kolar, M.; Pecina, A.; Hobza, P.; Fanfrlik, J. *Chempluschem* **2013**, *78*, 921.

6 Pecina, A.; Meier, R.; Fanfrlik, J.; Lepsik, M.; Rezac, J.; Hobza, P.; Baldauf, C. *Chem Commun* **2016**, *52*, 3312.
7 Fanfrlik, J.; Ruiz, F. X.; Kadlcikova, A.; Rezac, J.; Cousido-Siah, A.; Mitschler, A.; Haldar, S.; Lepsik, M.; Kolar, M. H.; Majer, P.; Podjarny, A. D.; Hobza, P. *ACS Chem Biol* **2015**, *10*, 1637.
8 Fanfrlik, J.; Brahmkshatriya, P. S.; Rezac, J.; Jilkova, A.; Horn, M.; Mares, M.; Hobza, P.; Lepsik, M. *J Phys Chem B* **2013**, *117*, 14973.
9 Rezacova P., C. P., Matejicek P., Lepsik M., Pokorna J., Grunner B., Konvalinka J. *In Boron Science – New Technologies and Applications*; CRC Press: New York, **2011**.
10 Lesnikowski, Z. *J. Expert Opin Drug Discov* **2016**, *11*, 569.
11 Scholz, M.; Hey-Hawkins, E. *Chem Rev* **2011**, *111*, 7035.
12 Issa, F.; Kassiou, M.; Rendina, L. M. *Chem Rev* **2011**, *111*, 5701.
13 Lesnikowski, Z. *J. J Med Chem* **2016**, *59*, 7738.
14 Lesnikowski, Z. J. *Collect Czech Chem Commun.* **2007**, *72*, 1646.
15 Tiwari, R.; Mahasenan, K.; Pavlovicz, R.; Li, C. L.; Tjarks, W. *J Chem Inf Model* **2009**, *49*, 1581.
16 Calvaresi, M.; Zerbetto, F. *J Chem Inf Model* **2011**, *51*, 1882.
17 Farras, P.; Juarez-Perez, E. J.; Lepsik, M.; Luque, R.; Nunez, R.; Teixidor, F. *Chem Soc Rev* **2012**, *41*, 3445.
18 Fanfrlik, J.; Lepsik, M.; Horinek, D.; Havlas, Z.; Hobza, P. *Chemphyschem* **2006**, *7*, 1100.
19 Hnyk, D; McKee, M.L. *Boron: The Fifth Element*; Springer International Publishing: Switzerland, **2015**; Vol. *20*.
20 Lepsik, M.; Srnec, M.; Hnyk, D.; Gruner, B.; Plesek, J.; Havlas, Z.; Rulisek, L. *Collect Czech Chem Commun* **2009**, *74*, 1.
21 Lepsik, M.; Srnec, M.; Plesek, J.; Budesinsky, M.; Klepetarova, B.; Hnyk, D.; Gruner, B.; Rulisek, L. *Inorg Chem* **2010**, *49*, 5040.
22 Muetterties, E. L. *Boron Hydride Chemistry*; Academic Press: New York, **1975**.
23 Sivaev, I. B.; Bregadze, V. I.; Sjoberg, S. *Collect Czech Chem Commun* **2002**, *67*, 679.
24 Grimes, R. N. *Carboranes*, 2nd Edition; Academic Press: London, **2011**.
25 Bregadze, V. I. *Chem Rev* **1992**, *92*, 209.
26 Hawthorne, M. F.; Young, D. C.; Wegner, P. A. *J Am Chem Soc* **1965**, *87*, 1818.
27 Buhl, M.; Hnyk, D.; Machacek, J. *Chem-Eur J* **2005**, *11*, 4109.
28 Lipscomb, W. N. *Boron Hydrides* Dover Publications: New York, **1963**.
29 Williams, R. E. *Chem Rev* **1992**, *92*, 177.
30 Chen, Z. F.; King, R. B. *Chem Rev* **2005**, *105*, 3613.
31 Desiraju, G. R.; Steiner, T. *The Weak Hydrogen Bond: In Structural Chemistry and Biology*; Oxford Univeristy Press: Oxford, **1999**.
32 Sedlak, R.; Fanfrlik, J.; Hnyk, D.; Hobza, P.; Lepsik, M. *J Phys Chem A* **2010**, *114*, 11304.
33 Custelcean, R.; Jackson, J. E. *Chem Rev* **2001**, *101*, 1963.
34 Belkova, N. V.; Shubina, E. S.; Epstein, L. M. *Acc Chem Res* **2005**, *38*, 624.
35 Metrangolo, P.; Murray, J. S.; Pilati, T.; Politzer, P.; Resnati, G.; Terraneo, G. *Cryst Growth Des* **2011**, *11*, 4238.
36 Lo, R.; Fanfrlik, J.; Lepsik, M.; Hobza, P. *Phys Chem Chem Phys* **2015**, *17*, 20814.
37 Pecina, A.; Lepsik, M.; Hnyk, D.; Hobza, P.; Fanfrlik, J. *J Phys Chem A* **2015**, *119*, 1388.
38 Fanfrlik, J.; Prada, A.; Padelkova, Z.; Pecina, A.; Machacek, J.; Lepsik, M.; Holub, J.; Ruzicka, A.; Hnyk, D.; Hobza, P. *Angew Chem Int Ed* **2014**, *53*, 10139.
39 Fanfrlik, J.; Holub, J.; Ruzickova, Z.; Rezac, J.; Lane, P.; Wann, D.; Hnyk, D.; Ruzicka, A.; Hobza, P. *Chemphyschem* **2016**, *17*, 3373.
40 Politzer, P.; Murray, J. S.; Clark, T. *Phys Chem Chem Phys* **2010**, *12*, 7748.

41 Grimme, S. *J Comput Chem* **2006**, *27*, 1787.
42 Jurecka, P.; Cerny, J.; Hobza, P.; Salahub, D. R. *J Comput Chem* **2007**, *28*, 555.
43 Rezac, J.; Hobza, P. *Chem Rev* **2016**, *116*, 5038.
44 Sedlak, R.; Janowski, T.; Pitonak, M.; Rezac, J.; Pulay, P.; Hobza, P. *J Chem Theory Comput* **2013**, *9*, 3364.
45 Grimme, S.; Waletzke, M. *Phys Chem Chem Phys* **2000**, *2*, 2075.
46 Ryde, U.; Soderhjelm, P. *Chem Rev* **2016**, *116*, 5520.
47 Antony, J.; Grimme, S. *J Comput Chem* **2012**, *33*, 1730.
48 Senn, H. M.; Thiel, W. *Angew Chem Int Ed* **2009**, *48*, 1198.
49 Stewart, J. J. P. *J Mol Model* **2009**, *15*, 765.
50 Rezac, J.; Fanfrlik, J.; Salahub, D.; Hobza, P. *J Chem Theory Comput* **2009**, *5*, 1749.
51 Rezac, J.; Hobza, P. *Chem Phys Lett* **2011**, *506*, 286.
52 Rezac, J.; Hobza, P. *J Chem Theory Comput* **2012**, *8*, 141.
53 Fanfrlik, J.; Kolar, M.; Kamlar, M.; Hurny, D.; Ruiz, F. X.; Cousido-Siah, A.; Mitschler, A.; Rezac, J.; Munusamy, E.; Lepsik, M.; Matejicek, P.; Vesely, J.; Podjarny, A.; Hobza, P. *ACS Chem Biol* **2013**, *8*, 2484.
54 Elstner, M.; Hobza, P.; Frauenheim, T.; Suhai, S.; Kaxiras, E. *J Chem Phys* **2001**, *114*, 5149.
55 Ponder, J. W.; Case, D. A. *Adv Protein Chem* **2003**, *66*, 27.
56 Wang, J. M.; Wolf, R. M.; Caldwell, J. W.; Kollman, P. A.; Case, D. A. *J Comput Chem* **2004**, *25*, 1157.
57 Casewit, C. J.; Colwell, K. S.; Rappe, A. K. *J Am Chem Soc* **1992**, *114*, 10046.
58 Bayly, C. I.; Cieplak, P.; Cornell, W. D.; Kollman, P. A. *J Phys Chem* **1993**, *97*, 10269.
59 Chevrot, G.; Schurhammer, R.; Wipff, G. *J Phys Chem B* **2006**, *110*, 9488.
60 Karki, K.; Gabel, D.; Roccatano, D. *Inorg Chem* **2012**, *51*, 4894.
61 Shirts, M. R.; Pitera, J. W.; Swope, W. C.; Pande, V. S. *J Chem Phys* **2003**, *119*, 5740.
62 Cramer, C. J.; Truhlar, D. G. *Chem Rev* **1999**, *99*, 2161.
63 Orozco, M.; Luque, F. J. *Chem Rev* **2000**, *100*, 4187.
64 Tomasi, J.; Mennucci, B.; Cammi, R. *Chem Rev* **2005**, *105*, 2999.
65 Klamt, A.; Schuurmann, G. *J Chem Soc-Perkin Trans 2* **1993**, 799.
66 Kolar, M.; Fanfrlik, J.; Lepsik, M.; Forti, F.; Luque, F. J.; Hobza, P. *J Phys Chem B* **2013**, *117*, 5950.
67 Kongsted, J.; Soderhjelm, P.; Ryde, U. *J Comput Aid Mol Des* **2009**, *23*, 395.
68 Lee, T. B.; Mckee, M. L. *Inorg Chem* **2011**, *50*, 11412.
69 Czodrowski, P.; Sotriffer, C. A.; Klebe, G. *J Mol Biol* **2007**, *367*, 1347.
70 Davalos, J. Z.; Gonzalez, J.; Ramos, R.; Hnyk, D.; Holub, J.; Santaballa, J. A.; Canle, M.; Oliva, J. M. *J Phys Chem A* **2014**, *118*, 2788.
71 Farras, P.; Teixidor, F.; Branchadell, V. *Inorg Chem* **2006**, *45*, 7947.
72 Cigler, P.; Kozisek, M.; Rezacova, P.; Brynda, J.; Otwinowski, Z.; Pokorna, J.; Plesek, J.; Gruner, B.; Doleckova-Maresova, L.; Masa, M.; Sedlacek, J.; Bodem, J.; Krausslich, H. G.; Kral, V.; Konvalinka, J. *Proc Natl Acad Sci U S A* **2005**, *102*, 15394.
73 Fanfrlik, J.; Brynda, J.; Rezac, J.; Hobza, P.; Lepsik, M. *J Phys Chem B* **2008**, *112*, 15094.
74 Kozisek, M.; Cigler, P.; Lepsik, M.; Fanfrlik, J.; Rezacova, P.; Brynda, J.; Pokorna, J.; Plesek, J.; Gruner, B.; Grantz Saskova, K.; Vaclavikova, J.; Kral, V.; Konvalinka, J. *J Med Chem* **2008**, *51*, 4839.
75 Rezacova, P.; Pokorna, J.; Brynda, J.; Kozisek, M.; Cigler, P.; Lepsik, M.; Fanfrlik, J.; Rezac, J.; Grantz Saskova, K.; Sieglova, I.; Plesek, J.; Sicha, V.; Gruner, B.; Oberwinkler, H.; Sedlacek, J.; Krausslich, H. G.; Hobza, P.; Kral, V.; Konvalinka, J. *J Med Chem* **2009**, *52*, 7132.

76 Reynolds, R. C.; Campbell, S. R.; Fairchild, R. G.; Kisliuk, R. L.; Micca, P. L.; Queener, S. F.; Riordan, J. M.; Sedwick, W. D.; Waud, W. R.; Leung, A. K. W.; Dixon, R. W.; Suling, W. J.; Borhani, D. W. *J Med Chem* **2007**, *50*, 3283.
77 Brynda J., M. P., Šícha V., Fábry M., Poncová K., Bakardiev M., Grüner B., Cígler P., Řezáčová P. *Angew Chem Int Ed* **2013**, *52*, 13760.
78 Mader, P.; Pecina, A.; Cigler, P.; Lepsik, M.; Sicha, V.; Hobza, P.; Gruner, B.; Fanfrlik, J.; Brynda, J.; Rezacova, P. *Biomed Res Int* **2014**, *8*, 389869.
79 Pecina, A.; Lepsik, M.; Rezac, J.; Brynda, J.; Mader, P.; Rezacova, P.; Hobza, P.; Fanfrlik, J. *J Phys Chem B* **2013**, *117*, 16096.
80 Fujii, S.; Masuno, H.; Taoda, Y.; Kano, A.; Wongmayura, A.; Nakabayashi, M.; Ito, N.; Shimizu, M.; Kawachi, E.; Hirano, T.; Endo, Y.; Tanatani, A.; Kagechika, H. *J Am Chem Soc* **2011**, *133*, 20933.
81 Zangi, R. *J. Phys. Chem. B* **2011**, *115*, 2303.
82 Neumann, W.; Frank, R.; Hey-Hawkins, E. *Dalton Trans* **2015**, *44*, 1748.

Part 2

Boron Compounds in Drug Delivery and Imaging

2.1

Closomers: An Icosahedral Platform for Drug Delivery

Satish S. Jalisatgi

International Institute of Nano and Molecular Medicine, School of Medicine, University of Missouri, Columbia, Missouri, USA

Dedicated to Professor M. Frederick Hawthorne

2.1.1 Introduction

Closomers are discrete spherical molecules with unique structural features. A closomer structure consists of an icosahedral polyhedral borane cluster core from which up to 12 radial arms carrying desired functionalities radiate outward. Closomers are similar to dendrimers in a few aspects but have some major differences. Dendrimers are highly branched, star-shaped molecules with a higher internal volume, while closomers tend to be more compact and rigid structures with a higher symmetry.

The chemical basis for the closomer structure is the discovery of hydroxylation of all 12 B–H vertices of the icosahedral polyhedral borane $[closo\text{-}B_{12}H_{12}]^{2-}$ with 30% hydrogen peroxide to give the *closo*-dodecahydroxy-dodecaborate $[closo\text{-}B_{12}(OH)_{12}]^{2-}$, **1**, in near-quantitative yield [1]. The reactivity of 12 B–OH groups in **1** is similar to a typical hydroxyl group of an alcohol, thus facilitating **1** to become a molecular scaffold where each of the 12 OH groups is conjugated to a substituent radial arm with desired pendant groups at generation zero. As a result, 12-fold ether, ester, carbonate, and carbamate closomer structures with diverse functionalities are currently available [1–10].

Recently, there has been renewed interest in combining the tissue specificity of targeting vectors, such as antibodies or peptides, with both diagnostic and therapeutic agents. Among a number of viable options, coupling targeting vectors to nanoparticle drug delivery platforms while simultaneously delivering a payload of both diagnostic and therapeutic agents has gained interest. Modifications of nanocarriers, such as liposomes, polymeric and metallic nanoparticles, micelles, and dendrimers, are normally used to enhance the delivery of such payloads [11,12]. Some of the modifications include increasing the stability and the ability to respond to local pathological stimuli such as pH and temperature changes. Closomers, with their ability to modify all

12 boron vertices, can be an attractive alternative to deliver a variety of biologically active agents. Additionally, the ability to selectively differentiate one vertex of the icosahedral core from the other 11 vertices allows attachment of multifunctional payloads. Closomer-based monodisperse multifunctional delivery systems are notably distinguishable from other drug delivery platforms. Although most nanoparticle formulations contain mixtures of sizes, shapes, and compositions, closomers are discrete molecules with precise structures. This uniform structure should be advantageous in biological applications because variations in structure might lead to inconsistent activity and side effects, such as immune reactions.

Figure 2.1.1 shows an example of 12-fold closomer species with various closomer components, a fluorescent dansyl payload anchored to the B_{12}^{-2} core through a triethylene glycol linker. The synthesis of such closomers is described further in Scheme 2.1.1.

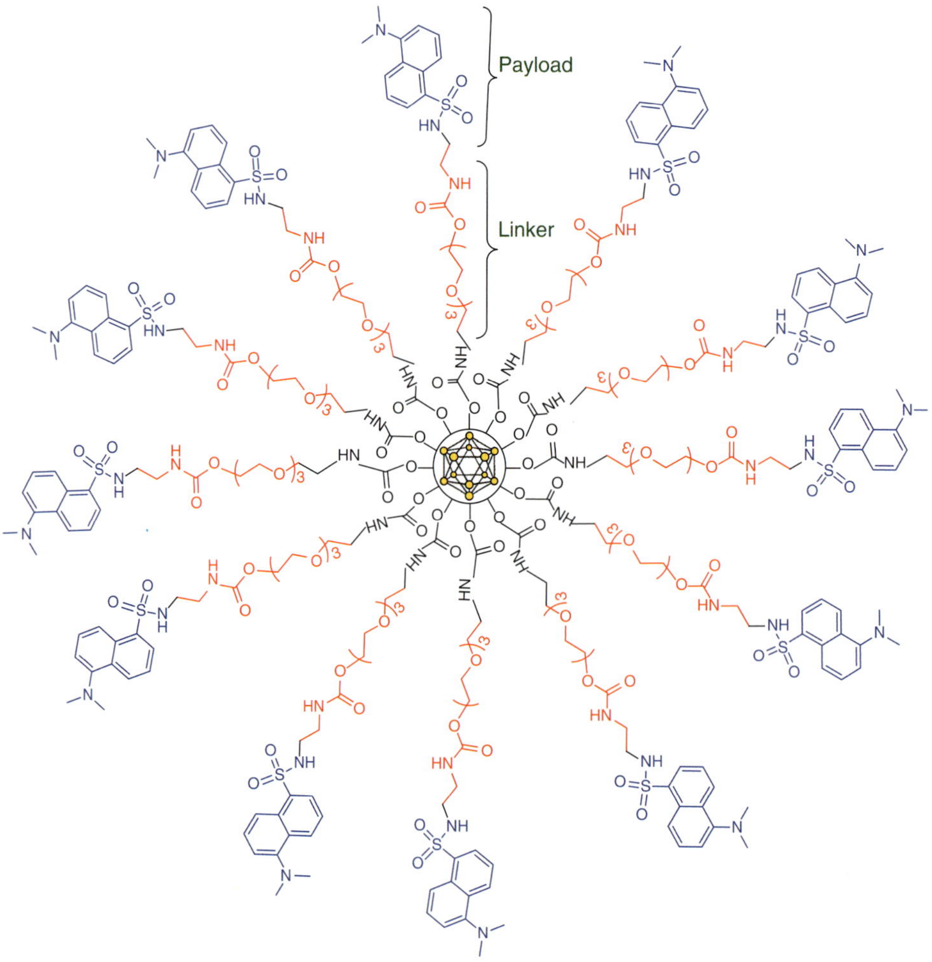

Figure 2.1.1 A representative closomer structure with 12 copies of the dansyl payload.

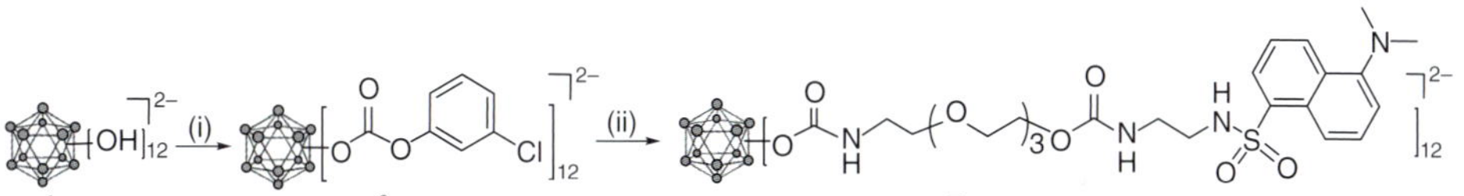

Scheme 2.1.1 Carbamate closomer synthesis. (i) m-Cl-Ph-OC(O)Cl (fivefold excess per vertex), pyridine (fivefold excess), acetonitrile, reflux; (ii) NH_2-$(CH_2)_2(CH_2O)_3OC(O)NH(CH_2)_2NH$-Dansyl, DMF, RT.

2.1.2 Synthesis and Chemistry of [closo-$B_{12}H_{12}$]$^{2-}$

After the discovery of the icosahedral dodecahydro-$closo$-dodecaborate anion [$closo$-$B_{12}H_{12}$]$^{2-}$ by Hawthorne et al. in 1960 [13], it has continued to fascinate researchers with its chemical properties and derivative chemistry. Currently, there are a number of methods to synthesize [$closo$-$B_{12}H_{12}$]$^{2-}$ [14,15], most popular being the pyrolysis of trimethylamine-borane with decaborane(14) in ultrasene at 190 °C [14,16]. Another convenient method for laboratory synthesis is the reaction of sodium borohydride with decaborane(14) in refluxing diglyme [17,18]. The [$closo$-$B_{12}H_{12}$]$^{2-}$ is a very stable anion. It is stable to strong acids and bases, and it does not react with aqueous sodium hydroxide or with 3 N hydrochloric acid at 95 °C. The cesium salt is stable up to 810 °C in an evacuated quartz tube [19]. The remarkable stability is due to the three-dimensional aromatic bonding in cage boron atoms [20,21]. The Na_2-[$closo$-$B_{12}H_{12}$] is nontoxic in rats with an LD_{50} of 7.5 g/kg body weight [22].

The substitution of hydrogen by electrophiles is a common reaction in the BH vertices in polyhedral boranes and carboranes. For example, the halogenation of icosahedral [$closo$-$B_{12}H_{12}$]$^{2-}$ proceeds readily, giving [$closo$-$B_{12}X_{12}$]$^{-2}$ (X: Cl, Br, and I) ions [23]. Other electrophiles have been successfully explored, although the introduction of 12 substituents on the icosahedral surface is rarely observed. The past chemistry of [$closo$-$B_{12}H_{12}$]$^{-2}$ has been extensively covered in a number of reviews. One of the earlier reviews on polyhedral boranes was published in 1968 in the form of a book by Muetterties and Knoth [24]. More recently, a couple of excellent reviews on polyhedral borane describe the synthesis of [$closo$-$B_{12}H_{12}$]$^{-2}$, its derivative chemistry, and its applications in great detail [25–27]. This chapter will focus on the chemistry of its dodecahydroxy derivative, [$closo$-$B_{12}(OH)_{12}$]$^{2-}$.

2.1.3 Hydroxylation of [closo-$B_{12}H_{12}$]$^{2-}$

The initial synthesis of monohydroxy derivative **2** was accomplished by refluxing [$closo$-$B_{12}H_{12}$]$^{2-}$ anion with 2-propanol in acidic condition followed by acid hydrolysis with concentrated hydrobromic acid [28]. Later, a number of approaches toward mono- and dihydroxy derivative were described [27,29–31]. Among these approaches, the reactions of Cs_2[$closo$-$B_{12}H_{12}$] with the concentrated aqueous solutions of H_2SO_4 at elevated

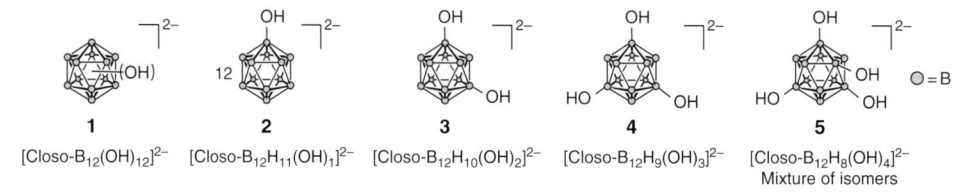

1 | **2** | **3** | **4** | **5**
[Closo-B$_{12}$(OH)$_{12}$]$^{2-}$ | [Closo-B$_{12}$H$_{11}$(OH)$_{1}$]$^{2-}$ | [Closo-B$_{12}$H$_{10}$(OH)$_{2}$]$^{2-}$ | [Closo-B$_{12}$H$_{9}$(OH)$_{3}$]$^{2-}$ | [Closo-B$_{12}$H$_{8}$(OH)$_{4}$]$^{2-}$ Mixture of isomers

Figure 2.1.2 Various hydroxy *closo*-dodecaborate anions.

temperatures and varied periods of time provide the monohydroxy, **2**; 1,7-dihydroxy, **3**; 1,7,9-trihydroxy, **4**; and 1,2,8,10-tetrahydroxy, **5**, derivatives (Figure 2.1.2) [32].

The synthesis of the *closo*-dodecahydroxy [*closo*-B$_{12}$(OH)$_{12}$]^{2} ion, **1**, though, required a stronger reagent. The synthesis of **1** was ultimately achieved by a reaction of Cs$_2$B$_{12}$H$_{12}$ with 30% hydrogen peroxide at the reflux temperature (approximately 105 °C). After several days, the reaction mixture provided dicesium **1** in a 95% yield [1,9]. Purification was achieved by recrystallization with hot water.

2.1.4 Ether and Ester Closomers

The reaction of **1** with the carboxylic acid anhydrides or acyl and aroyl halides in the presence of a tertiary amine provided the corresponding dodecacarboxylate esters in moderate yields [6,8]. The reaction time ranged from one day to more than a week, depending on the reaction conditions and reagents. The esterification reactions were monitored by observing the ^{11}B NMR (nuclear magnetic resonance) spectrum of a sample of the reaction mixture. The completion of the reaction was indicated by the appearance of a single sharp resonance arising from 12 equivalent B-vertices. The products were purified by column chromatography using a suitable closomer salt, which was soluble in the organic solvents after removing excess acylation reagents with a reactive amine resin [6]. Characterization was achieved by ^{1}H and ^{11}B NMR spectra, mass spectrometry, and X-ray diffraction studies. An example of a 12-fold ester closomer, [*closo*-B$_{12}$(OCOCH$_3$)$_{12}$]$^{2-}$, with 12 acetoxy groups is shown in Figure 2.1.3. Similarly, a reaction of **1** with alkyl and aralkyl halides produced 12-fold ether closomers [2,33]. These ether closomers have unique redox properties [33,34], which enable them to exist in 26-electron dianionic, 25-electron radical monoanionic, and 24-electron diamagnetic neutral species. The interesting electron-deficient radical monoanionic species is purple in color, while the diamagnetic neutral species is yellow. These electron-deficient radical anionic and diamagnetic closomers are designated as *hypercloso* clusters. They are partly stabilized by back-bonding of 12 ether substituents to the electron-deficient B$_{12}$ cage [2,35]. One such example, [*hypercloso*-B$_{12}$(OBn)$_{12}$], is shown in Figure 2.1.3. The single crystal X-ray diffraction studies of dodecabenzyl ether systems provide fascinating insight into the bonding characteristic of cage boron atoms in *hypercloso* clusters. The dianionic [*closo*-B$_{12}$(OBn)$_{12}$]$^{2-}$ and radical anionic species [*closo*-B$_{12}$(OBn)$_{12}$]$^{1-}$ have I_h symmetry, while the neutral *hypercloso*-B$_{12}$(OBn)$_{12}$ have a D_3d symmetry due to Jahn–Teller distortion [36]. As a result, the B–B distance in the cage boron atoms of two triangular faces facing each other is longer than the corresponding B–B distance of the dianionic species. In contrast, the B–O bond lengths associated with oxygen atoms attached to these six boron atoms are shortened

(a)   (b)

[closo-$B_{12}(OCOCH_3)_{12}]^{2-}$ hypercloso-$B_{12}(OCH_2Ph)_{12}$

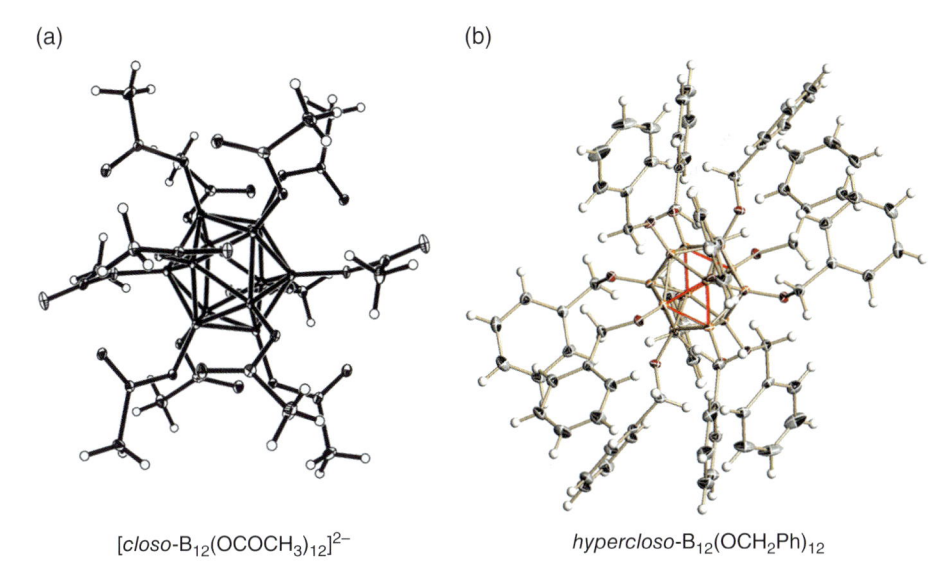

Figure 2.1.3 Solid-state structures of (a) *closo* dodecaacetate ester closomer and (b) *hypercloso* dodecabenzyl ether closomer; the B–B bonds shown in red are elongated due to Jahn–Teller distortion.

compared to those observed in the dianionic species. This can be attributed to the backbonding of the ether oxygen electron to the electron-deficient B_{12} cage. The fascinating redox properties of the ether closomers can be exploited for biochemical purposes [34].

The chemistry of ether closomers was recently expanded by constructing a PAMAM-type dendritic closomer [37]. This monodisperse dendritic closomer was comparable in size, weight, and number of terminal groups to a fourth-generation PAMAM dendrimer [38]. This closomer was able to form a stable complex with doxorubicin, a chemotherapeutic agent, although the encapsulation efficiency was slightly less than that of the conventional G4 PAMAM dendrimer. This can be attributed to the compactness and rigidity of closomers, which result in lower internal volume available for drug encapsulation.

2.1.5 Carbonate and Carbamate Closomers

The use of carboxylic acid ester derivatives as practical closomer radial arms involves several obstacles; most notably, the isolation and purification procedures for dianionic closomer ester products have been achieved in the presence of excess acylation reagents (three- to fivefold excess per vertex). Among the various methods that are available for the synthesis of 12-fold ester closomers, using acid anhydride resulted in virtually no formation of side products, and the 12-fold esters were relatively easy to isolate and purify [6]. Although the synthesis of 12-fold ester closomers with short-chain radial was standardized, there remained a need for a convenient and efficient process to synthesize multigram quantities of closomers with functional radial arms. In this regard, the reaction of **1** with an excess of substituted phenyl chloroformates provided the corresponding carbonates in high yields [5]. One such example is shown in Scheme 2.1.1. The reaction of **1** with m-chlorophenyl chloroformate gave the carbonate **6** in approximately 70% yield. Compared to the ester closomer synthesis using acid chlorides and anhydrides, the synthesis of **6** shown in Scheme 2.1.1 required approximately 24 h of reflux to achieve the 12-fold carbonate closomer in a yield of approximately 70%. Typical purification involves selective crystallization of the product from the excess chloroformate used in the reaction, followed by column chromatography on alumina. Further reaction of the 12-fold carbonate closomer **6** with 60 mole equivalents of a primary amine resulted in 12-fold carbamate closomers in excellent yields.

The reaction of **6** with 60 moles excess of the primary amine containing a fluorescent dansyl group gave the corresponding carbamate **7** in 82% yield. A number of carbamate closomers containing a variety of functional groups, such as –COOR (R: protecting group), –NH$_2$R (R: protecting group), –N$_3$, and –alkyne, were synthesized in yields ranging from 70 to 80% [5]. The ability to synthesize carbamate closomer species on a large scale with a high yield has greatly increased the attractiveness of the closomer platform for drug delivery purposes.

2.1.6 Azido Closomers

Since its discovery in 2001 by Sharpless and coworkers [39], the Cu(I)-catalyzed [3 + 2] cyclo-addition of an alkyne and an azide to generate five-membered 1,2,3-triazole rings, coined "click chemistry," has found immense applications in the design of dendrimers, polymers, and other molecules, with applications in the material sciences and biology [40]. These reactions are both regio- and chemoselective and provide an opportunity to design specific macromolecular structures with a 1,2,3-triazole ring. The triazole ring has the inherent advantage of being stable toward hydrolytic cleavage compared to the peptide linkage [41].

Blending the click chemistry module with the existing closomer chemistry offers immense opportunities for the formation of novel drug molecules and imaging entities. The 12-fold azido ester substituted closomers, which were synthesized using the route shown in Scheme 2.1.2, were further reacted with a functionalized terminal alkyne moiety to generate 12-fold 1,2,3-triazole rings attached to the parent B_{12}^{2-} scaffold. The synthetic route provided a mild, stereospecific, highly efficient, and environmentally friendly procedure for synthesizing highly functional closomer entities [3,5].

This chemistry was exploited in the development of the high-performance multi-nuclear T_1 Gd(III)-based magnetic resonance (MR) contrast agent **11**, shown in Scheme 2.1.2, in which two 12-fold DTTA-Gd(III) linked closomers were synthesized by convergent methods [42]. The functionalization of the closomer B_{12}^{2-} core was achieved by reacting **1** with a chloroacetic anhydride to provide the chloro terminated closomer **8**. This closomer was then reacted with sodium azide to provide the corresponding azide **9** in a good yield. The DTTA fragment **10** was synthesized in a multistep synthesis by *bis*-alkylation of NH$_2$ terminated alkynyl triethylene glycol. Using Sharpless click chemistry, the CuI mediated conjugation of **10** with **9** and the further deprotection of the ^{t}butyl groups followed by Gd complexation provided the final closomer MR contrast agent **11** in a good yield. This high-performance magnetic resonance image (MRI) contrast agent was water-soluble and carried 12 Gd^{3+} ions, which were chelated tightly

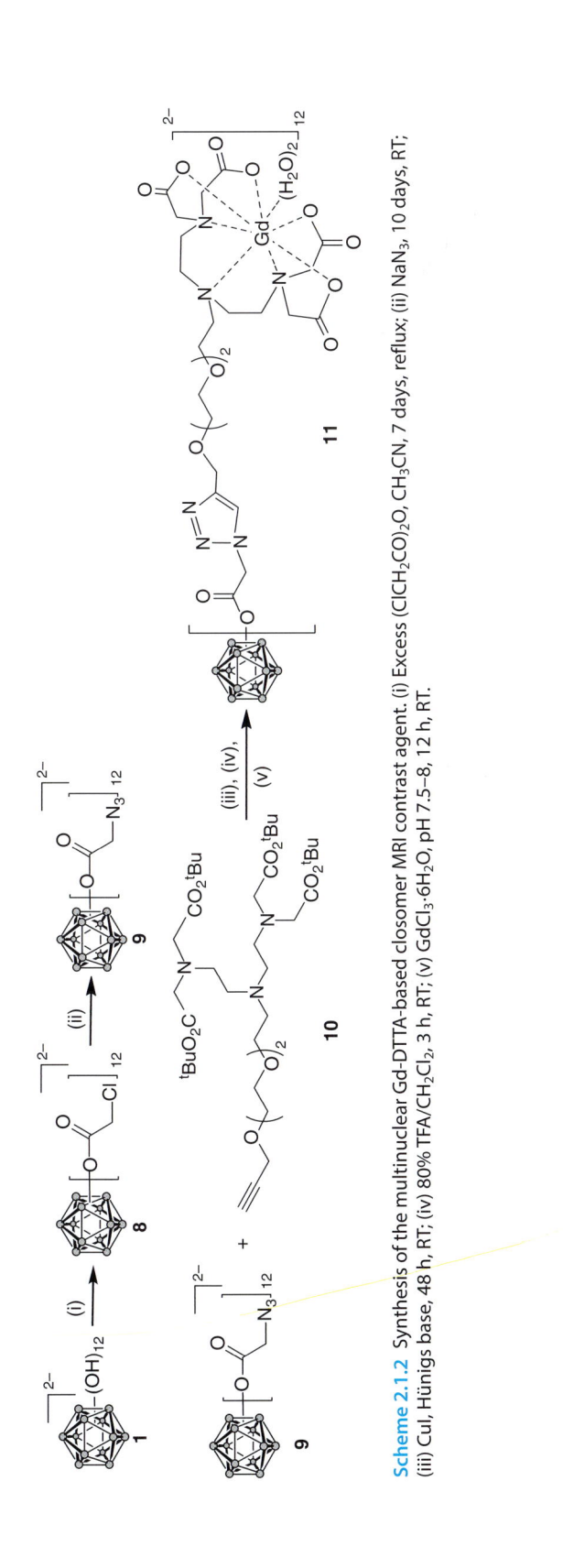

Scheme 2.1.2 Synthesis of the multinuclear Gd-DTTA-based closomer MRI contrast agent. (i) Excess (ClCH$_2$CO)$_2$O, CH$_3$CN, 7 days, reflux; (ii) NaN$_3$, 10 days, RT; (iii) CuI, Hünigs base, 48 h, RT; (iv) 80% TFA/CH$_2$Cl$_2$, 3 h, RT; (v) GdCl$_3$·6H$_2$O, pH 7.5–8, 12 h, RT.

in a sterically confined space. This unique configuration resulted in an exceptionally high per Gd^{3+} relaxivity value of 13.8 mM^{-1}s^{-1} compared to the clinically used contrast agent, Omniscan (4.2 mM^{-1}s^{-1}). The *in vivo* contrast image study of **11** in mice bearing human PC3 prostate cancer xenografts showed an enhanced contrast in tumors and kidneys compared to clinically used Omniscan, even at one-seventh of the safe clinical dose of 0.04 mmol/kg, without any apparent toxicity. Interestingly, the contrast-enhancing effect of **11** persisted for more than one hour. This might be useful for increasing both the signal-to-noise ratio and/or the image resolution during MRI procedures [42]. Later, this approach was adopted to attach 12 copies of Gd^{3+}-DOTA chelate to the B_{12}^{2-} core [43].

2.1.7 Methods of Vertex Differentiation for Multifunctional Closomers

The syntheses of the above-described closomers with 12 identical radial arms (also known as monomodal closomers) have greatly enhanced the understanding of the synthesis, purification, and linker chain optimization. However, the utility of these monomodal closomer species is limited by their inability to carry more than a single type of passenger species. The ability to accommodate discrete numbers of two or more types of passenger species is required to build nanostructures with targeting and diagnostic and/or therapeutic payloads as a part of a single closomer package. Thus, methods for differentiating closomer vertices must be developed to attain the full use of the closomer structure. This can be achieved by differentiating a single vertex of the closomer B_{12}^{2-} core prior to either global hydroxylation or selectively reacting one hydroxyl group in **1**. One such approach is shown in Figure 2.1.4, where a branched linker arm or arms with different reactive end groups can be attached to the bi-vertex-differentiated B_{12}^{2-} core.

2.1.7.1 Vertex Differentiation by Selective Derivatization of [*closo*-$B_{12}(OH)_{12}$]$^{2-}$, 1

This method involves the reaction of **1** with one mole equivalent of alkyl or aralalkyl halide to provide a monoether substituted closomer, such as **14**, as shown in Scheme 2.1.3. The chemistry of ether-based linkages in vertex-differentiated closomers was derived

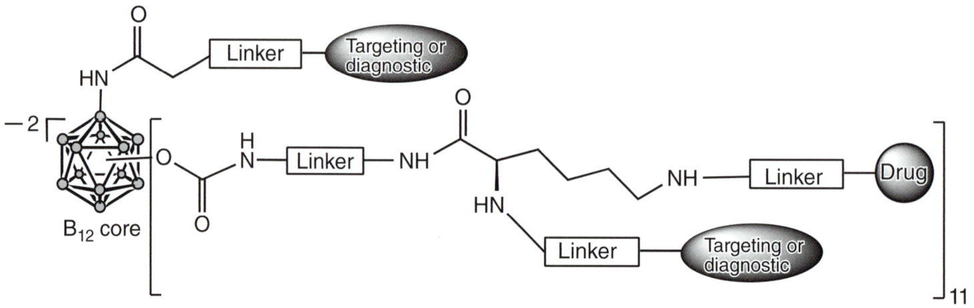

Figure 2.1.4 Trifunctionalized closomer delivery system.

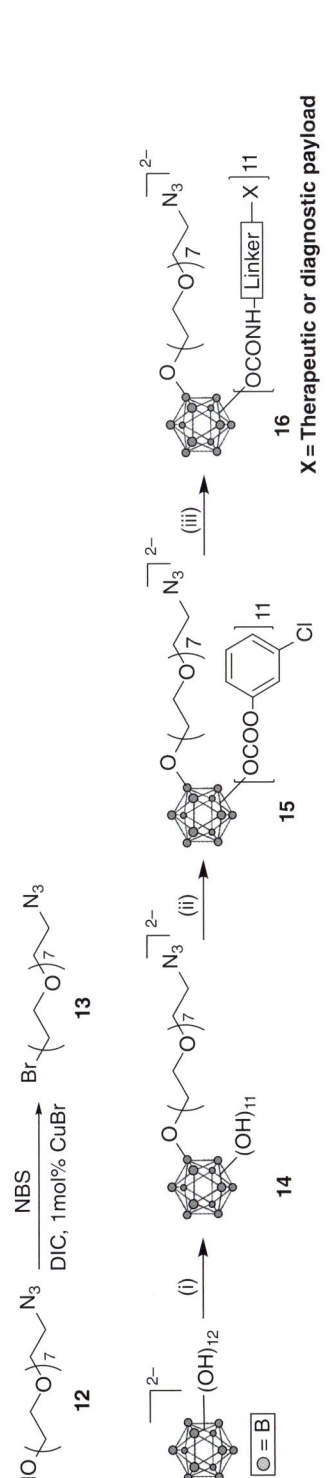

Scheme 2.1.3 Single vertex differentiation based on a monoether functionalized closomer. (i) Large excess of **1**, DIPA, CH$_3$CN; (ii) m-Cl-C$_6$H$_4$-OCOCl, pyridine, CH$_3$CN; (iii) H$_2$N-Linker-X (X: passenger payload).

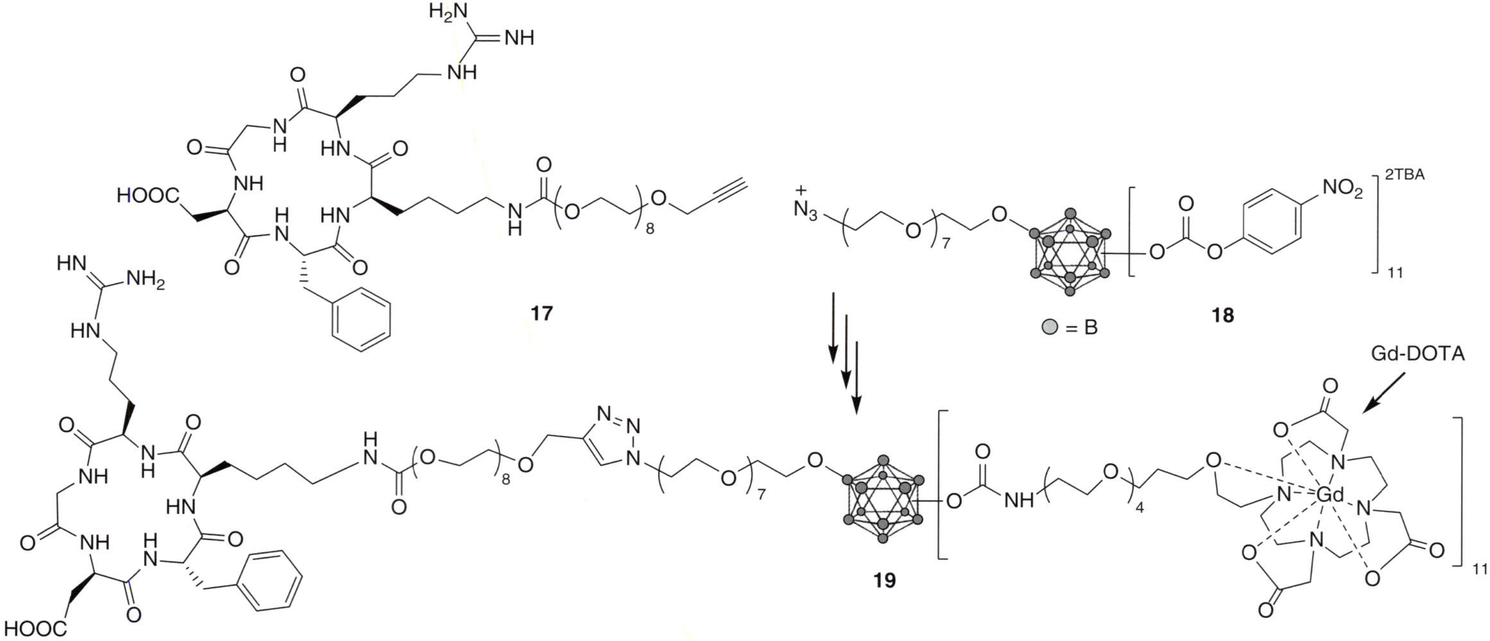

Scheme 2.1.4 Synthesis of a $\alpha_v\beta_3$ integrin targeted closomer MRI contrast agent.

from the well-studied dodecaether chemistry [2,34]. Ether closomers are readily formed by reacting **1** with alkyl, aralkyl halides, and tosylates in the presence of a sterically hindered tertiary amine. Polyethylene glycol (PEG) bromide as a source of alkyl halide provides monoether **14**, which is shown in Scheme 2.1.3, in a moderate yield of 35%. The type of PEG linker that is chosen depends on the desired use of the terminal functionality. Readily available groups include –OH, –NH$_2$, –COOH, and –N$_3$. For example, the commercially available azido octaethylene glycol **12** was treated with N-bromosuccinimide in the presence of *N,N*-diisopropylcarbodiimide and 1% CuBr to provide the corresponding bromo azide **13**. The conjugation of **1** with **13** was carried out using a large excess of **1** to ensure the formation of a monopegelated closomer. After the reaction, the product **14** was purified by size exclusion chromatography. This monopegylated ether **14** could then be converted to a type **16** carbamate via intermediate carbonate **15**. The azido function of the key bifunctional intermediate **15** can be used later in copper(I)-catalyzed 1,3 dipolar cyclization (Sharpless click chemistry) or Staudinger ligation for peptide conjugation, or can be reduced to the free amine for further conjugation to the desired biomolecule [44,45].

Based on this strategy, a targeted version of the Gd^{3+}-chelated closomer was synthesized, and the differentiation of a single vertex in the B$_{12}^{2-}$ cage was achieved by the monoetherification of a single B–OH vertex of [*closo*-B$_{12}$(OH)$_{12}$]$^{2-}$ and the functionalization of the remaining 11 vertices to ester or carbonate groups [42,44]. This differentiated single vertex permitted the attachment of a $\alpha_v\beta_3$ integrin-targeted cRGD peptide to the *closo*-B$_{12}^{2-}$ core via a long PEG linker. The remaining 11 vertices carried 11 Gd^{3+}-DOTA chelates through a PEG linker via carbamate linkages (Scheme 2.1.4). This unique MRI contrast agent, **19**, exhibited a higher relaxivity per Gd than the commercially available small molecule contrast agent Omniscan in PBS at a field strength of 7 tesla. The *in vitro* cell-binding experiments and *in vivo* MRI studies of tumor-bearing mice demonstrated the ability of this closomer-based contrast agent to selectively target $\alpha_v\beta_3$ integrin-expressing cells. Binding studies showed that the Gd content was threefold higher in cells expressing the $\alpha_v\beta_3$ integrin receptors compared to that in cells in which the expression of this receptor was absent. Serial T_1-weighted MRI of mice possessing human PC3 prostate cancer xenografts showed that a higher and more persistent contrast enhancement of tumors was achieved with the closomer contrast agent than that achieved with Omniscan. T_1 relaxivity studies on these MRI contrast agents showed marked improvements in these species. These MRI closomer structures are nearly spherical and are equipped with rigid linker arms terminating with the paramagnetic Gd-DOTA complex. The rigidity of these structures is hypothesized to reduce the rate of rotation, which will, in turn, favor increased relaxivity. The 12 Gd-DOTA chelates at the termini of the linker arms should undergo a metal-bound water exchange at a rate comparable to that of the parent Gd-DOTA complex.

2.1.7.2 Vertex Differentiation by Functionalizing the B$_{12}^{2-}$ Core Prior to the Cage Hydroxylation

Although the monoether derivative of [*closo*-B$_{12}$(OH)$_{12}$]$^{2-}$ shown in Scheme 2.1.3 is useful for the construction of targeted delivery systems, its synthesis and purification are tedious, and the yields are rather low at only 35% after multiple purifications by size-exclusion chromatography steps [45]. Therefore, a more efficient and convenient

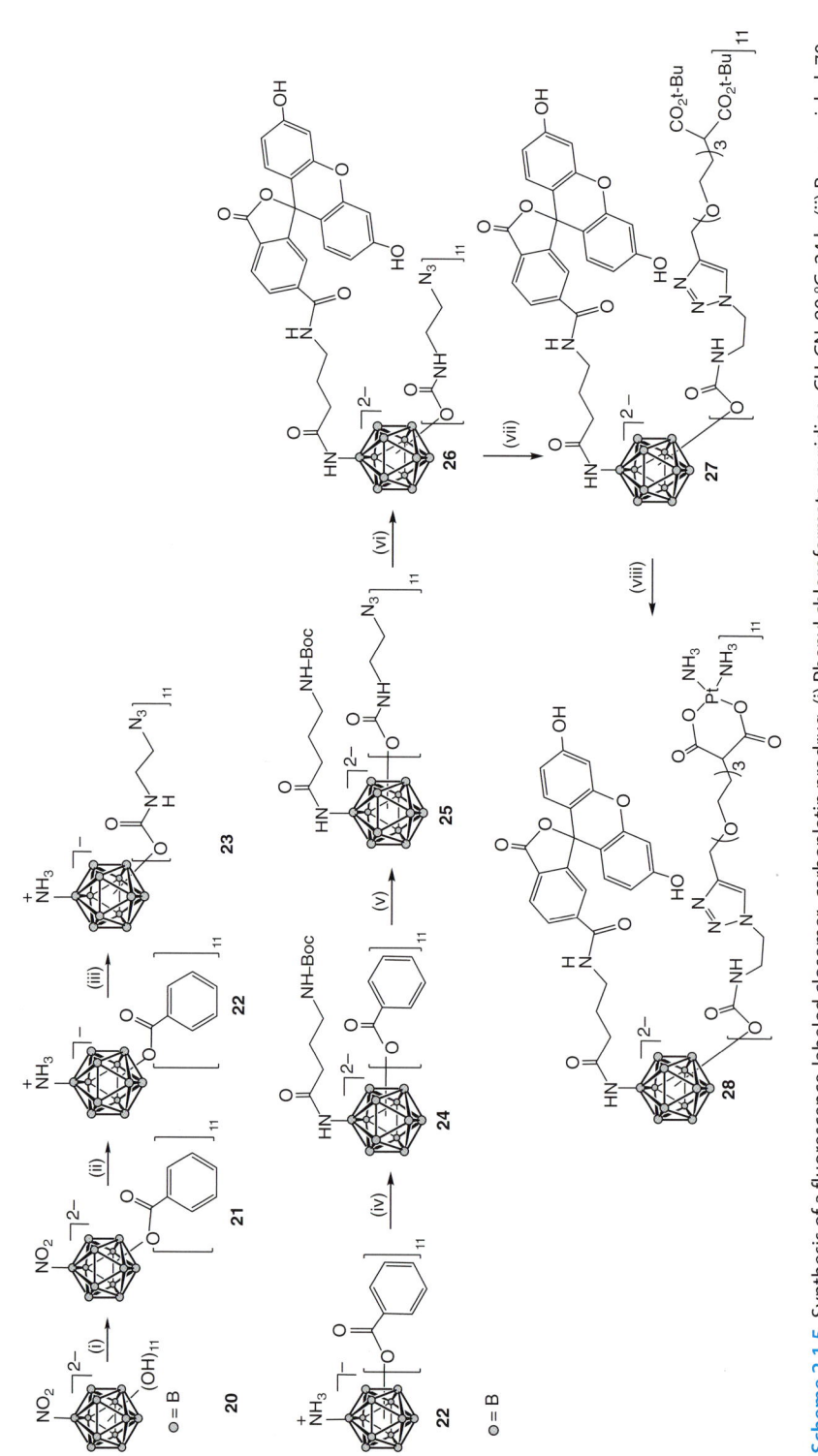

Scheme 2.1.5 Synthesis of a fluorescence-labeled closomer–carboplatin prodrug. (i) Phenyl chloroformate, pyridine, CH₃CN, 80 °C, 24 h; (ii) Raney nickel, 70 bar H₂, MeOH, 40 °C, 24 h; (iii) 2-azidoethanamine, CH₃CN, 40 °C, 72 h; (iv) ᵗBoc-GABA-NHS ester, DBU, CH₃CN, rt, 24 h, 86%; (v) 2-azidoethanamine, CH₃CN, 40 °C, 72 h, 76%; (vi) 80% TFA, 40 min, 98%, 5(6)-carboxyfluorescein succinimidyl ester, Et₃N, CH₂Cl₂, 66%; (vii) CuSO₄×5H₂O, sodium ascorbate, TBTA, MeOH/CH₂Cl₂ rt, 72 h, 52%; (viii) TFA, rt, 16 h, 97%, 1 M NaOH, pH 7, [Pt(NH₃)₂(H₂O)₂](NO₃)₂, H₂O, rt, 16 h, 46%.

method for generating vertex-differentiated closomers was investigated. First, it was necessary to identify a $[closo\text{-}B_{12}H_{11}(X)_1]^{2-}$ derivative capable of withstanding the harsh initial hydrogen peroxide treatment that is required for the conversion of the B–H vertices of $[closo\text{-}B_{12}H_{12}]^{2-}$ to the B–OH vertices of $[closo\text{-}B_{12}(OH)_{12}]^{2-}$.

One of the most promising candidates was the previously reported monosubstituted derivative $[closo\text{-}B_{12}H_{11}NH_3]^-$, whose synthesis was first reported in 1964 [46]. The synthesis of $[closo\text{-}B_{12}H_{11}NH_3]^-$ is typically achieved by the reaction of $[closo\text{-}B_{12}H_{12}]^{2-}$ with hydroxylamine-O-sulfonic acid. Under the conditions necessary for the hydroxylation of the B–H vertices, the amino group of $[closo\text{-}B_{12}H_{11}NH_3]^-$ is oxidized to form the nitro derivative $[closo\text{-}B_{12}(OH)_{11}NO_2]^{2-}$. The nitro group is then easily reduced to the amino form to permit further derivatization. This approach was exploited for the synthesis of a carboplatin–closomer drug delivery system, **28**, for which 11 carboplatin prodrugs were conjugated to 11 vertices of the B_{12}^{2-} core via carbamate linkages, and the remaining B–NH$_3$ vertex was conjugated to a fluorescein molecule via an amide linkage (Scheme 2.1.5). The intranucleus localization of this fluorescein-conjugated platinated closomer in A459 small-cell lung cancer cells was confirmed by fluorescence microscopy. An *in vitro* MTT cytotoxicity assay of a non-fluorescent analog of **28** showed comparable activity to that of carboplatin against a platinum-sensitive A459 lung cancer cell line, but showed a marked enhancement when tested against a platinum-resistant SK-OV-3 ovarian cancer cell line [47].

The vertex differentiation strategy described in Scheme 2.1.5 can also be utilized to develop trifunctional closomer systems that are capable of carrying therapeutic, diagnostic, and cell-targeting functions.

The branched linker shown in Figure 2.1.4 is based on lysine, which has two amino groups with dissimilar reactivities that can be conjugated to different functionalities. Using this approach, the trifunctional closomer **31** was recently synthesized, in which 11 copies of the chemotherapeutic drug chlorambucil were conjugated to the B_{12}^{2-} core along with 11 copies of glucosamine, which targets the GLUT1 receptor and a lone fluorescent sulforhodamine-B moiety (Scheme 2.1.6). Chlorambucil has poor bioavailability, low selectivity for DNA, and a relatively high drug clearance rate. Conjugation to nanocarriers, such as water-soluble closomers, could increase its bioavailability, resulting in an enhanced efficacy.

The synthesis of the trifunctional closomer **31** shown in Scheme 2.1.6 was accomplished by converting the monodifferentiated closomer **14** (shown in Scheme 2.1.3) to the azido ester **29**. It has 11 azido groups that are available for a click reaction and a lone sulforhodamine-B fluorophore for a diagnostic function. This azido ester closomer was reacted with an excess of the lysine-based branched linker **30** containing an alkyne end group. The lysine-based branched linker **30** containing both the chlorambucil drug and the glucosamine group was synthesized off-board using standard organic transformations and purified by standard separation techniques. The click reaction of the azido ester closomer **29** with linker **30** provided high yields of closomer **31**. The purification of closomer **31** was achieved by size exclusion column chromatography on a lipophilic Sephadex LH-20 column (GE Healthcare Life Sciences) and dialysis using a 3000 MW cutoff membrane. The purity was confirmed by size exclusion HPLC (high-performance liquid chromatography). This trimodal closomer **31** has a mass of 12250 Daltons, is monodisperse, and is one of the first examples of a closomer system in which all three therapeutic, diagnostic, and targeting functions assembled on the B_{12}^{2-} core in a precise manner [48].

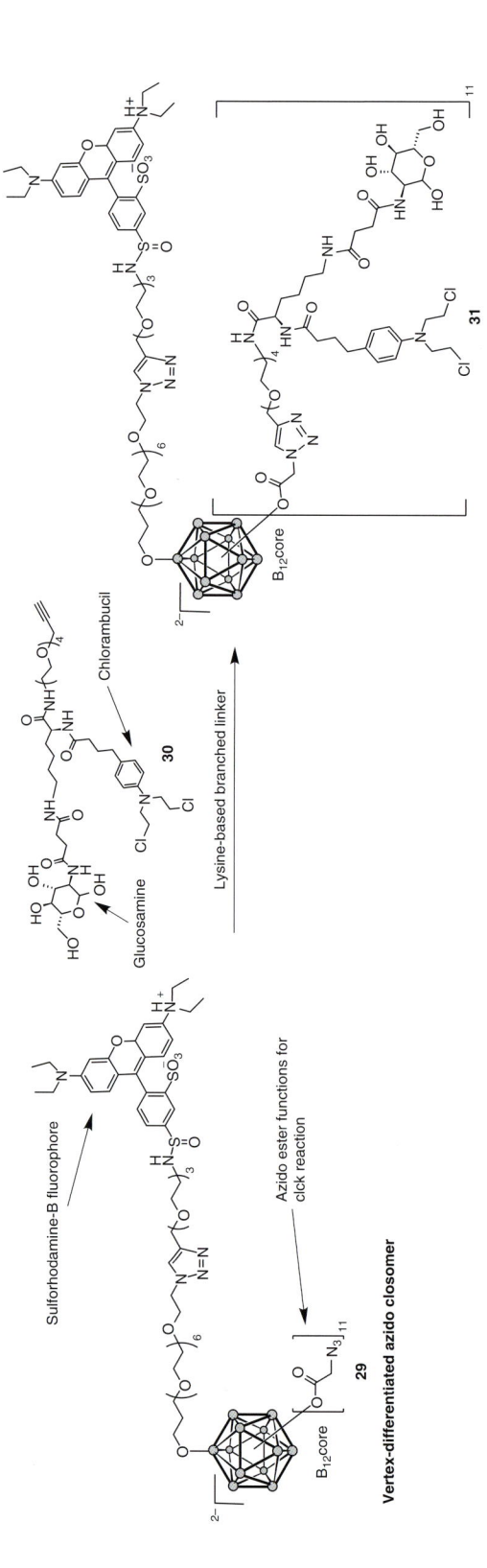

Scheme 2.1.6 Convergent synthesis of the trifunctional closomer drug delivery system.

The cytotoxic effect of the trifunctional chlorambucil–closomer **31** (CDDS-**31**) was tested in Jurkat cells, a human T-cell leukemia cell line, using the MTT assay. CDDS-**31** significantly inhibited the proliferation of the Jurkat cells by more than tenfold of that observed in free CLB. Several research groups have shown that increasing the number of glucosamine units per molecule increases the uptake of the therapeutic agent to the tumor tissue [49]. The closomer delivery system is uniquely suited to carry multiple copies of glucosamine to increase the cellular uptake of payloads.

2.1.8 Conclusions

Closomers are unique monodisperse nanomolecular species that are capable of carrying a number of payloads. The broad-ranging derivative chemistry of 12 hydroxyl groups of $[closo\text{-}B_{12}(OH)_{12}]^{2-}$ that results in 12-fold ether, ester, carbonate, and carbamate derivatives is useful in the closomer synthesis aimed toward particular application. Other notable advantages include the compact and globular structure, a precise synthetic approach providing a very high structural control, a spherical shape with variable polyvalences, a high organic and aqueous solubility as desired, and a noncrystalline nature. These advantages will translate into a higher loading of payload per closomer unit with a high degree of bioavailability, increased therapeutic efficacy, and reduced clearance through the reticular endothelial system.

References

1. Bayer, M.J., and Hawthorne, M.F. 2004. An improved method for the synthesis of $[closo\text{-}B_{12}(OH)_{12}]^{-2}$. *Inorg. Chem.*, vol. 43, no. 6, pp. 2018–2020.
2. Farha, O.K., Julius, R.L., Lee, M.W., Huertas, R.E., Knobler, C.B., and Hawthorne, M.F. 2005. Synthesis of stable dodecaalkoxy derivatives of hypercloso-$B_{12}H_{12}$. *J. Am. Chem. Soc.*, vol. 127, no. 51, pp. 18243–18251.
3. Goswami, L.N., Chakravarty, S., Lee, M.W., Jr., Jalisatgi, S.S., and Hawthorne, M.F. 2011. Extensions of the icosahedral closomer structure by using azide-alkyne click reactions. *Angew. Chem. Int. Ed. Engl.*, vol. 50, no. 20, pp. 4689–4691.
4. Hawthorne, M.F. 2003. New discoveries at the interface of boron and carbon chemistries. *Pure Appl. Chem.*, vol. 75, no. 9, pp. 1157–1164.
5. Jalisatgi, S.S., Kulkarni, V.S., Tang, B., Houston, Z.H., Lee, M.W., and Hawthorne, M.F. 2011. A convenient route to diversely substituted icosahedral closomer nanoscaffolds. *J. Am. Chem. Soc.*, vol. 133, pp. 12382–12385.
6. Li, T., Jalisatgi, S.S., Bayer, M.J., Maderna, A., Khan, S.I., and Hawthorne, M.F. 2005. Organic syntheses on an icosahedral borane surface: closomer structures with twelvefold functionality. *J. Am. Chem. Soc.*, vol. 127, no. 50, pp. 17832–17841.
7. Ma, L., Hamdi, J., Wong, F., and Hawthorne, M.F. 2006. Closomers of high boron content: synthesis, characterization, and potential application as unimolecular nanoparticle delivery vehicles for boron neutron capture therapy. *Inorg. Chem.*, vol. 45, no. 1, pp. 278–285.
8. Maderna, A., Knobler, C.B., and Hawthorne, M.F. 2001. Twelvefold functionalization of an icosahedral surface by total esterification of $[B_{12}(OH)_{12}]^{2-}$: 12(12)-closomers. *Angew. Chem. Int. Ed. Eng.*, vol. 40, no. 9, pp. 1662–1664.

9 Peymann, T., Knobler, C.B., Khan, S.I., and Hawthorne, M.F. 2001b. Dodecahydroxy-closo-dodecaborate(2−). *J. Am. Chem. Soc.*, vol. 123, no. 10, pp. 2182–2185.

10 Thomas, J., and Hawthorne, M.F. 2001. Dodeca(carboranyl)-substituted closomers: toward unimolecular nanoparticles as delivery vehicles for BNCT. *Chem. Commun.*, no. 18, pp. 1884–1885.

11 Fernandez-Fernandez, A., Manchanda, R., and McGoron, A.J. 2011. Theranostic applications of nanomaterials in cancer: drug delivery, image-guided therapy, and multifunctional platforms. *Appl. Biochem Biotech.*, vol. 165, nos. 7–8, pp. 1628–1651.

12 Parat, A., Bordeianu, C., Dib, H., Garofalo, A., Walter, A., Bégin-Colin, S., and Felder-Flesch, D. 2015. Dendrimer–nanoparticle conjugates in nanomedicine. *Nanomedicine*, vol. 10, no. 6, pp. 977–992.

13 Pitochelli, A.R., and Hawthorne, M.F. 1960. Isolation of the icosahedral $B_{12}H_{12}^{2-}$ ion. *J. Am. Chem. Soc.*, vol. 82, pp. 3228–3230.

14 Miller, H.C., Miller, N.E., and Muetterties, E.L. 1963. Synthesis of polyhedral boranes. *J. Am. Chem. Soc.*, vol. 85, no. 23, pp. 3885–3886.

15 Miller, H.C., Miller, N.E., and Muetterties, E.L. 1964. Chemistry of boranes. *XX. Syntheses of polyhedral boranes. Inorg. Chem.*, vol. 3, no. 10, pp. 1456–1463.

16 Miller, H.C., Muetterties, E.L., Boone, J.L., Garrett, P., and Hawthorne, M.F. 1967. Borane anions. In E.L. Muetterties (ed.), Inorganic Syntheses, vol. 10. John Wiley & Sons, New York, pp. 81–91.

17 Adams, R.M., Siedle, A.R., and Grant, J. 1964. Convenient preparation of the dodecahydrododecaborate ion. *Inorg. Chem.*, vol. 3, no. 3, p. 461.

18 Kuznetsov, N.T., and Klimchuk, G.S. 1971. Mixed dodecahydro-closo-dodecaborate chlorides of rubidium and cesium. *Russ. J. Inorg. Chem.*, vol. 16, no. 4, p. 645.

19 Muetterties, E.L., Balthis, J.H., Chia, Y.T., Knoth, W.H., and Miller, H.C. 1964. Chemistry of boranes. VIII. Salts and acids of $B_{10}H_{10}$-2 and $B_{12}H_{12}^{-2}$. *Inorg. Chem.*, vol. 3, no. 3, pp. 444–451.

20 Aihara, J. 1978. Three-dimensional aromaticity of polyhedral boranes. *J. Am. Chem. Soc.*, vol. 100, no. 11, pp. 3339–3342.

21 King, R.B. 2001. Three-dimensional aromaticity in polyhedral boranes and related molecules. *Chem. Rev.*, vol. 101, no. 5, pp. 1119–1152.

22 Sweet, W.H., Soloway, A.H., and Wright, R.L. 1962. Evaluation of boron compounds for use in neutron capture therapy of brain tumors. II. Studies in man. *J. Pharmacol. Exp. Thernn*, vol. 137, pp. 263–266.

23 Knoth, W.H., Miller, H.C., Sauer, J.C., Balthis, J.H., Chia, Y.T., and Muetterties, E.L. 1964. Chemistry of boranes. IX. Halogenation of $B_{10}H_{10}^{-2}$ and $B_{12}H_{12}^{-2}$. *Inorg. Chem.*, vol. 3, no. 2, pp. 159–167.

24 Muetterties, E.L., and Knoth, W.H. 1968, Polyhedral boranes. Marcel Dekker, New York.

25 Bregadze, V., and Sivaev, I. 2011. Polyhedral boron compounds for BNCT. In: Boron Science. CRC Press, Boca Raton, FL, pp. 181–208.

26 Sivaev, I.B., Bregadze, V.I., and Kuznetsov, N.T. 2002. Derivatives of the closododecaborate anion and their application in medicine. *Russ. Chem. Bull.*, vol. 51, no. 8, pp. 1362–1374.

27 Sivaev, I.B., Bregadze, V.I., and Sjöberg, S. 2002. Chemistry of closo-dodecaborate anion $[B_{12}H_{12}]^{2-}$: a review. *Collect. Czech. Chem. Commun.*, vol. 67, no. 6, pp. 679–727.

28. Knoth, W.H., Sauer, J.C., England, D.C., Hertler, W.R., and Muetterties, E.L. 1964. Chemistry of boranes. XIX.1 Derivative chemistry of $B_{10}H_{10}^{-2}$ and $B_{12}H_{12}^{-2}$. *J. Am. Chem. Soc.*, vol. 86, no. 19, pp. 3973–3983.
29. Antsyshkina, A.S., Sadikov, G.G., Gorobinskii, L.B., Chernyavskii, A.S., Solntsev, K.A., Sergienko, V.S., and Kuznetsov, N.T. 2001. Random disordering in hydroxo derivatives of closo-$[B_{12}H_{12}]^{2-}$ anion: crystal structures of $Cs_2[B_{12}H_{11}(OH)]\cdot CH_3C(O)OH$ and $(PPh_4)_2[B_{12}H_{11}(OH)]\cdot xCH_3C(O)OH\cdot yH_2O$. *Russ. J. Inorg. Chem.*, vol. 46, pp. 1323–1332.
30. Bechtold, R., and Kaczmarczyk, A. 1974. Coupled products from low-temperature decomposition of hydronium dodecahydrododecaborate(2−). *J. Am. Chem. Soc.*, vol. 96, no. 18, pp. 5953–5954.
31. Krause, U., and Preetz, W. 1995. Darstellung und spektroskopische Charakterisierung von Carboxylatododecaboraten. *Z. Anorg. Allg. Chem.*, vol. 621, no. 4, pp. 516–524.
32. Peymann, T., Knobler, C.B., and Hawthorne, M.F. 2000. A study of the sequential acid-catalyzed hydroxylation of dodecahydro-closo-dodecaborate(2−). *Inorg. Chem.*, vol. 39, no. 6, pp. 1163–1170.
33. Peymann, T., Knobler, C.B., Khan, S.I., and Hawthorne, M.F. 2001a. Dodeca(benzyloxy) dodecaborane, $B_{12}(OCH_2Ph)_{12}$: a stable derivative of hypercloso-$B_{12}H_{12}$. *Angew. Chem. Int. Ed.*, vol. 40, no. 9, pp. 1664–1667.
34. Lee, M.W., Farha, O.K., Hawthorne, M.F., and Hansch, C.H. 2007. Alkoxy derivatives of dodecaborate: discrete nanomolecular ions with tunable pseudometallic properties. *Angew. Chem. Int. Ed. Engl.*, vol. 46, no. 17, pp. 3018–3022.
35. McKee, M.L., Wang, Z.-X., and Schleyer, P.v.R. 2000. Ab initio study of the hypercloso boron hydrides BnHn and BnHn−: exceptional stability of neutral $B_{13}H_{13}$. *J. Am. Chem. Soc.*, vol. 122, no. 19, pp. 4781–4793.
36. Fujimori, M., and Kimura, K. 1997. Ground and excited states of an icosahedral $B_{12}H_{12}$ cluster simulating the B_{12} cluster in β-rhombohedral boron. *J. Solid State Chem.*, vol. 133, no. 1, pp. 178–181.
37. Pushechnikov, A., Jalisatgi, S.S., and Hawthorne, M.F. 2013. Dendritic closomers: novel spherical hybrid dendrimers. *Chem. Commun. (Cambridge, UK)*, vol. 49, no. 34, pp. 3579–3581.
38. Tomalia, D.A., Naylor, A.M., and Goddard, W.A. 1990. Starburst dendrimers: molecular-level control of size, shape, surface chemistry, topology, and flexibility from atoms to macroscopic matter. *Angew. Chem. Int. Ed. Engl.*, vol. 29, no. 2, pp. 138–175.
39. Rostovtsev, V.V., Green, L.G., Fokin, V.V., and Sharpless, K.B. 2002. A stepwise Huisgen cycloaddition process: copper(I)-catalyzed regioselective "ligation" of azides and terminal alkynes. *Angew. Chem. Int. Ed.*, vol. 41, no. 14, pp. 2596–2599.
40. Moses, J.E., and Moorhouse, A.D. 2007. The growing applications of click chemistry. *Chem. Soc. Rev.*, vol. 36, no. 8, pp. 1249–1262.
41. Appendino, G., Bacchiega, S., Minassi, A., Cascio, M.G., De Petrocellis, L., and Di Marzo, V. 2007. The 1,2,3-triazole ring as a peptido- and olefinomimetic element: discovery of click vanilloids and cannabinoids. *Angew. Chem. Int. Ed. Eng.*, vol. 46, no. 48, pp. 9312–9315.
42. Goswami, L.N., Ma, L., Chakravarty, S., Cai, Q., Jalisatgi, S.S., and Hawthorne, M.F. 2013. Discrete nanomolecular polyhedral borane scaffold supporting multiple gadolinium(III) complexes as a high performance MRI contrast agent. *Inorg. Chem.*, vol. 52, no. 4, pp. 1694–1700.

43 Goswami, L.N., Ma, L., Kueffer, P.J., Jalisatgi, S.S., and Hawthorne, M.F. 2013. Synthesis and relaxivity studies of a DOTA-based nanomolecular chelator assembly supported by an icosahedral closo-B_{12}^{2-}-core for MRI: a click chemistry approach. *Molecules*, vol. 18, pp. 9034–9048.

44 Goswami, L.N., Houston, Z.H., Sarma, S.J., Jalisatgi, S.S., and Hawthorne, M.F. 2013. Efficient synthesis of diverse heterobifunctionalized clickable oligo(ethylene glycol) linkers: potential applications in bioconjugation and targeted drug delivery. *Org. Biomol. Chem.*, vol. 11, pp. 1116–1126.

45 Goswami, L.N., Houston, Z.H., Sarma, S.J., Li, H., Jalisatgi, S.S., and Hawthorne, M.F. 2012. Synthesis of vertex-differentiated icosahedral closo-boranes: polyfunctional scaffolds for targeted drug delivery. *J. Org. Chem.*, vol. 77, no. 24, pp. 11333–11338.

46 Hertler, W.R., and Raasch, M.S. 1964. Chemistry of boranes. XIV. Amination of $B_{10}H_{10}^{-2}$ and $B_{12}H_{12}^{-2}$ with hydroxylamine-O-sulfonic acid. *J. Am. Chem. Soc.*, vol. 86, no. 18, pp. 3661–3668.

47 Bondarev, O., Khan, A.A., Tu, X., Sevryugina, Y.V., Jalisatgi, S.S., and Hawthorne, M.F. 2013. Synthesis of [closo-$B_{12}(OH)_{11}NH3$]$^-$: a new heterobifunctional dodecaborane scaffold for drug delivery applications. *J. Am. Chem. Soc.*, vol. 135, no. 35, pp. 13204–13211.

48 Sarma, S.J., Khan, A.A., Goswami, L.N., Jalisatgi, S.S., and Hawthorne, M.F. 2016. A trimodal closomer drug-delivery system tailored with tracing and targeting capabilities. *Chem. Eur. J.*, vol. 22, no. 36, pp. 12715–12723.

49 Korotcov, A., Ye, Y., Chen, Y., Zhang, F., Huang, S., Lin, S., Sridhar, R., Achilefu, S., and Wang, P. 2012. Glucosamine-linked near-infrared fluorescent probes for imaging of solid tumor xenografts. *Mol. Imaging Biol.*, vol. 14, no. 4, pp. 443–451.

2.2

Cobaltabisdicarbollide-based Synthetic Vesicles: From Biological Interaction to *in vivo* Imaging

Clara Viñas Teixidor,[1] Francesc Teixidor,[1] and Adrian J. Harwood[2]

[1] Institut de Ciència de Materials de Barcelona (ICMAB-CSIC), Campus UAB, Bellaterra, Spain
[2] School of Biosciences and Neuroscience and Mental Health Research Institute, Cardiff University, Cardiff, UK

2.2.1 Introduction

Boron-containing compounds are rarely seen in nature, but they can be synthesized in the laboratory to create complex three-dimensional structures of multiple boron atoms. These large boron hydrides, or boranes, can form polyhedral boron clusters, and among some of the most elaborate of these species are the metallabisdicarbollides, discovered by Hawthorne in 1965 [1]. In these, a transition metal, such as iron, nickel, or cobalt, is sandwiched between two anionic η^5-carboranyl ligands ($[C_2B_9H_{11}]^{2-}$) to produce molecules with a net negative charge dispersed over the whole molecule (Figure 2.2.1a). Interestingly, their weakly polarized B–H and C–H bonds promote non-electrostatic intermolecular interactions, making them simultaneously hydrophobic and hydrophilic, and hence soluble in both water and oils. Recently, cobaltabisdicarbollide $[3,3'\text{-}Co(1,2\text{-}C_2B_9H_{11})_2]^-$, known as COSAN, has been shown to form monolayer nanovesicles and membrane-like structures in aqueous solution [2]. Here, we discuss how these artificial membranes form, what happens when they meet the lipid bilayer membranes commonly found in biology, and the consequences of these interactions for living cells.

Biological membranes are a defining feature of cells, and they create the barriers between intracellular compartments and the external environment. They form by the self-assembly of lipid molecules into bilayers with their hydrophilic head groups facing the aqueous environment and a water-free interior through interaction of their hydrophobic tails (Figure 2.2.1b) [3]. Biological membranes contain a diversity of different lipid components, which increase membrane complexity and modify the physicochemical membrane properties, such as membrane thickness, lateral diffusion rate, and curvature [4,5]. As bio-membranes are effective barriers to movement of most molecules and ions, passage across them is generally controlled by transport and channel proteins, or by resculpting membranes via vesicle intermediates. As a consequence, a wide range of the proteins control molecular flux across bio-membranes, resulting in an extensive capacity for selective exchange between different cell compartments or the external environment.

Boron-Based Compounds: Potential and Emerging Applications in Medicine, First Edition.
Edited by Evamarie Hey-Hawkins and Clara Viñas Teixidor.
© 2018 John Wiley & Sons Ltd. Published 2018 by John Wiley & Sons Ltd.

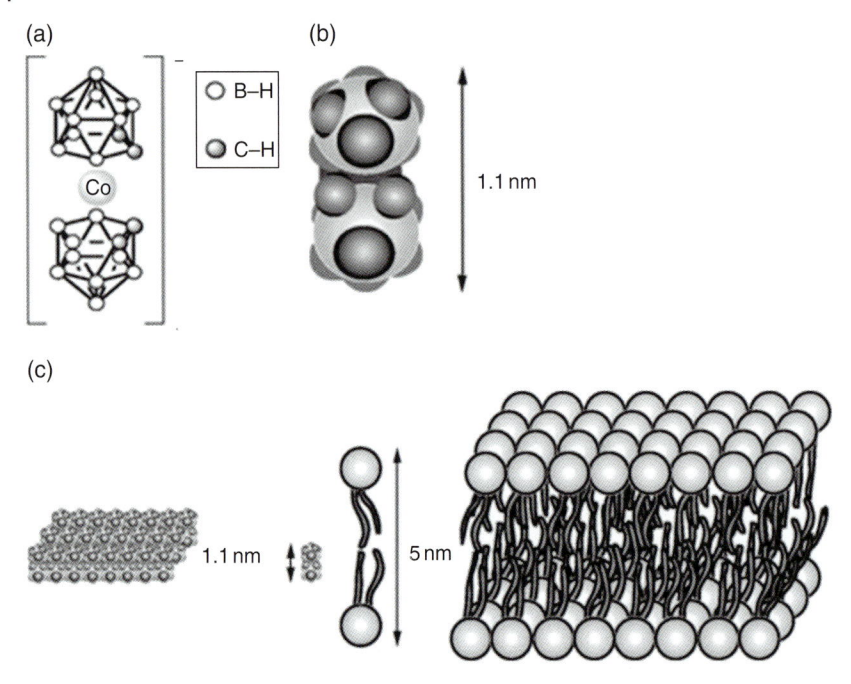

Figure 2.2.1 (a) Chemical structure of cobaltabisdicarbollide [3,3′-Co(1,2-$C_2B_9H_{11}$)$_2$] COSAN. (b) Scale diagram of COSAN. (c) A schematic lipid bilayer to show relative differences in membrane thickness.

These barrier properties create a problem for delivery of drugs and other biomolecules into cells. Artificial membranes and bilayer organic vesicles, known as liposomes, can be created from synthetic lipid molecules of similar amphiphilic structure, and are one means to promote drug delivery. Alternatively, bio-membranes can be breached by disruption with surfactants, which again possess similar amphiphilic properties, creating nonselective biocides. However, neither of these approaches is without problems. In this chapter, we describe the effects of the alternative membrane system formed from COSAN and its derivatives, examining the basis for its ability to form membranes, its unusual physicochemical properties, and its interactions with bio-membranes and living cells. These novel properties reveal an unexpected new biology at the dynamic interface between inorganic, synthetic membranes and naturally occurring biological membranes.

2.2.2 A Synthetic Membrane System

Pioneering studies of Bauduin and coworkers [2] found that H[COSAN] forms monolayer vesicles of 40–50 nm diameter and a constant wall thickness of 1.16 nm in the µM–18.6 mM range in water. Surface second harmonic generation (SHG) and surface tension methods lead to the same conclusion about the molecular orientation of [COSAN]$^-$ anions at the water surface: COSAN forms monolayer vesicles with the COSAN orthogonally aligned to the plane of the vesicle membranes. In addition, small COSAN micelles of approximately 14 molecules can form at a much higher concentration of 18.6 mM [2,6]. In comparison to lipid bilayers, COSAN vesicles are about half the size of the smallest liposomes, and their membranes are approximately 20% of the thickness (Figure 2.2.1b). These studies suggest that COSAN molecules align orthogonally to the plane of the membrane, and must be held together by a balance between electrostatic and hydrophobic forces. The mechanism of self-assembly appears to be very different from those of lipid bilayers. In general, self-assembly of this nature is likely to arise through an enthalpy-driven surface effect. One suggestion is that this arises by formation of B−H···H−C dihydrogen bonds [7]. However, a recent study has shown that the $[B_{21}H_{18}]^-$ anion that contains no C−H bonds also forms vesicles [8]. A detailed mechanism is still to be established, but appears to be related to molecular shape and size, which may exert their effects via changes to energetics of the local water molecules.

2.2.3 Crossing Lipid Bilayers

The interactions of COSAN with lipid bilayer membranes were initially investigated in cell-free systems using synthetic membranes and monitoring their ability to interact and transit from one side of the membrane to the other (Figure 2.2.2a) [9]. Application of COSAN to one side of a membrane formed from the synthetic lipid 1,2-dioleoyl-*sn*-glycero-3-phosphocholine (DOPC) resulted in a steady negative ionic current across the membrane (Figure 2.2.2b). This occurs without application of a transmembrane voltage, indicating that COSAN directly crosses the membrane without a driving force. As membrane electrical capacitance stays constant throughout the measurements, the membrane remains intact and COSAN molecules pass through it without creating pores or membrane disruption. Neither lipid composition nor membrane electrostatic properties are major factors determining COSAN transport rate, and no significant differences were seen for model membranes of neutral lipid composition that mimic either prokaryotic (1,2-diphytanoyl-*sn*-glycero-3-phosphocholine [DPhPC]) or eukaryotic (DOPC) cell membranes. These electrophysiological measurements fit with the results seen for direct monitoring of COSAN translocation by inductively coupled plasma mass spectrometry (ICP-MS; Figure 2.2.2c). As an anion, COSAN would not be expected to directly travel through a lipid bilayer due to its high Born energy, calculated as ~80 kJ/mol, or more than 30 kT [9], and it seems that its unusual combination of hydrophilic and lipophilic properties overcomes the energy barrier presented by lipid bilayers, enabling it to cross directly through lipid bilayer membranes. COSAN's lipid bilayer transiting property occurs with zero-order kinetics, where rate is independent of the initial concentration. This suggests a "tunneling effect," where once engaged at the aqueous membrane interface, COSAN molecules translocate through the lipid phase without resistance or pore formation to exit at the receiving interface.

The interactions of COSAN with lipid bilayer membranes can be directly visualized by cryo–transmission electron microscopy (cryoTEM) studies of mixtures of monolayer COSAN nanovesicles and bilayer liposomes in solution [9]. In mixtures of H[COSAN]:liposome 1:4 (v:v), an equivalent COSAN concentration of 4 mM, COSAN vesicles fuse with the liposome membrane (Figure 2.2.3a). At lower liposome ratios of 1:3 (v:v), two or more liposome units often become connected via COSAN vesicles. At the interface between COSAN vesicles and liposomes, there is a dramatic morphology change to form a planar lamellar microstructure (Figure 2.2.3b). These lamellar forms of COSAN appear to resolve from the membrane into the liposome center, recovering their spherical vesicle

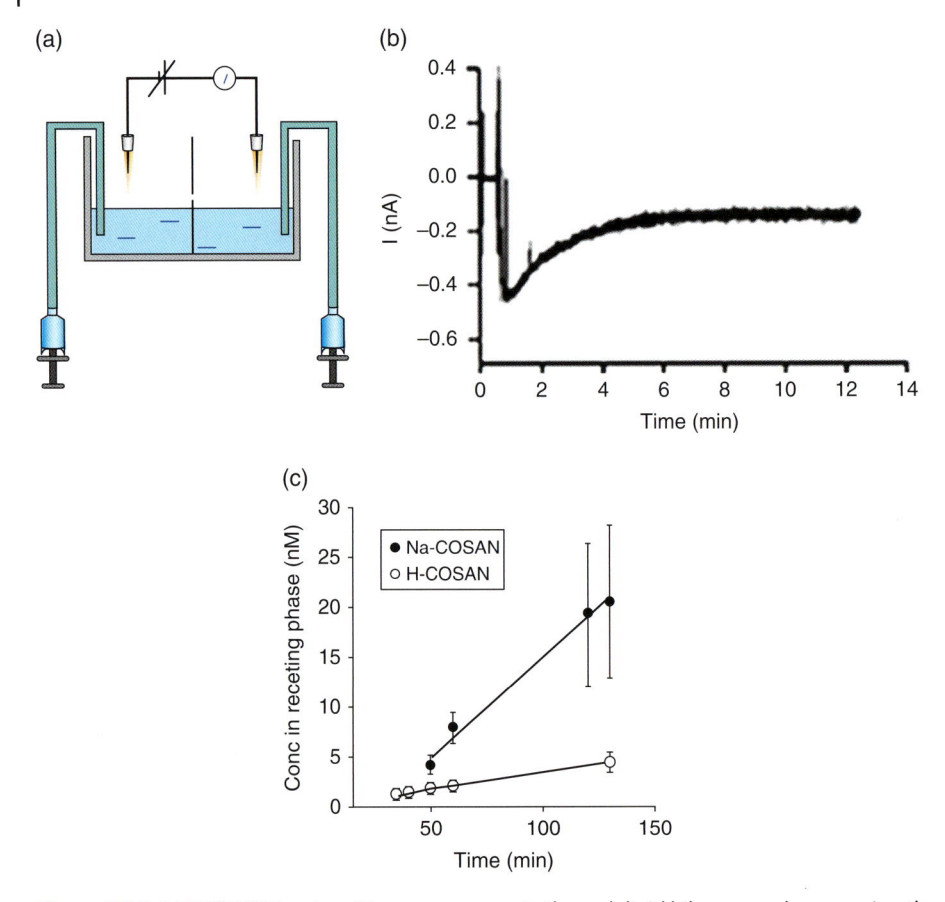

Figure 2.2.2 (a) [COSAN]⁻ ion transition measurements through lipid bilayer membranes using the Montal Mueller technique. (b) A typical current recording against time obtained when COSAN is applied to one side of a planar bilayer formed from DOPC. Current initiates as a strong, but transient, depolarization before stabilizing as a continuous, steady negative current across the membrane. (c) The negative current is accompanied by a steady flow of COSAN across a neutral planar membrane, as measured by ICP-MS. Commencing with the application of 100 µM COSAN to one side, permeation rates across the membrane vary according to the counterion of Na$^+$ or H$^+$. All show zero-order kinetics. Error bars are standard deviations resulting from ten independent experiments [9].

morphology (Figure 2.2.3b, arrows). As these liposome–COSAN double vesicles are often larger than the initial liposome population, it suggests that the interaction between COSAN vesicles and liposomes may cause fusion of multiple liposomes.

2.2.4 Visualization of COSAN within Cells

To investigate whether COSAN can enter cells as well as liposomes, micro-Raman spectroscopy was used to track COSAN via its distinctive vibrational peak at 2570 cm^{-1} due to its B–H bonds (Figure 2.2.4a) [10]. COSAN entry into HEK293 mammalian cells from culture media that contain 25 mM and 2 mM (Figure 2.2.4a and 2.2.4b, respectively)

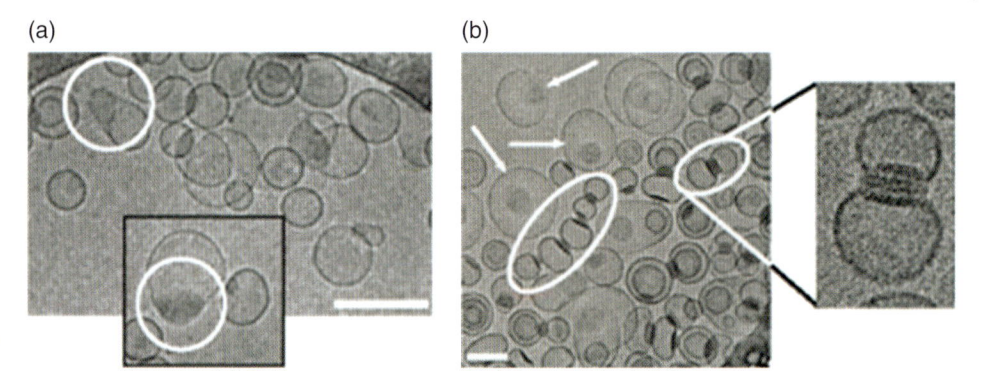

Figure 2.2.3 (a) CryoTEM image of COSAN:liposomes (1:4) suspended in vitreous ice at 15,000× magnification. Circles highlight fusion between COSAN vesicles and liposomes. Scale bar: 200 nm. (b) CryoTEM image of COSAN:liposomes (1:3) suspended in vitreous ice at 20,000× magnification. The circles highlight the joining of two or more liposome units linked by COSAN. Arrows indicate the complete penetration of the COSAN inside the liposome and the recovery of the monolayer vesicle form. Scale bar: 200 nm. Insert shows 50,000× magnification of the planar multilayer morphology of COSAN at the interface of two liposomes [9].

was extremely fast, occurring with a half maximal rate of tens of seconds (unpublished observations), but reached concentrations within the cell that were much higher than the external concentration, apparently entering against the concentration gradient [11]. If COSAN was then removed from the medium, it initially remained within the cell, but then was gradually lost (Figure 2.2.4a). This indicates that its entry rate is much greater than its exit rate, hence its accumulation. To date, no upper limit has been established for COSAN accumulation within cells. Rather than being homogeneously distributed throughout the cell, Raman spectral images of treated cells indicated a preferential accumulation of COSAN in the cytoplasm, although it was not excluded from the nucleus (Figure 2.2.4b). There appeared to be local subcellular COSAN accumulation within the cytoplasm. The identity of these regions is currently unclear; however, areas of high B–H signal match the areas of high C–H content (Figure 2.2.4b), indicating that COSAN-rich areas are associated with cell components, and are unlikely to exclude major proportions of the cytoplasm contents. Significantly, there was no substantial accumulation in the plasma membrane, consistent with the biophysical analysis of artificial membranes.

2.2.5 COSAN Interactions with Living Cells

The effect of COSAN accumulation has been investigated in a range of cells. COSAN treatment has been tested on a range of mammalian cells: HEK293, HeLa, 3 T3-4, and the lymphoblastoid cell line THP1 (Figure 2.2.5) [11]. In all concentrations examined, COSAN had no immediate effect on cell viability, with cells showing no signs of membrane disruption. However, prolonged exposure (>5 hours) blocked cell proliferation with half maximal effective dose concentration (ED_{50}) values of 99–157 µM. Specific effects varied with growth state and cell type. In confluent monolayers, cells took on a rounded appearance, but remained adhered to the substrate. Proliferating HEK293 cells

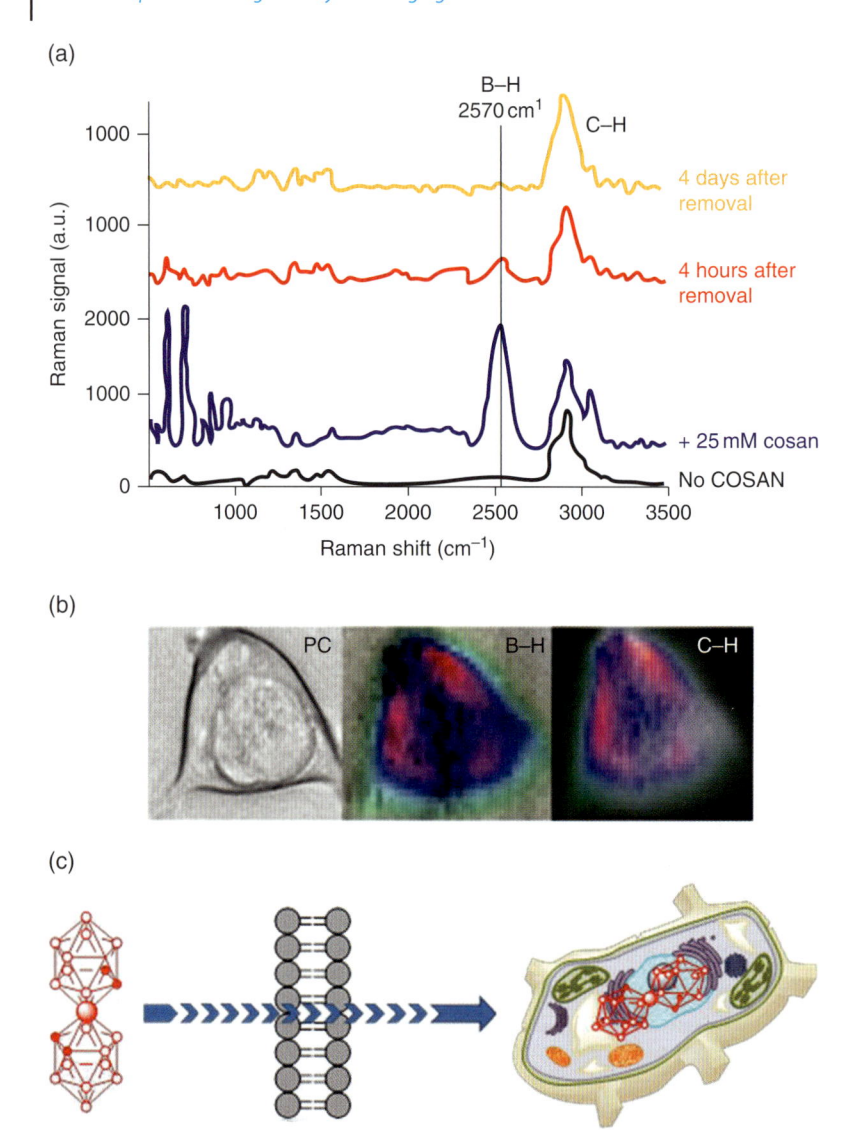

Figure 2.2.4 (a) Spectral fingerprints of HEK293 cells treated with 25 mM COSAN for 1 hour, followed by measurements of the same cells 4 hours and 4 days after the compounds have been removed. (b) Cellular imaging of HEK293 cell treated with 2 mM COSAN. Images show phase contrast image (PC) and Raman chemical images at 2570 cm^{-1} (B–H peak) and 2950 cm^{-1} (C–H peak). Pink zones in the Raman images show COSAN accumulation inside the cell [10]. (c) A simple schematic representation of the COSAN's uptake by the cells. *Source*: http://pubs.rsc.org/-/content/articlehtml/2014/cc/c3cc49658a. Licensed under CC 3.0.

began to bleb after 8 hours of treatment, having the appearance of cells entering apoptosis [12], with large numbers of cells dying at 24 hours. This effect could be suppressed in part by Ivachtin, a caspase 3 inhibitor. In contrast, HeLa cells took on an unusual highly vacuolated appearance after 24 hours (Figure 2.2.5). Importantly, in all cases when COSAN was only applied for 5 hours and then removed, cells fully recovered, indicating COSAN is not directly cytotoxic.

	No COSAN	COSAN	Wash
HEK293			
HeLa			
Dictyostelium			

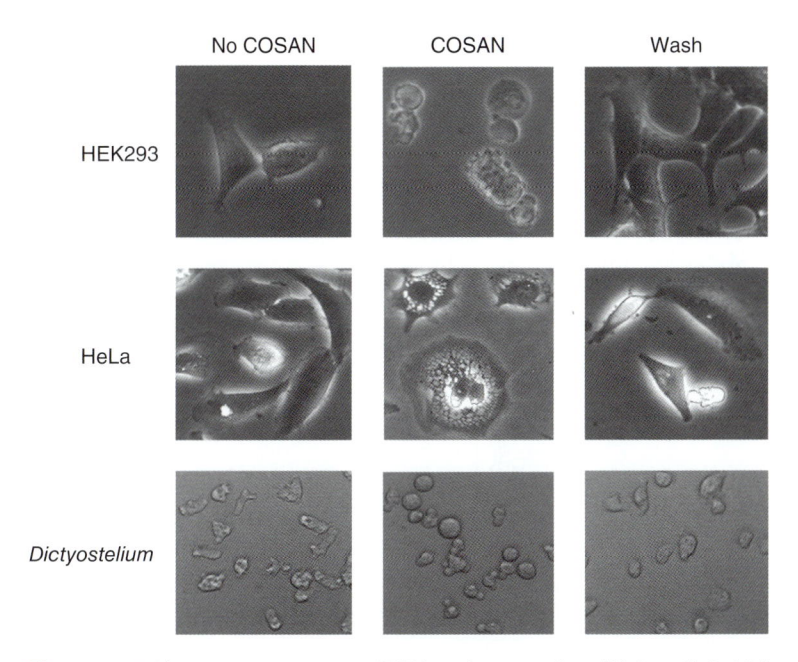

Figure 2.2.5 Phase contrast images of HEK293 (top panel) and HeLa cells (middle panel) grown in the presence or absence of 200 μM COSAN for 24 hours. Wash shows cells grown in the presence of 200 μM COSAN for 5 hours, before the cells were washed and replated in COSAN-free medium. HEK293 cells show a blebbing morphology often associated with apoptotic cells. HeLa cells show an unusual, highly vacuolated morphology within the perinuclear cytoplasm. *Dictyostelium* cells were photographed before treatment, 30 min following addition of 10 μM COSAN, and 2 hours after COSAN was removed [11].

COSAN also affects cell growth of non-mammalian cells. *Dictyostelium discoideum* is a eukaryotic amoeba with a close evolutionary relationship to animal cells [13]. These cells lack caspase-mediated apoptosis [14] and hence do not have the longer term confounding effects seen with HEK293 cells. *Dictyostelium* cells are in fact much more sensitive to COSAN, with a ED_{50} of 2.6 μM, and total proliferation arrest occurring at 4 μM (Figure 2.2.5). Under phase contrast microscopy, treated cells again retained the refractile appearance of living cells. Consistent with a lack of cytotoxicity, cells resumed division following COSAN removal. *Dictyostelium* cells remained viable after long periods in COSAN, and to date, we have found no limit for *Dictyostelium*, with cultures remaining viable for more than 1 month in COSAN and even after prolonged exposure to 500 μM of COSAN, almost 200 times the ED_{50} value.

2.2.6 Enhancing Cellular Effects of COSAN

Dictyostelium cells have been used to screen a range of several COSAN derivatives (Figure 2.2.6) [11]. Substitution of the central metal cobalt with iron $[3,3'\text{-Fe}(1,2\text{-}C_2B_9H_{11})_2]^-$ (FESAN) made no major difference to compound potency. However, methylation of COSAN, or cross-linking two COSAN clusters via a polyethylene glycol (PEG) chain, decreased potency. In contrast, addition of iodine to make $[3,3'\text{-Co}(8\text{-I-}1,2\text{-}C_2B_9H_{10})_2]^-$

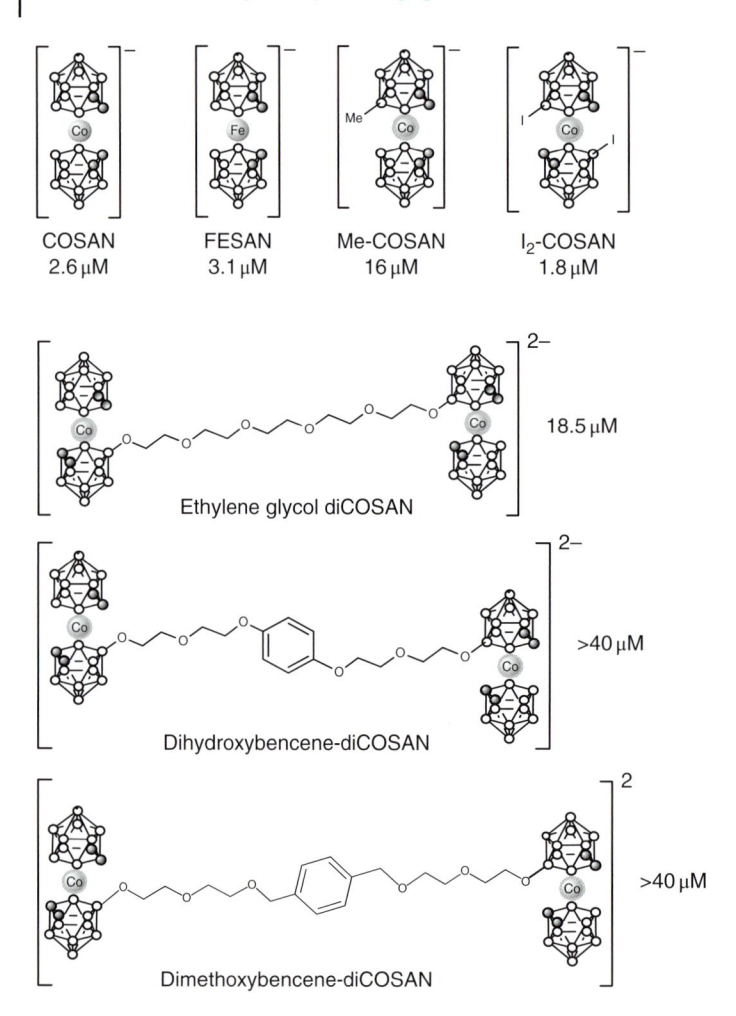

Figure 2.2.6 A chemical series based on [COSAN]⁻ reveals increased potency of [I$_2$-COSAN]⁻ on living cells. The ED$_{50}$ values on *Dictyostelium* are given for each compound.

(I$_2$-COSAN) increased potency. Recovery of *Dictyostelium* cells after I$_2$-COSAN treatment took longer than COSAN and was dose-dependent, suggesting a stronger interaction with its cellular targets. When tested on mammalian cell cultures, again I$_2$-COSAN was more potent than COSAN. Previously, we found no, or very little, effect of COSAN on the growth of bacteria. However, I$_2$-COSAN arrested growth of the two bacterial species tested, *Escherichia coli* and *Klebsiella pneumoniae*. Interestingly, we found that *E. coli* was significantly more sensitive to I$_2$-COSAN than *K. pneumoniae*. For both bacterial species, growth arrest was reversed by I$_2$-COSAN removal or dilution to below its ED$_{50}$ value.

Although both molecules possessed the ability to accumulate against the concentration gradient, I$_2$-COSAN reaches an intracellular concentration 3–4 times that of COSAN, although both have the same, slow exit rate [11]. I$_2$-COSAN was able to form vesicle structures of similar size and appearance of COSAN vesicles, arguing against this difference in behavior arising from major structural differences in solution [15]. Interestingly, increased I_2-COSAN potency was not due to a higher permeation rate, and in fact it has a lower overall permeation rate across planar lipid bilayers than seen for COSAN. However, measurements of partition coefficients between octanol and water revealed that I_2-COSAN has a significantly higher lipophilicity than COSAN [11]. This suggests a possible accumulation mechanism limited by COSAN's ability to interact with hydrophobic surfaces or environments. As COSAN does not accumulate within membranes, such as the plasma membrane, its interactions may be due more to surface effects than intrinsic interactions within the interior of hydrophobic lipid bilayers.

2.2.7 Tracking the *in vivo* Distribution of I-COSAN

The *in vitro* cell culture studies indicate that, in general, COSAN and its derivatives accumulate within cells. This raises the question of how COSAN-based molecules would distribute and accumulate in organs and tissues. This is particularly important, as many *o*-, *m*-, and *p-closo*-carborane derivatives have been proposed for use in boron neutron capture therapy (BNCT), a potential new cancer treatment based on neutron irradiation of ^{10}B and its conversion to ^{11}B, with a subsequent local release of an α-particle and a high-energy ^{7}Li atom [16–18]. Halogenated carborane derivatives have been used to prepare compounds for application in BNCT as well as in radio-imaging [19,20] arising from their ability to be radiolabeled with a range of medical isotopes. Even though halogenated *nido*-carboranes have potential relevance to important topics such as radioiodine carrier [21,22] and as BNCT reagents [23], relatively few B-iodinated [7,8-*nido*-C_2B_9]$^-$ derivatives are known. Several reviews on the radioisotopes' incorporation in organoboranes and boron clusters have recently appeared [24]. Based on this knowledge, $[3,3'$-Co(8-I-1,2-$C_2B_9H_{10}$)(8'-R-1',2'-$C_2B_9H_{10}$)]$^-$ [R = H, $C_6H_5COO(CH_2CH_2O)_2$] (I-COSAN and I-PEG-COSAN, respectively) compounds were radiolabeled with either ^{125}I (γ-radiation emitter) or ^{124}I (positron emitter) via palladium-catalyzed isotopic exchange reaction (Figure 2.2.7); these labeled COSAN derivatives were administered to mice, and their biodistribution was investigated [25].

Following ^{125}I-COSAN intravenous administration of 100 μL (25 ± 5 μCi) that was injected to mice (*n* = 3 per compound) through the catheterized tail vein, tissue samples taken at 10, 30, and 120 minutes were measured for radioactive accumulation. There was a rapid accumulation of I-COSAN in lungs and liver over 2 hours, matched by a decrease in blood content over the same time period (Figure 2.2.7). There was a lesser accumulation in heart, kidneys, spleen, and stomach over the same time course. Independently, ^{124}I-COSAN was administered to mice (*n* = 3 per compound) and then followed using positron emission tomography (PET), and again accumulation was observed in lung, heart, kidneys, liver, and stomach. Furthermore, synthesis of a new bifunctional (iodine and PEG) I-PEG-COSAN derivative, which would allow further functional derivatization, was tested and found to have the same distribution as I-COSAN. These results demonstrate that, as suggested from *in vitro* experiments, COSAN and its derivatives can accumulate in tissues within the body, offering a means to both deliver and track compounds for BCNT. Interestingly, *in vivo* studies did not show accumulation of radioactivity in the thyroid gland, suggesting the stability of both ^{124}I-COSAN and ^{124}I-PEG-COSAN derivatives. This general radiolabeling strategy, which can be applied in the

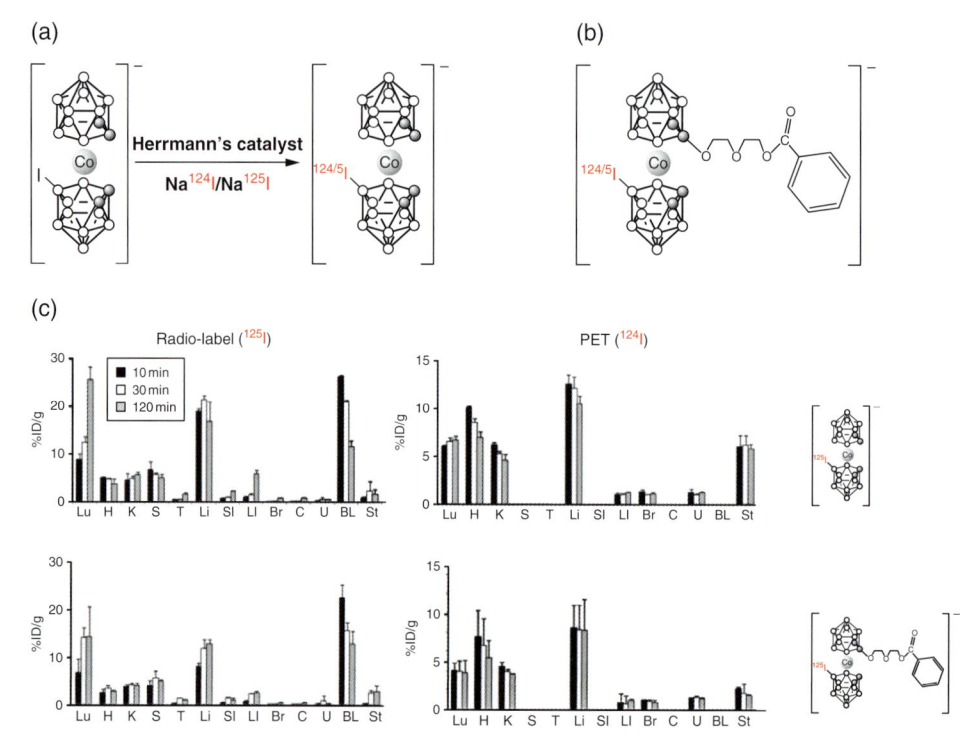

Figure 2.2.7 (a) COSAN was labeled with either ^{125}I (gamma emitter) or ^{124}I (positron emitter; not shown) via palladium-catalyzed isotopic exchange reaction. (b) Structure of iodine-labeled and PEGylated COSAN. (c) Accumulation in different organs of [^{125}I-COSAN]$^-$ / [^{124}I-COSAN]$^-$ (top) and [^{125}I-PEG-COSAN]$^-$/[^{124}I-PEG-COSAN]$^-$ (bottom). Left: Accumulation of ^{125}I-radiolabeled compounds were measured following dissection of organs and gamma counting at selected time points (10, 30, and 120 min.) after administration. Results are expressed as percentage of injected dose (%ID) per gram of tissue. Right: Accumulation of ^{124}I-labeled compounds were visualized by PET-C. Results are expressed as percentage of injected dose (%ID) per cubic centimeter of tissue. Mean ± standard deviation values are presented ($n = 3$). Herrmann's catalyst is a highly efficient palladacycle catalyst that corresponds to trans-di-μ-acetatobis[2-[bis(2-methylphenyl)phosphino]benzyl]dipalladium. Lu: Lungs; H: heart; K: kidneys; S: spleen; T: testicles; L: liver; SI: small intestine; LI: large intestine; Br: brain; C: cerebellum; U: urine; BL: blood; St: stomach [25].

future to COSAN derivatives bearing a wide range of functionalities, might be applicable to targeted cobaltabisdicarbollides able to selectively accumulate in tumors. Hence, this method may become an invaluable, widely applied tool for the fast and accurate evaluation of new COSAN-based BNCT drug candidates in animal tumor models. Due to the non-invasive nature of PET imaging, potential translation into the clinical setting to predict therapeutic efficacy on a patient-by-patient basis also can be foreseen.

2.2.8 Discussion and Potential Applications

This body of work demonstrates a set of novel interactions between living cells and COSAN or related molecules. This arises from COSAN's unusual physicochemical properties that enable it to easily cross membranes, and can be enhanced by chemical or structural modification, such as seen for the halogenated I_2-COSAN. Importantly, COSAN entry into cells neither disrupts the plasma membrane nor is immediately cytotoxic, consistent with its ability to cross membranes without disrupting the lipid bilayer.

For most eukaryotic and prokaryotic cell cultures, cells remain viable with prolonged COSAN treatment, but exhibit altered biology, such as cytostatic effects on cell growth and proliferation. In particular, many of these effects are reversible when COSAN is washed from the cells. Interestingly, irreversible effects can be seen in some cases, such as apoptosis seen in proliferating HEK293, indicating that COSAN within the cell may interact with cell-signaling processes to elicit change in cell behavior. The nature of these interactions is currently unknown. There is a differential sensitivity between cell state and cell species, so that proliferating cells may be more sensitive to cell death following prolonged COSAN treatment, compared to quiescent cells within tissue. If this is translated to the *in vivo* context, then it may be possible to target rapidly growing cells, such as in cancer treatment. *Dictyostelium* amoebae show a much higher sensitivity to COSAN and its derivatives. If this higher sensitivity to COSAN were also present in other unicellular eukaryotes, it may offer an amoebicidal agent to target protozoan and amoebozoan pathogens.

With our current imaging techniques, we cannot determine whether COSAN and lipid membranes inside cells mix to form hybrid membranes, although this seems less likely from the observed biophysical interactions seen in cell-free, artificial systems. Associations between membrane fluidity, cell proliferation arrest, and apoptosis have been seen with other small molecules, such as methyl jasmonate [26], and it is possible that all the biological effects we observe could arise due to interaction between the two membrane systems. Alternatively, COSAN may mediate its effects due to protein interactions either by enzyme inhibition via active site binding, such as the modified COSAN molecules that inhibit enzymes [27–29], or more general interference by binding to protein surfaces. More detailed analysis is required to resolve these mechanistic possibilities.

A valuable feature of the inorganic boron-based COSAN membranes reported here is that they are not present in nature, and no degradation or modification of COSAN or I_2-COSAN is observed in the cells that have currently been investigated. This biologically inert feature may offer new opportunities for drug design and molecular delivery systems. Boron clusters can be coupled to bioactive molecules, such as inhibitors, siRNAs, peptides [30], and protein ligands [31,32], and have the potential to carry these molecules across the plasma membrane and into cells. COSAN vesicles could also be used to encapsulate water-soluble or hydrophilic compound; however, it is currently unclear how the topology of the COSAN vesicles changes as they merge with lipid bilayer membranes and whether this would deliver into the cell or disperse on the cell surface.

As COSAN can accumulate to high concentrations in living cells, this may enhance its potential therapeutic effects. In particular, it may have advantages for BNCT in the treatment of cancer, where the transition from bench to bed (even in the preclinical setting) has only occasionally been approached. The main reason for this is a lack of techniques able to monitor accumulation of boron in the tumors and their surrounding tissues *in vivo* and in real time. Successful application of BNCT needs careful timing of neutron irradiation to coincide with the optimal ratio of boron accumulation in tumor–nontumor tissue, and hence needs a good pharmacokinetic tracer. Non-invasive PET imaging serves this purpose well, allowing screening for maximum therapeutic efficacy on a patient-by-patient basis. PET labeling of the anionic COSAN cluster has advantages over neutral icosahedral *closo*-carboranyl clusters in terms of water solubility, fast halogenation reaction, and its capacity to be multi-decorated by incorporation of distinct functional groups at the different vertices. When coupled to the high density of boron atoms in COSAN and its derivatives, and their propensity for intracellular accumulation, this should facilitate translation of BNCT into the clinical setting.

A more innovative application may be the potential of COSAN-based vesicles as a nanoscale platform for cell modification and re-engineering. In an analogous manner to biological membranes and vesicles, self-assembling COSAN-based nanovesicles introduced into cells could offer a surface on which to control molecule interactions with cellular components, or to assemble noncellular molecules for novel physical or chemical interactions. Controlling interactions on such synthetic membranes could be used to regulate chemical reactions, elicit signaling, or control molecule release. As COSAN and other borane clusters can be coupled to other bioactive molecules, they could be used to assemble complex functionalities and regulate molecular behavior on synthetic surfaces within cells. In this way, we propose that COSAN-based synthetic nanovesicles could be used to redesign and engineer new hybrid cells with novel biological properties. The ability to modify COSAN to change its potency and differential effects on cell types presents a wide range of possible permutations to explore.

2.2.9 Summary

Taken together, these observations demonstrate highly unusual properties of COSAN enabling it to both generate membrane-like structures and vesicles and rapidly transit lipid bilayer membranes to accumulate within cells without affecting membrane integrity. Once inside the cell, COSAN is not directly cytotoxic, but elicits biological effects such as cytostatic arrest of cell growth, apoptosis, or membrane rearrangements. For most cells, this is reversible, as cells can recover following its removal. Specific biological effects, however, vary with cell type and conditions, and we conclude that they arise due to COSAN interaction with cellular components and not changes to the plasma membrane.

These results reveal an unexpected biology at the interface of biological and synthetic membranes. This particular property offers new opportunities for cancer therapy, drug design, and molecular delivery systems. More excitingly, COSAN-based nanovesicles may offer new possibilities for nanoscale platforms to directly introduce new functionality and use cell re-engineering to create synthetic hybrid cells with novel biological properties.

Appendix: Abbreviations

BNCT	boron neutron capture therapy
COSAN	$[3,3'\text{-}Co(1,2\text{-}C_2B_9H_{11})_2]^-$
cryoTEM	cryo–transmission electron microscopy
DOPC	1,2-dioleoyl-*sn*-glycero-3-phosphocholine
DPhPC	1,2-diphytanoyl-*sn*-glycero-3-phosphocholine
ED_{50}	half maximal effective dose concentration
FESAN	$[3,3'\text{-}Fe(1,2\text{-}C_2B_9H_{11})_2]^-$
I-COSAN	$[3,3'\text{-}Co(8\text{-}I\text{-}1,2\text{-}C_2B_9H_{10})(1'\!,2\text{-}C_2B_9H_{11})]^-$
I_2-COSAN	$[3,3'\text{-}Co(8\text{-}I\text{-}1,2\text{-}C_2B_9H_{10})_2]^-$
ICP-MS	inductively coupled plasma mass spectrometry
I-PEG-COSAN	$[3,3'\text{-}Co(8\text{-}I\text{-}1,2\text{-}C_2B_9H_{10})(8'\text{-}R\text{-}1'\!,2\text{-}C_2B_9H_{10})]^-$

Acknowledgments

Our own contributions to this area of research would not have been possible without the enthusiastic work of our coworkers, our postdoctoral associates, and graduate project students whose names appear in the references. Financial support during the writing of this chapter has come from Generalitat de Catalunya (2014/SGR/149) and Ministerio de Economía y Competitividad (CTQ2013-44670-R). ICMAB-CSIC acknowledges financial support from MINECO through the Severo Ochoa Program for Centers of Excellence in R&D (SEV-2015-0496).

References

1. Hawthorne, M.F., Young, D.C., and Wegner, P.A. Carbametallic boron hydride derivatives. I. Apparent analogs of ferrocene and ferricinium ion. J Am Chem Soc 1965; 87:1818–1819.
2. Bauduin, P., Prevost, S., Farras, P., Teixidor, F., Diat, O., and Zemb, T. A theta-shaped amphiphilic cobaltabisdicarbollide anion: transition from monolayer vesicles to micelles. Angew Chem Int Ed Engl 2011; 50:5298–5300.
3. Singer, S.J. A fluid lipid-globular protein mosaic model of membrane structure. Ann N Y Acad Sci 1972; 195:16–23.
4. McMahon, H.T., and Gallop, J.L. Membrane curvature and mechanisms of dynamic cell membrane remodelling. Nature 2005; 438:590–596.
5. Antonny, B. Mechanisms of membrane curvature sensing. Annu Rev Biochem 2011; 80:101–123.
6. Gassin, P.-M., Girard, L., Martin-Gassin, G., Brusselle, D., Jonchère, A., Diat, O., Viñas, C., Teixidor, F., and Bauduin, P. Surface activity and molecular organization of metallacarboranes at the air–water interface revealed by nonlinear optics. Langmuir 2015; 31:2297–2303.
7. Viñas, C., Tarrés, M., González-Cardoso, P., Farràs, P., Bauduin, P., and Teixidor, F. Surfactant behaviour of metallacarboranes: a study based on the electrolysis of water. Dalton Trans 2014; 43:5062–5068.
8. Dordovic, V., Tosner, Z., Uchman, M., Zhigunov, A., Reza, M., Ruokolainen, J., Pramanik, G., Cígler, P., Kalíkova, K., Gradzielski, M., and Matějíček, P. Stealth amphiphiles: self-assembly of polyhedral boron clusters. Langmuir 2016; 32:6713–6722.
9. Verdia-Baguena, C., Alcaraz, A., Aguilella, V.M., Cioran, A.M., Tachikawa, S., Nakamura, H., Teixidor, F., and Viñas, C. Amphiphilic COSAN and I2-COSAN crossing synthetic lipid membranes: planar bilayers and liposomes. Chem Commun 2014; 50:6700–6703.
10. Tarres, M., Canetta, E., Viñas, C., Teixidor, F., and Harwood, A.J. Imaging in living cells using nuB-H Raman spectroscopy: monitoring COSAN uptake. Chem Commun 2014; 50:3370–3372.
11. Tarres, M., Canetta, E., Paul, E., Forbes, J., Azzouni, K., Viñas, C., Teixidor, F., and Harwood, A.J. Biological interaction of living cells with COSAN-based synthetic vesicles. Sci Rep 2015; 5:7804.

12 Ellis, R.E., Yuan, J.Y., and Horvitz, H.R. Mechanisms and functions of cell death. Annu Rev Cell Biol 1991; 7:663–698.
13 Kessin, R.H. Dictyostelium: Evolution, Cell Biology, and the Development of Multicellularity. Cambridge University Press, Cambridge, 2001.
14 Olie, R.A., Durrieu, F., Cornillon, S., Loughran, G., Gross, J., Earnshaw, W.C., and Golstein, P. Apparent caspase independence of programmed cell death in *Dictyostelium*. Curr Biol 1998; 8:955–958.
15 Brusselle, D., Bauduin, P., Girard, L., Zaulet, A., Viñas, C., Teixidor, F., Ly, I., and Diat. O. Lyotropic lamellar phase formed from monolayered θ-shaped carborane-cage amphiphiles. Angew Chem Int Ed 2013; 52:12114–12118.
16 Soloway, A.H., Tjarks, W., Barnum, B.A., Rong, F.G., Barth, R.F., Codogni, I.M., and Wilson, J.G. The chemistry of neutron capture therapy. Chem Rev 1998; 98:1515–1562.
17 Luderer, M.J., de la Puente, P., and Azab, A.K. Advancements in tumor targeting strategies for boron neutron capture therapy. Pharm Res 2015; 32:2824–2836.
18 Gahbauer, R., Gupta, N., Blue, T., Goodman, J., Barth, R., Grecula, J., Soloway, A.H., Sauerwein, W., and Wambersie, A. Boron neutron capture therapy: principles and potential. Cancer Res 1998, 150:183–209.
19 Hawthorne, M.F., and Maderna, A. Applications of radiolabeled boron clusters to the diagnosis and treatment of cancer. Chem Rev 1999; 99:3421–3434.
20 Tolmachev, V., and Sjöberg. S. Polyhedral boron compounds as potential linkers for attachment of radiohalogens to targeting proteins and peptides: a review. Collect Czech Chem Commun 2002; 67:913–935.
21 El-Zaria, M.E., Janzen, N., Blacker, M., and Valliant, J.F. Synthesis, characterisation, and biodistribution of radioiodinated C-hydroxy-carboranes. Chem Eur J 2012; 18:11071–11078.
22 Green, A.E.C., Parker, S.K., and Valliant, J.F. Synthesis and screening of bifunctional radiolabelled carborane-carbohydrate derivatives. J Organometal Chem 2009; 694:1736–1746.
23 Mizusawa, E.A., Thompson, M.R., and Hawthorne, M.F. Synthesis and antibody-labeling studies with the *p*-isothiocyanatobenzene derivatives of 1,2-dicarba-*closo*-dodecarborane(12) and the dodecahydro-7,8-dicarba-*nido*-undecaborate(-1) ion for neutron-capture therapy of human cancer: crystal and molecular structure of $Cs^+[nido$-7-(p-C_6H_4NCS)-9-I-7,8-$C_2B_9H_{11}]^-$. Inorg Chem 1985; 24:1911–1916.
24 Hosmane, N.S. (Ed.). Boron Science: New Technologies and Applications. CRC Press, Boca Raton, FL, 2012.
25 Gona, K.B., Zaulet, A., Gómez-Vallejo, V., Teixidor, F., Llop J., and Viñas, C. COSAN as a molecular imaging platform: synthesis and "in vivo" imaging. Chem Commun 2014; 50:11415–11417.
26 Yeruva, L., Elegbede, J.A., and Carper, S.W. Methyl jasmonate decreases membrane fluidity and induces apoptosis through tumor necrosis factor receptor 1 in breast cancer cells. Anti-Cancer Drugs 2008; 19:766–776.
27 Cigler, P., Kozísek, M,., Rezácová, P., Brynda, J., Otwinowski, Z., Pokorná, J., Plesek, J., Grüner, B., Dolecková-Maresová, L., Mása, M., Sedlácek, J., Bodem, J., Kräusslich, H.G., Král, V., and Konvalinka, J. From nonpeptide toward noncarbon protease inhibitors: metallacarboranes as specific and potent inhibitors of HIV protease. Proc Natl Acad Sci USA 2005; 102:15394–15399.

28 Rezacova, P., Pokorna, J., Brynda, J., Kozisek, M., Cigler, P., Lepsik, M., Fanfrlik, J., Rezac, J, Saskova, K.G., Sieglova, I., Plesek, J., Sicha, V., Gruner, B., Oberwinkler, H., Sedlacek, J., Krausslich, H.G., Hobza, P., Kral, V., and Konvalinka, J. Design of HIV protease inhibitors based on inorganic polyhedral metallacarboranes. J Med Chem 2009; 52:7132–7141.

29 Farras, P, Juarez-Perez, E.J., Lepsik, M., Luque, R., Nunez, R., and Teixidor, F. Metallacarboranes and their interactions: theoretical insights and their applicability. Chem Rev 2012; 41:3445–3463.

30 Frank, R., Ahrens, V.M., Boehnke, S., Beck-Sickinger, A.G., and Hey-Hawkins, E. Charge-compensated metallacarborane building blocks for conjugation with peptides. ChemBioChem 2016; 17(4):308–317.

31 Kwiatkowska, A., Sobczak, M., Mikolajczyk, B, Janczak, S., Olejniczak, A.B., Sochacki, M., Lesnikowski, Z.J., and Nawrot, B. siRNAs modified with boron cluster and their physicochemical and biological characterization. Bioconjugate Chemistry 2013; 24:1017–1026.

32 Capala, J., Barth, R.F., Bendayan, M., Lauzon, M., Adams, D.M., Soloway, A.H., Fenstermaker, R.A., and Carlsson, J. Boronated epidermal growth factor as a potential targeting agent for boron neutron capture therapy of brain tumors. Bioconjugate Chemistry 1996; 7: 7–15.

2.3

Boronic Acid–Based Sensors for Determination of Sugars

Igor B. Sivaev and Vladimir I. Bregadze

A.N. Nesmeyanov Institute of Organoelement Compounds, Russian Academy of Sciences, Moscow, Russia

2.3.1 Introduction

Diabetes is a chronic disease that has devastating human, social, and economic consequences. It is one of the most notorious saccharide-related diseases. Diabetes occurs either when the pancreas does not produce enough insulin, a hormone that regulates glucose level in blood (type I diabetes), or when the body cannot effectively use the insulin it produces (type II diabetes). Diabetes is associated with chronic ill health, disability, and premature mortality. From a physiological perspective, the debilitating long-term complications include heart disease, blindness, kidney failure, stroke, and nerve damage leading to amputation. Both the number of cases and the prevalence of diabetes have been steadily increasing over the past few decades. According to the World Health Organization, 422 million adults were living with diabetes in 2014 compared to 108 million in 1980. The prevalence of diabetes has nearly doubled since 1980, rising from 4.7% to 8.5% in the adult population. Diabetes caused 1.5 million deaths in 2012. Higher than optimal blood glucose caused an additional 2.2 million deaths, by increasing the risks of cardiovascular and other diseases [1]. Therefore, diabetes presents one of the largest health challenges to face us in the twenty-first century.

To date, although there is no means to cure or prevent diabetes, blood glucose monitoring and appropriate medication adjusting diabetic blood sugar levels to maintain them within tight boundaries dramatically reduce the health risks faced by diabetics. Therefore, tight control of blood glucose is the most important goal in the treatment of diabetes. This will limit the long-term consequences of the chronic disease, including damage to the heart, eyes, kidneys, nerves, and other organs caused by high glucose concentrations (i.e., hyperglycemia). The availability of affordable home blood glucose monitoring has revolutionized the quality of life experienced by diabetics. Ideally, one would like to have a continuous, real-time, and non-invasive monitoring method such as glucose-sensing contact lenses or implantable sensor devices. However, currently the most commonly used method for measuring blood glucose levels is a non-continuous and invasive approach that samples blood from a finger followed by *in vitro*

Boron-Based Compounds: Potential and Emerging Applications in Medicine, First Edition.
Edited by Evamarie Hey-Hawkins and Clara Viñas Teixidor.
© 2018 John Wiley & Sons Ltd. Published 2018 by John Wiley & Sons Ltd.

glucose concentration determination using a test strip and a meter. The majority of home blood glucose monitoring tools relies on the glucose oxidase enzyme (GOx) [2] and demonstrates good sensitivity [2–4]. However, there are some inherent limitations with an enzymatic approach. The systems have to be stored appropriately, they are specific for only a few saccharides, and in most cases they become unstable under harsh conditions. For this reason, much work has been focused on the development of non-enzyme sensors with the capacity to monitor saccharides under a broad range of environmental conditions, and thus allow access to more widespread diagnostic applications.

A powerful approach for detecting glucose in fluids is its complexation with boronic acids. This method is superior to enzyme-based sensors because it is not affected by factors that affect enzyme activity, such as environmental pH and temperature. In contrast, boronic acid–mediated detection is based on equilibrium thermodynamics and does not require special treatment of the sensor to maintain its structural integrity.

A number of excellent reviews covering the use of boronic acid in design of saccharide receptors were published during the last 15 years [5–16]. Therefore, the purpose of this chapter isn't to give a comprehensive review of the subject, but rather to demonstrate its place in the medicinal chemistry of boron compounds.

2.3.2 Interactions of Boronic Acids with Carbohydrates

The selective recognition of glucose over other saccharides presents a curious challenge. With glucose, blood also contains some amounts of other carbohydrates such as fructose, galactose, and mannose (Figure 2.3.1). D-Glucose is the major carbohydrate as compared to D-fructose. The concentration of glucose in blood is around 5 mM, which is much larger than the concentration of fructose even after a fructose-rich meal (<0.1 mM) [12]. However, D-fructose generally demonstrates stronger binding affinity to aryl monoboronic acids in comparison to D-glucose (discussed further in this chapter). The stronger binding of D-fructose can be explained by the different binding mode: D-fructose exhibits the tridentate binding mode, whereas D-glucose exhibits bidentate binding. Moreover, in aqueous solution, cleavage of the

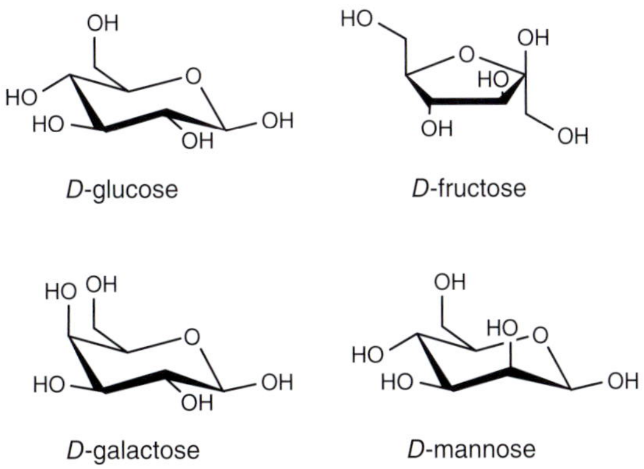

Figure 2.3.1 Common naturally occurring monosaccharides (β-forms).

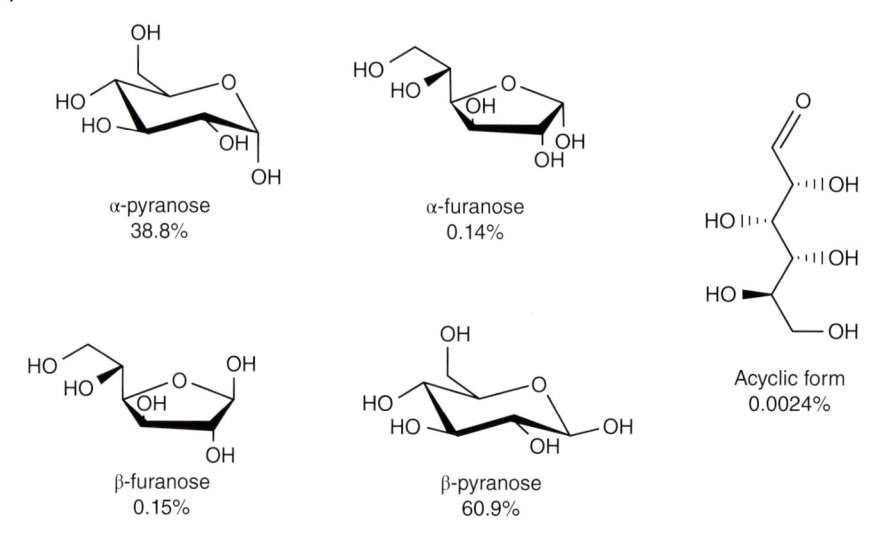

Figure 2.3.2 D-Glucose in various forms with percentage composition at equilibrium in D_2O at 27 °C (for acyclic form; equilibrated at 37 °C).

hemiacetal ring causes interconversion between the pyranose and furanose ring forms of glucose, via an acyclic intermediate, with inversion of configuration at the anomeric center equilibrating the α- and β-anomers [17] (Figure 2.3.2). Therefore, saccharides are difficult to differentiate from each other.

The first hint of the "marriage" between boron and polyols was detected by Jean-Baptiste Biot in his seminal studies on optical rotation. In 1832, he noted that the rotation of tartaric acid changed in the presence of boric acid [18]. It would be a century later before interaction of boron acids (boric, boronic, and borinic) and monosaccharides was studied in detail. In 1910–1930s, Jacob Böeseken and coworkers elucidated the absolute structural configuration of a series of carbohydrates based on change of their acidity and conductivity upon the ester formation with boric acid [19]. Given the significance of boric acid in the determination of saccharide configurations, it is perhaps surprising that the same properties were not observed in boronic acids until the 1950s, when Kuivila et al. postulated the formation of a cyclic boronic ester on phenylboronic acid and mannitol analogous to the one known to form between boric acid and polyols [20]. The next were Lorand and Edwards, who studied reversible formation of boronic esters on the interaction of phenylboronic acid with polyols in water [21]. By measuring the complexation equilibrium between phenylboronic acid and several simple diols (ethylene glycol, catechol) and common monosaccharides (glucose, fructose, mannose, galactose) using the pH depression method, ester formation was shown to be more favorable in solutions of high pH where the boronate ion exists in high concentrations. This study also confirmed the Lewis acid behavior of boronic acids and the tetracoordinate structure of their conjugate base (i.e., the hydroxyboronate anion). Another conclusion is that free boronic acids have lower Lewis acid strengths than their neutral complexes with 1,2-diols. For example, the pK_a of phenylboronic acid decreases from 8.9 to 6.8 and 4.5 upon the formation of cyclic esters with glucose and fructose, respectively [22]. The equilibria involved in phenylboronate binding of a diol are conventionally summarized as a set of coupled equilibria (Scheme 2.3.1).

Scheme 2.3.1

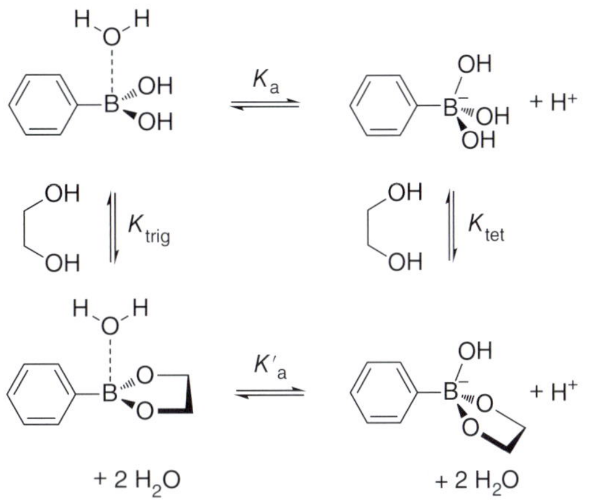

In aqueous solution, phenylboronic acid reacts with water to form the boronate anion plus a hydrated proton, thereby defining an acidity constant K_a (pK_a 8.90 in water at 25 °C). The formation of a diol boronate complex, defined by K_{tet}, formally liberates 2 equiv. of water, but this stoichiometric factor is usually ignored as a constant in dilute aqueous solution. The magnitude of log K_{tet} varies with the diol, ranging from about 3.8 for fructose to about 1.2 for simple diols such as ethylene glycol. Phenylboronic acid could also bind diols to form a trigonal complex (K_{trig}), and this species would itself act as an acid according to K'_a. It is observed that $K_{tet} > K_{trig}$. For instance, the logarithms of these constants for phenylboronic acid binding fructose in 0.5 M NaCl water are log $K_{tet} = 3.8$ and log $K_{trig} < -1.4$. This difference in the value of the binding constant between K_{tet} and K_{trig} is typical, with differences of up to approximately five orders of magnitude being commonplace. It is known that the neutral boronic acid becomes more acidic upon binding (i.e., $pK'_a > pK_a$); in other words, the boronic ester is more acidic than the boronic acid [22].

Formation of a saccharide complex at neutral pH is essential for practical development of the boronic acid–based sensors. Since pK'_a is 1–2 units lower than pK_a, the boronic acid–saccharide complex will only exist in significant amounts at neutral pH if the pK_a of the boronic acid itself is ≤7. Since the pK_a of phenylboronic acid is 8.9, it requires basic aqueous conditions to form strong complexes. Therefore, to realize strong binding at neutral pH, the pK_a of boronic acid should be lowered. This can be achieved by introduction of electron-withdrawing groups into the phenyl ring; for example, 4-nitrophenylboronic and 3-carboxymethyl-5-nitrophenylboronic acids have pK_a of 7.23 and 6.74, respectively [23]. However, such modification of phenylboronic acid is not simple and requires high synthetic efforts. Another approach is based on the discovery that the interaction between a boronic acid and neighboring aminomethyl substituent lowers the pK_a of the boronic acid [24]. The nuclear magnetic resonance (NMR) studies confirmed a strong but kinetically labile B–N interaction in ortho-(N,N-dimethylaminomethyl)benzene boronic acid derivatives. This interaction expands the pH range within which a tetrahedral boron atom exists. Accordingly, a strong binding between boronic acid and diols even at neutral pH can be provided and applied to carbohydrate sensing.

Table 2.3.1 The binding constants of phenylboronic acid and polyols in water at 25 °C

Polyol	K_{tet}, M^{-1}*	K_{eq}, M^{-1}**
1,3-propanediol	0.88	
Ethylene glycol	2.80	
D-glucose	110	4.16
D-mannose	170	13
D-galactose	280	15
D-fructose	4400	160
Sorbitol		370
Catechol	18,000	830

* Measured by the pH-depression method (25 °C) [21].
** Measured by the ARS competition method at pH 7.4 [25].

The very important point to be considered in boronic acid–carbohydrate binding is the predisposition of boronic acid to interact with different types of diols. In 1959, Lorand and Edwards published the first comprehensive study of the binding constants between various diol-containing compounds and phenylboronic acid using the pH-depression method [21]. The study demonstrated that different diols have different affinity for the boronic acid group and lower the acidity of the boron species to different degrees (Table 2.1.1). The numbers reported, although referred to as binding constants, are very different from the binding constants of other monoboronic acids determined later using spectroscopic methods. A careful analysis of the situation indicates that this discrepancy is because of a lack of clear definition of the term "binding constants." The binding constants determined using the pH-depression method are K_{tet} instead of K_{eq}. This was because the pH-depression method assumed that the boronic ester did not exist (or existed in negligible amount). Therefore, the K_{trig} part of the equation in Scheme 2.3.1 was omitted. In 2002, Springsteen and Wang published the first systematic study of the K_{eq} between phenylboronic acid and various diol-containing compounds and described the relation among K_{trig}, K_{tet}, and K_{eq} as K_{eq} = % acid form × $K_{eq\text{-trig}}$ + % ester form × $K_{eq\text{-tet}}$ [25]. These values are much lower than the K_{tet} values determined using the pH-depression method, but have the same trend (Table 2.3.1). Moreover, it is the case that the trends established are inherent in all monoboronic acids.

The high binding affinity of D-fructose and sorbitol in comparison with other compounds can be explained by the formation of the 2,3,6-tridentate complex [26], whereas D-glucose forms a tridentate complex with phenylboronic acid only in alkaline solution [27].

It should be noted that not only phenylboronic acids themselves are able to bind different sugars but some of their derivatives such as benzoxaborole can do it as well. This field was underestimated until 2006, when the exceptional sugar-binding properties of the parent benzoxaborole at physiological conditions were described [28,29]. Benzoxaboroles display higher Lewis acidity than the corresponding phenylboronic acids, which is explained by the ring strain generated in the five-membered heterocyclic

Scheme 2.3.2

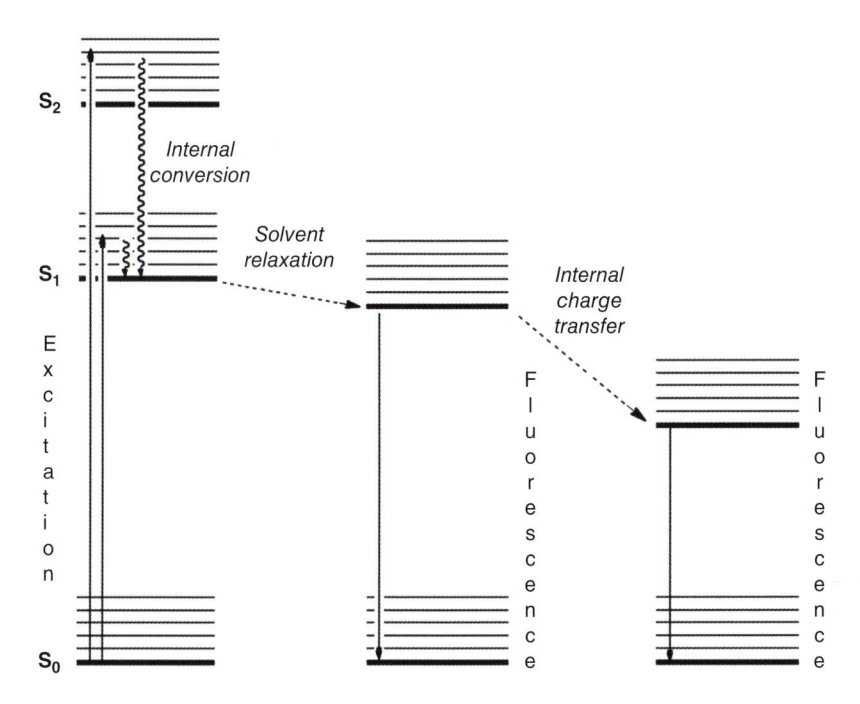

Figure 2.3.3 The dual fluorescence of an intramolecular charge transfer fluorophore.

ring that can be reduced upon addition of nucleophiles generating tetrahedral species (Scheme 2.3.2) [30]. It is a remarkable feature as the acid–base interactions play a crucial role in sugar sensing by phenylboronic acids.

2.3.3 Fluorescence Carbohydrate Sensors

Fluorescence sensors for carbohydrate detection are of particular practical interest because fluorescence demonstrates an exceptionally high sensitivity of detection (typically 10^{-6} M) offsetting the synthetic costs of such sensors. Another great advantage of fluorescent sensors is their very fast (sub-millisecond) response time. The sensor efficacy is described by three main parameters – sensitivity, selectivity, and fluorescence response value – and all these depend to a great extent on the fluorophore design.

Fluorescence is the emission of light by a substance that has absorbed light or other electromagnetic radiation. Electronic excitation of a fluorophore in its singlet ground state S_0, by an incident photon of sufficient energy, promotes one of the fluorophore electrons into a level of higher energy. The excited energy level initially populated will be a singlet state (such as S_1, S_2, or higher). For most molecules in solution, the subsequent relaxation of the molecule energy through collisional deactivation with solvent molecules will be rapid, resulting in the first singlet excited state S_1. From this S_1 state, fluorophores will dissipate their remaining energy as light. The emission of a photon from this locally excited state (LE) will result in an emission wavelength corresponding to the difference in energy between the initial and final electronic energy levels occupied.

The excitation of a fluorophore with a dipole moment in the ground state generally results in growth of its dipole moment. In solutions, where polar solvent molecules are free to re-orientate themselves and maximize favorable dipole–dipole interactions, the energy of the excited state can be lowered, resulting in the difference between positions of the band maxima of the absorption and emission spectra (Stokes shift). In most cases, the emitted light has a longer wavelength, and therefore lower energy, than the absorbed radiation (Figure 2.3.3).

Most of the literature on fluorescent sensors can be divided into three main groups depending on their principle design – sensors based on intramolecular charge transfer (ICT), sensors exploiting photoinduced electron transfer (PET), and sensors exploiting fluorescence resonance energy transfer (FRET).

2.3.3.1 Intramolecular Charge Transfer Sensors

The ICT-based sensors have a rather simple design and consist of two main parts – fluorophore and the receptor (boronic acid) – which is attached directly to fluorophore with a significant orbital overlap for electronic coupling. Alterations in the electron distribution of the sensor result in changes to both emission wavelength and fluorescence intensity. The first fluorescent sensor for carbohydrate detection was reported by Yoon and Czarnik in 1992 and comprised anthracene with attached boronic acid fragment (Scheme 2.3.3) [31]. On addition of saccharide, the intensity of the fluorescence emission of 2-anthrylboronic acid was reduced by ~30%. The change in fluorescence emission intensity was ascribed to the change of electronic effect of the substituent. At pH 7.4 (phosphate buffer), the neutral sp^2-hybridized boronic acid displays a strong fluorescence emission, whereas its sp^3-hybridized negatively charged complex with fructose displays a reduction in the intensity of fluorescence emission. The observed

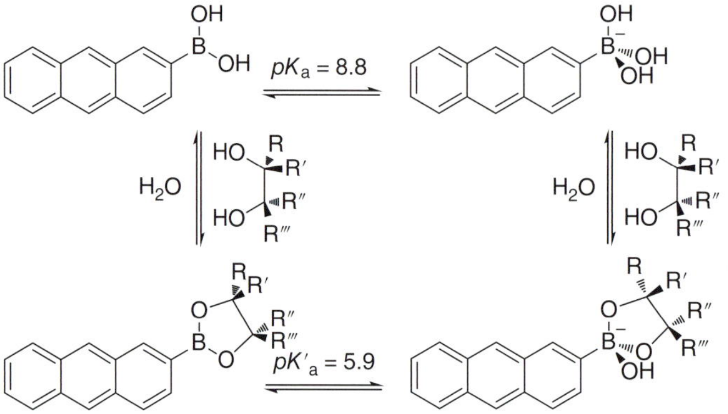

Scheme 2.3.3

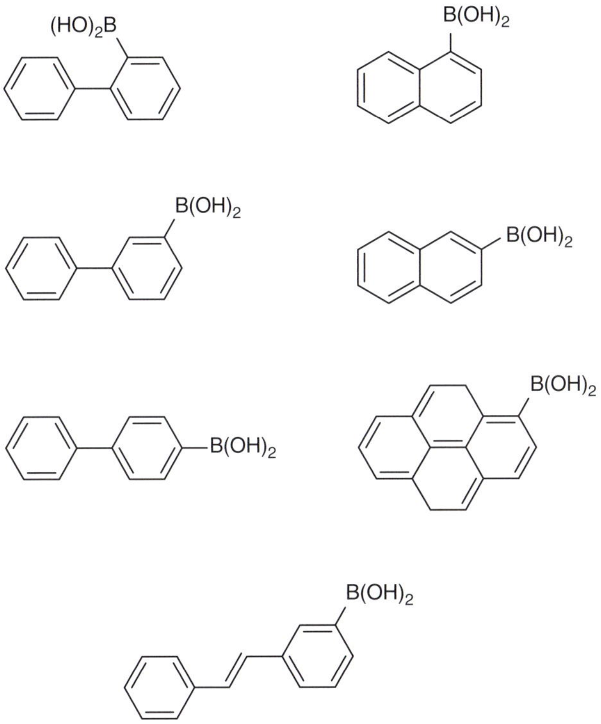

Chart 2.3.1

stability constant (K_{obs}) for the complex of 2-anthrylboronic acid with D-fructose at pH 7.4 in DMSO/water (1:99 v/v) was 270 M^{-1}; however, the fluorescence response was rather small.

Later, the sensing ability of a series of other simple arylboronic acids (Chart 2.3.1) against D-glucose was investigated [32,33].

Based on understanding of the simple fluorophore–boronic acid systems as ICT sensors, the fluorescence response can be significantly increased by proper design of fluorophore. In many respects, the ICT can be considered as an extension of the solvent relaxation mechanism. In general, molecules that exhibit intramolecular electron transfer are conjugate organic π-systems with acceptor (A) and donor (D) subunits linked by a formally single bond. The photoexcitation of D–A molecules is followed by an electron transfer from donor to acceptor. Excitation of a D–A system induces the motion of an electron from one orbital to another. If the initial and final orbital are separated in space, the electronic transition is accompanied by an almost instantaneous change in the dipole moment of the D–A system in the same molecule. When an electron-donating group is conjugated to an electron-withdrawing group, the dipole moment can be increased substantially, enhancing the favorable dipole–dipole interactions of the fluorophore molecule with the surrounding solvent molecules. This process of environmentally dependent stabilization of the S_1 state can result in dramatic changes in the wavelength of the emission band (bathochromic shift) and as such has found great application for fluorescence sensing.

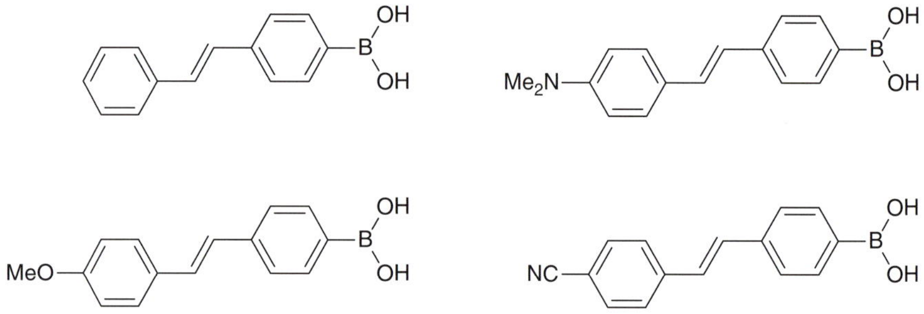

Chart 2.3.2

The best known example of the ICT system is *para*-N,N-dimethylaminobenzonitrile (DMABN). In nonpolar solvents, DMABN displays a "normal" fluorescence response (i.e., that characteristic of a benzene derivative in its LE state). In polar solvents, a second emission band of longer wavelength emerged. The relative intensity of the long-wavelength ICT band was found to grow with decreasing intensity of the short-wavelength LE band as a function of the increasing solvent polarity [34]. The solvent role in this process was demonstrated by the absence of ICT in crystalline DMABN [35].

To clarify the fluorescence mechanism in boronic acid derivatives, DiCesare and Lakowicz examined an interaction of saccharides with a series of stilbene boronic acid derivatives (Chart 2.3.2) [36]. It was demonstrated that the neutral sp^2-hybridized boronic acid is an electron-withdrawing group, whereas the anionic sp^3-hybridized boronic acid acts as an electron-donating group.

In these instances, when the moiety at the 4′-position and the hybridization at boron conspired to produce a system with a donor and an acceptor linked through the conjugated stilbene scaffold, ICT takes place, lowering the excited state energy. This influence was elegantly exemplified by the examination of two diametrically opposed systems.

In the case of 4′-dimethylaminostilbene-4-boronic acid, the electron-donating dimethylamino moiety is the donor group. When boron is sp^2 hybridized, and therefore an acceptor, excited-state ICT can occur between the amino donor and boron acceptor, red-shifting the emission wavelength of the sp^2 species. On the rehybridization of boron to sp^3, its acceptor properties are lost. This leads to a loss of the ICT effect in the excited state of the sp^3 species and shifts the emission wavelength of the fluorophore to higher energy. The inability of the sp^3-hybridized species to lower the energy of its excited state by a mechanism available to the sp^2-hybridized ones causes strong blue shift of the emission band with an increase in its intensity (Scheme 2.3.4).

The fluorescent response to the loss of the electron-withdrawing properties of boron on conversion from sp^2 to sp^3 hybridization was verified by increasing the pH of the sensor solution (from 6.0 to 12.0) and by the addition of saccharide to the sensor solution buffered at pH 8.0 ($pK_a' = 6.61$, pH = 8.0, $pK_a = 9.14$), both titrations yielding the same results.

Conversely, in 4′-cyanostilbene-4-boronic acid, the electron-withdrawing cyano moiety is the acceptor group. When boron is sp^2 hybridized and therefore also an acceptor, no excited-state ICT is feasible. On the rehybridization of boron to its sp^3 form, the boron becomes a donor group that allows the ICT between the boron donor and the

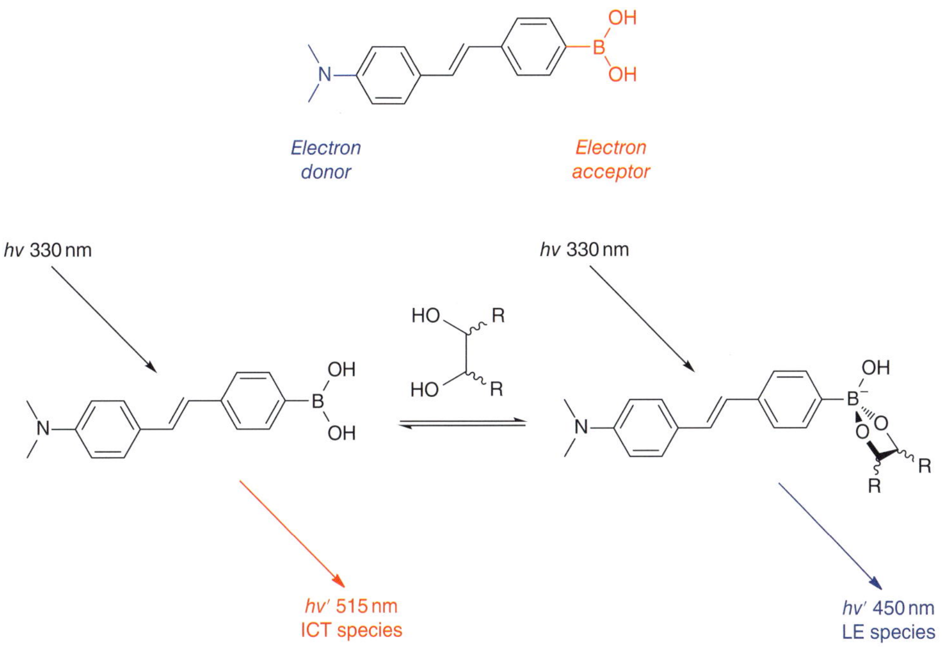

Scheme 2.3.4

cyano acceptor, resulting in red shift of the emission wavelength with a decrease in the emission intensity. Reversing the roles of the donor and acceptor groups for the boronic acid and 4′-substituent produces changes in emission wavelength and intensity for both sensors that are very similar in magnitude, but occur toward opposite ends of the electromagnetic spectrum (Scheme 2.3.5).

The fluorescence response to the revival of the electron-donating properties of boron on conversion from sp^2 to sp^3 hybridization was again verified by increasing the pH of the sensor solution (from 6.0 to 12.0) and by the addition of saccharide to the sensor solution buffered at pH 8.0 (pK_a' = 5.84, pH = 8.0, pK_a = 8.17), both titrations yielding the same results.

It should be noted that such simple modification of the fluorophore results in significant variations in stability of the boronic acid–carbohydrate complexes. Thus, the observed stability constants (K_{obs}) for the 4′-cyanostilbene-4-boronic acid sensor are 290 and 91 M^{-1} with D-fructose and D-glucose, respectively; and, for the 4′-methoxy-, 4′-dimethylamino-, and 4′-cyano derivatives, these values are 1000 and 23 M^{-1}, 400 and 10 M^{-1}, and 1500 and 55 M^{-1}, respectively, in 2:1 (v/v) methanol/water at pH 8.0 (phosphate buffer) [36].

Later, a range of ICT sensors containing other spacers between the dimethylamine and boric acid moieties were synthesized, and their carbohydrate-sensing ability was determined [37–39] (Chart 2.3.3).

A series of isomeric dimethylamine derivatives of naphthylboronic acids (Chart 2.3.4) were shown to have attractive fluorescence characteristics. Thus, 4-(dimethylamino) naphthalene-1-boronic acid produces a 41-fold emission intensity increase with D-fructose at pH 7.4 (phosphate buffer); the observed stability constants K_{obs} are 207 M^{-1}

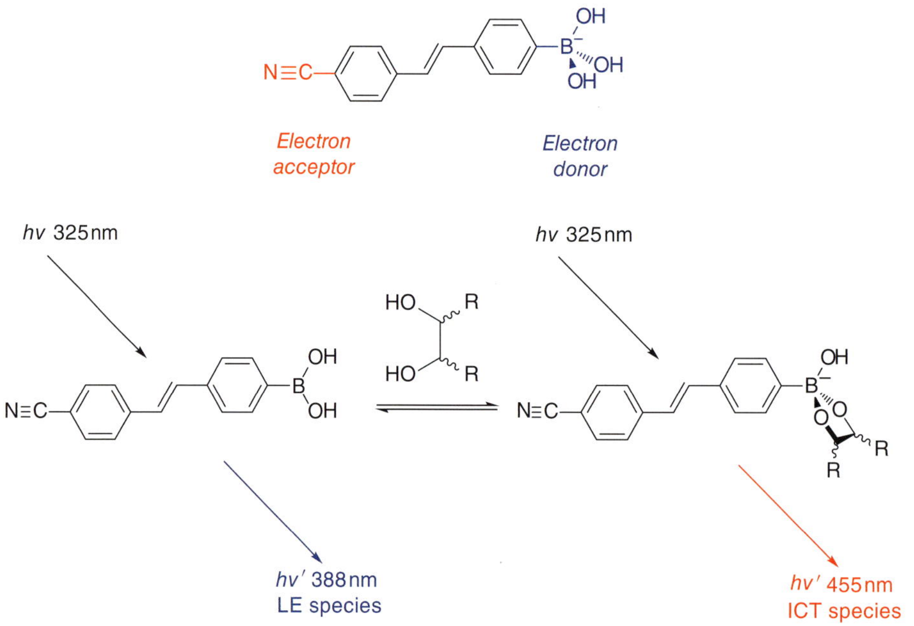

Scheme 2.3.5

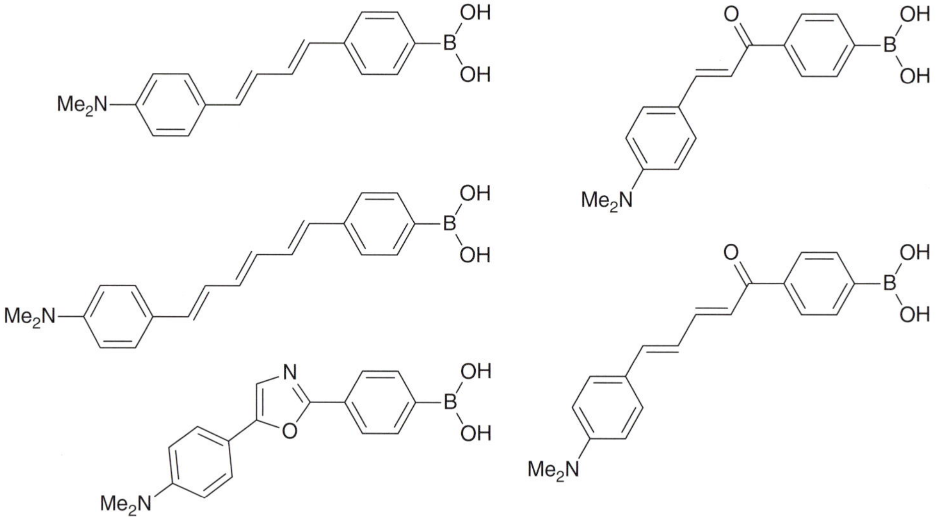

Chart 2.3.3

for D-fructose and 4.0 M^{-1} for D-glucose [40]. The isomeric 5-(dimethylamino)naphthalene-1-boronic acid is a ratiometric sensor that demonstrates large fluorescence intensity changes at two wavelengths – 433 nm (36-fold increase in intensity) and 133 nm (61% decrease in intensity) with D-fructose at pH 7.4 (phosphate buffer); the observed stability constants K_{obs} are 311 M^{-1} for D-fructose and 3.6 M^{-1} for D-glucose [41]. Another isomeric naphthalene, 6-(dimethylamino)naphthalene-2-boronic acid, demonstrates an

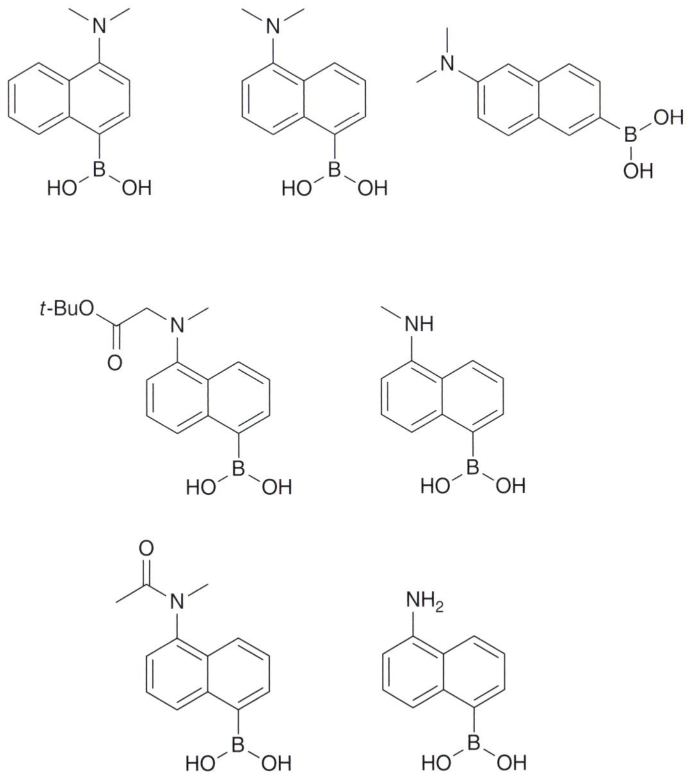

Chart 2.3.4

80% decrease in fluorescence on D-fructose addition at pH 7.4 (phosphate buffer); the observed stability constants K_{obs} are $120\,M^{-1}$ for D-fructose and $2.4\,M^{-1}$ for D-glucose [42]. A series of 5-(dimethylamino)naphthalene-1-boronic acid analogs with different substituents on the aniline group of the naphthalene ring were synthesized, and their binding with saccharides was studied. It was revealed that the substitution pattern on the aniline nitrogen atom has a significant effect on the fluorescence properties of these compounds. 5-(*tert*-Butoxy-carbonylmethylmethylamino)naphthalene-1-boronic acid shows ratiometric fluorescence changes upon binding of a sugar. 5-(Methylamino)naphthalene-1- and 5-aminonaphthalene-1-boronic acids show dramatic fluorescence increases upon binding of a sugar, so they are very good off-on fluorescence sensors for sugars. In addition to the quantifiable fluorescence property changes upon sugar addition, the fluorescence color changes in these compounds are also visible to the naked eye. Therefore, they can also be used as color sensors. In contrast, the aniline group in 5-(acetylmethylamino)naphthalene-1-boronic acid is masked as an amide, making it only weakly fluorescent since the lone-pair electrons of the nitrogen atom in the amide group are not readily available due to resonance stabilization [43].

A series of quinoline and indole-based boronic acid were synthesized, and their carbohydrate-sensing properties were studied (Chart 2.3.5) [44–48].

4,4-Difluoro-4-bora-3a,4a-diaza-*s*-indacene (BODIPY) is another attractive fluorophore for design of boronic-based fluorescence sensors due to its high quantum yield,

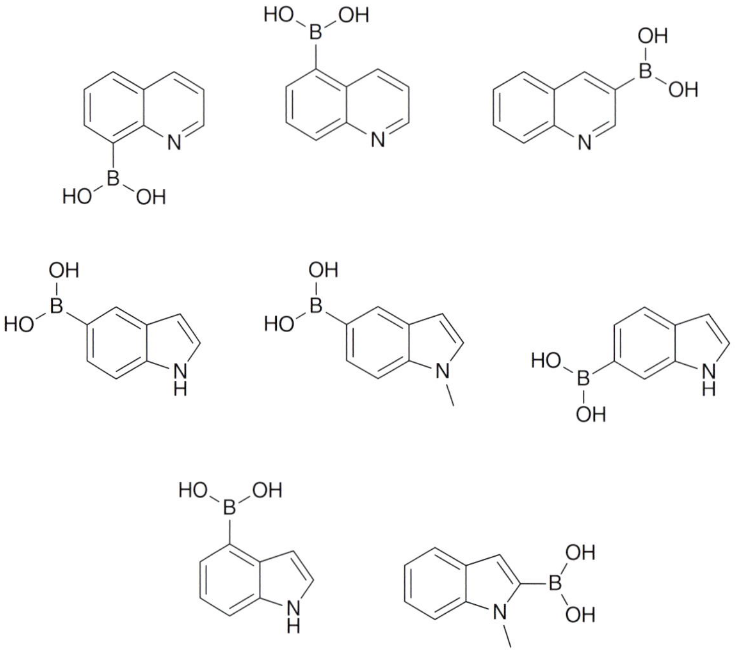

Chart 2.3.5

tunable fluorescence characteristics, high photostability, and narrow emission bandwidth [45]. Recently, a series of BODIPY-containing phenylboronic acids were synthesized, and their carbohydrate-sensing properties were studied (Chart 2.3.6) [50–53]. It was demonstrated that both the fluorophore wavelength tuning and effective fluorescence response can be achieved by attaching auxochromic substituents to the 5-position of the BODIPY core. For example, the introduction of substituent in 8-(4-$(HO)_2BC_6H_4$)-1,3,5,7-Me_4-BODIPY produces strong red shift of the emission maximum from 502 to 580 nm and nearly fivefold increase of the emission intensity (Figure 2.3.4) [52]. Similarly, the introduction of substituent in 1-(4-$(HO)_2BC_6H_4CH=CH$)-3-Me-BODIPY results in a nearly 20-fold increase of the fluorescence response to incubation with D-fructose (Figure 2.3.5) [53].

One of the boronic acids synthesized, so-called Fructose Orange, exhibited a remarkable 24-fold fluorescence enhancement (the quantum yield changed from 0.01 to 0.27 in the presence of 200 mM fructose) with an excellent selectivity among a large collection of 24 saccharides (Figure 2.3.6), and it was successfully applied for determination of fructose in Coca-Cola [53].

2.3.3.2 Photoinduced Electron Transfer Sensors

The next class of fluorescence systems for saccharides to be discussed is PET sensors, which are probably the most important class of boronic acid–based fluorescence sensors. Most PET sensors consist of a fluorophore linked to an amine moiety via a methylene spacer. PET, which takes place from amino groups to aromatic hydrocarbons, causes fluorescence quenching of the latter. When the amino group strongly interacts with a

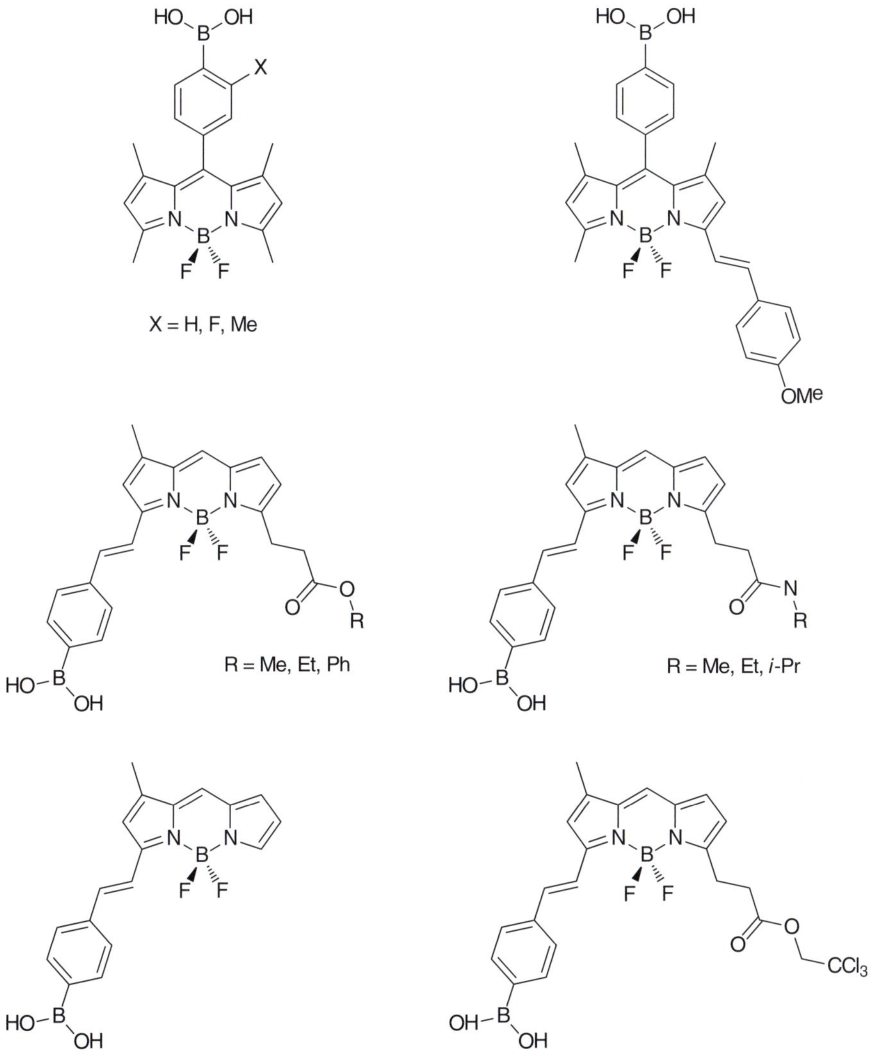

Chart 2.3.6

boric moiety, electron transfer is hindered and a very large enhancement of the fluorescence is observed [54]. Figure 2.3.7 illustrates the mechanism in terms of molecular orbitals. On excitation of the fluorophore, an electron of the highest occupied molecular orbital (HOMO) is promoted to the lowest unoccupied molecular orbital (LUMO), which enables PET from the HOMO of the donor (free amine) to that of the fluorophore, causing fluorescence quenching (OFF). Upon sugar binding, the redox potential of the donor is raised so that the relevant HOMO becomes lower in energy than that of the fluorophore, and consequently, PET is no longer possible and the fluorescence quenching is suppressed. In other words, the fluorescence intensity is enhanced upon cation binding (ON). This system can be interpreted as an "off-on" molecular switch.

9-{[N-Methyl-N-(*ortho*-boronobenzyl)amino]methyl}anthracene, shown in Figure 2.3.7, is the first rationally designed fluorescent PET sensor that displays the same inherent

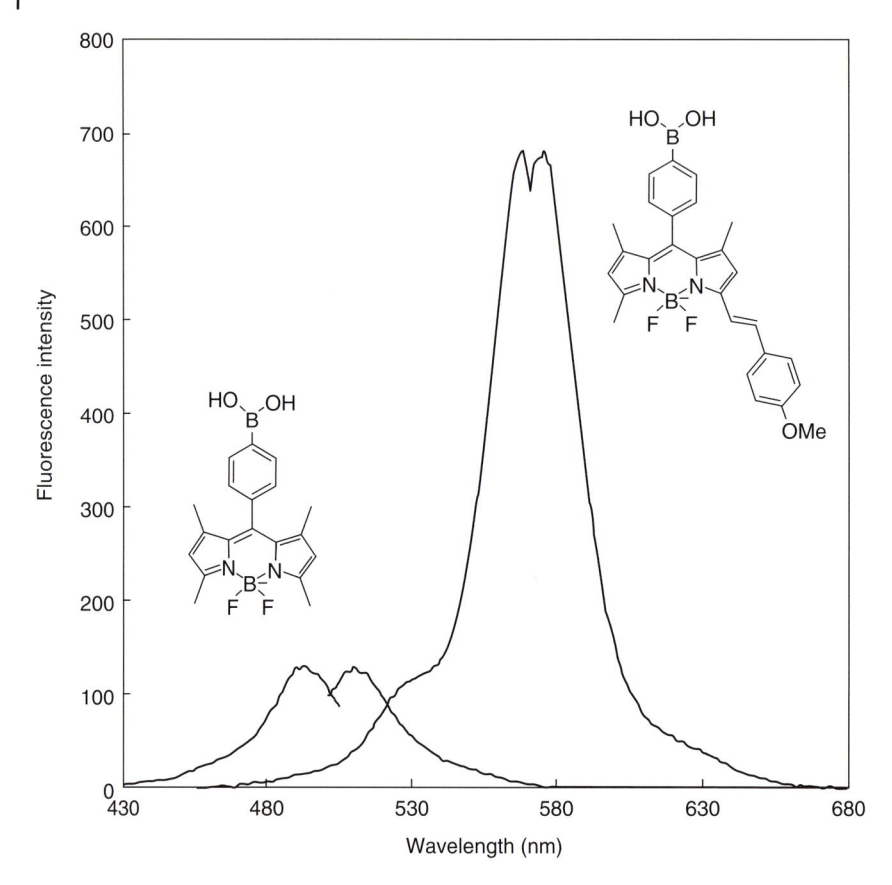

Figure 2.3.4 Excitation and emission fluorescence spectra of BODIPY-containing boronic acids in ethanol. *Source*: Adapted with permission from ref. [52]. Copyright (2012) American Chemical Society.

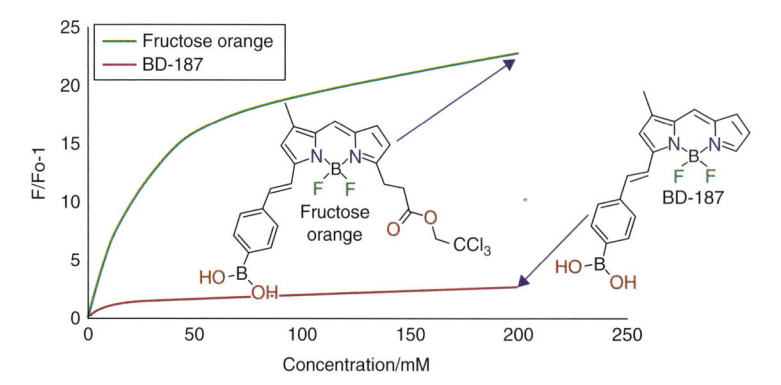

Figure 2.3.5 Fluorescence responses of BODIPY-containing boronic acids after incubation with D-fructose. *Source*: Reprinted with permission from ref. [53]. Copyright (2012) American Chemical Society.

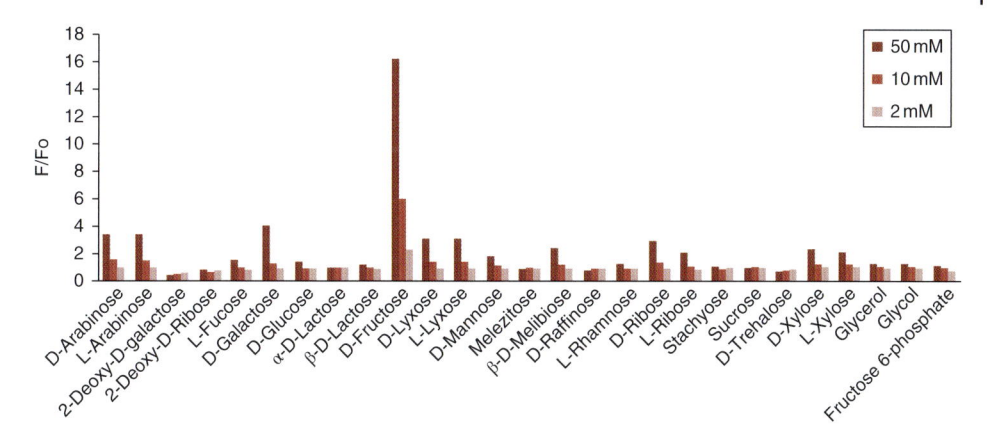

Figure 2.3.6 Selectivity of Fructose Orange against 24 different sugars, glycerol, ethylene glycol, and fructose 6-phosphate at different concentrations in HEPES buffer (pH: 7.4). *Source*: Reprinted with permission from ref. [53]. Copyright (2012) American Chemical Society.

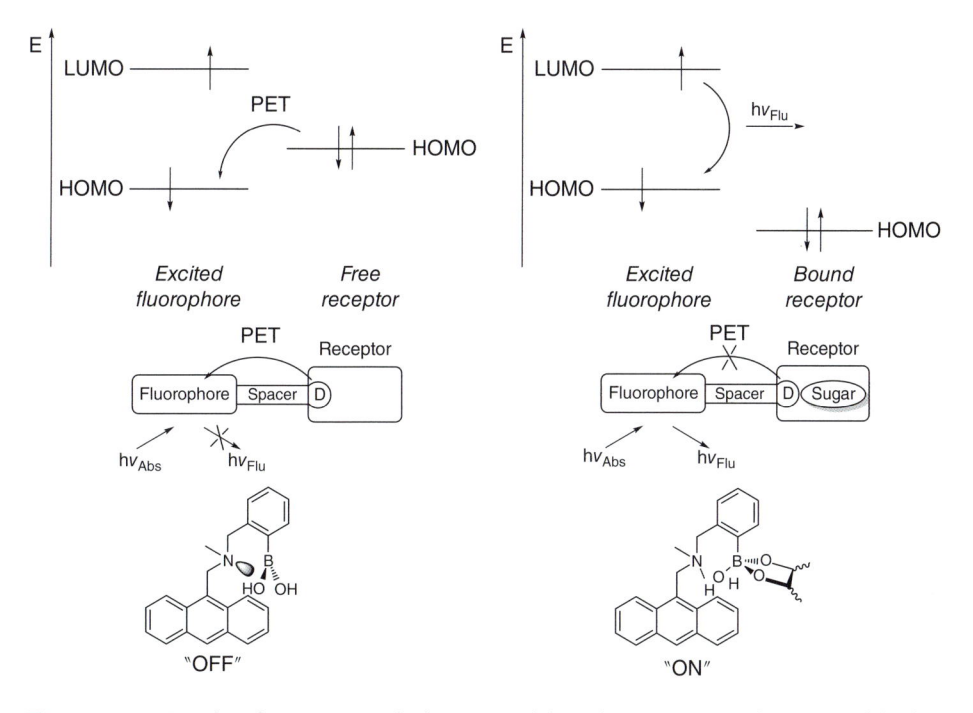

Figure 2.3.7 Principles of sugar sensing by boronic acid–based PET sensors, and structure of the first rationally designed fluorescent PET sensor.

trend in selectivity for saccharides as other monoboronic acids (the observed stability constants K_{obs} are $1000\,M^{-1}$ for D-fructose and $60\,M^{-1}$ for D-glucose in 33.3 wt.% methanol–water at pH 7.8) [55,56]. This simple "off-on" PET system was modified with the introduction of a second boronic group (Chart 2.3.7). The modification proved successful, and fortuitously the spacing of the two boronic acid groups provided an effective binding pocket for *D*-glucose. The complexation of *D*-glucose occurred with a 1:1 stoichiometry with the saccharide binding to form a macrocyclic ring. While the inherent selectivity of monoboronic acids is for *D*-fructose, in this compound the stabilization derived from the rigid macrocyclic ring produces a *D*-glucose selective system (the observed stability constants K_{obs} are $320\,M^{-1}$ with *D*-fructose and $4000\,M^{-1}$ with *D*-glucose in 33.3 wt.% methanol–water at pH 7.8) [56,57].

Later, the same design was used for synthesis of chiral sensors (Chart 2.3.8), which were found to demonstrate very high chemo- and enantioselectivity for sugar acids, such as tartaric, glucaric, gluconic, and glucuronic acids, as well as for six-carbon sugar alcohols, but do not bind strongly with five- or four-carbon sugar alcohols or monosaccharides [58,59].

The interaction between the amine nitrogen and boric acid moiety plays a pivotal role in signaling the binding event. For some time, the formation of a direct bond between nitrogen and boron was assumed to be responsible for the fluorescence enhancement seen when boronic acids bound diols. This interpretation does, however, raise certain questions.

The X-ray study of the diboronic acid sensor revealed that in the unbound receptor, the geometry in boron is trigonal planar. This is important as the absence of deviation from planarity implies that there is no direct N–B Lewis base–Lewis acid bond in boron. For the unbound receptor, the hydrogen atoms of the boronic acid were located and participate in the BO–H...N hydrogen bonds with the nitrogen atoms of the receptor framework (the N...O distances are 2.764 and 2.759 Å) [58]. The formation of such intramolecular BO–H...N bonds forming seven-membered rings was found to be typical for *ortho*-aminomethylphenylboronic acids [60–65]. When the receptor was bound to tartaric acid and the complex was crystallized from a methanol–dichloromethane

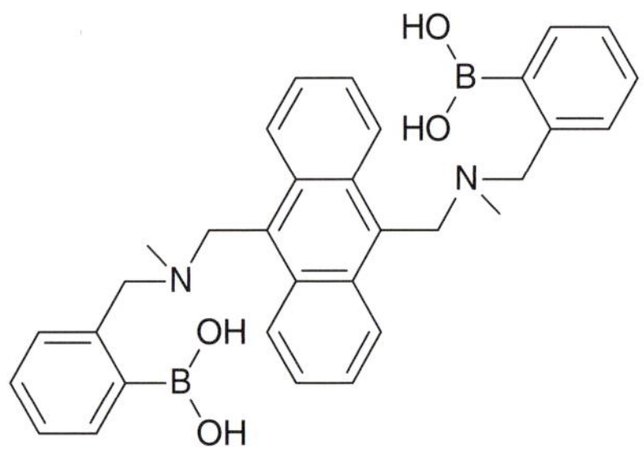

Chart 2.3.7

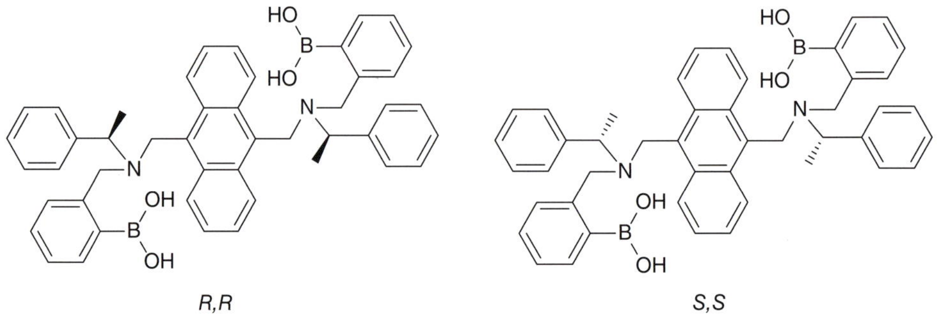

Chart 2.3.8

solution, the tetragonal geometry of the boron atom with methanol bound through its oxygen atom to the boron center and also hydrogen bonds to the nitrogen atom of the receptor framework (the N...O distances are 2.655 Å and 2.693 Å) were found. However, the hydrogen atom was not located, and it was impossible to determine whether that hydrogen was closer to the nitrogen or the oxygen. The position of the hydrogen is a subtle issue, but it defines whether the solvent inserted structure should be best considered as a neutral complexed boronic acid or whether the methanol is closer to being fully dissociated, creating a zwitterionic structure (Figure 2.3.8) [58].

The X-ray study of closely related complexes of *ortho-N,N*-tetraethyleneaminomethyl phenylboronic acid (Figure 2.3.9) [66] and 9-{[*N*-Methyl-*N*-(*ortho*-boronobenzyl)

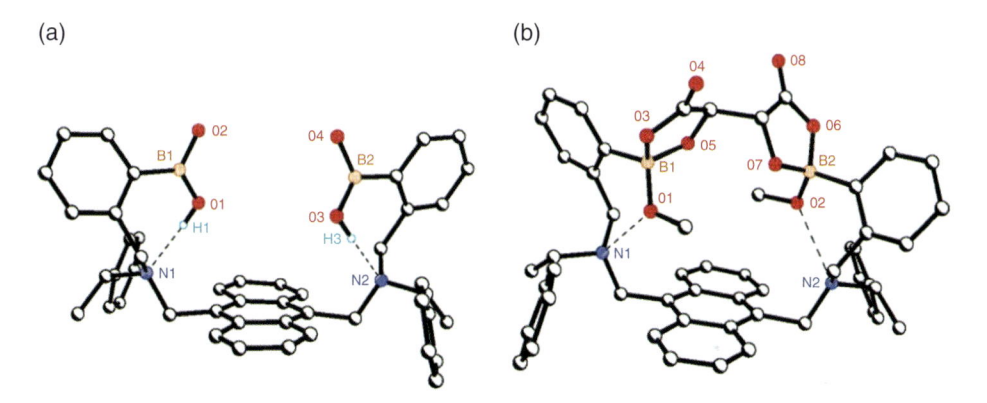

Figure 2.3.8 (a) X-ray structures of the R,R-chiral diboronic acid, and (b) the S,S-chiral diboronic acid–tartrate complex. *Source*: Reprinted with permission from ref. [58]. Copyright (2004) American Chemical Society.

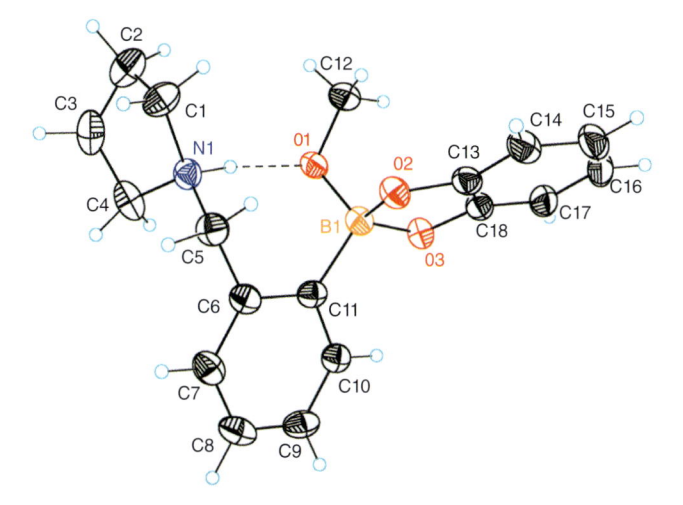

Figure 2.3.9 X-ray structure of the complex of *ortho-N,N*-tetramethyleneaminomethyl phenylboronic acid with catechol and methanol. *Source*: Reprinted with permission from ref. [66]. Copyright (2009) American Chemical Society.

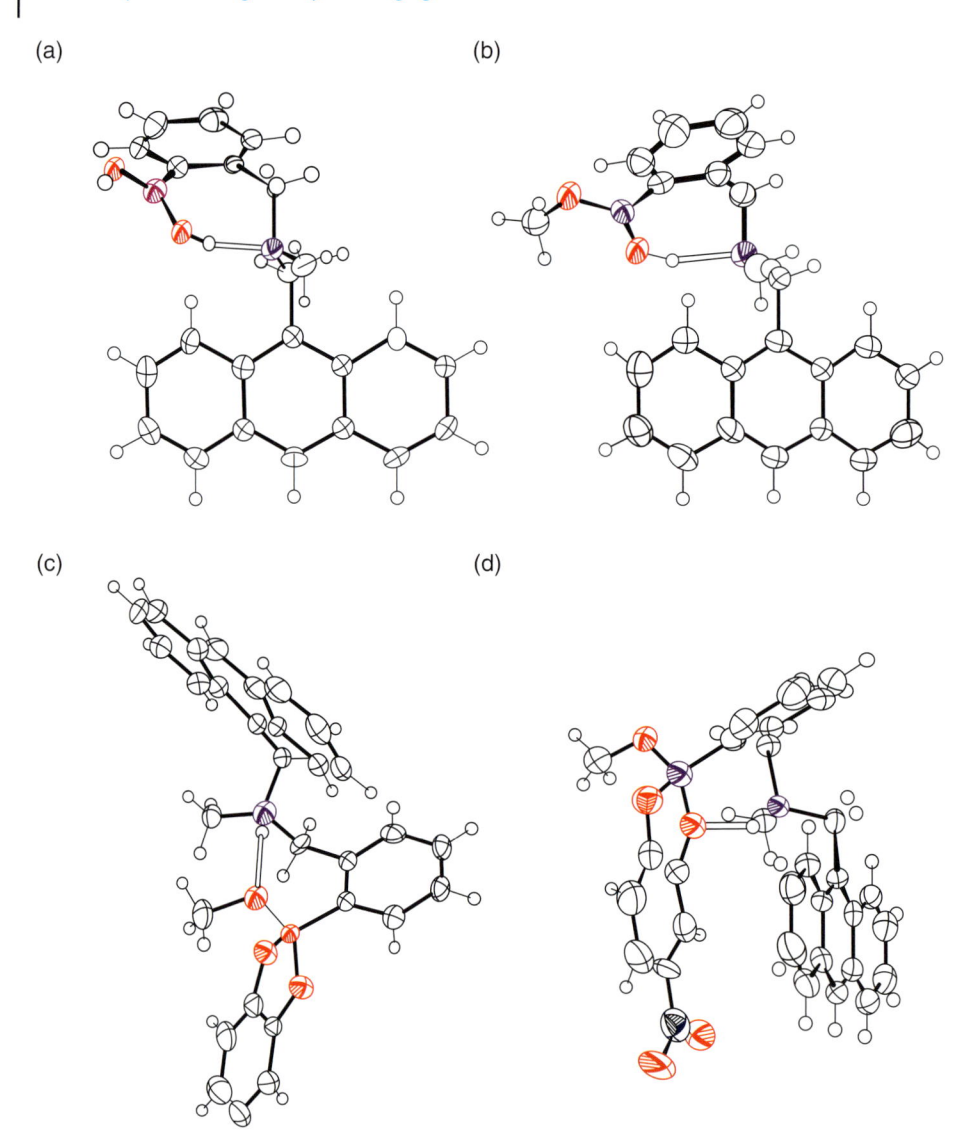

Figure 2.3.10 X-ray structures of 9-{[N-Methyl-N-(*ortho*-boronobenzyl)amino]methyl}anthracene crystallized from (a) dichloromethane and (b) methanol, and its complexes with (c) catechol and (d) 4-nitrocatechol. *Source*: Reprinted with permission from ref. [63]. Copyright (2009) the Royal Society of Chemistry.

amino]methyl}-anthracene (Figure 2.3.10) [63] with catechol and methanol revealed formation of the zwitterionic structure with the protonated nitrogen atom.

It should be noted that solvent insertion into the B–N bond is not a new idea. The similar structure was postulated more than 50 years ago for complexes of the *cis*-1, 2-cyclopentanediol ester of 8-quinolineboronic acid with solvent water and phenol molecules bridging the nitrogen and boron centers based on their infrared (IR) spectra (Chart 2.3.9) [67].

Chart 2.3.9

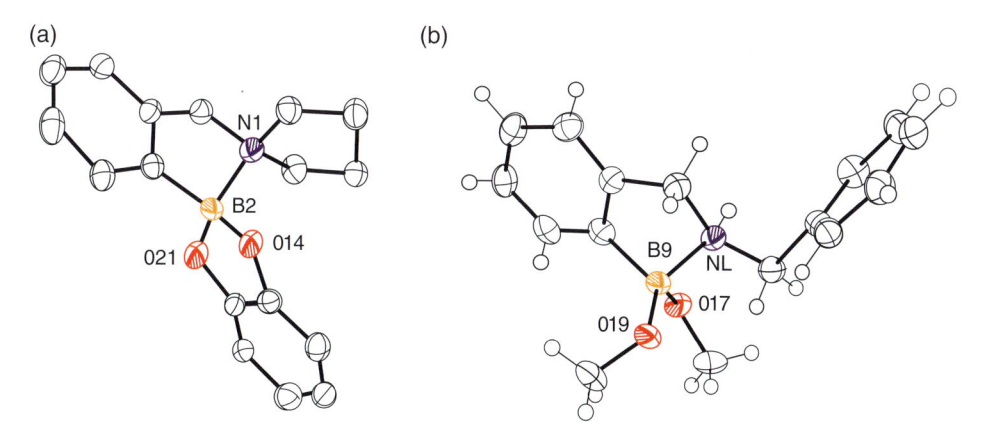

Figure 2.3.11 X-ray structures of (a) catechol *ortho*-(*N,N*-tetramethyleneaminomethyl) phenylboronate and (b) dimethyl *ortho*-(*N*-benzylaminomethyl)phenylboronate. *Source*: Reprinted with permission from ref. [68]. Copyright (2006) American Chemical Society.

On the other hand, the detailed ^{11}B NMR and X-ray studies of the N–B interactions in *ortho*-(*N,N*-dialkylaminomethyl) arylboronate systems demonstrated that in an aprotic solvent, the N–B dative bond is usually present (Figure 2.3.11). However, in protic media, solvent insertion of the N–B occurs to afford a hydrogen-bonded zwitterionic species [66,68].

One of the fundamental differences between the PET sensors and the above-described ICT sensors is the presence of a spacer between the receptor (boric acid group) and the fluorophore. This makes it possible to use such sensors in a modular design approach, when one or more blocks may be changed to achieve the desired result. One the most interesting examples is adjusting the spacer length between two boronic acid receptors to achieve selectivity to *D*-glucose (Chart 2.3.10) [69].

The sensor with a flexible six-carbon linker provides the optimal selectivity for *D*-glucose over other saccharides and exhibits the highest observed stability constant K_{obs} within these systems (Figure 2.3.12). This is in agreement with the observed selectivity of other sensors with pyrene and anthracene fluorophores, which also have linkers containing six carbon atoms [56,70].

Another impressive example is a library of diboronic acid sensors with varying spacers, where a series of 26 compounds with different flexible and rigid spacers was synthesized and their binding with various carbohydrates was studied (Chart 2.3.11) [71–73].

Chart 2.3.10

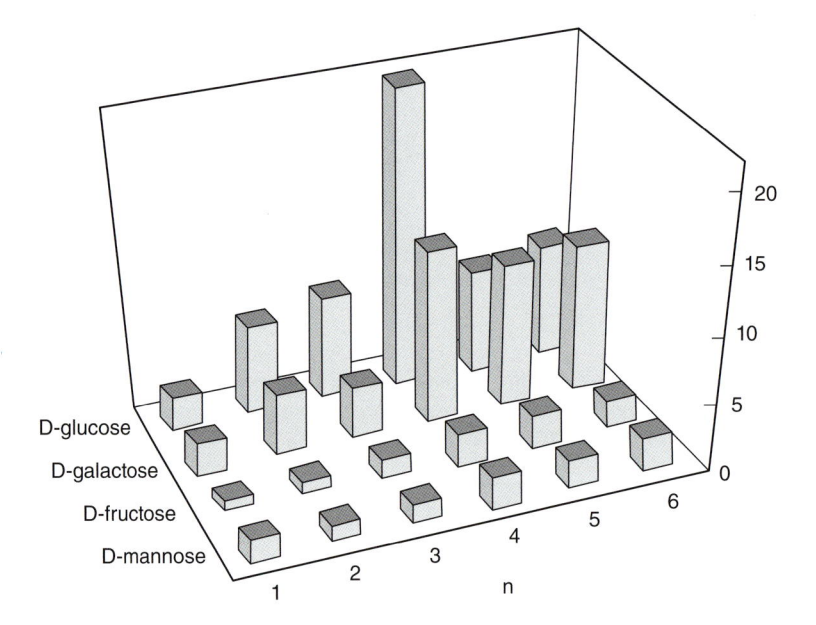

Figure 2.3.12 Relative stability constants of diboronic acid versus monoboronic acid with carbohydrates. *Source*: Adopted with permission from ref. [69]. Copyright (2002) the Royal Society of Chemistry.

Chart 2.3.11

Chart 2.3.12

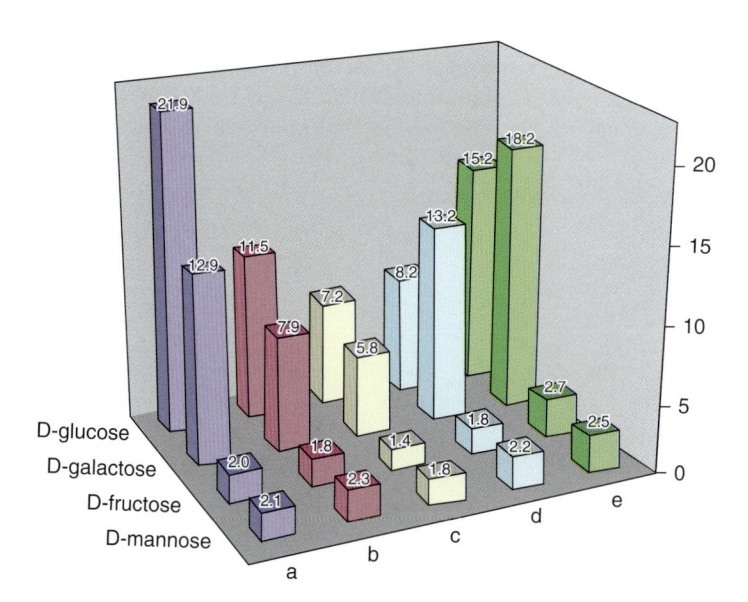

Figure 2.3.13 Relative stability constants of diboronic acid versus the corresponding monoboronic acids with carbohydrates. *Source*: Reprinted with permission from Ref. [71]. Copyright (2003) Elsevier.

Another example is the fluorophore selection for the sensors with given design (Chart 2.3.12). It was found that the relative stability constants' complexes with D-fructose and D-mannose do not depend markedly on the fluorophore choice, whereas the relative stability constants' complexes with D-glucose and D-galactose display a saddle-like trend from pyrene to 2-naphthalene (Figure 2.3.13) [74].

In general, the modular approach enables almost unlimited possibilities of synthesis of very complex boronic acid–based PET sensors with different numbers and locations of receptors and fluorophores.

2.3.3.3 Fluorescence Resonance Energy Transfer Sensors

Another type of fluorescent sensors for carbohydrate detection is FRET sensors. FRET involves the nonradiative energy transfer from a fluorescent donor molecule to an acceptor molecule in close proximity, which is usually caused by dipole–dipole interactions. As long as the donor and acceptor remain in close proximity, no emission occurs. Events that trigger displacement of the donor from the acceptor will result in a regain of emission. The rate of energy transfer in dipole–dipole interactions is inversely proportional to R^6, where R is the distance between the donor and acceptor. This means that even small changes in distance can be measured by FRET (i.e., Angstrom level). Therefore, FRET is commonly applied in biology as a powerful research tool to measure distance and detect molecular interactions at the cellular level [75,76].

The typical FRET sugar sensor consists of two components – a fluorescent reporter and a boronic acid-containing quencher/receptor. Examples of such FRET systems are the series of viologen-based sugar sensors developed by Singaram and coworkers. This simple example includes the commercially available dye 8-hydroxy-1,3,6-pyrenetrisulfonic acid trisodium salt (HPTS or pyranine) as the fluorescent reporter and the boronic acid–substituted viologen 4,4'-N,N-bis(benzyl-2-boronic acid)bipyridinium dibromide (o-BBV^{2+}) as the quencher/receptor molecule (Scheme 2.3.6). These sensors are capable of operating in aqueous solution at pH 7.4 and are highly sensitive to D-glucose in the physiological range. The perfect summary of this study was published in 2016 [77].

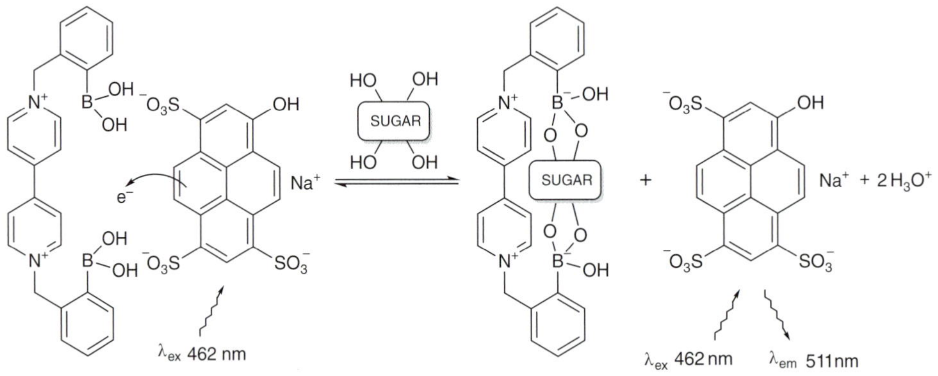

Scheme 2.3.6

2.3.4 Colorimetric Sensors

Colorimetric sensors for carbohydrates are of particular interest in a practical sense. If a system with a large color change can be developed, it could be incorporated into a diagnostic test paper for *D*-glucose, similar to universal indicator paper for pH. Such a system would make it possible to measure *D*-glucose concentrations without special equipment, and this would be of particular benefit to diabetics in developing countries [78].

There are fewer colorimetric sugar sensors based on boronic acids compared to the number of fluorescence sensors. The first boronic acid azo dyes were prepared at the end of the 1940s for use in boron neutron capture therapy [79,80]. However, only in the 1990s were some boronic acid–appended azobenzene derivatives synthesized for sugar sensing. Koumoto *et al.* proposed to use the boronic acid–amine interaction for molecular design of an intermolecular sensing system for saccharides. *m*-Nitrophenylboronic acid interacts with the pyridine nitrogen of 4-(4-dimethylaminophenylazo)pyridine in methanol and changes its color from yellow to orange. Added saccharides form complexes with the boronic acid and enhance the acidity of the boronic acid group. As a result, the boron–nitrogen interaction becomes stronger, and the intensified intramolecular charge-transfer band changes the solution color to red (Chart 2.3.13) [81].

This is a pioneering work using intermolecular interactions between boronic acid and dyes. The concept of the combination of two molecules has become a growing trend. Many researchers used dyes containing catechol structures, like alizarin red S and pyrocatechol violet. These dyes form a cyclic ester with boronic acid, which accompanies a change in the color or fluorescence of the dyes. The sugar addition induced the displacement of dyes from the cyclic ester, which results in a recovery of the original signal of the dyes [82–84]. These combinations of boronic acids and dyes will offer some interesting applications.

Ideally, a simple system using one constituent would be suitable for practical glucose sensing to avoid interferences in complex biological fluids. Ward *et al.* reported a series of dyes with a basic skeleton (Chart 2.3.14a) [85,86]. In particular, the azo dye with an electron-withdrawing nitro group shows a large color change upon sugar addition in aqueous MeOH at pH 11. The wavelength shifted circa 55 nm to a shorter wavelength upon sugar complexation, corresponding to the color change from purple to red. They proposed that a key structure of the signaling mechanism is an intramolecular B–N interaction between the boronic acid moiety and the nitrogen of the aniline moiety. This concept was applied to the dye with tricyanovinyl dye, which shows a color change

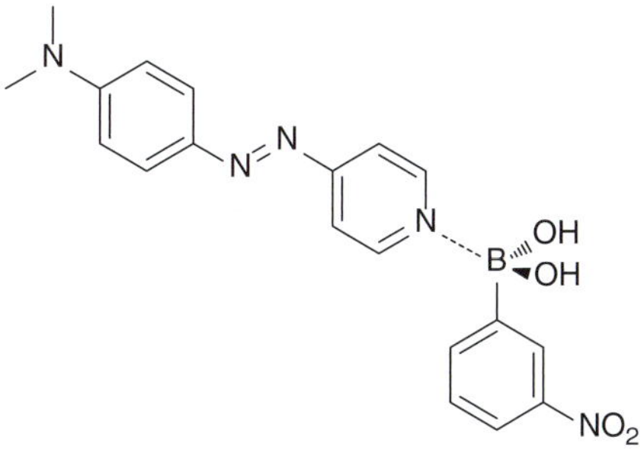

Chart 2.3.13

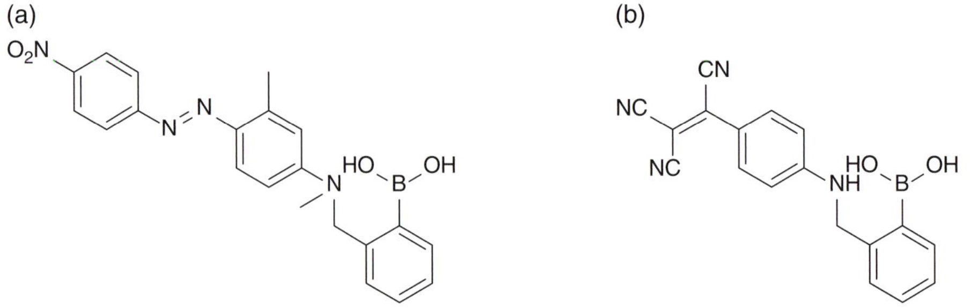

(a)

(b)

Chart 2.3.14

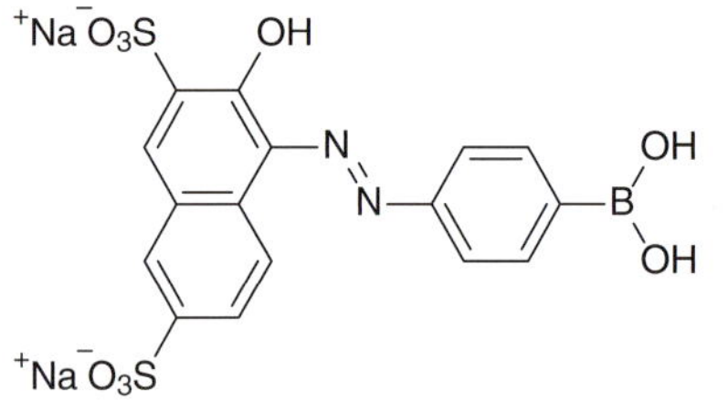

Chart 2.3.15

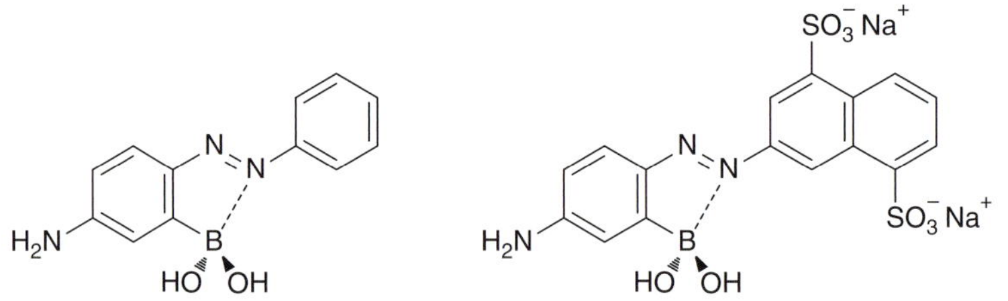

Chart 2.3.16

in a neutral aqueous solution (Chart 2.3.14b). In aqueous MeOH at pH 8.2, the addition of sugar induced a shape change in the absorption spectrum, which is recognized as a color change from purple to pink [87].

DiCesare and Lakowicz have demonstrated that azo dye containing naphthyl group (Chart 2.3.15) works as a visible color sensor in a complete water system at pH 7.0. The sugar addition induces red shift, which results in a change from orange to a purple-reddish color. They proposed that this color change is due to the conformational change of the boron atom from the neutral sp^2 form to the anionic sp^3 form [88].

In order to fabricate sugar sensors that show a significant color change, Egawa et al. developed a strategy to arrange a boronic acid group adjacent to a chromophore by introducing a boronic acid group to the *ortho*-position of the azo group. Some *ortho*-boronic acid substituted azobenzenes were successfully synthesized with diazo-coupling reactions (Chart 2.3.16) [89,90]. The absorption maximum at 505 nm in aqueous methanol shows significant blue shift to 386 nm on D-fructose addition (Figure 2.3.14) [90]. In order to improve the solubility in water, two sulfonyl groups were introduced to the azo dye [89].

Figure 2.3.14 UV-visible absorption spectra of the azo dye boronic acid in the absence and presence of *D*-fructose, measured in a methanol–water solution (1:1, v/v) at pH 10.0 (CHES buffer). *Source*: Reprinted with permission from ref. [90]. Copyright (2010) The Chemical Society of Japan.

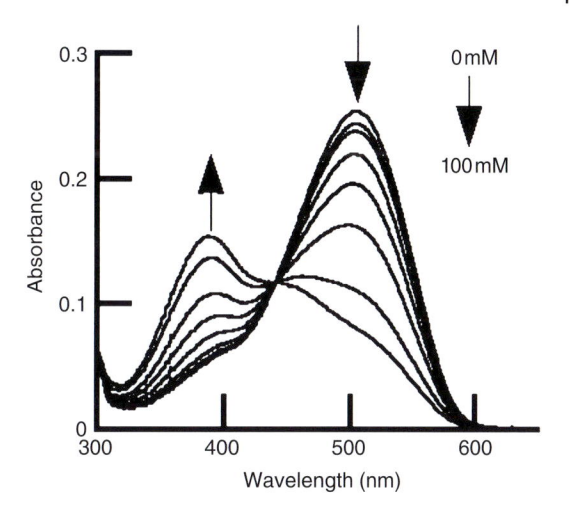

The ^{15}N NMR study demonstrated the existence of a B–N dative bond between boronic acid and azo groups [90]. The B–N dative bond causes significant red shift of the absorption maximum, and it is cleaved upon sugar addition, which results in a significant color change.

2.3.5 Conclusions

Saccharides play a significant role in the metabolic pathways of living organisms; therefore, identification of biologically important sugars (e.g., glucose, fructose, and galactose) and determination of their concentration are necessary in a variety of medical contexts. The boronic acids have the advantage of a reversible and fast equilibrium interaction with monosaccharides. In addition, the boronic acid group can be incorporated in many different systems, giving large possibilities for the development of analytic devices for the recognition and detection of sugars. Despite its long and rich history, understanding of boron acid interactions with carbohydrates continues to increase in the twenty-first century. Established in the past 25 years, sensing systems and the fundamental knowledge accumulated in the process of their development permit the design of more complex and sophisticated structures, thus enabling enhanced selectivity toward a wider range of sugar-like targets. Although further effort is still ongoing to address some common issues, such as selectivity and biocompatibility, it is reasonable to believe that future development of boronic acid–based saccharide sensors will overcome these problems and play an active role in biological research, clinical diagnosis, and even treatment of diseases.

References

1 World Health Organization (WHO), *Global Report on Diabetes*, WHO, Geneva, **2016**. http://apps.who.int/iris/bitstream/10665/204871/1/9789241565257_eng.pdf
2 A.L. Galant, R.C. Kaufman, J.D. Wilson, Glucose: Detection and analysis. *Food Chem.*, **2015**, 188, 149–160.

3 K.R. Ervin, E.J. Kiser, Issues and implications in the selection of blood glucose monitoring technologies. *Diabetes Technology & Therapeutics*, **1999**, 1, 3–11.
4 H.-C. Wang, A.-R. Lee, Recent developments in blood glucose sensors. *J. Food Drug Anal.*, **2015**, 23, 191–200.
5 T.D. James, S. Shinkai, Artificial receptors as chemosensors for carbohydrates. In: *Host-Guest Chemistry: Topics in Current Chemistry*, vol. 218, ed. S. Penadés, Springer, Berlin, **2002**, 159–200.
6 W. Wang, X. Gao, B. Wang, Boronic acid-based sensors. *Curr. Org. Chem.*, **2002**, 6, 1285–1317.
7 S. Striegler, Selective carbohydrate recognition by synthetic receptors in aqueous solution. *Curr. Org. Chem.*, **2003**, 7, 81–102.
8 H. Fang, G. Kaur, B. Wang, Progress in boronic acid-based fluorescent glucose sensors. *J. Fluoresc.*, **2004**, 14, 481–489.
9 T.D. James, Boronic acid-based receptors and sensors for saccharides. In: *Boronic Acids*, ed. D.G. Hall, Wiley-VCH, Weinheim, **2005**, 441–479.
10 J. Yan, H. Fang, B. Wang, Boronolectins and fluorescent boronolectins: An examination of the detailed chemistry issues important for the design. *Med. Res. Rev.*, **2005**, 25, 490–520.
11 T.D. James, M.D. Phillips, S. Shinkai, *Boronic Acids in Saccharide Recognition*, RSC Publishing, Cambridge, **2006**.
12 J.S. Hansen, J.B. Christensen, J.F. Petersen, T. Hoeg-Jensen, J.C. Norrild, Arylboronic acids: A diabetic eye on glucose sensing. *Sensors & Actuators B*, **2012**, 161, 45–79.
13 X. Wu, Z. Li, X.-X. Chen, J.S. Fossey, T.D. James, Y.-B. Jiang, Selective sensing of saccharides using simple boronic acids and their aggregates. *Chem. Soc. Rev.*, **2013**, 42, 8032–8048.
14 X. Sun, T.D. James, Glucose sensing in supramolecular chemistry. *Chem. Rev.*, **2015**, 115, 8001–8037.
15 W. Zhai, X. Sun, T.D. James, J.S. Fossey, Boronic acid-based carbohydrate sensing. *Chem. Asian J.*, **2015**, 10, 1836–1848.
16 X. Sun, W. Zhai, J. S. Fossey, T.D. James, Boronic acids for fluorescence imaging of carbohydrates. *Chem. Commun.*, **2016**, 52, 3456–3469.
17 S.J. Angyal, The composition of reducing sugars in solution: Current aspects. *Adv. Carbohydr. Chem. Biochem.*, **1991**, 49, 19–35.
18 T.M. Lowry, *Optical Rotary Power*, Longmans, Green, and Co., London, **1935**, 21.
19 J. Böeseken, The use of boric acid for the determination of the configuration of carbohydrates. *Adv. Carbohydr. Chem.*, **1949**, 4, 189–210.
20 H.G. Kuivila, A.H. Keough, E.J. Soboczenski, Areneboronates from diols and polyols. *J. Org. Chem.*, **1954**, 19, 780–783.
21 J.P. Lorand, J.O. Edwards, Polyol complexes and structure of the benzeneboronate ion. *J. Org. Chem.*, **1959**, 24, 769–774.
22 L.I. Bosch, T.M. Fylesb, T.D. James, Binary and ternary phenylboronic acid complexes with saccharides and Lewis bases. *Tetrahedron*, **2004**, 60, 11175–11190.
23 M.A. Martínez-Aguirre, R. Villamil-Ramos, J.A. Guerrero-Alvarez, A.K. Yatsimirsky, Substituent effects and pH profiles for stability constants of arylboronic acid diol esters. *Inorg. Chem.*, **2013**, 78, 4674–4684.
24 T. Burgemeister, R. Grobe-Einsler, R. Grotstollen, A. Mannschreck, G. Wulff, Fast thermal breaking and formation of a B-N bond in 2-(aminomethyl) benzeneboronates. *Chem. Ber.*, **1981**, 114, 3403–3411.

25 G. Springsteen, B. Wang, A detailed examination of boronic acid-diol complexation. *Tetrahedron*, **2002**, 58, 5291–5300.

26 J.C. Norrild, H. Eggert, Boronic acids as fructose sensors. Structure determination of the complexes involved using ^{1}J(CC) coupling constants. *J. Chem. Soc., Perkin Trans. 2*, **1996**, 2583–2588.

27 J.C. Norrild, H. Eggert, Evidence for mono- and bisdentate boronate complexes of glucose in the furanose form. Application of ^{1}J(C-C) coupling constants as a structural probe. *J. Am. Chem. Soc.*, **1995**, 117, 1479–1484.

28 M. Dowlut, D.G. Hall, An improved class of sugar-binding boronic acids, soluble and capable of complexing glycosides in neutral water. *J. Am. Chem. Soc.*, **2006**, 128, 4226–4227.

29 M. Bérubé, M. Dowlut, D.G. Hall, Benzoboroxoles as efficient glycopyranoside-binding agents in physiological conditions: Structure and selectivity of complex formation. *J. Org. Chem.*, **2008**, 73, 6471–6479.

30 A. Adamczyk-Wozniak, K.M. Borys, A. Sporzynski, Recent developments in the chemistry and biological applications of benzoxaboroles. *Chem. Rev.*, **2015**, 5224–5247.

31 J. Yoon, A.W. Czarnik, Fluorescent chemosensors of carbohydrates. A means of chemically communicating the binding of polyols in water based on chelation-enhanced quenching. *J. Am. Chem. Soc.*, **1992**, 114, 5874–5875.

32 H. Suenaga, M. Mikami, K.R.A.S. Sandanayake, S. Shinkai, Screening of fluorescent boronic acids for sugar sensing which show a large fluorescence change. *Tetrahedron Lett.*, **1995**, 36, 4825–4828.

33 H. Suenaga, H. Yamamoto, S. Shinkai, Screening of boronic acids for strong inhibition of the hydrolytic activity of α-chymotrypsin and for sugar sensing associated with a large fluorescence change. *Pure Appl. Chem.*, **1996**, 68, 2179–2186.

34 Z.R. Grabowski, K. Rotkiewicz, W. Rettig, Structural changes accompanying intramolecular electron transfer: Focus on twisted intramolecular charge-transfer states and structures. *Chem. Rev.*, **2003**, 103, 3899–4032.

35 A. Demeter, K.A. Zachariasse, Fluorescence of crystalline 4-(dimethylamino) benzonitrile. Absence of dual fluorescence and observation of single-exponential fluorescence decays. *Chem. Phys. Lett.*, **2003**, 380, 699–703.

36 N. DiCesare, J.R. Lakowicz, Spectral properties of fluorophores combining the boronic acid group with electron donor or withdrawing groups. Implication in the development of fluorescence probes for saccharides. *J. Phys. Chem. A*, **2001**, 105, 6834–6840.

37 N. Di Cesare, J.R. Lakowicz, Wavelength-ratiometric probes for saccharides based on donor-acceptor diphenylpolyenes. *J. Photochem. Photobiol. A: Chem.*, **2001**, 143, 39–47.

38 N. DiCesare, J.R. Lakowicz, Chalcone-analogue fluorescent probes for saccharides signaling using the boronic acid group. *Tetrahedron Lett.*, **2002**, 43, 2615–2618.

39 N. DiCesare, J.R. Lakowicz, A new highly fluorescent probe for monosaccharides based on a donor-acceptor diphenyloxazole. *Chem. Commun.*, **2001**, 2022–2023.

40 X. Gao, Y. Zhang, B. Wang, New boronic acid fluorescent reporter compounds. 2. A naphthalene-based on-off sensor functional at physiological pH. *Org. Lett.*, **2003**, 5, 4615–4618.

41 X. Gao, Y. Zhang, B. Wang, Naphthalene-based water-soluble fluorescent boronic acid isomers suitable for ratiometric and off-on sensing of saccharides at physiological pH. *New J. Chem.*, **2005**, 29, 579–586.

42 X. Gao, Y. Zhang, B. Wang, A highly fluorescent water-soluble boronic acid reporter for saccharide sensing that shows ratiometric UV changes and significant fluorescence changes. *Tetrahedron*, **2005**, 61, 9111–9117.

43 Y. Zhang, X. Gao, K. Hardcastle, B. Wang, Water-soluble fluorescent boronic acid compounds for saccharide sensing: Substituent effects on their fluorescence properties. *Chem. Eur. J.*, **2006**, 12, 1377–1384.

44 W. Yang, J. Yan, G. Springsteen, S. Deeter, B. Wang, A novel type of fluorescent boronic acid that shows large fluorescence intensity change upon binding with a carbohydrate in aqueous solution at physiological pH. *Bioorg. Med. Chem. Lett.*, **2003**, 13, 1019–1022.

45 W. Yang, L. Lin, B. Wang, A novel type of boronic acid fluorescent reporter compound for sugar recognition. *Tetrahedron Lett.*, **2005**, 46, 7981–7984.

46 Q.J. Shen, W.J. Jin, Chemical and photophysical mechanism of fluorescence enhancement of 3-quinolineboronic acid upon change of pH and binding with carbohydrates. *Luminescence*, **2011**, 26, 494–499.

47 Y. Nagai, K. Kobayashi, H. Toi, Y. Aoyama, Stabilization of sugar-boronic esters of indolylboronic acid in water via sugar-indole interaction: A notable selectivity in oligosaccharides. *Bull. Chem. Soc. Jpn.*, **1993**, 66, 2965–2971.

48 J. Wang, S. Jin, N. Lin, B. Wang, Fluorescent indolylboronic acids that are useful reporters for the synthesis of boronolectins. *Chem. Biol. Drug. Des.*, **2006**, 67, 137–144.

49 A. Loudet, K. Burgess, BODIPY dyes and their derivatives: Syntheses and spectroscopic properties. *Chem. Rev.*, **2007**, 107, 4891–4932.

50 J.S. Hansen, J.F. Petersena, T. Hoeg-Jensen, J.B. Christensen, Buffer and sugar concentration dependent fluorescence response of a BODIPY-based aryl monoboronic acid sensor. *Tetrahedron Lett.*, **2012**, 53, 5852–5855.

51 J.S. Hansen, M. Ficker, J.F. Petersen, J.B. Christensen, T. Hoeg-Jensen, *ortho*-Substituted fluorescent aryl monoboronic acid displays physiological binding of *D*-glucose. *Tetrahedron Lett.*, **2013**, 54, 1849–1852.

52 J. Zhai, T. Pan, J. Zhu, Y. Xu, J. Chen, Y. Xie, Y. Qin, Boronic acid functionalized boron dipyrromethene fluorescent probes: Preparation, characterization, and saccharides sensing applications. *Anal. Chem.*, **2012**, 84, 10214–10220.

53 D. Zhai, S.-C. Lee, M. Vendrell, L.P. Leong, Y.-T. Chang, Synthesis of a novel BODIPY library and its application in the discovery of a fructose sensor. *ACS Comb. Sci.*, **2012**, 14, 81–84.

54 K. Kubo, PET sensors. In: *Advanced Concepts in Fluorescence Sensing. Part A: Small Molecule Sensing, Topics in Fluorescence Spectroscopy*, vol. 9, Springer, Berlin, **2005**, 219–247.

55 T.D. James, K.R.A.S. Sandanayake, S. Shinkai, Novel photoinduced electron-transfer sensor for saccharides based on the interaction of boronic acid and amine. *J. Chem. Soc., Chem. Commun.*, **1994**, 477–478.

56 T.D. James, K.R.A.S. Sandanayake, R. Iguchi, S. Shinkai, Novel saccharide-photoinduced electron transfer sensors based on the interaction of boronic acid and amine. *J. Am. Chem. Soc.*, **1995**, 117, 8982–8987.

57 T.D. James, K.R.A.S. Sandanayake, S. Shinkai, A glucose-selective molecular fluorescence sensor. *Angew. Chem. Int. Ed.*, **1994**, 33, 2207–2209.

58 J. Zhao, M.G. Davidson, M.F. Mahon, G. Kociok-Köhn, T.D. James, An enantioselective fluorescent sensor for sugar acids. *J. Am. Chem. Soc.*, **2004**, 126, 16179–16186.

59 J. Zhao, T.D. James, Chemoselective and enantioselective fluorescent recognition of sugar alcohols by a bisboronic acid receptor. *J. Mater. Chem.*, **2005**, 15, 2896–2901.

60 S.W. Coghlan, R.L. Giles, J.A.K. Howard, L.G.F. Patrick, M.R. Probert, G.E. Smith, A. Whiting, Synthesis and structure of potential Lewis acid–Lewis base bifunctional catalysts: 2-*N,N*-Diisopropylaminophenylboronate derivatives. *J. Organomet. Chem.*, **2005**, 690, 4784–4793.

61 K. Arnold, A.S. Batsanov, B. Davies, A. Whiting, Synthesis, evaluation and application of novel bifunctional *N,N*-diisopropylbenzylamineboronic acid catalysts for direct amide formation between carboxylic acids and amines. *Green Chem.*, **2008**, 10, 124–134.

62 A. Adamczyk-Wozniak, Z. Brzozka, M.K. Cyranski, A. Filipowicz-Szymanska, P. Klimentowska, A. Zubrowska, K. Zukowski, A. Sporzynski, *ortho*-(Aminomethyl)phenylboronic acids – synthesis, structure and sugar receptor activity. *Appl. Organomet. Chem.*, **2008**, 22, 427–432.

63 L. Zhang, J.A. Kerszulis, R.J. Clark, T. Ye, L. Zhu, Catechol boronate formation and its electrochemical oxidation. *Chem. Commun.*, **2009**, 2151–2153.

64 A. Adamczyk-Woźniak, I. Madura, A. Pawełko, A. Sporzyński, A. Żubrowska, J. Żyła, Amination-reduction reaction as simple protocol for potential boronic molecular receptors. Insight in supramolecular structure directed by weak interactions. *Cent. Eur. J. Chem.*, **2011**, 9, 199–205.

65 A. Adamczyk-Woźniak, M.K. Cyranski, B.T. Fraczak, A. Lewandowska, I.D. Madura, A. Sporzyński, Imino- and aminomethylphenylboronic acids. *Tetrahedron*, **2012**, 68, 3761–3767.

66 B.E. Collins, S. Sorey, A.E. Hargrove, S.H. Shabbir, V.M. Lynch, E.V. Anslyn, Probing intramolecular B-N interactions in *ortho*-aminomethyl arylboronic acids. *J. Org. Chem.*, **2009**, 74, 4055–4060.

67 J.D. Morrison, R.L. Letsinger, Organoboron compounds. XVIII. Bifunctional binding of water by the *cis*-1,2-cyclopentanediol ester of 8-quinolineboronic acid. *J. Org. Chem.*, **1964**, 29, 3405–3407.

68 L. Zhu, S.H. Shabbir, M. Gray, V.M. Lynch, S. Sorey, E.V. Anslyn, A structural investigation of the N-B Interaction in an *o*-(N,N-dialkylaminomethyl)arylboronate system. *J. Am. Chem. Soc.*, **2006**, 128, 1222–1232.

69 S. Arimori, M.L. Bell, C.S. Oh, K.A. Frimat, T.D. James, Modular fluorescence sensors for saccharides. *J. Chem. Soc., Perkin Trans.1*, **2002**, 803–808.

70 K.R.A.S. Sandanayake, T.D. James, S. Shinkai, Two dimensional photoinduced electron transfer (PET) fluorescence sensor for saccharides. *Chem. Lett.*, **1995**, 24, 503–504.

71 W. Yang, S. Gao, X. Gao, V.V.R. Karnati, W. Ni, B. Wang, W.B. Hooks, J. Carson, B. Weston, Diboronic acids as fluorescent probes for cells expressing sialyl Lewis X. *Bioorg. Med. Chem. Lett.*, **2002**, 12, 2175–2177.

72 V.V. Karnati, X. Gao, S. Gao, W. Yang, W. Ni, S. Sankar, B. Wang, A glucose-selective fluorescence sensor based on boronic acid-diol recognition. *Bioorg. Med. Chem. Lett.*, **2002**, 12, 3373–3377.

73 W. Yang, H. Fan, X. Gao, S. Gao, V.V.R. Karnati, W. Ni, W.B. Hooks, J. Carson, B. Weston, B. Wang, The first fluorescent diboronic acid sensor specific for hepatocellular carcinoma cells expressing sialyl lewis X. *Chem. Biol.*, **2004**, 11, 439–448.

74 S. Arimori, G.A. Consiglio, M.D. Phillips, T.D. James, Tuning saccharide selectivity in modular fluorescent sensors. *Tetrahedron Lett.*, **2003**, 44, 4789–4792.

75 R.B. Sekar, A. Periasamy, Fluorescence resonance energy transfer (FRET) microscopy imaging of live cell protein localizations. *J. Cell Biol.*, **2003**, 160, 629–633.
76 D. Shrestha, A. Jenei, P. Nagy, G. Vereb, J. Szöllősi, Understanding FRET as a research tool for cellular studies. *Int. J. Mol. Sci.*, **2015**, 16, 6718–6756.
77 A. Resendeza, R.A. Wesslinga, B. Singaram, Boronic acid functionalized viologens as saccharide sensors, In: *Boron: Sensing, Synthesis and Supramolecular Self-Assembly*, ed. M. Li, J.S. Fossey, T.D. James, RSC Publishing, Cambridge, **2016**, 128–181.
78 Y. Egawa, R. Miki, T. Seki, Colorimetric sugar sensing using boronic acid-substituted azobenzenes. *Materials*, **2014**, 7, 1201–1220.
79 H.R. Snyder, S.L. Meisel, The synthesis of azo boronic acids. II. Dyes from tetrazotized benzidine-2,2'-diboronic acid. *J. Am. Chem. Soc.*, **1948**, 70, 774–776.
80 H.R. Snyder, C. Weaver, The preparation of some azo boronic acids. *J. Am. Chem. Soc.*, **1948**, 70, 232–234.
81 K. Koumoto, M. Takeuchi, S. Shinkai, Design of a visualized sugar sensing system utilizing a boronic acid-azopyridine interaction. *Supramol. Chem.*, **1998**, 9, 203–210.
82 G. Springsteen, B. Wang, Alizarin red S. as a general optical reporter for studying the binding of boronic acids with carbohydrates. *Chem. Commun.*, **2001**, 1608–1609.
83 S. Boduroglu, J.M. El Khoury, D.V. Reddy, P.L. Rinaldi, J. Hu, A colorimetric titration method for quantification of millimolar glucose in a pH 7.4 aqueous phosphate buffer. *Bioorg. Med. Chem. Lett.*, **2005**, 15, 3974–3977.
84 W.M.J. Ma, M.P.P. Morais, F. D'Hooge, J.M.H. van den Elsen, J.P.L. Cox, T.D. James, J.S. Fossey, Dye displacement assay for saccharide detection with boronate hydrogels. *Chem. Commun.*, **2009**, 532–534.
85 C.J. Ward, P.R. Ashton, T.D. James, P. Patel, A molecular colour sensor for monosaccharides. *Chem. Commun.*, **2000**, 229–230.
86 C.J. Ward, P. Patel, T.D. James, Boronic acid appended azo dyes-colour sensors for saccharides. *J. Chem. Soc., Perkin Trans.*, **2002**, 462–470.
87 C.J. Ward, P. Patel, T.D. James, Molecular color sensors for monosaccharides. *Org. Lett.*, **2002**, 4, 477–479.
88 N. DiCesare, J.R. Lakowicz, New color chemosensors for monosaccharides based on azo dyes. *Org. Lett.*, **2001**, 3, 3891–3893.
89 Y. Egawa, R. Gotoh, S. Niina, J. Anzai, *ortho*-Azo substituted phenylboronic acids for colorimetric sugar sensors. *Bioorg. Med. Chem. Lett.*, **2007**, 17, 3789–3792.
90 Y. Egawa, Y. Tanaka, R. Gotoh, S. Niina, Y. Kojima, N. Shimomura, H. Nakagawa, T. Seki, J. Anzai, Nitrogen-15 NMR spectroscopy of sugar sensor with B-N interaction as a key regulator of colorimetric signals. *Chem. Lett.*, **2010**, 39, 1188–1189.

2.4

Boron Compounds in Molecular Imaging

Bhaskar C. Das,[1-3],* Devi Prasan Ojha,[2] Sasmita Das,[1] and Todd Evans[3],*

[1] Departments of Medicine and Pharmacological Sciences, Icahn School of Medicine at Mount Sinai, New York, NY, USA
[2] Departments of Medicine and Pharmacological Sciences, Icahn School of Medicine at Mount Sinai, New York, NY, USA
[3] Department of Surgery, Weill Cornell Medical College of Cornell University, New York, NY, USA
* Corresponding authors.

2.4.1 Introduction

Molecular imaging (MI) is a pharmacodynamics strategy to visualize specific molecules *in vivo* using high-affinity probes that target specific molecular sites in a living system. These probes are specifically designed to overcome biological delivery barriers (vascular, interstitial, cell membrane). Small molecular probes currently in use can be broadly classified as radiopharmaceuticals, paramagnetic or fluorescent materials, or bubble-based agents. They are being used as physiological and molecular markers in a number of applications, monitoring cell trafficking, apoptosis, angiogenesis, cellular metabolism, and drug development studies. Probes may also be used for therapeutic interventions. Imaging technologies have had an important impact on many aspects of healthcare delivery. Physicians routinely rely on medical images for diagnostic and prognostic purposes, and scientists develop and apply new imaging strategies to non-invasively track pathophysiological processes and for development of novel therapeutic strategies.

Imaging was selected by the National Academy of Engineering as one of the 20 greatest engineering achievements of the twentieth century [1]. The increasing cost of healthcare places a substantial economic burden on society, resulting in a shift in emphasis from the treatment to the prevention of diseases. Moreover, individualized disease prevention, risk stratification, and therapy could lead to improvements and greater efficiency in healthcare delivery. With the advent of genomics, proteomics, and technological advances, modern targeted molecular imaging strategies are replacing the traditional anatomical or physiological approaches to the detection, evaluation, and monitoring of many diseases and their treatment. Molecular imaging using biologically targeted markers can provide unique insight into genetic

Boron-Based Compounds: Potential and Emerging Applications in Medicine, First Edition.
Edited by Evamarie Hey-Hawkins and Clara Viñas Teixidor.
© 2018 John Wiley & Sons Ltd. Published 2018 by John Wiley & Sons Ltd.

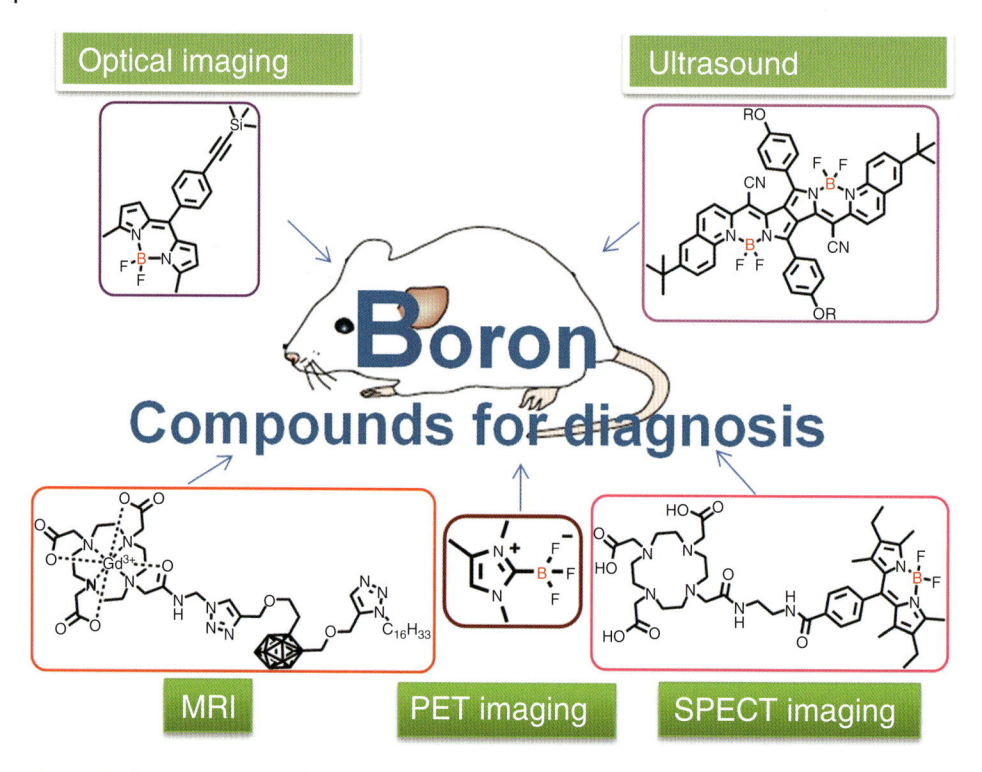

Figure 2.4.1 Boron-based molecular imaging probes.

and cellular processes and allow for disease evaluation and management as well as individualized therapy [2,3]. In this book chapter, we will discuss selected boron molecule-based molecular imaging agents for the evaluation of various disease areas (Figure 2.4.1) [4,5].

Historically, the earliest direct imaging approaches involved native radiolabeled monoclonal antibodies, which were later replaced by engineered antibodies and peptides or peptidomimetic probes for imaging of cell-specific antigens [6]. With direct molecular imaging, the magnitude of probe uptake and localization are directly related to the targeted molecule; however, this approach requires a customized probe for every target of interest. To overcome this limitation, indirect molecular imaging strategies, such as reporter gene technology, have been developed. This concept is complex and consists of multiple steps. The first step involves introduction of a reporter gene driven by a constitutive, inducible, or tissue-specific promoter into the cell nuclei by various methods, including viral and nonviral vectors. In the second step, the reporter gene transferred into the cell nuclei undergoes transcription followed by translation of messenger RNA (mRNA), which results in production of a reporter protein. This protein is then detected by its interactions with a complementary reporter probe radiolabeled with a Single-Photon Emission computed Tomography (SPECT) or Positron Emission Tomography (PET) isotope. Both direct and indirect molecular imaging procedures are complementary to each other for present patient care.

2.4.2 Molecular Imaging in Biomedical Research

Molecular imaging techniques can be performed in the intact organism with sufficient spatial and temporal resolution for studying biological processes *in vivo*. Furthermore, it allows a repetitive, noninvasive, uniform, and relatively automated study of the same living subject using identical or alternative biological imaging assays at different time points, thus harnessing the statistical power of longitudinal studies, and reducing the number of animals required and the cost. Therefore, molecular imaging may be used for early detection, characterization, and "real-time" monitoring of diseases as well as for investigating the efficacy of drugs. Furthermore, molecular imaging is a branch of medical imaging science that aims to detect, localize, and monitor critical molecular processes in cells, tissue, and living organisms using highly sensitive instrumentation and contrast strategies (Figure 2.4.2).

2.4.3 Molecular Imaging Modalities

Presently, there is a consensus among experts in the field of molecular imaging that the rapidly evolving imaging deals with three distinct imaging areas, specifically imaging of molecular biomarkers, single-cell imaging, and imaging therapeutics. The advances and increasing use of molecular imaging that have occurred over the last decade would not have been possible without the concurrent evolution in imaging technology. These advances include the development of new high-resolution, high-sensitivity imaging systems and novel imaging modalities. These collectively provide a selection of operational

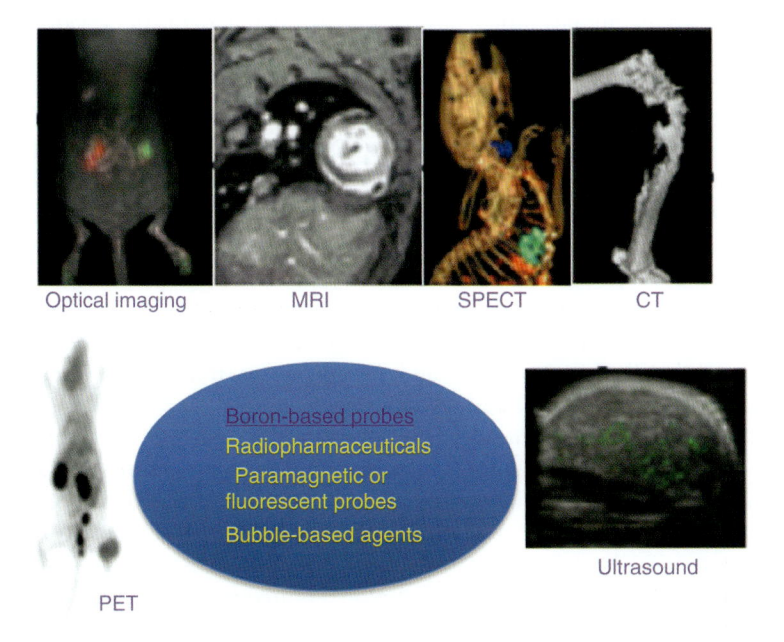

Figure 2.4.2 Overview of molecular probes in imaging.

parameters for various clinical imaging modalities and present both potential advantages and disadvantages for targeted molecular imaging.

A diverse range of imaging techniques, or modalities, is now available (Table 2.4.1). Magnetic Resonance Imaging (MRI), PET, Computed Tomography (CT), and SPECT are all used routinely to scan patients in hospitals; smaller, less expensive versions have been produced for animal research. These techniques can penetrate deep into tissue, and sources of distortion are relatively few and largely understood. They are often used to survey whole bodies for disease and to do cross-sectional imaging, particularly for research on deep-seated organs. Each modality has pros and cons.

Although a number of modalities have been developed, only a few are available for broad application in molecular imaging. The most sensitive molecular imaging techniques are the radionuclide-based PET and SPECT imaging modalities. PET or SPECT has the sensitivity needed to visualize most interactions between physiological targets and ligands such as neurotransmitters and brain receptors. Radionuclide-based imaging modalities are able to determine concentrations of specific biomolecules as low as in the picomolar range. It must be stressed here, however, that the choice of a certain imaging modality – whether MRI, ultrasound, or PET – depends primarily on the specific question to be addressed.

The nuclear approaches include SPECT and PET, which are particularly well suited for *in vivo* molecular imaging because of their high sensitivity and acceptable spatial resolution, and the availability of instrumentation and molecular probes [7]. Both SPECT and PET imaging have become standard approaches for physiological imaging in patients. Technological advances and the availability of preclinical small-animal micro-SPECT and micro-PET imaging systems have led to the emergence of novel imaging strategies with the potential to translate to clinical practice. Both SPECT and PET imaging strategies have advantages and disadvantages based on the underlying physical and chemical differences between the two techniques, which have been described in detail in previous reviews [9–12]. The selection of imaging modality should, therefore, be based on the properties of the biological system under evaluation, the availability of targeted molecular probes, and the accessibility of SPECT or PET instrumentation. Besides these techniques, as we discussed in this chapter, MRI, X-ray, CT, ultrasound, and light-based methods (endoscopy and optical coherence tomography [OCT]) are also used routinely for patient care. Research modalities include various light microscopy techniques (confocal, multiphoton, total internal reflection, and super-resolution fluorescence microscopy), electron microscopy, mass spectrometry imaging, fluorescence tomography, bioluminescence, variations of OCT, and optoacoustic imaging.

Although clinical imaging and research microscopy are often isolated from one another, it can be argued that their combination and integration are not only informative but also essential to discovering new biology and interpreting clinical datasets in which signals invariably originate from hundreds to thousands of cells per voxel. No single modality can be considered the best molecular strategy for whole-animal imaging under all circumstances. A multimodality approach is the preferred way to explore biomedical imaging. Instrument companies developed PET–SPECT, PET–CT–SPECT, PET–MRI, and so on to advance this field (see the Appendix). Taking this into consideration, academic scientists and instrumentation companies are developing multimodality probes and hybrid instrumentation technology to further expand the molecular imaging research field in the context of our modern healthcare system.

Table 2.4.1 Different molecular imaging modalities

Technique	Labels	Signal	Pros	Cons	Sensitivity	Resolution
PET	Radiolabeled nuclei	Positrons from radionuclides	Highly sensitive	High cost, detects only one radionuclide; operates through radioactive nuclei	10^{-15}	1–2 mm
SPECT	Radiolabeled nuclei	γ-rays	Different radionuclides can be distinguished, so more processes can be imaged at once	Requires radioactivity, high cost	10^{-14}	1–2 mm
CT	None	X-rays	Fast, cross-sectional images	High cost, poor resolution of soft tissues	10^{-6}	50 m
MRI	Can use isotope-labeled molecular tracers	Alterations in magnetic fields	Harmless, high resolution of soft tissues	Cannot follow many labels, high cost	10^{-9}–10^{-6}	50 µm
Optical	Genetically engineered proteins and bioluminescent and fluorescently labeled probes	Light, particularly in the infrared	Easy, nondamaging technique readily adapted to study specific molecular events, low cost	Poor depth penetration	10^{-12}	1–2 mm
Photoacoustic	Probes that absorb light and create sound signals	Sound	Better depth resolution than light, low cost	Information processing and machines still being optimized	10^{-12}	50 µm
Ultrasound	Microbubbles, which can be combined with targeted contrast agents	Sound	Quick, harmless, low cost	Poor image contrast, works poorly in air-containing organs	10^{-8}	50 µm

2.4.4 Boron Compounds in Molecular Imaging

Chemical tools are increasingly important in both clinical and research imaging because they can add molecular and cellular specificity and/or enhance physiological data extraction. Additionally, chemical imaging agents have two major advantages over fluorescent proteins, although the two are often used complementarily: chemical tools enable imaging in humans and in mice obviate the need for genetically engineered reporters. A considerable number of imaging agents have been developed over the last decade (Molecular Imaging and Contrast Agent Database [MICAD]), and some agents are commercially available or even FDA-approved [8–10]. Besides this, nanoparticles are particularly promising because they tend to accumulate in innate immunocytes, which are often "first responders" in pathologic processes. Furthermore, nanoparticles have unique pharmacokinetics: that is, they circulate longer, are not immediately cleared renally, and can be targeted to specific organs, cells, or proteins. Magnetic nanoparticles (MNPs), which are detected by MRI, are perhaps the best studied nanoparticle type. Ferumoxytol, for example, is a US Food and Drug Administration (FDA)-approved nanomaterial for iron replacement in treating anemia but has been used to enhance MRI; when tagged with fluorochromes, ferumoxytol also doubles as a magnetic resonance (MR) and/or optical imaging agent [13]. Quantum dots have been essential in certain microscopic imaging experiments, especially in conjunction with environmentally sensitive particles, targeted particles, and short-wave infrared particles that can be detected much deeper in tissue [14]. Labeled antibodies and antibody fragments have long been used for targeted imaging, and the introduction of long-lived imaging isotopes (^{89}Zr, ^{68}Ga, ^{64}Cu, and ^{124}I) has resulted in some spectacular clinical results. Newer alpaca-derived antibody fragments currently being developed offer several advantages over traditional antibodies. Specifically, single-chain camelid antibody fragments lack an Fc portion and are much smaller (~15 kDa) than immunoglobulins (~150 kDa), "diabody" antibody derivatives (~60 kDa), Fab fragments (~50 kDa), or single-chain variable fragments (ScFvs; ~25 kDa). Other important chemical imaging tools now in routine use include a large number of isotopes (Table 2.4.2), fluorochromes, or metal-labeled small molecules. There are a number of hyperpolarized C13 metabolites being developed for metabolic MRI.

Boron-based compounds in imaging are playing an increasingly important role in preclinical target evaluation through noninvasive clinical cancer diagnosis. This can be attributed to the ease of synthesis of boron-based precursors for 18-F radiolabeling, which can be produced in many different ways. The most common procedure is the cold fluorination of phenylboronic acid using an excess of KHF_2 in solution, with recrystallization of the resulting precipitate giving a relatively stable potassium phenyltrifluoroborate salt [15].

When a protein scaffold interacts with the boron atom in the chromophore through a bound water molecule, this converts the boron atom to be sp^3-hybridized, which further modulates the reactivity of the protein scaffold with reactive oxygen and nitrogen species (ROS and RNS, respectively). Although other surrounding residues in the protein may also affect the chemoselectivity (e.g., by modulating the bond angle or length of the new B–O bond), the identified sp^3-hybridized hydrated boronate itself is the determining factor for the unprecedented chemoselectivity of the scaffold. This boron chemistry in a folded non-native protein will facilitate the future development of highly

Table 2.4.2 Common isotopes used in nuclear medicine

Isotope	Symbol	Z	$T_{1/2}$	Decay
Imaging				
Fluorine-18	^{18}F	9	109.77 m	β^+
Gallium-67	^{67}Ga	31	3.26 d	EC
Indium-111	^{111}In	49	2.80 d	EC
Iodine-123	^{123}I	53	13.3 h	EC
Iodine-131	^{131}I	53	8.02 d	β^-
Krypton-81m	^{81m}Kr	36	13.1 s	IT
Nitrogen-13	^{13}N	7	9.97 m	β^+
Rubidium-82	^{82}Rb	37	1.27 m	β^+
Technetium-99m	^{99m}Tc	43	6.01 h	IT
Thallium-201	^{201}Tl	81	3.04 d	EC
Xenon-133	^{133}Xe	54	5.24 d	β^-
Yttrium-90	^{90}Y	39	2.67 d	β^-

Decay: mode of decay; EC: electron capture; photons: principle photon energies in kilo-electron volts (keV); $T_{1/2}$: half-life; Z: atomic number, the number of protons.

selective arylboronate-based sensors and inspire the tuning of chemical or biochemical reactions in protein scaffolds for a multitude of applications.

PET is revolutionizing our ability to visualize *in vivo* targets for target validation and personalized medicine. Of several classes of imaging agents, peptides afford high affinity and high specificity to distinguish pathologically distinct cell types by the presence of specific molecular targets. Of various available PET isotopes, [^{18}F]-fluoride ion is preferred because of its excellent nuclear properties and on-demand production in hospitals at Curie levels. However, the short half-life of ^{18}F and its lack of reactivity in water continue to challenge peptide labeling. Hence, peptides are often conjugated to a metal chelator for late-stage, one-step labeling [16].

In seeking new methods, one approach would be to eschew C–F bond formation in favor of B–F bond formation, hypothesizing that an organotrifluoroborate, which can be readily prepared in water, would afford late-stage, one-step labeling (Figure 2.4.3). Two concerns befall all radiotracers: (1) radiolabeling must be kinetically rapid and sufficiently favorable thermodynamically to ensure reasonable yields, and (2) the tracer must be stable *in vivo*. To address these issues, we [4] and others studied the synthesis of aryl trifluoroborates (ArBF$_3^-$), wherein three fluoride ions condense with an arylboronic acid (Figure 2.4.4).

Pinacolate esters and borimidines are also fluoridated, either directly or following solvolysis to give the boronic acid (Figure 2.4.5). Selective receptor-targeting radiopeptides have emerged as an important class of radiopharmaceuticals for molecular imaging and therapy of tumors that overexpress peptide receptors on the cell membrane. After such peptides labeled with γ-emitting radionuclides bind to their receptors, they allow clinicians to visualize receptor-expressing tumors noninvasively. Peptides labeled with particle emitters could also eradicate receptor-expressing tumors. The first attempt to

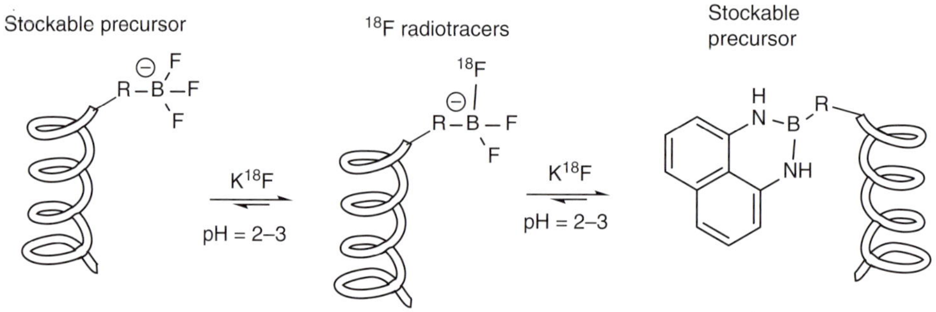

Figure 2.4.3 Boron-based ^{18}F-radiotracers.

$$ArB(OH)_2 + 2HF + KF \rightleftharpoons ArBF_3^-K^+ + 2H_2O$$

Figure 2.4.4 Reaction pathways for $ArBF_3^-$ stability.

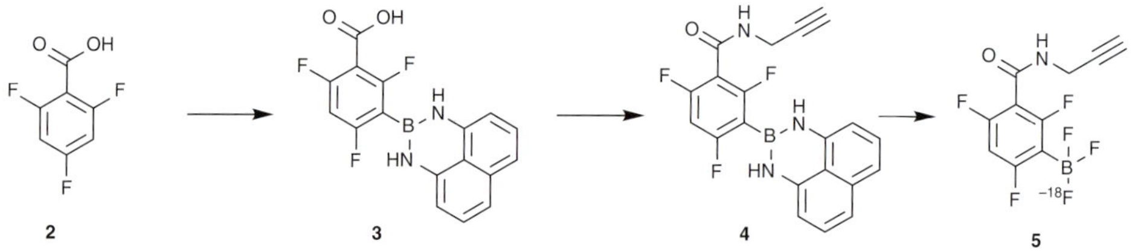

Figure 2.4.5 One-pot, two-step labeling to synthesize an [^{18}F] compound.

utilize a [18/19F] organotrifluoroborate group in an analog of an existing radiotracer was carried out with the synthesis of a multimodal PET/NIRF tracer, [^{18}F]BOMB, activated with a maleimide for subsequent conjugation with tilmanocept (a commercial agent that is labeled with 99mTc and used intraoperatively for lymphatic mapping). The radiofluorination of the benzopinacolate conjugation with tilmanocept was shown to give better results *in vivo* (mouse) than the clinically used 99mTc-tilmanocept. Recently, two methods have been described for radiolabeling a cyclic RGD (widely known to bind to the $\alpha_V\beta_3$ receptor overexpressed in several cancer cell lines) using B–F bonds [17].

The method following the characterized route starts with the biomolecule conjugated to the arylboronic ester **4**, followed by direct one-step radiolabeling with the same aqueous [$^{18/19}$F] KHF$_2$ and acidic conditions described for 1 h at room temperature to provide **5** [18]. However, although RGD will withstand acidic conditions, the strategy is not appropriate for other interesting and more sensitive biomolecules. The prior radiolabeling of a prosthetic group allows a larger range of reaction conditions, and efficient attachment to the peptide can be rapidly achieved under mild conditions. The innovative approach published by the Perrin group developed the first one-pot two-step method for radiofluorinations of different biomolecules through the B–F approach by using a 1,8-diaminonaphthalene protected borimidine with an alkyne arm as a prosthetic group for indirect radiolabeling [19].

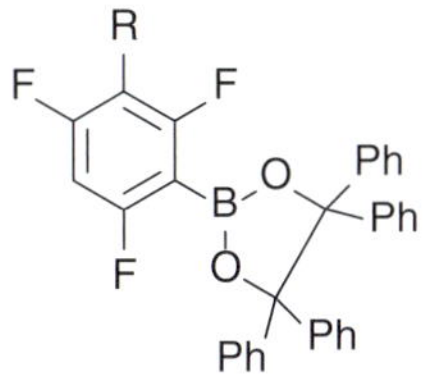
6 ref. 19a

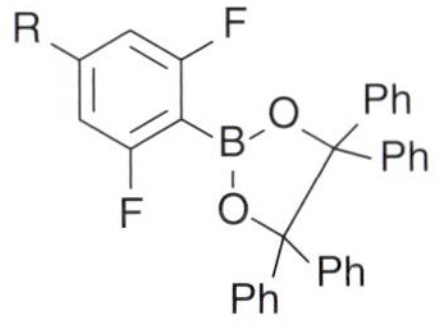
7 ref. 19b

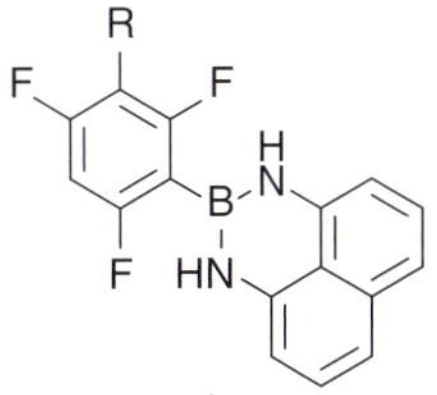
8 ref. 19c

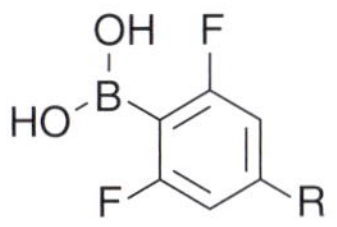
9 ref. 19d

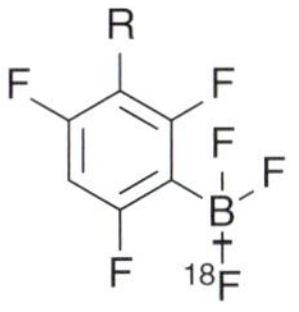
10 ref. 20a

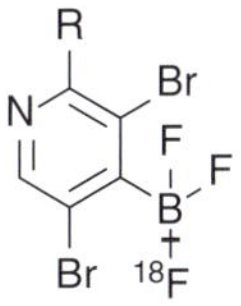
11 ref. 20b

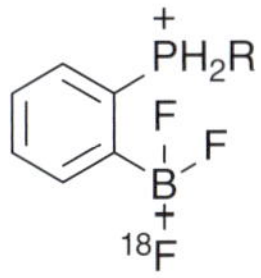
12 ref. 20c

Figure 2.4.6 Different B–^{18}F bond containing probes.

Despite significant efforts to improve radiofluorination in boron precursors, some issues still persist (Figure 2.4.6). Generally, reactions with [^{18}F] fluoride are slow and radiolabeling yields are low, meaning that effective purification methods are needed, which also contributes to the total synthesis time of a radiotracer leading to lower final specific activity (compounds **6–12**). This could be attenuated by producing very high concentrations of [^{18}F] fluoride on a microliter scale, and by using nanomolar amounts of reagents, which may be feasible in microfluidic synthesis systems [19,20].

2.4.5 Boron-Based Imaging Probes

2.4.5.1 Boron-Based Optical Probes

Living organisms produce hydrogen peroxide (H_2O_2) to kill invading pathogens and for cellular signaling, but aberrant generation of this Reactive Oxygen Species (ROS) is a hallmark of oxidative stress and inflammation in aging, injury, and disease. The effects of H_2O_2 on the overall health of living animals remain elusive, in part owing to a dearth of methods for studying this transient small molecule *in vivo*. H_2O_2 is also emerging as a newly recognized messenger in cellular signal transduction [21]. However, a substantial challenge in elucidating its diverse roles in complex biological environments is the lack of methods for probing this reactive oxygen metabolite in living systems with molecular specificity. To address these issues, many boron-based optical probes have been developed and used as novel agents for studying the biology of H_2O_2 and monitoring the transient transition at different levels of development.

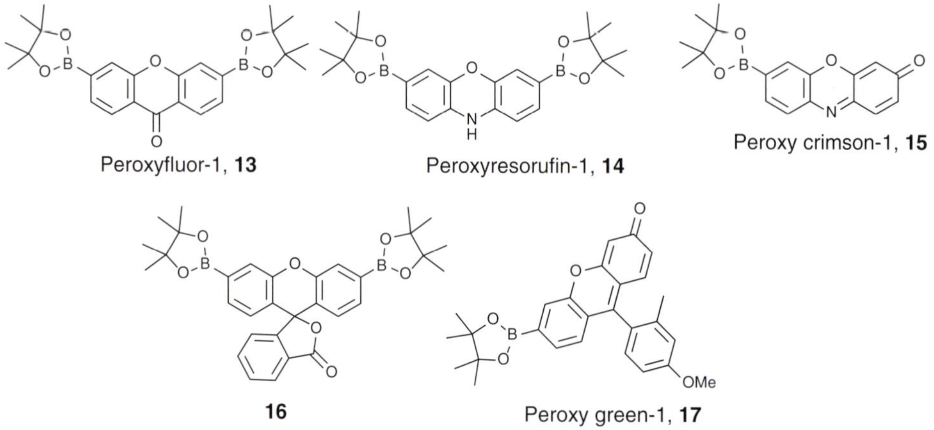

Peroxyfluor-1, **13**

Peroxyresorufin-1, **14**

Peroxy crimson-1, **15**

16

Peroxy green-1, **17**

Figure 2.4.7 Boronate fluorescent probes visualizing endogenous H_2O_2.

For example, the synthesis and application of Peroxy Green 1 (PG1, **17**), Peroxy Crimson 1 (PC1, **15**), and three new fluorescent probes **13**, **14**, and **16** show high selectivity for H_2O_2 and are capable of visualizing endogenous H_2O_2 produced in living cells by growth factor stimulation, including the first direct imaging of peroxide produced for brain cell signaling (Figure 2.4.7). The combined features of ROS selectivity, sensitivity to signaling levels of H_2O_2, and live-cell compatibility presage many new opportunities for PG1, PC1, and related synthetic reagents for exploring the physiological roles of H_2O_2 in living systems with molecular imaging.

A new family of fluorescent probes was developed with varying emission colors for selectively imaging H_2O_2 generated at physiological cell-signaling levels. This structurally homologous series of fluorescein- and rhodol-based reporters relies on a chemospecific boronate-to-phenol switch to respond to H_2O_2 over a panel of biologically relevant ROS with tunable excitation and emission maxima and sensitivity to endogenously produced H_2O_2 signals, as shown by studies in RAW264 macrophages during the phagocytic respiratory burst and A431 cells in response to EGF stimulation [22]. These reagents are used as new H_2O_2-specific probes, for example Peroxy Orange 1 (PO1, **22**), in conjunction with the green-fluorescent highly reactive oxygen species (hROS) probe APF. This dual-probe approach allows for selective discrimination between changes in H_2O_2 and hypochlorous acid (HOCl) levels in live RAW264 macrophages [23]. Moreover, when macrophages labeled with both PO1 and APF were stimulated to induce an immune response, three distinct types of phagosomes were detected: those that generated mainly hROS, those that produced mainly H_2O_2, and those that possessed both types of ROS (Figure 2.4.8). The ability to monitor multiple ROS fluxes simultaneously using a palette of differently colored fluorescent probes opens new opportunities to disentangle the complex contributions of oxidation biology to living systems by molecular imaging.

Another advancement was the development of peroxy caged luciferin-1 (PCL1) (Figure 2.4.9), a chemoselective bioluminescent probe for real-time detection of H_2O_2 within living animals. PCL1 (**24**) is a boronic acid–caged firefly luciferin molecule that selectively reacts with H_2O_2 to release firefly luciferin, which triggers a bioluminescent

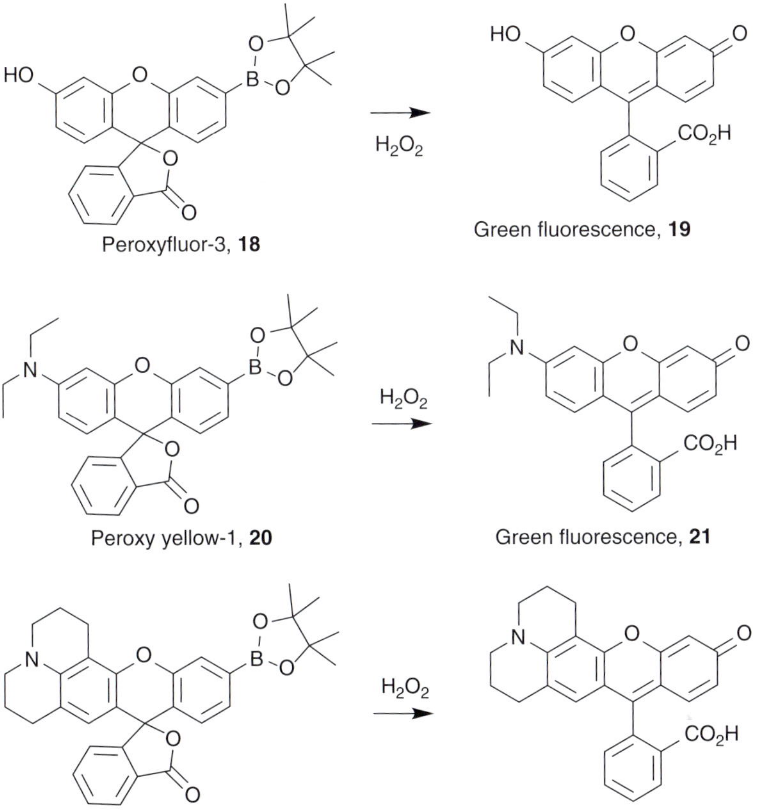

Figure 2.4.8 Reaction conditions for monoboronate fluorescent probes in the presence of H_2O_2.

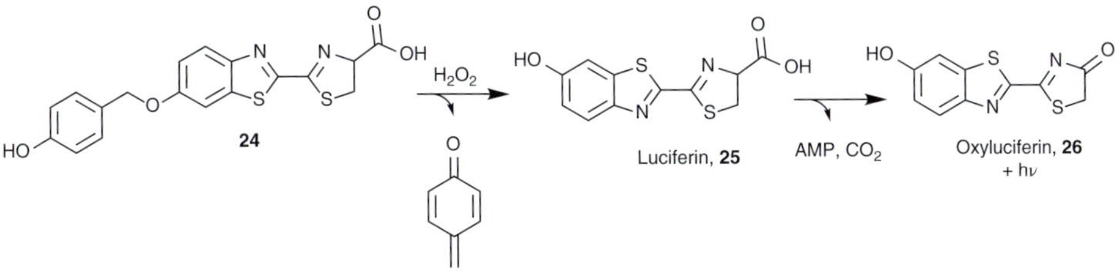

Figure 2.4.9 Visualizing changes in H_2O_2 levels in living cells by bioluminescent imaging.

response in the presence of firefly luciferase. The high sensitivity and selectivity of PCL1 for H_2O_2, combined with the favorable properties of bioluminescence for *in vivo* imaging, afford a unique technology for real-time detection of basal levels of H_2O_2 generated in healthy, living mice. Moreover, efficacy was demonstrated of PCL1 for monitoring physiological fluctuations in H_2O_2 levels by directly imaging elevations in H_2O_2 within testosterone-stimulated tumor xenografts *in vivo*. The ability to chemoselectively monitor H_2O_2 fluxes in real time in living animals offers opportunities to dissect H_2O_2's disparate contributions to health, aging, and disease [24].

In addition to using boron-based optical probes to detect hROS, they can also be used as fluorescence probes for labeling, and tracking of immune cells *in vitro* and *in vivo* is also appealing because they rapidly diffuse, and are noninvasive, nontoxic, and relatively sensitive.

2.4.5.2 Boron-Based Nuclear Probes

For *in vivo* imaging of boron compounds, radiolabeled derivatives are of particular interest since their bio-distribution can be easily monitored by using SPECT and PET, depending on the radionuclide employed. Although the amount of radioactivity used in radio-immunoimaging experiments is extremely small in comparison to the large number of boron atoms required for efficient Boron Neutron Capture Therapy (BNCT), the labeling of boron compounds with radionuclides enables their direct detection in biological systems during studies of bio-distribution and pharmacokinetics. This is important in the evaluation of boron systems especially designed for enhanced tumor selectivity with regard to BNCT, and it also offers the synthesis of new inorganic tumor-imaging agents that display high chemical stability of the radiolabel. This is of special importance since enzymatic cleavage *in vivo* resulting in loss of the radiolabel is often a problem. Initially, labeling reactions of boron clusters with iodine-131, iodine-125, astatine-211, tritium, and cobalt-57 were presented. The use of fluorine-18, a positron emitter detectable with PET, is only reported for studies with BPA.

The iodination of *nido*-7,8-$C_2B_9H_{12}^-$ (**27**) was first reported by Hawthorne *et al.* in 1965 [25]. The reaction of the potassium salt of this anion with elemental iodine in absolute ethanol afforded the mono-iodinated product (**28**) in 75% yield (Figure 2.4.10). Characterization was accomplished by elemental analyses and ^{11}B NMR. The reaction should be regarded as an electrophilic substitution at the open face of the *nido*-7,8-$C_2B_9H_{12}^-$ ion. Because of the pattern of the ^{11}B NMR spectrum, the authors assigned the position of the boron bearing the iodine substituent as adjacent to one of the carbon atoms. In a later study, the same authors reinvestigated the iodination reaction of 7-(4-C_6H_4NCS)-*nido*-7,8-$C_2B_9H_{11}^-$ employing sodium iodide and chloramine-T as an oxidizing agent [26]. The *p*-isothiocyanatophenyl substituent was present for later linkage of the iodinated *nido* anion to a protein amino group, thus accomplishing conjugate labeling.

Tritium is a β-emitter with a half-life of 12.323 years [27]. Because of its long half-life, this radionuclide is not suited for clinical applications. Nevertheless, it is used for *in vitro* and *in vivo* studies with animals, since tritium is easily introduced into molecules (e.g., via hydrogen or proton exchange reactions using tritium gas, HTO, or T_2O) [28]. Carboranes (**29**) can be tritiated by prior deprotonation of the CH vertex with n-BuLi, followed by quenching the resulting anion with T_2O (Figure 2.4.11).

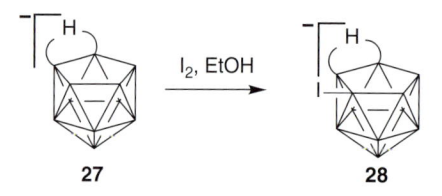

Figure 2.4.10 Iodination of *nido*-carborane anion.

Figure 2.4.11 Tritium is introduced to *closo*-carboranes.

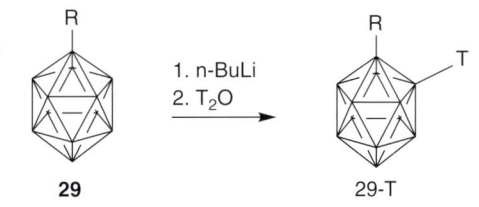

Figure 2.4.12 Venus flytrap complex (VFC) with radio cobalt (^{57}Co) conjugated to a monoclonal antibody.

In addition to radiolabeling using radioiodine, *nido*-carborane dianions form extremely stable complexes with cobalt and other transition metals and are potentially of great interest as carriers for radiometals. This can be seen in the so-called Venus flytrap complex (VFC) with ^{57}Co as the γ-photon-emitting radiometal. The structure of this complex was determined in an X-ray diffraction study using isotopically normal cobalt.

Conjugation of the mixture of ^{57}Co-labeled VFC isomers **30** (*dl* and *meso*) to the anti-CEA monoclonal antibody, T84.66, could be accomplished via its N-hydroxysulfosuccinimide ester (Figure 2.4.12). In this manner, an average of 0.03–0.05 molecules of VFC reagent per molecule of antibody was conjugated in accord with the VFC concept of providing an extraordinarily stable imaging agent rather than a therapeutic agent for BNCT. The VFC conjugate retained >90% immunoreactivity, was stable in serum (more than 7 days), and demonstrated excellent localization in LS174T tumor xenografts during *in vivo* studies with nude mice. In the latter study, the pharmacokinetics of ^{57}Co VFC-T84.66 were compared to those of T84.66 monoclonal antibody (MAb) conjugated with either DTPA or its benzylisothiocyanate derivative (BzDTPA) labeled with ^{111}In. The whole-body half-life for VFC-T84.66 was less than for either DTPA-T84.66 or BzDTPAT84.66. The blood clearance rate was similar for all three radioimmunoconjugates. Hepatic uptake of the radiolabel was rapid and remained constant for 7 days for both DTPA radioimmunoconjugates. For VFC, however, the initially observed liver radioactivity decreased rapidly to about 10% of its original value, suggesting a possible role for VFC radioimmunoconjugates in the imaging and β-therapy of liver metastases [29,30].

Recently, PET imaging has found its use expanding into pharmacology, and clinically in neurology, cardiology, and particularly oncology [31–33]. Early on, [^{18}F]-deoxyglucose (FDG), [^{18}F]-thymidine, and [^{18}F]-misonidazole provided images based on heightened metabolic flux or hypoxia characteristic of many cancers. Yet, because

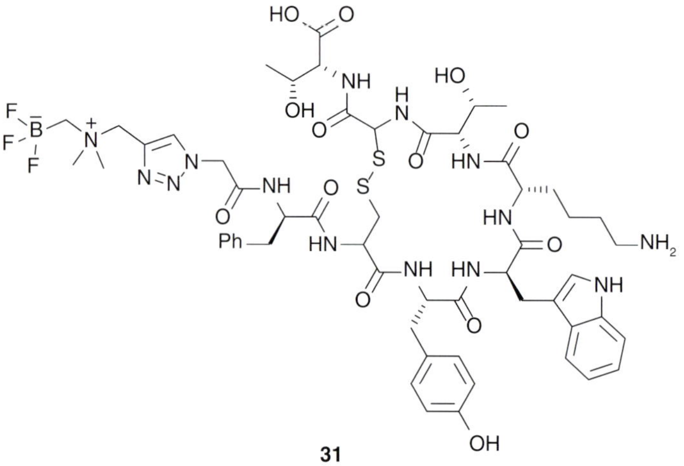

Figure 2.4.13 [^{18}F]-AMBF$_3^-$ octreotate as a probe for blocked and unblocked mice.

cancers are increasingly characterized by the presence of distinct extracellular targets, new target-specific imaging agents are needed to guide diagnosis and treatment. While new ^{18}F-labeling methods for small molecules serve diverse needs, various groups have focused on larger molecules such as peptides, which predictably exhibit high affinity and specificity for many targets. Advances in proteomics and combinatorial screens have provided peptidic tracers that distinguish pathologically distinct molecular targets, which is impossible to achieve with FDG. Examples of peptide tracers include octreotate (**31**), bombesin, and RGD (Figure 2.4.13) [34].

A one-step aqueous ^{18}F-labeling method, which can be applied to peptides to provide functional *in vivo* images, has been a longstanding challenge in PET imaging. Over the past few years, several groups developed rapid and mild radiolabeling methods based on the aqueous radiosynthesis of *in vivo* stable aryltrifluoroborate (ArBF$_3^-$) conjugates. Recent access to production levels of ^{18}F-fluoride led to a fluorescent-^{18}F-ArBF$_3^-$ **32** at unprecedentedly high specific activities of 15 C$_i$/μmol, which opened new avenues for developing boron-based PET imaging agents (three ^{18}F atoms introduced at a time to boron atoms rather than one ^{18}F atom to other elements). However, extending this method to labeling peptides as imaging agents has been exploited in RGD labeling (Figure 2.4.14). Isotope exchange on a clinically useful ^{18}F-ArBF$_3^-$ radiotracer leads to excellent radiochemical yields and exceptionally high specific activities, while the anionic nature of the aryltrifluoroborate prosthetic results in very rapid clearance. Since rapid clearance of the radioactive tracer is generally desirable for tracer development, these results suggest new directions for varying linker arm composition to slightly retard clearance rather enhancement [35].

Bradykinin B1 receptor (B1R) is involved in pain and inflammation pathways and is upregulated in inflamed tissues and cancer. Due to its minimal expression in healthy tissues, B1R is an attractive target for the development of therapeutic agents to treat inflammation, chronic pain, and cancer [36]. Perrin's group synthesized and compared two ^{18}F-labeled peptides derived from potent B1R antagonists B9858 (**33**) and

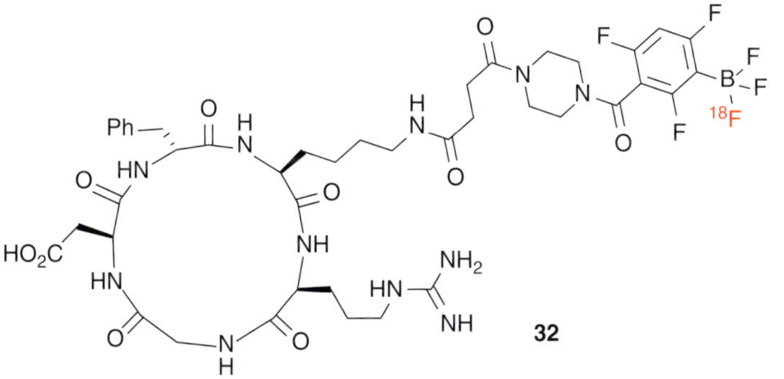

Figure 2.4.14 Structure of RGD-^{18}F-ArBF$_3^-$ with very high specific activity.

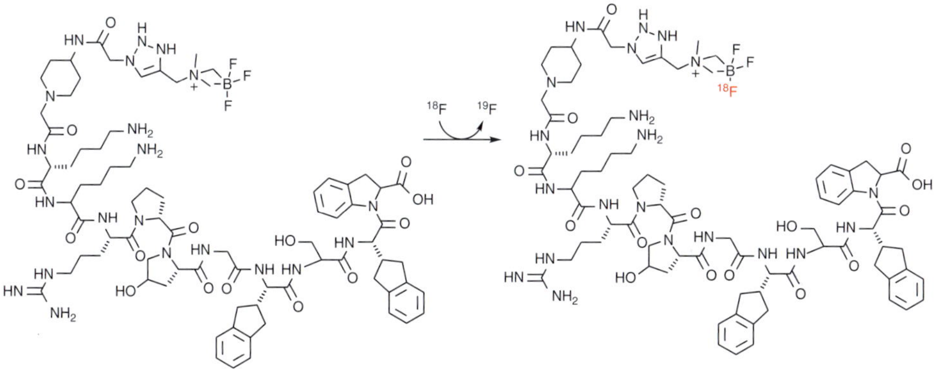

Figure 2.4.15 ^{18}F-exchange to synthesize AmBF$_3^-$ B9858.

B9958 (Figure 2.4.15) for imaging B1R expression with PET. These results indicate that ^{18}F-AmBF$_3^-$ B9858 and ^{18}F-AmBF^{3-} B9958 are promising agents for the *in vivo* imaging of B1R expression with PET. While boron-based PET imaging agent development programs have only recently begun, there appears to be a great future for developing – B^{18}F$_3$-based PET imaging agents for different areas.

2.4.5.3 Boron-Based MRI Probes

In vivo cell tracking has become essential for enhancing the efficacy of cell therapy. Improvements in noninvasive imaging strategies have led to the acquisition of detailed information on the transplanted cells for monitoring cellular therapeutics and assessing the immune microenvironment. Although imaging techniques such as CT, PET, ultrasound, and MRI have been commonly used for medical applications for cell therapy, MRI is the most attractive modality owing to its safety and high resolution. Recently, cell tracking using MRI has been extensively utilized to assess biological processes by modulating and monitoring immunotherapeutic cells using immunotherapies [37].

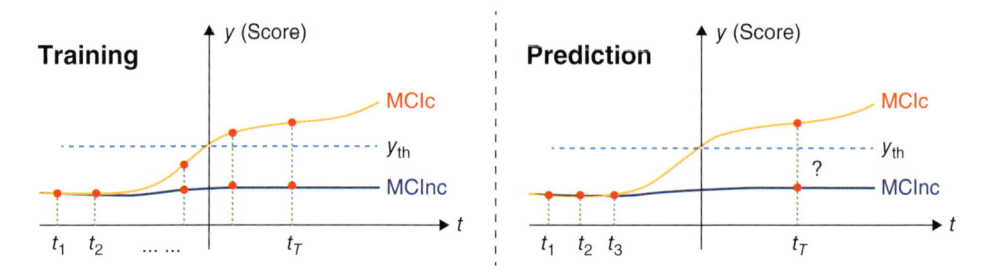

Figure 2.4.16 Longitudinal data to predict AD conversion.

In particular, T1-weighted MRI, which can show hyper-intense areas and produce bright positive contrast against background tissue, has been known as a suitable method for labeling and tracking of target cells [38].

Accurate prediction of Alzheimer's disease (AD) is important for the early diagnosis and treatment of this condition. Mild cognitive impairment (MCI) is an early stage of AD. Therefore, patients with MCI who are at high risk of fully developing AD should be identified to accurately predict AD. However, the relationship between brain images and AD is difficult to construct because of the complex characteristics of neuroimaging data. To address this problem, Perry *et al.* presented a longitudinal measurement of MCI brain images and a hierarchical classification method for AD prediction. Longitudinal images obtained from individuals with MCI were investigated to acquire important information on the longitudinal changes, which can be used to classify MCI subjects as either MCI conversion (MCIc) or MCI non-conversion (MCInc) individuals. Recent studies showed that MRI can contribute significant progress to understand the neural changes related to AD and other diseases. Moreover, MRI data provide some brain structure information; this information can be used to identify the anatomical differences between populations of AD patients and normal controls, and to assist in the diagnosis and evaluation of MCI progression (Figure 2.4.16) [39]. Generally, most MRI-based classification methods consist of two major steps: (1) feature extraction and selection, and (2) classifier learning. Basing on the type of features extracted from MRI, the MCInc/MCIc classification methods can be divided into three categories: the voxel-based approach, the vertex-based approach, and the region of interest (ROI)-based approach.

Very recently, Hanaoka's group developed boron-based targeted MRI contrast agents and used these for detection of atherosclerotic plaques [39b].

2.4.5.4 Boron-Based Molecular Probes for Disease

A fluorescent probe, HKGreen-2 (**35**), has been developed based on a specific reaction between ketone and peroxynitrite (ONOO−). This probe is highly sensitive and selective for the detection of peroxynitrite not only in abiotic but also in biological systems. With this probe, the Liu and Ahrenes groups successfully detected peroxynitrite generated in murine macrophage cells activated by phorbol 12-myristate 13-acetate (PMA), interferon-γ (IFN-γ), and lipopolysaccharide (LPS) (Figure 2.4.17). This new probe will be a useful tool for studying the roles of peroxynitrite in biological processes [40].

In recent years, boron-dipyrromethene (BODIPY) dyes have gained great popularity as fluorescent markers and sensors, owing to their very sharp emissions, high fluorescent quantum yields, good photostability, and insensitivity to pH. The typical BODIPY compounds emit strong green light [41,42]. Modification of its core to achieve red shifts has attracted considerable interest, for long-wavelength emission can avoid the interference of inherent biological fluorescence in the short-wavelength region. The extended conjugation length via introducing phenylethene groups on the 2,6 positions obtained strong red-emissive derivatives. BODIPY derivatives have the disadvantage of very small Stokes shifts, which lead to self-quenching and measurement errors by excitation light and scattering light. Attaching an electron donor to the core to form a donor–acceptor system is an efficient way to increase the Stokes shifts due to their very strong Intramolecular Charge Transfer (ICT) characteristics. However, these D–A (Donor-Acceptor) type BODIPY dyes exhibit significant fluorescence quenching in polar solvents, which is a limitation for biological applications. Liu and others have designed a novel donor–acceptor type BF complex and its isomer, derived from naphthalimide. Unlike the known D–A BODIPY derivatives, these analogs have a proper ICT effect, so they show both very large Stokes shifts (absorption at 424 nm and emission at 620 nm) and high fluorescence quantum yields in polar solvents [43].

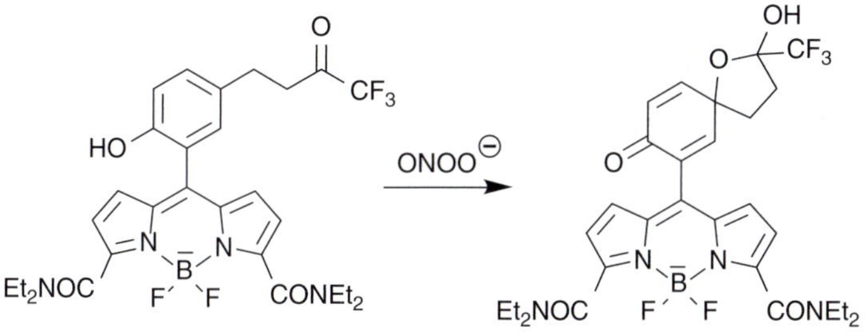

Figure 2.4.17 Oxidation of HKGreen (**34**) with peroxynitrite.

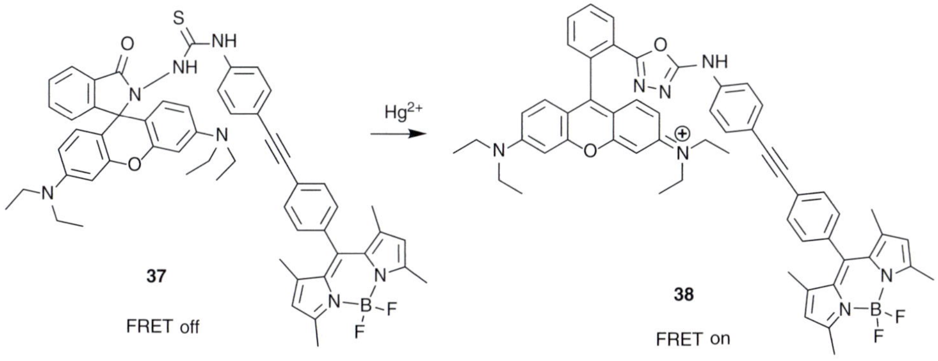

Figure 2.4.18 The structure and binding mode of a BODIPY–rhodamine chemodosimeter.

A FRET sensor for mercury ions (as shown in Figure 2.4.18) whose sensitivity reached the parts-per-billion (ppb) scale [44] has been developed to popularize this strategy in the development of more ratiometric sensors for various analytes. The ICT-based ratiometric sensors encounter two problems influencing their accuracy: (1) binding of the target ions results in remarkable shifts of sensor absorption maxima, and the difference in efficiency among multiple excitation wavelengths may be a potential origin of inaccuracy; and (2) relatively broad fluorescence spectra for ICT fluorophores, before and after binding target ions, have a high degree of overlap, which makes it difficult to accurately determine the ratio of the two fluorescence peaks. Therefore, this boron-based sensor **37** will be a general strategy to use as a novel probe for FRET sensors.

Molander and coworkers were the first to synthesize a aryltrifluoroborate functionalized with an azide (Figure 2.4.19) as an attractive tool for prosthetic group conjugation owing to it being a simple process for peptide labeling (**40**). Formation of the benzyl azide **40** proceeds in near-quantitative isolated yields with a range of substitution patterns, and reaction with a terminal alkyne was attempted under a variety of conditions (varying solvent, temperature, and reaction time). It was found that the reaction proceeded to completion in dimethyl sulfoxide (DMSO) after 1 h at 80 °C, and any other solvents used were largely ineffective owing to the lack of solubility of the starting materials. If the starting material used is not capable of withstanding the elevated temperature, the reaction can be performed at room temperature, giving an 82% yield after 48 h. Modification of the functional groups present on the aryl alkyne showed little impact on the yield, with all reactions attempted yielding 85–100% conversion. Formation of aryl azides has also been demonstrated using other aryltrifluoroborate reagents, offering an alternative procedure for the formation of "click" precursors while keeping aryltrifluoroborates (**41**) intact [45].

Two closely related phenyl selenyl–based BODIPY turn-on fluorescent probes for the detection of hypochlorous acid (HOCl) were synthesized for studies in chemical biology; emission intensity is modulated by a photoinduced electron-transfer (PET) process. Probe **42** intrinsically shows a negligible background signal; however, after reaction with HOCl, chemical oxidation of selenium forecloses the PET process, which evokes a significant increase in fluorescence intensity. The fluorescence intensity of probes **42** and **43** with HOCl involves an ~18 and ~50-fold enhancement compared with the respective responses from other ROS and RNS, and low detection limits (30.9 nm for 1 and 4.5 nm for 2). Both probes show a very fast response with HOCl; emission intensity reached a maximum within 1 s. These probes show high selectivity for HOCl, as confirmed by confocal microscopy imaging when testing with RAW264.7 and MCF-7 cells (Figure 2.4.20) [46].

Fluorescent probes with larger Stokes shifts in the far-visible and near-infrared spectral region (600–900 nm) are superior for cellular imaging and biological analyses due

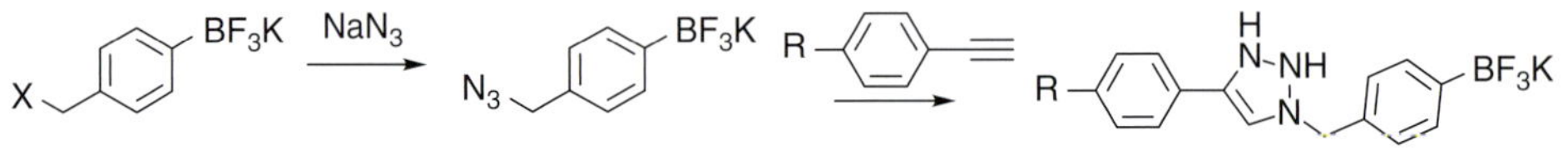

Figure 2.4.19 Application of click reaction to synthesize boron probes.

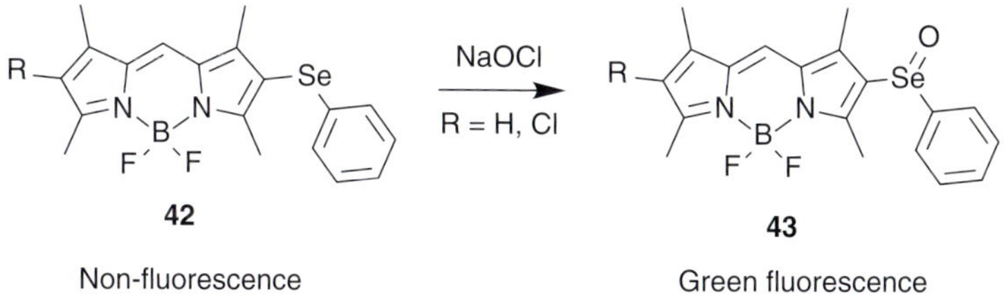

Figure 2.4.20 Detection of NaOCl using boron probes.

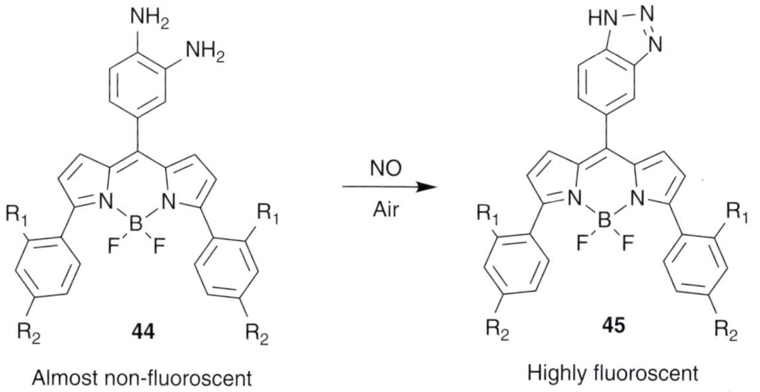

Figure 2.4.21 Synthesis routes and reaction mechanism of o-MOPB and p-MOPB with NO.

to avoiding light-scattering interference, reducing auto-fluorescence from biological samples, and encouraging deeper tissue penetration for *in vivo* imaging. Wang's group (Figure 2.4.21) synthesized two *bis*-methoxyphenyl-BODIPY fluorescent probes **44** for the detection of nitric oxide (NO) [47]. Under physiological conditions, these probes can react with NO to form the corresponding triazoles **45** with 250- and 70-fold turn-on fluorescence emitting at 590 and 620 nm, respectively. Moreover, the triazole forms of these probes have large Stokes shifts of 38 nm, in contrast to 10 nm of existing BODIPY probes for NO. Excellent selectivity has been observed against other ROS and RNS, ascorbic acid, and biological matrix. After the evaluation of MTT assay, new fluorescent probes have been successfully applied to fluorescence imaging of NO released from RAW 264.7 macrophages by co-stimulation of LPS and interferon-γ. The experimental results indicate that fluorescent probes can be powerful candidates for fluorescence imaging of NO due to the low background interference and high detection sensitivity.

The platinum drugs cisplatin, carboplatin, and oxaliplatin are highly utilized in the clinic and as a consequence have been extensively studied in the laboratory setting, sometimes by generating fluorophore-tagged analogs. Several groups synthesized Pt(II) complexes containing ethane-1,2-diamine ligands linked to a BODIPY fluorophore, and compared their biological activity with previously reported Pt(II) complexes conjugated to carboxyfluorescein and carboxyfluorescein diacetate (Figure 2.4.22). The cytotoxicity and DNA damage capacity of Pt–fluorophore

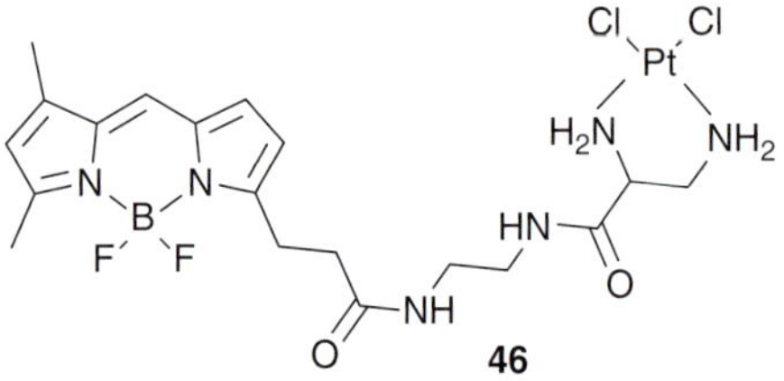

Figure 2.4.22 Pt–BODIPY as imaging agents.

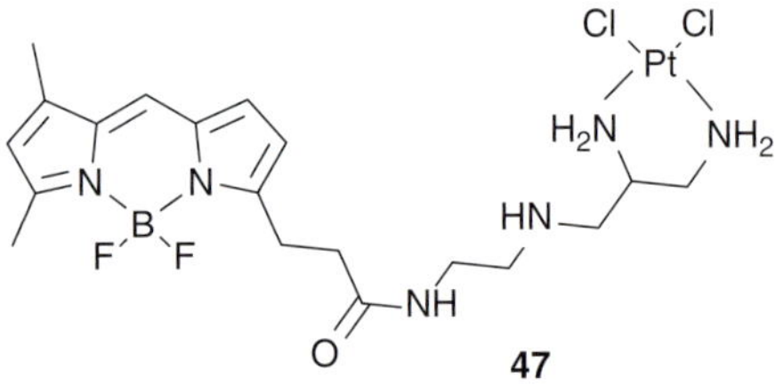

complexes **46** and **47** were compared to those of cisplatin, and the Pt–BODIPY complexes were found to be more cytotoxic with reduced cytotoxicity in cisplatin-resistant cells. Microscopy revealed a predominately cytosolic localization, with nuclear distribution at higher concentrations. Spheroids grown from parent and resistant cells revealed penetration of Pt–BODIPY into spheroids, and retention of the cisplatin-resistant spheroid phenotype. While most activity profiles were retained for the Pt–BODIPY complexes, accumulation in resistant cells was only slightly affected, suggesting that some aspects of Pt–fluorophore cellular pharmacology deviate from cisplatin [48,49].

Recently, elegant molecular designs have been developed to take full advantage of the unique high detection sensitivity of both optical and nuclear imaging methods. One possible approach seeks to fuse the two imaging systems into one molecule (a monomolecular multimodality imaging agent [MOMIA]) in order to ensure the same bio-distribution of the two probes [50]. Despite the high sensitivity and complementary nature of radionuclear and optical methods, combinations of these two imaging modalities are rare. This is attributable to a variety of reasons, one of them being the difficulty to synthesize and characterize species capable of this type of bimodal imaging [51–53]. More precisely, the properties of DOTA have been explored as MRI contrast agents, or for labeling biomolecules using metal radioisotopes for both diagnosis (SPECT or PET) and therapeutic purposes [54–58].

2.4.6 Future Perspectives

There is no doubt that molecular imaging will play a major role in biomedical research, not only for diagnosis of diseases before pathophysiological changes are irreversible but also in biomarker discovery, in new target-to-hit and hit-to-lead development, and in preclinical trials to establish the toxicity profile of new agents and finally delivery and prognosis of diseases. Molecular imaging will play a major role in identifying new biomarkers for stem cell biology, precision medicine, and single-cell detection. The success of these technologies depends on easy synthesis of probes and pharmacological agents with novel instruments and micromolar-scale resolution for detection. In this context, future boron-based molecular probes will play crucial roles as theranostic agents (both therapeutic and diagnostic) [59–61].

Appendix: Companies Offering Imaging Instruments and Reagents

Company	Web address
Agilent Technologies	http://www.agilent.com/
BD Biosciences	http://www.bdbiosciences.com/
Berthold Technologies	http://www.berthold.com/
Bioscan	http://www.bioscan.com/
Bruker Biospin	http://www.bruker-biospin.com/
Caliper Life Sciences/Xenogen	http://www.caliperls.com/
Carestream Molecular Imaging	http://www.carestreamhealth.com/
Chroma Technology	http://www.chroma.com/
Clontech	http://www.clontech.com/
Cri	http://www.cri-inc.com/
Evrogen	http://www.evrogen.com/
GE Healthcare	http://www.gehealthcare.com/
Hamamatsu	http://www.hamamatsu.com/
Hitachi Medical Systems	http://www.hitachimed.com/
Improvision	http://www.improvision.com/
Invitrogen	http://www.invitrogen.com/
LaVision	http://www.lavision.de/
LI-COR	http://www.licor.com/
Lightools Research	http://www.lightools.com/
M2M imaging	http://www.m2mimaging.com/
Mauna Kea Technologies	http://www.maunakeatech.com/
Millennium	http://www.mlnm.com/
MRVision	http://www.mrvision.com/
New England Biolabs	http://www.neb.com/
Nikon Instruments	http://www.nikon.com/
Philips Healthcare	http://www.medical.philips.com/
Promega	http://www.promega.com/
Semrock	http://www.semrock.com/
Siemens Molecular Imaging	http://www.medical.siemens.com/
SkyScan Micro CT	http://www.skyscan.be/
UVP	http://www.uvp.com/
Varian Magnetic Resonance	http://www.varianinc.com/
Visage Imaging	http://www.visageimaging.com/
VisEn Medical	http://www.visenmedical.com/

References

1 National Academy of Engineering. *Greatest Engineering Achievements of the 20th Century* [online], www.greatachievements.org (2009).
2 (a) Allport J.R., and Weissleder, R. In vivo imaging of gene and cell therapies. *Exp. Hematol.*, **29**, 1237–1246 (2001). (b) Sabbah, P., Foerenbach, H., Dutertre, G., Nioche, C., and DeBreuille, O. Multimodal anatomic, functional, and metabolic brain imaging for

tumor resection. *Clin Imaging* **26**, 6–12 (2002). (c) Li, C., Wang, W., Wu, Q., Ke, S., Houston, J., Sevick-Muraca, E., Dong, L., Chow, D., Charnsangavej, C., and Gelovani, J.G. Dual optical and nuclear imaging in human melanoma xenografts using a single targeted imaging probe. *Nucl. Med. Biol.* **33**, 349–358 (2006).

3 (a) Lin, K., Zheng, W., and Huang, Z. Integrated autofluorescence endoscopic imaging and point-wise spectroscopy for real-time *in-vivo* tissue measurements. *J. Biomed. Opt.* **14**, 0405071–0405073 (2010). (b) Wang, W., Ke, S., Kwon, S., Yallampalli, S., Cameron, A.G., Adams, K.E., Mawad, M.E., and Sevick-Muraca, E.M. A new optical and nuclear dual-labeled imaging agent targeting interleukin 11 receptor alpha-chain. *Bioconjug. Chem.* **18**, 397–402 (2007). (c) Culver, J., Akers, W., and Achilefu, S. Multimodality molecular imaging with combined optical and SPECT/PET modalities. *J. Nucl. Med.* **49**, 169–172 (2008).

4 (a) Das, B.C., Thapa, P., Karki, R., Schinke, C., Das, S., Kambhampati, S., Banerjee, S.K., Van Veldhuizen, P., Verma, A., Weiss, L.M., and Evans, T. Boron chemicals in diagnosis and therapeutics. *Future Med. Chem.* **5**, 653–676 (2013). (b) Das, B.C., Smith, M.E., and Kalpana, G.V. Design, synthesis of novel peptidomimetic derivatives of 4-HPR for rhabdoid tumors. *Bioorg. Med. Chem. Lett.* **18**, 4177–4180 (2008). (c) Solingapuram, Sai K.K., Das, B., Sattiraju, A., Almaguel, F., Craft, S., and Mintz, A. Radiosynthesis and initial *in vitro* evaluations of [^{18}F]KBM-1 as a potential RAR-a imaging agent. *J. Nucl. Med.* **57**, 1077 (2016).

5 (a) Burke, B.P., Clemente, G.S., and Archibald, S.J. Boron–^{18}F containing positron emission tomography radiotracers: advances and opportunities. *Contrast Media Mol. Imaging.* **10**, 96–110 (2015). (b) Liu, Z., Pourghiasian, M., Benard, F., Pan, J., Lin, K.-S., and Perrin, D.M. Preclinical evaluation of a high-affinity ^{18}F-trifluoroborate octreotate derivative for somatostatin receptor imaging. *J. Nucl. Med.* **55**, 1499–1505 (2014).

6 (a) Dannoon, S., Ganguly, T., Cahaya, H., Geruntho, J.J., Galliher, M.S., Beyer, S.K., Choy, C.J., Hopkins, M.R., Regan, M., Blecha, J.E., Skultetyova, L., Drake, C.R., Jivan, S., Barinka, C., Jones, E.F., Berkman, C.E., and VanBrocklin, H.F. Structure–activity relationship of ^{18}F-labeled phosphoramidate peptidomimetic prostate-specific membrane antigen (PSMA)-targeted inhibitor analogues for PET imaging of prostate cancer. *J. Med. Chem.* **59**, 5684–5694 (2016). (b) Soodgupta, D., Zhou, H., Beaino, W., Lu, L., Rettig, M., Snee, M., Skeath, J., DiPersio, J.F., Akers, W.J., Laforest, R., Anderson, C.J., Tomasson, M.H., and Shokeen, M. Ex vivo and in vivo evaluation of overexpressed VLA-4 in multiple myeloma using LLP2A imaging agents. *J. Nucl. Med.* **57**, 640–645 (2016).

7 Dobrucki, L.W., and Sinusas, A.J. PET and SPECT in cardiovascular molecular imaging. *Nat. Rev. Cardiol.* **7**, 38–47 (2010).

8 Chopra, A., *et al.* Molecular imaging and contrast agent database (MICAD): evolution and progress. *Mol. Imaging Biol.* **14**, 4–13 (2012).

9 Gaglia, J.L., *et al.* Noninvasive mapping of pancreatic inflammation in recent-onset type-1 diabetes patients. *Proc. Natl. Acad. Sci. USA* **112**, 2139–2144 (2015).

10 (a) Weissleder, R., Nahrendorf, M., and Pittet, M.J. Imaging macrophages with nanoparticles. *Nat. Mater.* **13**, 125–138 (2014). (b) Marti-Climent, J.M., Collantes, M., Jauregui-Osoro, M., Quincoces, G., Prieto, E., Bilbao, I., *et al.* Radiation dosimetry and biodistribution in non-human primates of the sodium/iodide PET ligand [^{18}F]-tetrafluoroborate. *EJNMMI Res.* **5**, 70 (2015).

11 Weeks, A.J., Jauregui-Osoro, M., Cleij, M., Blower, J.E., Ballinger, J.R., and Blower, P.J. Evaluation of [^{18}F]-tetrafluoroborate as a potential PET imaging agent for the human

sodium/iodide symporter in a new colon carcinoma cell line, HCT116, expressing hNIS. *Nucl. Med. Commun.* **32**, 98–105 (2011).

12 Louie, A. Multimodality imaging probes: design and challenges. *Chem. Rev.* **110**, 3146–3195 (2010).

13 Fu, W., Wojtkiewicz, G., Weissleder, R., Benoist, C., and Mathis, D. Early window of diabetes determinism in NOD mice, dependent on the complement receptor CRIg, identified by noninvasive imaging. *Nat. Immunol.* **13**, 361–368 (2012).

14 (a) Rao, J., Dragulescu-Andrasi, A., and Yao, H. Fluorescence imaging in vivo: recent advances. *Curr. Opin. Biotechnol.* **18**, 17 (2007). (b) Lemon, C.M., *et al.* Metabolic tumor profiling with pH, oxygen, and glucose chemosensors on a quantum dot scaffold. *Inorg. Chem.* **53**, 1900–1915 (2014). (c) Kim, S., *et al.* Near-infrared fluorescent type II quantum dots for sentinel lymph node mapping. *Nat. Biotechnol.* **22**, 93–97 (2004).

15 Vedejs, E., Chapman, R.W., Fields, S.C., Lin, S., and Schrimpf, M.R. Conversion of arylboronic acids into potassium aryltrifluoroborates: convenient precursors of arylboron difluoride Lewis acids. *J. Org. Chem.* **60**, 3020–3027 (1995).

16 (a) Tsien, R.Y. Imagining imaging's future. *Nat. Rev. Mol. Cell Biol.* **4**, SS16–SS21 (2003). (b) Aina, O.H., Liu, R.W., Sutcliffe, J.L., Marik, J., Pan, C.X., and Lam, K.S. From combinatorial chemistry to cancer-targeting peptides. *Mol. Pharmaceutics* **4**, 631–651 (2007). (c) Cai, L.S., Lu, S.Y., and Pike, V.W. Chemistry with F-18 fluoride ion. *Eur. J. Org. Chem.* 2008, 2853–2873 (**2008**).

17 (a) Ting. R., Aguilera, T.A., Crisp, J.L., Hall, D.J., Eckelman, W.C., Vera, D.R., and Tsien, R.Y. Fast ^{18}F labeling of a near-infrared fluorophore enables positron emission tomography and optical imaging of sentinel lymph. *Bioconjug. Chem.* **21**, 1811–1819 (2010). (b) Trembleau, L., Simpson, M., Cheyne, R.W., Escofet, I., Appleyard, M.V.C.A.L., Murray, K., Sharp, S., Thompson, A.M., and Smith, T.A.D. Development of ^{18}F-fluorinatable dendrons and their application to cancer cell targeting. *New J. Chem.* **35**, 2496–2502 (2011). (c) Li, Y., Guo, J., Tang, S., Lang, L., Chen, X., and Perrin, D.M. One-step and one-pot-two-step radiosynthesis of cyclo-RGD-^{18}F-aryltrifluoroborate conjugates for functional imaging. *Am. J. Nucl. Med. Mol. Imag.* **3**, 44–56 (2013). (d) Liu, S., Park, R., Conti, P.S., and Li, Z. Kit like ^{18}F labeling method for synthesis of RGD peptide-based PET probes. *Am. J. Nucl. Med. Mol. Imag.* **3**, 97–101 (2013).

18 (a) Li, Y., Liu, Z., Harwig, C.W., Pourghiasian, M., Lau, J., Lin, K.-S., Schaffer, P., Benard, F., and Perrin, D.M. ^{18}F-click labeling of a bombesin antagonist with an alkyne-^{18}F-ArBF3–: in vivo PET imaging of tumors expressing the GRP-receptor *Am. J. Nucl. Med. Mol. Imag.* **3**, 57–70 (2013). (b) Liu, Z., Li, Y., Lozada, J., Schaffer, P., Adam, M.J., Ruth, T.J., and Perrin, D.M. Stoichiometric leverage: rapid 18 F-aryltrifluoroborate radiosynthesis at high specific activity for click conjugation. *Angew. Chem. Int. Ed.* **52**, 2303–2307 (2013).

19 (a) Ting, R., Harwig, C., auf dem Keller, U., McCormick, S., Austin, P., Overall, C.M., Adam, M.J., Ruth, T.J., and Perrin. D.M. Toward [^{18}F]-labeled aryltrifluoroborate radiotracers: in vivo positron emission tomography imaging of stable aryltrifluoroborate clearance in mice. *J. Am. Chem. Soc.* **130**, 12045–12055 (2008). (b) Smith, T.A.D., Simpson, M., Cheyne, R., and Trembleau, L. *Appl. Radiat. Isot.* **69**, 1395–1400 (2011). (c) See 7(a). (c)Li, Y., Schaffer, P., and Perrin, D.M. Dual isotope labeling: conjugation of 32P-oligonucleotides with ^{18}F-aryltrifluoroborate via copper(I) catalyzed cycloaddition. *Bioorg. Med. Chem. Lett.* **23**, 6313–6316 (2013).

20 (a) Liu, Z., Li, Y., Lozada, J., Pan, J., Lin, K.-S., Schaffer, P., and Perrin, D.M. Rapid, one-step, high yielding ^{18}F-labeling of an aryltrifluoroborate bioconjugate by isotope exchange at very high specific activity. *J. Labelled Comp. Rad.* **55**, 491–496 (2012). (b) Liu, Z., Hundal-Jabal, N., Wong, M., Yapp, D., Lin, K.S., Benard, F., and Perrin, D.M. A new ^{18}F-heteroaryltrifluoroborate radio-prosthetic with greatly enhanced stability that is labelled by ^{18}F–^{19}F-isotope exchange in good yield at high specific activity. *MedChemComm.* **5**, 171–179 (2014). (c) Li, Z., Chansaenpak, K., Liu, S., Wade, C.R., Conti, P.S., and Gabbai, F.P. Harvesting ^{18}F-fluoride ions in water via direct ^{18}F–^{19}F isotopic exchange: radiofluorination of zwitterionic aryltrifluoroborates and in vivo stability studies. *MedChemComm.* **3**, 1305–1308 (2012).

21 (a) Miller, E.W., Albers, A.E., Pralle, A., Isacoff, E.Y., and Chang, C.J. Boronate-based fluorescent probes for imaging cellular hydrogen peroxide. *J. Am. Chem. Soc.* **127**, 16652–16659 (2005). (b) Miller, E.W., Tulyathan, O., Isacoff, E.Y., and Chang, C.J. Molecular imaging of hydrogen peroxide produced for cell signaling. *Nat. Chem. Biol.* **3**, 263–267 (2007).

22 Dickinson, B.C., Huynh, C., and Chang, C.J. A palette of fluorescent probes with varying emission colors for imaging hydrogen peroxide signaling in living cells. *J. Am. Chem. Soc.* **132**, 5906–5915 (2010).

23 Chang, M.C.Y., Pralle, A., Isacoff, E.Y., and Chang, C.J. A selective, cell-permeable optical probe for hydrogen peroxide in living cells. *J. Am. Chem. Soc.* **126**, 15392–15393 (2004).

24 Van de Bittnera, G.C., Dubikovskayaa, E.A., Bertozzi, C.R., and Chang, C.J. In vivo imaging of hydrogen peroxide production in a murine tumor model with a chemoselective bioluminescent reporter. *Proc. Natl. Acad. Sci. USA*. **107**, 21316–21321 (2010).

25 Olsen, F.P., and Hawthorne, M.F. Halodicarbaundecaborate(II) ions. *Inorg. Chem.* **4**, 1839 (1965).

26 (a) Varadarajan, A., Sharkey, R.M., Goldenberg, D.M., and Hawthorne, M.F. Conjugation of phenyl isothiocyanate derivatives of carborane to antitumor antibody and in vivo localization of conjugates in nude mice. *Bioconjug. Chem.* **2**, 102 (1991). (b) Mizusawa, E.A., Thompson, M.R., and Hawthorne, M.F. *Inorg. Chem.* **24**, 1911 (1985).

27 Stephenson, G.J., Jr., and Goldman, T. A possible solution to the tritium endpoint problem. *Phys. Lett.* **440B**, 89 (1998).

28 Mizusawa, E., Dahlman, H.L., Bennet, S.J., Goldenberg, D.M., and Hawthorne, M.F. Neutron-capture therapy of human cancer: in vitro results on the preparation of boron-labeled antibodies to carcinoembryonic antigen. *Proc. Natl. Acad. Sci. USA*. **79**, 3011 (1982).

29 Filer, C., Hurt, S., and Wan, Y.-P. In *Receptor Pharmacology and Function*, Williams, M., Glennon, R.A., Timmermans, P.B.M.W.M. (Eds.), New York, Marcel Dekker, 105 (1989).

30 Beatty, B.G., Paxton, R.J., Hawthorne, M.F., Williams, L.E., Rickard-Dickson, K.J., Do, T., Shively, J.E., and Beatty, J.D. Pharmacokinetics of an anti-carcinoembryonic antigen monoclonal antibody conjugated to a bifunctional transition metal carborane complex (venus flytrap cluster) in tumor-bearing mice. *J. Nucl. Med.* **34**, 1294 (1993).

31 Haubner, R., *et al.* Glycosylated RGD-containing peptides: tracer for tumor targeting and angiogenesis imaging with improved biokinetics. *J. Nucl. Med.* **42**, 326–336 (2001).

32 Pourghiasian, M., Liu, Z.B., Pan, J.H., Zhang, Z.X., Colpo, N., Lin, K.S., Perrin, D.M., and Benard, F. ^{18}F-AmBF3-MJ9: a novel radiofluorinated bombesin derivative for prostate cancer imaging. *Bioorg. Med. Chem.* **23**, 1500–1506 (2015).

33 Liu, Z., Radtke, M.A., Wong, M.Q., Lin, K.-S., Yapp, D.T., and Perrin, D.M. Dual mode fluorescent ^{18}F-PET tracers: efficient modular synthesis of rhodamine-[cRGD]2-[^{18}F]-organotrifluoroborate, rapid, and high yielding one-step ^{18}F-labeling at high specific activity, and correlated in vivo PET imaging and ex vivo fluorescence. *Bioconjug Chem.* **25**, 1951–1962 (2014).

34 Liu, Z.B., Pourghiasian, M., Radtke, M.A., Lau, J., Pan, J.H., Dias, G.M., Yapp, D., Lin, K.S., Benard, F., and Perrin, D.M. An organotrifluoroborate for broadly applicable one-step ^{18}F-labeling. *Angew. Chem. Int. Ed.* **53**, 11876–11880 (2014).

35 Liu, Z., Li, Y., Lozada, J., Wong, M.Q., Greene, J., Lin, K.S., Yapp, D., and Perrin, D.M. Kit-like ^{18}F-labeling of RGD-19F-arytrifluroborate in high yield and at extraordinarily high specific activity with preliminary *in vivo* tumor imaging. *Nucl. Med. Biol.* **40**, 841–849 (2013).

36 Liu, Z., Amouroux, G., Zhang, Z., Pan, J., Hundal-Jabal, N., Colpo, N., Lau, J., Perrin, D.M., Bénard, F., and Lin, K. ^{18}F-trifluoroborate derivatives of [Des-Arg10] kallidin for imaging bradykinin B1 receptor expression with positron emission tomography. *Mol. Pharmaceutics.* **12**, 974–982 (2015).

37 (a) Lu, J., Feng, F., and Jin, Z. *Curr. Pharm. Biotechnol.* **14**, 714–722 (2013). (b) Ahrens, E.T., and Bulte, J.W. Tracking immune cells in vivo using magnetic resonance imaging. *Nat. Rev. Immunol.* **13**, 755–763 (2013).

38 (a) Tseng, C.L., Shih, I.L., Stobinski, L., and Lin, F.H. *Biomaterials.* **31**, 5427–5435 (2010). (b) Kim, T., Momin, E., Choi, J., Yuan, K., Zaidi, H., Kim, J., Park, M., Lee, N., McMahon, M.T., Quinones-Hinojosa, A., Bulte, J.W., Hyeon, T., and Gilad, A.A. Mesoporous silica-coated hollow manganese oxide nanoparticles as positive T1 contrast agents for labeling and MRI tracking of adipose-derived mesenchymal stem cells. *J. Am. Chem. Soc.* **133**, 2955–2961 (2011).

39 (a) Huang, M., Yang, W., Feng, Q., and Chen, W. Longitudinal measurement and hierarchical classification framework for the prediction of Alzheimer's disease. *Scientific Reports* **7**, 39880 (2016). (b) Iwaki, S., Hokamura, K., Ogawa, M., Takehara, Y., Muramatsu, Y., Yamane, T., Hirabayashi, K., Morimoto, Y., Hagisawa, K., Nakahara, K., Mineno, T., Terai, T., Komatsu, T., Ueno, T., Tamura, K., Adachi, Y., Hirata, Y., Arita, M., Arai, H., Umemura, K., Nagano, T., and Hanaoka, K. A design strategy for small molecule based targeted MRI contrast agents: their application for detection of atherosclerotic plaques. *Org. Biomol. Chem.* **12**(43), 8611–8618 (2014).

40 Sun, Z.-N., Wang, H.-L., Liu, F.-Q., Chen, Y., Tam, P.K.H., and Yang, D. BODIPY-based fluorescent probe for peroxynitrite detection and imaging in living cells. *Org. Lett.* **11**, 1887–1894 (2009).

41 Zhang, X., Xiao, Y., and Qian, X. A ratiometric fluorescent probe based on FRET for imaging Hg^{2+} Ions in living cells. *Angew. Chem., Int. Ed.* **47**, 8025–8029 (2008).

42 Lee, M.H., Kim, H.J., Yoon, S., Park, N., and Kim, J.S. Resonance energy transfer approach and a new ratiometric probe for Hg^{2+} in aqueous media and living organism. *Org. Lett.* **11**, 2740–2743 (2009).

43 Shang, G.Q., Gao, X., Chen, M.X., Zheng, H., and Xu, G. A novel Hg2+ selective ratiometric fluorescent chemodosimeter based on an intramolecular FRET mechanism. *J. Fluoresc.* **18**, 1187 (2008).

44 (a) Zhou, Z.G., Yu, M.X., Yang, H., Huang, K.W., Li, F.Y., Yi, T., and Huang, C.H. FRET-based sensor for imaging chromium(III) in living cells. *Chem. Commun.* 3387–3389 (2008).

45 (a) Molander, G.A., and Figueroa, R. Synthesis of unsaturated organotrifluoroborates via Wittig and Horner–Wadsworth–Emmons olefination. *J. Org. Chem.* **71**, 6135–6140 (2006). (b) Cho, Y.A., Kim, D.-S., Ahn, H.R., Canturk, B., Molander, G.A., and Ham, J. Preparation of potassium azidoaryltrifluoroborates and their cross-coupling with aryl halides. *Org. Lett.* **11**, 4330–4333 (2009).

46 (a) Mulay, S.V., Choi, M., Jang, Y.J., Kim, Y., Jon, S., and Churchill, D.G. Enhanced fluorescence turn-on imaging of hypochlorous acid in living immune and cancer cells. *Chem. Eur. J.* **22**, 9642–9648 (2016).

47 (a) Chen, J.-B., Zhang, H.-X., Guo, X.-F., Wang, H., and Zhang, H.-S. "Off–on" red-emitting fluorescent probes with large Stokes shifts for nitric oxide imaging in living cells. *Anal. Bioanal. Chem.* **405**, 7447–7456 (2013). (b) Wang, H., et al. Highly sensitive determination of nitric oxide in biologic samples by a near-infrared BODIPY-based fluorescent probe coupled with high-performance liquid chromatography. *Talanta.* **116**, 335–342 (2013).

48 Rocca, J.D., Liu, D., and Lin, W. Nanoscale metal-organic frameworks for biomedical imaging and drug delivery. *Acc. Chem. Res.* **44**, 957–968 (2011).

49 Pushie, M.J., Pickering, I.J., Korbas, M., Hackett, M.J., and George, G.N. Elemental and chemically specific X-ray fluorescence imaging of biological systems. *Chem. Rev.* **114**, 8499–8541 (2014).

50 Li, C., Wang, W., Wu, Q., Ke, S., Houston, J., Sevick-Muraca, E., Dong, L., Chow, D., Charnsangavej, C., and Gelovani, J.G. Dual optical and nuclear imaging in human melanoma xenografts using a single targeted imaging probe. *Nucl. Med. Biol.* **33**, 349–358 (2006).

51 Lee, H., Akers, W.J., Cheney, P.P., Edwards, W.B., Liang, K., Culver, J.P., and Achilefu, S., Complementary optical and nuclear imaging of caspase-3 activity using combined activatable and radio-labeled multimodality molecular probe. *J. Biomed. Opt.* **14**, 0405071–0405073 (2009).

52 Wang, W., Ke, S., Kwon, S., Yallampalli, S., Cameron, A.G., Adams, K.E., Mawad, M.E., and Sevick-Muraca, E.M. A new optical and nuclear dual-labeled imaging agent targeting interleukin 11 receptor alpha-chain. *Bioconjugate Chem.* **18**, 397–402 (2007).

53 Culver, J., Akers, W., and Achilefu, S. Multimodality molecular imaging with combined optical and SPECT/PET modalities. *J. Nucl. Med.* **49**, 169–172 (2008).

54 (a) Caravan, P. Strategies for increasing the sensitivity of gadolinium based MRI contrast agents *Chem. Soc. Rev.* **35**, 512–523 (2006). (b) Caravan, P., Ellison, J.J., Mc Murry, T.J., and Lauffer, R.B. Gadolinium(III) chelates as MRI contrast agents: structure, dynamics, and applications. *Chem. Rev.* **99**, 2293–2352 (1999).

55 (a) Hermann, P., Kotek, J., Kubicek, V., and Lukes, I. Gadolinium(iii) complexes as MRI contrast agents: ligand design and properties of the complexes. *Dalton Trans.* 3027–3047 (2008). (b) Merbach, A.E., and Toth, E. *The Chemistry of Contrast Agents in Medical Magnetic Resonance Imaging*, Wiley, Chichester, 2001.

56 Azad, B.B., Rota, V.A., Breadner, D., Dhanvantari, S., and Luyt, L.G. Design, synthesis and in vitro characterization of glucagon-like peptide-1 derivatives for pancreatic beta cell imaging by SPECT. *Bioorg. Med. Chem.* **18**, 1265–1272 (2010).

57 Velikyan, I., Sundin, A., Eriksson, B., Lundqvist, H., Sörensen, J., Bergström, M., and Långström, B. In vivo binding of [68Ga]-DOTATOC to somatostatin receptors in neuroendocrine tumours—impact of peptide mass. *Nucl. Med. Biol.* **37**, 265–275 (2010).
58 Cordier, D., Forrer, F., Bruchertseifer, F., Morgenstern, A., Apostolidis, C., Good, S., Müller-Brand, J., Mäcke, H., Reubi, J.C., and Merlo, A. Targeted alpha-radionuclide therapy of functionally critically located gliomas with 213Bi-DOTA-[Thi8,Met(O2)11]-substance P: a pilot trial. *Eur. J. Nucl. Med. Mol. Imaging.* **37**, 1335–1344 (2010).
59 Auletta, L., Gramanzini, M., Gargiulo, S., Albanese, S., Salvatore, M., and Greco, A. Advances in multimodal molecular imaging. *Q. J. Nucl. Med. Mol. Imaging.* **61**(1), 19–32 (2017).
60 Rieffel, J., Chitgupi, U., and Lovell, J.F. Recent advances in higher-order, multimodal, biomedical imaging agents. *Small.* **11**(35), 4445–4461 (2015).
61 Same, S., Aghanejad, A., Akbari Nakhjavani, S., Barar, J., and Omidi, Y. Radiolabeled theranostics: magnetic and gold nanoparticles. *Bioimpacts.* **6**(3), 169–181 (2016).

2.5

Radiolabeling Strategies for Boron Clusters: Toward Fast Development and Efficient Assessment of BNCT Drug Candidates

Kiran B. Gona,[1,2,3] Vanessa Gómez-Vallejo,[1] Irina Manea,[4] Jonas Malmquist,[5] Jacek Koziorowski,[5] and Jordi Llop[1,]*

[1] Radiochemistry and Nuclear Imaging Group, CIC biomaGUNE, San Sebastian, Guipúzcoa, Spain
[2] Cardiovascular Molecular Imaging Laboratory, Section of Cardiovascular Medicine and Yale Cardiovascular Research Center, Yale University School of Medicine, New Haven, Connecticut, USA
[3] Veterans Affairs Connecticut Healthcare System, West Haven, Connecticut, USA
[4] Colentina Clinical Hospital, Bucuresti, Romania
[5] Department of Radiology and Department of Medical and Health Sciences, Linköping University, Linköping, Sweden
*Corresponding author. Radiochemistry and Nuclear Imaging Group, CIC biomaGUNE, San Sebastian, Guipúzcoa, Spain.
Email: jllop@cicbiomagune.es

2.5.1 Boron Neutron Capture Therapy

2.5.1.1 Boron: The Element

Boron is a chemical element placed in Group XIII in the Periodic Table, with symbol B, atomic number 5, and molecular weight of 10.811 g/mol. It can be found in two allotropic forms: amorphous boron (brown) and crystalline boron (black). Boron has two stable isotopes: ^{11}B and ^{10}B, with natural abundances of 80.1% and 19.9%, respectively.

2.5.1.2 The Principle behind Boron Neutron Capture Therapy (BNCT)

BNCT, first described by Locher 80 years ago [1], is a binary therapeutic approach based on the ability of the non-radioactive nuclide boron-10 (^{10}B) to capture thermal neutrons, which results in the $^{10}B(n, \alpha, \gamma)^{7}Li$ nuclear reaction. Alpha particles and ^{7}Li recoil ions have high linear energy transfer (LET) properties and path lengths in the range of 4 to 10 μm. If selective accumulation of ^{10}B atoms is achieved in cancer cells, the ions produced as a consequence of neutron irradiation trigger cell death, while sparing healthy surrounding tissue (Figure 2.5.1).

Strategies to accumulate boron atoms in cancer cells are usually based in the attachment of a boron-rich moiety to a targeting unit. To date, abnormal metabolism and the overexpression of membrane receptors have been exploited to accumulate ^{10}B nuclei in cancer cells. For example, small boron molecules have been used to prepare carbohydrates [2–8], amino acids and peptides [9–17], nucleic acid derivatives [18–25], and immunoconjugates [26–29]. More recently, nanometer-sized drug delivery systems

Boron-Based Compounds: Potential and Emerging Applications in Medicine, First Edition.
Edited by Evamarie Hey-Hawkins and Clara Viñas Teixidor.
© 2018 John Wiley & Sons Ltd. Published 2018 by John Wiley & Sons Ltd.

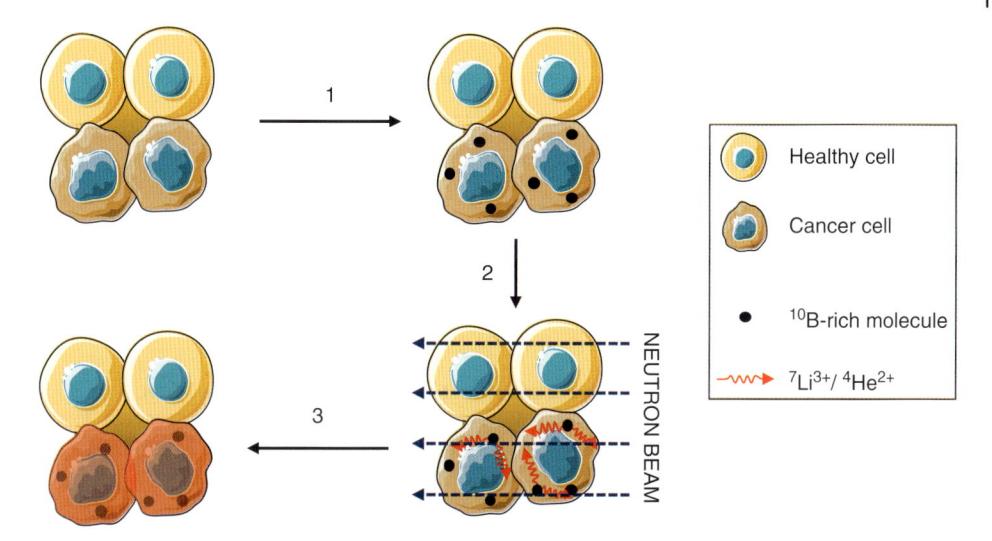

Figure 2.5.1 Schematic representation of the principle behind BNCT: if ^{10}B atoms preferentially accumulate in cancer cells (1), subsequent neutron irradiation (2) produces the rapid nuclear reaction ^{10}B(n, α, γ)^{7}Li. α-particles and ^{7}Li ions have high linear energy transfer, triggering cell damage and death (3) while sparing healthy surrounding cells.

have gained attention [30–32], since nanomedicines preferentially accumulate in cancer tissue due to the well-known enhanced permeability and retention (EPR) effect. This effect is based on the presence of leaky vasculature in the vicinity of tumors, whose endothelium is fenestrated with gaps between 100 and 780 nm in size. This, together with a deficient lymphatic drainage, results in a passive accumulation of nanomedicines in tumor tissue [33].

The successful application of BNCT has been hindered by a number of problems, not least the unease around conducting clinical irradiations in nuclear research reactors. Apart from this, which might be mitigated by the installation of accelerator-based neutron sources, the main limitation of BNCT is the need to develop drugs that are able to deposit a sufficient number of ^{10}B atoms specifically in tumor cells or tissues. Two main parameters need to be considered during development of new BNCT drug candidates: first, the boron concentration in the tumor should be >20–35 µg ^{10}B/g, although lower amounts might be efficient if the boron atoms are located close to the cell nucleus; second, tumor-to-normal tissue (TtT) and tumor-to-blood (TtB) ratios should be greater than five to prevent damage to healthy tissue in the path of the neutron beam [34]. Thus, there is the need to determine, *in vivo* and in a noninvasive way, the accumulation of boron in the tumor, surrounding tissue, and blood as a function of time, in order to (1) define the optimal time window for the application of neutron irradiation, in which the therapeutic effect is optimal while side effects and destruction of the surrounding tissue are minimized (Figure 2.5.2); and (2) predict the therapeutic efficacy of the treatment on a patient-by-patient basis while enabling dose adjustment (personalized medicine).

Currently, different alternatives are available to determine the concentration of boron in tissues, including inductively coupled plasma–atomic emission spectroscopy (ICP-AES), high-resolution alpha autoradiography, neutron capture radiography, laser post-ionization secondary neutral mass spectrometry (laser-SNMS), and electron

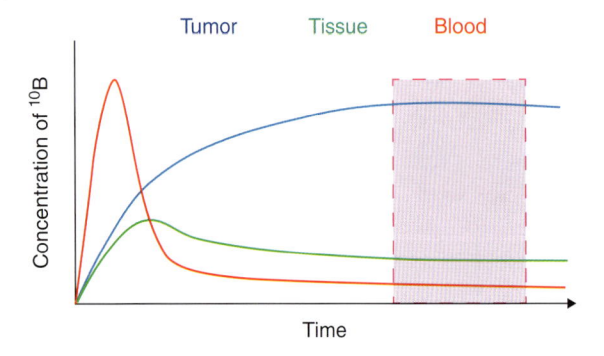

Figure 2.5.2 Fictitious curves representing the concentration of ^{10}B in the tumor (blue line), surrounding tissue (green line), and blood (red line), after administration of a BNCT drug candidate. Determination of these curves enables the identification of the optimal time window for applying neutron irradiation (in purple) and predicting the therapeutic efficacy.

energy loss spectroscopy (EELS) [35]. However, all these techniques are performed *ex vivo* in small tissue samples, usually surgically removed from the patient or experimental animal. Prompt γ-ray spectroscopy, based on γ-ray spectroscopy following neutron capture in ^{10}B, is a fast method to determine the average ^{10}B content in a macroscopic sample [36]. Even though this technique opens the possibility to perform *in vivo* γ-ray spectroscopy of the subject during treatment [37,38], information about boron concentration can be obtained only in small volumes (typically a few cubic centimeters), and the temporal resolution is suboptimal (a few minutes). An alternative method to determine, *in vivo* and noninvasively, the concentration of boron in an organ or tissue is by magnetic resonance imaging (MRI). Both natural boron isotopes ^{11}B and ^{10}B are detectable by nuclear magnetic resonance spectroscopy (NMR) and consequently are useful for MRI. Considering the natural abundance and the nuclear properties, ^{11}B displays a higher sensitivity and better spectral resolution than ^{10}B. However, the ^{10}B T$_2$ is longer than ^{11}B T$_2$ for the same molecular site, which might be an advantage for the detection of ^{10}B. In 2001, Bendel and coworkers published the first *in vivo* MRI images of ^{10}B-enriched sodium borocaptate (BSH) [39], after intravenous injection in mice with implanted M2R melanoma xenografts. The boron concentration could be determined at realistic tissue concentration levels, with an acceptable spatial resolution and in reasonable imaging times. The same group demonstrated a few years later the feasibility of using ^{1}H magnetic resonance spectroscopy (MRS) and magnetic resonance spectroscopic imaging (MRSI) to follow the distribution of boronophenylalanine (BPA) *in vivo* [40]. Despite the promising results, implementation of this technology is still challenging in the clinical setting, the quantification of the images is not straightforward, and the sensitivity is suboptimal. For a recent revision on the application of magnetic resonance to *in vivo* mapping of boron compounds, refer to Ref. [41].

In this context, both the evaluation of newly developed BNCT drug candidates and the determination of the optimal time window for neutron irradiation in clinical therapy would benefit from a technique capable to determine – noninvasively, *in vivo*, with high sensitivity, in real time, and at the whole-body level – the concentration of boron in organs, tissues (including the tumor), and blood. This can be achieved by using nuclear imaging techniques, such as positron emission tomography (PET) and single-photon emission computerized tomography (SPECT), which are described in this chapter.

2.5.2 Boron Clusters

As mentioned in Section 2.5.1.2, one of the major limitations for BNCT to become a routine clinical practice is the need to develop drugs that are able to deposit a sufficient number of ^{10}B atoms specifically (or preferentially) in tumor cells or tissues. In order to maximize the amount of boron in tumor cells, two strategies can be followed: (1) accumulate a high number of molecules containing just one or a few number of boron atoms each, or (2) accumulate a lower number of molecules containing a huge amount of boron atoms each. Two clear examples of these strategies are p-BPA, which contains one boron atom per molecule (Figure 2.5.3), and sodium borocaptate, which contains 12 boron atoms per molecule (BSH; Figure 2.5.3); both currently are approved for clinical trials and used as stand-alone BNCT drugs or in combination [42].

Like carbon, boron has a high self-catenation capacity. However, whereas carbon tends to have four connections when it self-catenates, boron tends to have more than four, producing compact polyhedral structures known as boron clusters. Because boron clusters accumulate several boron atoms in one structure, they are *a priori* ideal candidates for the preparation of BNCT drug candidates.

Depending on their chemical composition, boron clusters can be classified as boranes (which contain boron and hydrogen atoms), carboranes (which contain boron, hydrogen, and carbon atoms), and metallacarboranes (which contain at least one metallic atom besides boron, hydrogen, and carbon). General boron cluster chemistry has been the topic of several review articles (see, e.g., Ref. [43]). Therefore, only basic information is provided in this chapter.

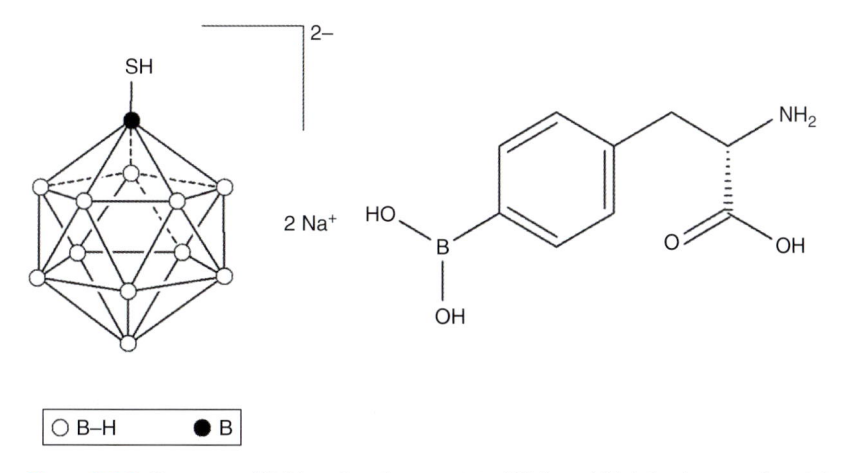

Figure 2.5.3 Structure of (left) sodium borocaptate (BSH), and (right) p-boronophenylalanine (BPA).

2.5.2.1 Boranes

Boranes contain only boron and hydrogen atoms in their structures. With the general formula B_xH_y, boranes are characterized by delocalized electron-deficient linkages. In other words, they contain too few valence electrons in their structure to be described in terms of two-center, two-electron (2c2e) bonds. Instead, they can be described in terms of three-center, two-electron (3c2e) bonds, which result in the formation of trigonal faces and hyper-coordination. The high connectivity of atoms compensates for the low electron density in the skeletal bonds. As a result, three-dimensional (3D) deltahedral shapes are favored in boron clusters (see the structure of BSH as an example: Figure 2.5.3), contrary to the formation of chains and rings observed in carbon-based molecules.

The simplest borane is BH_3, which is unstable and dimerizes to form diborane. Boranes with >5 boron atoms form deltahedral clusters, which can be neutral or anionic. The geometries of boron clusters are described as: (1) deltahedra, or (2) those same deltahedra with one or more missing vertices. The classification is based on the polyhedral skeletal electron pair theory, originally formulated by Kenneth Wade [44]. Because this theory is also applicable to carboranes (see Section 2.5.2.2), a brief discussion is included next.

2.5.2.2 Carboranes

Carboranes are clusters containing boron, hydrogen, and carbon atoms. Like boranes, they form polyhedral clusters with a high coordination number in their skeletal atoms (generally between 5 and 6), and also can be classified as deltahedra, or deltahedra with missing vertices, according to Wade's rules. Carboranes have the general formula $[C_nB_mH_{n+m+p}]^{x-}$, where n is the number of carbon atoms in the cluster, m is the number of boron atoms in the cluster, p is the number of bridging hydrogen atoms, and x is the net charge (notice that boranes are also covered by this general formula, with $n = 0$). By applying Wade's rules, the total number of skeletal electrons as a function of the occupied vertices can be $2n + 2$ (so-called *closo-* compounds), $2n + 4$ (*nido-* compounds), $2n + 6$ (*arachno-* compounds), or $2n + 8$ (*hypho-* compounds). In all clusters, each B–H vertex contributes two electrons to the cluster, while each C–H vertex contributes three electrons.

The most widely investigated carboranes are the extremely stable dicarba-*closo*-dodecaboranes (commonly known as carboranes) [45], which are icosahedral *closo*-carboranes with the general formula $C_2B_{10}H_{12}$ (see [46,47]). Their rich derivative chemistry makes dicarba-*closo*-dodecaboranes suitable building blocks with application for the preparation of macro-molecular and supramolecular entities [48,49], nonlinear optics [50–52], and medicinal chemistry, particularly in the context of BNCT and as hydrophobic pharmacophores of bioactive molecules [53–62].

Carboranes exist as *ortho-*, *meta-*, and *para-* isomers (*o-*, *m-*, and *p*-carborane, respectively), which differ in the relative positions of the carbon atoms in the cluster (Figure 2.5.4). The carbon and boron atoms in the carborane structures are hexacoordinate and result in a nearly icosahedral geometry.

The chemistry of carboranes is rich and unique, because the carbon and boron vertices in the cluster have orthogonal reactivity. The CH groups are weakly acidic with

○ B–H   ● C–H

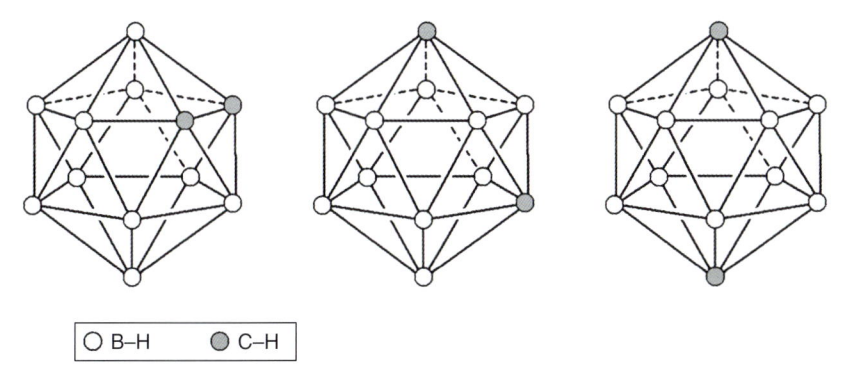

Figure 2.5.4 (Left to right) Structures of o-, m-, and p-carborane.

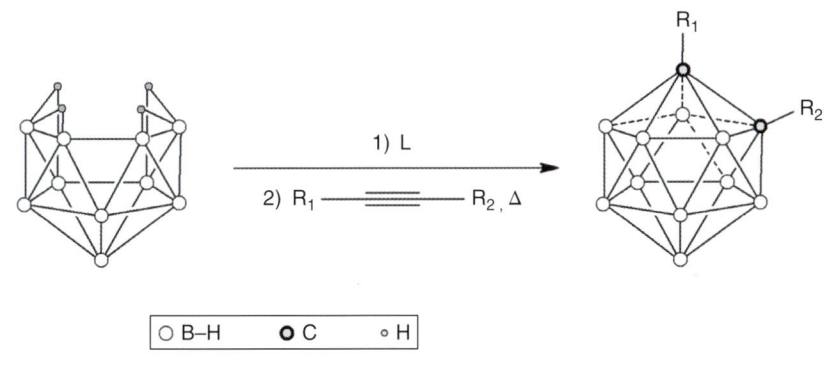

○ B–H   ● C   ○ H

Figure 2.5.5 Strategy for the preparation of C_c-substituted o-carboranes by reaction of decaborane with alkynes.

pKa values of 22.0, 25.6, and 26.8 for *ortho-*, *meta-*, and *para-* isomers, respectively [63]. These carbon atoms can be readily deprotanated generating nucleophiles, whereas the boron vertices can be derivatized by using reactive electrophiles. It is therefore possible to regioselectively synthesize a variety of C- and/or B-substituted carborane derivatives without the need for complex protecting group strategies. Additionally, o-carborane derivatives can be prepared following a straightforward strategy based on the reaction of decaborane ($B_{10}H_{14}$) with a Lewis base in order to form a reactive complex ($B_{10}H_{12}L_2$) [64], which is further reacted under conventional heating with an alkyne to yield the corresponding o-carborane derivative (Figure 2.5.5). After the pioneering work by Heying et al. [65], this methodology has been widely applied to the preparation of mono- and di-C_c-substituted o-carboranes (see Ref. [66] for an extensive review).

Very recently, the preparation of C_c-substituted carboranes has been approached by addition of functionalized alkynes (including mono- and polyfunctional internal and terminal alkynes) to $B_{10}H_{12}(CH_3CN)_2$ in the presence of a series of Cu, Ag, and Au salts. Excellent results were obtained when $AgNO_3$ was used in catalytic amounts [67]. Also very recently, the preparation of C,B-substituted-o-carborane derivatives was reported by reaction of 1,3-dehydro-o-carborane (prepared from 3-iodo-1-lithio-o-carborane) with alkynes via Pd/Ni-cocatalyzed [2+2+2] cycloaddition reaction [68].

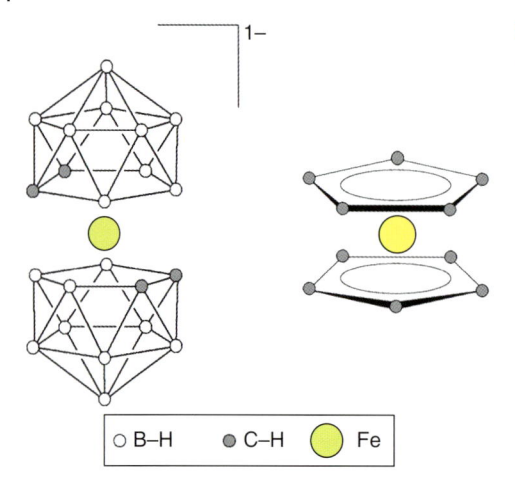

Figure 2.5.6 Structure of (left) [(3,3'-Fe(1,2-$C_2B_9H_{11}$)$_2$] and (right) ferrocene.

2.5.2.3 Metallocarboranes

Metallocarboranes can be thought as heteroboranes with at least one metal atom present in or associated with the cage framework. The first metallocarborane was prepared in 1965 by Hawthorne and coworkers [69]. This sandwich-type complex had a central iron atom that was complexed to two dicarbollide units, in a similar fashion to ferrocene (Figure 2.5.6). After the discovery of the first metallocarborane, many mixed-sandwich complexes involving both the dicarbollide dianion $(C_2B_9H_{11})^{2-}$ and carbonyl or cyclopentadienyl ligands were prepared and reported, as well as complexes with lower carboranes and incorporating other transition metals (e.g., cobalt, copper, gallium, palladium, manganese, molybdenum, nickel, rhenium, and tungsten) [70,71]. For a recent review, see Ref. [72].

2.5.3 Nuclear Imaging: Definition and History

Molecular imaging can be defined as a set of techniques that allow the noninvasive visualization of cell function and the monitoring of molecular processes in living organisms. They differ from traditional imaging techniques because biomarkers are used, enabling the investigation of biological, physiological, and medical processes *in vivo* and in a wide variety of animal species, ranging from small rodents to humans. Nuclear imaging techniques, which comprise PET, SPECT, and planar γ-camera imaging, are considered molecular imaging techniques. These techniques offer the possibility to detect functional changes occurring inside the body. Since the functional changes take place before the structural ones, their detection enables earlier diagnosis and evaluation of the response to treatment, and therefore a much better and more cost-effective chance to cure diseases.

The history of nuclear imaging started at the end of the nineteenth century, when Henry Bequerel discovered radioactivity (1896) and Marie Curie discovered radium (1898). The next important development was made by George de Hevesy, who studied the chemical processes in biological systems using radioactive tracers. Blumgart and

Yens, in 1927, made the first human study using aqueous mixture of ^{214}Pb and ^{214}Bi to measure the transit time of the blood flow from one arm to the other [73]. In 1930, the cyclotron was invented by Ernest O. Lawrence and the production of new artificial radionuclides became possible, thus opening new avenues for the investigation of complex functions in living systems. The development of the technology permitted in 1950 the first images of the radionuclide's distribution. The rectilinear scanner was made in 1951 by Benedict Cassen, and in 1958 the Anger γ-camera arrived and the single-photon imaging system started to be worldwide. Approximately at the same time, the imaging properties of positron emitters were described.

Until 1960, the nuclear imaging field was dominated by iodine-131 (^{131}I) until the discovery of technetium-99 m (^{99m}Tc) in 1964, which launched a big development of nuclear medicine imaging and remains to date – due to its imaging properties, easy labeling, and facile production – the most used radionuclide. In recent decades, the increase of positron-emitting radionuclides with better imaging properties and improved methods opened a wide area of research and diagnostic applications. Nuclear imaging techniques have recently emerged as powerful tools that can provide valuable information far beyond diagnostic applications, by attaching a radioactive atom (radionuclide) to a molecule or macromolecule (e.g., a small drug, peptide, or protein) that can be tracked noninvasively after administration to a living organism; hence, the spatiotemporal distribution of the labeled entity can be resolved with excellent sensitivity and good resolution, to ultimately provide quantitative pharmacokinetic information. Such information can be used in the context of drug development, for example to determine the ADME (absorption, distribution, metabolism, and elimination) profile, to anticipate potential toxicological effects, or to aid in the process of optimal dose determination.

2.5.3.1 Radioactivity: The Basis of Nuclear Imaging

Nuclear imaging uses radioactive substances to obtain images. The radioactivity refers to the processes that take place spontaneously in the nucleus of an atom that is in an unstable state because it has an excess of energy. Emitting radiation is the natural and only way for an unstable atom to turn into a stable atom, in a process called radioactive decay. All the substances that emit radiation are called radioactive substances, and their atoms are radionuclides.

Radioactive decay is a stochastic process in which the probability of an atom to undergo radioactive decay is constant over time. The number of radionuclides, N, is a discrete variable (a natural number) – but for any physical sample, N is so large that it can be treated as a continuous variable. The radioactive decay formula is expressed as shown in Equation 2.5.1:

$$N(t) = N_0 e^{-\lambda t} \qquad (2.5.1)$$

where N_0 is the initial number of radioactive atoms, $N(t)$ is the number of radioactive atoms at time = t, λ is the decay constant ($\ln 2/T_{1/2}$), and t is time. $T_{1/2}$ is the half-life of the radioisotope, and corresponds to the period of time required to decrease the number of radioactive atoms to one half of the starting value. The mechanism of emitting radiation is closely linked to the atom's energy levels and can have different forms: α-particles, β-particles (electrons and positrons), γ-rays, conversion electrons (internal conversion), and neutrinos (electron capture). The set of properties of radioactive decay (the way of radioactive decay, the emission type, the level of transition energy, and the lifetime of the radionuclide before decaying) is characteristic for each radionuclide. In the International System of Units (SI), the activity is expressed in becquerels (Bq), 1 Bq corresponding to 1 disintegration per second. The amount of radioactivity handled in animal or human imaging experiments is usually in the MBq range.

Positron or γ-emitters (both are γ-emitters ultimately, resulting in the emission of γ-rays, *vide infra*) are ideally suited to be tracked *in vivo*. Due to their high penetration capacity, γ-rays are capable of escaping from the organism under investigation without suffering significant attenuation or scattering, and hence can be detected using detectors placed outside the organism body. This is the basis of PET, SPECT, and planar γ-camera imaging (*vide infra*). For therapeutic applications, the use of radionuclides presenting a high LET (i.e., with the capacity to create localized damage) is more convenient. β-particles (electrons) have a longer range in tissues, and are an attractive choice for targeting solid, heterogeneous, or large-volume tumors. α-particles have the highest LET, and this makes them highly efficient and specific for treatment of residual, microscopic, or small-volume tumors.

2.5.3.2 Single-Photon Emission Computerized Tomography

SPECT relies on the detection of γ-rays originated during radioactive decay of single-photon emitting radionuclides, which decay by (β⁻, γ) emission, electron capture (EC, γ) emission, or isometric transition (IT). Commonly used single-photon emitters present half-lives of several hours and have photon energies in the range of 100–300 KeV (Table 2.5.1), ^{99m}Tc being the most frequently used radionuclide. The main emission energy of this radionuclide (141 keV) is very suitable for a wide number of investigations in many medical and research fields, and the absence of any α- or β-emission turns it into a favorable radionuclide regarding patient radiation safety.

The typical configuration of SPECT scanners consists of at least one γ-ray detection module and one collimator. The core of the detection module is normally a scintillation crystal that, when excited by a γ-ray, absorbs its energy and re-emits it in the form of a flash of light. This flash is subsequently detected by a photo-electronic system that records its location in the crystal and its intensity, which is proportional to the energy

Table 2.5.1 Typical single-photon emitters (with half-life and energy)

Isotope	Half-life	γ Energy (γ fraction)
^{99m}Tc	6.02 h	141 KeV (89%)
^{123}I	13.22 h	159 KeV (83%)
^{111}In	2.80 d	171 (91%), 245 KeV (94%)
^{67}Ga	3.26 d	93 (39%), 185 (21%), 300 (17%), 394 (5%)
^{131}I	8.02 d	364 KeV (82%)

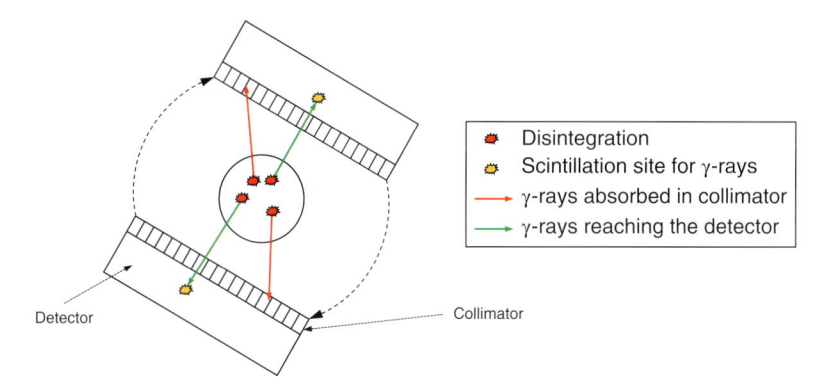

Figure 2.5.7 Schematic representation of the detection of photons using SPECT. Only a fraction (~0.001%) of the emitted γ-rays reach the detectors, while others are absorbed in the collimator.

of the incidental γ-ray. The collimator is formed by parallel lead walls that select the proper γ-rays for the investigation, usually used at a perpendicular angle on the region of interest. In such a way, the collimator forms a projected image of the radioisotope distribution on the surface of the scintillation crystal.

The first scintillation cameras were equipped with one static head detector. With these systems, a planar projection of the subject under investigation was obtained (planar γ-imaging). Due to the superposition of the different planar layers, the resulting image of the radiotracer's distribution had a very low contrast, and no in-depth information could be obtained. The next step was to have more acquisitions gained from multiple positions of the detector head around the imaged subject. By providing 3D information, this modality led to better contrast and higher resolution. Incorporation of two or more head systems further improved the quality of the images, while enabling the use of lower amounts of radioactivity (with the consequent reduction in patients' dose exposure) and scanning time. Figure 2.5.7 shows a schematic representation of a SPECT system with two detector heads that rotate around the imaged subject.

2.5.3.3 Positron Emission Tomography

The positron, the electron's anti-particle, is a positively charged electron with a very short path. After radioactive decay, the emitted positron loses its energy very quickly as a result of collisions with the surrounding matter atoms. This path, which depends on the energy of the emitted positron (Table 2.5.2), is not longer than a few millimeters in water or body tissues. After losing all its energy ("thermalizing"), the positron reacts with a nearby, negatively charged electron, and the masses of the two particles transform into two γ-rays through a process called annihilation. The two γ-rays, which have energies of 511 keV (the energy corresponding to the rest mass of a β-particle) each, are emitted simultaneously and 180° apart.

This unique feature of γ-rays generated by positron–electron annihilation is exploited by coincidence detectors, which are placed around the subject under investigation. Two photons, detected almost simultaneously by a pair of detectors, will be assumed to arise from a single annihilation, and the direction of the incident rays will be determined as

Table 2.5.2 Typical positron emitters (with half-life and E_{max} positron)

Isotope	Half-life	β^+ Energy
^{18}F	109.8 min	0.63 MeV
^{11}C	20.4 min	0.96 MeV
^{13}N	9.97 min	1.20 MeV
^{15}O	122 sec	1.73 MeV
^{68}Ga	67.6 min	1.89 MeV

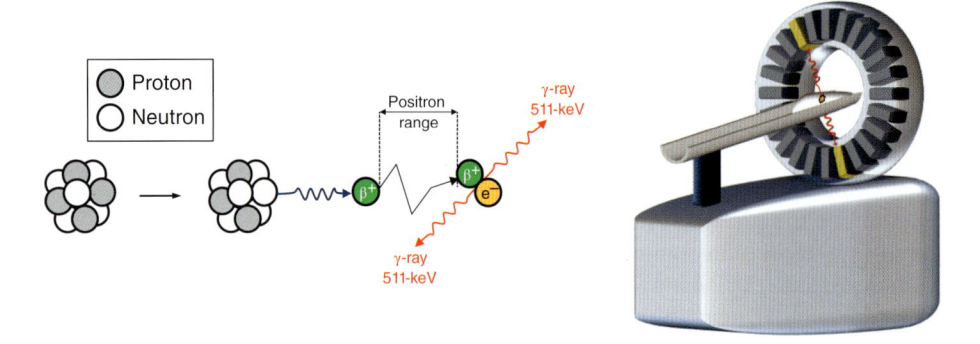

Figure 2.5.8 (Left) Schematic representation of the annihilation process of one positron and one electron, with subsequent emission of two γ-rays. (Right) Representation of a PET camera. The two photons emitted after the annihilation process are detected simultaneously by two detectors on the ring, placed around the subject under investigation.

the line passing through the two detectors. Hence, PET imaging requires back-to-back detection heads (typically, PET scanners consist of a stationary array of full-ring detectors) and coincidence detection circuitry. Importantly, PET scanners do not need collimators: the detection of the annihilation coincidences is known as *electronic* collimation, as opposed to the *physical* collimation implemented in SPECT scanners. A schematic representation of a PET scanner is shown in Figure 2.5.8.

In both modalities, SPECT and PET, a set of projections at different angular positions around the subject under investigation are acquired. Information on the concentration of radioactivity within the imaged subject is gained after reconstruction of the images. A full description of the mathematical principles behind image reconstruction falls beyond the scope of this chapter. However, it is worth mentioning that reconstruction algorithms can be roughly divided into analytical (e.g., filtered-back projection [FBP]) and (statistical) iterative (e.g., ordered subset expectation maximization [OSEM]) methods. Both have advantages and disadvantages, and their use will depend on many factors, including the nature of the information required from the images. Typically, iterative reconstruction methods provide images with improved spatial resolution and better signal-to-noise ratios at the expense of increased computational time and some unpredictability in the final outcome [74].

2.5.3.4 Multimodal Imaging

Nuclear imaging techniques such as PET and SPECT offer information about the spatiotemporal distribution of the radionuclide within the organism under investigation, although they do not provide any anatomical information. Because of this, nuclear imaging techniques are often combined with other imaging modalities that can provide accurate morphological information, which is essential to unequivocally determine the location of the radioactivity. Computerized tomography (CT) (3D X-ray imaging) is the most commonly used. Indeed, most PET and SPECT scanners currently available exist only as hybrid systems (PET-CT or SPECT-CT) at both the clinical and preclinical levels. The CT image is used for anatomical localization of the radioactive signal, but also to create the attenuation map for appropriate correction during image reconstruction (for more details, please refer to Ref. [75]). In recent years, hybrid PET-MRI systems have been developed. Contrary to CT, MRI provides excellent resolution and high contrast in soft tissue; unfortunately, determination of the attenuation map is still a challenge, and hence this hybrid technology is still not fully established [76].

2.5.3.5 Nuclear Imaging and the Need for Radiolabeling

As previously mentioned, nuclear imaging techniques rely on the administration of compounds labeled with radioactive isotopes that enable external, noninvasive detection. Nuclear imaging investigations are typically performed using a trace amount of the radioactive substance, namely the radiotracer or radiopharmaceutical.

In this context, it is clear that the first step in the application of nuclear imaging techniques is the preparation of the radiotracer, by attachment of the radionuclide to a biologically active molecule. Thanks to the virtually unlimited possibilities and techniques for radiolabeling and the emergence of free and open sources of scientific information such as Google Scholar, Wikipedia, interest groups (e.g., LinkedIn and ResearchGate), and free or open-access journals (e.g., www.intechopen.com and www.springeropen.com), this section encourages the reader to visit these sources. For the open and free references, the keyword or phrase will be held by quotation marks (e.g., "radiolabeling"). Alternative references will be also provided here when appropriate.

The first step in radiolabeling is the addition of a radionuclide to a biomolecule in such a way that: it (1) does not alter the pharmacological properties in an adverse way, and (2) remains bound to the biomolecule under *in vivo* conditions. For the first criterion, unless a stable atom of the biomolecule is exchanged for a radionuclide using, for example, "isotopic exchange" or by the introduction of a "leaving group" that is replaced by a radioactive isotope of the original atom, some evaluation of the radiopharmaceutical has to be performed to ensure that the biological properties are not modifying as a result of the radiolabeling. If the target for the radiopharmaceutical is a receptor, a ligand binding assay [77] such as a Scatchard plot or Lindmo assay (also known as Lindmo analysis and the Lindmo method) can be performed, the first suitable for both small molecules and monoclonal antibodies (mAbs) and the latter to measure the "immunoreactive fraction" of radiolabeled mAbs. For *in vivo* studies of radiotracer binding (i.e., to study the pharmacokinetics of a radiopharmaceutical that binds, e.g., to an enzyme or a receptor), graphical methods such as a Logan plot [78], which is used for reversibly binding, and a Patlak plot [79,80], which is used for irreversibly binding radiopharmaceuticals, can be used. For receptor studies, it is also important to have a controlled *specific activity* (or amount of radioactivity per unit mass) [81] in order not to saturate the target with the non-radioactive analog of the radiopharmaceutical; compare carrier-free, no carrier added, and carrier added. For nonsaturable systems (e.g., blood flow and metabolism), the specific activity is generally not of importance. For a metabolic tracer, an *in vitro* cell assay is a good starting point, followed by small-animal imaging. For the second criterion (the stability of the radiopharmaceutical), the ultimate answer will be given by *in vivo* experiments but should be initiated by some *ex vivo* experiments like incubating the radiopharmaceutical in blood plasma, blood, or liver homogenate at 37 °C, and taking samples for analysis at different time points to see the stability over the time of the radiopharmaceutical.

In certain cases, the radiopharmaceutical is metabolized *in vivo*, and because the camera cannot distinguish the different metabolites but only sees the radiation, radiometabolite analysis together with a compartmental model can be used to correct the image data for the radiometabolites.

If the radiopharmaceutical passes the *in vitro* and *ex vivo* tests but shows poor results *in vivo*, further investigation is needed. For protein-based radiopharmaceuticals, the isoelectric point may be compared with the original pharmaceutical. For small-molecule radiopharmaceuticals, the *distribution constant* (also known as distribution coefficient and partition ratio) K_D can be measured; this should preferably be performed using high-performance liquid chromatography (HPLC), as small amounts of impurities may give significant errors when using the octanol–saline extraction method. As an example, if an ^{18}F-labeled radiopharmaceutical has a (true) K_D of 4, but it contains 1% of free ^{18}F-fluoride, which has a K_D of −2, the (extraction) measured K_D will thus be 2 (i.e., two orders of magnitude lower than the true K_D).

The final challenge for a radiopharmaceutical is the interspecies differences. A radiopharmaceutical that works well in a mouse model may fail in the human – and vice versa. Even though there are some tools to compensate for the pharmacokinetic and pharmacodynamic differences between different species (allometric scaling, Dedrick plot, and Kleiber's law) [82], it may be difficult to predict the outcome.

2.5.4 Radiolabeling of Boron Clusters

Boron clusters are suitable tools for the preparation of BNCT drug candidates. As mentioned in this chapter, incorporation of a radiolabel to the drug candidate enables subsequent investigation of the pharmacokinetic properties, and hence the prediction of therapeutic efficacy and the determination of the optimal time window for application of neutron irradiation. Generally, boron clusters are not used as stand-alone chemical entities, but are attached to targeting moieties that lead to preferential accumulation of the boron atoms in the tumor tissue. Obviously, in this scenario the radiolabel can be incorporated either in the boron cluster or somewhere else in the molecule (e.g., in the targeting moiety).

One important aspect of nuclear imaging techniques that needs to be taken into consideration is that only the location of the radiolabel is monitored. In other words, if the molecule under investigation is metabolized, the labeled metabolite will be tracked. Because the determination of the location of the boron-rich residue is the most relevant information when BNCT drug candidates are investigated, strategies for the incorporation of the radiolabel in the boron cluster are anticipated to be the preferred alternative and will be mainly considered in this chapter. Examples covering other scenarios will be provided if appropriate.

Radiolabeling of boron clusters can be achieved mainly by radiohalogenation and radiometallation.

2.5.4.1 Radiohalogenation

Halogens have very well-known chemistry, and their most commonly used radioisotopes cover a wide range of half-lives, that is, from 109.8 min (^{18}F) to almost 60 days (^{125}I); additionally, radiohalogens with different emission modes (e.g., almost 100% positron emission for ^{18}F; 100% electron capture with subsequent emission of low-energy γ-rays for ^{125}I; and electron and γ emission for ^{131}I) can be easily obtained either from accelerators or from commercial suppliers, enabling a wide variety of *in vivo* and *in vitro* applications. Consequently, radiohalogens are one of the first options to be considered when approaching the radiolabeling of molecules, and boron clusters have not been an exception.

2.5.4.1.1 Radioastatination of Boron Clusters

Astatine-211 (^{211}At) is a 7.2-h half-lived α-emitter that offers potential advantages for targeted α-particle therapy. However, its use is constrained by its limited availability: it can be produced efficiently only from natural bismuth via the ^{209}Bi(α,2n)^{211}At nuclear reaction, and there are only a few cyclotrons in the world with appropriate particle acceleration characteristics to achieve this nuclear reaction [83]. Additionally, as an α-emitter, ^{211}At is not suitable for *in vivo* imaging, and hence radiolabeling of boron clusters with this radionuclide will be only briefly discussed here.

As an α-emitter, radiolabeling of monoclonal antibodies (mAbs) applicable to targeted radiation therapy has been approached in several studies [84,85]. However, direct astatination of mAbs is often problematic. First, direct reaction of electrophilic ^{211}At with tyrosine moieties of mAbs does not work well [86]; and, second, the labeled molecules are not stable *in vivo* [87]. To circumvent this problem, deactivated aryl compounds (e.g., benzoates) have been investigated as pendant groups for attaching ^{211}At to carrier molecules [88]. Despite deastatination not being a problem for mAbs labeled with astatobenzoate derivatives when slowly metabolized intact mAbs and their F(ab')$_2$ fragments are used [88,89], deastatination occurs rapidly when smaller mAb fragments are used. Hence, alternative radiolabeling routes needed to be developed.

Because boron–halogen bonds are generally stronger than carbon–halogen bonds, it was hypothesized that astatinated carborane derivatives might be appropriate pendant groups for the radiolabeling of mAbs and their fragments. To prove this hypothesis, the *in vivo* stability of a series of astatinated benzamide derivatives, astatinated *nido*-carborane derivatives, and compounds containing two *nido*-carboranyl moieties, also referred to as Venus flytrap complexes (VFCs), was compared [90]. Radioastatination was carried out by reaction of the corresponding *nido*-carboranyl compound with Na[^{211}At]At solution in aqueous acidic media (methanol was added to improve solubility when needed) in the presence of Chloramine-T. The reaction was quenched with sodium metabisulfite, and the labeled compound was purified by HPLC. The *in vivo* studies demonstrated that most of the small molecules labeled with [211]At were found to release free astatide, with the exception of astatinated VFC-containing compounds, which were found to be quite stable to *in vivo* deastatination.

After this work, Wilbur's group started a series of studies in the pursuit of the ideal [211]At-labeling agent, which should fulfill the following features: high labeling efficiency, high *in vivo* stability, and does not alter the biodistribution of the labeled entity. With that aim, protein-reactive derivatives containing *nido*-carborane, bis-*nido*-carborane (VFP), and 2-nonahydro-*closo*-decaborate(2-) derivatives [91]; maleimido-*closo*-decaborate(2-) derivatives [92]; and *closo*-decaborate moieties functionalized with acid-cleavable hydrazone linkers [93] were prepared, conjugated to mAbs, and tested. Biodistribution studies were conducted in rodents using the radioiodinated mAbs as controls. In a more recent study, the biodistribution of an intact antibody conjugated with a maleimido-*closo*-decaborate(2-) reagent via sulfhydryl groups was compared to the same antibody conjugated with an isothiocyanato-*closo*-decaborate(2-) reagent via lysine residues [94]. Considerably higher renal uptake was observed for the antibody conjugated via sulfhydryl groups, suggesting that further optimization in the labeling approach is still required.

2.5.4.1.2 Radioiodination of Boron Clusters

Radioiodination of boron clusters has been carried out following mainly two different approaches: (1) oxidative radioiodination in the presence of oxidizing agents, and (2) isotopic exchange under catalytic conditions. In isotopic exchange reactions, the final radiolabeled molecule and the non-labeled precursor have exactly the same chemical structure. Hence, their separation is not possible, and low specific activity values (amount of radioactivity per unit mass) are obtained. In oxidative radioiodination, the labeled molecule and the non-labeled precursor have different chemical structures; hence, the radiotracer can be isolated and higher specific activities can be achieved. The most commonly used iodine radioisotopes are ^{124}I, ^{125}I, and ^{131}I. Besides having the advantage of a stronger bond strength, ~380 kJ/mol (B–I) versus ~210 kJ/mol (C–I) boron clusters can be iodinated in a wider pH range than the "classical" tyrosine iodination method [95].

The radioiodination of different neutral and anionic boron clusters with potential application in the biomedical arena using the methodologies discussed here has been reported in the literature, and an extensive review was published by Agarwal *et al.* [96]. Hence, only a few illustrative examples will be presented here, in addition to recent advances not covered in the review [96].

Radioiodination of simple neutral clusters has been carried out mainly by Pd-catalyzed isotope exchange. For example, the synthesis of 2-^{125}I-*p*-carborane could be achieved in >80% labeling yields by reaction of 2-iodo-*p*-carborane with Na[^{125}I]I (sodium hydroxide solution) using tris(dibenzylideneacetone)-dipalladium (Pd$_2$(dba)$_3$) as the catalyst and tetra-butylammonium hydrogen sulfate to quench the hydroxide anion [97]. A different catalyst, trans-bis(acetato)bis[o-(di-o-tolylphosphino)benzyl] dipalladium(II) (Herrmann's catalyst), was successfully used to conduct the same reaction with similar radiochemical yields without the addition of the quencher; this process was extended to the preparation of other radioiodinated boron clusters, including *o*-carborane and *m*-carborane derivatives with incorporation yields of the

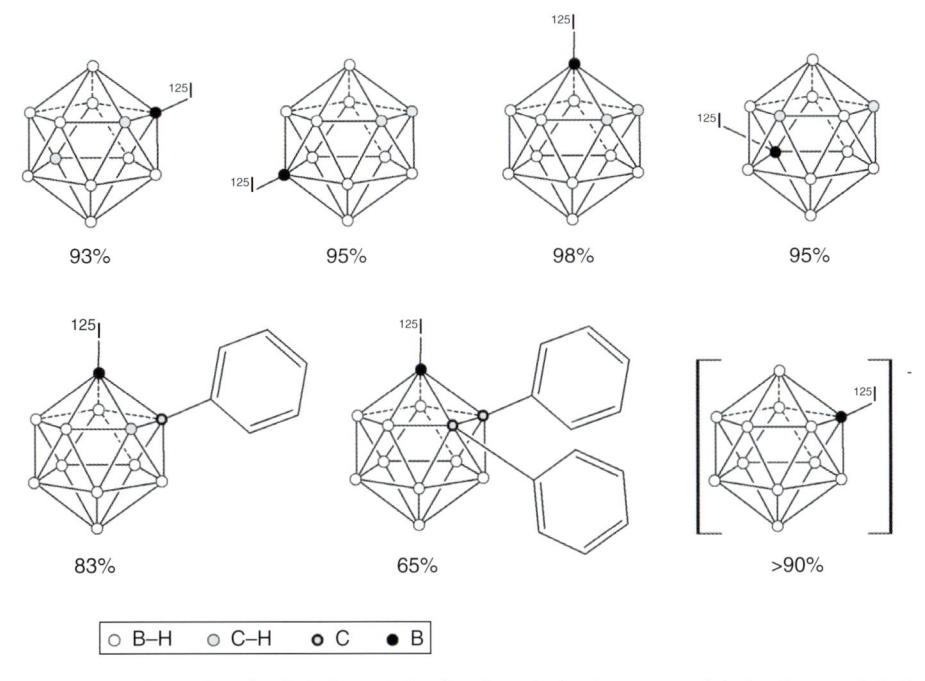

Figure 2.5.9 Examples of radioiodinated dicarba-*closo*-dodecaboranes and dodecaborates. Labeling yields are included under the chemical structure. Radiolabeled o-carborane derivatives using a completely different approach were recently reported by Gona et al. (see Figure 2.5.10) [101].

radionuclide between 65 and 98%, depending on the substituents present in the carborane cage (Figure 2.5.9) [98]. Using a similar strategy, radioiodinated dodecaborates were synthesized using copper sulfate as the iodine exchange catalyst in both organic [99] and aqueous media (Figure 2.5.9) [100].

The strategy relies on the preparation of radiolabeled decaborane using ^{125}I and catalytic isotopic exchange followed by ring closure using different alkynes in a one-pot, one step reaction. The methodology could be extended to the preparation of ^{131}I-labeled analogs.

Also very recently, a parallel strategy was applied to the radioiodination of a new bifunctional derivative of [3,3'-Co(8-I-1,2-C$_2$B$_9$H$_{10}$)(1',2'-C$_2$B$_9$H$_{11}$)], commonly known as COSAN, incorporating a polyethylene glycol (PEG) arm and one iodine atom (Figure 2.5.11) [102]. The radiolabeling was successful using both ^{125}I and ^{124}I. The latter enabled the determination of the biodistribution pattern of the radiolabeled cobaltabisdicarbollide species using PET-CT combined with dissection and γ counting (Figure 2.5.12).

Radioiodination under oxidative conditions has been conducted in a wide collection of boron clusters. Wilburg et al. synthesized the radioiodinated monocarbon carborane derivatives (Figure 2.5.13) using N-chlorosuccinimide as the oxidizing agent [103]. Valliant et al. synthesized several mono- and bifunctional cage-radioiodinated derivatives of *nido-o*-carborane using Chloramine-T or Iodogen® as the oxidizing agents [104,105]. Boron clusters (7,8-dicarba-*nido*-undecaborate(1-) and *closo*-dodecaborate(1-) derivatives) containing the isothiocyanate group for subsequent conjugation to biomolecules via thiourea bonds have also been labeled with ^{125}I under

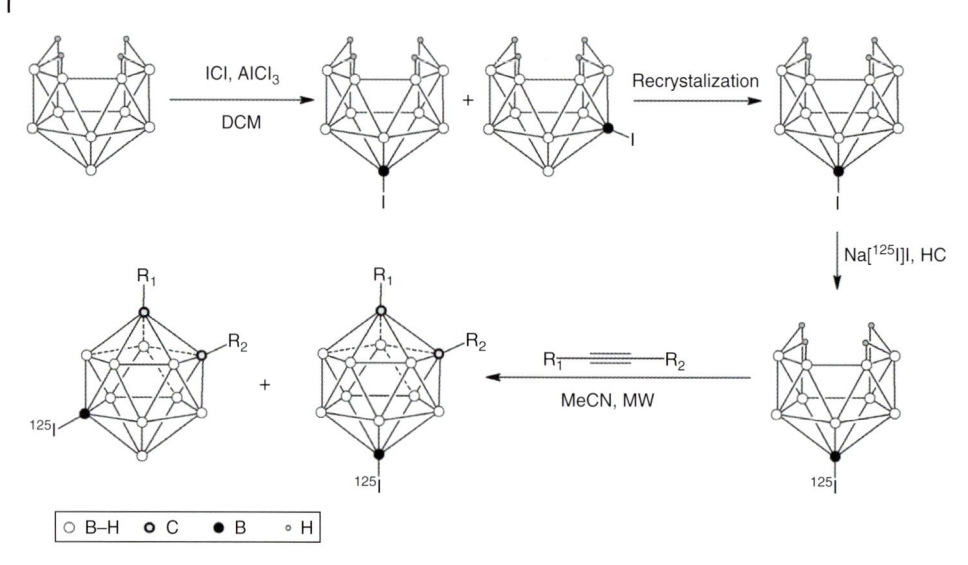

Figure 2.5.10 Synthesis and purification of 1-iododecaborane, radiolabeling by catalytic isotopic exchange using ^{125}I, and subsequent reaction for the preparation of ^{125}I-labeled o-carborane derivatives using microwave (MW) heating and acetonitrile (MeCN) as both the solvent and the Lewis base (one-pot, one-step reaction). The same strategy can be used for the preparation of ^{131}I-labeled analogs. HC: Herrmann's catalyst.

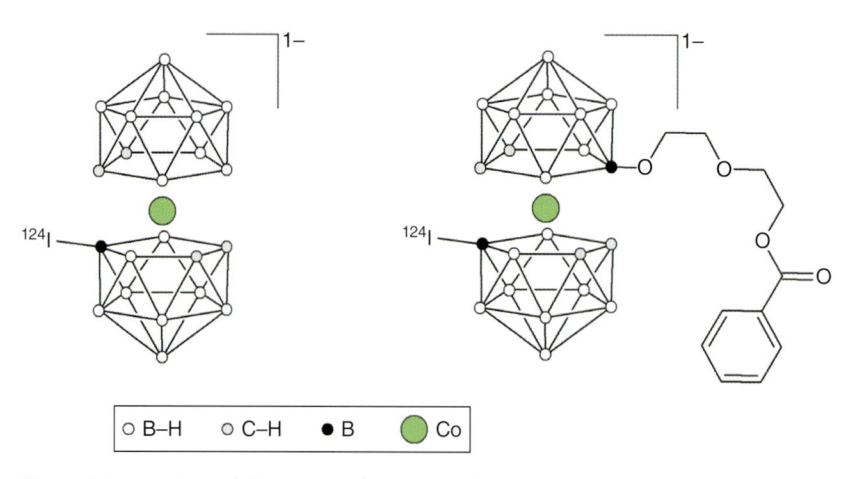

Figure 2.5.11 Iodinated derivatives of COSAN. (Left) [3,3'-Co(8-I-1,2-C$_2$B$_9$H$_{10}$)(1',2'-C$_2$B$_9$H$_{11}$)]; (right) [3,3'-Co(8-I-1,2-C$_2$B$_9$H$_{10}$)(8'-(OCH$_2$CH$_2$)$_2$COOC$_6$H$_5$-1',2'-C$_2$B$_9$H$_{10}$)]$^-$.

oxidative conditions [106–109]. Taking advantage of the presence of the isothiocyanate group, the boron clusters were attached to antibodies, and some of these conjugates were identified as promising BNCT agents.

2.5.4.1.3 Radiofluorination of Boron Clusters

Fluorine-18 (^{18}F) is an attractive alternative to radioiodine. It can be produced in large quantities in biomedical cyclotrons, it has a reasonably long half-life (109.7 min), and it has a small positron range, enabling the acquisition of high-resolution and quantifiable

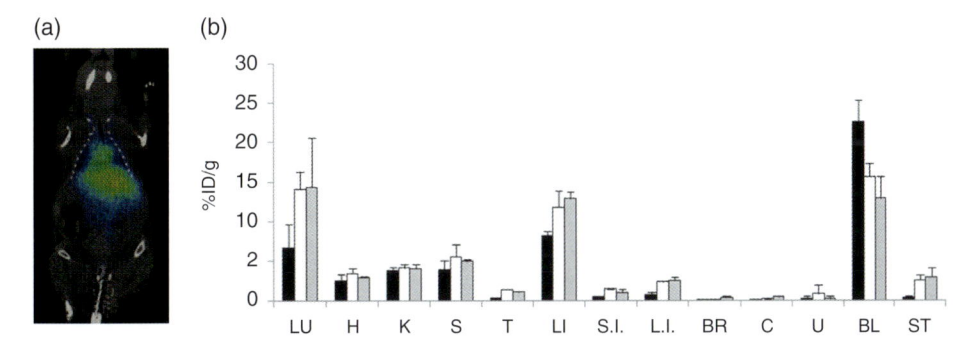

Figure 2.5.12 (a) PET coronal projection (co-registered with CT image) resulting from averaged images obtained after administration of [3,3'-Co(8-^{125}I-1,2-C$_2$B$_9$H$_{10}$)(8'-(OCH$_2$CH$_2$)$_2$COOC$_6$H$_5$-1',2'-C$_2$B$_9$H$_{10}$)]$^-$; (b) biodistribution of [3,3'-Co(8-^{125}I-1,2-C$_2$B$_9$H$_{10}$)(8'-(OCH$_2$CH$_2$)$_2$COOC$_6$H$_5$-1',2'-C$_2$B$_9$H$_{10}$)]$^-$ in mice tissues using the dissection method. Radioactivity is expressed as the percentage of the injected dose (ID) per gram of tissue. LU: Lungs; H: heart; K: kidneys; S: spleen; T: testicles; L: liver; S.I.: small intestine; L.I.: large intestine; BR: brain; C: cerebellum; U: urine; BL: blood; ST: stomach.

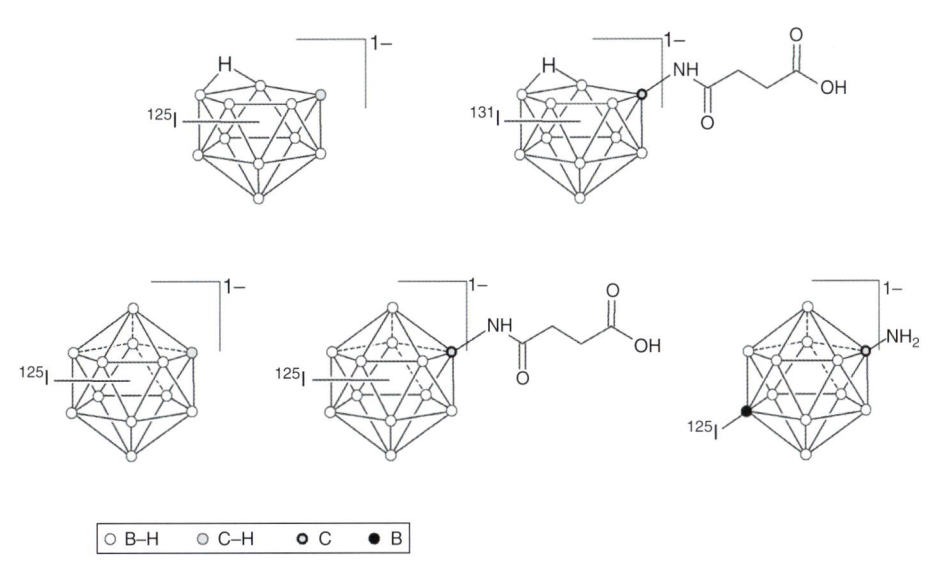

Figure 2.5.13 Examples of radioiodinated boron clusters synthesized by radioiodination under oxidative conditions.

PET images. Indeed, fluorine-18 is the most widely used positron emitter in the clinical field, thanks to the widespread use of the well-known glucose metabolism marker 2-deoxy-2-[^{18}F]-fluoro-D-glucose ([^{18}F]FDG).

To date, only the radiofluorination of *ortho*-carborane has been reported. The first methodology was based on direct reaction of cyclotron-produced [^{18}F]F$_2$ with o-carborane to yield 9-fluoro-o-carborane in 21% radiochemical yield [110]. However, the production of [^{18}F]$_2$ is challenging. Because of this, alternative routes using the labeling agent [^{18}F]F$^-$, which can be produced at a multi-gigabecquerel level, are

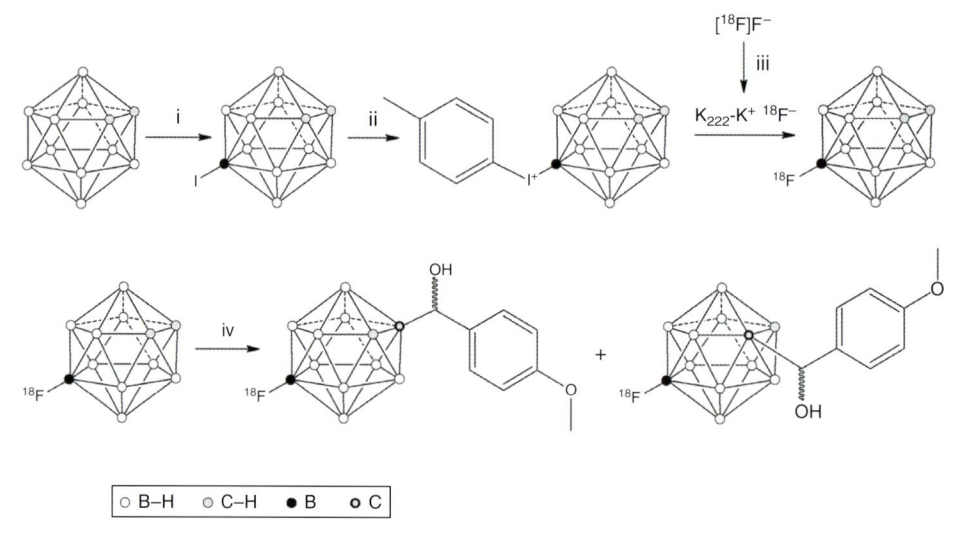

Figure 2.5.14 (Top) Synthesis of 9-[^{18}F]fluoro-o-carborane. (i) ICl–AlCl$_3$ in CH$_2$Cl$_2$, reflux, 4 h; (ii) NaBO$_3$–Ac$_2$O, RT, 90 min, then toluene–H$_2$SO$_4$(c), RT, 16h; (iii) Kryptofix® 222–K$_2$CO$_3$, solvent. (Bottom) Syntheses of ^{18}F-labeled C$_c$-substituted o-carborane by reaction of 9-[^{18}F]fluoro-o-carborane with 4-methoxybenzaldehyde (MBA). (iv) n-BuLi, THF, Δ; then MBA in THF. RT: Room temperature; THF: tetrahydrofuran.

preferred. Recently, the mono-[^{18}F]fluorination of o-carborane via nucleophilic substitution using cyclotron-produced [^{18}F]F$^-$ as the labeling agent and a carboranyl iodonium salt was reported [111]. The 9-[^{18}F]fluoro-o-carborane could be prepared in good radiochemical yields (44%, decay corrected). Further C$_c$ derivatization of the ^{18}F-labeled carborane could be achieved by treatment with n-BuLi and reaction with aldehydes (Figure 2.5.14).

2.5.4.1.4 Radiobromination of Boron Clusters

Bromine has several radioisotopes with good properties for *in vivo* imaging. Bromine-75 (^{75}Br, T$_{1/2}$ = 1.6 h) and Bromine-76 (^{76}Br, T$_{1/2}$ = 16 h) are positron emitters, while Bromine-77 (T$_{1/2}$ = 57 h) may be used for single-photon imaging. The versatility of the nuclear properties of the bromine isotopes makes radiobromine an attractive label for the assessment of BNCT drug candidates. As in the case of radioiodination, oxidative halogenations and halogen exchange strategies have been developed.

Tolmachev *et al.* demonstrated that *closo*-dodecaborate can be radiobrominated using a mixture of cyclotron-produced ^{77}Br, ^{76}Br, and ^{82}Br (the latter being a β-emitter with a half-life of 35.3 h) via oxidative halogenations using Chloramine-T as the oxidizing agent. Radiochemical yields close to 90% could be obtained [112]. Using the same labeling strategy, [^{76}Br]undecahydro-bromo-7,8-dicarba-*nido*-undecaborate(1-) was prepared by bromination of the 7-(p-isothiocyanato-phenyl)dodecahydro-7,8-dicarba-*nido*-undecaborate(1-) ion in almost quantitative radiochemical yields. The labeled boron cluster was further conjugated to the monoclonal humanized anti-HER2 antibody trastuzumab (Herceptin®) without intermediate purification in a one-pot reaction [113].

Halogen exchange has been employed, for example, for the radiobromination of iodo-*closo*-carboranes. Winberg *et al.* prepared the radiobrominated analogs of 3-iodo-*ortho*-carborane, 9-iodo-*ortho*-carborane, 9-iodo-*meta*-carborane, and 2-iodo-*para*-carborane using ^{76}Br and palladium-catalyzed (Herrmann's catalyst) halogen exchange in toluene. Excellent radiochemical yields could be obtained when a reaction time of 40 min was used [114].

2.5.4.2 Radiometallation

Radiometals have been widely used to label a wide variety of biomolecules, including proteins and antibodies. The most commonly used approach for the incorporation of radiometals consists of using bifunctional chelators (BFCs), which contain one functional group and a metal-binding moiety function. The first allows for anchoring the BFC to the biomolecule, while the latter enables the binding of the metallic radionuclide via formation of a complex. To note, radiolabeling using radiometals and BFCs usually results in the modification of the molecule under investigation. Hence, the effects of radiolabeling on the biological properties of the biomolecules need careful investigation. Ligand systems used for metal complexation include derivatives of diethylenetriaminepentaacetic acid (DTPA), ethylenediaminetetraacetic acid (EDTA), and cryptands [115]. Other important metal-chelating ligands are TETA (1,4,8,11-tetraazacyclo-tetradecane-1,4,8,11-tetrayltetraacetic acid) [116,117], which complexes copper with improved *in vitro* and *in vivo* stability, and DOTA (1,4,7,10-tetraazacyclododecane-1,4,7,10-tetrayltetraacetic acid) [118], which has been employed as a ligand system for ^{90}Y, ^{67}Ga/^{68}Ga, and ^{111}In. Stable ^{90}Y and ^{67}Ga complexes are formed with 1,4,7-triazacyclononane-1,4,7-triyltriacetic acid (NOTA) [119,120].

Boron cluster derivatives have been successfully labeled using the above-mentioned method. For example, mercapto-*closo*-undecahydrododecaborate (BSH) fused with a short arginine peptide (1R, 2R, or 3R) was prepared and fused with DOTA (see Figure 2.5.15) for subsequent incorporation of the positron emitter ^{64}Cu [121].

The above-mentioned strategy can be applied to any biomolecule, including those containing boron clusters. However, because the label is formally not incorporated in the boron cluster but in the biomolecules themselves, these will not be further considered in this section.

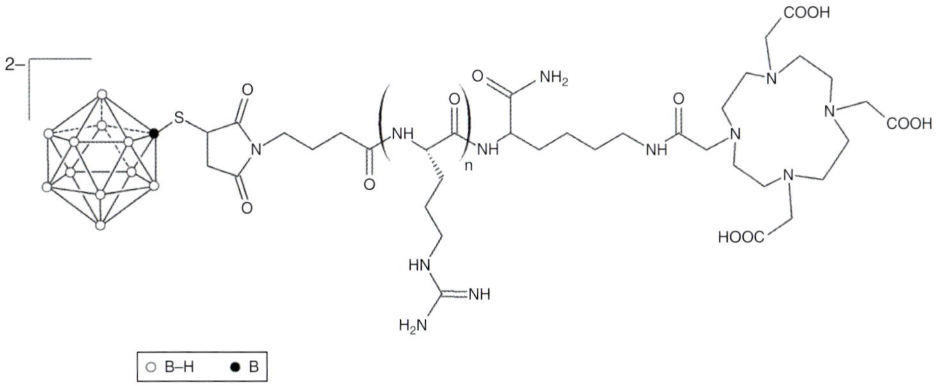

Figure 2.5.15 Mercapto-*closo*-undecahydrododecaborate (BSH) fused with a short arginine peptide (n = 1, 2, or 3) and linked to a DOTA chelator to enable complexation with the positron emitter ^{64}Cu.

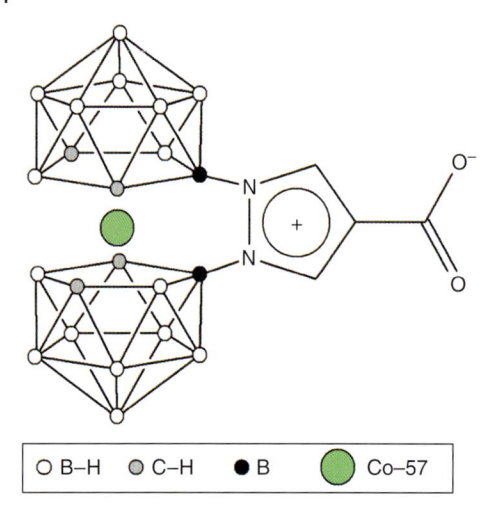

Figure 2.5.16 Cobaltacarborane with a pyrazole ring substituted with a carboxylate function as a handle for subsequent coupling. Green circle: ^{57}Co.

The capacity of boron clusters to form metallacarboranes has been exploited to develop radiolabeled boron cluster derivatives. In one of the first examples, Hawthorne and coworkers described the use of a carborane derivative as a chelating agent for radiometals, leading to so-called VFCs (Figure 2.5.16) [122]. This consists of two *nido*-carboranyl ligands hinged together through a bifunctional pyrazole molecule, which provides an attachment point for a carboxyl group employed to bond the radioactive cluster to an amino group of a biomolecule.

Cobalt-57 has a half-life of 271.7 days and decays by electron capture with subsequent emission of γ-rays. However, its long half-life is quite inappropriate to approach *in vivo* investigations. The same complex could be prepared, for example, with Cobalt-55, which has a half-life of 17.5 h, a positron branching ratio of 77%, and an average positron energy of 570 keV [123], enabling imaging using PET.

Labeled metallacarboranes have also been prepared using ^{99m}Tc. The method, developed by Valliant and coworkers [124], enabled the preparation of ^{99m}Tc-labeled metallocarboranes in water under mild reaction conditions, by reaction of *nido*-carborane ligands with $[Re(CO)_3Br_3]^{2-}$ in the presence of aqueous potassium fluoride to give the corresponding η^5-$Re(CO)_3$–carborane complexes (Figure 2.5.17). The resulting complexes were isolated in 80% radiochemical yield following a simple solid-phase (Sep-Pak®) purification process, and showed high stability even when exposed to cysteine and histidine challenges.

This approach entails using fluoride ions to prevent the decomposition of the technetium starting materials over the prolonged reaction times and elevated temperatures needed to form the desired complexes. However, only modest product yields could be achieved for carboranes containing sterically demanding substituents. Microwave-assisted heating increased the reaction rate substantially, and occasionally the yield of the reactions compared to those of conventional heating methods [125]. Importantly, targeting vectors can be incorporated to these complexes at a carbon or boron atom using a number of robust synthetic strategies [126,127]. Indeed, a ten-year period of refining the linkers has resulted in several derivatives [124,125,128,129].

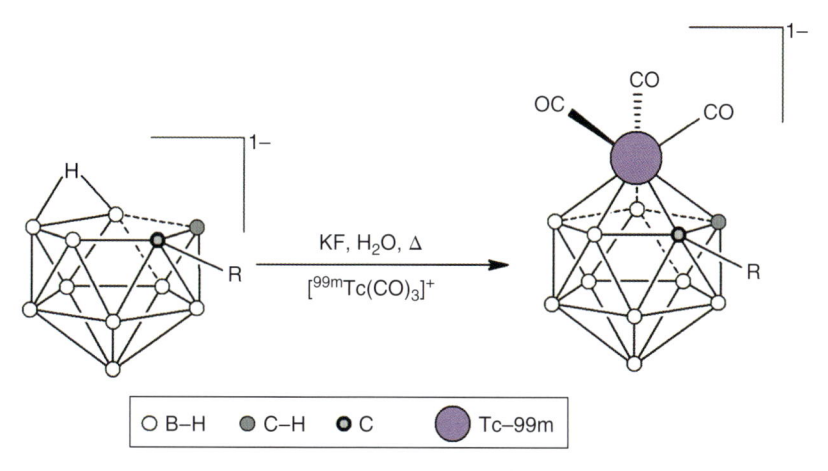

Figure 2.5.17 Reaction scheme for the formation of ^{99m}Tc-labeled metallacarboranes.

2.5.4.2.1 Radiolabeling using other Radionuclides

Carbon-11 (^{11}C) is a very convenient PET isotope, because it enables the preparation of labeled species that are identical to the non-labeled molecules by simple substitution of a stable carbon atom by an ^{11}C atom. It decays 100% by positron emission, it has a reasonably low β^+ energy (Eβ_{max} = 0.96 MeV), and hence its use began to receive increasing attention in the 1960s due to the widespread installation of cyclotrons. The main drawback of ^{11}C is its short half-life (20.4 min), which limits its application especially for the investigation of molecules with slow pharmacokinetics and long biological half-lives.

Carbon-11 (^{11}C) has not yet been attached directly in or to a carborane vertex. A few [^{11}C]methyl labeling examples of carboranes have been done via a linker on the CH vertex to mimic raclopride [130]. Carbon-11 has also been incorporated to carboranes by attachment to the amino residue of amino-*m*-carboranes [54], to prepare two analogs of 2-(4-aminophenyl)benzothiazole in which the phenyl ring was substituted by a *m*-carborane cage. Both compounds were radiolabeled with carbon-11 using [^{11}C]CH$_3$OTf as the labeling agent (Figure 2.5.18).

Tritium (^{3}H, T) is a very useful low-energy β^- emitter. Its applicability *in vivo* is rather limited, because the electrons emitted during radioactive decay lack the sufficient penetration capacity to escape from the subject under investigation, and hence external detection is not feasible. However, it is extremely useful for both *in vitro* and *ex vivo* experiments, and is often used in the development of radioligands for PET and SPECT, as it can provide a good estimation regarding the affinity of the ligands for the specific targets. To date, radiolabeling of boron clusters with tritium has been conducted by deprotonation of one of the CH vertices with a strong base and subsequent quenching of the resulting anion with T$_2$O using tritiated water [131,132]. A tritiodehalogenation from a halogenated carborane has not yet been shown in the literature.

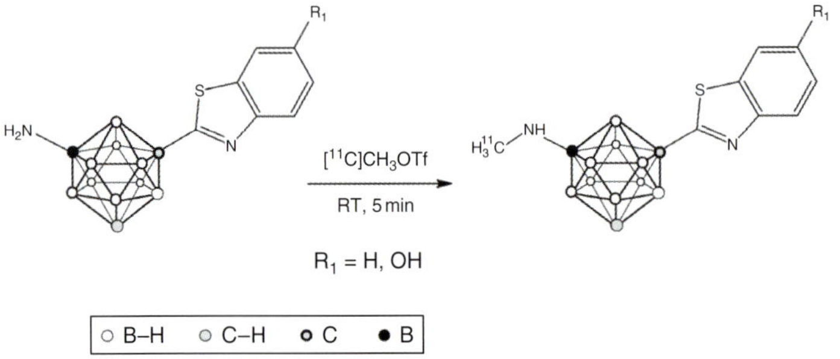

Figure 2.5.18 Reaction scheme for the preparation of ^{11}C-labeled carboranyl benzothiazoles.

2.5.5 The use of Radiolabeling in BNCT Drug Development: Illustrative Examples

Despite the promise of nuclear imaging in the process of drug development, the relatively high cost of the technique and the lack of research-dedicated installations have limited its use in the development of BNCT drugs, and especially in the context of boron clusters. Beyond boron clusters, PET has been used to estimate the accumulation of BPA-fr (p-boronophenylalanine labeled with ^{10}B and conjugated with fructose, commonly used in therapeutic clinical practice) in tumors [133]. The measurement of [^{18}F]FBPA accumulation is made about 1 hour after [^{18}F]FBPA administration [134–137]. However, there are several limitations of FBPA PET in predicting BPA-fr accumulation in the tumors and normal tissues [138]. First, the chemical structure of FBPA differs from that of BPA-fr, and hence different pharmacokinetic behavior might be expected; second, [^{18}F]FBPA PET is performed at a tracer dose of the radiotracer, while therapeutic doses of ^{10}BPA are administered (approximately 500 mg/kg) for BNCT; and, finally, the administration pattern used in both cases is different (a single bolus for [^{18}F]FBPA, and a slow bolus intravenous injection followed by drip infusion during neutron irradiation for BPA-fr). Hence, the predictive value of FBPA PET for BPA-fr accumulation in the tumor and normal tissues remains unclear.

In the context of boron clusters, BNCT drug candidates have been evaluated at the preclinical level using radiolabeling followed by *in vivo* imaging.

Conjugation of ^{57}Co-labeled VFCs to a mAb could be accomplished, providing an extraordinarily stable imaging agent rather than a therapeutic agent for BNCT [139]. The VFC conjugate retained >90% immunoreactivity, was stable in serum for more than 7 days, and demonstrated excellent localization in tumor xenografts during *in vivo* studies with nude mice.

BSH fused with a short arginine peptide (3R) was labeled with copper-64, and the drug's pharmacokinetics were evaluated using PET in U87DEGFR tumor-bearing mice [121]. The results were compared to ^{64}Cu-labeled BSH-DOTA as the control. BSH-3R showed a high uptake in the tumor area on PET imaging (see Figure 2.5.19). It was concluded that BSH-3R is the ideal boron compound for clinical use during BNCT and that in developing this compound for clinical use, the BSH-3R PET probe is essential for pharmacokinetic imaging.

Several studies were performed with mAbs conjugated with a series of discrete, precisely synthesized oligomeric reagents (trailers), each of which contains a fixed number of boron atoms up to 200 (Figure 2.5.20) [140]. The *nido*-carborane moieties in

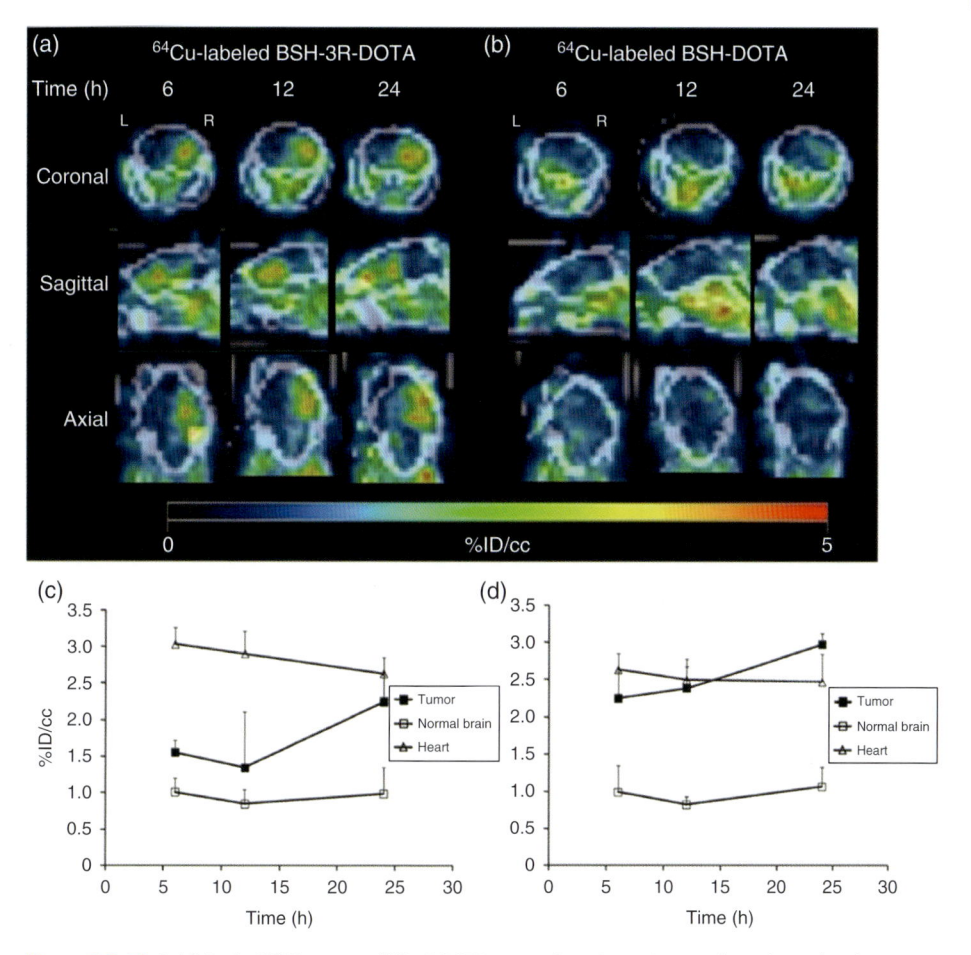

Figure 2.5.19 (a,b) Brain PET images of U87DEGFR tumor-bearing mice at 6 h, 12 h, and 24 h post injection of (a) ^{64}Cu-labeled BSH-3R-DOTA and (b) ^{64}Cu-labeled BSH-DOTA. (c,d) Quantification analysis by PET images of radioactivity accumulation in the tumor, normal brain, and heart at 6 h, 12 h, and 24 h post injection of (c) ^{64}Cu-labeled BSH-DOTA and (d) ^{64}Cu-labeled BSH-3R-DOTA. Source: Reprinted with permission from Ref. [121].

these trailers were radioiodinated [141]. These studies demonstrate that biospecific antibodies prepared with an anti-*nido*-carborane anion specificity can selectively localize a *nido*-carborane anion structure to a tumor cell surface. This is of particular importance, since the binding of a boron-rich oligomeric phosphate diester (trailer) composed of 10 haptenic *nido* cages to an anti-*nido*-mAb could be demonstrated.

The typical biodistribution pattern for [^{18}F]FDG was observed (Figure 2.5.22): high accumulation was detected in the kidneys and the bladder due to elimination mainly via urine, while significant uptake was also observed in the heart and the brain, organs that are known to consume a significant amount of glucose. For both models (PANC-1 and A549), the tumors could be clearly visualized on the images. The two labeled COSAN analogs showed a similar biodistribution pattern, irrespective of the tumor model: long residence time of the labeled species in the blood pool; significant uptake in the liver,

Figure 2.5.20 Example of an oligomeric *nido*-carborane phosphate diester derivative. Recently, two [124]I-labeled COSAN analogs (see Figure 2.5.21) have been prepared by some of the authors using palladium0catalyzed iodine exchange, and subsequently evaluated in PANC-1 (human pancreatic carcinoma) and A549 (carcinomic human alveolar basal epithelial cell line) xerograft mouse models [142,143]. [^{18}F]FDG PET studies were conducted in the same animals to assess tumor metabolism.

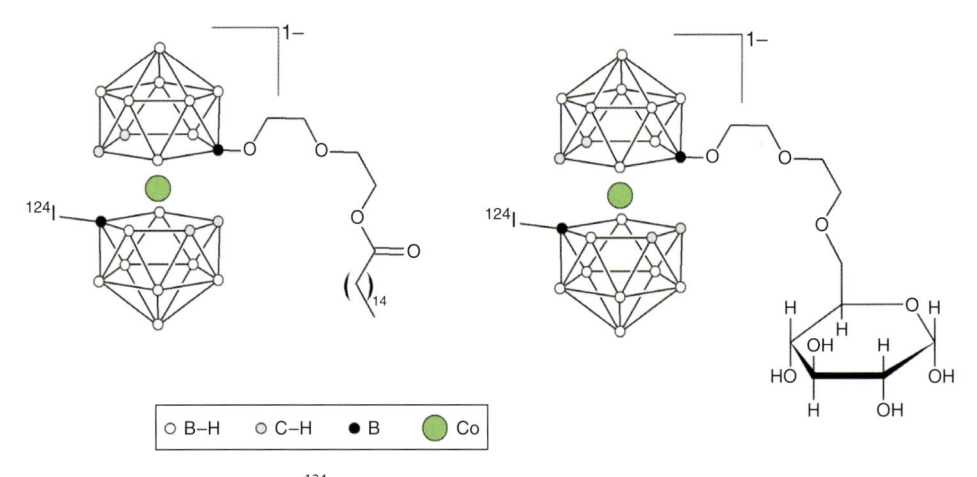

Figure 2.5.21 Structure of the [124]I-labeled COSAN analogs.

the kidneys, and the lungs; and lower but detectable uptake in the tumor (approximately 1%ID/g, irrespective of the compound and the tumor model). Despite the low accumulation found in the tumor, radiolabeling followed by nuclear imaging proved efficient to determine the concentration of boron in the tumor as a function of time.

Biodistribution studies of iodinated C-hydroxy-*nido*-carborane ligands (e.g., carboranol and salborin, which are *closo*-carborane analogs of phenol and salicyclic acid) in BALB/c mice were performed by El-Zaria et al. (Figure 2.5.23) [144]. Low accumulation of the labeled compounds was observed in the liver and intestines, which are sites where labeled carboranes typically localize. The labeled cluster bearing hydroxy and carboxylic acid groups on the two carbon vertices demonstrated preferential

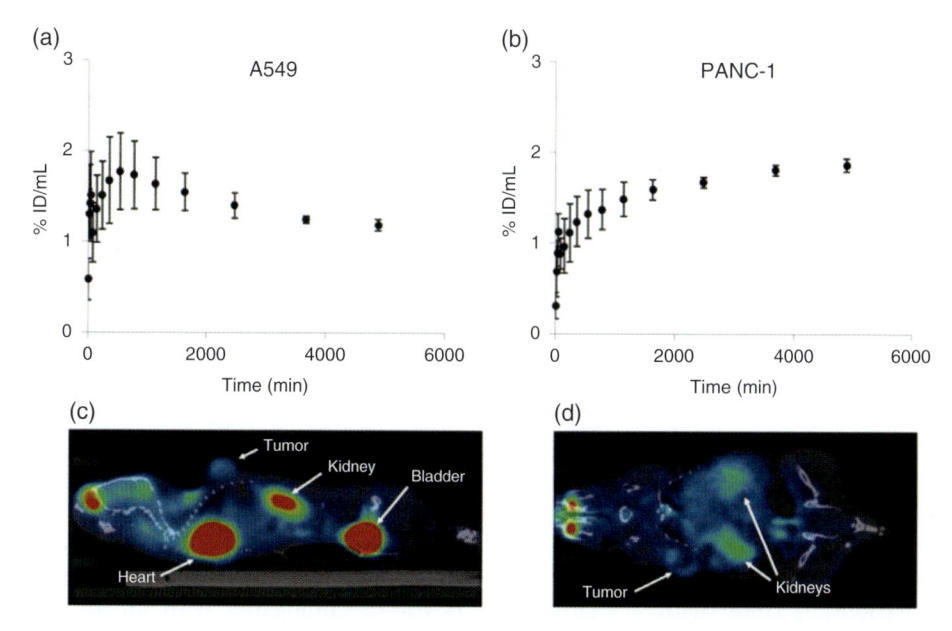

Figure 2.5.22 Biodistribution of [^{18}F]FDG in subcutaneous PANC-1 (human pancreatic carcinoma) and A549 (carcinomic human alveolar basal epithelial cell line) xerograft mouse models. Time–activity curves in the tumor, expressed as the percentage of the injected dose (%ID) per cubic centimeter of tissue, are shown in (a) and (b). PET sagittal (c) and coronal (d) slices resulting from averaged images obtained after administration of [^{18}F]FDG in mice bearing A549 (c) and PANC-1 (d) tumors are shown; optimal viewpoint and slides have been selected for correct visualization of the tumor.

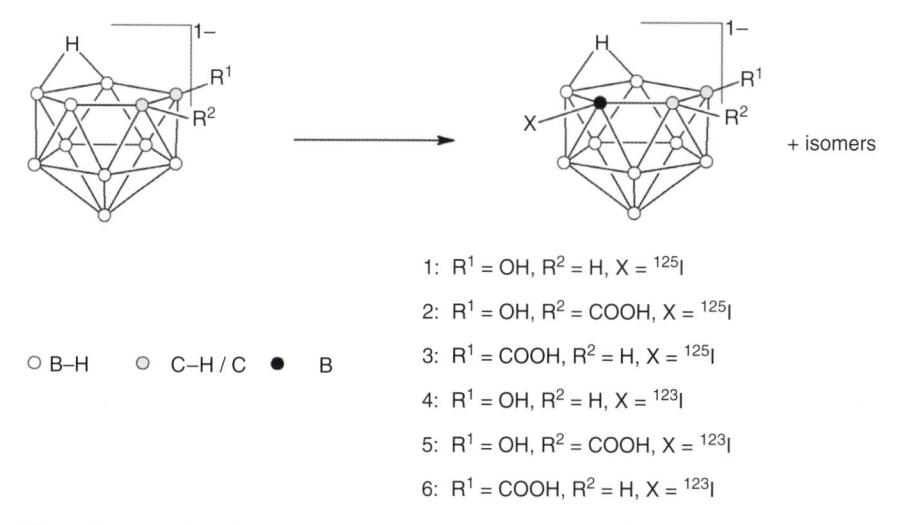

○ B–H   ○ C–H / C   ● B

1: R^1 = OH, R^2 = H, X = ^{125}I
2: R^1 = OH, R^2 = COOH, X = ^{125}I
3: R^1 = COOH, R^2 = H, X = ^{125}I
4: R^1 = OH, R^2 = H, X = ^{123}I
5: R^1 = OH, R^2 = COOH, X = ^{123}I
6: R^1 = COOH, R^2 = H, X = ^{123}I

Figure 2.5.23 Radioiodinated 7-hydroxy-nido-carboranol (I-nido-carboranol, **1** and **4**), nido-salborin (I-nido-salborin, **2** and **5**), and 7,8-dicarba-nido-undecaborate-7-carboxylic acid (**3** and **6**).

clearance through the kidneys and low thyroid uptake. This compound had substantially reduced nonspecific binding compared to the deshydroxy analog, making it an attractive bifunctional ligand for preparing targeted molecular imaging and therapy agents.

Radiolabeling at the targeting moiety has also been applied to the determination of the pharmacokinetic properties or drug candidates containing boron clusters. In an investigation on the use of an ^{211}At-labeled anti-CD45 mAb as a replacement of total body irradiation in conditioning regimens designed to decrease the toxicity of hematopoietic cell transplantation (HCT), dose-escalation studies were conducted in dogs using ^{211}At-labeled anticanine CD45 mAb, CA12.10C12, conjugated with a maleimido-closodecaborate(2-) derivative. Unacceptable renal toxicity was noted in the dogs receiving doses in the 0.27–0.62 mCi/kg range. This result was not anticipated, as no toxicity had been noted in prior biodistribution and toxicity studies conducted in mice. Studies were conducted to understand the cause of the renal toxicity and to find a way to circumvent it. A dog biodistribution study was conducted with ^{123}I-labeled CA12.10C12 that had been conjugated with the maleimido-closodecaborate(2-) derivative. The biodistribution data showed that tenfold higher kidney concentrations were obtained with the maleimido conjugate than had been obtained in a previous biodistribution study with ^{123}I-labeled CA12.10C12 conjugated with an amine-reactive phenylisothiocyanato-CHX-A″ derivative. The difference in kidney concentrations observed in dogs for the two conjugation approaches led to an investigation of the reagents. Size exclusion HPLC (SE-HPLC) analyses showed that the purity of the CA12.10C12 conjugated via reduced disulfides was lower than that obtained with amine-reactive conjugation reagents, and nonreducing SDS-PAGE (sodium dodecyl sulfate–polyacrylamide gel electrophoresis) analyses indicated protein fragments were present in the disulfide reduced conjugate. Amine-reactive decaborate (phenylisothiocyanate) was prepared for use in the studies. A biodistribution was conducted with co-administered ^{125}I- and ^{211}At-labeled CA12.10C10 conjugated with the amine-reactive decaborate. In that study, lower kidney concentrations were obtained for both radionuclides than had been obtained in the earlier study of the same antibody conjugated with the maleimido-closodecaborate(2-) derivative after reduction of disulfide bonds [94].

Very recently, labeled carborane–tetrazine derivatives suitable to be used in pretargeting strategies involving bioorthogonal inverse electron demand Diels–Alder chemistry have been reported by Genady et al. [145]. The carborane derivative 1,12-dicarba-*closo*-dodecaborate-1-(4-(1,2,4,5-tetrazin-3-yl)phenyl)methanamide could be synthesized, converted to the corresponding *nido*-cluster, and successfully radiolabeled with ^{125}I and ^{123}I. The labeled tetrazines showed good stability *in vitro* and reacted fast with (E)-cyclooct-4-enol (TCO) and TCO-functionalized antibodies. The ^{123}I-labeled compound was used to assess the biodistribution in healthy mice using SPECT-CT. Accumulation was observed in the liver from short times up to 24 hours after administration. Interestingly, no accumulation of radioactivity was observed in the thyroid gland, suggesting high radiochemical stability of the labeled species. Hence, the described carborane–tetrazine derivatives show promise for future application for pretargeting studies, using bioorthogonal coupling chemistry with a range of different TCO-functionalized vectors, including tumor-targeting antibodies.

2.5.6 Conclusion and Future Perspectives

Boron neutron capture therapy has been for decades a promising technology to treat cancers lacking efficient alternative treatments. However, after 80 years since it was described for the first time, BNCT has not been established as a routine clinical practice. This is due to a combination of reasons; one of them is the need to develop drugs capable to accumulate a sufficient number of boron atoms in the target tissue. In this context, the use of boron clusters is anticipated to be advantageous, because they accumulate a certain number of boron atoms in one single chemical structure. Besides the need of developing the appropriate drugs, it is clear that gaining access to a technology capable of determining the accumulation of boron atoms in the tumor, surrounding tissue, and blood *in vivo*, noninvasively, at the whole-body level, and in real time would be a great advantage, both during the process of drug development and also to predict therapeutic efficacy on a patient-by-patient basis (personalized medicine) and to define the optimal time window for neutron irradiation.

Nuclear imaging techniques, based on the radiolabeling of the molecule under investigation with positron or γ emitters, fulfill the conditions described here. After incorporation of the radionuclide in the target molecules and administration into the patient (or subject under investigation) using different administration routes (usually intravenous), 3D-tomographic images can be obtained, enabling the quantification of the spatiotemporal distribution of the labeled species within the organism. Certain aspects have to be taken into consideration during radiolabeling and nuclear imaging: (1) the chemical structure of the drug and the labeled molecule should be identical, or at least they should present equivalent biological activity and pharmacokinetic properties; and (2) the labeled molecule should be stable *in vivo*: if the radionuclide detaches or the radiotracers undergo metabolism, only the location of the label (or the labeled fragment) will be tracked, leading to wrong results and misinterpretation of the data.

With the widespread installation of biomedical cyclotrons around the world and the continuous advances in the development of more sensitive imaging cameras, it is expected that nuclear imaging will play a pivotal role in the near future in the process of drug development, including the development of new BNCT drug candidates. Additionally, the combination of nuclear imaging with neutron sources as hybrid systems could help in providing more efficient and personalized treatment to patients.

References

1 Locher, G.L. (1936) Biological effects and therapeutic possibilities of neutrons. Am. J. Roentgenol., **36**, 1–13.
2 Tietze, L.F., et al. (2001) *ortho*-Carboranyl glycosides for the treatment of cancer by boron neutron capture therapy. Bioorg. Med. Chem., **9**(7), 1747–1752.
3 Maurer, J.L., et al. (1988) Hydrophilically augmented glycosyl carborane derivatives for incorporation in antibody conjugation reagents. Organometallics, **7**(12), 2519–2524.
4 Giovenzana, G.B., et al. (1999) Synthesis of carboranyl derivatives of alkynyl glycosides as potential BNCT agents. Tetrahedron, **55**(49), 14123–14136.
5 Tjarks, W., et al. (1992) Synthesis and *in vitro* evaluation of boronated uridine and glucose derivatives for boron neutron capture therapy. J. Med. Chem., **35**(9), 1628–1633.

6 Campo, F., et al. (2012) Synthesis of complex glycosylated carboranes for BNCT. Synlett, (**1**), 120–122.
7 Jacobsson, M., et al. (2008) Xylose as a carrier for boron containing compounds. Bioorg. Med. Chem. Lett., **18**(7), 2451–2454.
8 Satapathy, R., et al. (2015) Glycoconjugates of polyhedral boron clusters. J. Organomet. Chem., **798**, 13–23.
9 Semioshkin, A., et al. (2007) Reactions of oxonium derivatives of $[B_{12}H_{12}]^{2-}$ with amines: synthesis and structure of novel B12-based ammonium salts and amino acids. J. Organomet. Chem., **692**(19), 4020–4028.
10 Morin, C., et al. (2004) Synthesis and evaluation of boronated lysine and bis(carboranylated) γ-amino acids as monomers for peptide assembly. Eur. J. Org. Chem., (**18**), 3828–3832.
11 Beletskaya, I.P., et al. (2004) New nonnatural α-amino acid derivatives with carboranyl fragments in α- and β-positions. Synlett, (**7**), 1247–1248.
12 Kusaka, S., et al. (2011) Synthesis of optically active dodecaborate-containing l-amino acids for BNCT. Appl. Radiat. Isot., **69**(12), 1768–1770.
13 Kabalka, G.W., et al. (2011) Boronated unnatural cyclic amino acids as potential delivery agents for neutron capture therapy. Appl. Radiat. Isot., **69**(12), 1778–1781.
14 Hattori, Y., et al. (2014) Synthesis and in vitro evaluation of thiododecaborated α, α- cycloalkylamino acids for the treatment of malignant brain tumors by boron neutron capture therapy. Amino Acids, **46**(12), 2715–2720.
15 Futamura, G., et al. (2017) Evaluation of a novel sodium borocaptate-containing unnatural amino acid as a boron delivery agent for neutron capture therapy of the F98 rat glioma. Radiation Oncology, **12**(1).
16 Michiue, H., et al. (2014) The acceleration of boron neutron capture therapy using multi-linked mercaptoundecahydrododecaborate (BSH) fused cell-penetrating peptide. Biomaterials, **35**(10), 3396–3405.
17 Matuszewski, M., et al. (2015) Nucleoside bearing boron clusters and their phosphoramidites-building blocks for modified oligonucleotide synthesis. New J. Chem., **39**(2), 1202–1221.
18 Tjarks, W. (2000) The use of boron clusters in the rational design of boronated nucleosides for neutron capture therapy of cancer. J. Organomet. Chem., **614–615**, 37–47.
19 Hasabelnaby, S., et al. (2012) Synthesis, chemical and enzymatic hydrolysis, and aqueous solubility of amino acid ester prodrugs of 3-carboranyl thymidine analogs for boron neutron capture therapy of brain tumors. Eur. J. Med. Chem., **55**, 325–334.
20 Semioshkin, A., et al. (2013) Synthesis of the first conjugates of 5-ethynyl-2′-deoxyuridine with *closo*-dodecaborate and cobalt-bis-dicarbollide boron clusters. Tetrahedron, **69**(37), 8034–8041.
21 Lesnikowski, Z.J. (2009) Nucleoside-boron cluster conjugates: beyond pyrimidine nucleosides and carboranes. J. Organomet. Chem., **694**(11), 1771–1775.
22 Agarwal, H.K., et al. (2015) Synthesis and evaluation of thymidine kinase 1-targeting carboranyl pyrimidine nucleoside analogs for boron neutron capture therapy of cancer. Eur. J. Med. Chem., **100**, 197–209.
23 Nizioł, J., et al. (2014) Synthesis, reactivity and biological activity of N(4)-boronated derivatives of 2′-deoxycytidine. Bioorg. Med. Chem., **22**(15), 3906–3912.
24 Nizioł, J., et al. (2015) Biological activity of N(4)-boronated derivatives of 2′-deoxycytidine, potential agents for boron-neutron capture therapy. Bioorg. Med. Chem., **23**(19), 6297–6304.

25 Khalil, A., et al. (2013) N3-substituted thymidine bioconjugates for cancer therapy and imaging. Future Medicinal Chemistry, **5**(6), 677–692.

26 Wellum, G.R., et al. (1982) Boron neutron capture radiation therapy of cerebral glimoas: an analysis of the possible use of boron-loaded tumor-specific antibodies for the selective concentration of boron in gliomas. Int. J. Radiat. Oncol. Biol. Phys., **8**(8), 1339–1345.

27 Ranadive, G.N., et al. (1993) A technique to prepare boronated B72.3 monoclonal antibody for boron neutron capture therapy. Nucl. Med. Biol., **20**(1), 1–6.

28 Novick, S., et al. (2002) Linkage of boronated polylysine to glycoside moieties of polyclonal antibody; boronated antibodies as potential delivery agents for neutron capture therapy. Nucl. Med. Biol., **29**(2), 159–167.

29 Yang, W., et al. (2009) Boron neutron capture therapy of EGFR or EGFRvIII positive gliomas using either boronated monoclonal antibodies or epidermal growth factor as molecular targeting agents. Appl. Radiat. Isot., **67**(7–8 Suppl.), S328–S331.

30 Tietze, R., et al. (2015) Boron containing magnetic nanoparticles for neutron capture therapy: an innovative approach for specifically targeting tumors. Appl. Radiat. Isot., **106**, 151–155.

31 Xiong, H., et al. (2015) Doxorubicin-loaded carborane-conjugated polymeric nanoparticles as delivery system for combination cancer therapy. Biomacromolecules, **16**(12), 3980–3988.

32 Cioran, A.M., et al. (2014) Preparation and characterization of Au nanoparticles capped with mercaptocarboranyl clusters. Dalton Trans., **43**(13), 5054–5061.

33 Ciofani, G., et al. (2009) Folate functionalized boron nitride nanotubes and their selective uptake by glioblastoma multiforme cells: Implications for their use as boron carriers in clinical boron neutron capture therapy. Nanoscale Res. Lett., **4**(2), 113–121.

34 Barth, R.F., et al. (1992) Boron neutron capture therapy for cancer: realities and prospects. Cancer, **70**(12), 2995–3007.

35 Wittig, A., et al. (2008) Boron analysis and boron imaging in biological materials for boron neutron capture therapy (BNCT). Crit. Rev. Oncol. Hematol., **68**(1), 66–90.

36 Kobayashi, T., et al. (1983) Microanalysis system of ppm-order ^{10}B concentrations in tissue for neutron capture therapy by prompt gamma-ray spectrometry. Nucl. Instr. Meth. Phys. Res., **204**(2–3), 525–531.

37 Verbakel, W.F.A.R., et al. (2003) Boron concentrations in brain during boron neutron capture therapy: in vivo measurements from the Phase I trial EORTC 11961 using a gamma-ray telescope. Int. J. Radiat. Oncol. Biol. Phys., **55**(3), 743–756.

38 Munck Af Rosenschöld, P.M., et al. (2001) Toward clinical application of prompt gamma spectroscopy for *in vivo* monitoring of boron uptake in boron neutron capture therapy. Med. Phys., **28**(5), 787–795.

39 Bendel, P., et al. (2001) In vivo imaging of the neutron capture therapy agent BSH in mice using ^{10}B MRI. Magn. Reson. Med., **46**(1), 13–17.

40 Bendel, P., et al. (2005) Noninvasive quantitative in vivo mapping and metabolism of boronophenylalanine (BPA) by nuclear magnetic resonance (NMR) spectroscopy and imaging. Radiat. Res., **164**(5), 680–687.

41 Bendel, P. (2012) Boron imaging: localized quantitative detection and imaging of boron by magnetic resonance. In: Sauerwein WAG, Wittig A, Moss R, Nakagawa Y (Editors). Neutron Capture Therapy: Principles and Applications. Berlin: Springer Science & Business Media, 2012, p. 213–223.

42 Yokoyama, K., et al. (2006) Pharmacokinetic study of BSH and BPA in simultaneous use for BNCT. J. Neurooncol., **78**(3), 227–232.

43 Bregadze, V.I., et al. (2006) Polyhedral boron compounds as potential diagnostic and therapeutic antitumor agents. Anticancer Agents Med. Chem., **6**(2), 75–109.

44 Wade, K. (1971) The structural significance of the number of skeletal bonding electron-pairs in carboranes, the higher boranes and borane anions, and various transition-metal carbonyl cluster compounds. J. Chem. Soc. Chem. Commun., (**15**), 792–793.

45 Jemmis, E.D. (1982) Overlap control and stability of polyhedral molecules. *closo*-carboranes. J. Am. Chem. Soc., **104**(25), 7017–7020.

46 Grimes, R.N. (2016) Icosahedral carboranes: 1,2-$C_2B_{10}H_{12}$. In: Carboranes, 3rd ed. New York: Academic Press, p. 283–502.

47 Grimes, R.N. (2016) Icosahedral carboranes: 1,7-$C_2B_{10}H_{12}$ and 1,12-$C_2B_{10}H_{12}$. In: Carboranes, 3rd ed. New York: Academic Press, p. 503–615.

48 Grimes, R.N. (1993) Carboranes, anti-crowns, and big wheels. Angew. Chem. Int. Ed., **32**(9), 1289–1290.

49 Wedge, T.J., et al. (2003) Multidentate carborane-containing Lewis acids and their chemistry: mercuracarborands. Coord. Chem. Rev., **240**(1–2), 111–128.

50 Murphy, D.M., et al. (1993) Synthesis of icosahedral carboranes for second-harmonic generation. Part 1. J. Mater. Chem., **3**(1), 67–76.

51 Murphy, D.M., et al. (1993) Synthesis of icosahedral carboranes for second-harmonic generation. Part 2. J. Mater. Chem., **3**(2), 139–148.

52 Kaszynski, P., et al. (2002) Investigations of electronic interactions between closo-boranes and triple-bonded substituents. Collect. Czech. Chem. Commun., **67**(7), 1061–1083.

53 Gona, K.B., et al. (2013) Synthesis of m-carboranyl amides via palladium-catalyzed carbonylation. Tetrahedron Lett., **54**(8), 941–944.

54 Gona, K.B., et al. (2015) Synthesis and C-radiolabelling of 2-carboranyl benzothiazoles. Molecules, **20**(5), 7495–7508.

55 Vázquez, N., et al. (2011) Synthesis of D2 receptor ligand analogs incorporating one dicarba-*closo*-dodecaborane unit. Tetrahedron Lett., **52**(5), 615–618.

56 Vázquez, N., et al. (2012) Synthesis of novel pyrazole derivatives incorporating one dicarba-*closo*-dodecaborane unit. Tetrahedron Lett., **53**(35), 4743–4746.

57 Endo, Y., et al. (1999) Potent estrogenic agonists bearing dicarba-*closo*-dodecaborane as a hydrophobic pharmacophore. J. Med. Chem., **42**(9), 1501–1504.

58 Ogawa, T., et al. (2006) *m*-Carborane bisphenol structure as a pharmacophore for selective estrogen receptor modulators. Bioorg. Med. Chem. Lett., **16**(15), 3943–3946.

59 Issa, F., et al. (2011) Boron in drug discovery: carboranes as unique pharmacophores in biologically active compounds. Chem. Rev., **111**(9), 5701–5722.

60 Frank, R., et al. (2015) Carbaboranes: more than just phenyl mimetics. Pure Appl. Chem., **87**(2), 163–171.

61 Scholz, M., et al. (2011) Carbaboranes as pharmacophores: properties, synthesis, and application strategies. Chem. Rev., **111**(11), 7035–7062.

62 Stadlbauer, S., et al. (2012) Imitation and modification of bioactive lead structures via integration of boron clusters. Pure Appl. Chem., **84**(11), 2289–2298.

63 Hermansson, K., et al. (1999) o-, m-, and p-carboranes and their anions: *ab initio* calculations of structures, electron affinities, and acidities. Inorg. Chem., **38**(26), 6039–6048.

64 Schaeffer, R. (1957) A new type of substituted borane. J. Am. Chem. Soc., **79**(4), 1006–1007.
65 Heying, T.L., et al. (1963) A new series of organoboranes. I. Carboranes from the reaction of decaborane with acetylenic compounds. Inorg. Chem., **2**(6), 1089–1092.
66 Bregadze, V.I. (1992) Dicarba-*closo*-dodecaboranes $C_2B_{10}H_{12}$ and their derivatives. Chem. Rev., **92**(2), 209–223.
67 Toppino, A., et al. (2013) High yielding preparation of dicarba-closo-dodecaboranes using a silver(I) mediated dehydrogenative alkyne-insertion reaction. Inorg. Chem., **52**(15), 8743–8749.
68 Qiu, Z., et al. (2010) Palladium/nickel-cocatalyzed cycloaddition of 1,3-dehydro-o-carborane with alkynes: facile synthesis of C,B-substituted carboranes. J. Am. Chem. Soc., **132**(45), 16085–16093.
69 Hawthorne, M.F., et al. (1965) Carbametallic boron hydride derivatives. I. Apparent analogs of ferrocene and ferricinium ion. J. Am. Chem. Soc., **87**(8), 1818–1819.
70 Warren Jr, L.F., et al. (1968) Metallocene analogs of copper, gold, and palladium derived from the (3)-1,2-dicarbollide ion. J. Am. Chem. Soc., **90**(18), 4823–4828.
71 Hawthorne, M.F., et al. (1965) Carborane analogues of cobalticinium ion. Chem. Commun., (**19**), 443–444.
72 Grimes, R.N. (2016) Metallacarboranes of the transition and lanthanide elements. In: Carboranes, 3rd ed. New York: Academic Press, p. 773–1014.
73 Blumgart, H.L., et al. Studies on the velocity of blood flow: I. The method utilized 1. J. Clin. Invest., **4**(1), 1–13.
74 Popota, F.D., et al. (2012) Comparison of the performance evaluation of the MicroPET R4 scanner according to NEMA standards NU 4-2008 and NU 2-2001. IEEE Trans. Nucl. Sci., **59**(5 Part 1), 1879–1886.
75 Ay, M.R., et al. (2007) Computed tomography based attenuation correction in PET/CT: principles, instrumentation, protocols, artifacts and future trends. Iran. J. Nucl. Med., **15**(28), 1–29.
76 Paulus, D.H., et al. (2016) Hybrid positron emission tomography/magnetic resonance imaging: challenges, methods, and state of the art of hardware component attenuation correction. Invest. Radiol., **51**(10), 624–634.
77 De Jong, L.A.A., et al. (2005) Receptor-ligand binding assays: technologies and applications. J. Chromatogr. B, **829**(1–2), 1–25.
78 Logan, J., et al. (1990) Graphical analysis of reversible radioligand binding from time-activity measurements applied to [N-^{11}C-methyl]-(−)-cocaine PET studies in human subjects. J. Cereb. Blood Flow Metab., **10**(5), 740–747.
79 Patlak, C.S., et al. (1983) Graphical evaluation of blood-to-brain transfer constants from multiple-time uptake data. J. Cereb. Blood Flow Metab., **3**(1), 1–7.
80 Patlak, C.S., et al. (1985) Graphical evaluation of blood-to-brain transfer constants from multiple-time uptake data: generalizations. J. Cereb. Blood Flow Metab., **5**(4), 584–590.
81 Gómez-Vallejo, V.G., Koziorowski, J., Llop, J. (2012) Specific activity of ^{11}C-labelled radiotracers: a big challenge for PET chemists. In: Hsieh C-H (Editor). Positron Emission Tomography: Current Clinical and Research Aspects. Rijeka, Croatia: Intech, p. 183–210.
82 Shingleton, A. (2010) Allometry: the study of biological scaling. Nature Education Knowledge, **3**(10).

83 Zalutsky, M.R., et al. (2011) Astatine-211: production and availability. Curr. Radiopharm., **4**(3), 177–185.

84 Goldenberg, D.M. (2002) Targeted therapy of cancer with radiolabeled antibodies. J. Nucl. Med., **43**(5), 693–713.

85 Von Mehren, M., et al. Monoclonal antibody therapy for cancer. Annu. Rev. Med., 2003, 343–369.

86 Vaughan, A.T.M., et al. (1977) The preparation of astatotyrosine. Int. J. Appl. Radiat. Isot., **28**(6), 595–598.

87 Visser, G.W.M., et al. (1981) The synthesis and in vitro stability of organic astatine compounds. J. Labelled Compd. Radiopharm., **18**(1–2), 127.

88 Zalutsky, M.R., et al. (1988) Astatination of proteins using an N-succinimidyl tri-n-butylstannyl benzoate intermediate. Int. J. Rad. Appl. Instrum., **39**(3), 227–232.

89 Wilbur, D.S., et al. (1993) Preparation and evaluation of para-[^{211}At]astatobenzoyl labeled anti-renal cell carcinoma antibody A6H F(ab′)$_2$. In vivo distribution comparison with para-[^{125}I]iodobenzoyl labeled A6H F(ab′)$_2$. Nucl. Med. Biol., **20**(8), 917–927.

90 Wilbur, D.S., et al. (2004) Reagents for astatination of biomolecules: comparison of the in vivo distribution and stability of some radioiodinated/astatinated benzamidyl and nido-carboranyl compounds. Bioconjug. Chem., **15**(1), 203–223.

91 Wilbur, D.S., et al. (2007) Reagents for astatination of biomolecules. 2. Conjugation of anionic boron cage pendant groups to a protein provides a method for direct labeling that is stable to in vivo deastatination. Bioconjug. Chem., **18**(4), 1226–1240.

92 Wilbur, D.S., et al. (2009) Reagents for astatination of biomolecules. 4. Comparison of maleimido-closo-decaborate(2-) and meta-[^{211}At]astatobenzoate conjugates for labeling anti-CD45 antibodies with [^{211}At]astatine. Bioconjug. Chem., **20**(10), 1983–1991.

93 Wilbur, D.S., et al. (2011) Reagents for astatination of biomolecules. 5. Evaluation of hydrazone linkers in ^{211}At- and ^{125}I-labeled closo-Decaborate(2-) conjugates of Fab as a means of decreasing kidney retention. Bioconjug. Chem., **22**(6), 1089–1102.

94 Wilbur, D.S., et al. (2012) Reagents for astatination of biomolecules. 6. An intact antibody conjugated with a maleimido-closo-decaborate(2-) reagent via sulfhydryl groups had considerably higher kidney concentrations than the same antibody conjugated with an isothiocyanato-closo-decaborate(2-) reagent via lysine amines. Bioconjug. Chem., **23**(3), 409–420.

95 Tolmachev, V., et al. (1999) Closo-dodecaborate(2-) as a linker for iodination of macromolecules: aspects on conjugation chemistry and biodistribution. Bioconjug. Chem., **10**(3), 338–345.

96 Agarwal, H.K., Hasabelnaby, S, Tiwari, R., Tjarks, W. (2012) Boron cluster (radio)halogenations in biomedical research. In: B Hosmane N (Editor). Boron Science: New Technologies and Applications. Boca Raton, FL: CRC Press, Taylor & Francis Group, p. 107–144.

97 Eriksson, L., et al. (2003) Feasibility of palladium-catalyzed isotopic exchange between sodium [^{125}I]I and 2-iodo-*para*-carborane. J. Labelled Compd. Radiopharm., **46**(7), 623–631.

98 Winberg, K.J., et al. (2003) High yield [^{125}I]iodide-labeling of iodinated carboranes by palladium-catalyzed isotopic exchange. J. Organomet. Chem., **680**(1–2), 188–192.

99 Korsakov, M.V., et al. (2003) Feasibility of isotopic exchange in the system [^{125}I]iodide-undecahydro-iodo-*closo*-dodecaborate(2-) anion. J. Radioanal. Nucl. Chem., **256**(1), 67–71.

100 Shchukin, E., et al. (2004) Copper-mediated isotopic exchange between [^{125}I]iodide and bis(triethylammonium) undecahydro-12-iodo-*closo*-dodecaborate in aqueous media. J. Radioanal. Nucl. Chem., **260**(2), 295–299.
101 Gona, K.B., et al. (2015) Straightforward synthesis of radioiodinated C-substituted *o*-carboranes: towards a versatile platform to enable the in vivo assessment of BNCT drug candidates. Dalton Trans., **44**(21), 9915–9920.
102 Gona, K.B., et al. (2014) COSAN as a molecular imaging platform: synthesis and "in vivo" imaging. Chem. Commun., **50**(77), 11415–11417.
103 Wilbur, D.S., et al. (2004) Synthesis, radioiodination, and biodistribution of some *nido*- and *closo*-monocarbon carborane derivatives. Nucl. Med. Biol., **31**(4), 523–530.
104 Green, A.E.C., et al. (2008) Carborane–carbohydrate derivatives: versatile platforms for developing targeted radiopharmaceuticals. Can. J. Chem., **86**(11), 1063–1069.
105 Green, A.E.C., et al. (2009) Synthesis and screening of bifunctional radiolabelled carborane-carbohydrate derivatives. J. Organomet. Chem., **694**(11), 1736–1746.
106 Mizusawa, E.A., et al. (1985) Synthesis and antibody-labeling studies with the p-isothiocyanatobenzene derivatives of 1,2-dicarba-closo-dodecaborane(12) and the dodecahydro-7,8-dicarba-nido-undecaborate(1-) ion for neutron-capture therapy of human cancer: crystal and molecular structure of Cs + [nido-7-(*p*-C$_6$H$_4$NCS)-9-I-7,8-C$_2$B$_9$H$_{11}$]. Inorg. Chem., **24**(12), 1911–1916.
107 Varadarajan, A., et al. (1991) Conjugation of phenyl isothiocyanate derivatives of carborane to antitumor antibody and in vivo localization of conjugates in nude mice. Bioconjug. Chem., **2**(2), 102–110.
108 Nestor, M., et al. (2003) Biodistribution of the chimeric monoclonal antibody U36 radioiodinated with a closo-dodecaborate-containing linker: comparison with other radioiodination methods. Bioconjug. Chem., **14**(4), 805–810.
109 Bruskin, A., et al. (2004) Radiobromination of monoclonal antibody using potassium [^{76}Br] (4 isothiocyanatobenzyl-ammonio)-bromo-decahydro-closo-dodecaborate (Bromo-DABI). Nucl. Med. Biol., **31**(2), 205–211.
110 Kabalka, G.W.L., Lee, S. K., Longford, C. P. D. (1996). J. Nucl. Med., **37**(S5), 43.
111 Gona, K.B., et al. (2013) [^{18}F]Fluorination of o-carborane via nucleophilic substitution: Towards a versatile platform for the preparation of ^{18}F-labelled BNCT drug candidates. Chem. Commun., **49**(98), 11491–11493.
112 Tolmachev, V., et al. (2002) Radiobromination of closo-dodecaborate anion: aspects of labelling chemistry in aqueous solution using chloramine-T. Radiochimica Acta, **90**(4), 229–235.
113 Winberg, K.J., et al. (2004) Radiobromination of anti-HER2/neu/ErbB-2 monoclonal antibody using the p-isothiocyanatobenzene derivative of the [76Br]undecahydro-bromo-7, 8-dicarba-nido-undecaborate(1-) ion. Nucl. Med. Biol., **31**(4), 425–433.
114 Winberg, K.J., et al. (2005) Radiobromination of closo-carboranes using palladium-catalyzed halogen exchange. J. Labelled Compd. Radiopharm., **48**(3), 195–202.
115 Brechbiel, M.W., et al. (1986) Synthesis of 1-(p-isothiocyanatobenzyl) derivatives of DTPA and EDTA. Antibody labeling and tumor-imaging studies. Inorg. Chem., **25**(16), 2772–2781.
116 Desphande, S.V., et al. (1988) Copper-67-labeled monoclonal antibody Lym-1, a potential radiopharmaceutical for cancer therapy: labeling and biodistribution in RAJI tumored mice. J. Nucl. Med., **29**(2), 217–225.

117 Moi, M.K., et al. (1985) Copper chelates as probes of biological systems: stable copper complexes with a macrocyclic bifunctional chelating agent. Anal. Biochem., **148**(1), 249–253.

118 Cox, J.P.L., et al. (1989) Synthesis of a kinetically stable yttrium-90 labelled macrocycle-antibody conjugate. J. Chem. Soc., Chem. Commun., (**12**), 797–798.

119 Craig, A.S., et al. (1989) Towards tumour imaging with indium-111 labelled macrocycle-antibody conjugates. J. Chem. Soc., Chem. Commun., (**12**), 794–796.

120 Craig, A.S., et al. (1989) Stability, Ga NMR, and crystal structure of a neutral gallium(III) complex of 1,4,7-triazacyclononanetriacetate: a potential radiopharmaceutical? J. Chem. Soc., Chem. Commun., (**23**), 1793–1794.

121 Iguchi, Y., et al. (2015) Tumor-specific delivery of BSH-3R for boron neutron capture therapy and positron emission tomography imaging in a mouse brain tumor model. Biomaterials, **56**, 10–17.

122 Paxton, R.J., et al. (1991) A transition metal complex (Venus flytrap cluster) for radioimmunodetection and radioimmunotherapy. Proc. Natl. Acad. Sci. USA, **88**(8), 3387–3391.

123 Mastren, T., et al. (2015) Cyclotron production of high specific activity ^{55}Co and in vivo evaluation of the stability of ^{55}Co metal-chelate-peptide complexes. Molecular Imaging, **14**(10), 526–532.

124 Sogbein, O.O., et al. (2004) Preparation of Re(I)- and ^{99m}Tc(I)-metallocarboranes in water under weakly basic reaction conditions. Inorg. Chem., **43**(10), 3032–3034.

125 Green, A.E.C., et al. (2006) Microwave-assisted synthesis of 3,1,2- and 2,1,8-Re(I) and ^{99m}Tc(I)-metallocarborane complexes. Inorg. Chem., **45**(15), 5727–5729.

126 Weller, A. (2011) Ligand design: the two faces of carboranes. Nature Chem., **3**(8), 577–578.

127 Ahrens, V.M., et al. (2011) Incorporation of *ortho*-carbaboranyl-Ne-modified L-lysine into neuropeptide Y receptor Y_1- and Y_2-selective analogues. J. Med. Chem., **54**(7), 2368–2377.

128 Sogbein, O.O., et al. (2005) Synthesis of ortho- and meta-Re(I)-metallocarboranes in water. Inorg. Chem., **44**(25), 9574–9584.

129 Louie, A.S., et al. (2011) Preparation, characterization, and screening of a high affinity organometallic probe for α-adrenergic receptors. J. Med. Chem., **54**(9), 3360–3367.

130 Gómez-Vallejo, V., et al. (2014) Synthesis and in vivo evaluation of ^{11}C-labeled (1,7-dicarba-closo-dodecaboran-1-yl)-N-{[(2S)-1-ethylpyrrolidin-2-yl]methyl} amide. J. Labelled Compd. Radiopharm., **57**(4), 209–214.

131 Schinazi, R.F., et al. (1994) Cellular pharmacology and biological activity of 5-carboranyl-2′-deoxyuridine. Int. J. Radiat. Oncol. Biol. Phys., **28**(5), 1113–1120.

132 Mizusawa, E., et al. (1982) Neutron-capture therapy of human cancer: in vitro results on the preparation of boron-labeled antibodies to carcinoembryonic antigen. Proc. Natl. Acad. Sci. USA, **79**(9 I), 3011–3014.

133 Ishiwata, K., et al. (1991) Synthesis and radiation dosimetry of 4-borono-2-[^{18}F] fluoro-D,L-phenylalanine: a target compound for PET and boron neutron capture therapy. Int. J. Rad. Appl. Instrum., **42**(4), 325–328.

134 Wang, L.W., et al. (2011) BNCT for locally recurrent head and neck cancer: preliminary clinical experience from a phase I/II trial at Tsing Hua Open-Pool Reactor. Appl. Radiat. Isot., **69**(12), 1803–1806.

135 Aihara, T., et al. (2006) First clinical case of boron neutron capture therapy for head and malignancies using ^{18}F-BPA PET. Head Neck, **28**(9), 850–855.

136 Ariyoshi, Y., *et al.* (2011) Fluorine-18-labeled boronophenylalanine positron emission tomography for oral cancers: qualitative and quantitative analyses of malignant tumors and normal structures in oral and maxillofacial regions. Oncology Letters, **2**(3), 423–427.

137 Kabalka, G.W., *et al.* (1997) Evaluation of fluorine-18-BPA-fructose for boron neutron capture treatment planning. J. Nucl. Med., **38**(11), 1762–1767.

138 Hanaoka, K., *et al.* (2014) FBPA PET in boron neutron capture therapy for cancer: prediction of B concentration in the tumor and normal tissue in a rat xenograft model. EJNMMI Research, **4**(1).

139 Beatty, B.G., *et al.* (1993) Pharmacokinetics of an anti-carcinoembryonic antigen monoclonal antibody conjugated to a bifunctional transition metal carborane complex (Venus flytrap cluster) in tumor-bearing mice. J. Nucl. Med., **34**(8), 1294–1302.

140 Pak, R.H., *et al.* (1995) Preparation and properties of nido-carborane-specific monoclonal antibodies for potential use in boron neutron capture therapy for cancer. Proc. Natl. Acad. Sci. USA, **92**(15), 6986–6990.

141 Primus, F.J., *et al.* (1996) Bispecific antibody mediated targeting of nido-carboranes to human colon carcinoma cells. Bioconjug. Chem., **7**(5), 532–535.

142 Zaulet, A. On the verge of bioorganic and inorganic chemistry: metallacarboranes in nanomedicine. Universitat Autònoma de Barcelona, 2015. http://www.tesisenred.net/handle/10803/323097

143 Gona, K.B. Exploring new labelling strategies for boronated compounds: towards fast development and efficient assessment of BNCT drug candidates. Universidad del pais Vasco (UPV-EHU), 2015. http://www.tesisenred.net/handle/10803/358847

144 El-Zaria, M.E., *et al.* (2012) Synthesis, characterisation, and biodistribution of radioiodinated C-hydroxy-carboranes. Chemistry, **18**(35), 11071–11078.

145 Genady, A.R., *et al.* (2015) Synthesis, characterization and radiolabeling of carborane-functionalized tetrazines for use in inverse electron demand Diels-Alder ligation reactions. J. Organomet. Chem., **791**, 204–213.

Part 3

Boron Compounds for Boron Neutron Capture Therapy

3.1

Twenty Years of Research on 3-Carboranyl Thymidine Analogs (3CTAs): A Critical Perspective

Werner Tjarks

Division of Medicinal Chemistry & Pharmacognosy, The Ohio State University, Columbus, Ohio, USA

3.1.1 Introduction

First efforts in the design, synthesis, and evaluation of boronated analogs of nucleic acids and their metabolic precursors for biomedical purposes can probably be traced back to 1959, when Maitlis *et al.* [1] synthesized a purine analog with a boron atom incorporated into the heterocyclic ring system for potential use in boron neutron capture therapy (BNCT). Since the early work of Maitlis *et al.* [1], numerous other boronated nucleobases, nucleosides, nucleotides, and oligonucleotides were synthesized and evaluated. In some cases, boron atoms were incorporated into the ring systems in fashions similar to the one described by Maitlis *et al.* [2,3]. In other cases, boron substituents, including *closo*-carboranes, *nido*-carboranes, metallocarboranes, dodecaborates, borane (RBH_3^-), cyanoborane, or the dihydroxyboryl group, were attached to the heterocyclic scaffolds [2–32]. In addition to BNCT, potential anti-inflammatory, anti-osteoporotic, general antineoplastic, antilipidemic, antiviral, purine receptor–modulating, antisense and siRNA-based gene-silencing, human neutrophil activity–modulating, electrochemical DNA-detecting, and human blood platelet function–modulating functions were discussed for these agents [2–32]. The focus of this chapter will be exclusively on 3-carboranyl thymidine analogs (3CTAs), a class of boronated nucleosides that was originally discovered by Anthony J. Lunato during his dissertation work under the guidance of Albert H. Soloway at The Ohio State University (OSU) Division of Medicinal Chemistry & Pharmacognosy in Columbus, Ohio, USA [33]. Subsequently, 3CTAs were further developed for BNCT of brain tumors by Werner Tjarks (OSU Division of Medicinal Chemistry & Pharmacognosy), Staffan Eriksson (Department of Anatomy, Physiology, and Biochemistry, Swedish University of Agricultural Sciences, Uppsala, Sweden), Rolf F. Barth (OSU Department of Pathology), Carol E. Cass (Department of Oncology, University of Alberta, Edmonton, Canada), and coworkers [4,5,34–50]. The review will begin with a short description of the basic BNCT concept and some characteristics of carborane clusters, as this will facilitate a better understanding of 3CTAs. This will be followed by a discussion of design concepts, synthetic strategies, structural aspects, enzymatic and metabolic properties,

Boron-Based Compounds: Potential and Emerging Applications in Medicine, First Edition.
Edited by Evamarie Hey-Hawkins and Clara Viñas Teixidor.
© 2018 John Wiley & Sons Ltd. Published 2018 by John Wiley & Sons Ltd.

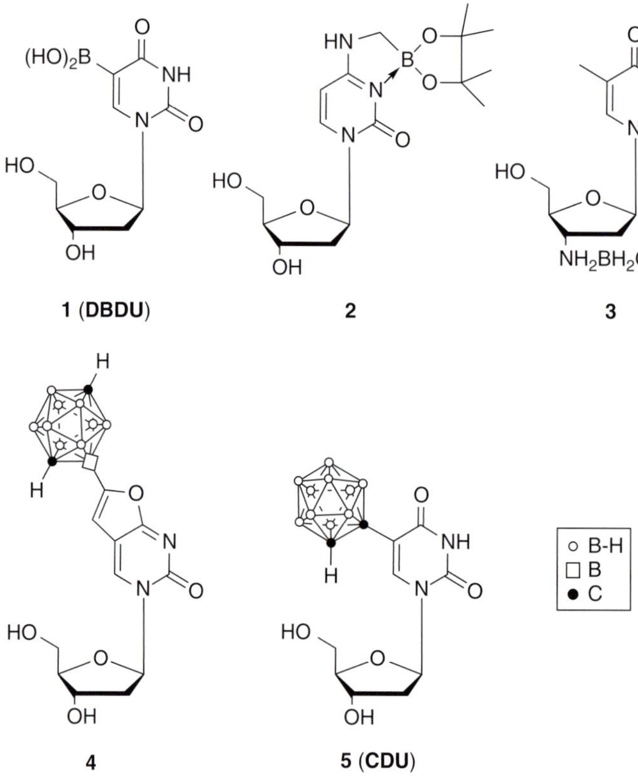

Figure 3.1.1 Representative structures of boronated nucleoside analogs that do not belong to the class of the 3CTAs.

cellular influx and efflux mechanisms, toxicological features, *in vitro* uptake, and *in vivo* biodistribution characteristics of 3CTAs. Finally, the results of preclinical BNCT studies with rodents and potential non-BNCT applications for 3CTAs will be addressed. Representative structures of boronated nucleoside analogs that do not belong to the class of the 3CTAs are shown in Figure 3.1.1. These compounds were extensively discussed in several recent reviews and original research papers [2–32], and will not be the subject of this review.

3.1.2 Boron Neutron Capture Therapy

BNCT is a method for the treatment of cancer that resembles radiation-sensitizing strategies in as far as it relies on both radiation and chemical components [51–54]. This modality is based on the selective accumulation of boron-10 (^{10}B) in tumor tissue, followed by irradiation with neutrons to produce high linear energy transfer (LET) lithium nuclei and helium nuclei (α-particles) [51]. The cytocidal effect of these particles is primarily caused by DNA double-strand breaks [51,55,56]. Approximately 10^9 ^{10}B atoms per tumor cells (~20 μg ^{10}B/g tumor cell) and tumor-to-normal tissue ratios of 3–4 to 1 are necessary for BNCT to be effective. Clinical BNCT trials were performed (for, e.g., brain, head and neck, liver, and skin cancers) using sodium mercaptoundecahydro-*closo*-dodecaborate (BSH) or boronophenylalanine (BPA) as boron delivery agents [51,54]. A pressing task of BNCT research remains the development of further improved tumor-selective boron delivery vehicles [51,57]. In addition to nucleobases, nucleosides, nucleotides, and oligonucleotides, preclinical boron delivery agents for BNCT include, for example, porphyrins and related structures [58–60], carbohydrates [61–63], amino acids [64], polyamines [65,66], monoclonal antibodies [67,68], and liposomes [69].

3.1.3 Carboranes

Carborane clusters, representative structures of which are shown in Figure 3.1.2, are widely used in the design and synthesis of pharmacologically active boron compounds [10,62,70–75]. The isomeric *closo*-carboranes (Figure 3.1.2a, 3.1.2b, and 3.1.2c1) are highly hydrophobic, whereas *nido*-carboranes (Figure 3.1.2c2) are negatively charged [76]. The latter can be obtained by reaction of fluoride anions or bases with the corresponding *closo*-counterparts [76]. Carboranes are in size comparable to a rotating phenyl group or adamantane [70], chemically highly versatile, and, except for the *closo*-to-*nido* conversion, very stable [76]. In addition, methods for the computer-aided design of carborane-containing agents have been developed [77]. Because of these features, carboranes have been frequently used in the design and synthesis of various bioactive compounds, including boron delivery vehicles for BNCT and pharmacologically active agents such as carbonic anhydrase inhibitors, HIV protease inhibitors, purinergic receptor antagonists, estrogen receptor modulators, nicotinamide phosphoryl transferase inhibitors, cyclooxygenase inhibitors, vitamin D receptor agonists, hypoxia-inducing factor (HIF) inhibitors, transthyretin dissociation inhibitors, antifolates, thrombin inhibitors, local anesthetics, retinoid receptor modulators, and androgen receptor modulators [10,62,71–75]. In addition, carboranes have been explored as prosthetic groups for radiohalogens in therapeutic and imaging applications [78].

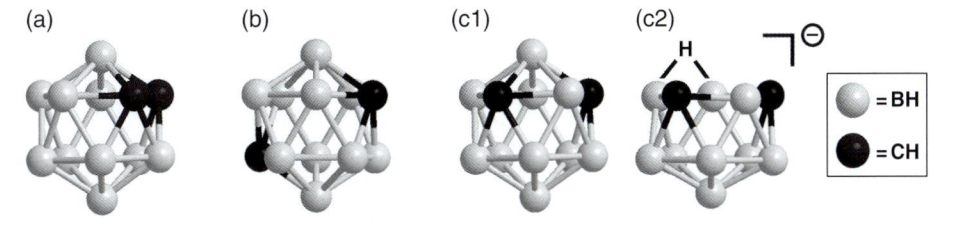

Figure 3.1.2 Structures of carboranes used in the synthesis of 3CTAs discussed in this review: (a) *closo*-o-carborane, (b) *closo*-p-carborane, (c1) *closo*-m-carborane, (c2) *nido*-m-carborane. Generated by Chem 3D, Perkin Elmer, Cambridge, MA, USA.

3.1.4 Rational Design of 3CTAs

Lunato et al. designed 3CTAs primarily based on the hypotheses that they are substrates of human thymidine kinase-1 (hTK1) and that binding to the enzyme leads to the formation of 3CTA-monophosphates [33]. Further conversation to the di- and triphosphates by other cellular kinases and subsequent polymerase-facilitated incorporation into DNA were considered as desirable at that time. However, the initial monophosphorylation presumably is sufficient to cause intracellular entrapment of 3CTAs because the acquired phosphate group is negatively charged under physiological conditions, which should prevent efflux by passive diffusion. In addition, nucleoside monophosphates are apparently not substrates of nucleoside transporters located in cell membranes [79]. This mechanism, which has been termed kinase-mediated trapping (KMT) [80], could be utilized for BNCT provided it is selective for tumor cells. Indeed, KMT can lead to entrapment of millimolar quantities of the phosphates of clinically used nucleoside analogs in tumor cells [81–86].

Since crystal structures of thymidine kinase-1 (TK1) were not available at that time, the original design of 3CTAs was based on two observations made in the purification of kinases and dehydrogenases by affinity chromatography [87–89]: (1) binding of the

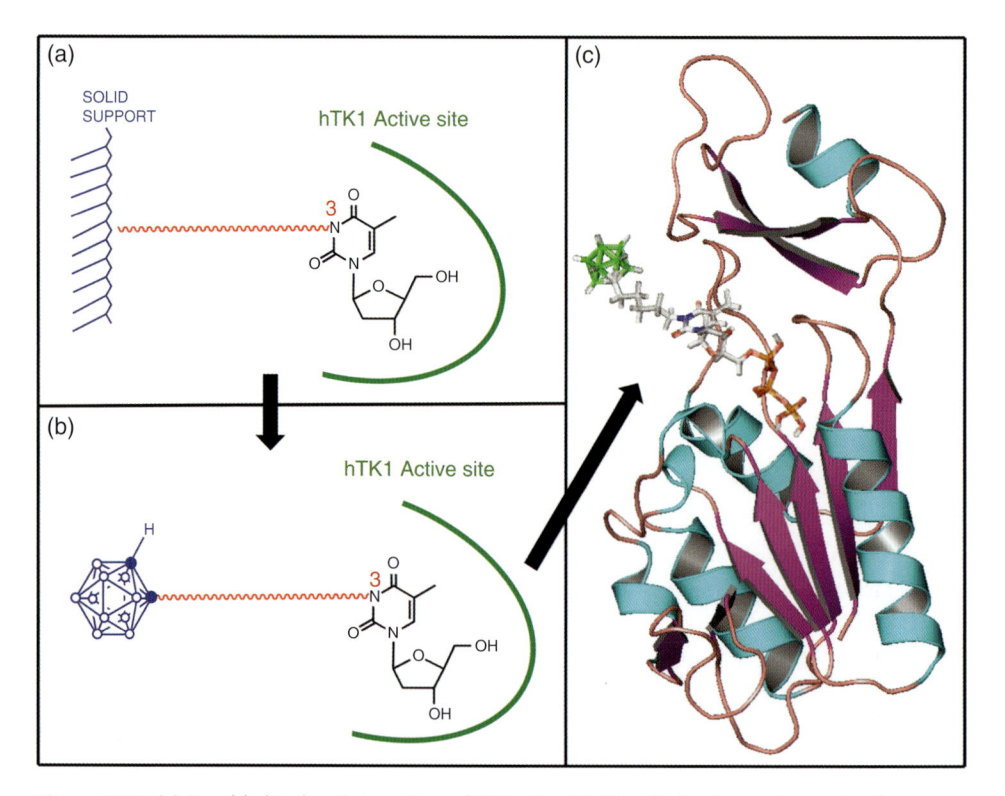

Figure 3.1.3 (a) Possible binding interaction of TK1 with dThd in affinity chromatography. (b) Affinity chromatography–based model developed for the design of 3CTAs. (c) Docking pose of **N5 (6)**-triphosphate into a homology model of TK1. *Source*: Adapted with permission from Ref. [47]. Copyright 2016 American Chemical Society. See also Figure 3.1.4 for the structure of **6 (N5)**, a first-generation 3CTA.

enzymes to the matrix improved when the nucleosides or nucleotides were tethered through a spacer to the solid support [87,88]; and (2) in the case of purifying TK1 from *Physarum polycephalum*, thymidine (dThd) was most likely bound covalently at the N3 position to the solid support (Figure 3.1.3a) [89]. It was hypothesized that in the case of 3CTAs, the bulky carborane cluster would take the role of the solid support and, propelled outside of the dThd binding pocket of TK1 by a spacer element, would not interfere with the binding of the dThd scaffold to the enzyme (Figure 3.1.3b).

Since the original design of 3CTAs, several crystal structures of so-called TK1-like enzymes, which include human TK1, vaccinia virus thymidine kinase (TK), and TKs from both gram-negative and gram-positive bacteria, have become available [90–98]. Human TK1 has significant sequence identities with vaccinia virus TK, gram-positive *Bacillus anthracis* (*Ba*) TK and *Staphylococcus aureus* (*Sa*) TK, and gram-negative *Thermotoga maritima* (*Tm*) TK of 70%, 42%, 41%, and 36%, respectively [99]. These kinases play essential roles in the biosynthetic salvage pathway of nucleotide synthesis [100], complementing the *de novo* pathway in particular in proliferating cells and bacteria [101–107]. In normal human cells, TK1 activity is high only in S-phase. In cancer cells, however, TK1 activity can remain high in other phases of the cell cycle as well [108–110]. This seems to contribute to the apparent selective accumulation of 3CTAs in tumor cells.

Figure 3.1.3c shows the docked structure of the triphosphate of **6** (**N5**), a first-generation 3CTA (Figure 3.1.4), in the substrate-binding pocket of a human TK1 homology model [47]. Tiwari *et al.* used the triphosphate of **6** in this docking study because it stabilized the docked pose of the dThd portion within the binding pocket [47]. The homology model was developed from an *apo* form of *Tm*TK [47,96]. The docking pose suggests that the carborane cluster is positioned toward the surface of TK1 and is exposed to the aqueous environment, thus supporting the original design concept for 3CTAs. However, the exact binding modes of 3CTAs to TK1 have not yet been confirmed by X-ray crystallography, and recent results of enzyme kinetic studies with second-generation 3CTAs, discussed in Section 3.1.6, challenge this binding model.

3.1.5 Synthesis and Initial Screening of 3CTAs as TK1 Substrates

3.1.5.1 First-Generation 3CTAs

Approximately 20 first-generation 3CTAs were synthesized and initially screened in phosphate transfer assays using recombinant TK1 (Figure 3.1.4) [37,48,49,111]. The obtained phosphorylation rates are expressed relative to dThd, the endogenous TK1 substrate. Such relative phosphorylation rates (rPRs) can be used routinely to establish raw structure–activity relationship (SAR) data for substrate characteristics of larger quantities of 3CTAs. However, rPRs do not provide accurate information on enzyme kinetic and enzyme inhibitor properties. Initially, the rPRs of **6** and **7** (*rac*-**N5-2OH**) were established together with those of the corresponding ethyl, propyl, butyl, hexyl, and heptyl homologs. There were no marked differences in rPRs between the compounds of this library. Overall, **6** and **7** appeared to be somewhat superior as TK1 substrates compared with their homologs, with rPRs of ~40% for both agents [111]. The presence of the highly lipophilic carborane cage [112] and the absence of ionizable functions

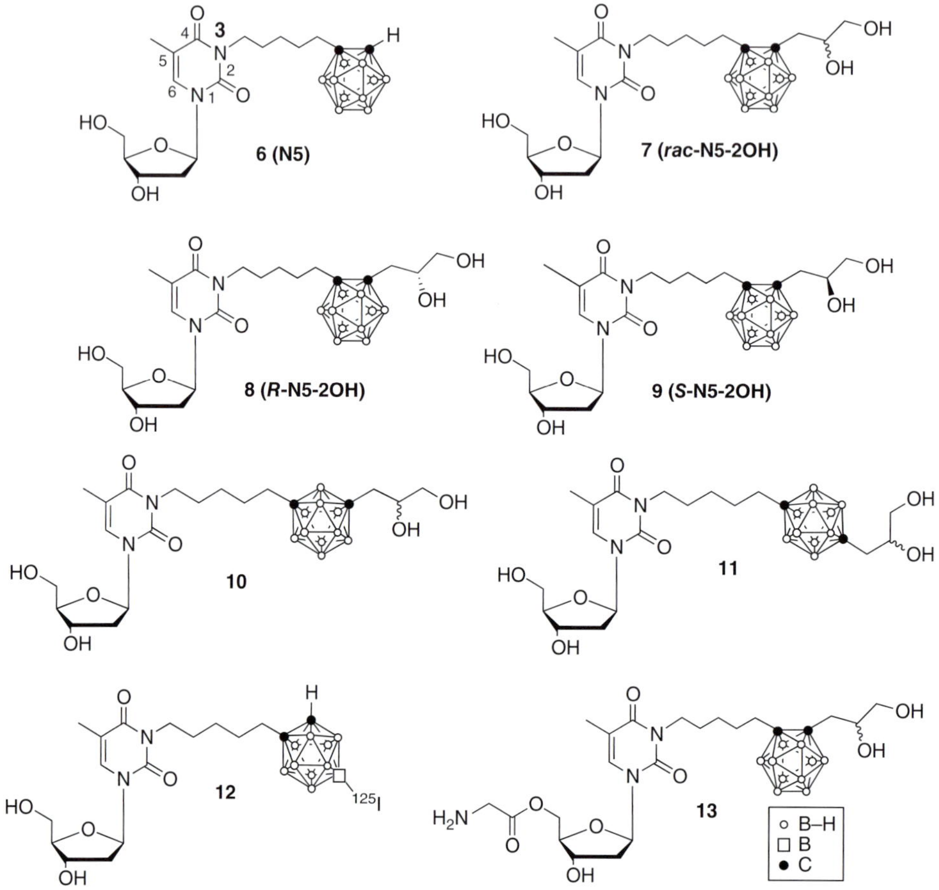

Figure 3.1.4 Representative structures of first-generation 3CTAs (**6–11**), (radio)halogenated 3CTAs (**12**), and amino acid ester prodrugs of 3CTAs (**13**). *rac*-**N5-2OH** (**7**) was prepared with normal boron distribution, ^{10}B-enriched, and in tritiated form.

render 3CTAs in general fairly water-insoluble [46]. Therefore, a dihydroxypropyl group was attached to the carborane cluster of **7** and its homologs in an attempt to increase water-solubility. Another consideration for the attachment of this group to the carborane cage was the presumable location of the cluster outside of the active site of TK1 in aqueous environment, as this potentially could reduce the disruption of the aqueous hydrogen bond network and consequently improve binding of **7** to TK1 [35]. Byun *et al.* synthesized the pure epimers (**8,9**) and structural isomers (**10,11**) of **7**, the latter containing either *m*-carborane (**10**) or *p*-carborane (**11**) [50]. The rPRs of **8–11** did not differ significantly from those of **7**, which was not unexpected assuming that, upon binding to TK1, the carborane cluster of the compounds are located outside of the active site of the enzyme [47]. A small library of four cage-halogenated and cage-radiohalogenated derivatives of **6**, represented by **12**, was synthesized by Tiwari *et al.* for potential use in tumor imaging and radiotherapy [47]. Compounds such as **12** potentially could also find use in the indirect real-time determination of tumor concentrations of 3CTAs in BNCT. As in the case of the epimers (**8,9**) and structural isomers (**10,11**) of **7**, the rPRs of cage-halogenated derivatives of **6** were comparable to those of **7**.

Another attempt to improve the water-solubility of 3CTAs was reported by Hasabelnaby et al. [46]. A library of 12 3′-monosubstituted-, 5′-monosubstituted-, and 3′,5′-disubstituted L-valine-, L-glutamate-, and glycine-ester prodrugs of **6** and/or **7** was synthesized. The structure of the 5′-glycine-ester prodrug of **7**, compound **13**, is shown in Figure 3.1.4. These prodrugs were specifically designed for BNCT of brain tumors following intracerebral administration. Thus, they must be readily cleavable in brain tissue. Overall, this prodrug strategy proved to be successful, as half-lives for the most promising candidates of the library in bovine cerebrospinal fluid, which has protein concentration similar to that of the interstitial fluid of the brain, ranged from 0.96 to 2.11 h and aqueous solubilty improved by up to 6600 times compared with the parental 3CTAs. These prodrug properties are potentially suitable for intracerebral administration.

3.1.5.2 Second-Generation 3CTAs

A significant limitation observed for both **6** and **7** was inadequate competition with endogenous dThd at the TK1–substrate binding site [41]. Similar problems were previously reported for 3′-deoxy-3′-[^{18}F]fluorothymidine ([^{18}F]FLT) in tumor-imaging applications and ganciclovir and zidovudine (AZT) in experimental suicide gene therapy [113–117]. This drawback as well as low water-solubility and moderate capacities of **6** and **7** to function as TK1 substrates were motivations for the design and synthesis of improved second-generation 3CTAs (Figures 3.1.5 and 3.1.6).

Hydrogen bond formation between a carbonyl oxygen of the peptide backbone in TK1-like enzymes and the hydrogen at the N3 position of dThd is crucial for substrate binding [90–98]. As discussed in Section 3.1.4, when attached to N3 of dThd via an alkyl linker, the bulky carborane cluster may be positioned toward the surface of the enzyme. However, such a carboranylalkyl substitution at N3 is also accompanied by the loss of hydrogen bonding, and it has been hypothesized that this could prevent a complete substrate-dependent *apo*-to-*holo* conformational switch of TK1 [5,35]. This could explain the insufficient TK1-inhibiting capacity of 3CTAs, since competing endogenous dThd may still have ample access to the substrate-binding site. Overall, about 50 second-generation 3CTAs were synthesized [35,36,39,40,48,49]. Many of the agents were designed based on the hypothesis that the introduction of hydrogen bond donors, and possibly even hydrogen bond acceptors, into the linker between carborane and dThd could restore hydrogen bond interactions, resulting in improved binding of 3CTAs to TK1. In addition, the introduction of hydrogen donor and acceptor groups could potentially also increase water solubility.

Representative second-generation 3CTAs with modified linker are shown in Figure 3.1.5. Hydrogen donor and acceptor groups include ether (**14**) [40], hydroxyl (**15**) [39], amidinyl (**16**) [36], guanidyl (**17**) [36], amino (**18**) [35], and amido (**19**) groups [35]. Unfortunately, none of the second-generation 3CTAs represented by compounds **14–19** had rPRs that were significantly superior to those of first-generation **6** and **7**, and many proved to be unstable during biological testing and also storage. The solubility of some of the amidinyl-modified 3CTAs (**16**) in phosphate-buffered saline (PBS) at pH 7.4 was up to 1900 times higher than that of **7**.

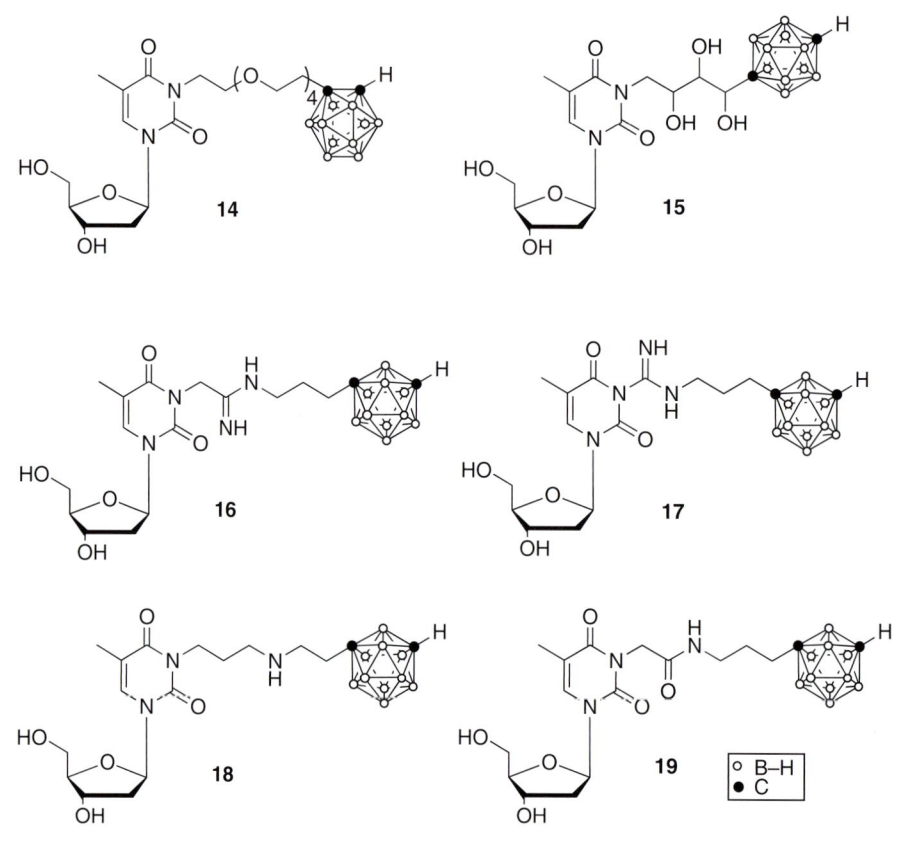

Figure 3.1.5 Representative second-generation 3CTAs with hydrogen donor–acceptor modifications in the spacer between the dThd scaffold and carborane cluster.

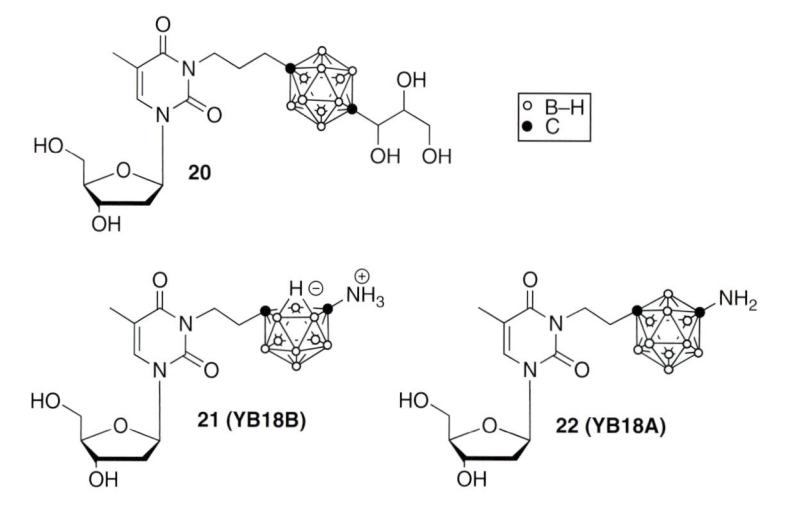

Figure 3.1.6 Representative second-generation 3CTAs with hydrophilic and ionic structural modifications at the carborane cluster.

Representative second-generation 3CTAs of another type are shown in Figure 3.1.6 [39,40,48,49]. In these structures, such as trihydroxypropyl (**20**) [39], ammonium (**21**) [49], or amino (**22**) [49], substituents were attached to the carborane cluster in an attempt to improve water solubility and TK1 substrate/inhibitor properties. The rPR values obtained for compound **22** (**YB18A**) in two independent experiments ranged from 72 to 79% [35,49]. This was about two times higher than the rPR values obtained for **7** in three independent experiments (~40%) [35,38,49].

3.1.6 Enzyme Kinetic and Inhibitory Studies

Studies were carried out to establish IC_{50} values and apparent k_{cat}/K_m values as a measure of the inhibition of TK1 by 3CTAs and the apparent catalytic efficiency of TK1 for 3CTAs, respectively. The k_{cat}/K_m values were expressed relative to those of dThd (rk_{cat}/K_m values). Among first- and second-generation 3CTAs represented by structures **1–20**, compound **7** emerged initially as the agent with the most favorable combination of IC_{50} values, rk_{cat}/K_m values, and physicochemical properties [35,36,38,39]. However, in a recent comparative experiment with **7** and **22**, IC_{50} values of 37.5 µM and 13.5 µM, respectively, and rk_{cat}/K_m values of 5.8 and 21.6%, respectively, were obtained, identifying **22** as a 3–4 times better inhibitor and substrate of TK1 than **7** [35]. This finding challenges the original hypothesis that the carborane cluster actually remains outside of the substrate-binding pocket of TK1 following binding of 3CTAs. Based on the enzymatic data obtained for **22**, it is conceivable that, depending on an appropriate spacer length between carborane cage and nucleoside scaffold in 3CTAs, the cluster is actually located within the substrate-binding pocket. Furthermore, a single amino substituent at the cluster, as in **22**, appears to interact with amino acid residues within the substrate-binding site, leading to markedly enhanced TK1 binding affinity and, thus, more effective competition with endogenous dThd.

3.1.7 Cell Culture Studies

Many of the 3CTA-related *in vitro* cytotoxicity, uptake, and retention studies were carried out with **6** and **7** and some of their cage-halogenated derivatives (**12**) in F98 glioma cells, RG2 glioma cells, wild-type TK1(+) L292 fibroblast cells, TK1(−) L929 cells, wild-type TK1(+) CCRF-CEM cells (human T lymphoblast-like cell line), and TK1(−) CCRF-CEM cells [41,42,45]. The TK1 levels in these cells were determined by both western blot analysis and/or quantifying phosphorylation of tritiated dThd, which was added to the cell culture medium. TK1 protein concentrations in these cell lines decreased in the following order: F98 >> RG2 > TK1(+) L929 >>> TK1(−) L929 [35,42,43]. The cytotoxicities of **6** and **7** were moderate in F98, RG2, TK1(+) L929, TK1(−) L929, TK1(+) CEM, and TK1(−) CEM cells with IC_{50} values ranging from 18 to >160 µM [41,42,45]. There was no apparent difference in toxicity between TK1(+) and TK1(−) cell lines. The uptake and/or retention of **6** and **7** and some of their cage-halogenated derivatives was consistently higher in L929 TK1(+) cells compared with their L929 TK1(−) counterparts [41,42,45,47]. On the other hand, there was no difference in uptake and retention between both cell lines for the clinical BNCT agent BPA, which is an amino acid derivative [42,47]. In the case of **7**, uptake of 92.0 μg B/10^9 cells following 24 h of incubation and 29% retention following subsequent incubation for 12 h in test-compound-free medium was observed in TK1(+) L929 cells [42]. The corresponding values for TK1(−) L929 cells were 9.3 μg uptake and 0% retention [42]. These data indicate that KMT may contribute to the uptake of **7** and other 3CTAs. Surprisingly, the uptake of **7** in F98 cells (25.7 μg) was 3.5 times lower than in TK1(+) L929 cells, although the TK1 content was about ~2 times higher [42,43]. This finding was supported by similar uptake ratios of 3.4 and 2.0 observed for **6** and the established TK1 substrate 5-bromo-2′-deoxyuridine (5-BrdUrd), respectively [42]. The retention of **7** in F98 cells was 0% [42]. These data indicated that other mechanisms, possibly including multidrug resistance-associated protein (MRP)-mediated efflux of 3CTA-monophosphates (Section 3.1.9), contribute to the cell-line-dependent differences of 3CTA uptake and retention.

3.1.8 Metabolic Studies

Compound **7** and other 3CTAs were not substrates of thymidine kinase-2 (TK2) [38], deoxycytidine kinase (dCK) [118], and catabolic dThd phosphorylase (TPase) [41]. In addition, the monophosphate of **7** (**7-MP**) was not a substrate of catabolic deoxynucleotidase-1 (dNT1) [41]. Detailed studies to explore the anabolic properties of 3CTAs were recently carried out with compounds **6** and **7** and their respective mono-, di-, and triphosphates [45]. For anabolic studies in cell culture, TK1(+) CEM, TK1(−) CEM, TK1(+) L929, and TK1(−) L929 cells were used [45]. The results are shown in Figure 3.1.7. The cellular anabolism of tritiated **7** was evaluated in comparison with that of tritiated dThd. In TK1(+) cells, high quantities of dThd monophosphate (dTMP) and moderate amounts of **7-MP** (monophosphate) were detected. In TK1(−) cells, monophosphates of both compounds were found in markedly lower concentrations. Small quantities of dThd diphosphate (dTDP), but not the diphosphate of **7** (**7-DP**), were found in TK1(+) cells. TK1(−) cells did not contain diphosphate forms of either compound. Interestingly, high quantities of **7** were retained in both TK1(+) and TK1(−) cells, whereas the measured concentrations of dThd were only moderate in both types of cells. The obtained results indicate that the presence of TK1 is necessary for the production of **7-MP** in cells but that further conversion to **7-DP** does not occur. However, high remaining concentrations of **7** in both TK1(+) and TK1(−) cells indicate that mechanisms other than KMT contribute significantly to the cellular uptake of 3CTAs such as **7**. These could include nonspecific binding to lipophilic cellular components, as **7** is a very lipophilic compound (log*P*: 2.09) [41].

Several phosphate transfer assays with nucleoside and nucleotide kinases were also carried out to explore possible anabolic pathways of **6** and **7** and to confirm the findings of the cell anabolism studies described in this chapter [45]. Neither uridine monophosphate–cytidine monophosphate kinase (UMP-CMPK) nor dThd monophosphate kinase (TMPK) catalyzed the formation of diphosphates when synthetic **6-MP** or **7-MP** were used as substrates, confirming the lack of **7-DP** formation observed in the anabolic cell culture experiments. Interestingly, nucleoside diphosphate kinase (NDPK) catalyzed the formation of **6-TP** (triphosphate) from **6-DP**, albeit with a markedly lower efficiency than for the formation of dThd triphosphate (dTTP) from dTDP. No incorporation of **6-TP** into oligonucleotide templates was observed when Klenow DNA polymerase I was used as the catalyzing enzyme. It was reported previously that attempts to incorporate

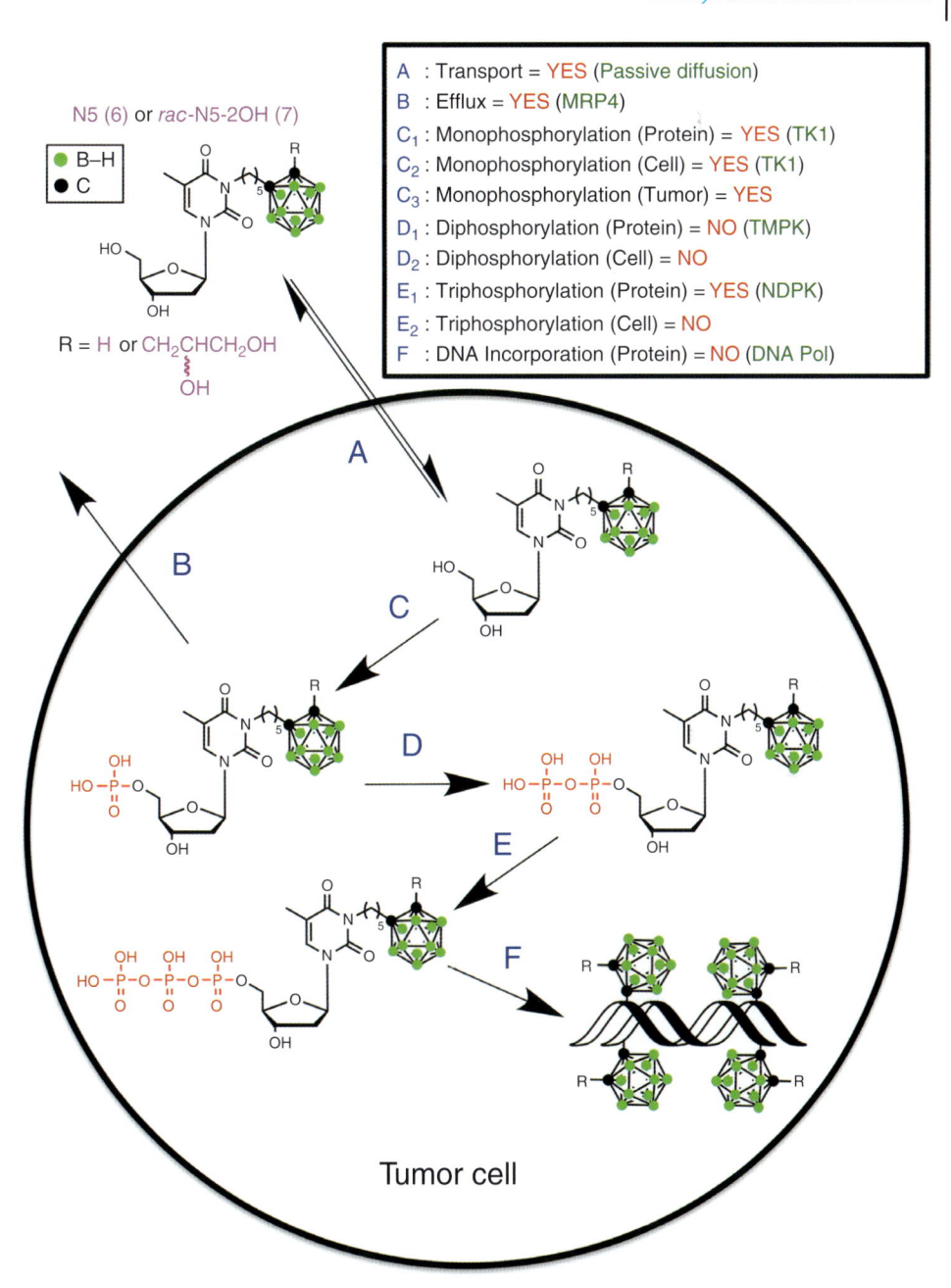

Figure 3.1.7 Potential tumor cell-related influx, efflux, and anabolic processes of 3CTAs. DNA Pol: DNA polymerase; NDPK: nucleoside diphosphate kinase; MRP4: multidrug resistance-associated protein-4; TK1: thymidine kinase-1; TMPK: thymidine monophosphate kinase.

3-methyl-dTTP with the help of DNA polymerase into DNA failed [119], which supports the results obtained for **6-TP**. The moderate *in vitro* toxicities observed for 3CTAs (Section 3.1.7) are probably due to the lack of DNA incorporation, as this is usually the major mechanism of action (MOA) of clinical chemotherapeutic anticancer and antiviral nucleoside analogs [79,120,121]. The results described here were corroborated by data from recent *in vitro* and *in vivo* metabolomic high-resolution mass spectrometry (HRMS) studies with **7** and **22** using RG2 glioma cells and rats implanted with intracerebral RG2 tumors [35]. Monophosphate forms of both **7** and **22** were detected both in cells and in tumor tissue along with the corresponding parental nucleoside analogs. Di- and triphosphates of **7** and **22** were not found. The obtained results suggest a unique intracellular metabolic profile of 3CTAs, which seems to be restricted to monophosphorylation by TK1.

3.1.9 Cellular Influx and Efflux Studies

Efficient transport through membranes into diseased cells is critical for all clinical chemotherapeutic nucleoside analogs [122]. Also, it is well established that MRPs, such as MRP4 and MRP5, can transport the monophosphates of nucleoside analogs, such as AZT, adefovir, ganciclovir, abacavir, and gemcitabine, out of cells [123–125]. Therefore, the cellular influx and efflux properties of **7** and **7-MP** were explored (Figure 3.1.7) [45]. Initially, *Saccharomyces cerevisiae* ("baker's yeast"), genetically altered to express the human equilibrative and concentrative nucleoside transporter hENT1, hENT2, hCNT1, hCNT2, or hCNT3, was used to study the effects of **7** on cellular nucleoside transporters. The result indicated that **7** only inhibited the transport of uridine (Urd) by hENT1. Compound **7** also inhibited the uptake of Urd into TK1(+) CCRF-CEM cells, which only express hENT1 [126]. Interestingly, however, the uptake of small quantities of tritiated **7** into TK1(+) CCRF-CEM cells was not significantly inhibited by either substantially larger quantities of non-tritiated ("cold") **7** or nitrobenzylthioinosine (NBMPR), a potent hENT1 inhibitor. These results indicated that **7** is an inhibitor of hENT1 but that its cellular uptake is mediated primarily by passive diffusion.

The effects of dipyridamole, indomethacin, and verapamil on the efflux of **7-MP** from TK1(+) CCRF-CEM cells were studied. Dipyridamole is an inhibitor of both MRP4 and MRP5, whereas indomethacin is a relatively selective inhibitor of MRP4 [123]. Verapamil is an inhibitor of P-glycoprotein (P-gp) [127]. Studies by Peng *et al.* have shown that MRP4 is a major efflux pump in (TK1+) CCRF-CEM cells [128]. The obtained results showed that both dipyridamole and indomethacin strongly inhibited the efflux of **7-MP** from TK1(+) CCRF-CEM cells, whereas verapamil affected the efflux only moderately. Thus, efflux inhibition of **7-MP** by the selective MRP4 inhibitor indomethacin implicates a role for this transporter in the cellular efflux of **7-MP**.

3.1.10 *In vivo* Uptake and Preclinical BNCT Studies

In vivo studies were carried out with **7**, **22**, and the propyl homolog of **22** using the intracerebral (i.c.) rat F98 and RG2 glioma models and both the subcutaneous (s.c.) and the i.c. murine TK1(+) and TK1(–) L929 tumor models. The RG2 and the F98 glioma have very similar histopathological appearances, but there are apparently similarities and differences in their known genomic and proteomic profiles [129]. Uptake and toxicity experiments were carried out with agents having normal boron distribution, whereas ^{10}B-enriched forms were used for BNCT studies [42,50]. Initial experiments showed that intraperitoneal (i.p.) injection of **7** into mice bearing TK1(+) L929 tumors resulted in undetectable quantities of boron within the tumor [42]. Therefore, direct intratumoral (i.t) injection and i.c. infusion were used for experiments with mice bearing s.c. L929 tumors and rodents implanted with i.c. L929, RG2, or F98 tumors, respectively.

Administration (i.t.) of **7** to mice bearing s.c. TK1(+) L929 and TK1(−) L929 tumors resulted in 22.8 and 8.4 µg B/g tumor tissue, respectively (~3:1 ratio), with a tumor-to-blood ratio of 47 for animals with the TK1(+) tumors [44]. A similar uptake ratio of ~3:1 was observed between i.c. TK1(+) L929 and TK1(−) L929 tumors (39.8 vs. 12.4 µg B/g tumor) following i.c. administration of **7** [42]. In another study using the s.c. murine TK1(+)–TK1(−) L929 tumor model, i.t. injection of the propyl homolog of **22** led to 21.7 and 7.0 µg B/g tumor, respectively (~3:1 ratio) [49]. These data were comparable with those obtained for **7** in the same study (29.8 vs. 12.1 µg B/g tumor), whereas no marked difference in boron uptake was observed between TK1(+) and TK1(−) tumors when BPA was injected intravenously (i.v.) (4.0 vs. 5.8 µg B/g tumor) [49]. Neutron irradiation of s.c. TK1(+) L929 tumor–bearing mice that had received **7** i.t. resulted in a ninefold decrease in tumor growth compared with TK1(+) L929 tumor–bearing mice that were irradiated with neutrons but did not receive **7**. Animals with TK1(−) L929 tumor that had received both neutron radiation and **7** showed tumor growth comparable to mice with TK1(+) L929 tumor that were only irradiated with neutrons without prior administration of **7** [44]. There was a marked decrease in the number of proliferating cells in TK1(+) L929 tumors following administration of **7** and neutron irradiation, as was shown by Ki67 antigen immunostaining [44].

Following i.c. administration of **7** into rats with i.c. RG2 and F98 glioma, 27.6 and 17.3 µg B/g tumor, respectively, were found. The corresponding tumor-to-blood ratios were 55 and 35, respectively. Although **7** was administered directly into the brain in solutions containing high amounts of dimethyl sulfoxide (DMSO), no apparent neurologic deficits and/or neuropathologic alterations could be observed [44]. Neutron irradiation of rats with RG2 tumors resulted in a survival time of 52.9 days, compared with 28.1 days for neutron-irradiated control animals that had not received **7** [44]. The corresponding survival times for rats with F98 tumors were 37.9 and 31.3 days, respectively [43]. BPA was used as a control agent in neutron irradiation experiments with both RG2 and F98 glioma-bearing rats, resulting in survival days of 35.9 and 36.7 days, respectively [43,44]. These experiments indicated that the RG2 gliomas accumulate higher quantities of compound **7** and are consequently more susceptible to neutron irradiation than F98 gliomas. This is surprising because the TK1 content in F98 glioma cells is about two times higher than in RG2 cells [35,42,43]. However, the results obtained for the *in vivo* uptake and preclinical BNCT studies seem to be consistent with the data obtained for F98 glioma cells *in vitro* (Section 3.1.7). This indicates that the uptake of 3CTAs into tumor cells and tumors does not solely depend on TK1 activity but also on other cell-specific metabolic and transport protein properties. In the case of i.c. administration, it is also conceivable that the observed tumor-selective uptake of **7** is in part due to simple precipitation at the injection site, which is located at the same coordinates as the tumor. The same could be the case for i.t. injections in s.c. tumors.

In a recent comparative *in vivo* biodistribution study with rats bearing i.c. RG2 gliomas, compounds **7** and **22** were administered i.c., resulting in concentrations of 8.9 µg and 32.5 µg B/g tumor, respectively [35]. This uptake ratio reflects approximately the difference in enzyme kinetic and inhibitory properties between both compounds, and it suggests that these factors contribute to the uptake of both compounds in tumor cells. However, it needs to be noted that the uptake observed for **7** in i.c. RG2 gliomas in this study was approximately three times lower than that previously reported for **7** in the same tumor model (27.6 µg B/g tumor) [44], indicating the potential for significant experimental variations in these technically challenging experiments. Preclinical BNCT studies have not yet been carried out with **22**.

3.1.11 Potential Non-BNCT Applications for 3CTAs

TK1 is an established cell proliferation marker, and the level of TK1 in serum has been used as a diagnostic and prognostic marker in cancer in both human and veterinary medicine [130–136]. Stavudine (d4T) and AZT are used as HIV/AIDS prodrugs [137], whereas FLT and 1-(2′-deoxy-2′-fluoro-1-β-D-arabinofuranosyl)-5-iodouracil (FIAU) are being explored as imaging agents for cancer and bacterial infections, respectively [138–141]. All of the agents are effectively 5′-monophosphorylated by TKs. Their MOAs are based on the formation of toxic triphosphate metabolites (AZT and d4T) or intracellular trapping of phosphate species (FLT and FIAU). Floxuridine (5-FdUrd) activation via TK1 occurs in a minor metabolic pathway, whereas its major MOA is thymidylate synthase (TS) inhibition [142,143]. All of these agents are derivatives of dThd and function as TK1 substrates rather than TK1 inhibitors. Several reports indicate that TK1-mediated dThd salvage has a major role in DNA repair *in vitro* following exposure of cells to radiation, DNA alkylating agents, or combined ribonucleotide reductase inhibition and radiation [144–151]. Recent studies have demonstrated that DNA damage caused by γ-radiation and various chemotherapeutics, including doxorubicin, etoposide, and hydroxyurea, resulted in increased TK1 expression, especially in p53-deficient or -defective HCT-116, SW480, and H1299 cells and, to a lesser extent, in p53-proficient HCT-116 and U2OS cells. TK1 expression proved to be nonessential for the growth of p53-deficient HCT-116 cells but was important in providing sufficient dTTP pools for DNA repair triggered by genotoxic insult of these cells. Consequently, small interfering RNA (siRNA)-facilitated TK1 knockout in p53-deficient HCT-116 caused an increase in cell death following DNA damage [152,153]. In addition, simultaneous siRNA-mediated silencing of TS and TK in p53-deficient HCT-116 cells caused significant increase of cell death by doxorubicin. Similar observations were made when TS-knockout p53-deficient HCT-116 cells were cultured in dThd-deprived medium [154]. Furthermore, TS inhibition in various cell lines caused increased TK1 activity and siRNA-mediated silencing of TK1, and it enhanced the activity of the antifolate pemetrexed [155–157]. These are very important findings, since it is known that in 50% of all human tumors, p53 is mutated or deleted, and that such alterations can lead to poor prognosis and resistance to chemotherapeutic agents [158–163].

Considering the importance of TK1 for DNA repair and cancer cell proliferation, in particular in p53-deficient tumors, it is surprising that TK1 has been utilized scarcely as a molecular target for selective inhibitors. Indeed, no pharmaceutically relevant selective inhibitors have been developed as yet that could complement inhibitors of the de novo nucleotide pathway (e.g., antifolates, 5-fluorouracil [5-FU], and 5-FdUrd) in anticancer chemotherapy. It should be noted that selective TK1 inhibitors per se are presumably nontoxic and should not contribute to undesired side effects in combination with other chemotherapeutic agents while concomitantly sensitizing, for example, p53-deficient tumor cells to the cytotoxic effects of the latter. Thus, there seems to be a need for the development of selective TK1 inhibitors.

The enzyme kinetic and inhibitory results obtained with **22** strongly indicate that the substrate-binding site of TK1 around the N3 position of dThd is fairly flexible. Depending on an appropriate spacer length between the carborane cage and nucleoside scaffold in 3CTAs, the cluster is actually located within the substrate-binding pocket. Furthermore, a single amino substituent at the cluster appeared to interact with amino acid residues within the substrate-binding site, leading to markedly enhanced TK1 binding affinity and, thus, more effective competition with dThd. This warrants the hypothesis that the identification of the detailed molecular interactions between 3CTAs and TK1 and the use of the carborane scaffold as a 3D platform for the introduction of additional or different substituents than the amino substituent in **22** can lead to third-generation 3CTAs with further enhanced TK1 inhibitor capacities that could eventually find use in the treatment of, for example, p53-deficient or -defective cancers. Indeed, such third-generation 3CTAs may even surpass the binding affinity of dThd to TK1.

Carborane clusters can be substituted via two basic routes: (1) reaction of C-metallated (usually Li^+) carboranes with appropriate reagents to introduce, for example, alkyl, aryl, amine, amide, nitrile, hydroxyl, sulfhydryl, carboxyl, and boronic acid substituents [76,164]; or (2) metal-catalyzed (usually palladium) cross-coupling reactions of B-halo-carboranes (often iodine) resulting in substitution patterns similar to the ones described in (1) [76]. Among the three isomeric *closo*-carboranes, B-iodination has been explored most widely for *o*-carborane. Various levels and combinations of iodination, including boron atoms 3, 4, 6–10, and 12, have been explored [47,78,165–173]. In the cases of *p*- and *m*-carborane, 2-monohalogenation and 9-mono/9,10-dihalogenation, respectively, have been reported widely [78].

In summary, there is ample opportunity to further improve the TK1 binding properties of 3CTAs, which may eventually lead to agents that can be used in BNCT but also as general TK1 inhibitors in combination with γ-radiation and chemotherapeutics, in particular inhibitors of the de novo nucleotide pathway for DNA synthesis.

3.1.12   Conclusion

3CTAs are promising boron delivery agents for BNCT, as KMT provides a solid theoretical, biochemical, and physiological basis for their tumor cell selective accumulation. Overall, about 70 3CTAs were synthesized over a period of 20 years. This is a fairly small number even for an academic environment. While it is, of course, possible that this is due to an incapability of the chemists involved in the project, it seems more likely that this indicates a lack of adaptability of carborane cluster–based chemistry to compound library production as compared with conventional carbon-based chemistry. Nevertheless, this compound library was developed following general drug design, discovery, and development strategies, and a wealth of information was obtained on *in vitro* and *in vivo* toxicities, tumor cell and tumor tissue accumulation, metabolic properties, cell membrane transport, and the potential to function as boron delivery agents for BNCT. Major drawbacks of current 3CTAs are that they do not seem to be suitable for systemic administration and that not all aspects of their tumor selectivity are fully understood. However, a second-generation 3CTA (**22**) was discovered that has markedly superior enzyme kinetic and inhibitory properties compared with all other 3CTAs, and a thorough biochemical and biological evaluation of this compound could shed more light on the potential of 3CTAs to serve as boron delivery agents for BNCT. Furthermore, the specific structural properties of **22** suggested that third-generation 3CTAs with even further improved pharmacological activity could be developed and that such compounds might even find applications beyond that of BNCT.

The involved investigators have terminated research on 3CTAs, and it is uncertain if there will be corporate, federal, or any other type of support in the USA that could move 3CTAs into a clinical setting. The same might be true for other investigational BNCT agents. There are meanwhile a few US Food and Drug Administration (FDA)-approved boronic acid derivatives (bortezomib, ixazomib citrate, and tavaborole) that draw activity from the fact that the boronic acid moiety in these structures forms strong tetravalent complexes with critical hydroxyl groups within enzyme active sites. This is a rare but effective trade among approved pharmacologically active agents [174–177]. The situation is different for carborane clusters, which are presumably not capable of forming reversible coordinative bonds with biomolecules, as in the case of boronic acids. Carboranes per se probably have to rely on their unique geometry, physicochemical properties, and synthetic versatility to become useful entities in drug design. It remains to be determined if these properties are indeed so unique that they can provide opportunities that conventional carbon-based drug design cannot.

Acknowledgments

The preparation of this book chapter was financially supported by The Ohio State University College of Pharmacy. The author thanks Keisuke Ishita for helping to prepare this document. Many thanks to Tony for his great idea.

References

1 Chissick SS, Dewar MJS, Maitlis PM. A boron-containing purine analog. J Am Chem Soc. 1959; 81:6329–6330.
2 Niziol J, Rode W, Ruman T. Boron nucleic acid bases, nucleosides and nucleotides. Mini Rev Org Chem. 2012; 9(4):418–425.
3 Soloway AH, Tjarks W, Barnum BA, Rong F-G, Barth RF, Codogni IM, et al. The chemistry of neutron capture therapy. Chem Rev. 1998; 98(4):1515–1562.
4 Tjarks W, Tiwari R, Byun Y, Narayanasamy S, Barth RF. Carboranyl thymidine analogues for neutron capture therapy. Chem Commun. 2007:4978–4991.
5 Khalil A, Ishita K, Ali T, Tjarks W. N3-substituted thymidine bioconjugates for cancer therapy and imaging. Future Med Chem. 2013; 5(6):677–692.
6 Martin AR, Vasseur J-J, Smietana M. Boron and nucleic acid chemistries: merging the best of both worlds. Chem Soc Rev. 2013; 42(13):5684–5713.
7 Tjarks W, Wang J, Chandra S, Ji W, Zhuo JC, Lunato AJ, et al. Boronated nucleosides for BNCT. KURRI-KR-54. 2000:157–158.

8 Lesnikowski ZJ. Nucleoside-boron cluster conjugates – beyond pyrimidine nucleosides and carboranes. J Organomet Chem. 2009; 694(11):1771–1775.
9 Goudgaon NM, El-Kattan GF, Schinazi RF. Boron containing pyrimidines, nucleosides, and oligonucleotides for neutron capture therapy. Nucleosides Nucleotides. 1994; 13(1–3):849–880.
10 Lesnikowski ZJ. Challenges and opportunities for the application of boron clusters in drug design. J Med Chem. 2016; 59(17):7738–7758.
11 Niziol J, Uram L, Szuster M, Sekula J, Ruman T. Biological activity of N(4)-boronated derivatives of 2′-deoxycytidine, potential agents for boron-neutron capture therapy. Bioorg Med Chem. 2015; 23(19):6297–6304.
12 Niziol J, Zielinski Z, Les A, Dabrowska M, Rode W, Ruman T. Synthesis, reactivity and biological activity of N(4)-boronated derivatives of 2′-deoxycytidine. Bioorg Med Chem. 2013; 22(15):3906–3912.
13 Semioshkin A, Laskova J, Ilinova A, Bregadze V, Lesnikowski ZJ. Reactions of oxonium derivatives of $[B_{12}H_{12}]^{2-}$ with sulphur nucleophiles. synthesis of novel B_{12}-based mercaptanes, sulfides and nucleosides. J Organomet Chem. 2011; 696(2):539–543.
14 Semioshkin A, Bregadze V, Godovikov I, Ilinova A, Lesnikowski ZJ, Lobanova I. A convenient approach towards boron cluster modifications with adenosine and 2′-deoxyadenosine. J Organomet Chem. 2011; 696(23):3750–3755.
15 Laskova J, Kozlova A, Bialek-Pietras M, Studzinska M, Paradowska E, Bregadze V, et al. Reactions of closo-dodecaborate amines. Towards novel bis-(closo-dodecaborates) and closo-dodecaborate conjugates with lipids and non-natural nucleosides. J Organomet Chem. 2016; 807:29–35.
16 Semioshkin A, Ilinova A, Lobanova I, Bregadze V, Paradowska E, Studzinska M, et al. Synthesis of the first conjugates of 5-ethynyl-2′-deoxyuridine with closo-dodecaborate and cobalt-bis-dicarbollide boron clusters. Tetrahedron. 2013; 69(37):8034–8041.
17 Ilinova A, Semioshkin A, Lobanova I, Bregadze VI, Mironov AF, Paradowska E, et al. Synthesis, cytotoxicity and antiviral activity studies of the conjugates of cobalt bis(1,2-dicarbollide)(-I) with 5-ethynyl-2′-deoxyuridine and its cyclic derivatives. Tetrahedron. 2014; 70(35):5704–5710.
18 Vyakaranam K, Hosmane NS. Novel carboranyl derivatives of nucleoside mono- and diphosphites and phosphonates: a synthetic investigation. Bioinorg Chem Appl. 2004; 2(1–2):31–42.
19 Adamska A, Rumijowska-Galewicz A, Ruszczynska A, Studzińska M, Jabłońska A, Paradowska E, et al. Anti-mycobacterial activity of thymine derivatives bearing boron clusters. Eur J Med Chem. 2016; 121:71–81.
20 Olejniczak AB, Adamska AM, Paradowska E, Studzinska M, Suski P, Lesnikowski ZJ. Modification of selected anti-HCMV drugs with lipophilic boron cluster modulator. Acta Pol Pharm. 2013; 70(3):489–504.
21 Bednarska K, Olejniczak AB, Wojtczak BA, Sulowska Z, Lesnikowski ZJ. Adenosine and 2′-deoxyadenosine modified with boron cluster pharmacophores as new classes of human blood platelet function modulators. ChemMedChem. 2010; 5(5):749–756.
22 Jelen F, Olejniczak AB, Kourilova A, Lesnikowski ZJ, Palecek E. Electrochemical DNA detection based on the polyhedral boron cluster label. Anal Chem. 2009; 81(2):840–844.

23. Bednarska K, Olejniczak AB, Piskala A, Klink M, Sulowska Z, Lesnikowski ZJ. Effect of adenosine modified with a boron cluster pharmacophore on reactive oxygen species production by human neutrophils. Bioorg Med Chem. 2012; 20(22):6621–6629.
24. Kwiatkowska A, Sobczak M, Mikolajczyk B, Janczak S, Olejniczak AB, Sochacki M, et al. siRNAs modified with boron cluster and their physicochemical and biological characterization. Bioconjug Chem. 2013; 24(6):1017–1026.
25. Olejniczak AB, Kierzek R, Wickstrom E, Lesnikowski ZJ. Synthesis, physicochemical and biochemical studies of anti-IRS-1 oligonucleotides containing carborane and/or metallacarborane modification. J Organomet Chem. 2013; 747:201–210.
26. Grabowska I, Stachyra A, Gora-Sochacka A, Sirko A, Olejniczak AB, Lesnikowski ZJ, et al. DNA probe modified with 3-iron bis(dicarbollide) for electrochemical determination of DNA sequence of avian influenza virus H5N1. Biosens Bioelectron. 2014; 51:170–176.
27. Bialek-Pietras M, Olejniczak AB, Paradowska E, Studzinska M, Suski P, Jablonska A, et al. Synthesis and in vitro antiviral activity of lipophilic pyrimidine nucleoside/carborane conjugates. J Organomet Chem. 2015; 798(1):99–105.
28. Wojtczak BA, Olejniczak AB, Wang L, Eriksson S, Lesnikowski ZJ. Phosphorylation of nucleoside-metallacarborane and carborane conjugates by nucleoside kinases. Nucleosides Nucleotides Nucleic Acids. 2013; 32(10):571–588.
29. Bednarska K, Olejniczak AB, Klink M, Sulowska Z, Lesnikowski ZJ. Modulation of human neutrophil activity by adenosine modified with a carborane pharmacophore. Bioorg Med Chem Lett. 2014; 24(14):3073–3078.
30. Matuszewski M, Kiliszek A, Rypniewski W, Lesnikowski ZJ, Olejniczak AB. Nucleoside bearing boron clusters and their phosphoramidites. building blocks for modified oligonucleotide synthesis. New J Chem. 2015; 39(2):1202–1221.
31. Janczak S, Olejniczak A, Balabanska S, Chmielewski MK, Lupu M, Vinas C, et al. Boron clusters as a platform for new materials: synthesis of functionalized o-carborane ($C_2B_{10}H_{12}$) derivatives incorporating DNA fragments. Chem Eur J. 2015; 21(43):15118–15122.
32. Wojtczak BA, Andrysiak A, Gruener B, Lesnikowski ZJ. "Chemical ligation": a versatile method for nucleoside modification with boron clusters. Chem Eur J. 2008; 14(34):10675–10682.
33. Lunato AJ. Part 1. Synthesis and in vitro phosphorylation of a homologous series of 5-s-alkylcarboranyl-2′-deoxyuridines and 3-N-alkylcarboranylthymidines for use in boron neutron capture therapy of brain cancer. Part 2. An investigation into the stereoselective formation of 3,5-di-o-(p-toluoyl)-2-deoxy-alpha-d-ribofuranosyl chloride (chlorination). PhD thesis. The Ohio State University; 1996.
34. Byun Y, Narayanasamy S, Johnsamuel J, Bandyopadhyaya AK, Tiwari R, Al-Madhoun AS, et al. 3-Carboranyl thymidine analogues (3CTAs) and other boronated nucleosides for boron neutron capture therapy. Anti-Cancer Agents Med Chem. 2006; 6(2):127–144.
35. Agarwal HK, Khalil A, Ishita K, Yang W, Nakkula RJ, Wu L-C, et al. Synthesis and evaluation of thymidine kinase 1-targeting carboranyl pyrimidine nucleoside analogs for boron neutron capture therapy of cancer. Eur J Med Chem. 2015; 100:197–209.
36. Agarwal HK, McElroy CA, Sjuvarsson E, Eriksson S, Darby MV, Tjarks W. Synthesis of N3-substituted carboranyl thymidine bioconjugates and their evaluation as substrates of recombinant human thymidine kinase 1. Eur J Med Chem. 2013; 60:456–468.

37 Lunato AJ, Wang J, Woollard JE, Anisuzzaman AKM, Ji W, Rong F-G, et al. Synthesis of 5-(carboranylalkylmercapto)-2′-deoxyuridines and 3-(carboranylalkyl)thymidines and their evaluation as substrates for human thymidine kinases 1 and 2. J Med Chem. 1999; 42(17):3378–3389.

38 Al-Madhoun AS, Johnsamuel J, Yan J, Ji W, Wang J, Zhuo J-C, et al. Synthesis of a small library of 3-(carboranylalkyl)thymidines and their biological evaluation as substrates for human thymidine kinases 1 and 2. J Med Chem. 2002; 45(18):4018–4028.

39 Narayanasamy S, Thirumamagal BTS, Johnsamuel J, Byun Y, Al-Madhoun AS, Usova E, et al. Hydrophilically enhanced 3-carboranyl thymidine analogues (3CTAs) for boron neutron capture therapy (BNCT) of cancer. Bioorg Med Chem. 2006;14(20):6886–6899.

40 Johnsamuel J, Lakhi N, Al-Madhoun AS, Byun Y, Yan J, Eriksson S, et al. Synthesis of ethyleneoxide modified 3-carboranyl thymidine analogues and evaluation of their biochemical, physicochemical, and structural properties. Bioorg Med Chem. 2004; 12(18):4769–4781.

41 Al-Madhoun AS, Johnsamuel J, Barth RF, Tjarks W, Eriksson S. Evaluation of human thymidine kinase 1 substrates as new candidates for boron neutron capture therapy. Cancer Res. 2004; 64(17):6280–6286.

42 Barth RF, Yang W, Al-Madhoun AS, Johnsamuel J, Byun Y, Chandra S, et al. Boron-containing nucleosides as potential delivery agents for neutron capture therapy of brain tumors. Cancer Res. 2004; 64(17):6287–6295.

43 Barth RF, Yang W, Nakkula RJ, Byun Y, Tjarks W, Wu LC, et al. Evaluation of TK1 targeting carboranyl thymidine analogs as potential delivery agents for neutron capture therapy of brain tumors. Appl Radiat Isot. 2015; 106:251–255.

44 Barth RF, Yang W, Wu G, Swindall M, Byun Y, Narayanasamy S, et al. Thymidine kinase 1 as a molecular target for boron neutron capture therapy of brain tumors. Proc Natl Acad Sci USA. 2008; 105(45):17493–17497.

45 Sjuvarsson E, Damaraju VL, Mowles D, Sawyer MB, Tiwari R, Agarwal HK, et al. Cellular influx, efflux, and anabolism of 3-carboranyl thymidine analogs: potential boron delivery agents for neutron capture therapy. J Pharmacol Exp Ther. 2013; 347(2):388–397.

46 Hasabelnaby S, Goudah A, Agarwal HK, abd Alla MS, Tjarks W. Synthesis, chemical and enzymatic hydrolysis, and aqueous solubility of amino acid ester prodrugs of 3-carboranyl thymidine analogs for boron neutron capture therapy of brain tumors. Eur J Med Chem. 2012; 55:325–334.

47 Tiwari R, Toppino A, Agarwal HK, Huo T, Byun Y, Gallucci J, et al. Synthesis, biological evaluation, and radioiodination of halogenated closo-carboranylthymidine analogues. Inorg Chem. 2012; 51(1):629–639.

48 Byun Y, Yan J, Al-Madhoun AS, Johnsamuel J, Yang W, Barth RF, et al. The synthesis and biochemical evaluation of thymidine analogs substituted with nido carborane at the N-3 position. Appl Radiat Isot. 2004; 61(5):1125–1130.

49 Byun Y, Yan J, Al-Madhoun AS, Johnsamuel J, Yang W, Barth RF, et al. Synthesis and biological evaluation of neutral and zwitterionic 3-carboranyl thymidine analogues for boron neutron capture therapy. J Med Chem. 2005; 48(4):1188–1198.

50 Byun Y, Thirumamagal BTS, Yang W, Eriksson S, Barth RF, Tjarks W. Preparation and biological evaluation of ^{10}B-enriched 3-[5-{2-(2,3-dihydroxyprop-1-yl)-o-carboran-1-yl} pentan-1-yl]thymidine (N5-2OH), a new boron delivery agent for boron neutron capture therapy of brain tumors. J Med Chem. 2006; 49(18):5513–5523.

51. Barth RF, Coderre JA, Vicente MGH, Blue TE. Boron neutron capture therapy of cancer: current status and future prospects. Clin Cancer Res. 2005; 11(11):3987–4002.
52. Hosmane NS, Maguire JM, Zhu, Y. Takagaki, M. Boron and gadolinium neutron capture therapy for cancer treatment. Singapore: World Scientific Publishing Company; 2012.
53. Barth RF, Vicente MGH, Harling OK, Kiger WS 3rd, Riley KJ, Binns PJ, et al. Current status of boron neutron capture therapy of high grade gliomas and recurrent head and neck cancer. Radiat Oncol. 2012; 7:146.
54. Moss RL. Critical review, with an optimistic outlook, on boron neutron capture therapy (BNCT). Appl Radiat Isot. 2015; 88:2–11.
55. Pouget J-P, Mather SJ. General aspects of the cellular response to low- and high-LET radiation. Eur J Nucl Med. 2001; 28(4):541–561.
56. Hall EJ. Radiobiology for the radiologist. 5th ed. Philadelphia: Lippincott Williams & Williams; 2000.
57. Barth RF, Joensuu H. Boron neutron capture therapy for the treatment of glioblastomas and extracranial tumours: as effective, more effective or less effective than photon irradiation? Radiother Oncol. 2007; 82(2):119–122.
58. Isaac MF, Kahl SB. Synthesis of ether- and carbon-linked polycarboranyl porphyrin dimers for cancer therapies. J Org Chem. 2003; 680:232–243.
59. Jori G, Soncin M, Friso E, Vicente MGH, Hao E, Miotto G, et al. A novel boronated-porphyrin as a radio-sensitizing agent for boron neutron capture therapy of tumors: in vitro and in vivo studies. Appl Radiat Isot. 2009; 67:S321–S324.
60. Renner MW, Miura M, Easson MW, Vicente MGH. Recent progress in the syntheses and biological evaluation of boronated porphyrins for boron neutron-capture therapy. Anticancer Agents Med Chem. 2006; 6(2):145–157.
61. Tietze LF, Griesbach U, Schuberth I, Bothe U, Marra A, Dondoni A. Novel carboranyl C-glycosides for the treatment of cancer by boron neutron capture therapy. Chemistry. 2003; 9(6):1296–1302.
62. Bregadze V, Semioshkin A, Sivaev I. Synthesis of conjugates of polyhedral boron compounds with tumor-seeking molecules for neutron capture therapy. Appl Radiat Isot. 2011; 69:1774–1777.
63. Orlova AV, Kononov LO. Synthesis of conjugates of polyhedral boron compounds with carbohydrates. Russ Chem Rev. 2009; 78:629–642.
64. Das BC, Das S, Li G, Bao W, Kabalka GW. Synthesis of a water soluble carborane containing amino acid as a potential therapeutic agent. Synlett. 2001(9): 1419–1420.
65. Cai J, Soloway AH, Barth RF, Adams DM, Hariharan JR, Wyzlic IM, et al. Boron-containing polyamines as DNA targeting agents for neutron capture therapy of brain tumors: synthesis and biological evaluation. J Med Chem. 1997; 40(24):3887–3896.
66. Zhuo J-C, Cai J, Soloway AH, Barth RF, Adams DM, Ji W, et al. Synthesis and biological evaluation of boron-containing polyamines as potential agents for neutron capture therapy of brain tumors. J Med Chem. 1999; 42(7):1282–1292.
67. Yang W, Barth RF, Wu G, Tjarks W, Binns P, Riley K. Boron neutron capture therapy of EGFR or EGFRvIII positive gliomas using either boronated monoclonal antibodies or epidermal growth factor as molecular targeting agents. Appl Radiat Isot. 2009; 67:S328–S331.
68. Yang W, Wu G, Barth RF, Swindall MR, Bandyopadhyaya AK, Tjarks W, et al. Molecular targeting and treatment of composite EGFR and EGFRvIII-positive gliomas using boronated monoclonal antibodies. Clin Cancer Res. 2008; 14:883–891.

69 Heber EM, Hawthorne MF, Kueffer PJ, Garabalino MA, Thorp SI, Pozzi ECC, *et al*. Therapeutic efficacy of boron neutron capture therapy mediated by boron-rich liposomes for oral cancer in the hamster cheek pouch model. Proc Natl Acad Sci USA. 2014;111(45):16077–16081.
70 Gabel D. Boron clusters in medicinal chemistry: perspectives and problems. Pure Appl Chem. 2015; 87(2):173–179.
71 Issa F, Kassiou M, Rendina LM. Boron in drug discovery: carboranes as unique pharmacophores in biologically active compounds. Chem Rev. 2011; 111:5701–5722.
72 Kahlert J, Austin CJD, Kassiou M, Rendina LM. The fifth element in drug design: boron in medicinal chemistry. Aust J Chem. 2013; 66(10):1118–1123.
73 Lesnikowski ZJ. Recent developments with boron as a platform for novel drug design. Expert Opin Drug Discov. 2016; 11(6): 569–578.
74 Ban HS, Nakamura H. Boron-based drug design. Chem Rec. 2015; 15(3):616–635.
75 Luderer MJ, de la Puente P, Azab AK. Advancements in tumor targeting strategies for boron neutron capture therapy. Pharm Res. 2015; 32(9):2824–2836.
76 Grimes NG. Carboranes. 2nd ed. Amsterdam: Elsevier; 2011.
77 Tiwari R, Mahasenan K, Pavlovicz R, Li C, Tjarks W. Carborane clusters in computational drug design: a comparative docking evaluation using AutoDock, FlexX, Glide, and Surflex. J Chem Inf Model. 2009; 49(6):1581–1589.
78 Agarwal HK, Hasabelnaby S, Tiwari R, Tjarks W. Boron cluster (radio) halogenation in biomedical research. In: Hosmane NS (Editor). Boron science. Boca Raton: CRC Press; 2012. p. 107–144.
79 Galmarini CM, Mackey JR, Dumontet C. Nucleoside analogues: mechanisms of drug resistance and reversal strategies. Leukemia. 2001; 15(6):875–890.
80 Aspland SE, Ballatore C, Castillo R, Desharnais J, Eustaquio T, Goelet P, *et al*. Kinase-mediated trapping of bi-functional conjugates of paclitaxel or vinblastine with thymidine in cancer cells. Bioorg Med Chem Lett. 2006; 16(19):5194–5198.
81 Grunewald R, Abbruzzese JL, Tarassoff P, Plunkett W. Saturation of 2',2'-difluorodeoxycytidine 5'-triphosphate accumulation by mononuclear cells during a phase I trial of gemcitabine. Cancer Chemother Pharmacol. 1991; 27(4):258–262.
82 Chabner BA. Cytidine analogues. In: Chabner BA, Longo DL (Editors). Cancer chemotherapy and biotherapy. 2nd ed. Philadelphia: Lippincott-Raven; 1996. p. 213–233.
83 Liliemark J. The clinical pharmacokinetics of cladribine. Clin Pharmacokinet. 1997; 32(2):120–131.
84 Plunkett W, Liliemark JO, Adams TM, Nowak B, Estey E, Kantarjian H, *et al*. Saturation of 1-beta-D-arabinofuranosylcytosine 5'-triphosphate accumulation in leukemia cells during high-dose 1-beta-D-arabinofuranosylcytosine therapy. Cancer Res. 1987; 47(11):3005–3011.
85 Frick LW, Nelson DJ, St. Clair MH, Furman PA, Krenitsky TA. Effects of 3'-azido-3'-deoxythymidine on the deoxynucleotide triphosphate pools of cultured human cells. Biochem Biophys Res Communs. 1988; 154(1):124–129.
86 Fridland A, Connelly MC, Ashmun R. Relationship of deoxynucleotide changes to inhibition of DNA synthesis induced by the antiretroviral agent 3'-azido-3'-deoxythymidine and release of its monophosphate by human lymphoid cells (CCRF-CEM). Mol Pharmacol. 1990; 37(5):665–670.
87 Lowe CR, Dean PDG. Affinity chromatography. New York: John Wiley & Sons; 1974.

88 Lowe CR, Harvey MJ, Craven DB, Dean PD. Some parameters relevant to affinity chromatography on immobilized nucleotides. Biochem J. 1973; 133(3):499–506.
89 Grobner P, Loidl P. Thymidine kinase. a novel affinity chromatography of the enzyme and its regulation by phosphorylation in physarum polycephalum. J Biol Chem. 1984; 259(12):8012–8014.
90 Sandrini MPB, Piskur J. Deoxyribonucleoside kinases: two enzyme families catalyze the same reaction. Trends Biochem Sci. 2005; 30(5):225–228.
91 Kosinska U, Carnrot C, Eriksson S, Wang L, Eklund H. Structure of the substrate complex of thymidine kinase from ureaplasma urealyticum and investigations of possible drug targets for the enzyme. FEBS J. 2005; 272(24):6365–6372.
92 Kosinska U, Carnrot C, Sandrini MP, Clausen AR, Wang L, Piskur J, et al. Structural studies of thymidine kinases from bacillus anthracis and bacillus cereus provide insights into quaternary structure and conformational changes upon substrate binding. FEBS J. 2007; 274(3):727–737.
93 Welin M, Kosinska U, Mikkelsen NE, Carnrot C, Zhu C, Wang L, et al. Structures of thymidine kinase 1 of human and mycoplasmic origin. Proc Natl Acad Sci USA. 2004; 101(52):17970–17975.
94 Segura-Pena D, Sekulic N, Ort S, Konrad M, Lavie A. Substrate-induced conformational changes in human UMP/CMP kinase. J Biol Chem. 2004; 279(32):33882–33889.
95 Segura-Pena D, Lutz S, Monnerjahn C, Konrad M, Lavie A. Binding of ATP to TK1-like enzymes is associated with a conformational change in the quaternary structure. J Mol Biol. 2007; 369(1):129–141.
96 Segura-Pena D, Lichter J, Trani M, Konrad M, Lavie A, Lutz S. Quaternary structure change as a mechanism for the regulation of thymidine kinase 1-like enzymes. Structure. 2007; 15(12):1555–1566.
97 Birringer MS, Claus MT, Folkers G, Kloer DP, Schulz GE, Scapozza L. Structure of a type II thymidine kinase with bound dTTP. FEBS L. 2005; 579(6):1376–1382.
98 El Omari K, Solaroli N, Karlsson A, Balzarini J, Stammers DK. Structure of vaccinia virus thymidine kinase in complex with dTTP: insights for drug design. BMC Struct Biol. 2006; 6(1):22.
99 Research Collaboratory for Structural Bioinformatics Protein Data Bank (RCSB PDB). Available from: http://www.rcsb.org/pdb/home/home.do
100 Arner ES, Eriksson S. Mammalian deoxyribonuclease kinases. Pharmacol Ther. 1995; 67(2):155–186.
101 Yusa T, Yamaguchi Y, Ohwada H, Hayashi Y, Kuroiwa N, Morita T, et al. Activity of the cytosolic isozyme of thymidine kinase in human primary lung tumors with reference to malignancy. Cancer Res. 1988; 48(17):5001–5006.
102 Derenzini M, Montanaro L, Trere D, Chilla A, Tazzari PL, Dall'Olio F, et al. Thymidylate synthase protein expression and activity are related to the cell proliferation rate in human cancer cell lines. Mol Pathol. 2002; 55(5):310–314.
103 Maehara Y, Moriguchi S, Emi Y, Watanabe A, Kohnoe S, Tsujitani S, et al. Comparison of pyrimidine nucleotide synthetic enzymes involved in 5-fluorouracil metabolism between human adenocarcinomas and squamous cell carcinomas. Cancer. 1990; 66(1):156–161.
104 Sandrini MPB, Clausen AR, Munch-Petersen B, Piskur J. Thymidine kinase diversity in bacteria. Nucleosides Nucleotides Nucleic Acids. 2006; 25(9–11):1153–1158.

105 Sandrini MPB, Clausen AR, On SLW, Aarestrup FM, Munch-Petersen B, Piskur J. Nucleoside analogues are activated by bacterial deoxyribonucleoside kinases in a species-specific manner. J Antimicrob Chemother. 2007; 60(3):510–520.
106 Sandrini MPB, Shannon O, Clausen AR, Bjorck L, Piskur J. Deoxyribonucleoside kinases activate nucleoside antibiotics in severely pathogenic bacteria. Antimicrob Agents Chemother. 2007; 51(8):2726–2732.
107 Saito H, Tomioka H. Thymidine kinase of bacteria: activity of the enzyme in actinomycetes and related organisms. J Gen Microbiol. 1984; 130(7):1863–1870.
108 Hengstschlaeger M, Knoefler M, Muellner EW, Ogris E, Wintersberger E, Wawra E. Different regulation of thymidine kinase during the cell cycle of normal versus DNA tumor virus-transformed cells. J Biol Chem. 1994; 269(19):13836–13842.
109 Hengstschlaeger M, Muellner EW, Wawra E. Thymidine kinase is expressed differently in transformed versus normal cells: a novel test for malignancy. Int J Oncol. 1994; 4(1):207–210.
110 Sherley JL, Kelly TJ. Regulation of human thymidine kinase during the cell cycle. J Biol Chem. 1988; 263(17):8350–8358.
111 Al-Madhoun AS, Ji W, Johnsamuel J, Yan J, Lakhi N, Byun J, et al. The biological evaluation of a small library of 3-(carboranylalkyl) thymidines as substrates for human thymidine kinases 1 and 2. In: Sauerwein W, Moss R, Wittig A (Editors). Research and development in neutron capture therapy. Bologna: Monduzi Editori; 2002. p. 89–93.
112 Endo Y, Iijima T, Yaguchi K, Kawachi E, Inoue N, Kagechika H, et al. Structure-activity study of retinoid agonists bearing substituted dicarba-closo-dodecaborane. relation between retinoidal activity and conformation of two aromatic nuclei. Bioorg Med Chem Lett. 2001; 11(10):1307–1311.
113 van Waarde A, Cobben DCP, Suurmeijer AJH, Maas B, Vaalburg W, de Vries EFJ, et al. Selectivity of ^{18}F-FLT and ^{18}F-FDG for differentiating tumor from inflammation in a rodent model. J Nucl Med. 2004; 45(4):695–700.
114 Zhang CC, Yan Z, Li W, Kuszpit K, Painter CL, Zhang Q, et al. [^{18}F]FLT-PET imaging does not always "light up" proliferating tumor cells. Clin Cancer Res. 2012; 18(5):1303–1312.
115 Shields AF. PET imaging of tumor growth: not as easy as it looks. Clin Cancer Res. 2012; 18(5):1189–1191.
116 Knecht W, Munch-Petersen B, Piskur J. Identification of residues involved in the specificity and regulation of the highly efficient multisubstrate deoxyribonucleoside kinase from drosophila melanogaster. J Mol Biol. 2000; 301(4):827–837.
117 Kokoris MS, Black ME. Characterization of herpes simplex virus type 1 thymidine kinase mutants engineered for improved ganciclovir or acyclovir activity. Protein Sci. 2002; 11(9):2267–2272.
118 Thirumamagal BT, Johnsamuel J, Cosquer GY, Byun Y, Yan J, Narayanasamy S, et al. Boronated thymidine analogues for boron neutron capture therapy. Nucleosides Nucleotides Nucleic Acids. 2006; 25(8):861–866.
119 Huff AC, Topal MD. DNA damage at thymine N-3 abolishes base-pairing capacity during DNA synthesis. J Biol Chem. 1987; 262(26):12843–12850.
120 Galmarini CM, Jordheim L, Dumontet C. Pyrimidine nucleoside analogs in cancer treatment. Expert Rev Anticancer Ther. 2003; 3(5):717–728.
121 Deville-Bonne D, El Amri C, Meyer P, Chen Y, Agrofoglio LA, Janin J. Human and viral nucleoside/nucleotide kinases involved in antiviral drug activation: structural and catalytic properties. Antiviral Res. 2010; 86(1):101–120.

122 Pastor-Anglada M, Perez-Torras S. Nucleoside transporter proteins as biomarkers of drug responsiveness and drug targets. Front Pharmacol. 6:1–14.
123 Ritter CA, Jedlitschky G, zu Schwabedissen HM, Grube M, Koeck K, Kroemer HK. Cellular export of drugs and signaling molecules by the ATP-binding cassette transporters MRP4 (ABCC4) and MRP5 (ABCC5). Drug Metab Rev. 2005; 37(1):253–278.
124 Pastor-Anglada M, Cano-Soldado P, Molina-Arcas M, Lostao MP, Larrayoz I, Martinez-Picado J, et al. Cell entry and export of nucleoside analogues. Virus Res. 2005; 107(2):151–164.
125 Borst P, Balzarini J, Ono N, Reid G, de Vries H, Wielinga P, et al. The potential impact of drug transporters on nucleoside-analog-based antiviral chemotherapy. Antiviral Res. 2004; 62(1):1–7.
126 Belt JA, Marina NM, Phelps DA, Crawford CR. Nucleoside transport in normal and neoplastic cells. Adv Enzyme Regul. 1993; 33:235–252.
127 Summers MA, Moore JL, McAuley JW. Use of verapamil as a potential P-glycoprotein inhibitor in a patient with refractory epilepsy. Ann Pharmacother 2004; 38:1631–1614.
128 Peng X-X, Shi Z, Damaraju VL, Huang X-C, Kruh GD, Wu H-C, et al. Up-regulation of MRP4 and down-regulation of influx transporters in human leukemic cells with acquired resistance to 6-mercaptopurine. Leuk Res. 2008; 32(5):799–809.
129 Barth Rolf F, Kaur B. Rat brain tumor models in experimental neuro-oncology: the C6, 9L, T9, RG2, F98, BT4C, RT-2 and CNS-1 gliomas. J Neurooncol. 2009; 94(3):299–312.
130 Nisman B, Allweis T, Kaduri L, Maly B, Gronowitz S, Hamburger T, et al. Serum thymidine kinase 1 activity in breast cancer. Cancer Biomark. 2010; 7(2):65–72.
131 Nisman B, Yutkin V, Nechushtan H, Gofrit Ofer N, Peretz T, Gronowitz S, et al. Circulating tumor M2 pyruvate kinase and thymidine kinase 1 are potential predictors for disease recurrence in renal cell carcinoma after nephrectomy. Urology. 2010; 76(2):513 e1–513 e6.
132 Li Z, Wang Y, He J, Ma J, Zhao L, Chen H, et al. Serological thymidine kinase 1 is a prognostic factor in oesophageal, cardial and lung carcinomas. Eur J Cancer Prev. 2010; 19(4):313–318.
133 Svobodova S, Topolcan O, Holubec L, Treska V, Sutnar A, Rupert K, et al. Prognostic importance of thymidine kinase in colorectal and breast cancer. Anticancer Res. 2007; 27(4A):1907–1909.
134 He Q, Fornander T, Johansson H, Johansson U, Hu GZ, Rutqvist L-E, et al. Thymidine kinase 1 in serum predicts increased risk of distant or loco-regional recurrence following surgery in patients with early breast cancer. Anticancer Res. 2006; 26(6C):4753–4759.
135 Alegre MM, Robison RA, O'Neill KL. Thymidine kinase 1 upregulation is an early event in breast tumor formation. J Oncol. 2012:575647, 5 p.
136 Sharif H, von Euler H, Westberg S, He E, Wang L, Eriksson S. A sensitive and kinetically defined radiochemical assay for canine and human serum thymidine kinase 1 (TK1) to monitor canine malignant lymphoma. Vet J. 2012; 194(1):40–47.
137 De Clercq E. Antiviral drugs in current clinical use. J Clin Virol. 2004; 30(2):115–133.
138 Grierson JR, Schwartz JL, Muzi M, Jordan R, Krohn KA. Metabolism of 3′-deoxy-3′-[F-18]fluorothymidine in proliferating A549 cells: validations for positron emission tomography. Nucl Med Biol. 2004; 31(7):829–837.

139 Plotnik DA, McLaughlin LJ, Chan J, Redmayne-Titley JN, Schwartz JL. The role of nucleoside/nucleotide transport and metabolism in the uptake and retention of 3′-fluoro-3′-deoxythymidine in human B-lymphoblast cells. Nucl Med Biol. 2011; 38(7):979–986.

140 Peterson KL, Reid WC, Freeman AF, Holland SM, Pettigrew RI, Gharib AM, et al. The use of ^{14}C-FIAU to predict bacterial thymidine kinase presence: implications for radiolabeled FIAU bacterial imaging. Nucl Med Biol. 2013; 40(5):638–642.

141 Bettegowda C, Foss CA, Cheong I, Wang Y, Diaz L, Agrawal N, et al. Imaging bacterial infections with radiolabeled 1-(2′-deoxy-2′-fluoro-beta-D-arabinofuranosyl)-5-iodouracil. Proc Natl Acad Sci USA. 2005; 102(4):1145–1150.

142 Longley DB, Harkin DP, Johnston PG. 5-Fluorouracil: mechanisms of action and clinical strategies. Nat Rev Cancer. 2003; 3(5):330–338.

143 Andersen S, Heine T, Sneve R, Koenig I, Krokan HE, Epe B, et al. Incorporation of dUMP into DNA is a major source of spontaneous DNA damage, while excision of uracil is not required for cytotoxicity of fluoropyrimidines in mouse embryonic fibroblasts. Carcinogenesis. 2005; 26(3):547–555.

144 Haveman J, Sigmond J, Van Bree C, Franken NA, Koedooder C, Peters GJ. Time course of enhanced activity of deoxycytidine kinase and thymidine kinase 1 and 2 in cultured human squamous lung carcinoma cells, SW-1573, induced by gamma-irradiation. Oncol Rep. 2006; 16(4):901–905.

145 McKenna PG, McKelvey VJ, Frew TL. Sensitivity to cell killing and the induction of cytogenetic damage following gamma irradiation in wild-type and thymidine kinase-deficient Friend mouse erythroleukemia cells. Mutat Res. 1988; 200(1–2):231–242.

146 Boothman DA, Davis TW, Sahijdak WM. Enhanced expression of thymidine kinase in human cells following ionizing radiation. Int J Radiat Oncol, Biol, Phys. 1994; 30(2):391–398.

147 Wei S, Ageron-Blanc A, Petridis F, Beaumatin J, Bonnet S, Luccioni C. Radiation-induced changes in nucleotide metabolism of two colon cancer cell lines with different radiosensitivities. Int J Radiat Biol. 1999; 75(8):1005–1013.

148 He Q, Skog S, Welander I, Tribukait B. X-irradiation effects on thymidine kinase (TK): I. TK1 and 2 in normal and malignant cells. Cell Prolif. 2002; 35(2):69–81.

149 Al-Nabulsi I, Takamiya Y, Voloshin Y, Dritschilo A, Martuza RL, Jorgensen TJ. Expression of thymidine kinase is essential to low dose radiation resistance of rat glioma cells. Cancer Res. 1994; 54(21):5614–5617.

150 Wakazono Y, Kubota M, Furusho K, Liu L, Gerson SL. Thymidine kinase deficient cells with decreased TTP pools are hypersensitive to DNA alkylating agents. Mutat Res, DNA Repair. 1996; 362(1):119–125.

151 Kunos CA, Ferris G, Pyatka N, Pink J, Radivoyevitch T. Deoxynucleoside salvage facilitates DNA repair during ribonucleotide reductase blockade in human cervical cancers. Radiat Res. 2011; 176(4):425–433.

152 Chen Y-L, Eriksson S, Chang Z-F. Regulation and functional contribution of thymidine kinase 1 in repair of DNA damage. J Biol Chem. 2010; 285(35):27327–27335.

153 Radivoyevitch T, Saunthararajah Y, Pink J, Ferris G, Lent I, Jackson M, et al. dNTP supply gene expression patterns after p53 loss. Cancers. 2012; 4:1212–1224.

154 Hu C-M, Chang Z-F. Synthetic lethality by lentiviral short hairpin RNA silencing of thymidylate kinase and doxorubicin in colon cancer cells regardless of the p53 status. Cancer Res. 2008; 68(8):2831–2840.

155 Pressacco J, Mitrovski B, Erlichman C, Hedley DW. Effects of thymidylate synthase inhibition on thymidine kinase activity and nucleoside transporter expression. Cancer Res. 1995; 55(7):1501–1508.

156 Lee SJ, Kim SY, Chung JH, Oh SJ, Ryu JS, Hong YS, et al. Induction of thymidine kinase 1 after 5-fluorouracil as a mechanism for 3′-deoxy-3′-[^{18}F]fluorothymidine flare. Biochem Pharmacol. 2010; 80(10):1528–1536.

157 Di Cresce C, Figueredo R, Ferguson PJ, Vincent MD, Koropatnick J. Combining small interfering RNAs targeting thymidylate synthase and thymidine kinase 1 or 2 sensitizes human tumor cells to 5-fluorodeoxyuridine and pemetrexed. J Pharmacol Exp Ther. 2011; 338(3):952–963.

158 Lowe SW, Bodis S, McClatchey A, Remington L, Ruley HE, Fisher DE, et al. p53 Status and the efficacy of cancer therapy in vivo. Science. 1994; 266(5186):807–810.

159 Tonini G, Fratto ME, Schiavon G. Molecular prognostic factors: clinical implications in patients with breast cancer. Cancer Ther. 2008;6:773–782.

160 Martinez-Rivera M, Siddik ZH. Resistance and gain-of-resistance phenotypes in cancers harboring wild-type p53. Biochem Pharmacol. 2012; 83(8):1049–1062.

161 Chen KG, Sikic BI. Molecular pathways: regulation and therapeutic implications of multidrug resistance. Clin Cancer Res. 2012; 18(7):1863–1869.

162 Chorawala MR, Oza PM, Shah GB. Mechanisms of anticancer drugs resistance: an overview. Int J Pharm Sci Drug Res. 2012; 4(1):1–9.

163 Lai D, Visser-Grieve S, Yang X. Tumour suppressor genes in chemotherapeutic drug response. Biosci Rep. 2012; 32(4):361–374.

164 Malan C, Morin C. Synthesis of unsymmetrical C-disubstituted para-carboranes: access to functionalized carboranyl-boronic acid and carboranol. Tetrahedron Lett. 1997; 38(37):6599–6602.

165 Safronov AV, Hawthorne MF. Novel synthesis of 3-iodo-ortho-carborane. Inorganica Chim Acta. 2011;375(1):308–310.

166 Yamazaki H, Ohta K, Endo Y. Regioselective synthesis of triiodo-o-carboranes and tetraiodo-o-carborane. Tetrahedron Lett. 2005; 46(17):3119–3122.

167 Barbera G, Vaca A, Teixidor F, Sillanpaa R, Kivekas R, Vinas C. Designed synthesis of new ortho-carborane derivatives: from mono- to polysubstituted frameworks. Inorg Chem. 2008; 47(16):7309–7316.

168 Himmelspach A, Finze M. Dicarba-closo-dodecaboranes with one and two ethynyl groups bonded to boron. Eur J Inorg Chem. 2010(13):2012–2024.

169 Sevryugina Y, Julius RL, Hawthorne MF. Novel approach to aminocarboranes by mild amidation of selected iodo-carboranes. Inorg Chem. 2010; 49(22):10627–10634.

170 Safronov AV, Sevryugina YV, Jalisatgi SS, Kennedy RD, Barnes CL, Hawthorne MF. Unfairly forgotten member of the iodocarborane family: synthesis and structural characterization of 8-iodo-1,2-dicarba-closo-dodecaborane, its precursors, and derivatives. Inorg Chem. 2012; 51(4):2629–2637.

171 Rudakov DA, Kurman PV, Potkin VI. Synthesis and deborination of polyhalo-substituted ortho-carboranes. Russ J Gen Chem. 2011; 81(6):1137–1142.

172 Ramachandran BM, Knobler CB, Hawthorne MF. Example of an intramolecular, non-classical C-H...pi hydrogen bonding interaction in an arylcarborane derivative. J Mol Struct. 2006; 785(1–3):167–169.

173 Ogawa T, Ohta K, Yoshimi T, Yamazaki H, Suzuki T, Ohta S, *et al.* m-Carborane bisphenol structure as a pharmacophore for selective estrogen receptor modulators. Bioorg Med Chem Lett. 2006; 16(15):3943–3946.

174 Richardson PG, Moreau P, Laubach JP, Gupta N, Hui A-M, Anderson KC, *et al.* The investigational proteasome inhibitor ixazomib for the treatment of multiple myeloma. Future Oncol. 2015; 11(8):1153–1168.

175 Gentile M, Offidani M, Vigna E, Corvatta L, Recchia AG, Morabito L, *et al.* Ixazomib for the treatment of multiple myeloma. Expert Opin Investig Drugs. 2015; 24(9):1287–1298.

176 Touchet S, Carreaux F, Carboni B, Bouillon A, Boucher J-L. Aminoboronic acids and esters: from synthetic challenges to the discovery of unique classes of enzyme inhibitors. Chem Soc Rev. 2011; 40(7):3895–3914.

177 Zhao H, Palencia A, Seiradake E, Ghaemi Z, Cusack S, Luthey-Schulten Z, *et al.* Analysis of the resistance mechanism of a benzoxaborole inhibitor reveals insight into the leucyl-tRNA synthetase editing mechanism. ACS Chem Biol. 2015; 10(10):2277–2285.

3.2

Recent Advances in Boron Delivery Agents for Boron Neutron Capture Therapy (BNCT)

Sunting Xuan and Maria da Graça H. Vicente*

Department of Chemistry, Louisiana State University, Baton Rouge, Louisiana, USA
*Corresponding author: Department of Chemistry, Louisiana State University, Baton Rouge, LA 70803, USA

3.2.1 Introduction

3.2.1.1 Mechanisms of BNCT

Boron neutron capture therapy (BNCT) is a promising binary anticancer therapy that selectively targets and destroys malignant tumor cells in the presence of healthy normal cells [1–5]. BNCT is based on the very high nuclear cross section of ^{10}B nuclei for neutron capture (3838 barns, 1 barn = 10^{-24} cm^2), a naturally occurring and stable boron isotope (19.9%). The irradiation of non-radioactive ^{10}B-containing tumors with low-energy thermal or epithermal neutrons produces excited ^{11}B nuclei, which spontaneously fission to give cytotoxic high linear energy transfer (high-LET) α and recoiling ^{7}Li particles, along with γ radiation, and approximately 2.4 MeV of kinetic energy (Figure 3.2.1). Low-energy epithermal neutrons are able to penetrate up to 8 cm into tissues to reach deep-located tumors. The generated high-LET particles have short path lengths of about 5–9 µM in tissue (about the diameter of a single cell), restricting the damage to ^{10}B-containing tumor cells. The biologically abundant nuclei ^{12}C (0.0034 barn), ^{1}H (0.33 barn), and ^{14}N (1.8 barn) show negligible interference with the ^{10}B (n,α)^{7}Li neutron capture reaction due to their much smaller nuclear cross sections in comparison with ^{10}B.

BNCT has been used clinically for several decades in the treatment of various types of tumors, including head and neck cancers, primary and metastatic melanomas, and high-grade brain tumors (e.g., glioblastoma multiforme [GBM]) [4,6,7]. Nuclear reactors available in the United States, Japan, China, Argentina, and several European countries have been used as the neutron sources in BNCT clinical trials. The first clinical trials took place in the United States, at the Brookhaven National Laboratory (BNL) and the Massachusetts Institute of Technology (MIT), approximately 50 years ago using thermal neutrons. However, epithermal neutrons are mainly used in current BNCT clinical trials around the world due to their higher tissue-penetrating ability.

$$^{10}_{5}B + ^{1}_{0}n \longrightarrow ^{11}_{5}B \longrightarrow ^{7}Li^{3+} + ^{4}He^{2+} + \gamma + 2.4\,MeV$$

Figure 3.2.1 The $^{10}B(n,\alpha)^{7}Li$ neutron capture and fission reactions [3].

3.2.1.2 General Criteria for BNCT Agents

Generally, an effective boron delivery drug for BNCT should display the following characteristics [1–5]: (1) deliver therapeutic amounts of ^{10}B to the tumor, estimated to be about 20 μg/g tumor or approximately 10^9 atoms/tumor cell; (2) high tumor-to–normal tissue and tumor-to-blood concentration ratios (>5); (3) low systemic toxicity; (4) rapid clearance from blood and normal tissues while persisting in tumor during the irradiation treatment; and (5) ready quantification of tissue-localized boron. In addition, for brain tumor treatment, the BNCT drugs should be able to permeate the blood–brain barrier (BBB), either by passive transmembrane diffusion or through targeting of transporters and receptors that are expressed at the BBB [8]. The passive diffusion of compounds across cellular membranes and the BBB depends on various factors, including the compounds' lipophilicity, molecular weight (MW), polar surface area, electrical charge, degree of ionization, and hydrogen bonding ability [8–10]. In general, low-molecular-weight (<500 Da) boron delivery agents with favorable lipophilicity (octanol–water partition coefficient, $\log P$ <5) show enhanced passive diffusion across cellular membranes and the BBB [9–12]. However, a major challenge in drug development for BNCT has been the selective delivery of therapeutic amounts of boron to tumor cells, with minimal toxicity to normal tissues. Another challenge in the treatment of brain tumors is the often-inefficient BBB permeability of the BNCT drugs.

Over the last few decades, thousands of boron carriers for BNCT have been designed and synthesized, but only a few have been tested in preclinical trials. Among these, many are able to deliver a higher amount of boron to tumors relative to the surrounding normal tissue, due to the *enhanced permeability and retention* (EPR) effect. However, the selectivity of boron accumulation is often low, which can cause undesired toxic effects to healthy tissues. To improve the tumor selectivity of BNCT agents, their encapsulation into delivery agents such as liposomes and their association with tumor-targeting moieties (e.g., peptides, folic acid, and antibodies) have been explored. In addition, several administration methodologies and combination therapies have been investigated. One very promising methodology that bypasses the BBB is convection-enhanced delivery (CED), a method for local drug infusion directly into the brain. CED has been shown to deliver high amounts of boron (>100 μg/g tumor) to intracerebral tumors, with very high tumor-to-blood and tumor-to-normal brain ratios [5,13]. More recently, another tumor treatment strategy called magnetic drug targeting (MDT) has also been investigated [14]. MDT uses an external magnetic field to direct boron-containing magnetic nanoparticles, administrated intraarterially, to tumor tissues. The MTD strategy is independent from biological recognition and passive accumulation, and allows selective accumulation of the drug at the tumor site.

3.2.1.3 Main Categories of BNCT Agents

The two most-used boron-containing drugs in clinical trials are the sodium salt of sulfhydryl boron hydride $Na_2B_{12}H_{11}SH$ (BSH) and the amino acid (*L*)-4-dihydroxyborylphenylalanine (BPA) (Figure 3.2.2, **1–2**) [4,6,7,15–18]. Although BSH and BPA show

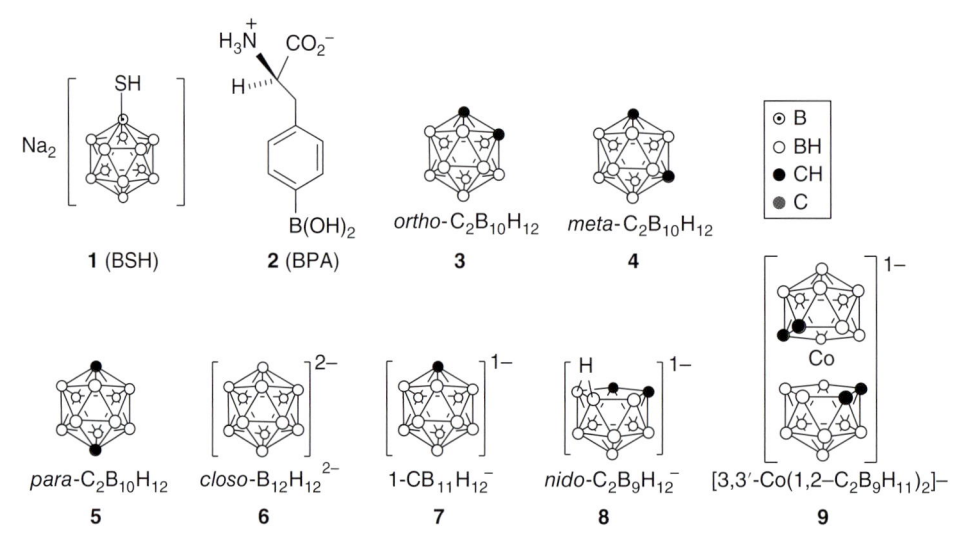

Figure 3.2.2 Structures of BSH, BPA, and common boron clusters currently used in BNCT drug development. The cluster atom representation is used throughout this chapter [5].

low toxicity and have recorded efficacy in clinical BNCT trials, they have only low selectivity and retention time in tumor [5,18,19]. Other promising classes of boronated compounds, including boronated amino acids, peptides, monoclonal antibodies (MAbs), nucleosides, lipids, carbohydrates, liposomes, porphyrin derivatives, boron dipyrromethenes (BODIPYs), and nanoparticles, have been synthesized and investigated as potential BNCT agents [3,5,13,20]. The preferred boron sources in these compounds are the neutral isomeric carboranes *ortho-*, *meta-*, and *para-*$C_2B_{10}H_{12}$; negatively charged *closo-*$B_{12}H_{12}^{2-}$; *closo-*$CB_{11}H_{12}^-$; the open-cage *nido-*$C_2B_9H_{12}^-$; and metallo-bis(dicarbollides) such as $[3,3'\text{-Co}(1,2\text{-}C_2B_9H_{11})_2]^-$ (Figure 3.2.2, **3–9**). This is because of their unique properties that include: (1) high boron content; (2) lipophilicity; (3) high photochemical, kinetic, and hydrolytic stabilities; and (4) ease of functionalization and attachment to targeting molecules [13,21]. Several reviews have been published on compound development for BNCT [2,3,5,13,20–30]. In this chapter, we review the synthesis and properties of the most promising classes of boron delivery agents, with emphasis on those reported in the last decade. Among the different classes of boron delivery agents, some of the most promising BNCT agents are nucleosides, amino acids, porphyrin derivatives, and tumor-targeted compounds that recognize a tumor-associated epitope, due to their generally lower toxicity and increased tumor specificity and uptake.

3.2.2 Amino Acids and Peptides

Considerations about the biochemical and metabolic differences between tumor and normal cells are important for achieving tumor selectivity with minimal damage to normal tissues. Amino acids, as precursors of specific biological components and/or substrates of certain enzymes in tumor cells, show favorable transport across plasma membranes of rapidly growing malignant cells, where they can be used for biological synthesis [2,5,20]. In addition, specific amino acid receptors (e.g., tyrosine receptor [31] and somatostatin receptor [32]) might be overexpressed in tumor cells. As a result, amino acids have the tendency to selectively accumulate and be temporarily retained in tumor cells [5,20,33]. The clinically used boronated amino acid BPA, which structurally resembles phenylalanine, despite its recorded low systemic toxicity, has the drawbacks of low water solubility and suboptimal tumor-targeting ability [5,19]. Therefore, BPA derivatives (e.g., BPA-fructose [34]) and other boronated amino acids have been developed with increased water solubility and enhanced uptake into tumor. Boronated naturally occurring amino acids (e.g., L-tyrosine and L-dopa [35]) have been of long-standing interest for selective targeting of tumor cells. The boron-containing compounds (**10–11**) of naturally occurring L-tyrosine and L-dopa amino acids, which are precursors of melanin and natural substrates of the tyrosinase enzyme, were shown to have higher selectivity toward melanin cells compared with BPA [35].

Several boronated unnatural amino acids are currently under development as potential boron delivery agents for BNCT (Figure 3.2.3). The reported BSH derivatives (**12–14**) exhibited low cytotoxicity, although slightly higher than that of BPA, toward the cell lines C6 (rat glioma), B16 (mouse melanoma), and SAS (human oral squamous). Compounds **12–14** all showed higher incorporation and higher killing effects than BPA in C6 cells [36,37]. More recently, boronated unnatural cyclic amino acids have received growing attention due to their enhanced tumor uptake and superior tumor retention times. Boronic acids containing cyclic amino acid moieties with *cis* structures (**15–16**), synthesized from hydantoin precursors, showed comparable boron uptake in B16 cells,

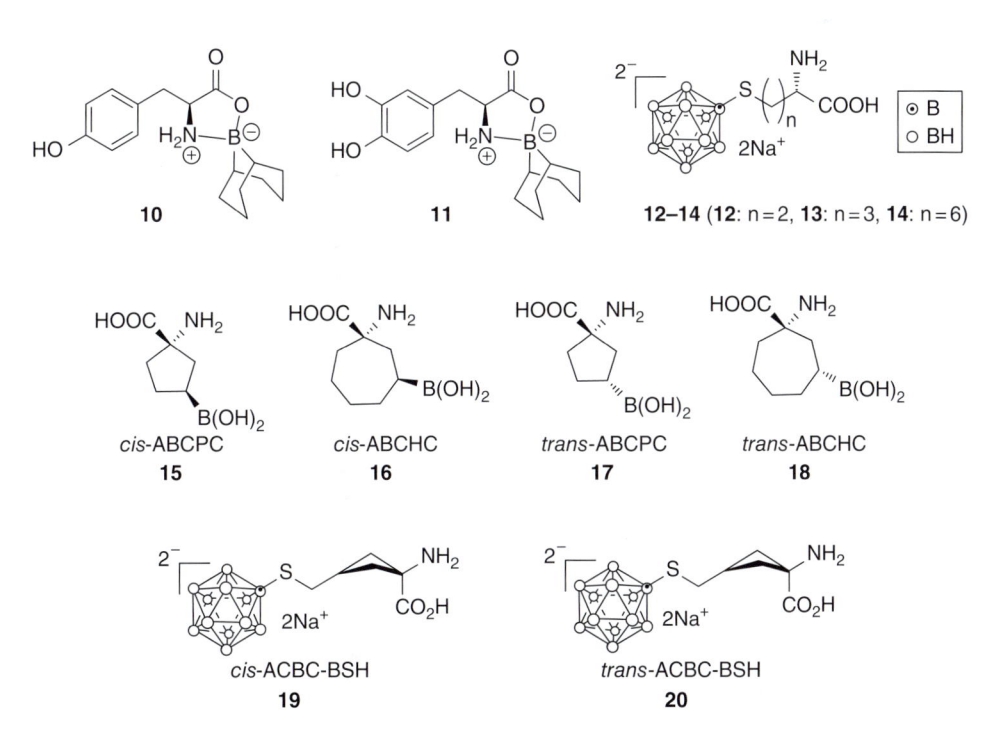

Figure 3.2.3 Structures of boronated amino acids [35–38,46].

but a superior tumor-to-blood boron concentration ratio even as a racemic mixture compared to that of the *trans* isomer (**17–18**) and BPA, in both *in vitro* and *in vivo* studies [38–41]. These types of unnatural amino acids have been shown to cross an intact BBB due to high upregulation of the L-type amino acid transporters (LATs) in high-grade gliomas [42,43]. In addition, there is evidence showing that these compounds cannot be metabolized, which may further enhance their tumor retention time in comparison with BPA [44]. The remarkable difference in tumor-targeting ability between the *cis* and *trans* isomers indicates that the pure enantiomers may provide even better tumor-targeting and localizing abilities. These properties render the boronated cyclic amino acids very promising BNCT agents, especially for the treatment of high-grade brain tumors due to their ability to cross the BBB. It has also been shown that unnatural amino acids bearing small 2,2-alkyl rings are selectively incorporated into tumor cells via LATs, and temporarily retained in tumor cells [45].

The BSH-containing cyclic amino acids (*cis*/*trans*) (**19–20**) were synthesized by condensation of (*cis*/*trans*)-hydantoin bearing a halogenated cycloalkyl group with S-(cyanoethyl)-BSH, followed by deprotection and hydrolysis of the hydantoin moieties in high yields. In contrast, the *trans* compound (**20**) in this study delivered a much larger amount of boron to C6 cells than the *cis* (**19**) analog, BPA, and the early reported linear derivative (**12**). Immunostaining microscopy indicated that **20** and **12** were incorporated into the C6 cell membranes and aggregated on the fringe of the cell nuclei, in contrast to BPA, which is evenly distributed within the cells with no particular regions of high concentration [36,46–48]. The unique cell microdistribution of these BSH derivatives may lead to more efficient damage to tumor cells upon neutron irradiation.

Peptides, composed of two or more amino acids, have the potential to selectively bind to receptors that are highly expressed in tumor tissues [49,50]. The neuropeptide Y (NPY) is a member of the pancreatic polypeptide family that binds to four Y-receptor subtypes (Y_1, Y_2, Y_4, and Y_5) [51,52]. The derivative [F^7,P^{34}]-NPY has been shown to be a breast cancer–selective ligand that binds to the Y_1-receptor subtype, which is found in 90% of breast cancer tissue and in all breast-cancer–derived metastases [51,53]. The [F^7,P^{34}]-NPY conjugated to *ortho*-carborane (**21**) exhibited nanomolar affinity and activity, and high uptake in MCF-7 breast adenocarcinoma cells [53].

Arg–Gly–Asp (RGD)-bearing peptides can bind to integrin αvβ3 specifically expressed on proliferating endothelial cells and tumor cells [54]. Cyclic RGD, conjugated to either BSH or *ortho*-carborane through suitable linkers (**22–27**) to control distance, orientation, flexibility, and physicochemical properties, was investigated (Figure 3.2.4) [55]. All the conjugates showed high *in vitro* integrin αvβ3 affinity, and the RGD–dimer–boron clusters (**25–27**) exhibited higher tumor uptake and slower clearance compared with the monomeric RGD conjugates (**22–24**) in an *in vivo* biodistribution study using mice inoculated subcutaneously with squamous cell carcinoma. In particular, **27** had significantly longer tumor retention time [55] and sensitized tumor cells more markedly [56] than BSH. As BSH has limited cell membrane permeability, a cell-penetrating peptide (CPP) (e.g., polyarginine) was attached to BSH to overcome this disadvantage [57]. The eight BSH-fused polyarginine (**28**) was shown to localize in the cell nucleus, and its *in vivo* administration to tumor-bearing mice showed significantly higher cancer-killing effects compared with 100-fold higher concentration of the BSH-administrated group (Figure 3.2.5) [57].

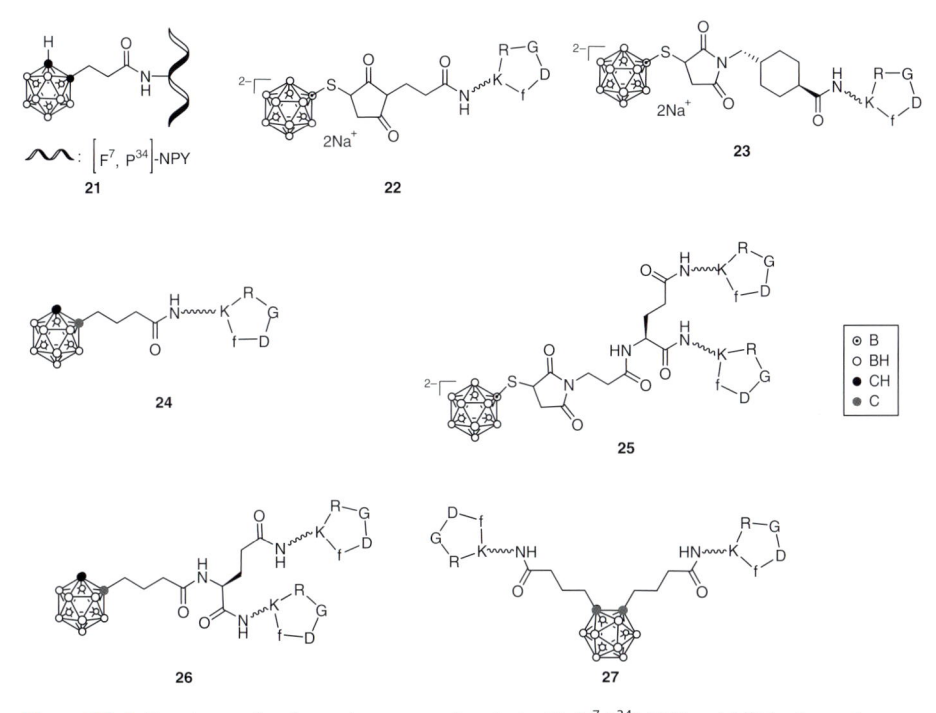

Figure 3.2.4 Structures of *ortho*-carborane conjugated with [F⁷,P³⁴]-NPY and BSH/*ortho*-carborane conjugated with cyclic RGD [53,55].

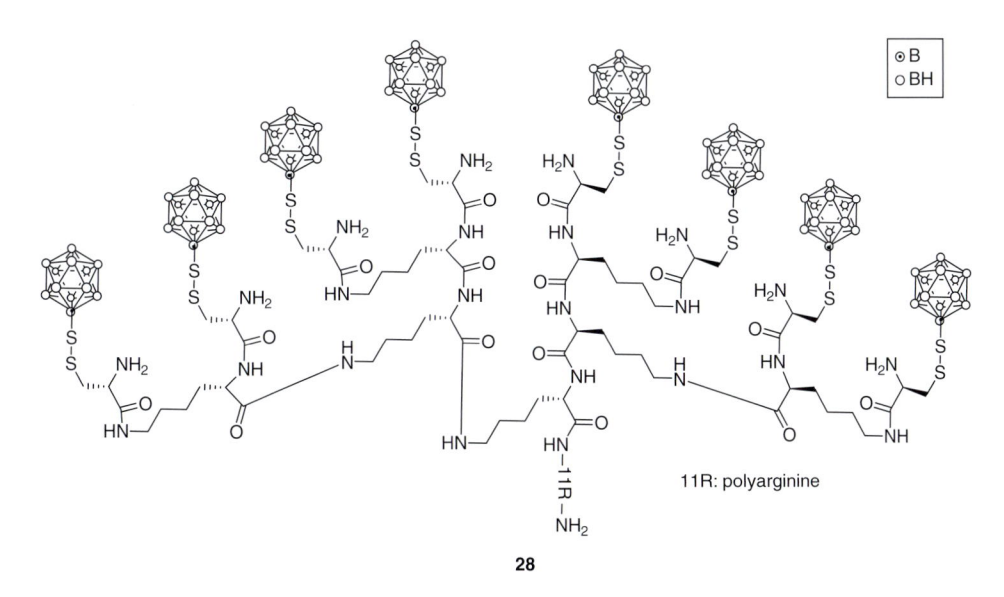

Figure 3.2.5 Structure of eight BSH-fused polyarginine [57].

3.2.3 Nucleosides

Boronated nucleosides, the biochemical precursors of nucleic acids, have been developed as boron delivery agents for BNCT, and investigated in multiple *in vivo* and *in vitro* biological studies that have led to structure–activity relationship (SAR) studies. The kinase enzyme, which is predominantly active in proliferating tumor cells, can catalyze the monophosphorylation of the corresponding boronated nucleosides [5,20,58–60]. This monophosphorylation pathway causes the selective accumulation of boronated nucleosides in close proximity to the cell nucleus by the kinase-mediated trapping (KMT) process [60–62]. The DNA-targeting property of boron-containing nucleosides could decrease the amount of boron required in cells for effective BNCT, since less boron is necessary when it localizes near or within the cell nucleus [5]. The recently reported boron-containing nucleosides, synthesized by nucleophilic substitution [59,63–65] or nucleophilic ring-opening [66] reactions of the corresponding nucleosides and boron compounds, are shown in Figure 3.2.6. 3-Carboranyl thymidine analogs (3-CTAs), such

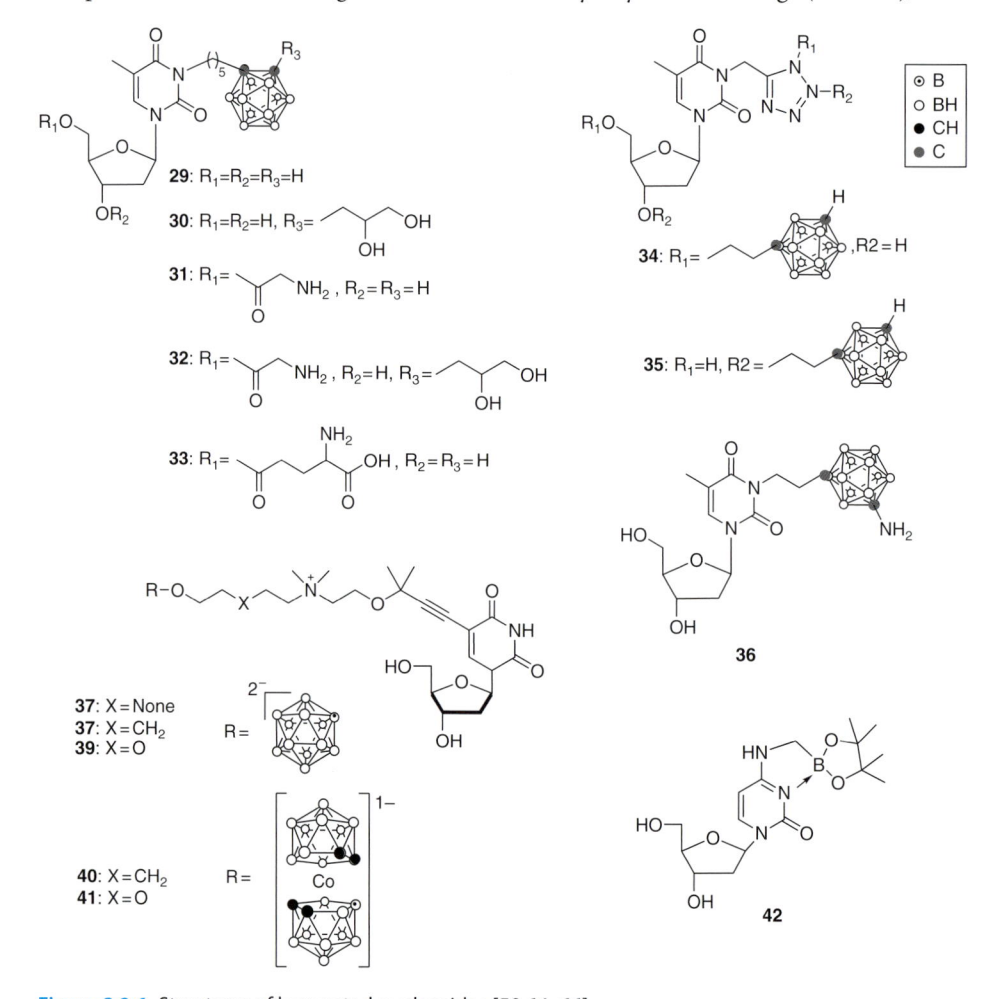

Figure 3.2.6 Structures of boronated nucleosides [59,64–66].

as N5 (**29**) and N5-2OH (**30**), were evaluated in preclinical cellular and *in vivo* studies. They exhibited selective *in vitro* uptake in TK1(+) versus TK1(−) cells, and favorable *in vivo* biodistribution and tumor uptake [60,67,68]. The major disadvantages of 3-CTAs are their generally low water solubility and limited ability to compete with endogenous thymidine (dTHd) at the substrate-binding site of human thymidine kinase (hTK1) [68]. The amino acid functionalized 3-CTAs exhibited much higher water solubility (48–6600 times) compared with their parental N5 and N5-2OH (**29–30**) compounds [59]; 30–70% DMSO was used to solubilize the latter compounds for intrathecal (i.t.) and intravenous (i.v.) administrations [67]. The 5′-glutamate ester and 5′-glycine ester functionalized compounds (**31–33**) were the most sensitive to chemical and enzymatic hydrolysis. The rapid cleavage of the ester bond can spare the 5′-deoxyribose for monophosphorylation to achieve tumor cell targeting, rendering them more promising candidates for BNCT compared with the valine-functionalized analogs. The linker between the dTHd scaffold and the carborane cluster at the N3-position of 3-CTAs was tuned to increase the water solubility and the ability for kinase enzyme catalyzing. The boronated tetrazolyl-type N3-substituted dTHd (**34–35**) was shown to be a better substrate for human hTK1 based on its enhanced water solubility (1.3–13 times) and the enzyme kinetics studies, in comparison with N5-2OH [63]. Compound **36** with an amino group directly attached to a *meta*-carborane cluster was about 3–4 times better as a substrate of hTK1 than N5-2OH [65]. These substrates of hTK1 may have higher intracellular trapping ability. The biological studies of the boronated nucleoside have not yet been reported. The boronated deoxyuridine derivatives containing $[B_{12}H_{12}]^{2-}$ and $[Co(C_2B_9H_{11})_2]^-$ were recently reported, and their cytotoxicity was evaluated in several cell lines (MRC-5, A549, LLC-MK2, L929, and Vero) [66]. The *closo*-dodecaborane conjugates (**37–39**) exhibited low cytotoxicity in all examined cell lines, whereas the Co-bisdicarbollide conjugates (**40–41**) showed relatively high toxicity ($IC_{50} < 31\,\mu M$) in MRC-5 and L929 cells, in agreement with the relatively higher toxicity of dTHd Co-bisdicarbollide conjugates [69]. The deoxyuridine boronated with pinacol (**42**) derivative was recently shown to be taken up in human colorectal adenocarcinoma C85 cells and subsequently undergo phosphorylation, as evidenced by the presence of *in vivo* synthesized mono-, di-, and triphosphates of **42** [64]. Boron nucleoside (**42**) showed higher toxicity toward C85 cells ($IC_{50} = 6.2\,mM$) [64] and SCC-15 cells ($IC_{50} = 16.4\,mM$) than to normal fibroblast cells ($IC_{50} = 49.5\,mM$) [70]. Compound **42** can insert into nucleic acids as a functional nucleotide derivative, as evidenced by mass spectrometry analysis of DNA from the SCC-15 cancer cells. The nuclear magnetic resonance (NMR) studies indicated a high degree of the dG–boron nucleoside pair, similar to the degree of the natural dG–dC base pair. These properties suggested **42** to be a very promising tumor-targeting agent for BNCT.

3.2.4 Antibodies

High-molecular-weight boronated delivery agents, such as monoclonal antibodies (mAbs) and their fragments, are very promising boron carriers for BNCT because of their high specificity for the target, due to their ability to recognize a tumor-associated epitope, such as the epidermal growth factor receptors (EGFRs) overexpressed in tumor tissues. In some cases, although they can selectively target tumors, the amount that reaches the tumor can be limited as a result of their rapid clearance by the reticuloendothelial system. In addition, the general non-internalizable characteristic of boronated mAbs limits their ability to cross capillary vascular endothelial cells, and might only lead to extracellular boron delivery [3]. Cetuximab (IMC-C225), the anti-EGFR mAb, has been approved for the treatment of recurrent EGFR(+) squamous cell carcinomas of the head and neck [71]. The cetuximab chemically linked with boronated polyamidoamine dendrimer (BD-C225) (Figure 3.2.7, **43**), which recognizes EGFR overexpressed in gliomas, has shown BNCT efficacy in EGFR gene-transfected rat glioma cells ($F98_{EGFR}$) [72,73]. The $F98_{EGFR}$ cells can rapidly take up this bioconjugate and showed 21% survival compared with 85% for the irradiated control. The mean tumor boron concentration received by combination of i.v. BPA and CED of BD-C225 ($87.9 \pm 16.5\,\mu g$ B/g) was higher than that using CED of BD-C225 alone, whereas the boron concentrations in blood and normal brain (<$0.5\,\mu g$ B/g) only reached the background limit of detection. Tumor-bearing rats that received both BD-C225 and BPA had longer mean survival times (MSTs) than the tumor-bearing rats that received

43: Cetuximab (IMC-C225)
44: L8A4

Polyamidoamine dendrimer

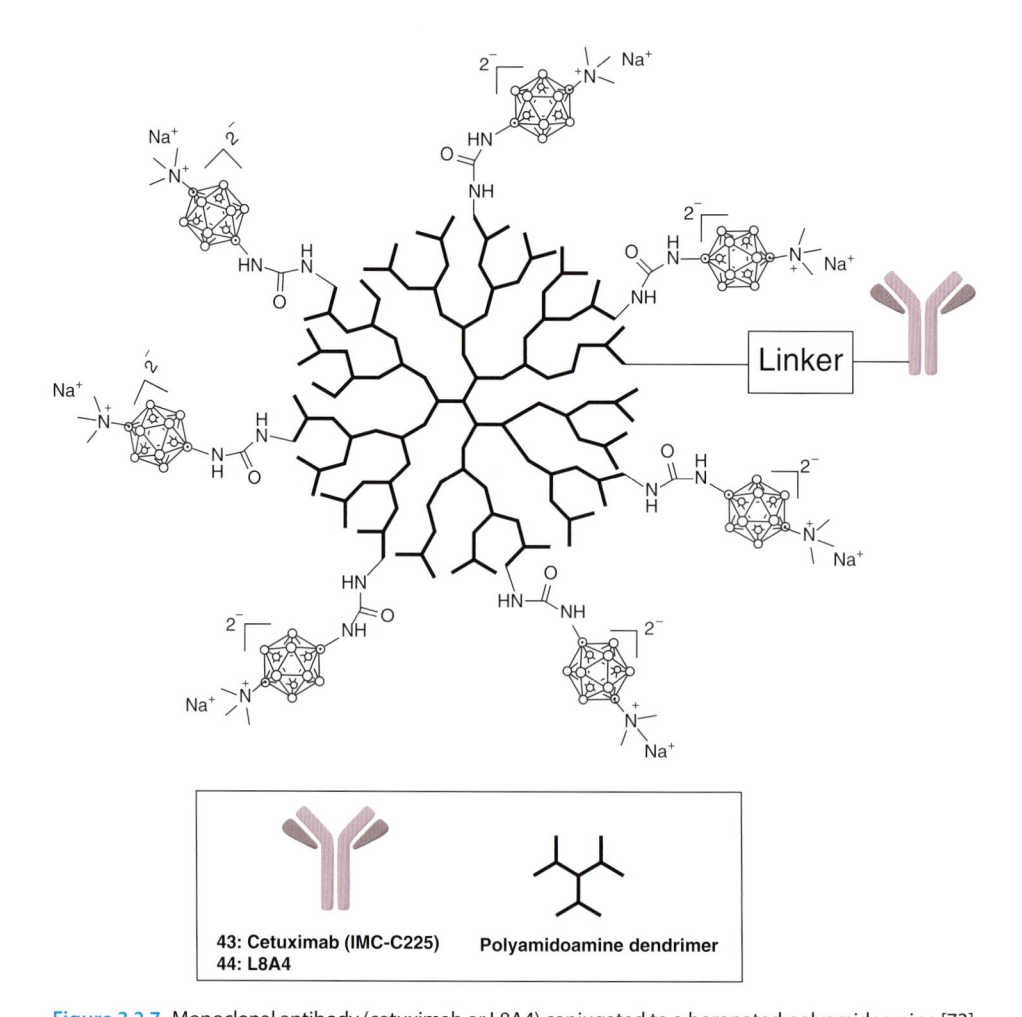

Figure 3.2.7 Monoclonal antibody (cetuximab or L8A4) conjugated to a boronated polyamidoamine [73].

BD-C225 alone after neutron irradiation. The boronated polyamidoamine dendrimer containing anti-EGFRvIII mAb (L8A4) (BD-L8A4) (Figure 3.2.7, **44**) was used in combination with BD-C225 to target rat gliomas containing both mutant EGFRvIII and wild-type EGFR [74]. The BNCT efficacy in rats bearing both $F98_{npEGFR}$ and $F98_{EGFR}$ gliomas was much higher using a combination of CED of BD-C225 and BD-L8A4 (24.4 μg B/g) than using CED of BD-C225 (13.8 μg B/g) or BD-L8A4 (12.3 μg B/g) alone. For the rats bearing $F98_{npEGFR}$ gliomas, the best survival data were obtained using CED of BD-L8A4 either alone or in combination with i.v. BPA [75]. These results show that due to the heterogeneity of receptor expression in brain tumors, a combination of boronated mAbs with another agent or agents, rather than a stand-alone boron delivery drug, should be used to achieve maximum BNCT efficacy.

3.2.5 Porphyrin Derivatives

Porphyrin derivatives, including porphyrins, chlorins, bacteriochlorins, corroles, and phthalocyanines, have emerged as particularly attractive boron carriers based on their excellent characteristics that include [5,13,21]: (1) high tumor cell uptake and retention, (2) the ability to deliver large amounts of boron intracellularly, (3) low dark toxicities, (4) high tumor-to-blood and tumor-to-normal tissue boron concentration ratios, (5) optical tumor detection and ease of quantification of tissue-localized boron, and (6) the possibility of using photodynamic therapy (PDT) as an adjuvant treatment for BNCT (two porphyrin-based macrocycles are US Food and Drug Administration [FDA]-approved as PDT photosensitizers) [13,21,76]. Several reviews have been published on the synthesis of boronated porphyrin derivatives and their application in BNCT [13,21,77–79]. In this section, we review the synthesis and properties of the most recently reported porphyrin derivatives.

3.2.5.1 Porphyrin Macrocycles

Boronated porphyrins are typically synthesized by post-functionalization of the parent macrocycle, for example through nucleophilic substitution [80–83], nucleophilic addition [84], nucleophilic ring-opening reaction [85–88], and amination [89], or by tetramerization of boron-substituted precursors [13,90–92].

BOPP (**45**) has been intensely investigated in both *in vitro* and *in vivo* biological studies [5,93–96] that revealed high tumor uptake, good localizing properties, and retention ability in tumor-bearing mice, with high tumor-to-blood and tumor-to-brain boron concentration ratios. The combined administration of BOPP and BPA was shown to significantly increase the tumor uptake compared with BOPP or BPA alone [94]. It was also reported that changing the delivery of BOPP from i.v. injection to CED significantly enhanced the boron concentration in tumor, as well as the tumor-to-brain and tumor-to-blood ratios [93]. The high tumor cell uptake of BOPP was shown to be related to the low-density lipoprotein (LDL) receptors, which are upregulated in many tumors, including gliomas [95,96]. The porphyrin conjugated to a peptide nuclear localization sequence (**46**) was shown to associate non-covalently with LDL, as evidenced by gel electrophoresis, indicating high potential for tumor-seeking and -targeting abilities (Figure 3.2.8) [95].

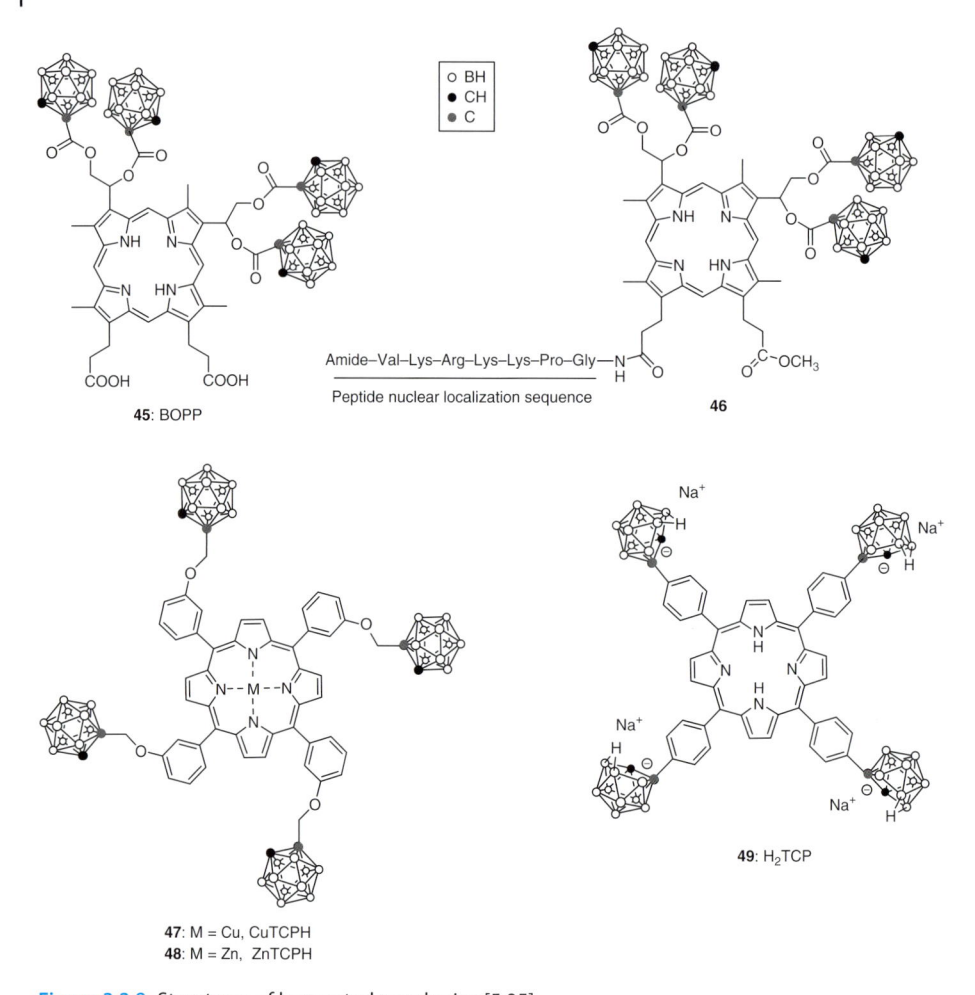

Figure 3.2.8 Structures of boronated porphyrins [5,95].

Among the metalated porphyrins, the Zn(II) and Cu(II) complexes have been the most investigated in *in vitro* and *in vivo* studies [5]. The Zn(II) complexes, for example ZnTCPH (**48**), generally enhance the fluorescence and photosensitizing properties of the macrocycle, while the Cu(II) complexes, such as (**47**), could be used for tumor imaging by subsequent labeling with ^{67}Cu for single-photon emission computed tomography (SPECT) or with ^{64}Cu for positron emission tomography (PET) [5]. A recent study showed that CuTCPH (**47**) and ZuTCPH (**48**) generated similar microdistributions of boron in various tissues of tumor-bearing mice upon intraperitoneal (i.p.) injection, with the liver and tumor taking up the largest amounts of boron [97].

Anionic carboranyl porphyrins are of particular interest for development as BNCT drugs due to their enhanced water solubility compared with the neutral analogs, and their normally lower toxicity and enhanced tumor uptake. Water-soluble H$_2$TCP (**49**) showed low dark toxicity (IC$_{50}$ > 100 µM) [91] and significant photosensitizing ability [91,98] toward murine melanotic melanoma cells (B16F1), making it possible to enhance anti-melanoma activity by combining BNCT with PDT treatment. The fluorescence microscopy observations showed that the porphyrin was largely localized intracellularly [91,98]. The *in vivo* study in mice bearing subcutaneous B16F1 tumors, via i.v. (~60 ppm ^{10}B) or i.t. (intratumoral) (~6 ppm ^{10}B) administration, indicated no detectable toxic effects, while the amount of H$_2$TCP (**49**) accumulated in the melanotic melanomas was sufficient to induce significant delay (5–6 days) in tumor growth after thermal neutron irradiation, compared with the control mice [98,99]. The H$_2$TCP (**49**) accumulated in low amounts in the skin and was cleared at a fast rate; therefore, it was unlikely to cause undesirable side effects, such as the onset of persistent skin photosensitivity as observed with first-generation PDT photosensitizers (e.g., Photofrin) [98]. To increase the boron content, more carboranyl clusters were attached to the porphyrin macrocycle. The water-soluble octa(*nido*-carboranyl)porphyrin H$_2$OCP (**50**) and its zinc(II) complex ZnOCP (**51**) showed low dark toxicity (IC$_{50}$ > 250 μM) toward V97 lung fibroblasts, and were readily taken up by human glioblastoma T98G cells, preferentially localizing in the lysosomes (Figure 3.2.9) [90]. Their cellular uptake, however, is slower and lower in comparison with the tetra(*nido*-carboranyl)porphyrin H$_2$TCP (**49**), possibly as a result of their higher tendency for aggregation in solutions. In addition, the free-base H$_2$OCP (**50**) showed higher cellular uptake in T98G cells compared with its zinc counterpart (**51**), in part due to the higher hydrophobic character and higher aggregation tendency of **50**.

Due to the facile synthesis of zwitterionic cobaltabis(dicarbollide) [3,3'-Co(8-C$_4$H$_8$O$_2$-1,2-C$_2$B$_9$H$_{10}$)(1',2'-C$_2$B$_9$H$_{11}$)], a series of porphyrins containing zwitterionic cobaltacarboranes (**52–62**) were synthesized by nucleophilic ring-opening of the dioxane

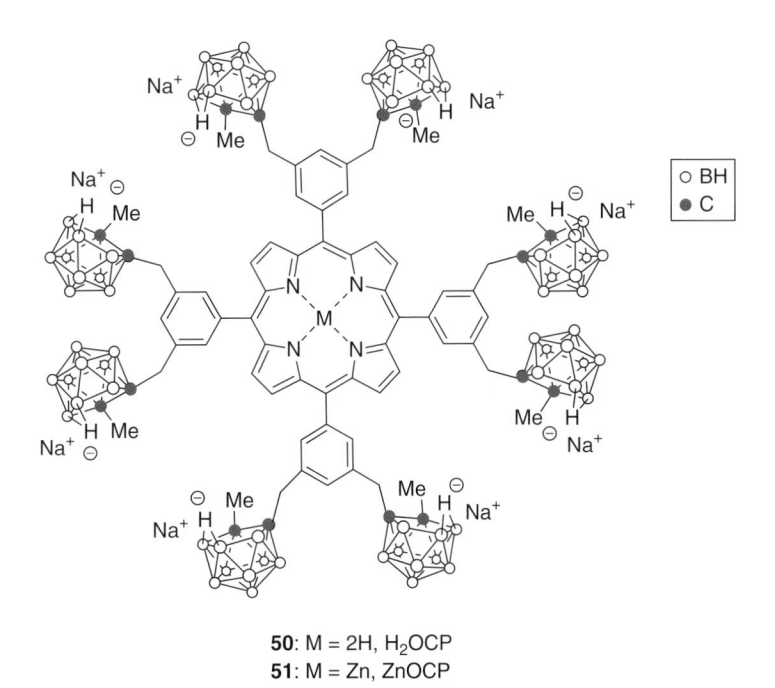

50: M = 2H, H$_2$OCP
51: M = Zn, ZnOCP

Figure 3.2.9 Structures of H$_2$OCP and ZnOCP [90].

ring. Their cellular uptake was highly dependent on the number of cobaltacarboranes on the porphyrin macrocycle, the charge distribution, as well as the lipophilic character and aggregation behavior (Figure 3.2.10) [85–88]. The cellular uptake in human HEp2 cells for compounds **52–56** increased as the number of cobaltacarborane moieties increased, with the porphyrin bearing four cobaltacarboranes (**56**) accumulating the most and the porphyrin bearing one cobaltacarborane (**52**) accumulating the least [85]. The position of the cobaltacarborane residues linked to the porphyrin macrocycle and the difference in charge distribution also affects the cellular uptake; compound **54** bearing two cobaltacarboranes on adjacent pyridyl rings accumulated to a higher extent than compound **53** with

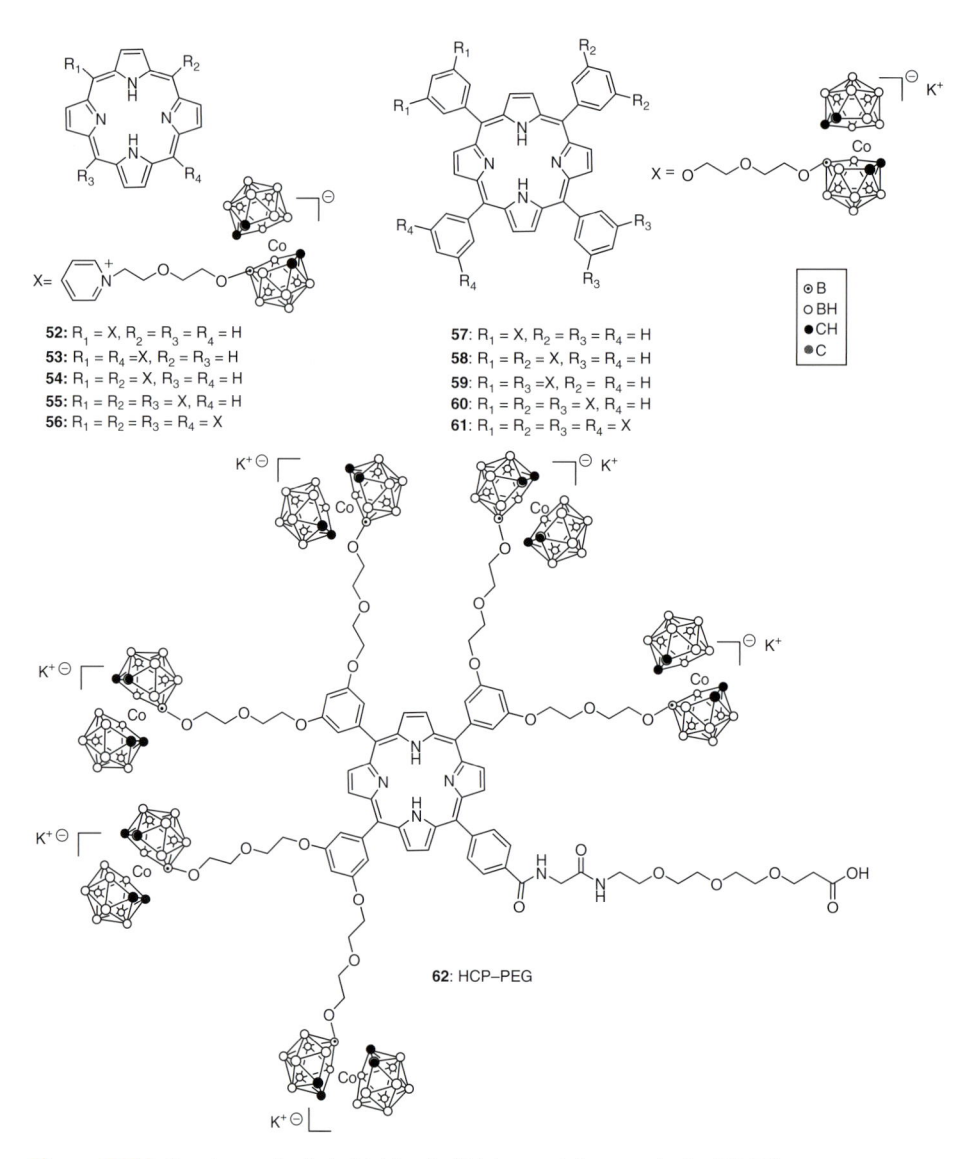

Figure 3.2.10 Structures of cobaltabis(dicarbollide)-containing porphyrins [85,87].

two cobaltacarboranes on opposite pyridyl rings [85]. More cobaltacarboranes were introduced into the porphyrin macrocycle to increase the loading of boron per molecule. The octa-cobaltacarboranyl porphyrin **61**, however, did not accumulate as much in HEp2 cells as porphyrin **58**, although it contained the largest amount of boron among the series of **57–61**, due to its poor solubility in aqueous solution and high tendency for aggregation [88]. Compounds **58** and **60** having cobaltacarboranes on adjacent positions showed higher cellular uptake after 24 h than compound **59** having cobaltacarboranes on opposite positions, the same trend as that of compounds **53** and **54** [88]. To increase the water solubility and to provide functionalization for further conjugation of the cobaltacarboranyl-porphyrin, a polyethylene glycol (PEG) segment was introduced to the macrocycle (**62**). The H_2OCP (**61**) and the pegylated HCP-PEG (**62**) were found to be nontoxic to human HEp2 cells up to concentrations of 50 μM (higher concentrations were not tested due to precipitation) and 400 μM, respectively [87]. Compounds **61** and **62** were also found to be non-phototoxic up to concentrations of 50 μM and 100 μM, respectively, limiting their use as adjuvant PDT sensitizers. The *in vivo* investigation in BALB/c mice revealed low toxicity for both compounds **61** and **62**, with the determined maximum tolerated dose (MTD) of 160 mg/kg for **61** and of 320 mg/kg for **62**. These compounds were relatively less toxic than other reported carboranyl porphyrins, including BOPP (**45**).

Fluorinated porphyrins have attracted considerable interest due to their enhanced biological properties (e.g., metabolism stability and cellular uptake) and potential use of ^{19}F-NMR and/or ^{18}F-PET in addition to porphyrin fluorescence for imaging [100–102]. A series of fluorinated carboranylporphyrins (**63–67**), both in the free-base form and as metal complexes, were synthesized by nucleophilic substitution of fluorinated porphyrins with lithiocarboranes, but their biological properties were not reported (Figure 3.2.11) [81]. To further increase tumor selectivity and uptake, boronated porphyrins incorporating a tumor-targeting moiety (e.g., peptides, linear and branched

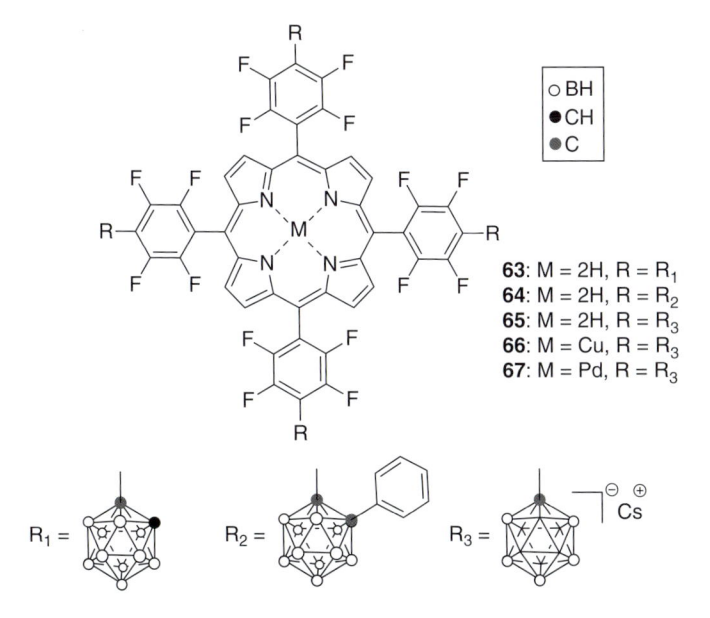

Figure 3.2.11 Fluorinated carboranylporphyrins [81].

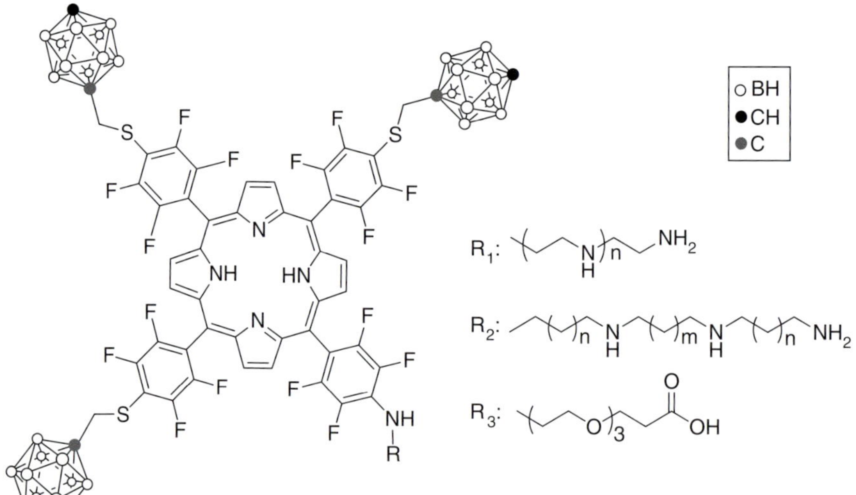

68: R = R₁, n = 2
69: R = R₁, n = 3
70: R = R₁, n = 4
71: R = R₂, n = 0, m = 1
72: R = R₂, n = 1, m = 1
73: R = R₂, n = 1, m = 2
74: R = R₂, n = 1, m = 0
75: R = R₃

Figure 3.2.12 Fluorinated carboranylporphyrins functionalized with polyamines and PEG [82,83].

polyamines, glucose, and arginine) were investigated (Figures 3.2.12 and 3.2.13) [82,83,103]. Recently, a series of fluorinated carboranylporphyrins conjugated to tumor-targeting moieties (**68–80**) were synthesized through sequential nucleophilic substitution reactions on the *p*-phenyl fluorides of *meso*-tetra-(pentafluorophenyl) porphyrin (TPPF) using thiol-terminated carborane [82,83]. The boron cluster can be attached either to the porphyrin macrocycle or to the tumor-targeting moiety, which results in different physicochemical and biological properties. In contrast to some carborane-containing derivatives of spermidine and spermine of significant toxicity (IC$_{50}$ < 25 µM in F98 rat glioma cells) [104], all compounds (**68–80**), including the ones bearing spermidine and spermine residues, showed low dark toxicity toward T98G cells (IC$_{50}$ > 250 µM) (Table 3.2.1), maybe as a result from the attachment of the carborane clusters to the porphyrin macrocycle rather than to the targeting molecule. All the compounds (**68–80**) preferentially localized in the ER, the Golgi, and the lysosomes of HEp2 cells, and their uptake into T98G glioma cells, varied significantly. The polyamine conjugates (**68–74**) showed faster and higher cellular uptake compared with the PEG conjugate (**75**) (Table 3.2.1), due to their positive charge (pKa of amine groups are in the range of 7–10) under physiological conditions that likely favor interactions with the negatively charged plasma membranes, and the upregulated polyamine transport system in tumor cells [105]. The extent of cellular uptake of polyamine conjugates (**68–74**) generally increases with the hydrophilicity of the conjugates (**73** > **72** > **69** ~ **70** > **71** > **74** > **68**) (Table 3.2.1). Among the series of **76–80**, the arginine conjugate

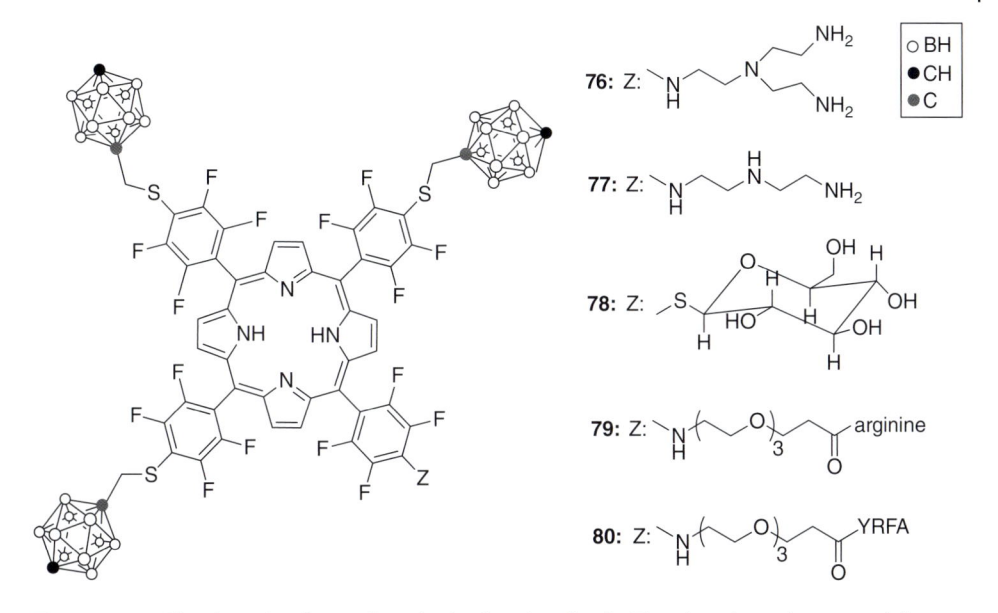

Figure 3.2.13 Fluorinated carboranylporphyrins functionalized with polyamines, glucose, arginine, and peptide [82,83].

Table 3.2.1 Cytotoxicity and uptake for porphyrin conjugates in T98G cells, and permeability coefficients (P) for porphyrin conjugates in hCMEC/D3 cells [82,83]

Compound	Dark toxicity IC$_{50}$ (μM)	T98G uptake at 24 h (nM/cell)	$P \times 10^{-6}$ (cm/s) in hCMEC/D3 cells
68	>400	0.0052	1.31 ± 0.08
69	>400	0.065	0.82 ± 0.06
70	>400	0.055	1.21 ± 0.04
71	>400	0.042	3.29 ± 0.03
72	>400	0.071	1.10 ± 0.06
73	>400	0.083	0.87 ± 0.08
74	>400	0.025	1.47 ± 0.05
75	296	0.0045	1.18 ± 0.03
76	>400	0.0072	2.32 ± 0.02
77	>400	0.0058	0.82 ± 0.02
78	>400	0.0068	0.62 ± 0.05
79	>400	0.013	1.44 ± 0.03
80	>400	0.0051	1.71 ± 0.06

exhibited especially higher uptake in T98G tumor cells at times >4 h, probably due to the unique ability of the guanidinium group of arginine to form bidentate hydrogen bonds with plasma membrane phosphates, favoring its internalization. The branched polyamine conjugate (**76**) accumulated to a higher extent in T98G cells than compound

77, likely due to its greater positive charge and more favorable charge distribution. All the conjugates displayed low dark toxicity and phototoxicity (IC$_{50}$ > 40 µM) in T98G cells, limiting their use as potential adjuvant PDT photosensitizers. All the conjugates also showed lower BBB permeability than lucifer yellow, which is a marker for low BBB permeability, including the glucose (**78**) and YRFA (**80**) conjugates. The reason for incorporation of glucose and opioid peptide YRFA was to target the glucose transporters and µ-opioid receptors, respectively, which are highly expressed in the brain capillary endothelial cells that form the BBB [106,107]. The observed low BBB permeabilities could be due to the high MW and high hydrophobicity of these conjugates.

3.2.5.2 Chlorin Macrocycles

Boronated chlorin and phthalocyanine derivatives are very promising to be used as dual BNCT and PDT sensitizers, as they absorb light in the red region of the spectrum (λ > 640 nm), which has deeper penetration through most tissues [5,21]. Several reviews have been published on the synthesis of boronated chlorins and their investigation as BNCT and PDT agents [5,13,21,108]. The most recently reported boronated chlorin derivatives are shown in Figures 3.2.14 through 3.2.18. These compounds are mainly synthesized by post-functionalization of the parent chlorin macrocycle with boron-substituted molecules through condensation reactions [109,110], nucleophilic substitution reactions [111,112], nucleophilic ring-opening reactions [113–116], Sonogashira couplings [117], and click reactions [118]. The boronated chlorins can also be synthesized by reduction of the corresponding boronated porphyrins [81].

TPFC (**81**) showed low dark toxicity toward HEp2 cells (IC$_{50}$ > 100 µM) and in tumor-bearing mice [119]. TPFC was shown to be readily taken up by melanotic melanoma B16F1 cells [119] and F98 rat glioma cells [112] in significant amounts (twice those achieved with BPA), showing extensive phototoxicity and BNCT efficiency upon light and neutron irradiation, respectively. The i.v. injected TPFC showed moderate tumor selectivity at relatively short times (<18 h) and fast clearance from tissues. Significant inhibition of tumor growth was observed upon light irradiation 3 h after injection, in comparison with the control mice [119]. The TPFC is not able to cross the BBB and therefore showed little accumulation in the brain, while the *in vivo* concomitant administration of BPA (i.v.) and TPFC (CED) greatly enhanced the BNCT efficiency compared with TPFC or BPA alone [112]. In contrast to TPFC, another two fluorinated carboranylchlorins (**82–83**) with cesium counterions showed good water solubility; however, their biological properties have not yet been reported [81].

The chlorin e$_6$–based boronated derivatives (**84–87**), bearing BSH or cobaltacarborane, were shown to accumulate in A549 human lung adenocarcinoma cells [117,118], although all the chlorin e$_6$–based boronated derivatives (**84–87**) had limited solubility in aqueous solutions. Compounds **88–90** were taken up by A549 cells efficiently, had relatively long retention times, and exhibited significant phototoxicity upon light irradiation [115,117]. The intracellular concentrations of **84** and **85**, bearing aromatic linkers, were below the therapeutic amounts for effective BNCT treatment [118]. This result is consistent with studies showing that flexible polyalkyl linkers between chlorin e$_6$ and the carboranyl group as in compounds (**88–90**) promote internalization of the conjugates, presumably due to enhanced anchoring of the conjugates onto the cellular membrane, while rigid aromatic linkers as in (**86–87**) disfavor anchoring [114].

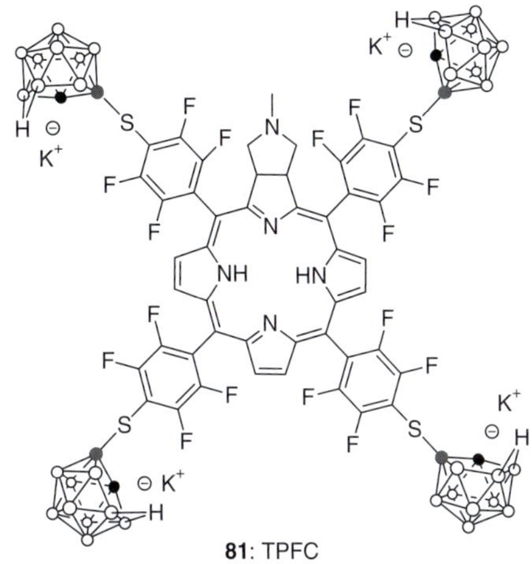

Figure 3.2.14 Structures of boronated chlorins [81,119].

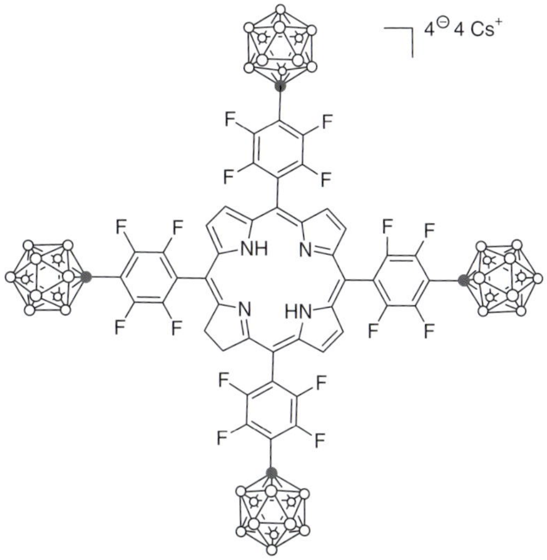

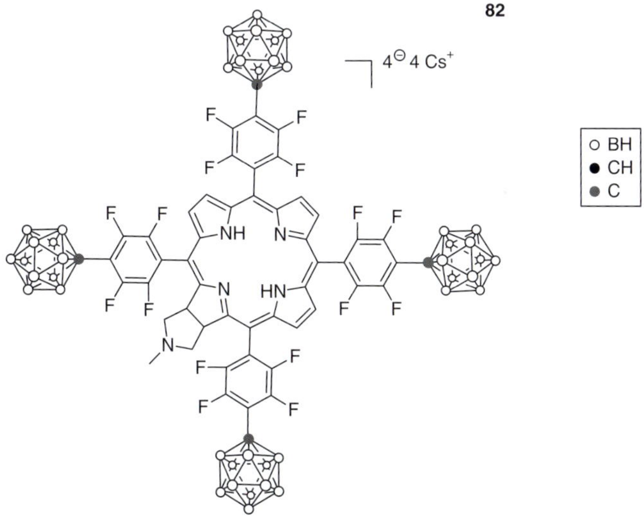

A compound bearing two cobaltacarborane moieties (**91**) showed lower cellular uptake than compounds **88–89** bearing a single cobaltacarborane group, maybe due to the enhanced negative charge of **91**. The *in vivo* biodistribution of **89–90** in tumor-bearing mice indicated that they accumulate preferentially in the liver and tumor, with a tumor-to-muscle ratio of approximately 3, 3 h after i.v. injection [115]. In contrast to **88**, **89**, and **91**, the recently reported compounds **93** and **95** bearing two BPA or BSH moieties showed one-fourth maximum accumulation time relative to that of their counterparts **92** and **94** bearing a single boron-containing moiety, in the tumor tissue of tumor-bearing mice. Furthermore, **93** and **95** showed rapid clearance from normal tissue within 24 h after i.v. injection [110,111]. Compounds **93** and **95** also showed significant tumor

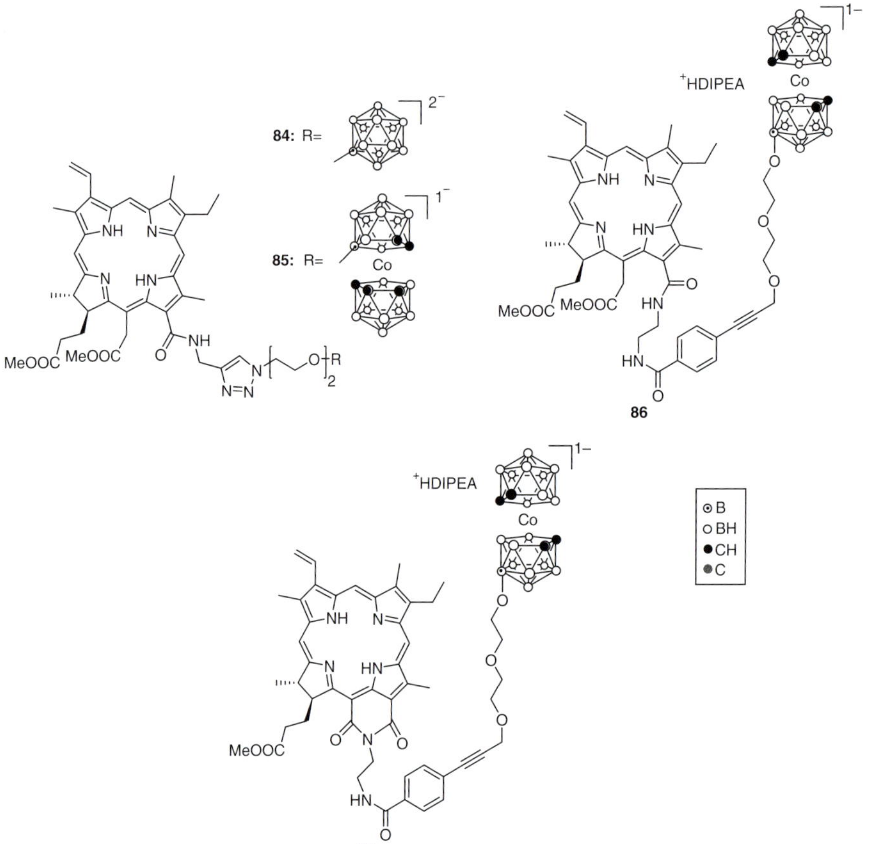

Figure 3.2.15 Structures of chlorin e_6–based boronated derivatives [117,118].

growth inhibition 3 h after injection, upon light irradiation (660 nm), suggesting **93** and **95** might be candidates for dual PDT and BNCT treatment. The ^{10}B concentration of **93** in the tumor at the maximum accumulation and the tumor-to-blood ratio, however, were below those required for effective BNCT [110].

3.2.5.3 Phthalocyanine Macrocycles

The most recently developed boronated phthalocyanines are shown in Figures 3.2.19 and 3.2.20. These phthalocyanines were synthesized either by cyclotetramerization of boronated precursors [120,121] or via nucleophilic substitution and nucleophilic ring-opening reactions on a preformed phthalocyanine macrocycle using boronated electrophiles and nucleophiles [122,123]. The A_3B-type Zn(II)-phthalocyanines (ZnPcs) containing one or two cobaltacarborane residues (**96–97**) showed good solubility in polar solvents, including methanol, acetone, DMF, and DMSO, and their quantum yields were in the range of 0.10–0.14. In addition, compounds **98** and **99** were shown to be efficient singlet oxygen generators [121–123]. The axially dicarborane substituted phthalocyanine (**98**) has higher singlet oxygen quantum yield (0.41 in DMSO) than its

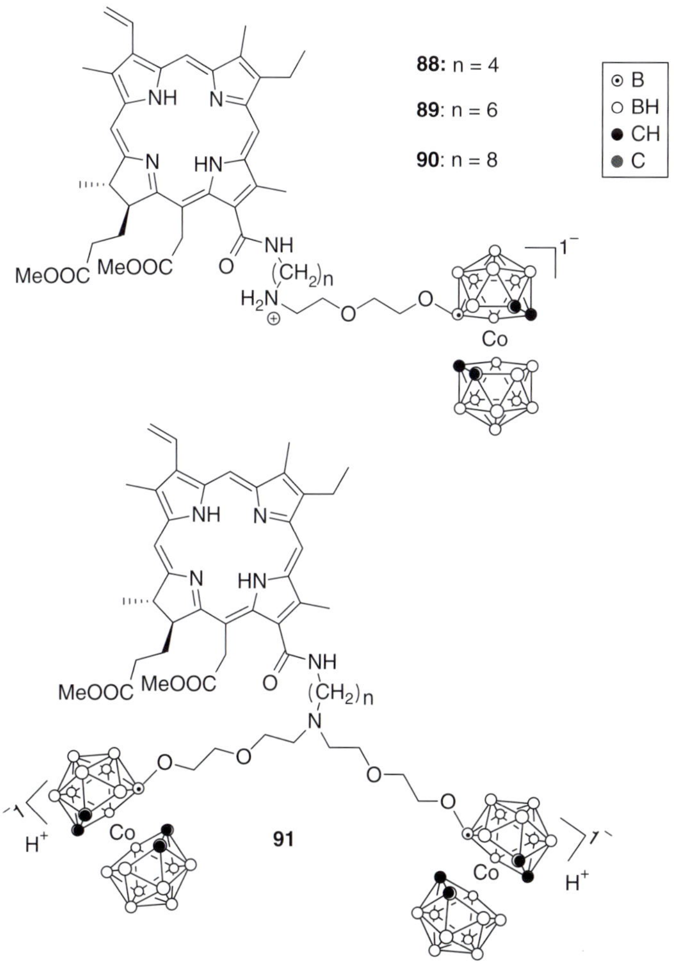

Figure 3.2.16 Structures of chlorin e_6–based boronated derivatives bearing cobaltabis(dicarbollide) moieties [114,115].

unsubstituted counterpart (0.15 in DMSO), likely due to partial protection of the bulky substituents from intermolecular quenching. Compound **98** was also more resistant to photochemical degradation compared with its unsubstituted analog. However, the potassium salts of these ZnPcs showed limited water solubility, in agreement with previous observations of potassium salt derivatives of cobaltacarboranes [124]. The biological properties of boronated phthalocyanines (**96–99**) have not yet been reported. On the other hand, the polyanionic *nido*-[ZnMCHESPc]Cs8 complex (**100**) showed relatively higher solubility in aqueous solutions, but still with significant tendency for aggregation [121]. The singlet oxygen quantum yield of **100** (0.24) was much lower than that of its neutral counterpart (0.63), but still high enough for photosensitization. The *in vitro* BNCT studies of **100** using neutron autoradiography indicated higher boron concentration compared with the control images.

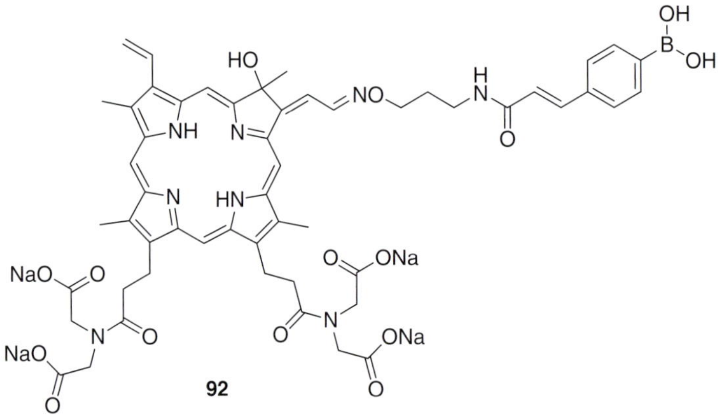

Figure 3.2.17 Structures of chlorin e₆–based boronated derivatives bearing BPA moieties [110].

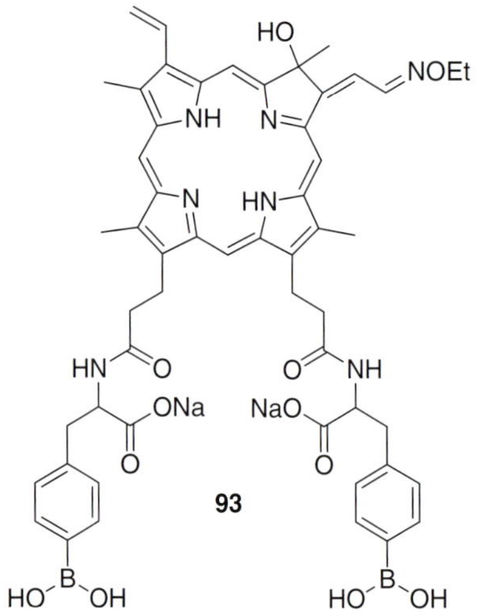

3.2.6 Boron Dipyrromethenes

BODIPYs, sometimes referred to as *semi-porphyrins*, have attracted considerable interest in recent years due to their various applications in imaging, biological labeling, drug delivery, sensing, and theranostics [125–129]. BODIPYs usually have very sharp Gaussian-shaped absorption and emission spectra, high molar extinction coefficients (usually $\varepsilon > 80{,}000\,M^{-1}\,cm^{-1}$), and high fluorescence quantum yields (commonly $\phi > 0.50$), which facilitate the detection of tissue-localized boron in BNCT. BODIPYs show relatively high photostability and biocompatibility, negligible sensitivity to solution pH and solvent polarity, and generally low cytotoxicity. In addition, BODIPYs are easy to synthesize and functionalize, allowing the fine-tuning of their physicochemical and biological properties. Despite their remarkable characteristics, studies on the potential use of BODIPYs as boron delivery agents are limited. Nakata et al. reported, for the first time, that BODIPYs (**101–103**) can cause destructive dynamic damage of

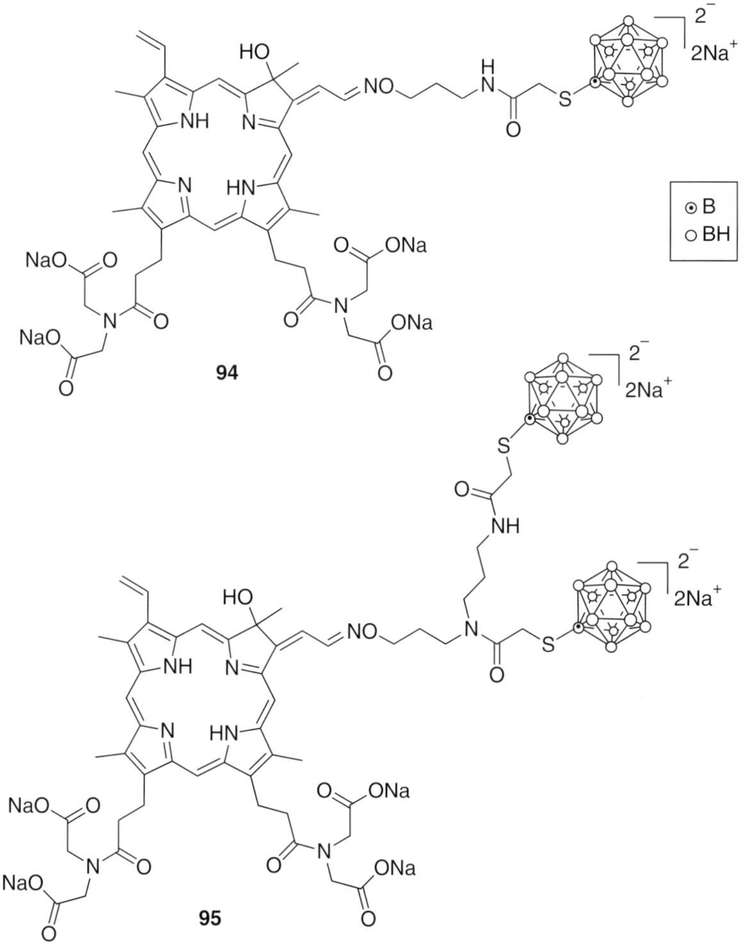

Figure 3.2.18 Structures of chlorin e₆–based boronated derivatives bearing BSH moieties [111].

BSA during thermal neutron irradiation, suggesting high potential of BODIPYs to be used as dynamic drugs for BNCT (Figure 3.2.21) [130,131].

Since BODIPYs have lower MW and generally higher solubility compared with porphyrin derivatives, they are of interest for investigation as BNCT agents. We recently reported, for the first time, two series of carboranyl BODIPYs and investigated their cytotoxicity, uptake into human glioma cells, and BBB permeability [130,131]. BODIPYs (**104–107**) bearing one or two *ortho*- or *para*-carborane clusters were synthesized by palladium(0)-catalyzed Suzuki cross-coupling reactions or by nucleophilic substitution reactions at the 2,6- or 8- positions, respectively, of the corresponding halogenated BODIPY precursors. The *in vitro* BBB permeability of the BODIPYs using hCMEC/D3 brain endothelial cells showed that only BODIPY **106** had higher permeability ($2.52 \pm 0.88 \times 10^{-5}$ cm/s) than lucifer yellow, likely due to its lower MW and hydrophobicity, as measured by the partition coefficients (log$P$) in 1-octanol-HEPES (Figure 3.2.22) [132,133]. To further investigate the relationship between the BBB permeability and the MW and hydrophobicity of

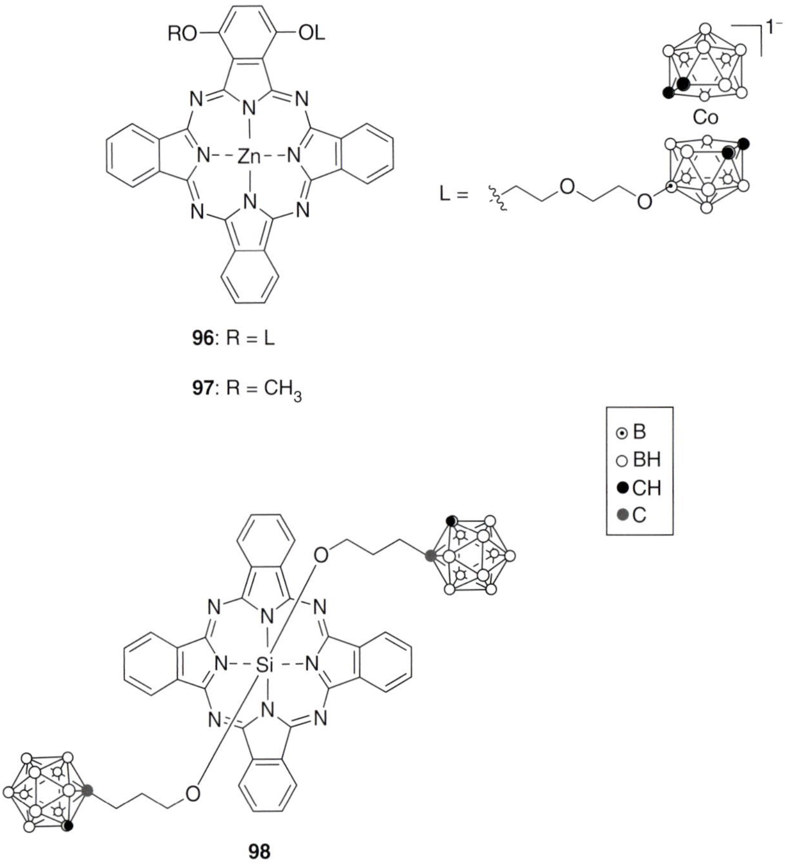

Figure 3.2.19 Structures of boronated phthalocyanines [120,122].

the BODIPYs, we most recently reported a series of seven carborane-containing BODIPYs (MW = 366–527 Da, log$P$ = 1.50–2.70) with *ortho*-carborane clusters at the 8- or 3/5-positions (**106, 108–113**), synthesized via nucleophilic substitution or palladium(0)-catalyzed Suzuki or Stille cross-coupling reactions. All the BODIPYs exhibited low dark toxicity and phototoxicity toward human glioma T98G cells. Their cellular uptake varied significantly, with **111** accumulating the most and **113** accumulating the least. All the BODIPYs (**106, 108–113**) showed higher BBB permeability than lucifer yellow, with **108** showing the highest value, as a result of its lower and most favorable MW (366 Da) and hydrophobicity (log$P$ = 1.50). The combination of amphiphilicity, high boron content, high tumor cellular uptake, low toxicity, and moderate BBB permeability makes BODIPYs promising BNCT agents for the treatment of brain tumors [133–135].

Carboranyl BODIPYs (e.g., **114**; Figures 3.2.23 and 3.2.24), synthesized using Sonogashira coupling reactions between BODIPYs containing aryl halogen groups and carboranes with terminal alkynyl groups or between BODIPYs containing terminal alkynyl groups and carboranes with aryl halogen groups, were recently reported [136–140]. All of those carboranyl BODIPYs, however, were investigated as fluorescent nanocars for single-molecule imaging, nonlinear materials, or low-energy photosensitizers, and were not yet evaluated as boron carriers for BNCT.

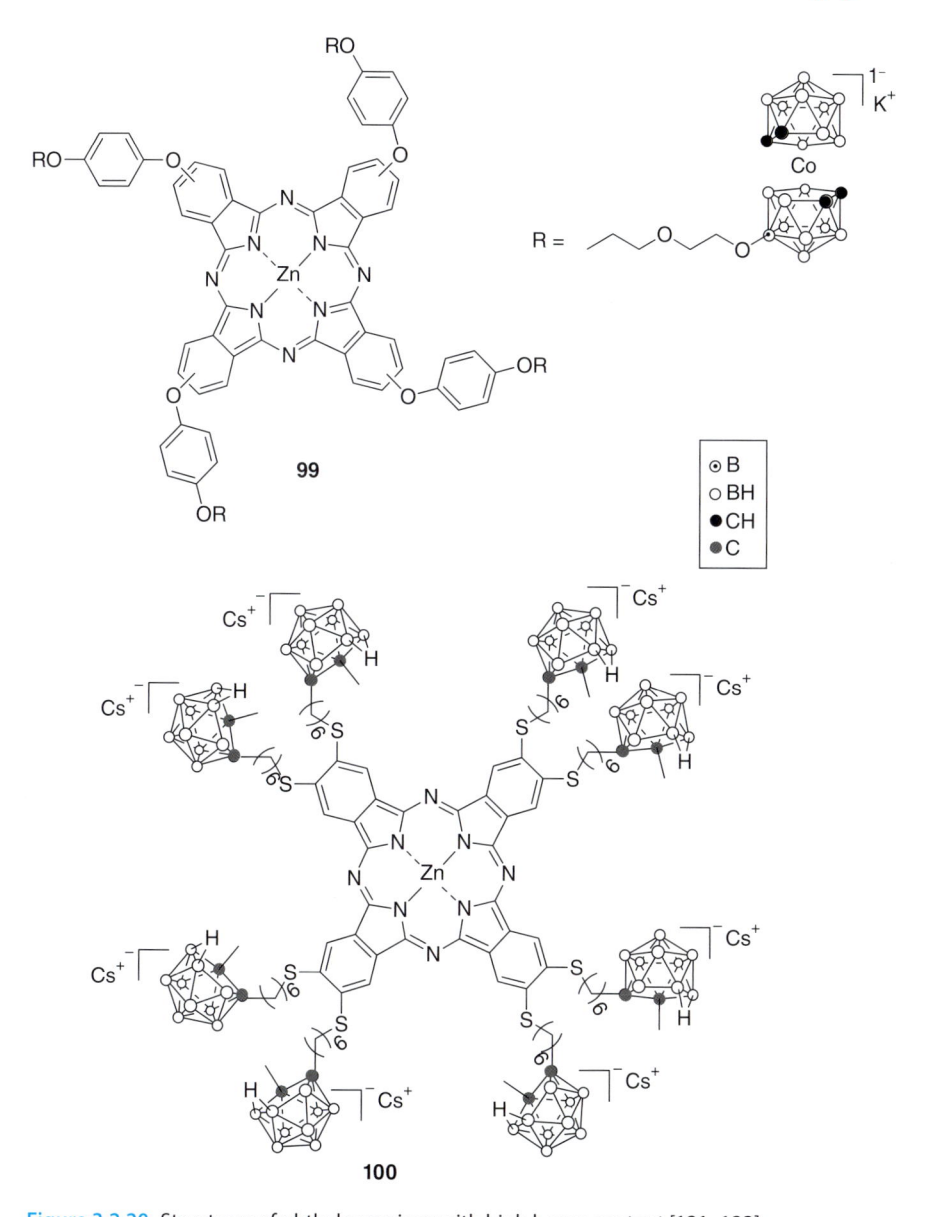

Figure 3.2.20 Structures of phthalocyanines with high boron content [121–123].

3.2.7 Liposomes

Liposomes are promising boron delivery agents for BNCT, due to their unique closed phospholipid bilayers that can encapsulate a large amount of boron drug(s) within the lipid bilayers and/or the aqueous core, depending on the lipophilicity of the drug. Another strategy is the incorporation of boron by using boron-functionalized lipids [5]. Liposomes can reduce the cytotoxicity of some drugs (e.g., doxorubicin), increase drug stability, and improve drug delivery [20,141]. Different types of liposomal boron carriers have been developed, including passive targeting liposomes based on the EPR effect

Figure 3.2.21 Structures of BODIPYs, which are effective for neutron dynamic therapy [131].

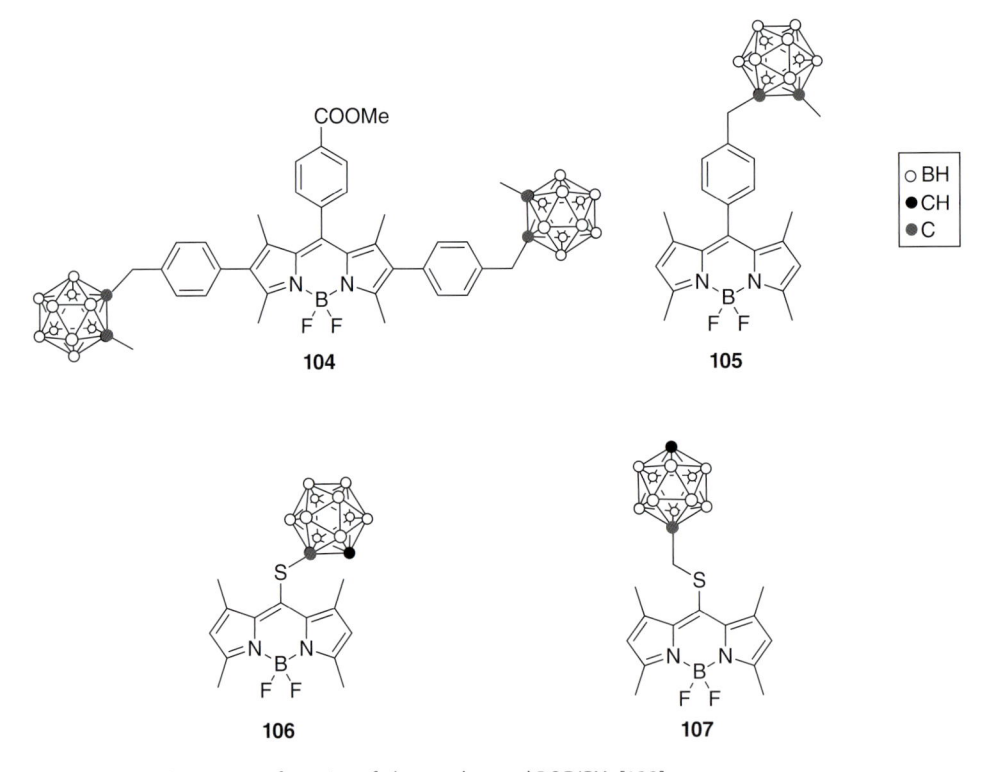

Figure 3.2.22 Structures of a series of *closo*-carboranyl BODIPYs [132].

[142–147] and/or active targeting liposomes incorporating tumor-selective moieties (e.g., folic acid [148], endothelial growth factor [149], and transferrin [150]). Several reviews have been published on the development of liposome-based boron delivery agents [5,20,29,30]. Recently, the DSBL (**115**)–25% PEG liposomes were shown in an *in vitro* study to be rapidly taken up by colon 26 cells into the cytoplasm via endocytosis, without degradation of the liposomes [146]. The liposomes retained in the cells caused low toxicity, but significant toxicity upon neutron irradiation. The *in vivo* study of DSBL–25% PEG liposomes in tumor-bearing mice revealed 22.7 ppm boron concentration (20 mg B/kg dose), and the tumor was significantly inhibited after neutron irradiation compared with the control mice. The liposomes were readily eliminated from the

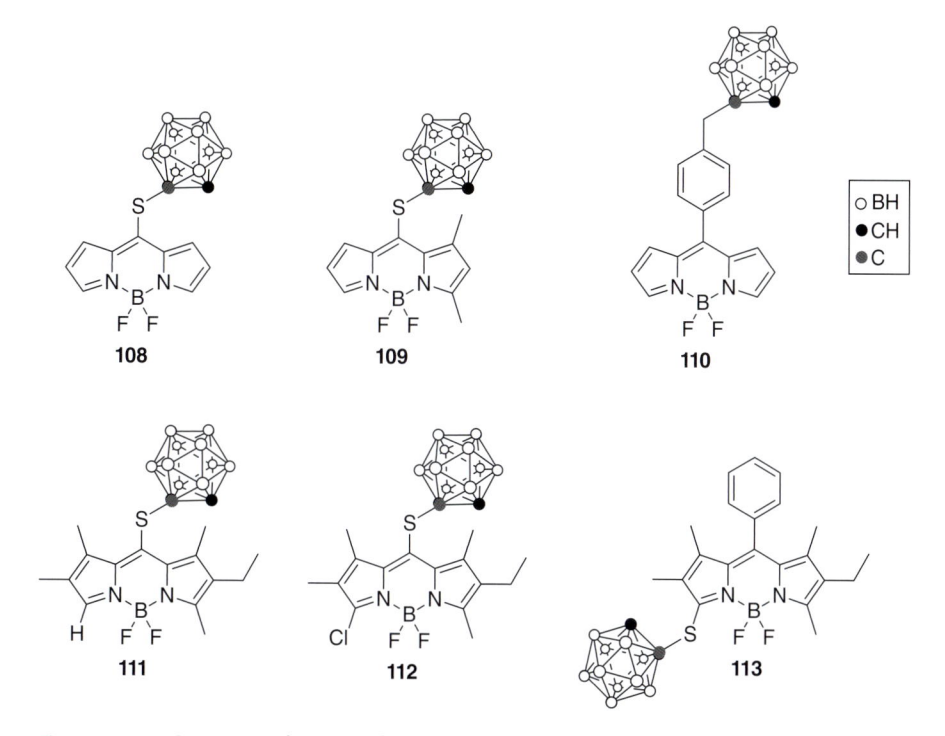

Figure 3.2.23 Structures of a series of carboranyl BODIPYs [134].

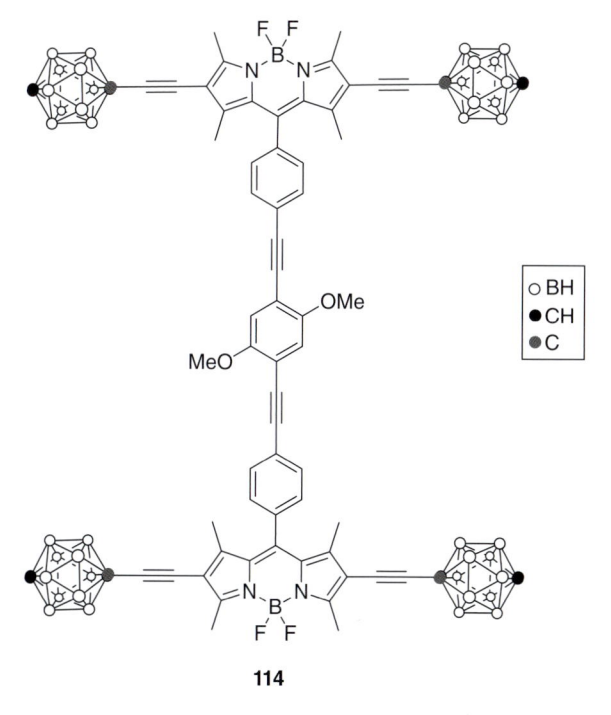

Figure 3.2.24 Structure of a carboranyl *bis*-BODIPY.

body without causing acute toxicity, in contrast to the *nido*-carborane lipid, which showed acute toxicity at 14 mg B/kg dose [150]. To increase the boron content of liposomes, BSH was encapsulated into DSBL (**115**) liposomes [144]. These BSH-encapsulating 10% DSBL liposomes displayed significant antitumor effect in colon 26 tumor-bearing mice; the tumor completely disappeared three weeks after thermal neutron irradiation following a 15 mg B/kg dose. To investigate the biodistribution of the boronated liposomes, fluorescently labeled *closo*-dodecaborate lipids (FL-SBL, **116**) were synthesized [142]. The *in vivo* imaging study of FL-SBL-labeled DSPS liposomes using colon 26 tumor-bearing mice showed selective delivery of boron liposomes to tumor tissue. Other dual MRI–BNCT probes were reported containing carboranyl cholesterol attached to Gd(III) (Gd-B-AC01, **117**), as well as to a folate receptor overexpressed in many tumors; these lipids selectively targeted tumors, and quantification of the boron concentration could be achieved by MRI (Figure 3.2.25) [148].

Cholesterol–metallacarborane (**118–120**) liposomes were recently reported to have effective incorporation into tumor tissues [143]. Up to 43.0 ppm (15 mg B/kg dose) of boron was achieved using the cholesterol conjugate-bearing cobalt (**118**) liposome, which is above the requirement for effective BNCT. In addition, liposomes carrying the hydrophilic polyhedral borane TAC as aqueous core and amphiphilic *nido*-carborane MAC in the bilayer were reported (Figure 3.2.26) [151,152]. TAC was shown to have long retention times in tumor as a result of covalent bond formation with nucleophiles present on endogenous intracellular proteins. The negative surface charge MAC was demonstrated to internalize via clathrin-mediated endocytosis. This liposomal system containing both TAC and MAC showed a high tumor-to-blood boron ratio and high boron concentration in tumor (5.68:1; 43 µg boron/g tumor) at 96 h after i.v. injection in mice bearing EMT6 tumors. Substantial tumor growth inhibition was achieved upon neutron irradiation. An important factor in the design and preparation of boron liposomes is the nature of the counter cation of negatively charged carboranes (e.g., *closo*-dodecaborates) [145,153]. It has been shown that the counter cations of *closo*-dodecaborates not only affect their encapsulation amount into liposomes and the liposome yield, but also significantly affect boron delivery into tumors. The use of spermidinium (spd) as a counter cation of *closo*-dodecaborates increased the yield of liposomes and favored a high boron-loading ratio in liposomes [145].

3.2.8 Nanoparticles

In recent years, nanoparticles (NPs) have attracted considerable interest for investigation as boron delivery agents for BNCT. Due to their high surface-to-volume ratio, a large amount of boron can be loaded into NPs while lowering the overall dose and toxicity. In addition, NPs generally increase the blood circulation time, and rely on the EPR effect for tumor accumulation. Several reviews on the design of boron-containing NPs and their potential as BNCT agents have been published [20,26,28,29]. Different strategies, including surface adsorption, encapsulation, or direct covalent linkage, can be used for incorporation of boron agents into NPs. In this section, we review the most recent examples of NP boron delivery agents.

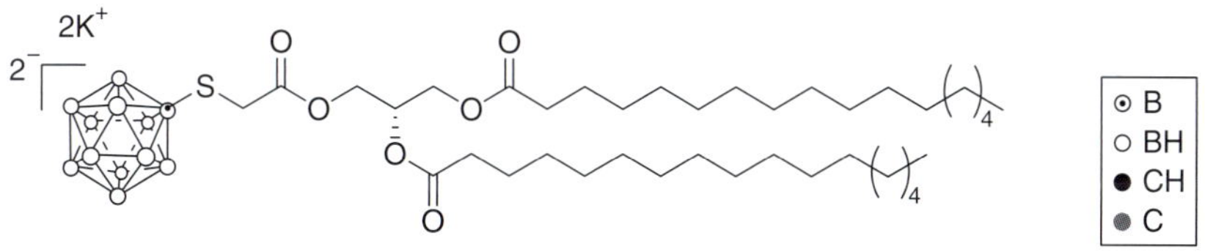

115: DSBL

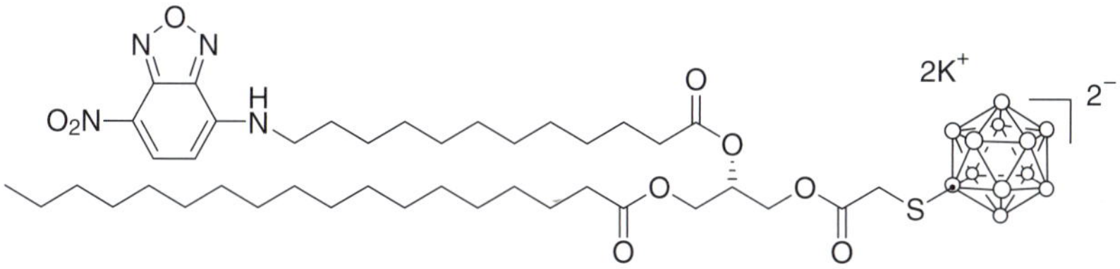

116: FL-SBL

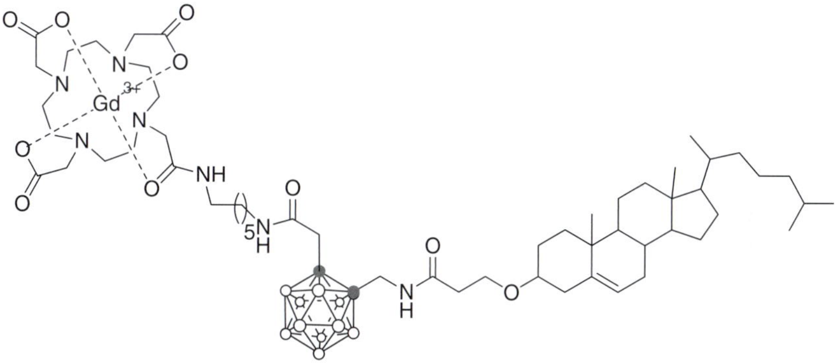

117: Gd-B-AC01

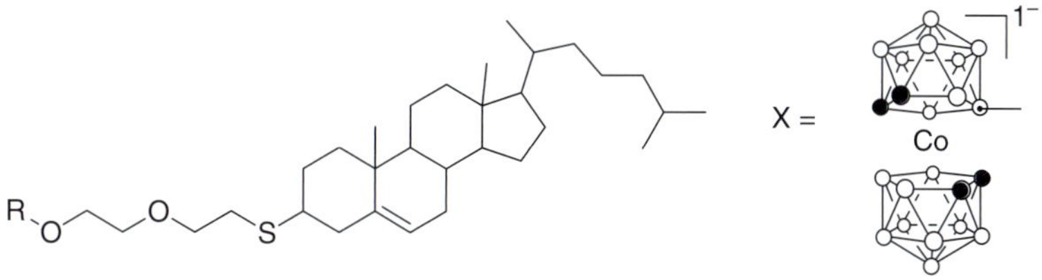

118: R = X, M = Co
119: R = X, M = Fe
120: R = X, M = Cr

Figure 3.2.25 Structures of boronated lipids [142,143,146,148].

The growth of BSH onto the surface of silicon nanowires (SiNWs), using an anodic bias process based on the negatively charged BSH, has been investigated [154]. The boron concentration loaded on SiNWs was shown to be six orders of magnitude higher than the boron concentration achieved in tumor tissue after i.v. administration of BSH

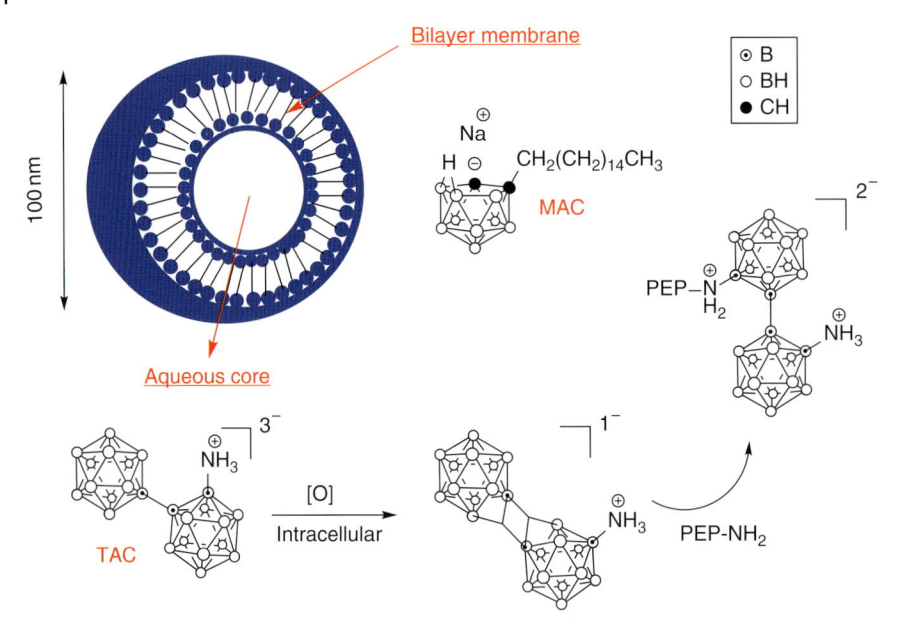

Figure 3.2.26 Liposomes carrying hydrophilic polyhedral borane TAC in the aqueous core and *nido*-carborane MAC in the bilayer [151].

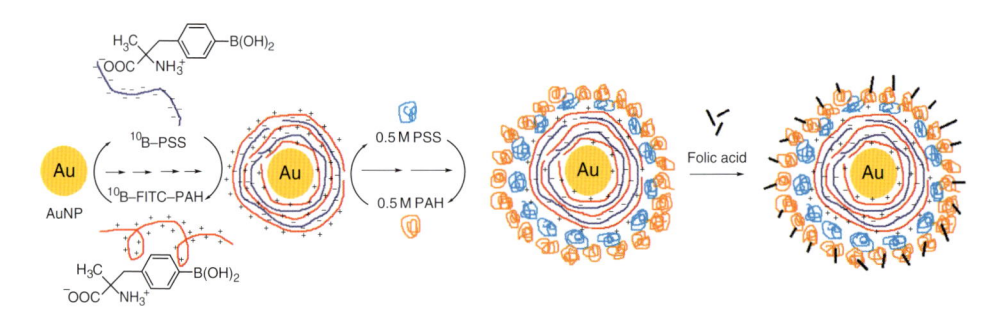

Figure 3.2.27 Schematic representation of the model design and synthesis of multifunctionalized Au NPs [156].

[155]. Multifunctionalized gold NPs containing a fluorescent dye (FITC), BPA, and folic acid were obtained by deposition of positively charged boronated poly-FITC allylamine hydrochloride (^{10}B-FITC-PAH) and negatively charged boronated polysodium 4-styrenesulfonate (^{10}B-PSS) onto gold NPs, layer by layer, followed by electrostatic conjugation of folic acid to the outermost layer (Figure 3.2.27) [156]. The folic acid incorporated was used as a tumor-targeting moiety to bind folic acid receptors overexpressed in tumor cells. These dual imaging–BNCT multilayer gold NPs exhibited significant uptake and favorable biocompatibility in human cancer cells.

Thiolate-stabilized gold NPs have facilitated boron incorporation into NPs due to their facile preparation. The mercaptocarborane was shown to be capped onto gold NPs through Au–thiolate ligand interactions [157,158]. These hydrophilic mercaptocarborane-capped gold NPs displayed high cellular uptake, accumulating inside membranes

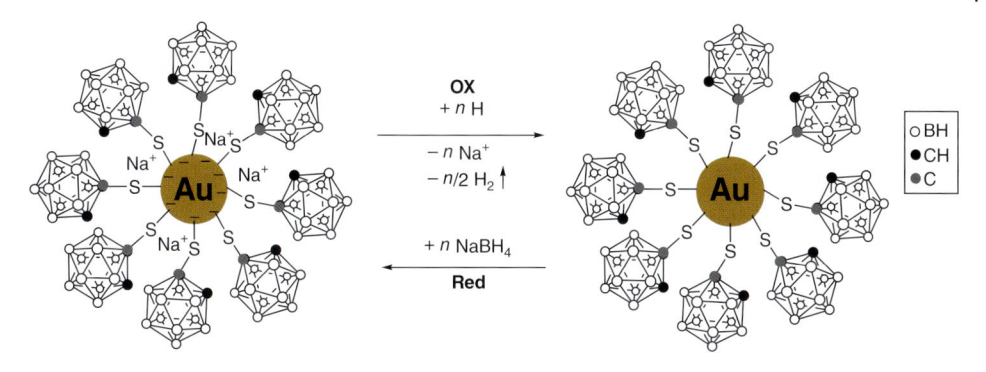

Figure 3.2.28 Mercaptocarborane-capped Au NPs [157].

due to the oxidation of the NPs and their subsequent inclusion into cell membranes (Figure 3.2.28). These NPs, however, showed pronounced toxicity toward HeLa cells. Carboranes and PEG moieties were incorporated onto gold NPs (Figure 3.2.29) through either Cu(I)-catalyzed azide–alkyne cycloaddition CuAAC ("click") reactions at the periphery of azido-terminated AuNPs or simply stabilization of AuNPs using a tris-carborane thiol dendron or a hybrid dendron containing both carborane and PEG resulting in water-soluble gold NPs [159]. The stability of these physically deposited boron moieties on NPs under physiological conditions, however, remains to be investigated.

Boron moieties can also be covalently incorporated into NPs to increase their stability. Covalently incorporated boron NPs can be achieved through a radical polymerization of dextran-bounded 3-acrylamidophenylboronic acid (APBA) (**121–122**) [160]. These NPs showed good stability in a wide pH range, high cellular uptake, and low toxicity in human gastric carcinoma MKN-28 cells ($IC_{50} > 200\,\mu g/mL$). Two biocompatible polymers, poly(2-(hydroxyethyl)methacrylate) (HEMA) and poly(2-(methacryloyloxy)ethyl succinate) (MES), were introduced to the surface of Si NPs (**123–124**) by surface-initiated atom transfer radical polymerization. The carboxylic acid and hydroxyl functional groups of the polymer chains were further functionalized with carboranyl clusters via condensation reactions to obtain covalently boronated Si NPs [161]. The NPs contained 13 and 18% boron atoms by weight, providing high amounts of boron for BNCT. The Si NPs can also be functionalized *in situ* with silylated-FITC via a Stöber method, followed by functionalization with decaborate–triethoxysilane precursor to produce dye-doped decaborate Si NPs (**125**) as dual imaging and BNCT agents [162]. However, the grafting of decaborate onto Si NPs is low (Figures 3.2.30 and 3.2.31).

The inexpensive and standard boron phosphate (BPO_4) is another candidate for boron delivery NPs. BPO_4 NPs conjugated with the tumor-targeting folic acid (**126**) were synthesized via condensation reactions between amine-terminated BPO_4 and folic acid [163]. These folic acid–functionalized BPO_4 NPs showed detectable cytotoxicity in several cell lines, although the toxicity is lower compared with the bare BPO_4 NPs.

Boron nitride nanotubes (BNNTs) are currently attracting wide attention due to their advantages over carbon nanotubes (CNTs), including biocompatibility and high thermal and chemical stabilities [164]. A dispersed solution combining BNNTs and DSPE-PEG2000 exhibited higher boron accumulation and higher BNCT antitumor effect in B16 melanoma cells compared with BSH, rendering it a promising BNCT agent [165].

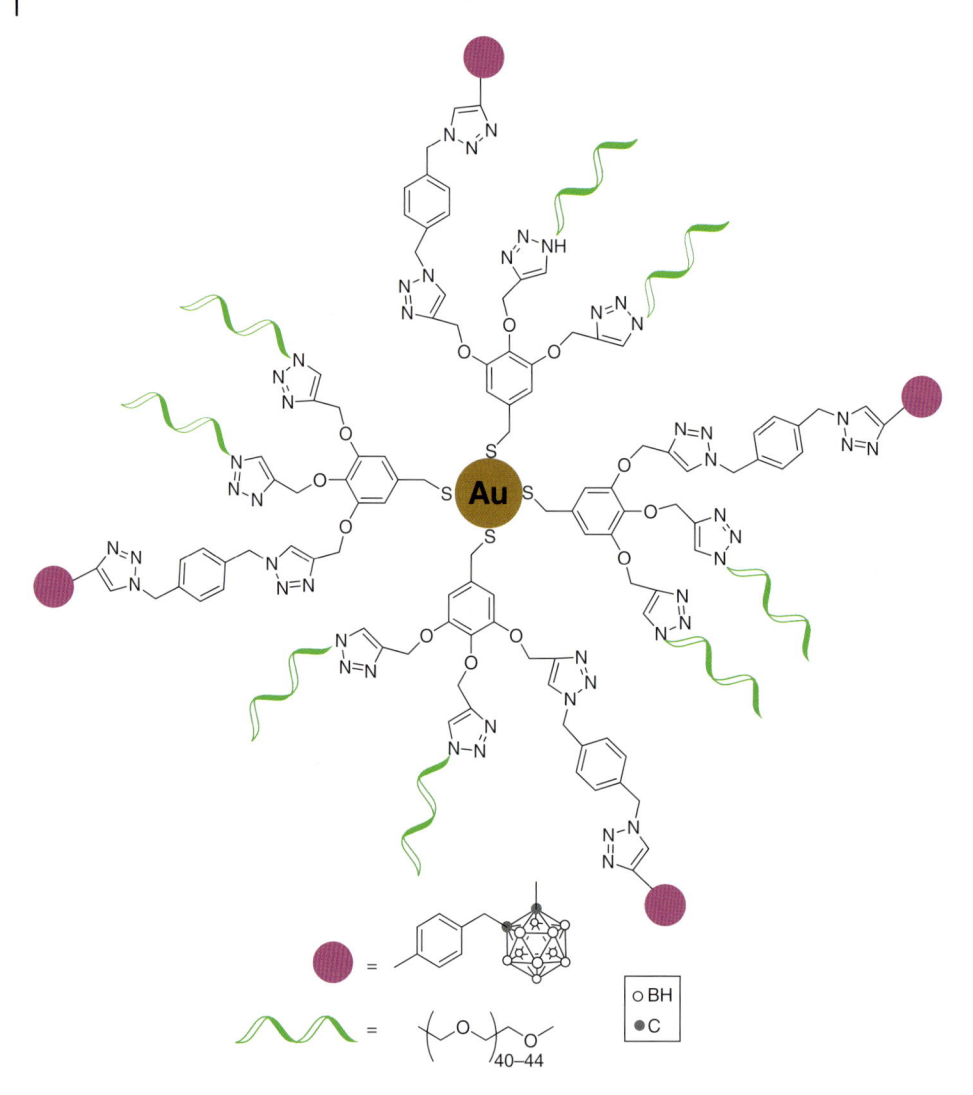

Figure 3.2.29 Au NPs incorporating *ortho*-carboranes and PEG groups [159].

Amphiphilic copolymer NPs have been of long-standing interest in drug delivery. Recently, a carborane-containing copolymer, poly(ethyleneglycol)-b-poly(L-lactide-co-2-methyl-2(2-dicarba-closo-dodecarborane)propyloxycarbonyl-propyne carbonate (PLMB) (**127**), was synthesized via reaction between decaborane and the side alkynyl groups, and self-assembled with doxorubicin (DOX) to form drug-loaded NPs, for potential binary use in BNCT and chemotherapy [166]. The *in vivo* neutron irradiation study of the DOX@PLMB NPs displayed higher tumor suppression efficiency than that of chemotherapy or BNCT alone. As the γ-ray emission from the boron nuclear reaction can cause adverse effects such as inflammation, a reactive oxygen species (ROS)-scavenging moiety along with a boron cluster were incorporated into the polymeric NPs to increase the efficacy of BNCT [167]. The multifunctional NPs were synthesized via

Figure 3.2.30 Dextran-bounded APBA and boronated Si NPs [160,161].

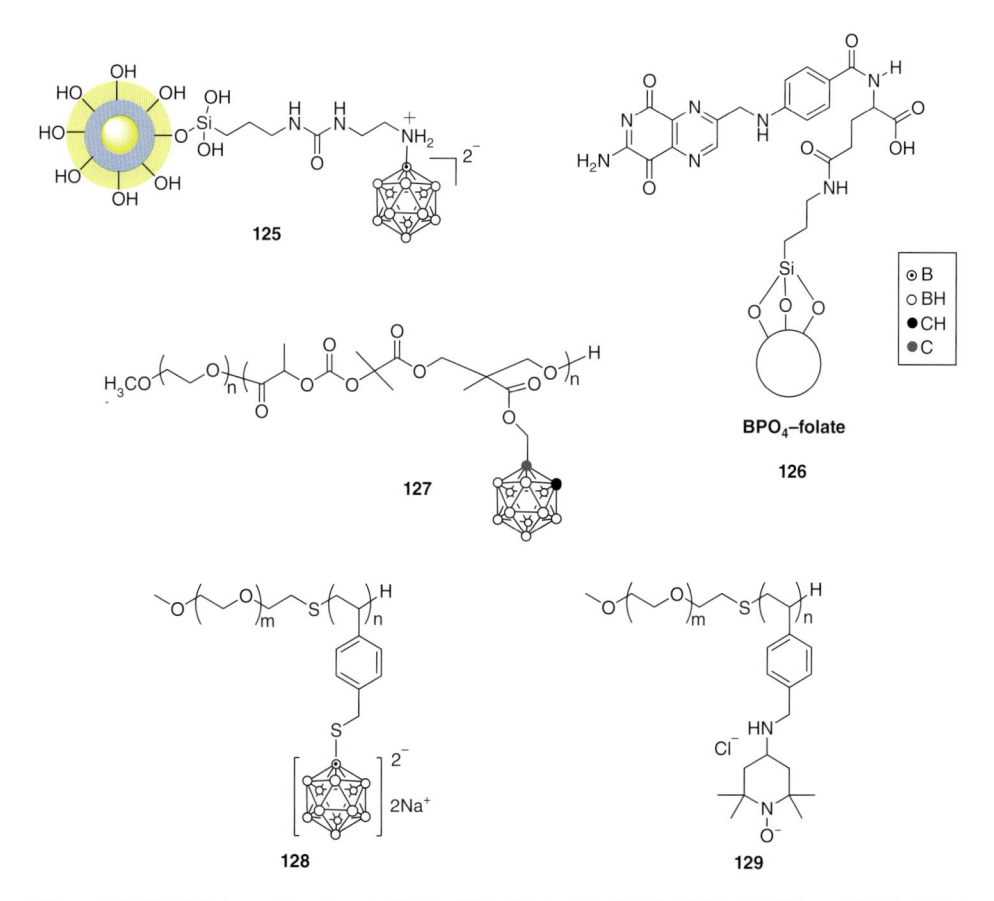

Figure 3.2.31 FITC-doped decaborate Si NPs, BPO$_4$–folate NPs, PLMB, PEG-b-PMBSH, and PEG-b-PMNT [162,163,166,167].

polyion complex formation of an anionic block copolymer (PEG-b-poly((*closo*-dodecaboranyl)thiomethylstyrene) (PEG-b-PMBSH) (**128**)) containing BSH clusters on the side chains and a cationic block copolymer (PEG-b-poly(4-(2,2,6,6-tetramethylpiperidine-N-oxyl)aminomethylstyrene) (PEG-b-PMNT) (**129**)) containing nitroxide radical moieties on the side chains. The resulting NPs showed low cytotoxicity *in vitro* toward cancer cells (C26), and threefold higher cellular uptake in C26 cells compared with normal cells (HAECs). Significant suppression of tumor growth, metastasis, and adverse effects caused by ROS was observed in tumor-bearing mice after neutron irradiation, although the boron accumulation in tumor was lower than the therapeutic amount. New boron-containing superparamagnetic iron oxide nanoparticles (SPIONs) were recently reported to potentially target tumor via an external applied magnetic field after intraarterial administration [14].

3.2.9 Conclusions

BNCT has the potential to be used as an effective therapy for the treatment of various tumors, especially high-grade gliomas and other difficult-to-treat malignancies for which chemo and radiation therapies are inefficient. Since the initial BNCT clinical trials over 50 years ago that used boronic acid as the boron carrier and thermal neutrons of limited tissue penetration, considerable advances have been achieved in BNCT drug development. Two clinically approved second-generation BNCT agents, BPA and BSH, have shown to be safe and efficacious in BNCT clinical trials, although the results are still far from ideal. Third-generation boron delivery agents of low toxicity, with enhanced tumor-seeking and tumor-localizing properties, and able to deliver therapeutic amounts of boron selectively into tumors, have been the focus of research in recent years. Different classes of BNCT agents currently under investigation are discussed in this chapter. Several of these third-generation agents have shown high tumor selectivity and uptake, with high tumor-to-normal tissue and tumor-to-blood boron concentration ratios, and are highly promising BNCT agents. In some cases, the boron carriers have the ability to bind to DNA and localize in close proximity to the cell nucleus, which enhances their killing effects and decreases the amount of boron required for effective BNCT. Additional preclinical and clinical trials on the most promising BNCT agents need to be carried out before the full potential of BNCT is realized. In addition, several studies indicate that the combination of different boron carriers, such as BSH and BPA, can lead to higher BNCT efficacy, compared with the use of a single boron delivery agent, due to the target of multiple tumor compartments. Therefore, the investigation and evaluation of combined boron delivery agents should constitute a future direction in BNCT research. In addition, for enhancement of BNCT efficacy, alternative administration methods, such as CED, should be further explored for the delivery of large amounts of boron into tumors with very high selectivity. Furthermore, to facilitate treatment planning and maximize the tumor-killing effect with minimal damage to normal tissues, the neutron irradiation treatment should be applied at the highest tumor-to-normal tissue and tumor-to-blood ratios. To achieve this goal, BNCT agents containing readily detectable moieties, such as a fluorescence label, or a PET, SPECT, or MRI agent, may play a prominent role in tracking and quantifying tissue-localized boron, treatment planning, and outcome. In addition, the combination of BNCT with surgical resection, chemotherapy, and/or radiation therapy could lead to superior clinical efficacy for tumor eradication, particularly for high-grade gliomas and other difficult-to-treat tumors for which there is currently no cure.

References

1 Hawthorne, M. F. The Role of Chemistry in the Development of Boron Neutron Capture Therapy of Cancer. *Angew. Chem. Int. Ed.* **1993**, *32* (7), 950–984.
2 Soloway, A. H.; Tjarks, W.; Barnum, B. A.; Rong, F.-G.; Barth, R. F.; Codogni, I. M.; Wilson, J. G. The Chemistry of Neutron Capture Therapy. *Chem. Rev.* **1998**, *98* (4), 1515–1562.
3 Barth, R. F.; Coderre, J. A.; Vicente, M. G. H.; Blue, T. E. Boron Neutron Capture Therapy of Cancer: Current Status and Future Prospects. *Clin. Cancer Res.* **2005**, *11* (11), 3987–4002.
4 Barth, R. F.; Vicente, M.; Harling, O. K.; Kiger, W.; Riley, K. J.; Binns, P. J.; Wagner, F. M.; Suzuki, M.; Aihara, T.; Kato, I.; Kawabata, S. Current Status of Boron Neutron Capture Therapy of High Grade Gliomas and Recurrent Head and Neck Cancer. *Radiat. Oncol.* **2012**, *7* (1), 1–21.
5 Sibrian-Vazquez, M.; Vicente, M. G. H. Boron Tumor-Delivery for BNCT: Recent Developments and Perspectives. In *Boron Science: New Technologies & Applications*, Hosmane, N. S. (ed.); CRC Press: Boca Raton, FL, **2011**; pp. 203–232.
6 Kankaanranta, L.; Seppälä, T.; Koivunoro, H.; Saarilahti, K.; Atula, T.; Collan, J.; Salli, E.; Kortesniemi, M.; Uusi-Simola, J.; Välimäki, P.; Mäkitie, A.; Seppänen, M.; Minn, H.; Revitzer, H.; Kouri, M.; Kotiluoto, P.; Seren, T.; Auterinen, I.; Savolainen, S.; Joensuu, H. Boron Neutron Capture Therapy in the Treatment of Locally Recurred Head-and-Neck Cancer: Final Analysis of a Phase I/II Trial. *Int. J. Radiat. Oncol. Biol. Phys.* **2012**, *82* (1), e67–e75.
7 Hopewell, J. W.; Gorlia, T.; Pellettieri, L.; Giusti, V.; H-Stenstam, B.; Sköld, K. Boron Neutron Capture Therapy for Newly Diagnosed Glioblastoma Multiforme: An Assessment of Clinical Potential. *Appl. Radiat. Isot.* **2011**, *69* (12), 1737–1740.
8 Banks, W. A. Characteristics of Compounds That Cross the Blood-Brain Barrier. *BMC Neurol.* **2009**, *9* (Suppl. 1), S3–S3.
9 Habgood, M. D.; Begley, D. J.; Abbott, N. J. Determinants of Passive Drug Entry into the Central Nervous System. *Cell. Mol. Neurobiol.* **2000**, *20* (2), 231–253.
10 Vries, H. E. d.; Kuiper, J.; Boer, A. G. d.; Berkel, T. J. C. V.; Breimer, D. D. The Blood-Brain Barrier in Neuroinflammatory Diseases. *Pharmacol. Rev.* **1997**, *49* (2), 143–156.
11 Pardridge, W. The Blood-Brain Barrier: Bottleneck in Brain Drug Development. *NeuroRx* **2005**, *2* (1), 3–14.
12 Palmer, A. M.; Alavijeh, M. S. Translational CNS Medicines Research. *Drug Discov. Today* **2012**, *17* (19–20), 1068–1078.
13 Bhupathiraju, N. V. S. D. K.; Vicente, M. G. H. Synthesis of Carborane-Containing Porphyrin Derivatives for the Boron Neutron Capture Therapy of Tumors. In *Applications of Porphyrinoids*, Paolesse, R. (ed.); Springer: Berlin, **2014**; pp. 31–52.
14 Tietze, R.; Unterweger, H.; Dürr, S.; Lyer, S.; Canella, L.; Kudejova, P.; Wagner, F. M.; Petry, W.; Taccardi, N.; Alexiou, C. Boron Containing Magnetic Nanoparticles for Neutron Capture Therapy: An Innovative Approach for Specifically Targeting Tumors. *Appl. Radiat. Isot.* **2015**, *106*, 151–155.

15 Kawabata, S.; Miyatake, S.-I.; Hiramatsu, R.; Hirota, Y.; Miyata, S.; Takekita, Y.; Kuroiwa, T.; Kirihata, M.; Sakurai, Y.; Maruhashi, A.; Ono, K. Phase II Clinical Study of Boron Neutron Capture Therapy Combined with X-Ray Radiotherapy/Temozolomide in Patients with Newly Diagnosed Glioblastoma Multiforme: Study Design and Current Status Report. *Appl. Radiat. Isot.* **2011**, *69* (12), 1796–1799.

16 Kankaanranta, L.; Seppälä, T.; Koivunoro, H.; Välimäki, P.; Beule, A.; Collan, J.; Kortesniemi, M.; Uusi-Simola, J.; Kotiluoto, P.; Auterinen, I.; Serèn, T.; Paetau, A.; Saarilahti, K.; Savolainen, S.; Joensuu, H. L-Boronophenylalanine-Mediated Boron Neutron Capture Therapy for Malignant Glioma Progressing after External Beam Radiation Therapy: A Phase I Study. *Int. J. Radiat. Oncol. Biol. Phys.* **2011**, *80* (2), 369–376.

17 Yamamoto, T.; Nakai, K.; Nariai, T.; Kumada, H.; Okumura, T.; Mizumoto, M.; Tsuboi, K.; Zaboronok, A.; Ishikawa, E.; Aiyama, H.; Endo, K.; Takada, T.; Yoshida, F.; Shibata, Y.; Matsumura, A. The Status of Tsukuba BNCT Trial: BPA-Based Boron Neutron Capture Therapy Combined with X-Ray Irradiation. *Appl. Radiat. Isot.* **2011**, *69* (12), 1817–1818.

18 Barth, R. F.; Joensuu, H. Boron Neutron Capture Therapy for the Treatment of Glioblastomas and Extracranial Tumours: As Effective, More Effective or Less Effective Than Photon Irradiation? *Radiother. Oncol.* **2007**, *82* (2), 119–122.

19 Ichikawa, H.; Taniguchi, E.; Fujimoto, T.; Fukumori, Y. Biodistribution of BPA and BSH after Single, Repeated and Simultaneous Administrations for Neutron-Capture Therapy of Cancer. *Appl. Radiat. Isot.* **2009**, *67* (7–8, Suppl.), S111–S114.

20 Luderer, M. J.; de la Puente, P.; Azab, A. K. Advancements in Tumor Targeting Strategies for Boron Neutron Capture Therapy. *Pharm. Res.* **2015**, *32* (9), 2824–2836.

21 Vicente, M. G. H.; Sibrian-Vazquez, M. Syntheses of Boronated Porphyrins and Their Application in BNCT. In *Handbook of Porphyrin Science: With Applications to Chemistry, Physics, Materials Science, Engineering, Biology and Medicine*, vol. 4; Kadish, K. M.; Smith, K. M.; Guilard, R. (eds.). World Scientific: Singapore, **2010**; pp. 191–248.

22 Bregadze, V.; Semioshkin, A.; Sivaev, I. Synthesis of Conjugates of Polyhedral Boron Compounds with Tumor-Seeking Molecules for Neutron Capture Therapy. *Appl. Radiat. Isot.* **2011**, *69* (12), 1774–1777.

23 Sivaev, I. B.; Bregadze, V. V. Polyhedral Boranes for Medical Applications: Current Status and Perspectives. *Eur. J. Inorg. Chem.* **2009**, *2009* (11), 1433–1450.

24 Hosmane, N. S.; Maguire, J. A.; Zhu, Y.; Takagaki, M. Major Neutron Capture Therapy (NCT) Drug Prototypes. In *Boron and Gadolinium Neutron Capture Therapy for Cancer Treatment*. World Scientific: Singapore, **2012**; pp. 41–98.

25 Ban, H. S.; Nakamura, H. Boron-Based Drug Design. *The Chemical Record* **2015**, *15* (3), 616–635.

26 Yinghuai, Z.; Cheng Yan, K.; Maguire, J. A.; Hosmane, N. S. Recent Developments in Boron Neutron Capture Therapy (BNCT) Driven by Nanotechnology. *Curr. Chem. Biol.* **2007**, *1* (2), 141–149.

27 Wang, J.; Wu, W.; Jiang, X. Nanoscaled Boron-Containing Delivery Systems and Therapeutic Agents for Cancer Treatment. *Nanomedicine* **2015**, *10* (7), 1149–1163.

28 Sumitani, S.; Nagasaki, Y. Boron Neutron Capture Therapy Assisted by Boron-Conjugated Nanoparticles. *Polym. J.* **2012**, *44* (6), 522–530.

29 Yinghuai, Z.; Yan, K. C.; Maguire, J. A.; Hosmane, N. S. Boron-Based Hybrid Nanostructures: Novel Applications of Modern Materials. In *Hybrid Nanomaterials*. John Wiley & Sons, Inc.: Hoboken, **2011**; pp. 181–198. doi:10.1002/9781118003497.ch6.

30 Nakamura, H. Liposomal Boron Delivery for Neutron Capture Therapy. *Methods Enzymol.* **2009**, *465*, 179–208.
31 Prota, G.; D'Ischia, M.; Mascagna, D. Melanogenesis as a Targeting Strategy against Metastatic Melanoma: A Reassessment. *Melanoma Res.* **1995**, *4* (6), 351–358.
32 Reubi, J. C. Peptide Receptors as Molecular Targets for Cancer Diagnosis and Therapy. *Endocr. Rev.* **2003**, *24* (4), 389–427.
33 Yoshimoto, M.; Kurihara, H.; Honda, N.; Kawai, K.; Ohe, K.; Fujii, H.; Itami, J.; Arai, Y. Predominant Contribution of L-Type Amino Acid Transporter to 4-Borono-2-18f-Fluoro-Phenylalanine Uptake in Human Glioblastoma Cells. *Nucl. Med. Biol.* **2013**, *40* (5), 625–629.
34 Nemoto, H.; Cai, J.; Iwamoto, S.; Yamamoto, Y. Synthesis and Biological Properties of Water-Soluble P-Boronophenylalanine Derivatives. Relationship between Water Solubility, Cytotoxicity, and Cellular Uptake. *J. Med. Chem.* **1995**, *38* (10), 1673–1678.
35 Bonjoch, J.; Drew, M. G. B.; González, A.; Greco, F.; Jawaid, S.; Osborn, H. M. I.; Williams, N. A. O.; Yaqoob, P. Synthesis and Evaluation of Novel Boron-Containing Complexes of Potential Use for the Selective Treatment of Malignant Melanoma. *J. Med. Chem.* **2008**, *51* (20), 6604–6608.
36 Hattori, Y.; Kusaka, S.; Mukumoto, M.; Uehara, K.; Asano, T.; Suzuki, M.; Masunaga, S.-I.; Ono, K.; Tanimori, S.; Kirihata, M. Biological Evaluation of Dodecaborate-Containing L-Amino Acids for Boron Neutron Capture Therapy. *J. Med. Chem.* **2012**, *55* (15), 6980–6984.
37 Kusaka, S.; Hattori, Y.; Uehara, K.; Asano, T.; Tanimori, S.; Kirihata, M. Synthesis of Optically Active Dodecaborate-Containing L-Amino Acids for BNCT. *Appl. Radiat. Isot.* **2011**, *69* (12), 1768–1770.
38 Barth, R. F.; Kabalka, G. W.; Yang, W.; Huo, T.; Nakkula, R. J.; Shaikh, A. L.; Haider, S. A.; Chandra, S. Evaluation of Unnatural Cyclic Amino Acids as Boron Delivery Agents for Treatment of Melanomas and Gliomas. *Appl. Radiat. Isot.* **2014**, *88*, 38–42.
39 Chandra, S.; Barth, R. F.; Haider, S. A.; Yang, W.; Huo, T.; Shaikh, A. L.; Kabalka, G. W. Biodistribution and Subcellular Localization of an Unnatural Boron-Containing Amino Acid (Cis-Abcpc) by Imaging Secondary Ion Mass Spectrometry for Neutron Capture Therapy of Melanomas and Gliomas. *PLoS One* **2013**, *8* (9), e75377.
40 Kabalka, G. W.; Shaikh, A. L.; Barth, R. F.; Huo, T.; Yang, W.; Gordnier, P. M.; Chandra, S. Boronated Unnatural Cyclic Amino Acids as Potential Delivery Agents for Neutron Capture Therapy. *Appl. Radiat. Isot.* **2011**, *69* (12), 1778–1781.
41 Kabalka, G. W.; Yao, M. L.; Marepally, S. R.; Chandra, S. Biological Evaluation of Boronated Unnatural Amino Acids as New Boron Carriers. *Appl. Radiat. Isot.* **2009**, *67* (7–8, Suppl.), S374–S379.
42 Aoyagi, M.; Agranoff, B. W.; Washburn, L. C.; Smith, Q. R. Blood-Brain Barrier Transport of 1-Aminocyclohexanecarboxylic Acid, a Nonmetabolizable Amino Acid for in Vivo Studies of Brain Transport. *J. Neurochem.* **1988**, *50* (4), 1220–1226.
43 Detta, A.; Cruickshank, G. S. L-Amino Acid Transporter-1 and Boronophenylalanine-Based Boron Neutron Capture Therapy of Human Brain Tumors. *Cancer Res.* **2009**, *69* (5), 2126–2132.
44 Svantesson, E.; Capala, J.; Markides, K. E.; Pettersson, J. Determination of Boron-Containing Compounds in Urine and Blood Plasma from Boron Neutron Capture Therapy Patients: The Importance of Using Coupled Techniques. *Anal. Chem.* **2002**, *74* (20), 5358–5363.

45 Rice, S. L.; Roney, C. A.; Daumar, P.; Lewis, J. S. The Next Generation of Positron Emission Tomography Radiopharmaceuticals in Oncology. *Semin. Nucl. Med.* **2011**, *41* (4), 265–282.

46 Hattori, Y.; Kusaka, S.; Mukumoto, M.; Ishimura, M.; Ohta, Y.; Takenaka, H.; Uehara, K.; Asano, T.; Suzuki, M.; Masunaga, S.-i.; Ono, K.; Tanimori, S.; Kirihata, M. Synthesis and in Vitro Evaluation of Thiododecaborated A, A- Cycloalkylamino Acids for the Treatment of Malignant Brain Tumors by Boron Neutron Capture Therapy. *Amino Acids* **2014**, *46* (12), 2715–2720.

47 Chandra, S.; Kabalka, G. W.; Nd, L. D.; Smith, D. R.; Coderre, J. A. Imaging of Fluorine and Boron from Fluorinated Boronophenylalanine in the Same Cell at Organelle Resolution by Correlative Ion Microscopy and Confocal Laser Scanning Microscopy. *Clin. Cancer. Res.* **2002**, *8* (8), 2675–2683.

48 Wittig, A.; Arlinghaus, H. F.; Kriegeskotte, C.; Moss, R. L.; Appelman, K.; Schmid, K. W.; Sauerwein, W. A. Laser Postionization Secondary Neutral Mass Spectrometry in Tissue: A Powerful Tool for Elemental and Molecular Imaging in the Development of Targeted Drugs. *Mol. Cancer Ther.* **2008**, *7* (7), 1763–1771.

49 Beck-Sickinger, A. G.; Khan, I. U. Targeted Tumor Diagnosis and Therapy with Peptide Hormones as Radiopharmaceuticals. *Anticancer Agents Med. Chem.* **2008**, *8* (2), 186–199.

50 Mezo, G. M. Marilena Receptor-Mediated Tumor Targeting Based on Peptide Hormones. *Expert Opin. Drug Deliv.* **2010**, *7* (1), 79–96.

51 Ahrens, V. M.; Frank, R.; Boehnke, S.; Schütz, C. L.; Hampel, G.; Iffland, D. S.; Bings, N. H.; Hey-Hawkins, E.; Beck-Sickinger, A. G. Receptor-Mediated Uptake of Boron-Rich Neuropeptide Y Analogues for Boron Neutron Capture Therapy. *ChemMedChem* **2015**, *10* (1), 164–172.

52 Böhme, I.; Stichel, J.; Walther, C.; Mörl, K.; Beck-Sickinger, A. G. Agonist Induced Receptor Internalization of Neuropeptide Y Receptor Subtypes Depends on Third Intracellular Loop and C-Terminus. *Cell. Signal.* **2008**, *20* (10), 1740–1749.

53 Ahrens, V. M.; Frank, R.; Stadlbauer, S.; Beck-Sickinger, A. G.; Hey-Hawkins, E. Incorporation of Ortho-Carbaboranyl-Nε-Modified L-Lysine into Neuropeptide Y Receptor Y1- and Y2-Selective Analogues. *J. Med. Chem.* **2011**, *54* (7), 2368–2377.

54 Jean, C.; Gravelle, P.; Fournie, J. J.; Laurent, G. Influence of Stress on Extracellular Matrix and Integrin Biology. *Oncogene* **2011**, *30* (24), 2697–2706.

55 Kimura, S.; Masunaga, S.-I.; Harada, T.; Kawamura, Y.; Ueda, S.; Okuda, K.; Nagasawa, H. Synthesis and Evaluation of Cyclic RGD-Boron Cluster Conjugates to Develop Tumor-Selective Boron Carriers for Boron Neutron Capture Therapy. *Biorg. Med. Chem.* **2011**, *19* (5), 1721–1728.

56 Masunaga, S.-I.; Sadaaki, K.; Harada, T.; Okuda, K.; Sakurai, Y.; Tanaka, H.; Suzuki, M.; Kondo, N.; Maruhashi, A.; Nagasawa, H. Evaluating the Usefulness of a Novel 10b-Carrier Conjugated with Cyclic RGD Peptide in Boron Neutron Capture Therapy. *World J. Oncol.* **2012**, *3* (3), 103–112.

57 Michiue, H.; Sakurai, Y.; Kondo, N.; Kitamatsu, M.; Bin, F.; Nakajima, K.; Hirota, Y.; Kawabata, S.; Nishiki, T.-I.; Ohmori, I.; Tomizawa, K.; Miyatake, S.-I.; Ono, K.; Matsui, H. The Acceleration of Boron Neutron Capture Therapy Using Multi-Linked Mercaptoundecahydrododecaborate (BSH) Fused Cell-Penetrating Peptide. *Biomaterials* **2014**, *35* (10), 3396–3405.

58 Arnér, E. S. J.; Eriksson, S. Mammalian Deoxyribonucleoside Kinases. *Pharmacol. Ther.* **1995**, *67* (2), 155–186.

59 Hasabelnaby, S.; Goudah, A.; Agarwal, H. K.; Abd alla, M. S. M.; Tjarks, W. Synthesis, Chemical and Enzymatic Hydrolysis, and Aqueous Solubility of Amino Acid Ester Prodrugs of 3-Carboranyl Thymidine Analogs for Boron Neutron Capture Therapy of Brain Tumors. *Eur. J. Med. Chem.* **2012**, *55*, 325–334.

60 Barth, R. F.; Yang, W.; Al-Madhoun, A. S.; Johnsamuel, J.; Byun, Y.; Chandra, S.; Smith, D. R.; Tjarks, W.; Eriksson, S. Boron-Containing Nucleosides as Potential Delivery Agents for Neutron Capture Therapy of Brain Tumors. *Cancer Res.* **2004**, *64* (17), 6287–6295.

61 Tjarks, W.; Tiwari, R.; Byun, Y.; Narayanasamy, S.; Barth, R. F. Carboranyl Thymidine Analogues for Neutron Capture Therapy. *Chem. Commun.* **2007**, (*47*), 4978–4991.

62 Aspland, S. E.; Ballatore, C.; Castillo, R.; Desharnais, J.; Eustaquio, T.; Goelet, P.; Guo, Z.; Li, Q.; Nelson, D.; Sun, C.; Castellino, A. J.; Newman, M. J. Kinase-Mediated Trapping of Bi-Functional Conjugates of Paclitaxel or Vinblastine with Thymidine in Cancer Cells. *Biorg. Med. Chem. Lett.* **2006**, *16* (19), 5194–5198.

63 Agarwal, H. K.; McElroy, C. A.; Sjuvarsson, E.; Eriksson, S.; Darby, M. V.; Tjarks, W. Synthesis of N3-Substituted Carboranyl Thymidine Bioconjugates and Their Evaluation as Substrates of Recombinant Human Thymidine Kinase 1. *Eur. J. Med. Chem.* **2013**, *60*, 456–468.

64 Nizioł, J.; Zieliński, Z.; Leś, A.; Dąbrowska, M.; Rode, W.; Ruman, T. Synthesis, Reactivity and Biological Activity of N(4)-Boronated Derivatives of 2′-Deoxycytidine. *Biorg. Med. Chem.* **2014**, *22* (15), 3906–3912.

65 Agarwal, H. K.; Khalil, A.; Ishita, K.; Yang, W.; Nakkula, R. J.; Wu, L.-C.; Ali, T.; Tiwari, R.; Byun, Y.; Barth, R. F.; Tjarks, W. Synthesis and Evaluation of Thymidine Kinase 1-Targeting Carboranyl Pyrimidine Nucleoside Analogs for Boron Neutron Capture Therapy of Cancer. *Eur. J. Med. Chem.* **2015**, *100*, 197–209.

66 Semioshkin, A.; Ilinova, A.; Lobanova, I.; Bregadze, V.; Paradowska, E.; Studzińska, M.; Jabłońska, A.; Lesnikowski, Z. J. Synthesis of the First Conjugates of 5-Ethynyl-2′-Deoxyuridine with Closo-Dodecaborate and Cobalt-Bis-Dicarbollide Boron Clusters. *Tetrahedron* **2013**, *69* (37), 8034–8041.

67 Al-Madhoun, A. S.; Johnsamuel, J.; Barth, R. F.; Tjarks, W.; Eriksson, S. Evaluation of Human Thymidine Kinase 1 Substrates as New Candidates for Boron Neutron Capture Therapy. *Cancer Res.* **2004**, *64* (17), 6280–6286.

68 Khalil, A.; Ishita, K.; Ali, T.; Tjarks, W. N3-Substituted Thymidine Bioconjugates for Cancer Therapy and Imaging. *Future Med. Chem.* **2013**, *5* (6), 10.4155/fmc.4113.4131.

69 Leśnikowski, Z. J.; Paradowska, E.; Olejniczak, A. B.; Studzińska, M.; Seekamp, P.; Schüßler, U.; Gabel, D.; Schinazi, R. F.; Plešek, J. Towards New Boron Carriers for Boron Neutron Capture Therapy: Metallacarboranes and Their Nucleoside Conjugates. *Biorg. Med. Chem.* **2005**, *13* (13), 4168–4175.

70 Nizioł, J.; Uram, Ł.; Szuster, M.; Sekuła, J.; Ruman, T. Biological Activity of N(4)-Boronated Derivatives of 2′-Deoxycytidine, Potential Agents for Boron-Neutron Capture Therapy. *Biorg. Med. Chem.* **2015**, *23* (19), 6297–6304.

71 Frieze, D. A.; McCune, J. S. Current Status of Cetuximab for the Treatment of Patients with Solid Tumors. *Ann. Pharmacother.* **2006**, *40* (2), 241–250.

72 Wu, G.; Yang, W.; Barth, R. F.; Kawabata, S.; Swindall, M.; Bandyopadhyaya, A. K.; Tjarks, W.; Khorsandi, B.; Blue, T. E.; Ferketich, A. K.; Yang, M.; Christoforidis, G. A.; Sferra, T. J.; Binns, P. J.; Riley, K. J.; Ciesielski, M. J.; Fenstermaker, R. A. Molecular Targeting and Treatment of an Epidermal Growth Factor Receptor–Positive Glioma Using Boronated Cetuximab. *Amer. Assoc. Cancer Res.* **2007**, *13* (4), 1260–1268.

73 Wu, G.; Barth, R. F.; Yang, W.; Chatterjee, M.; Tjarks, W.; Ciesielski, M. J.; Fenstermaker, R. A. Site-Specific Conjugation of Boron-Containing Dendrimers to Anti-EGF Receptor Monoclonal Antibody Cetuximab (IMC-C225) and Its Evaluation as a Potential Delivery Agent for Neutron Capture Therapy. *Bioconj. Chem.* **2004**, *15* (1), 185–194.

74 Yang, W.; Wu, G.; Barth, R. F.; Swindall, M. R.; Bandyopadhyaya, A. K.; Tjarks, W.; Tordoff, K.; Moeschberger, M.; Sferra, T. J.; Binns, P. J.; Riley, K. J.; Ciesielski, M. J.; Fenstermaker, R. A.; Wikstrand, C. J. Molecular Targeting and Treatment of Composite EGFR and EGFRVIII-Positive Gliomas Using Boronated Monoclonal Antibodies. *Amer. Assoc. Cancer Res.* **2008**, *14* (3), 883–891.

75 Yang, W.; Barth, R. F.; Wu, G.; Tjarks, W.; Binns, P.; Riley, K. Boron Neutron Capture Therapy of EGFR or EGFRVIII Positive Gliomas Using Either Boronated Monoclonal Antibodies or Epidermal Growth Factor as Molecular Targeting Agents. *Appl. Radiat. Isot.* **2009**, *67* (7–8, Suppl.), S328–S331.

76 Ackroyd, R.; Kelty, C.; Brown, N.; Reed, M. The History of Photodetection and Photodynamic Therapy. *Photochem. Photobiol.* **2001**, *74* (5), 656–669.

77 Giuntini, F.; Boyle, R.; Sibrian-Vazquez, M.; Vicente, M. G. H. Porphyrin Conjugates for Cancer Therapy. In *Handbook of Porphyrin Science*, vol. 27. World Scientific: Singapore, **2013**; pp. 303–416.

78 Renner, M. W.; Miura, M.; Easson, M. W.; Vicente, M. G. H. Recent Progress in the Syntheses and Biological Evaluation of Boronated Porphyrins for Boron Neutron-Capture Therapy. *Anticancer Agents Med. Chem.* **2006**, *6* (2), 145–157.

79 Pietrangeli, D.; Rosa, A.; Ristori, S.; Salvati, A.; Altieri, S.; Ricciardi, G. Carboranyl-Porphyrazines and Derivatives for Boron Neutron Capture Therapy: From Synthesis to in Vitro Tests. *Coord. Chem. Rev.* **2013**, *257* (15–16), 2213–2231.

80 Koo, M.-S.; Ozawa, T.; Santos, R. A.; Lamborn, K. R.; Bollen, A. W.; Deen, D. F.; Kahl, S. B. Synthesis and Comparative Toxicology of a Series of Polyhedral Borane Anion-Substituted Tetraphenyl Porphyrins. *J. Med. Chem.* **2007**, *50* (4), 820–827.

81 Olshevskaya, V. A.; Zaitsev, A. V.; Sigan, A. L.; Kononova, E. G.; Petrovskiĭ, P. V.; Chkanikov, N. D.; Kalinin, V. N. Synthesis of Boronated Porphyrins and Chlorins by Regioselective Substitution for Fluorine in Pentafluorophenylporphyrins on Treatment with Lithiocarboranes. *Dokl. Chem.* **2010**, *435* (2), 334–338.

82 Bhupathiraju, N. V. S. D. K.; Vicente, M. G. H. Synthesis and Cellular Studies of Polyamine Conjugates of a Mercaptomethyl–Carboranylporphyrin. *Biorg. Med. Chem.* **2013**, *21* (2), 485–495.

83 Bhupathiraju, N. V. S. D. K.; Hu, X.; Zhou, Z.; Fronczek, F. R.; Couraud, P.-O.; Romero, I. A.; Weksler, B.; Vicente, M. G. H. Synthesis and In Vitro Evaluation of BBB Permeability, Tumor Cell Uptake, and Cytotoxicity of a Series of Carboranylporphyrin Conjugates. *J. Med. Chem.* **2014**, *57* (15), 6718–6728.

84 Ol'shevskaya, V. A.; Savchenko, A. N.; Shtil, A. A.; Cheong, C. S.; Kalinin, V. N. Boronated Porphyrins and Chlorins as Potential Anticancer Drugs. *Bull. Korean Chem. Soc.* **2007**, *28*, 1910.

85 Hao, E.; Jensen, T. J.; Courtney, B. H.; Vicente, M. G. H. Synthesis and Cellular Studies of Porphyrin – Cobaltacarborane Conjugates. *Bioconj. Chem.* **2005**, *16* (6), 1495–1502.

86 Hao, E.; Vicente, M. G. H. Expeditious Synthesis of Porphyrin-Cobaltacarborane Conjugates. *Chem. Commun.* **2005**, (*10*), 1306–1308.

87 Bhupathiraju, N. V. S. D. K.; Gottumukkala, V.; Hao, E.; Hu, X.; Fronczek, F. R.; Baker, D. G.; Wakamatsu, N.; Vicente, M. G. H. Synthesis and Toxicity of

Cobaltabisdicarbollide-Containing Porphyrins of High Boron Content. *J. Porphyr. Phthalocyanin.* **2012**, *15* (9–10), 973–983.

88 Hao, E.; Sibrian-Vazquez, M.; Serem, W.; Garno, J. C.; Fronczek, F. R.; Vicente, M. G. H. Synthesis, Aggregation and Cellular Investigations of Porphyrin–Cobaltacarborane Conjugates. *Chem. Eur. J.* **2007**, *13* (32), 9035–9042.

89 Genady, A. R. Synthesis and Characterization of a Novel Functionalized Azanonaborane Cluster for Boron Neutron Capture Therapy. *Org. Biomol. Chem.* **2005**, *3* (11), 2102–2108.

90 Gottumukkala, V.; Luguya, R.; Fronczek, F. R.; Vicente, M. G. H. Synthesis and Cellular Studies of an Octa-Anionic 5,10,15,20-Tetra[3,5-(Nido-Carboranylmethyl)Phenyl] Porphyrin (H_2OCP) for Application in BNCT. *Biorg. Med. Chem.* **2005**, *13* (13), 1633–1640.

91 Fabris, C.; Vicente, M. G. H.; Hao, E.; Friso, E.; Borsetto, L.; Jori, G.; Miotto, G.; Colautti, P.; Moro, D.; Esposito, J.; Ferretti, A.; Rossi, C. R.; Nitti, D.; Sotti, G.; Soncin, M. Tumour-Localizing and -Photosensitising Properties of Meso-Tetra(4-Nido-Carboranylphenyl)Porphyrin (H_2TCP). *J. Photochem. Photobiol. B: Biol.* **2007**, *89* (2–3), 131–138.

92 Ristori, S.; Salvati, A.; Martini, G.; Spalla, O.; Pietrangeli, D.; Rosa, A.; Ricciardi, G. Synthesis and Liposome Insertion of a New Poly(Carboranylalkylthio)Porphyrazine to Improve Potentiality in Multiple-Approach Cancer Therapy. *J. Am. Chem. Soc.* **2007**, *129* (10), 2728–2729.

93 Ozawa, T.; Afzal, J.; Lamborn, K. R.; Bollen, A. W.; Bauer, W. F.; Koo, M. S.; Kahl, S. B.; Deen, D. F. Toxicity, Biodistribution, and Convection-Enhanced Delivery of the Boronated Porphyrin BOPP in the 9l Intracerebral Rat Glioma Model. *Med. Phys.* **2005**, *63* (1), 247–252.

94 Dagrosa, M. A.; Viaggi, M.; Rebagliati, R. J.; Batistoni, D.; Kahl, S. B.; Juvenal, G. J.; Pisarev, M. A. Biodistribution of Boron Compounds in an Animal Model of Human Undifferentiated Thyroid Cancer for Boron Neutron Capture Therapy. *Mol. Pharm.* **2005**, *2* (2), 151–156.

95 Dozzo, P.; Koo, M.-S.; Berger, S.; Forte, T. M.; Kahl, S. B. Synthesis, Characterization, and Plasma Lipoprotein Association of a Nucleus-Targeted Boronated Porphyrin. *J. Med. Chem.* **2005**, *48* (2), 357–359.

96 Novick, S.; Laster, B.; Quastel, M. R. Positive Cooperativity in the Cellular Uptake of a Boronated Porphyrin. *Intl. J. Biochem. Cell Biol.* **2006**, *38* (8), 1374–1381.

97 Smilowitz, H. M.; Slatkin, D. N.; Micca, P. L.; Miura, M. Microlocalization of Lipophilic Porphyrins: Non-Toxic Enhancers of Boron Neutron-Capture Therapy. *Int. J. Radiat. Biol.* **2013**, *89* (8), 611–617.

98 Jori, G.; Soncin, M.; Friso, E.; Vicente, M. G. H.; Hao, E.; Miotto, G.; Colautti, P.; Moro, D.; Esposito, J.; Rosi, G. A Novel Boronated-Porphyrin as a Radio-Sensitizing Agent for Boron Neutron Capture Therapy of Tumours: In Vitro and in Vivo Studies. *Appl. Radiat. Isot.* **2009**, *67* (7–8), S321–S324.

99 Soncin, M.; Friso, E.; Jori, G.; Hao, E.; Vicente, M. G. H.; Miotto, G.; Colautti, P.; Moro, D.; Esposito, J.; Rosi, G. Tumor-Localizing and Radiosensitizing Properties of Meso-Tetra (4-Nido-Carboranylphenyl)Porphyrin (H_2TCP). *J. Porphyr. Phthalocyan.* **2008**, *12* (7), 866–873.

100 Purser, S.; Moore, P. R.; Swallow, S.; Gouverneur, V. Fluorine in Medicinal Chemistry. *Chem. Soc. Rev.* **2008**, *37* (2), 320–330.

101 Gryshuk, A.; Chen, Y.; Goswami, L. N.; Pandey, S.; Missert, J. R.; Ohulchanskyy, T.; Potter, W.; Prasad, P. N.; Oseroff, A.; Pandey, R. K. Structure–Activity Relationship among Purpurinimides and Bacteriopurpurinimides: Trifluoromethyl Substituent Enhanced the Photosensitizing Efficacy. *J. Med. Chem.* **2007**, *50* (8), 1754–1767.

102 Ko, Y.-J.; Yun, K.-J.; Kang, M.-S.; Park, J.; Lee, K.-T.; Park, S. B.; Shin, J.-H. Synthesis and in Vitro Photodynamic Activities of Water-Soluble Fluorinated Tetrapyridylporphyrins as Tumor Photosensitizers. *Biorg. Med. Chem. Lett.* **2007**, *17* (10), 2789–2794.

103 Bhupathiraju, N. V. S. D. K.; Rizvi, W.; Batteas, J. D.; Drain, C. M. Fluorinated Porphyrinoids as Efficient Platforms for New Photonic Materials, Sensors, and Therapeutics. *Org. Biomol. Chem.* **2016**, *14* (2), 389–408.

104 Cai, J.; Soloway, A. H.; Barth, R. F.; Adams, D. M.; Hariharan, J. R.; Wyzlic, I. M.; Radcliffe, K. Boron-Containing Polyamines as DNA Targeting Agents for Neutron Capture Therapy of Brain Tumors: Synthesis and Biological Evaluation. *J. Med. Chem.* **1997**, *40* (24), 3887–3896.

105 Cullis, P. M.; Green, R. E.; Merson-Davies, L.; Travis, N. Probing the Mechanism of Transport and Compartmentalisation of Polyamines in Mammalian Cells. *Chem. Biol.* **1999**, *6* (10), 717–729.

106 Zhang, M.; Zhang, Z.; Blessington, D.; Li, H.; Busch, T. M.; Madrak, V.; Miles, J.; Chance, B.; Glickson, J. D.; Zheng, G. Pyropheophorbide 2-Deoxyglucosamide: A New Photosensitizer Targeting Glucose Transporters. *Bioconj. Chem.* **2003**, *14* (4), 709–714.

107 Deeken, J. F.; Löscher, W. The Blood-Brain Barrier and Cancer: Transporters, Treatment, and Trojan Horses. *Amer. Assoc. Cancer Res.* **2007**, *13* (6), 1663–1674.

108 Mironov, A. F.; Grin, M. A. Synthesis of Chlorin and Bacteriochlorin Conjugates for Photodynamic and Boron Neutron Capture Therapy. *J. Porphyr. Phthalocyanin.* **2008**, *12* (11), 1163–1172.

109 Grin, M. A.; Semioshkin, A. A.; Titeev, R. A.; Nizhnik, E. A.; Grebenyuk, J. N.; Mironov, A. F.; Bregadze, V. I. Synthesis of a Cycloimide Bacteriochlorin P Conjugate with the Closo-Dodecaborate Anion. *Mendeleev Commun.* **2007**, *17* (1), 14–15.

110 Asano, R.; Nagami, A.; Fukumoto, Y.; Miura, K.; Yazama, F.; Ito, H.; Sakata, I.; Tai, A. Synthesis and Biological Evaluation of New Boron-Containing Chlorin Derivatives as Agents for Both Photodynamic Therapy and Boron Neutron Capture Therapy of Cancer. *Biorg. Med. Chem. Lett.* **2014**, *24* (5), 1339–1343.

111 Asano, R.; Nagami, A.; Fukumoto, Y.; Miura, K.; Yazama, F.; Ito, H.; Sakata, I.; Tai, A. Synthesis and Biological Evaluation of New BSH-Conjugated Chlorin Derivatives as Agents for Both Photodynamic Therapy and Boron Neutron Capture Therapy of Cancer. *J. Photochem. Photobiol. B: Biol.* **2014**, *140*, 140–149.

112 Hiramatsu, R.; Kawabata, S.; Tanaka, H.; Sakurai, Y.; Suzuki, M.; Ono, K.; Miyatake, S.-I.; Kuroiwa, T.; Hao, E.; Vicente, M. G. H. Tetrakis(P-Carboranylthio-Tetrafluorophenyl)Chlorin (Tpfc): Application for Photodynamic Therapy and Boron Neutron Capture Therapy. *J. Pharm. Sci.* **2015**, *104* (3), 962–970.

113 Grin, M. A.; Titeev, R. A.; Brittal, D. I.; Ulybina, O. V.; Tsiprovskiy, A. G.; Berzina, M. Y.; Lobanova, I. A.; Sivaev, I. B.; Bregadze, V. I.; Mironov, A. F. New Conjugates of Cobalt Bis(Dicarbollide) with Chlorophyll a Derivatives. *Mendeleev Commun.* **2011**, *21* (2), 84–86.

114 Efremenko, A. V.; Ignatova, A. A.; Grin, M. A.; Sivaev, I. B.; Mironov, A. F.; Bregadze, V. I.; Feofanov, A. V. Chlorin e_6 Fused with a Cobalt-Bis(Dicarbollide) Nanoparticle

Provides Efficient Boron Delivery and Photoinduced Cytotoxicity in Cancer Cells. *Photochem. Photobiol. Sci.* **2014**, *13* (1), 92–102.

115 Volovetskiy, A. B.; Shilyagina, N. Y.; Dudenkova, V. V.; Pasynkova, S. O.; Ignatova, A. A.; Mironov, A. F.; Grin, M. A.; Bregadze, V. I.; Feofanov, A. V.; Balalaeva, I. V.; Maslennikova, A. V. Study of the Tissue Distribution of Potential Boron Neutron-Capture Therapy Agents Based on Conjugates of Chlorin e_6 Aminoamide Derivatives with Boron Nanoparticles. *Biophysics* **2016**, *61* (1), 133–138.

116 Efremenko, A. V.; Ignatova, A. A.; Borsheva, A. A.; Grin, M. A.; Bregadze, V. I.; Sivaev, I. B.; Mironov, A. F.; Feofanov, A. V. Cobalt Bis(Dicarbollide) versus Closo-Dodecaborate in Boronated Chlorin e_6 Conjugates: Implications for Photodynamic and Boron-Neutron Capture Therapy. *Photochem. Photobiol. Sci.* **2012**, *11* (4), 645–652.

117 Grin, M. A.; Titeev, R. A.; Brittal, D. I.; Chestnova, A. V.; Feofanov, A. V.; Lobanova, I. A.; Sivaev, I. B.; Bregadze, V. I.; Mironov, A. F. Synthesis of Cobalt Bis(Dicarbollide) Conjugates with Natural Chlorins by the Sonogashira Reaction. *Russ. Chem. Bull.* **2010**, *59* (1), 219–224.

118 Bregadze, V. I.; Semioshkin, A. A.; Las'kova, J. N.; Berzina, M. Y.; Lobanova, I. A.; Sivaev, I. B.; Grin, M. A.; Titeev, R. A.; Brittal, D. I.; Ulybina, O. V.; Chestnova, A. V.; Ignatova, A. A.; Feofanov, A. V.; Mironov, A. F. Novel Types of Boronated Chlorin e_6 Conjugates via 'Click Chemistry'. *Appl. Organomet. Chem.* **2009**, *23* (9), 370–374.

119 Hao, E.; Friso, E.; Miotto, G.; Jori, G.; Soncin, M.; Fabris, C.; Sibrian-Vazquez, M.; Vicente, M. G. H. Synthesis and Biological Investigations of Tetrakis(P-Carboranylthio-Tetrafluorophenyl)Chlorin (TPFC). *Org. Biomol. Chem.* **2008**, *6* (20), 3732–3740.

120 Li, H.; Fronczek, F. R.; Vicente, M. G. H. Synthesis and Properties of Cobaltacarborane-Functionalized Zn(II)-Phthalocyanines. *Tetrahedron Lett.* **2008**, *49* (33), 4828–4830.

121 Pietrangeli, D.; Rosa, A.; Pepe, A.; Altieri, S.; Bortolussi, S.; Postuma, I.; Protti, N.; Ferrari, C.; Cansolino, L.; Clerici, A. M.; Viola, E.; Donzello, M. P.; Ricciardi, G. Water-Soluble Carboranyl-Phthalocyanines for BNCT: Synthesis, Characterization, and In Vitro Tests of the Zn(II)-Nido-Carboranyl-Hexylthiophthalocyanine. *Dalton Trans.* **2015**, *44* (24), 11021–11028.

122 Atmaca, G. Y.; Dizman, C.; Eren, T.; Erdoğmuş, A. Novel Axially Carborane-Cage Substituted Silicon Phthalocyanine Photosensitizer; Synthesis, Characterization and Photophysicochemical Properties. *Spectrochim. Acta, Pt. A: Mol. Biomol. Spectrosc.* **2015**, *137*, 244–249.

123 Nar, I.; Gül, A.; Sivaev, I. B.; Hamuryudan, E. Cobaltacarborane Functionalized Phthalocyanines: Synthesis, Photophysical, Electrochemical and Spectroelectrochemical Properties. *Synth. Met.* **2015**, *210* (Pt. B), 376–385.

124 Shmal'ko, A. V.; Efremenko, A. V.; Ignatova, A. A.; Sivaev, I. B.; Feofanov, A. V.; Hamuryudan, E.; Gül, A.; Kovalenko, L. V.; Qi, S.; Bregadze, V. I. Synthesis and in Vitro Study of New Highly Boronated Phthalocyanine. *J. Porphyr. Phthalocyanin.* **2014**, *18* (10–11), 960–966.

125 Loudet, A.; Burgess, K. BODIPY Dyes and Their Derivatives: Syntheses and Spectroscopic Properties. *Chem. Rev.* **2007**, *107* (11), 4891–4932.

126 Boens, N.; Leen, V.; Dehaen, W. Fluorescent Indicators Based on BODIPY. *Chem. Soc. Rev.* **2012**, *41* (3), 1130–1172.

127 Lu, H.; Mack, J.; Yang, Y.; Shen, Z. Structural Modification Strategies for the Rational Design of Red/Nir Region BODIPYs. *Chem. Soc. Rev.* **2014**, *43* (13), 4778–4823.

128 Yuan, L.; Lin, W.; Zheng, K.; He, L.; Huang, W. Far-Red to near Infrared Analyte-Responsive Fluorescent Probes Based on Organic Fluorophore Platforms for Fluorescence Imaging. *Chem. Soc. Rev.* **2013**, *42* (2), 622–661.

129 Kowada, T.; Maeda, H.; Kikuchi, K. BODIPY-Based Probes for the Fluorescence Imaging of Biomolecules in Living Cells. *Chem. Soc. Rev.* **2015**, *44* (14), 4953–4972.

130 Nakata, E.; Koizumi, M.; Yamashita, Y.; Onaka, K.; Sakurai, Y.; Kondo, N.; Ono, K.; Uto, Y.; Hori, H. Design, Synthesis and Destructive Dynamic Effects of BODIPY-Containing and Curcuminoid Boron Tracedrugs for Neutron Dynamic Therapy. *Anticancer Res.* **2011**, *31* (7), 2477–2481.

131 Nakata, E.; Koizumi, M.; Yamashita, Y.; Uto, Y.; Hori, H. Boron Tracedrug: Design, Synthesis, and Pharmacological Activity of Phenolic BODIPY-Containing Antioxidants as Traceable Next-Generation Drug Model. In *Oxygen Transport to Tissue Xxxiii*; Wolf, M.; Bucher, U. H.; Rudin, M.; Van Huffel, S.; Wolf, U.; Bruley, F. D.; Harrison, K. D. (eds.). Springer: New York, **2012**; pp. 301–306.

132 Gibbs, J. H.; Wang, H.; Bhupathiraju, N. V. S. D. K.; Fronczek, F. R.; Smith, K. M.; Vicente, M. G. H. Synthesis and Properties of a Series of Carboranyl-BODIPYs. *J. Organomet. Chem.* **2015**, *798*, Part 1, 209–213.

133 Wang, H.; Vicente, M. G. H.; Fronczek, F. R.; Smith, K. M. Synthesis and Transformations of 5-Chloro-2,2′-Dipyrrins and Their Boron Complexes, 8-Chloro-BODIPYs. *Chem. Eur. J.* **2014**, *20* (17), 5064–5074.

134 Xuan, S.; Zhao, N.; Zhou, Z.; Fronczek, F. R.; Vicente, M. G. H. Synthesis and in Vitro Studies of a Series of Carborane-Containing Boron Dipyrromethenes (BODIPYs). *J. Med. Chem.* **2016**, *59* (5), 2109–2117.

135 Zhao, N.; Vicente, M. G. H.; Fronczek, F. R.; Smith, K. M. Synthesis of 3,8-Dichloro-6-Ethyl-1,2,5,7-Tetramethyl–BODIPY from an Asymmetric Dipyrroketone and Reactivity Studies at the 3,5,8-Positions. *Chem. Eur. J.* **2015**, *21* (16), 6181–6192.

136 Godoy, J.; Vives, G.; Tour, J. M. Synthesis of Highly Fluorescent BODIPY-Based Nanocars. *Org. Lett.* **2010**, *12* (7), 1464–1467.

137 Khatua, S.; Godoy, J.; Tour, J. M.; Link, S. Influence of the Substrate on the Mobility of Individual Nanocars. *J. Phys. Chem. Lett.* **2010**, *1* (22), 3288–3291.

138 Ziessel, R.; Ulrich, G.; Olivier, J. H.; Bura, T.; Sutter, A. Carborane-BODIPY Scaffolds for through Space Energy Transfer. *Chem. Commun.* **2010**, *46* (42), 7978–7980.

139 Hablot, D.; Sutter, A.; Retailleau, P.; Ziessel, R. Unsymmetrical P-Carborane Backbone as a Linker for Donor–Acceptor Dyads. *Chem. Eur. J.* **2012**, *18* (7), 1890–1895.

140 Chu, P.-L. E.; Wang, L.-Y.; Khatua, S.; Kolomeisky, A. B.; Link, S.; Tour, J. M. Synthesis and Single-Molecule Imaging of Highly Mobile Adamantane-Wheeled Nanocars. *ACS Nano* **2013**, *7* (1), 35–41.

141 Wicki, A.; Witzigmann, D.; Balasubramanian, V.; Huwyler, J. Nanomedicine in Cancer Therapy: Challenges, Opportunities, and Clinical Applications. *J. Controlled Release* **2015**, *200*, 138–157.

142 Nakamura, H.; Ueda, N.; Ban, H. S.; Ueno, M.; Tachikawa, S. Design and Synthesis of Fluorescence-Labeled Closo-Dodecaborate Lipid: Its Liposome Formation and in Vivo Imaging Targeting of Tumors for Boron Neutron Capture Therapy. *Org. Biomol. Chem.* **2012**, *10* (7), 1374–1380.

143 Białek-Pietras, M.; Olejniczak, A. B.; Tachikawa, S.; Nakamura, H.; Leśnikowski, Z. J. Towards New Boron Carriers for Boron Neutron Capture Therapy: Metallacarboranes Bearing Cobalt, Iron and Chromium and Their Cholesterol Conjugates. *Biorg. Med. Chem.* **2013**, *21* (5), 1136–1142.

144 Koganei, H.; Ueno, M.; Tachikawa, S.; Tasaki, L.; Ban, H. S.; Suzuki, M.; Shiraishi, K.; Kawano, K.; Yokoyama, M.; Maitani, Y.; Ono, K.; Nakamura, H. Development of High Boron Content Liposomes and Their Promising Antitumor Effect for Neutron Capture Therapy of Cancers. *Bioconj. Chem.* **2013**, *24* (1), 124–132.

145 Tachikawa, S.; Miyoshi, T.; Koganei, H.; El-Zaria, M. E.; Vinas, C.; Suzuki, M.; Ono, K.; Nakamura, H. Spermidinium Closo-Dodecaborate-Encapsulating Liposomes as Efficient Boron Delivery Vehicles for Neutron Capture Therapy. *Chem. Commun.* **2014**, *50* (82), 12325–12328.

146 Ueno, M.; Ban, H. S.; Nakai, K.; Inomata, R.; Kaneda, Y.; Matsumura, A.; Nakamura, H. Dodecaborate Lipid Liposomes as New Vehicles for Boron Delivery System of Neutron Capture Therapy. *Biorg. Med. Chem.* **2010**, *18* (9), 3059–3065.

147 Gifford, I.; Vreeland, W.; Grdanovska, S.; Burgett, E.; Kalinich, J.; Vergara, V.; Wang, C. K. C.; Maimon, E.; Poster, D.; Al-Sheikhly, M. Liposome-Based Delivery of a Boron-Containing Cholesteryl Ester for High-LET Particle-Induced Damage of Prostate Cancer Cells: A Boron Neutron Capture Therapy Study. *Int. J. Radiat. Biol.* **2014**, *90* (6), 480–485.

148 Alberti, D.; Toppino, A.; Geninatti Crich, S.; Meraldi, C.; Prandi, C.; Protti, N.; Bortolussi, S.; Altieri, S.; Aime, S.; Deagostino, A. Synthesis of a Carborane-Containing Cholesterol Derivative and Evaluation as a Potential Dual Agent for MRI/BNCT Applications. *Org. Biomol. Chem.* **2014**, *12* (15), 2457–2467.

149 Zhao, X. B.; Bandyopadhyaya, A. K.; Johnsamuel, J.; Tiwari, R.; Golightly, D. W.; Patel, V.; Jehning, B. T.; Backer, M. V.; Barth, R. F.; Lee, R. J.; Backer, J. M.; Tjarks, W. Receptor-Targeted Liposomal Delivery of Boron-Containing Cholesterol Mimics for Boron Neutron Capture Therapy (BNCT). *Bioconj. Chem.* **2006**, *17* (5), 1141–1150.

150 Miyajima, Y.; Nakamura, H.; Kuwata, Y.; Lee, J.-D.; Masunaga, S.; Ono, K.; Maruyama, K. Transferrin-Loaded nido-Carborane Liposomes: Tumor-Targeting Boron Delivery System for Neutron Capture Therapy. *Bioconj. Chem.* **2006**, *17* (5), 1314–1320.

151 Kueffer, P. J.; Maitz, C. A.; Khan, A. A.; Schuster, S. A.; Shlyakhtina, N. I.; Jalisatgi, S. S.; Brockman, J. D.; Nigg, D. W.; Hawthorne, M. F. Boron Neutron Capture Therapy Demonstrated in Mice Bearing EMT6 Tumors Following Selective Delivery of Boron by Rationally Designed Liposomes. *Proc. Natl. Acad. Sci.* **2013**, *110* (16), 6512–6517.

152 Heber, E. M.; Hawthorne, M. F.; Kueffer, P. J.; Garabalino, M. A.; Thorp, S. I.; Pozzi, E. C. C.; Hughes, A. M.; Maitz, C. A.; Jalisatgi, S. S.; Nigg, D. W.; Curotto, P.; Trivillin, V. A.; Schwint, A. E. Therapeutic Efficacy of Boron Neutron Capture Therapy Mediated by Boron-Rich Liposomes for Oral Cancer in the Hamster Cheek Pouch Model. *Proc. Natl. Acad. Sci.* **2014**, *111* (45), 16077–16081.

153 Theodoropoulos, D.; Rova, A.; Smith, J. R.; Barbu, E.; Calabrese, G.; Vizirianakis, I. S.; Tsibouklis, J.; Fatouros, D. G. Towards Boron Neutron Capture Therapy: The Formulation and Preliminary In Vitro Evaluation of Liposomal Vehicles for the Therapeutic Delivery of the Dequalinium Salt of bis-nido-Carborane. *Biorg. Med. Chem. Lett.* **2013**, *23* (22), 6161–6166.

154 Jiang, K.; Coffer, J. L.; Gillen, J. G.; Brewer, T. M. Incorporation of Cesium Borocaptate onto Silicon Nanowires as a Delivery Vehicle for Boron Neutron Capture Therapy. *Chem. Mater.* **2010**, *22* (2), 279–281.

155 Gibson, C. R.; Staubus, A. E.; Barth, R. F.; Yang, W.; Ferketich, A. K.; Moeschberger, M. M. Pharmacokinetics of Sodium Borocaptate: A Critical Assessment of Dosing Paradigms for Boron Neutron Capture Therapy. *J. Neurooncol.* **2003**, *62* (1), 157–169.

156 Mandal, S.; Bakeine, G. J.; Krol, S.; Ferrari, C.; Clerici, A. M.; Zonta, C.; Cansolino, L.; Ballarini, F.; Bortolussi, S.; Stella, S.; Protti, N.; Bruschi, P.; Altieri, S. Design, Development and Characterization of Multi-Functionalized Gold Nanoparticles for Biodetection and Targeted Boron Delivery in BNCT Applications. *Appl. Radiat. Isot.* **2011**, *69* (12), 1692–1697.

157 Cioran, A. M.; Musteti, A. D.; Teixidor, F.; Krpetić, Ž.; Prior, I. A.; He, Q.; Kiely, C. J.; Brust, M.; Viñas, C. Mercaptocarborane-Capped Gold Nanoparticles: Electron Pools and Ion Traps with Switchable Hydrophilicity. *J. Am. Chem. Soc.* **2012**, *134* (1), 212–221.

158 M. Cioran, A.; Teixidor, F.; Krpetic, Z.; Brust, M.; Vinas, C. Preparation and Characterization of Au Nanoparticles Capped with Mercaptocarboranyl Clusters. *Dalton Trans.* **2014**, *43* (13), 5054–5061.

159 Li, N.; Zhao, P.; Salmon, L.; Ruiz, J.; Zabawa, M.; Hosmane, N. S.; Astruc, D. "Click" Star-Shaped and Dendritic Pegylated Gold Nanoparticle-Carborane Assemblies. *Inorg. Chem.* **2013**, *52* (19), 11146–11155.

160 Zhang, L.; Lin, Y.; Wang, J.; Yao, W.; Wu, W.; Jiang, X. A Facile Strategy for Constructing Boron-Rich Polymer Nanoparticles via a Boronic Acid-Related Reaction. *Macromol. Rapid Commun.* **2011**, *32* (6), 534–539.

161 Brozek, E. M.; Mollard, A. H.; Zharov, I. Silica Nanoparticles Carrying Boron-Containing Polymer Brushes. *J. Nanopart. Res.* **2014**, *16* (5), 1–12.

162 Abi-Ghaida, F.; Clément, S.; Safa, A.; Naoufal, D.; Mehdi, A. Multifunctional Silica Nanoparticles Modified via Silylated-Decaborate Precursors. *J. Nanomater.* **2015**, *2015*, 8.

163 Achilli, C.; Grandi, S.; Ciana, A.; Guidetti, G. F.; Malara, A.; Abbonante, V.; Cansolino, L.; Tomasi, C.; Balduini, A.; Fagnoni, M.; Merli, D.; Mustarelli, P.; Canobbio, I.; Balduini, C.; Minetti, G. Biocompatibility of Functionalized Boron Phosphate (BPO4) Nanoparticles for Boron Neutron Capture Therapy (BNCT) Application. *Nanomed. Nanotechnol. Biol. Med.* **2014**, *10* (3), 589–597.

164 Golberg, D.; Bando, Y.; Huang, Y.; Terao, T.; Mitome, M.; Tang, C.; Zhi, C. Boron Nitride Nanotubes and Nanosheets. *ACS Nano* **2010**, *4* (6), 2979–2993.

165 Nakamura, H.; Koganei, H.; Miyoshi, T.; Sakurai, Y.; Ono, K.; Suzuki, M. Antitumor Effect of Boron Nitride Nanotubes in Combination with Thermal Neutron Irradiation on BNCT. *Biorg. Med. Chem. Lett.* **2015**, *25* (2), 172–174.

166 Xiong, H.; Zhou, D.; Qi, Y.; Zhang, Z.; Xie, Z.; Chen, X.; Jing, X.; Meng, F.; Huang, Y. Doxorubicin-Loaded Carborane-Conjugated Polymeric Nanoparticles as Delivery System for Combination Cancer Therapy. *Biomacromolecules* **2015**, *16* (12), 3980–3988.

167 Gao, Z.; Horiguchi, Y.; Nakai, K.; Matsumura, A.; Suzuki, M.; Ono, K.; Nagasaki, Y. Use of Boron Cluster-Containing Redox Nanoparticles with ROS Scavenging Ability in Boron Neutron Capture Therapy to Achieve High Therapeutic Efficiency and Low Adverse Effects. *Biomaterials* **2016**, *104*, 201–212.

3.3

Carborane Derivatives of Porphyrins and Chlorins for Photodynamic and Boron Neutron Capture Therapies: Synthetic Strategies

Valentina A. Ol'shevskaya,[1] Andrei V. Zaitsev,[1] and Alexander A. Shtil[2]

[1] A.N. Nesmeyanov Institute of Organoelement Compounds, Russian Academy of Sciences, Moscow, Russia
[2] Blokhin National Medical Research Center of Oncology, Moscow, Russia

3.3.1 Introduction

Porphyrins, a broad class of tetrapyrrolic macrocycles, have found numerous applications in diverse fields ranging from catalysts [1] to electronic materials [2], biomimetic models for photosynthesis [3,4], and medicine [5]. Medicinal use of porphyrins is based largely on their ability to preferentially accumulate in metabolically active (particularly, malignant) cells, being a good tool for antitumor photodynamic therapy (PDT) and drug delivery [6]. Porphyrin-based compounds have gained momentum in PDT of cancer. PDT employs the capability of a photosensitizer (PS) compound to produce reactive oxygen species (ROS) such as singlet oxygen ($^1O_2$) and free radicals (e.g., •OH, •HO_2, and H_2O_2) upon activation with light, thereby evoking damage to proteins, nucleic acids, membranes, and other cell components [7,8]. Therefore, the PS is a critical factor in PDT. An ideal PS should demonstrate high efficacy in light-activated ROS generation and low toxicity in the dark; a long-wavelength absorption (>600 nm) is preferable for deeper tissue penetration [9]. The structure of the porphyrin macrocycle allows for the conjugation of a variety of functionally different chemical moieties. These modifications can profoundly improve physical, chemical, and biological properties of the molecule, including amphiphilicity, bioavailability, metabolic stability, and antitumor potency. In particular, the addition of boron polyhedra to the periphery of the tetrapyrrolic macrocycle makes the conjugates applicable in boron neutron capture therapy (BNCT). This therapeutic modality is based on selective absorption of the non-radioactive ^{10}B isotope in the tumor, followed by irradiation with low-energy thermal neutrons. The $^{10}B(n,\alpha)^7Li$ reaction produces high-energy α-particles and a residual ^7Li nucleus (energy of 200 and 350 keV μm^{-1}, respectively) whose trafficking at a distance comparable with one cell diameter causes lethal damage of ^{10}B-enriched tumor cells. Consequently, the dual therapeutic potential of boronated porphyrins is linked to their applicability as PSs for PDT and as radiosensitizers for BNCT [6,10].

Some porphyrin-containing pharmaceuticals, such as Photogem [11] or the closely related Photofrin [12], are used in the clinic as PSs for PDT. The second generation of chlorin-based PSs for PDT is under development [13–17]. Chlorins are more promising PSs because they absorb light in the red spectral region (λ = 650–660 nm). This property provides deeper penetration of tissue photodamage compared with that achievable with porphyrins. Furthermore, light energy used for excitation of chlorins is less absorbable by surrounding tissues. Overall, the synthesis of boronated porphyrins and chlorins makes it possible to obtain compounds with properties that are advantageous for both BNCT and PDT: preferential accumulation in neoplasms, phototoxicity, and the ability to generate local radioactive reactions and cytotoxic particles generated by thermal neutrons.

To develop efficient compounds for PDT and BNCT, many new types of derivatives of the tetrapyrrole macrocycle with polyhedral boron units within one molecule have been prepared over the past 20 years [18–21]. These compounds have been tested in PDT and BNCT; *in vitro* and cell culture studies included (1) the ability to produce ROS, (2) dark toxicity, (3) tumor selectivity, (4) interaction with biomacromolecules in the cell (intracellular distribution), and (5) fluorescent properties.

Researchers around the world [22–24] as well as our group [25] continue to develop the synthetic approaches to boronated porphyrins and chlorins as candidate agents for PDT and BNCT. This review especially focuses on recent results on boronated analogs of synthetic porphyrins and chlorins with high boron content and boronated analogs of chlorin e_6.

3.3.2 Recent Synthetic Routes to Carboranyl-Substituted Derivatives of 5,10,15,20-Tetraphenylporphyrin

The last few decades have shown considerable advances in the design and study of synthetic porphyrins [26]. Reasons for this interest mainly lie in the key role played by the unique 18π-electron (aromatic) macrocycle in biological systems containing hemes.

We developed a series of carboranylporphyrin congeners based on functional derivatives of 5,10,15,20-tetraphenylporphyrin and neutral and anionic derivatives of icosahedral carboranes. Our choice of 5,10,15,20-tetraphenylporphyrin as a platform for the preparation of boronated derivatives was dictated by: (1) availability of this compound, (2) suitability for introducing various functional groups, and (3) ease of obtaining final products with a reasonable yield and purity. Based on the available 5-(4-aminophenyl)-10,15,20-triphenyl-porphyrin (**1**) [27], 5,10,15,20-tetrakis(4-aminophenyl)porphyrin (**2**) [28], and substituted carboranes (triflates, epoxides, or isocyanates), a simple, one-stage synthesis of boronated porphyrins with hydrolytically resistant spacer groups was developed [29].

Alkylation of amino groups in **1,2** with 1-trifluoromethane sulfonylmethyl-*o*-carborane [30] and 1-trifluoromethanesulfonylmethyl-1-carba-*closo*-dodecaborate cesium [15] in the presence of sodium acetate in acetonitrile produced monocarborane- and tetracarborane-substituted porphyrins **3–6** in quantitative yields (Figure 3.3.1). Importantly, **4** and **6** (which contain the anionic *closo*-1-carba-monocarbon carborane group) were soluble in water, which makes these compounds prospective for practical use.

Structurally close compounds **7,8**, in which three *o*-carborane polyhedra are linked to the porphyrin macrocycle through the aminomethylene group [31], have been synthesized by sequential transformations of 5,10,15-tris(4-nitrophenyl)-20-phenylporphyrin

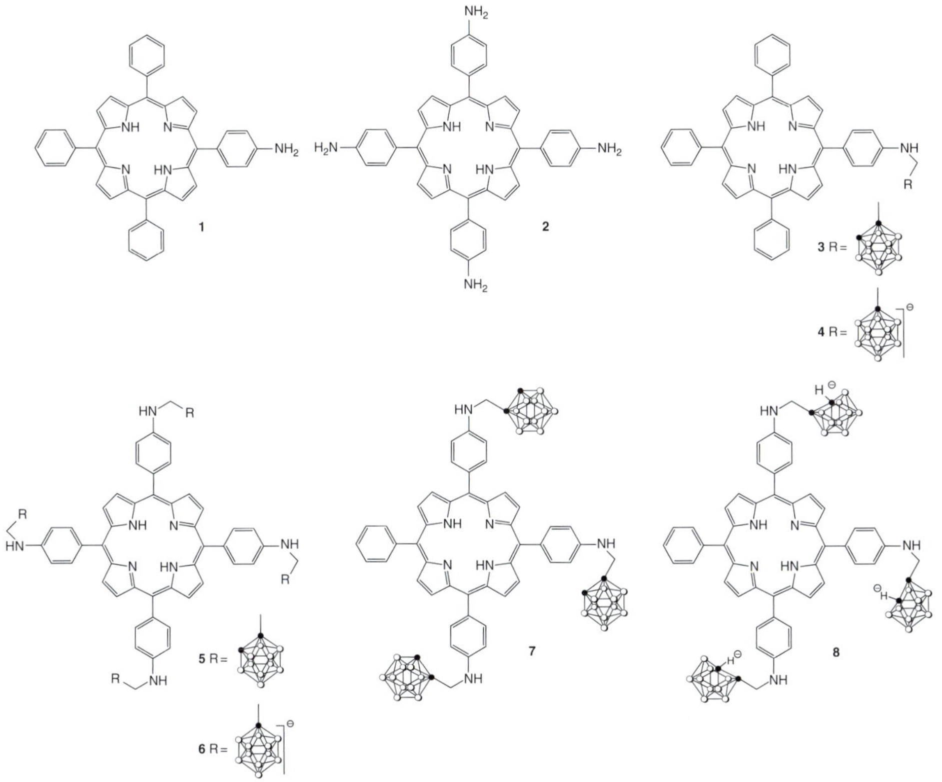

Figure 3.3.1 Carboranylporphyrins containing the aminomethylene bond.

into the respective amino derivative after reduction in SnCl$_2$–HCl. Then, the amino derivative reacted with 1-formyl-o-carborane; ultimately, the prepared imino derivative was reduced with NaBH$_4$ to produce 5,10,15-tris[(carboranylaminomethyl) phenyl]-20-phenylporphyrin.

Using this method, structurally similar compounds can be synthesized in one step with high yields.

Another approach to direct introduction of boron polyhedra into aminoporphyrins is based on the reaction of aminolysis of 1,2-carboranylepoxides [32,33]. Aminolysis of epoxides is an efficient route for preparation of β-aminoalcohols, important intermediates in the synthesis of biologically active compounds. To functionalize the amino groups in **1** and **2** with carborane epoxides [34], the catalytic activity of zinc chloride was used in the epoxide ring opening with nucleophiles [35]. However, zinc chloride might react with porphyrins to form metal complexes. Therefore, first Zn and Pd complexes of aminoporphyrins **1,2** [36,37] were obtained, and these metal complexes were put into the reactions with carborane epoxides. The catalytic opening of the epoxide ring in carboranes by amino groups of metal containing aminoporphyrins easily occurred in boiling acetonitrile or tetrahydrofuran (THF) in the presence of 1–3 mol% ZnCl$_2$, resulting in boronated metal porphyrins **9–16** (Figure 3.3.2). In these compounds, the aminoalcohol spacer linked the carborane polyhedron with the porphyrin macrocycle. All compounds were obtained in high yields.

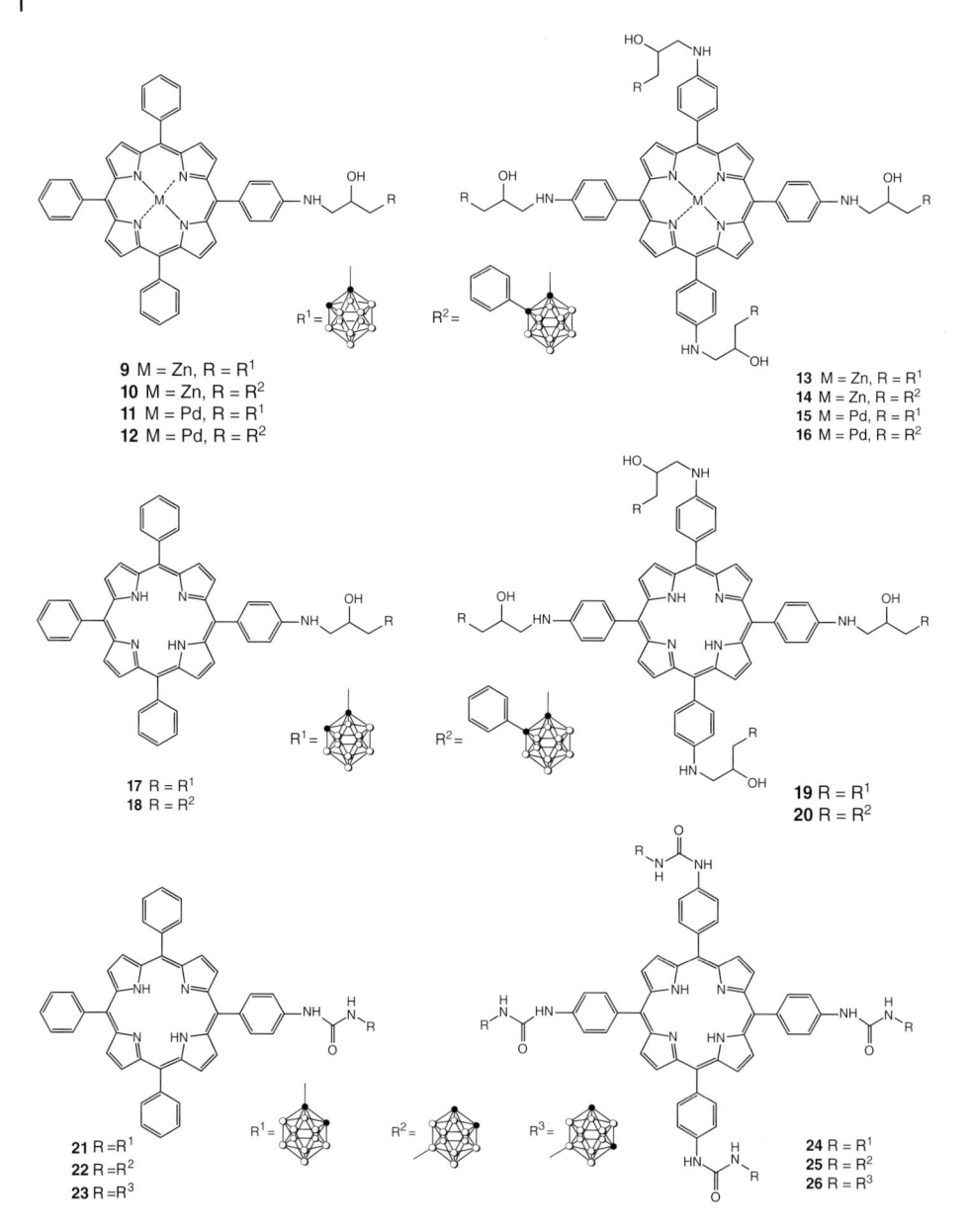

Figure 3.3.2 Carboranylporphyrins with aminoalcohol and amide bonds.

Demetallation of the products of aminolysis of **9,10,13,14** with trifluoroacetic acid in methylene chloride [38] allowed to obtain boronated β-aminoalcohol derivatives of porphyrins **17–20** in good yields (Figure 3.3.2). Interestingly, only one regio-isomer was formed in this reaction. It seems likely that the attack by the amine nucleophile takes place exclusively at the less hindered position, namely, at the methylene group of the epoxyde ring. Compounds **9–20** possess an asymmetric carbon atom; presumably, two enantiomeres can be formed.

Also, synthesis of boronated porphyrins with amide bonds was developed [29]. The latter bonds (that are more resistant to hydrolysis than ester bonds) are present in proteins and in numerous other natural products as well as in drugs.

Reactions of **1,2** with 1-isocyanato-*o*-carborane [39] or 9-isocyanato-*o*- and 9-isocyanato-*m*-carboranes [40] in which the isocyanate group is hydrolytically stable resulted in boron containing conjugates **21–26** (Figure 3.3.2). The position of the isocyanate group in the carborane significantly influenced the rate of the reaction and the yields of carboranylporphyrins **21–26**. As expected, the ability of 1-isocyanato-*o*-carborane to react with amino groups in **1,2** was higher than that of 9-isocyanato-*o*- and 9-isocyanato-*m*-carboranes. This difference can be explained by a non-uniform distribution of electron density on skeletal atoms of carborane polyhedra [41]. This difference gives rise to various electronic effects of carboranyl groups. Therefore, the same functional groups linked to the skeletal atoms would show differential activity. In compound **21**, the N=C=O group is linked to the electron-withdrawing 1-*o*-carboranyl group ($\sigma i = +0.38$); whereas in **22,23**, the N=C=O groups are located at electron-donating 9-*o*- ($\sigma i = -0.23$) and 9-*m*- ($\sigma i = -0.12$) carboranyl groups [42,43], making the nucleophilic attack of **1,2** more difficult. One may anticipate that **21–26** will fit the requirements for BNCT, since the conjugation of boron polyhedra to dendrimers via the amide bonds of similar nature has been shown to produce efficient agents [44,45].

Another type of carboranylporphyrins with amide bonds was synthesized [46] from a precursor prepared by acylation of the amino group in **1** with succinic and maleic anhydrides. This procedure produced porphyrin conjugates containing substituted succinic **27** and maleic **28** acid monoamides as pharmacophore groups (Figure 3.3.3).

Functionalization of the carboxylic group in **27** activated by TBTU with amines in ethyl acetate, in particular propargylamine or 3-amino-*o*-carborane, gave asymmetric porphyrin diamides **29,30**. Boronation of compound **29** containing a terminal triple bond was performed by copper-catalyzed 1,3-dipolar [2+3]-cycloaddition reaction with 1-azidomethyl-*o*-carborane to give porphyrin **31** with heterocyclic 1,2,3-triazole spacer group (Figure 3.3.3). The reaction was carried out in a CH_2Cl_2–H_2O system at room temperature using $Cu(OAc)_2$ $2H_2O$–sodium ascorbate as a catalyst.

The porphyrin **28** contains two reactive sites, that is, the carboxylic group and the activated C=C double bond, to which the carborane polyhedron can be added. The reaction of the carboxylic group (TBTU activation) of **28** with 3-amino-*o*-carborane in ethyl acetate at 20 °C in the presence of DIPEA resulted in the formation of asymmetric carborane diamide **32** in 90% yield. To obtain stable compounds for biological assays, addition reactions to the activated double bond in **28** with functionally substituted carboranes were carried out. Compound **28** reacted easily with 9-mercapto-*m*-carborane [47] or 1-azidomethyl-*o*-carborane [48] to give corresponding carboranylthio- (**33**) and carboranyltriazoline (**34**) derivatives, respectively, in high yields. Note that the reaction with 9-mercapto-*m*-carborane occurred as a regioselective nucleophilic addition to the double bond, the initial attack of the carborane S-nucleophile being directed at the β-carbon atom relative to the carboxylic group of **28**. The reaction of diamide **32** with 9-mercapto-*m*-carborane by refluxing in chloroform yielded compound **35** with two carborane polyhedra (Figure 3.3.3).

Dehydration of porphyrin **28** under the action of Ac$_2$O resulted in the formation of 5-[4-(*N*-maleimido)phenyl]-10,15,20-triphenylporphyrin **36**. Substituted maleimides are extensively studied due to the activated double bond prone to various chemical transformations. Porphyrins containing the maleimide moiety are convenient substrates for introducing carborane clusters because they can behave as dipolarophiles in the reaction of 1,3-dipolar cycloaddition or as dienophiles in the Diels–Alder reaction, as well as acceptors in the Michael reaction with S-, N-, and O-nucleophiles. Porphyrin **36** readily underwent Michael addition with 9-mercapto-*m*-carborane, 3-amino-*o*-carborane, and 3-hydroxy-*o*-carborane to form corresponding succinimide-substituted carboranylporphyrins **37–39** in high (70–85%) yields [49]. The reaction of 3-hydroxy-*o*-carborane with porphyrin **36** was carried out in the presence of cerium ammonium nitrate (CAN) catalyst. The reduced reactivity of this carborane in Michael addition

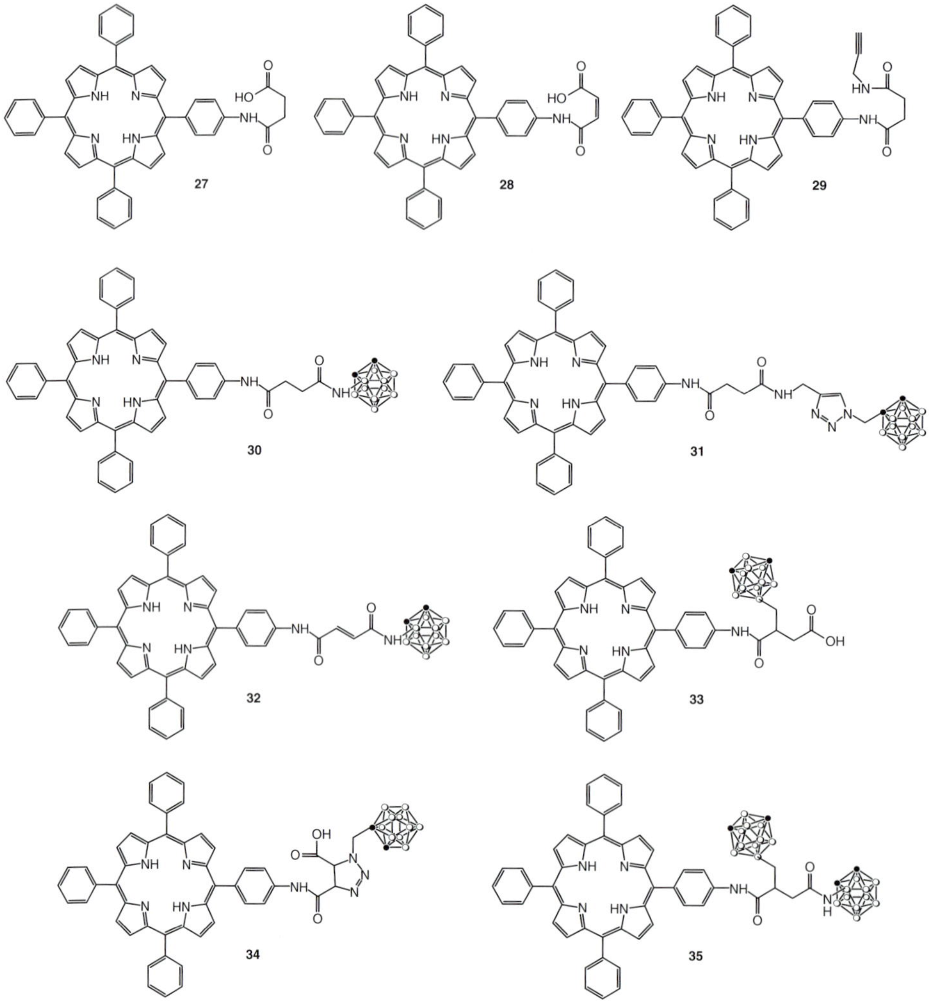

Figure 3.3.3 Types of boronated porphyrins prepared from succinic and maleic aminoporphyrin derivatives.

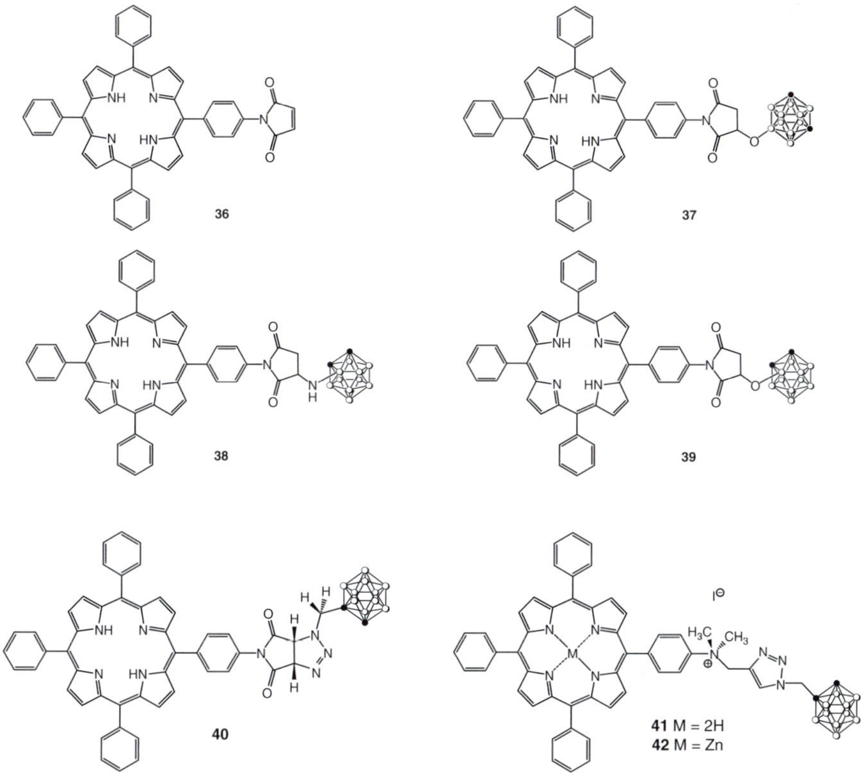

Figure 3.3.3 (Continued)

is likely to be caused by the electron-withdrawing character of 3-*o*-carboranyl group ($\sigma_i = 0.11$), which decreases the nucleophilicity of the oxygen atom. It should be noted that the carborane nucleophiles did not cause recyclization of the maleimide ring in porphyrin **36** in the course of Michael addition. The reaction of the double bond in the maleimide moiety of **36** with 1-azidomethyl-*o*-carborane led to the formation of carboranylporphyrin **40** containing bicyclic pyrrolidinotriazoline fragment (95% yield). The resulting carboranylporphyrins showed an increased affinity to tumor cells that allows their use as potential medicines for BNCT and PDT [46].

Cationic boronated porphyrins **41,42**, potential DNA ligands, were prepared [50] by the reaction of mono-substituted aminopropargyl derivatives of **1** and its zinc complex with 1-azidomethyl-*o*-carborane [48], and subsequent quaternization of the amino group in carboranyltriazolyl aminoporphyrins with methyl iodide under reflux in CHCl$_3$–MeCN (Figure 3.3.3).

Biological characteristics of porphyrins **28**, **32**, and **35** were assessed by measuring the dark toxicity and the ability to cause light-inducible cell death (HCT116 human colon cancer cell line). The dark cytotoxicity was determined in the MTT test [51] after cell exposure to **28**, **32**, or **35** for 72 h. A 50% inhibition of cell growth was observed at rather high concentrations, the IC$_{50}$ values being >12.5 µM for **37**, >10 µM for **32**, and >25 µM for **35**. The low dark cytotoxicity enables the use of these compounds in anticancer PDT. The experimental PDT was carried out after 30–60 min incubation of cells with 1 µM of **28**, **32**, or **35**, and removal of the compounds from the culture medium [52]. Illumination of cells loaded with **28** triggered necrosis of 100% cells within 5–7 min of light exposure. Necrotic death was detected by typical morphology and the inclusion of propidium iodide detectable by microscopy or flow cytometry [52]. After illumination of cells loaded with compound **32**, the rate of death was slower, and the morphology of damaged cells was indicative of apoptosis; the cells were rounded and gradually detached from the plastic. Within 24 h post illumination, ~90% of cells were floating; however, the morphological patterns of death differed from those of light-induced necrosis. Finally, illumination of cells incubated with **35** evoked a much slower damage: after 24–48 h, many cells changed insignificantly and remained attached to the substrate.

Thus, compounds **28**, **32**, and **35** showed no dark toxicity. In striking contrast, the action of light on the cells loaded with non-boronated **28** caused rapid (within minutes) necrosis. In the presence of mono-substituted carboranylporphyrin **32**, cell death developed slower via a mechanism other than PDT-induced necrosis. In the presence of dicarboranylporphyrin **35**, cell damage was even less pronounced. These results showed that the boron polyhedra differentially modulate the properties of porphyrins in the cells. The addition of one single boron polyhedron to the macrocycle can switch the death mechanism from necrosis to apoptosis, which is important for selection of the therapeutic protocol. The addition of the second boron polyhedron can reduce the photoactivity, probably due to retarded transport into the cells or by other reasons that remain to be elucidated to justify future strategies of design of boronated porphyrins as therapeutic agents.

3.3.3 Synthesis of Carborane Containing Porphyrins and Chlorins from Pentafluorophenyl-Substituted Porphyrin

In recent years, the synthesis of fluorine-containing products attracted considerable attention in medical chemistry. Indeed, ~20–25% of drug substances developed by the pharmaceutical industry contain fluorine atoms. Introduction of fluorine into the drug molecules is known to enhance metabolic stability and binding to target proteins, as well as to improve bioavailability [53]. Fluorinated porphyrins are suitable and well-known PSs with reasonable selectivity for cancer cells owing to their versatile photophysics, electron-withdrawing properties, and resistance to oxidative degradation. Selectivity and cell death–inducing potency depend on functionalization of the porphyrin's periphery. Therefore, development of selective and efficient fluorinated PSs is underway and remains a challenge. The use of fluorinated porphyrins and chlorins can provide optimal lipophilic characteristics of the reaction products. In combination with carborane clusters, this would furnish products suitable for both BNCT and PDT. We developed facile one-step methods for synthesis of fluorine-containing neutral and anionic boronated porphyrins and chlorins [54]. Fluorine nucleophilic substitution in the *para*-position of phenyl rings of 5,10,15,20-tetrakis(pentafluorophenyl)porphyrin and its metal complexes (Cu and Pd) [55] with neutral and anionic lithiocarboranes such as cesium 1-lithio-1-carba-*closo*-dodecaborate, 1-lithio-*o*-carborane, and 1-lithio-2-phenyl-*o*-carborane [56] resulted in fluorinated carboranylporphyrins **43–47** (Figure 3.3.4).

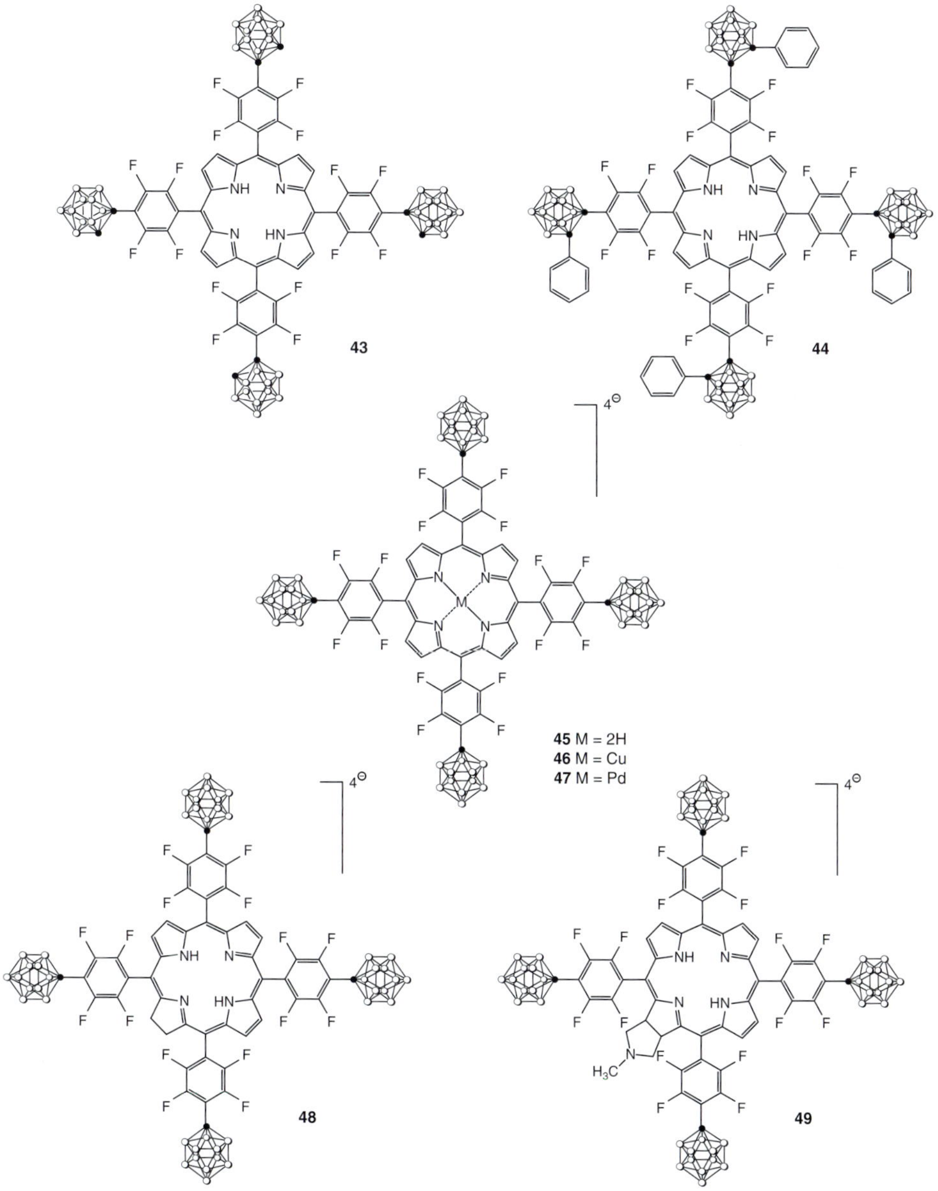

Figure 3.3.4 Fluorinated porphyrin and chlorin conjugates with *closo*-carborane polyhedra.

Reactions were carried out in THF and proceeded with retention of the *closo*-structure of the carborane skeleton to produce **43–47** in 65–85% yield. To increase the phototoxic effect, the boronated porphyrin **40** was transformed to water-soluble chlorins **48,49** by reduction with *p*-toluenesulfonylhydrazide (for **48**) [57] or azomethine ylide complex (for **49**) obtained from N-methylglycine and formaldehyde [58] (Figure 3.3.4).

Some biologically relevant properties of tetraanionic chlorin **48** were studied. This compound can permeate through a bilayer lipid membrane [59]. Next, being initially designed as radiosensitizers for BNCT, the fluorinated porphyrins, chlorins, and their metal complexes appeared to be more active than their non-boronated analogs for PDT in cell culture and in laboratory animals. These compounds accumulate in the cytoplasm and induce rapid necrosis upon light activation. The mechanism of cell death is the photodamage of biomacromolecules, primarily in the plasma membrane and organelles [60].

The fluorinated carboranylchlorin **48** showed a significantly bigger accumulation in B16 mouse melanoma and C6 rat glioma cells (peak at 36 h) than non-fluorinated carboranylchlorin or fluorinated chlorin (no boron). Compound **48** demonstrated favorable characteristics such as low toxicity in mice (IC_{50} = 161 mg/kg), no significant organ toxicity, and a prolonged circulation in the blood. Compound **48** was negligibly cytotoxic in the dark; in contrast, rapid photodamage was detectable within the initial 5–10 min of light illumination of wild-type and drug-resistant tumor cells loaded with low micromolar concentrations of **48**. Cell death was associated with the entry of propidium iodide into the nuclei, a hallmark of necrosis. Importantly, **48** was more potent than its single element modified derivatives in photodestruction of subcutaneously transplanted C6 xenografts in Balb/c-nu/nu mice. Finally, **48** sensitized C6 xenografts to thermal neutrons (BNCT). Altogether, the simplified synthesis, good water solubility, favorable pharmacological properties, and high potency in cell culture and *in vivo* make the novel fluorinated carboranylchlorin **48** a prospective drug candidate [60].

Next, neutral and anionic fluorinated carboranylporphyrins **50–56** containing *closo*-carborane polyhedra were synthesized from 5,10,15,20-tetrakis(pentafluorophenyl)porphyrin and its metal complexes (Zn, Cu, and Pd) and *closo*-mercaptocarboranes (9-mercapto-*o*-carborane, 9-mercapto-*m*-carborane, 1-mercapto-7-isopropyl-*m*-carborane, and cesium 1-mercapto-1-carba-*closo*-dodecaborate) using NaOAc as a base (Figure 3.3.5) [25]. Usually a nucleophilic substitution of fluorine is carried out in DMF in the presence of

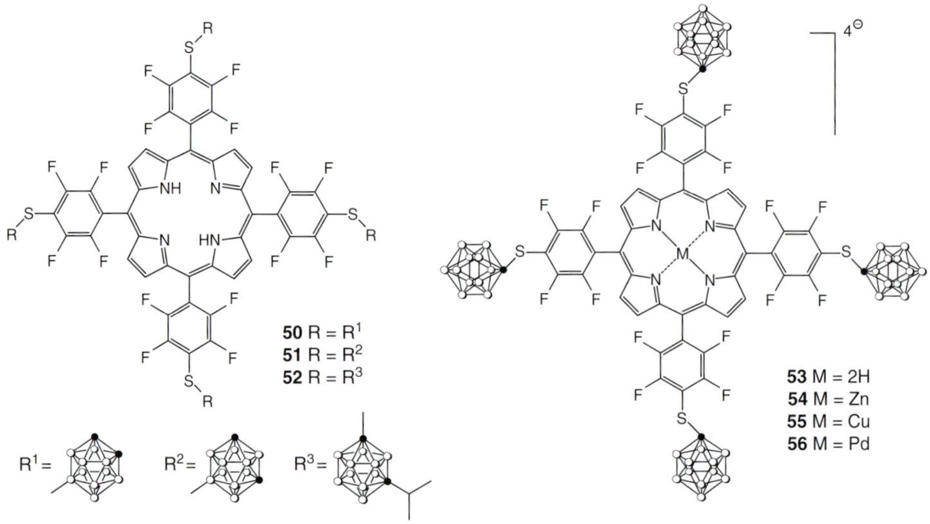

Figure 3.3.5 Fluorinated porphyrin conjugates with neutral and anionic *closo*-carboranylthio-substituents.

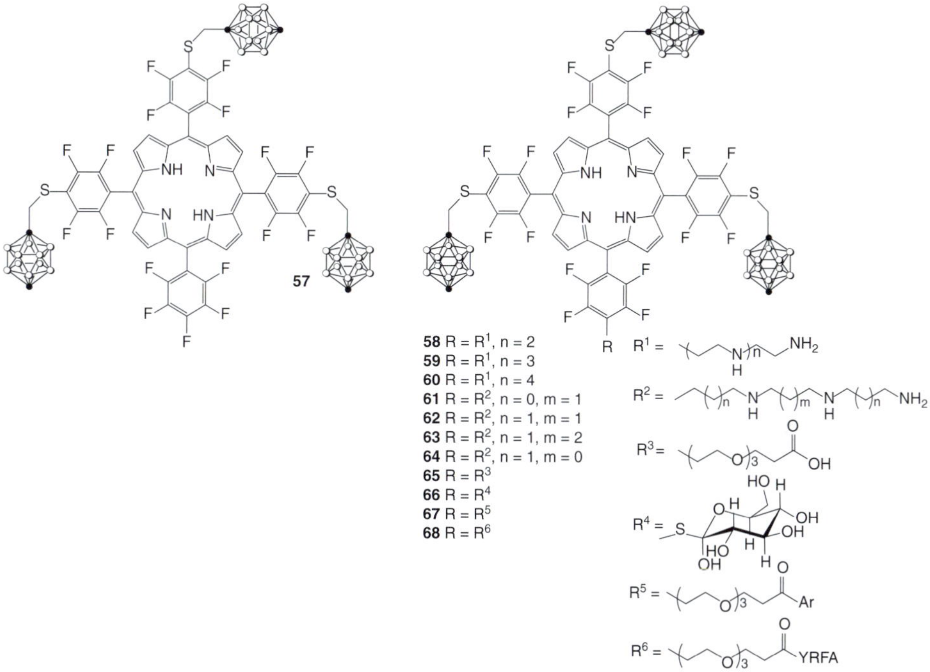

Figure 3.3.6 Polyamine, peptide, and glucose conjugates of fluorinated p-carboranylmethylthio-porphyrin.

K$_2$CO$_3$, leading to deboronation of the carborane polyhedron into a *nido*-dicarbaundecaborate anion [14].

Studies of accumulation of carboranylporphyrins **53–56** in HCT116 colon carcinoma cells showed that maximum concentrations were reached by 120 min of cell exposure, followed by a slow efflux. The dark cytotoxicity was low: a 50% cell growth inhibition was achieved only after a continuous 72 h cell exposure with 5–10 μM of tested compounds. In contrast, upon light illumination, compounds **53–56** at submicromolar concentrations caused a necrotic-like cell death within 3–15 min.

Fluorinated carboranylporphyrin **57** containing three *closo-p*-carboranylmethylthio substituents was prepared by the reaction of 5,10,15,20-tetrakis(pentafluorophenyl) porphyrin with 1-mercaptomethyl-*p*-carborane in the presence of K$_2$CO$_3$ in DMF at room temperature [22]. The subsequent conjugation of **57** with polyamines, glucose, and amino acids afforded the target polyamine–porphyrin conjugates **58–65**, glucose–porphyrin (**66**) and arginine–porphyrin (**67**) conjugates, and peptide Tyr-D-Arg-Phe-β-Ala (YRFA)–porphyrin conjugate (**68**) (Figure 3.3.6) investigated as boron delivery agents for BNCT [23]. Conjugates **58–68** were investigated for blood–brain barrier (BBB) permeability in human hCMEC/D3 brain endothelial cells, and their cytotoxicity for and uptake by human glioma T98G cells. For comparison, a symmetric tetra[(*p*-carboranylmethylthio)tetrafluorophenyl]porphyrin was also synthesized, and its crystal structure was obtained. All conjugates were efficiently taken up by T98G cells, particularly the cationic polyamine and arginine conjugates, and localized in various organelles including mitochondria and lysosomes. All compounds showed relatively low *in vitro*

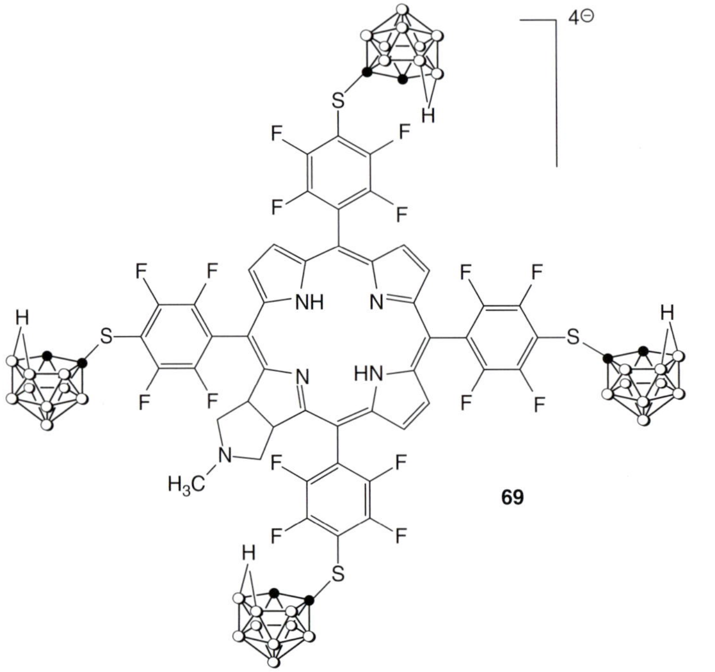

Figure 3.3.7 Tetrakis(p-carboranylthio-tetrafluorophenyl)chlorin (TPFC).

BBB permeability due to their higher molecular weight, hydrophobicity, and tendency for aggregation in solutions. Among this series, the branched polyamine and YRFA conjugates showed the highest permeability coefficients, while the glucose conjugate showed the least.

Fluorinated carboranylchlorin **69** containing four *nido*-carborane substituents (Figure 3.3.7) was prepared by conversion (CH_3NHCH_2COOH, paraformaldehyde, and toluene) of 5,10,15,20-tetrakis(pentafluorophenyl)porphyrin into chlorin [14]. The latter compound reacted with an excess of 1-mercapto-*o*-carborane and K_2CO_3 in DMF at room temperature to produce **69** in 69% yield. This compound was nontoxic in the dark but showed good photosensitizing ability both in cell culture and *in vivo* despite its relatively low quantum yield of singlet oxygen. In particular, TPFC exhibited significant PS activity against highly pigmented melanotic melanoma transplants in mice.

Illumination of **69** in a dimethyl sulfoxide (DMSO) solution by red light induced a very modest decrease in the intensity of visible absorption bands of chlorin (<10% decrease after 30 min illumination). The formation of new absorption bands was not detected even after prolonged illumination. The high degree of photostability displayed by tetrakis(*p*-carboranylthio-tetrafluorophenyl)chlorin (TPFC) is rather unusual since most derivatives of tetrapyrrole are known to undergo extensive photobleaching upon exposure to visible light [61]. These observations imply that the concentration of TPFC in the illuminated tissue remains at reasonably large levels throughout light exposure, thereby ensuring an efficient absorption of the incident light during the phototherapeutic procedure. On the other hand, the singlet oxygen quantum yield for TPFC was found to be 0.1, a significantly lower value compared with circa 0.5 typical for most PSs [62]. The high photochemical stability and comparatively low quantum yield of singlet oxygen observed for TPFC are probably consequences of the presence of four carborane cages, as previously observed for other carborane-containing molecules [63].

Compound **69** demonstrated a low dark toxicity in HEp2 human carcinoma cells at concentrations up to 100 µM within 24 h. *In vivo* studies showed that intravenously injected TPFC exhibited some selectivity of tumor targeting at a relatively short time after administration. The maximum ratio (2.5-fold) of PS concentration in melanoma versus peritumoral skin was registered 3 h post injection. All mice subjected to PDT remained tumor-free for 5 days, while the untreated or mock-treated mice (illumination in the absence of PS) displayed a fairly rapid tumor growth during this time interval. This compound was also studied in both PDT and BNCT [62,64], demonstrating a high phototoxic effect and equal or superior performance compared with boronated phenylalanine (BPA) upon neutron irradiation. However, *in vivo* TPFC did not show as high an antitumor effect as did BPA.

Thus, data on photo- and radiosensitizing activities of fluorinated carboranylporphyrins and chlorins obtained in animal models suggest that the modification of tetrapyrrole compounds with fluorine atoms and boron polyhedra could be regarded as a novel promising direction in organoelement chemistry of antitumor compounds. Modifications at the periphery of the porphyrin macrocycle using different substituents including boron clusters can change the properties of the entire molecule. As a result, modified PS with desirable lipophilic, hydrophilic, and amphiphilic properties can be obtained [65–69]. These characteristics are key factors that determine the mechanisms of tumor cell death upon photo- and radiotherapy.

3.3.4 Carborane Containing Derivatives of Chlorins: New Properties for PDT and Beyond

Currently, the PSs of various classes are used in the clinic or are at different stages of clinical trials [12,70]. Among them, the chlorophyll derivatives such as natural chlorins e_6, bacteriochlorins, and cycloimides are most prospective due to their intense absorption in the red and nearest-infrared spectral region, so their phototherapeutic "window" (660–800 nm) opens a new possibility for tumor diagnosis and treatment. However, chlorins and bacteriochlorins themselves have limited application as PSs due to high hydrophobicity, low chemical and light stability, and moderate selectivity of accumulation in tumor cells. This leads to unwanted side effects, including photodamage of surrounding healthy tissues. From a biological viewpoint, an enhancement of cellular uptake of PSs in malignant cells can improve the photocytotoxicity.

We have demonstrated a significantly higher efficacy of a water-soluble boronated derivative of natural protohemin IX (compared with boron-free protohemin IX) in PDT of sarcoma-bearing rats [71]. This finding paved the road to the conjugation of boron polyhedra to the chlorin macrocycle, a modification aimed initially at compounds for BNCT, to obtain agents more efficient in PDT than their non-boronated analogs. Thus, boronation not merely retains the PS properties of tetrapyrrole-containing compounds. Rather, the porphyrin–boron conjugates can possess an enhanced therapeutic potency.

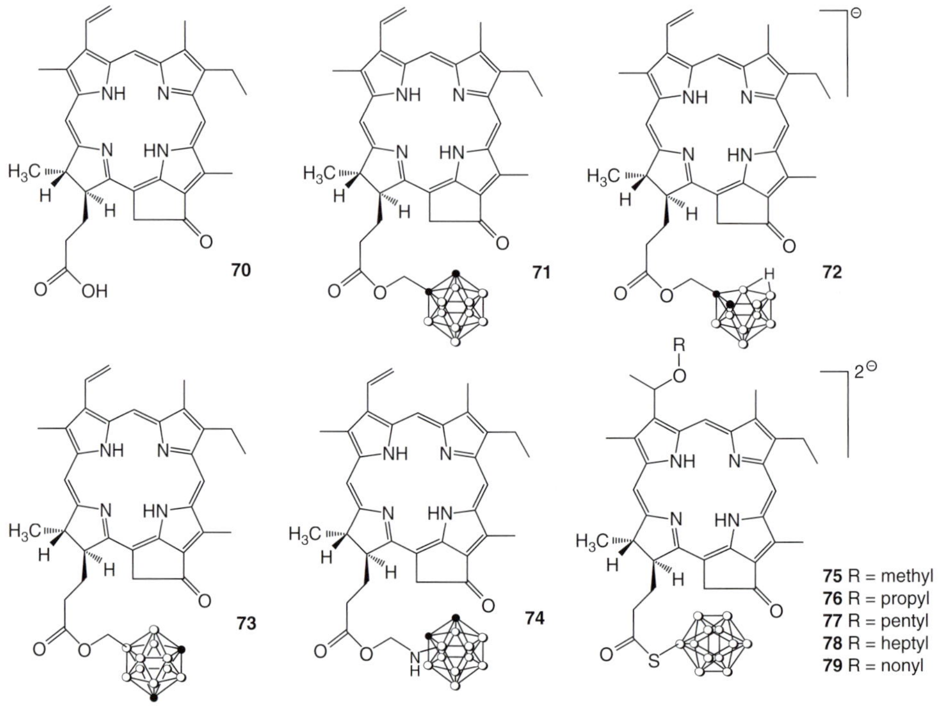

Figure 3.3.8 Boronated derivatives of pyropheophorbide *a*.

In particular, the design of boronated chlorophyll derivatives (pyropheophorbide *a*, pheophorbide *a*, chlorin e$_6$, bacteriochlorin, and cycloimide) enables to use them as therapeutic agents for PDT and BNCT [72,73], using the efficiency and advantages of each method and minimizing disadvantages.

3.3.4.1 Carborane Containing Derivatives of Pyropheophorbide *a* and Pheophorbide *a*

Pyropheophorbide *a* **70** was functionalized [74] with carborane polyhedra via the esterification of its carboxypropyl group with 1-hydroxymethyl-*o*-carborane and 9-hydroxymethyl-*m*-carborane or via amidation with 3-amino-*o*-carborane to give compounds **71–74** (Figure 3.3.8). Reactions were performed exploring the mixed anhydride method using di-*tert*-butyl pyrocarbonate in pyridine as a condensation agent. In case of 1-hydroxymethyl-*o*-carborane, equal amounts of two substances containing carboranes in *closo-* (**71**) and *nido-* (**72**) forms were obtained. From 9-hydroxymethyl-*m*-carborane, a single *closo*-carboranylated derivative (**73**) was obtained in 86% yield. The reaction of 3-amino-*o*-carborane with **70** yielded the corresponding amide derivative (**74**) in 56% yield. Electron absorption spectra of **70–74** were of the chlorin type, having identical position of bands in the visible spectral region. In the spectra of all compounds, the extinction coefficients were smaller than in the spectrum of the initial pyropheophorbide *a*. Testing of **70–74** against nonmalignant cells (human skin fibroblasts) and tumor cells (HCT116 colon carcinoma) demonstrated little to no dark cytotoxicity.

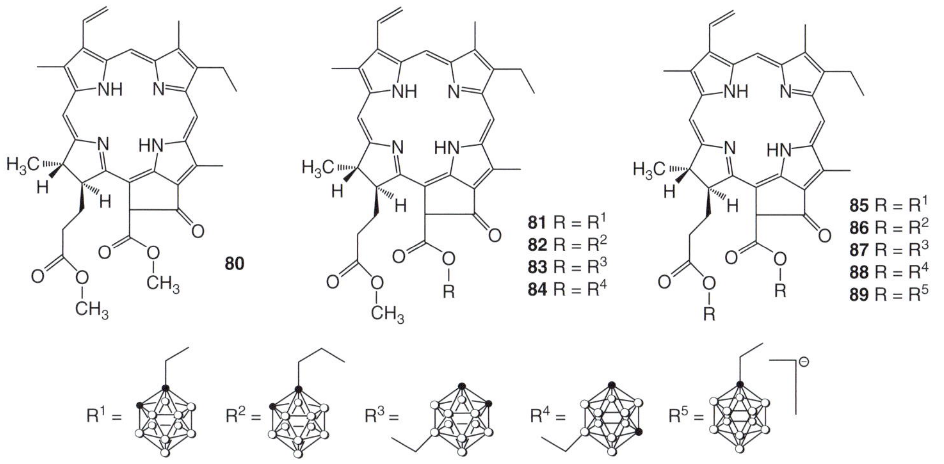

Figure 3.3.9 Boronated derivatives of methylpheophorbide a.

Some minor toxicity was detectable for the pyropheophorbide *a* derivative **72** in which the carborane polyhedron was in the *nido* form. However, given limited water-solubility of **71** and **72**, one may expect that boronated derivatives of pyropheophorbide should undergo further structural optimization to be applicable in PDT and BNCT.

Thioester derivatives **75–79** of pyropheophorbide *a* with hydrophilic *closo*-dodecaborate dianion have been prepared using oxalyl chloride and $[Me_4N]_2[B_{12}H_{11}SH]$ [75]. The compounds were clearly water-soluble when converted by ion exchange chromatography to the sodium salts from the tetramethylammonium salts used in the synthesis. Compounds **75–79** readily accumulated in V79 Chinese hamster cells (except the nonyl derivative), localized mainly in mitochondria, and showed a moderate dark cytotoxicity.

Methylpheophorbide *a* **80** is a promising compound for biomedical applications because of its reduced hydrophobicity compared to pyropheophorbide *a*. The carborane analogs **81–89** of methylpheophorbide *a* were obtained [76] via a classical organic chemistry reaction, trans-esterification, which makes it possible to obtain carboxylic esters under mild conditions in high yields (Figure 3.3.9).

Trans-esterification of one methoxycarbonyl group with carborane alcohols such as 1-hydroxymethyl-*o*-carborane, 1-hydroxyethyl-*o*-carborane, 9-hydroxymethyl-*o*-carborane, 9-hydroxymethyl-*m*-carborane, and cesium 1-hydroxymethyl-*closo*-monocarbadodecaborate occurred smoothly using I_2 in refluxing toluene or 2-chloro-1-methylpyridinium iodide and DMAP to give 13(2)-carborane esters **81–84** in 80% yield. Introduction of two carborane polyhedra into **80** via trans-esterification of 13(2)- and 17(3)-methoxycarbonyl groups was performed with the distannoxane triflate $[Bu_2Sn(OH)(OTf)]_2$; the conjugates **85–89** were obtained in 65–80% yield.

Using spectroscopic methods, Golovina et al. [77] demonstrated that methylpheophorbide *a* forms complexes with serum albumin and low-density lipoproteins (LDLs), whereas two diboronated derivatives, (13(2),17(3)-[di(*o*-carboran-1-yl)methoxycarbonyl] pheophorbide *a*) (**85**) and (13(2),17(3)-[di(1-carba-*closo*-dodecaboran-1-yl) methoxycarbonyl] pheophorbide *a*) (**89**), were capable of binding to LDL but not to albumin. Molecular modeling showed a mode of interaction of **75** with the amino acid residues in the albumin's hemin binding site. In contrast, for diboronated derivatives **85,89**, such interactions are sterically hindered by boron polyhedra, in line with experimentally determined lack of complex formation with albumin. These data strongly suggest that LDL might be the preferred carrier for polycarborane containing methylpheophorbide *a* derivatives.

3.3.4.2 Carborane Containing Derivatives of Chlorin e_6

The chlorin moiety in **90** is advantageous as a pharmacophore since it is readily available, relatively low toxic in the dark, and prone to simple conjugation chemistry. It has a Q-band absorption at ~660 nm and a high quantum yield of generated singlet oxygen. Therefore, **90** is highly promising for modification with boron polyhedra.

For the preparation of boronated derivatives of chlorin e_6 with the amide bond, the aminochlorin **91** was prepared by the nucleophilic opening of the exocyclic ring of methylpheophorbide *a* with ethylenediamine. Subsequent acylation of the free amino group in **91** with succinic anhydride and amidation of the resulting carboxylic group with 3-amino-*o*-carborane using DCC in CH_2Cl_2–Py mixture resulted in the boronated chlorin **92**. Hydrolysis of the ester group in **92** at the 17(3)-position with 70% TFA and amidation of the carboxylic group with 3-amino-*o*-carborane under similar conditions afforded chlorin **93** with two carborane clusters at the periphery of the chlorin macrocycle [78]. The aminochlorin **91** can be readily alkylated with 1-trifluoromethanesulfonylmethyl-*o*-carborane [79] or 1-trifluoromethanesulfonylmethyl-1-carba-*closo*-dodecaborate cesium [15] in THF to give corresponding neutral **94** and anionic **95** boronated chlorins in 22–80% yield. Furthermore, the metal complexes **96–101** (Zn(II), Pd(II), or Sn(IV)) of boronated chlorins **94–95** were prepared (Figure 3.3.10) [15].

Compound **94** formed complexes with serum albumin, a major porphyrin carrier [79]. The binding constant of these complexes was fourfold bigger than the respective value for complexes of albumin with boron-free aminochlorin **91**. Compound **94** potently sensitized rat fibroblasts to illumination with monochromatic red light: >98% of cells were necrotic by 24 h post illumination with 1 µM of **94**. This compound demonstrated a high efficacy in *in vivo* PDT of rat M-1 sarcoma. After illumination with 25 mg/kg of **94**, the residual tumors were significantly smaller than in animals subjected to PDT with an equal concentration of boron-free aminochlorin **91**. No signs of general toxicity were detectable after PDT with **94**. Thus, boronation can enhance the potency of chlorins in PDT, in part due to an increased binding to albumin. These data expand the therapeutic applicability of boronated chlorins beyond BNCT; the agents emerge as efficacious PSs and radiosensitizers.

Antimicrobial photodynamic inactivation is widely considered as an alternative to antibiotic chemotherapy of infective diseases, thereby attracting attention to the design of novel PSs. Compound **95** appeared to be much more potent than the starting compound **91** against *Bacillus subtilis*, *Staphylococcus aureus*, and *Mycobacterium* sp. Confocal fluorescence spectroscopy and membrane leakage experiments indicated that bacterial cell death upon photodynamic treatment with **95** is caused by loss of cell membrane integrity. The enhanced photobactericidal activity was attributed to the increased accumulation of **95** by bacterial cells, as evaluated by centrifugation and fluorescence correlation spectroscopy. Gram-negative bacteria were rather resistant to antimicrobial photodynamic inactivation by **95**. Unlike **91**, compound **95** showed

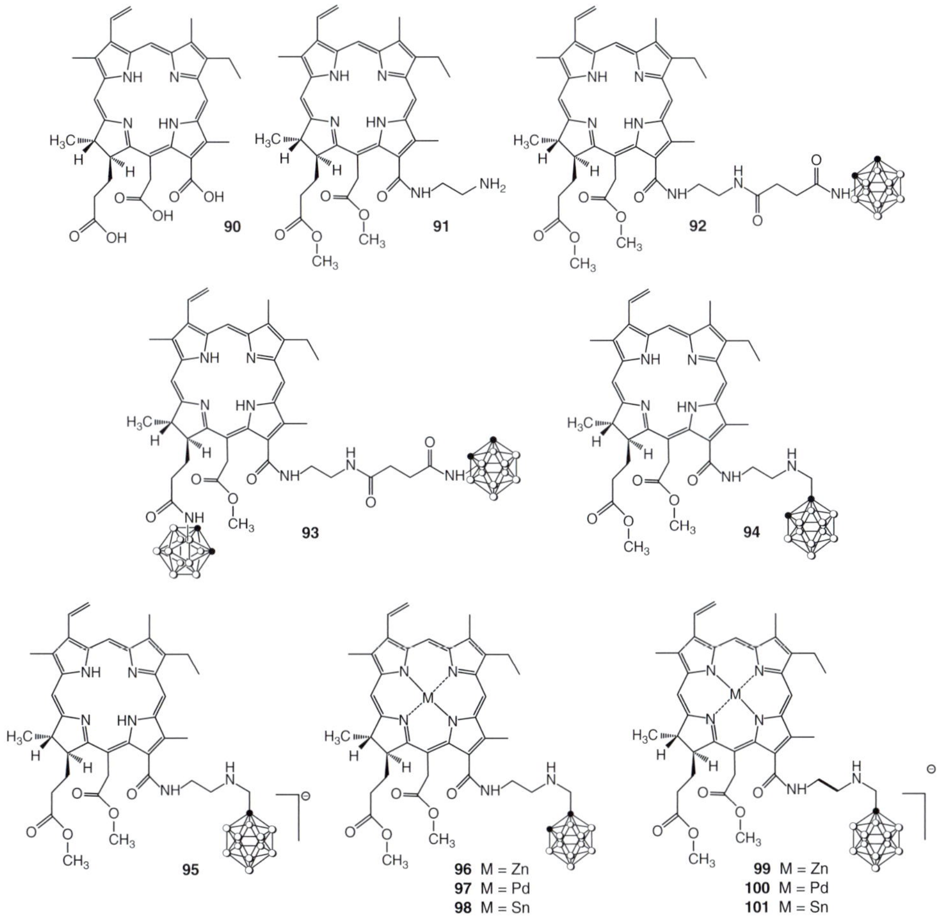

Figure 3.3.10 Carborane-substituted derivatives of chlorin e$_6$.

higher (compared to the wild-type strain) dark toxicity for *Escherichia coli* Δ*tolC* mutant deficient in TolC-requiring multidrug transporters [80].

Compound **95** evoked no dark toxicity and demonstrated a significantly higher photosensitizing efficacy than chlorin e$_6$ against transplanted aggressive tumors such as B16 melanoma and M-1 sarcoma. Illumination with red light of tumor cells loaded with 0.1 μM of **95** caused rapid (within the initial minutes) necrosis as determined by propidium iodide staining. The laser confocal microscopy–assisted analysis of cell death revealed the following order of events: prior to illumination, **95** accumulated in Golgi cysternae, in endoplasmic reticulum, and in some (but not all) lysosomes. In response to light, the ROS burst was concomitant with the drop of mitochondrial transmembrane electric potential, the dramatic changes of mitochondrial shape, and the loss of integrity of mitochondria and lysosomes. Within 3 min post illumination, the plasma membrane became permeable for propidium iodide [52].

Compounds **91** and **95** were one order of magnitude more potent than chlorin e$_6$ in photodamage of artificial liposomes as monitored in a dye release assay. The latter effect depended on the content of non-saturated lipids; in liposomes consisting of saturated lipids, no photodamage was detectable. The increased therapeutic efficacy of **95** compared to **91** was attributed to a striking difference in the ability of these PSs to traverse the hydrophobic membrane interior, as evidenced by measurements of voltage jump–induced relaxation of transmembrane current on planar lipid bilayers [52].

Compound **95** is an efficient PS in PDT of rat M-1 sarcoma [81]. At doses 2.5–10 mg/kg intraperitoneally, laser energy density of 150 and 300 J/cm^2, and power density of 0.25 and 0.42 W/cm^2, complete tumor regression was achieved.

PDT of B16 melanoma was the most efficient when **95** was administered at 10 mg/kg, laser radiation of 150 J/cm^2, and power density of 0.42 W/cm^2. With these parameters, PDT of B16 melanoma with **95** prevented the emergence of lung metastases in mice at least by day 21 post treatment.

Based on the water-soluble chlorin derivative **102** [82], disodium mercaptoundecahydro-*closo*-dodecaborate (BSH)-conjugated chlorin derivatives **103–106** (Figure 3.3.11) as agents for both antitumor PDT and BNCT were synthesized [24]. The *in vivo* biodistribution and clearance of **103–106** were investigated in tumor-bearing mice. Compounds

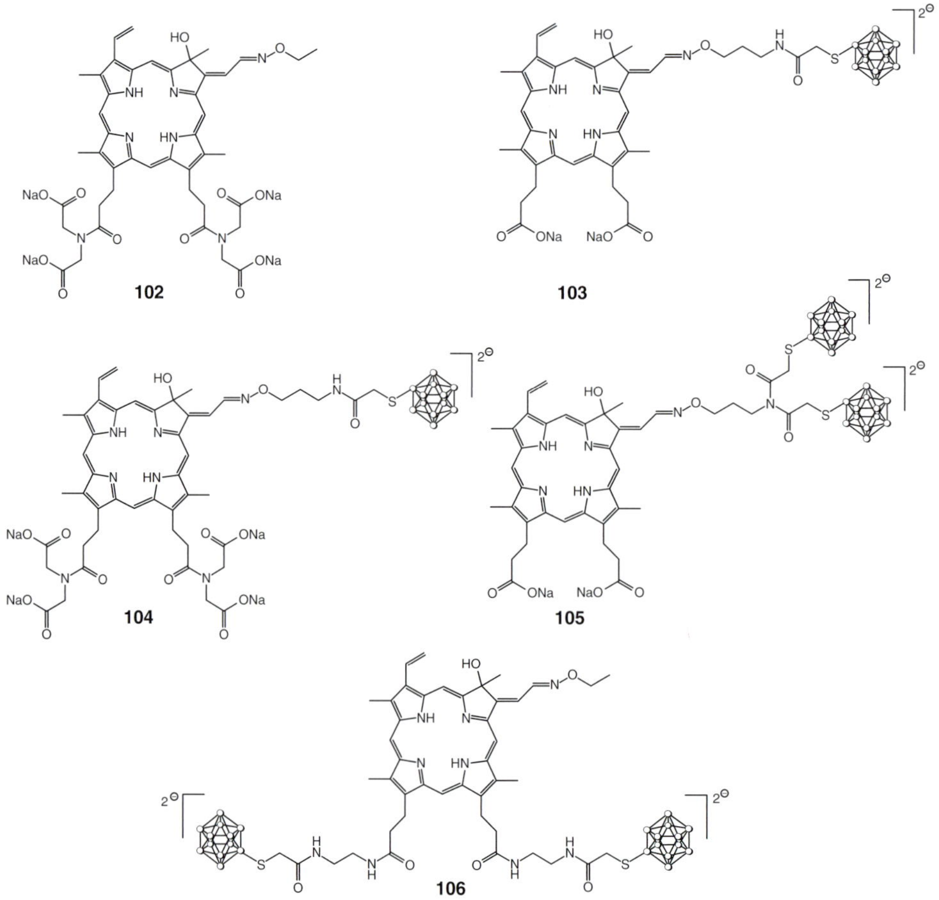

Figure 3.3.11 Chlorin derivatives prepared from BSH-containing protoporphyrin IX dimethyl ester.

104,105 showed good tumor-selective accumulation among the four derivatives. The time to maximum accumulation of **105** in the tumor was one-fourth of that for **104**, whereas clearance of **105** from normal tissues was similar to that of **104**. The *in vivo* therapeutic efficacy of PDT using **105**, which has twice as many boron atoms compared to **104**, was evaluated by measuring tumor growth rates after 660 nm light-emitting diode illumination 6 h post injection of **105**. Tumor growth was significantly inhibited by PDT with **105**, suggesting that **105** is a good candidate for both PDT and BNCT.

The carborane derivative of aminochlorin **91** (compound **107**) was prepared by reaction with *p*-iodobenzoyl chloride followed by palladium-catalyzed Sonogashira coupling reaction with cobalt bis(dicarbollide) derivative containing a terminal acetylene group (Figure 3.3.12) [83]. Ring-opening reactions of aminochlorin precursor with the zwitterionic 1,4-dioxane derivative of cobalt bis(dicarbollide), [3,3'-Co(8-$C_4H_8O_2$-1,2-$C_2B_9H_{10}$)(1',2'-$C_2B_9H_{11}$), resulted in formation of boronated chlorins **108–111** in 60% yield [16,84]. Chlorin e_6 conjugated with two cobalt bis(dicarbollide) clusters (compound **109**) accumulated in A549 human lung adenocarcinoma cells, stained the cytoplasm diffusely (largely lysosomes), and showed no discernible dark cytotoxicity. The conjugate **109** has a high

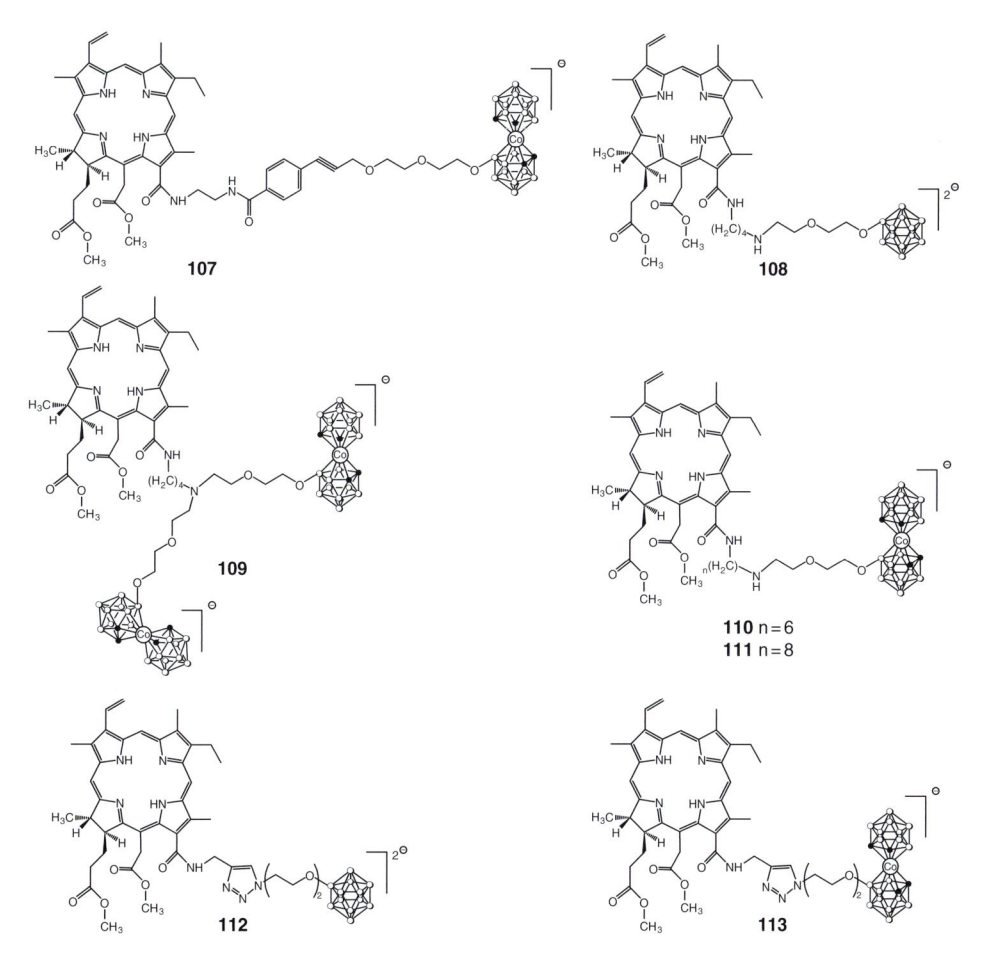

Figure 3.3.12 BSH- and cobalt bis(dicarbollide) conjugates of chlorin e_6.

quantum yield of singlet oxygen generation but possesses no photoinduced cytotoxicity [84]. The presence of cobalt complexes in **109** is supposed to be a reason for the observed antioxidative effect, but an exact mechanism of this phenomenon is unclear.

The study of **110,111** showed that the boron-containing conjugates of chlorin e_6 have no significant differences in accumulation and biodistribution. The conjugates, to a certain extent, meet the requirements imposed on BNCT agents. In particular, the compounds demonstrated relatively selective accumulation in the tumor, with a tumor versus normal tissue concentration ratio of ~3:1 [85].

Synthesis of boronated chlorins **112,113** containing the triazole spacer involved nucleophilic opening of the exocyclic ring in methylpheophorbide *a* **75** with propargyl amine, followed by treatment of the terminal acetylene group with *closo*-dodecaborate and cobalt bis(dicarbollide) azide derivatives in a CuI–Et$_3$N–MeCN system. These boronated conjugates accumulated in A549 human lung adenocarcinoma cells, but their intracellular concentration was insufficient for therapeutic efficacy [86].

3.3.4.3 Carborane Containing Derivatives of Purpurin-18 and Bacteriopurpurinimide

The compounds of this class strongly absorb light at the $\lambda=700$–770 nm region. This unique property opens new therapeutic opportunities due to deeper light penetration and increased tumor photodamage. Boronated ester conjugates of purpurin-18 (**114**) were obtained by the reaction of Boc-activated 17(3)-carboxylic group of **114** with 1-hydroxymethyl-*o*-carborane or 1-hydroxymethyl-*closo*-1-monocarbadodecaborate cesium in 53%

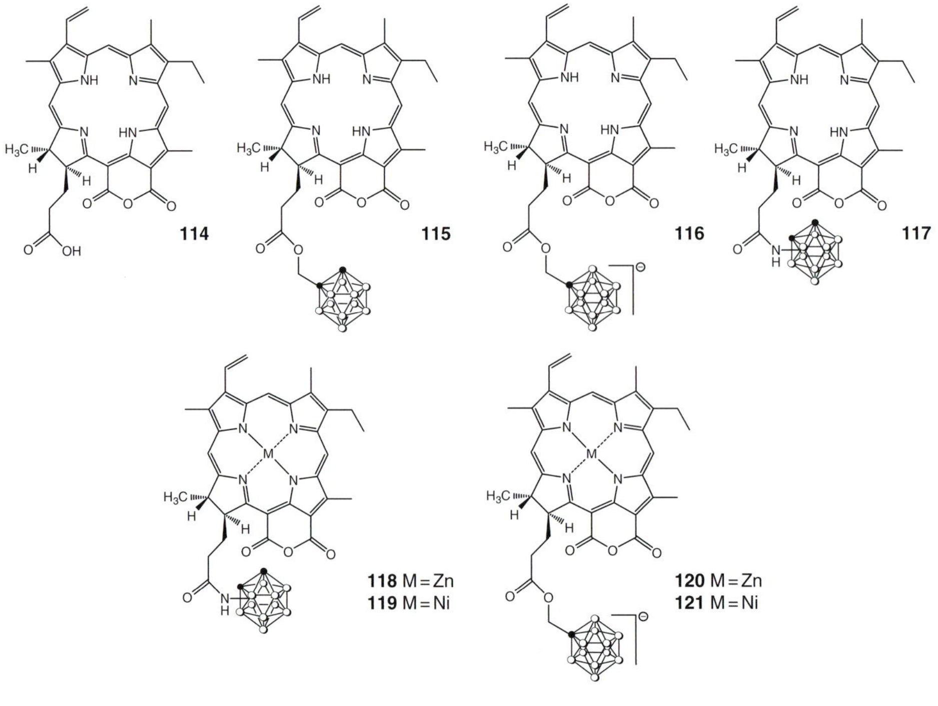

Figure 3.3.13 Boronated ester and amide conjugates of purpurin-18.

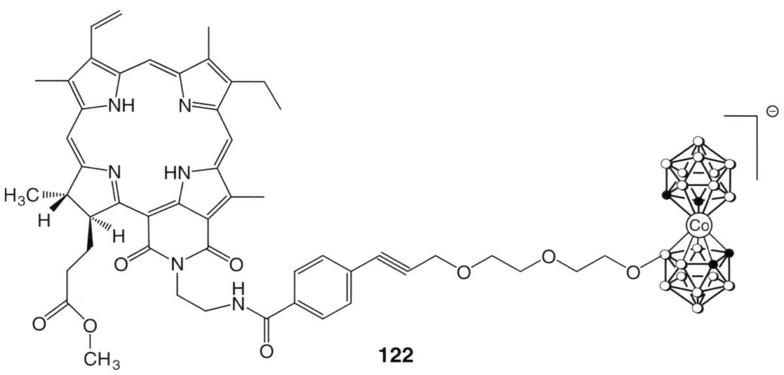

Figure 3.3.14 Boronated conjugates of N-aminobacteriopurpurinimide.

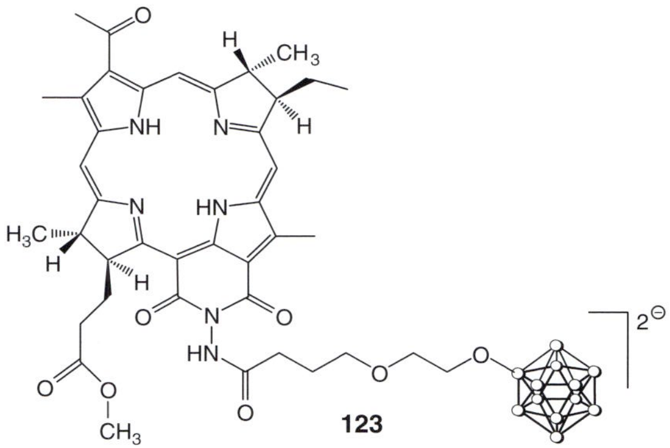

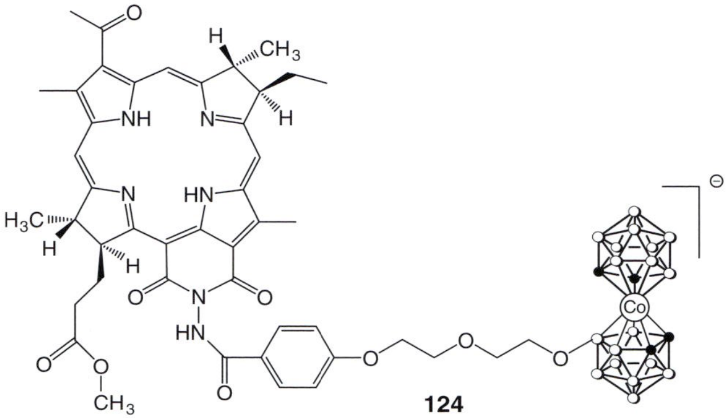

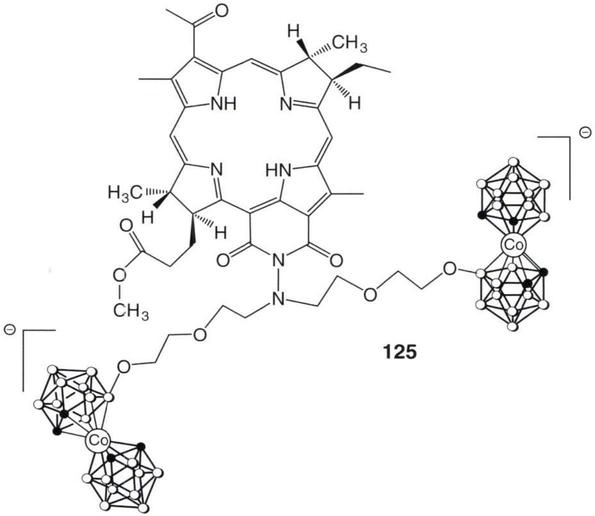

(**115**) and 63% (**116**) yields, respectively. Similarly, the reaction of **114** with 3-amino-*o*-carborane afforded the amide derivative **117** in 68% yield [87]. Using an alternative strategy, the boronated Zn and Ni complexes **118–121** were prepared in 46–93% yield from the corresponding metallopurpurins via esterification or amidation of 17(3)-carboxylic group with 3-amino-*o*-carborane and 1-hydroxymethyl-*closo*-1-monocarbadodecaborate cesium (Figure 3.3.13). These compounds formed high-affinity complexes with serum albumin. The dark toxicity of new compounds was tested against the K562 chronic myelogenous leukemia cell line. Compounds **114–117** did not influence cell proliferation rate or cell viability at concentrations <10 µM. Neutral carboranylpurpurin **115** was least toxic (IC_{50} >25 µM) compared to 12 µM for the starting purpurin **114**; therefore, introduction of the boron polyhedron decreases the dark toxicity of **114**. For studying PS activity of the class of purpurin derivatives, compound **115** was chosen. Red light illumination of HCT116 colon carcinoma cells loaded with 10 µM of **115** induced death after 20 min of exposure.

Synthesis of boronated conjugates of N-aminobacteriopurpurinimide was carried out via the nucleophilic ring-opening reaction of the oxonium derivative of cobalt bis(dicarbollide) [88,89] with the amino group of the macrocycle. As a result, the conjugates **122–125** containing two cobalt bis(dicarbollide) units were obtained in a high yield (Figure 3.3.14) [90].

3.3.5 Conclusion

We herein summarized current knowledge on simple and efficient synthetic routes leading to new tumor-selective derivatives of tetrapyrrole compounds via the functionalization of the macrocycles with various carborane clusters. Many of these compounds were evaluated in cell culture and in animal models for application in PDT and BNCT. More studies definitely are needed to evaluate important factors of therapeutic efficacy such as intratumoral accumulation, organ distribution, *in vivo* toxicity, and so on. Still, among an array of compounds synthesized to date, one can select the most potent and applicable for further in-depth mechanistic studies and provisionally clinical trials. Importantly, the addition of carborane clusters can further increase the sensitizing potency of porphyrins and chlorins not only to thermal neutrons but to light as well. A special attention to a balance between dark cytotoxicity and photocytotoxicity in the case of agents with potential dual PDT–BNCT therapeutic applications must be taken into consideration, and technologies allowing for the safe treatment of patients must be developed. Overall, the synthetic approaches described in this chapter yield multifunctional conjugates useful for diagnosis, drug delivery, and treatment in oncology.

Acknowledgments

The authors are grateful to V. Kalinin (A.N. Nesmeyanov Institute of Organoelement Compounds, Russian Academy of Sciences, Moscow) for the long-term supervision of the projects in carborane chemistry, M. Moisenovich, Yu. Antonenko, T. Rokitskaya (Moscow State University), N. Miyoshi (University of Fukui, Japan), and members of the Laboratory of Tumor Cell Death at Blokhin National Medical Research Center of Oncology, Moscow, for experiments and discussions.

References

1. Liu W, Groves JT. Manganese porphyrins catalyze selective C–H bond halogenations. *J Am Chem Soc* 2010; 132: 12847–12849.
2. Drain CM, Varotto A, Radivojevic I. Self-organized porphyrinic materials. *Chem Rev* 2009; 109: 1630–1658.
3. Aratani N, Kim D, Osuka A. Discrete cyclic porphyrin arrays as artificial light-harvesting antenna. *Acc Chem Res* 2009; 42: 1922–1934.
4. Balaban TS. Tailoring porphyrins and chlorins for self-assembly in biomimetic artificial antenna systems. *Acc Chem Res* 2005; 38: 612–623.
5. Denis TGS, Huang YY, Hamblin MR. Cyclic tetrapyrroles in photodynamic therapy: the chemistry of porphyrins and related compounds in medicine. *Handbook of Porphyrin Science* 2014; 27: 256–303.
6. Soloway AH, Tjarks W, Barnum BA, Rong FG, Barth RF, Codogni IM, Wilson JG. The chemistry of neutron capture therapy. *Chem Rev* 1998; 98: 1515–1562.
7. Pass HI. Photodynamic therapy in oncology: mechanisms and clinical use. *J Natl Cancer Inst* 1993; 85: 443–456.
8. Dolmans DE, Fukumura D, Jain RK. Photodynamic therapy for cancer. *Nat Rev Cancer* 2003; 3: 380–387.
9. Baker NR. Chlorophyll fluorescence: a probe of photosynthesis in vivo. *Annu Rev Plant Biol* 2008; 59: 89–113.
10. Ol'shevskaya VA, Zaytsev AV, Savchenko AN, Shtil AA, Cheong CS, Kalinin VN. Boronated porphyrins and chlorins as potential anticancer drugs. *Bull Korean Chem Soc* 2007; 28: 1910–1916.
11. Mironov AF, Nokel AY. Pharmacological agent "photoheme" for photodynamic cancer therapy. Russian Patent No. 2128993, 1999.
12. Mody TD. Pharmaceutical development and medical applications of porphyrin-type macrocycles. *J Porph Phthalocyan* 2000; 4: 362–367.
13. Luguya R, Fronczek FR, Smith KM, Vicente MGH. Synthesis of novel carboranylchlorins with dual application in boron neutron capture therapy (BNCT) and photodynamic therapy (PDT). *Appl Radiat Isot* 2004; 61: 1117–1123.
14. Hao E, Friso E, Miotto G, Jori G, Soncin M, Fabris C, Sibrian-Vazquez M, Vicente MGH. Synthesis and biological investigations of tetrakis(p-carboranylthio-tetrafluorophenyl)chlorin (TPFC). *Org Biomol Chem* 2008; 6: 3732–3740.
15. Ol'shevskaya VA, Savchenko AN, Zaitsev AV, Kononova EG, Petrovskii PV, Ramonova AA, Tatarskiy VV, Uvarov OV, Moisenovich MM, Kalinin VN, Shtil AA. Novel metal complexes of boronated chlorin e_6 for photodynamic therapy. *J Organomet Chem* 2009; 694: 1632–1637.
16. Grin MA, Titeev RA, Brittal DI, Ulybina OV, Tsiprovskiy AG, Berzina MY, Lobanova IA, Sivaev IB, Bregadze VI, Mironov AF. New conjugates of cobalt bis(dicarbollide) with chlorophyll *a* derivatives. *Mendeleev Commun* 2011; 21: 84–86.
17. Asano R, Nagami A, Fukumoto Y, Miura K, Yazama F, Ito H, Sakata I, Tai A. Synthesis and biological evaluation of new boron-containing chlorin derivatives as agents for both photodynamic therapy and boron neutron capture therapy of cancer. *Bioorg Med Chem Lett* 2014; 24: 1339–1343.
18. Chen W, Mehta SC, Lu DR. Selective boron drug delivery to brain tumors for boron neutron capture therapy. *Adv Drug Delivery Rev* 1997; 26: 231–247.

19 Pietrangeli D, Rosa A, Ristori S, Salvati A, Altieri S, Ricciardi G. Carboranyl-porphyrazines and derivatives for boron neutron capture therapy: from synthesis to in vitro tests. *Coord Chem Rev* 2013; 257: 2213–2231.

20 Grimes RN. Carboranes in medicine. In: *Carboranes*, 3rd ed. Elsevier: Amsterdam, 2016, 945–984.

21 Giuntini F, Boyle R, Sibrian-Vazquez M, Vicente MGH. Porphyrin conjugates for cancer therapy. *Handbook of Porphyrin Science* 2014; 27: 304–416.

22 Bhupathiraju NVSDK, Vicente MGH. Synthesis and cellular studies of polyamine conjugates of a mercaptomethyl–carboranylporphyrin. *Bioorg Med Chem* 2013; 21: 485–495.

23 Bhupathiraju NVSDK, Hu X, Zhou Z, Fronczek FR, Couraud PO, Romero IA, Weksler B, Vicente MGH. Synthesis and in vitro evaluation of BBB permeability, tumor cell uptake, and cytotoxicity of a series of carboranylporphyrin conjugates. *J Med Chem* 2014; 57: 6718–6728.

24 Asano R, Nagami A, Fukumoto Y, Miura K, Yazama F, Ito H, Sakata I, Tai A. Synthesis and biological evaluation of new BSH-conjugated chlorin derivatives as agents for both photodynamic therapy and boron neutron capture therapy of cancer. *J Photochem Photobiol: B: Biology* 2014; 140: 140–149.

25 Ol'shevskaya VA, Zaitsev AV, Kalinin VN, Shtil AA. Synthesis and antitumor activity of novel tetrakis[4-(*closo*-carboranylthio)tetrafluorophenyl]porphyrins. *Russ Chem Bull* 2014; 63: 2383–2387.

26 Kadish KM, Smith KM, Guilard R (Eds.). The Porphyrin Handbook. Academic Press, San Diego, 2000 (vols. 1–10), 2003 (vols. 11–20).

27 Kruper WJ, Chamberlin TA, Kochanny M. Regiospecific aryl nitration of meso-substituted tetraarylporphyrins: a simple route to bifunctional porphyrins. *J Org Chem Soc* 1989; 54: 2753–2756.

28 Semeikin AS, Koifman OI, Berezin BD. Synthesis of tetraphenylporphins with active groups in the phenyl rings. 1. Preparation of tetrakis(4-aminophenyl)porphin. *Chem Heterocycl Comp* 1982; 18: 1046–1047.

29 Ol'shevskaya VA, Zaitsev AV, Dutikova YV, Luzgina VN, Kononova EG, Petrovsky PV, Kalinin VN. A one step synthesis of boronated *meso*-tetraphenylporphyrins. *Macroheterocycles* 2009; 2: 221–227.

30 Kalinin VN, Ol'shevskaya VA, Rys EG, Tyutyunov AA, Starikova ZA, Korlyukov AA, Sung DD, Ponomaryov AB, Petrovskii PV, Hey-Hawkins E. The first carborane triflates: synthesis and reactivity of 1-trifluoromethanesulfonylmethyl- and 1,2-bis(trifluoromethanesulfonylmethyl)-o-carborane. *Dalton Trans* 2005; 5: 903–908.

31 Luguya R, Jaquinod L, Fronczek FR, Vicente MGH, Smith KM. Synthesis and reactions of meso-(p-nitrophenyl)porphyrins. *Tetrahedron* 2004; 60: 2757–2763.

32 Zakharkin LI, Kenzhetaeva VD, Zhigareva GG. Action of nucleophilic reagents on 3-(R-o-carboranyl)-1,2-epoxypropanes. *Russ Chem Bull* 1975; 24: 521–526.

33 Zakharkin LI, Zhigareva GG, Kenzhetaeva VD. Some reactions of 1-epoxyisopropyl-o-carborane. *Russ Chem Bull* 1975; 24: 527–529.

34 Kenzhetaeva VD, Kazantsev AV, Zakharkin LI. Synthesis and some transformations of 3-(R-o-carboranyl)-1,2-epoxypropanes. *Zh Obsch Khim* 1976; 46: 340–348 (in Russian).

35 Pachon LD, Games P, van Brussel JJM, Reedijk J. Zinc-catalyzed aminolysis of epoxides. *Tetrahedron Lett* 2003; 44: 6025–6027.

36 Fleischer EB, Shachter AM. Coordination oligomers and a coordination polymer of zinc tetraarylporphyrins. *Inorg Chem* 1991; 30: 3763–3769.

37 Adler AD, Longo FR, Kampas F, Kim JJ. On the preparation of metalloporphyrins. *Inorg Nucl Chem* 1970; 32: 2443–2445.
38 Miura M, Gabel D, Oenbrink G, Fairchied RG. Preparation of carboranyl porphyrins for boron neutron capture therapy. *Tetrahedron Lett* 1990; 31: 2247–2250.
39 Kalinin VN, Astakhin AV, Kazantsev AV, Zakharkin LI. Synthesis of 1-isocyanatocarboranes. *Russ Chem Bull* 1984; 33: 1508–1510.
40 Ol'shevskaya VA, Zaitsev AV, Ayub R, Petrovskii PV, Kononova EG, Tatarskii VV, Shtil AA, Kalinin VN. New 9-isocyanato-*o*- and 9-isocyanato-*m*-carboranes: synthesis and chemical and biological properties. *Doklady Chem* 2005; 405: 230–234.
41 Kalinin VN, Ol'shevskaya VA. Some aspects of the chemical behavior of icosahedral carboranes. *Russ Chem Bull* 2008; 57: 815–836.
42 Zakharkin LI, Kalinin VN, Snyakin AP, Kvasov BA. Effect of solvents on the electronic properties of 1-*o*-, 3-*o*- and 1-*m*-carboranyl groups. *J Organomet Chem* 1969; 18: 19–26.
43 Zakharkin LI, Kovredov AI, Ol'shevskaya VA. Synthesis of 9-(fluorophenyl)-*o*-, 9-(fluorophenyl)-*m*-, and 2-(fluorophenyl)-*p*-carboranes and determination of electronic effects of 9-*o*-, 9-*m*-, and 2-*p*-carboranyl groups. *Russ Chem Bull* 1981; 30: 1775–1777.
44 Barth RF, Adams DM, Soloway AH, Alam F, Darby MV. Boronated starburst dendrimer-monoclonal antibody immunoconjugates: evaluation as a potential delivery system for neutron capture therapy. *Bioconjug Chem* 1994; 5: 58–66.
45 Liu L, Barth RF, Adams DM, Soloway AH, Reisfeld RA. Critical evaluation of bispecific antibodies as targeting agents for boron neutron capture therapy of brain tumors. *Anticancer Res* 1996; 16: 2581–2588.
46 Ol'shevskaya VA, Luzgina VN, Kurakina YA, Makarenkov AV, Petrovskii PV, Kononova EG, Mironov AF, Shtil AA, Kalinin VN. Synthesis and antitumor properties of carborane conjugates of 5-(4-aminophenyl)-10,15,20-triphenylporphyrin. *Doklady Chem* 2012; 443: 91–96.
47 Plesek J, Janousek Z, Hermanek S. Chemistry of 9-mercapto-1,7-dicarba-*closo*-dodecaborane. *Collect Czech Chem Commun* 1978; 43: 1332–1338.
48 Ol'shevskaya VA, Makarenkov AV, Kononova EG, Petrovskii PV, Verbitskii EV, Rusinov GL, Kalinin VN, Charushin VN. 1,3-Dipolar cycloaddition of [(o-carboran-1-yl)methyl] azide to alkynes. *Doklady Chem* 2010; 434: 245–248.
49 Ol'shevskaya VA, Makarenkov AV, Korotkova NS, Kononova EG, Konovalova NV, Kalinin VN. Synthesis of carborane conjugates based on the maleimide derivative of 5,10,15,20-tetraphenylporphyrin. *Doklady Chem* 2014; 458: 165–168.
50 Ol'shevskaya VA, Korotkova NS, Makarenkov AV, Luzgina VN, Kalinin VN. Synthesis of cationic boronated porphyrins by modification of amino group of 5-(4'-aminophenyl)-10,15,20-triphenylporphyrin. *Vest Lobachevsky State Univ. Nizhni Novgorod* 2013; 1: 118–123.
51 Shchekotikhin AE, Glazunova VA, Luzikov YN, Buyanov VN, Susova OY, Shtil AA, Preobrazhenskaya MN. Synthesis and structure–activity relationship studies of 4,11-diaminonaphtho[2,3-*f*]indole-5,10-diones. *Bioorg Med Chem* 2006; 14: 5241–5251.
52 Moisenovich MM, Ol'shevskaya VA, Rokitskaya TI, Ramonova AA, Nikitina RG, Savchenko AN, Tatarskiy VV, Kaplan MA, Kalinin VN, Kotova EA, Uvarov OV, Agapov II, Antonenko YN, Shtil AA. Novel photosensitizers trigger rapid death of malignant human cells and rodent tumor transplants via lipid photodamage and membrane permeabilization. *PLoS One* 2010; 5: e12717.

53 Purser S, Moore PR, Swallow S, Gouverneur V. Fluorine in medicinal chemistry. *Chem Soc Rev* 2008; 37: 320–330.
54 Olshevskaya VA, Zaitsev AV, Sigan AL, Kononova EG, Petrovskii PV, Chkanikov ND, Kalinin VN. Synthesis of boronated porphyrins and chlorins by regioselective substitution for fluorine in pentafluorophenylporphyrins on treatment with lithiocarboranes. *Doklady Chem* 2010; 435: 334–338.
55 Tome JPC, Neves MGPMS, Tome AC, Cavaleiro JAS, Mendonc AF, Pegado IN, Duarte R, Valdeira ML. Synthesis of glycoporphyrin derivatives and their antiviral activity against herpes simplex virus types 1 and 2. *Bioorg Med Chem* 2005; 13: 3878–3888.
56 Knoth WH. $B_{10}H_{12}CNH_3$, B_9H_9CH-, $B_{11}H_{11}CH$-, and metallomonocarboranes. *Inorg Chem* 1971; 10: 598–603.
57 Banfi S, Caruso E, Caprioli S, Mazzagatti L, Canti G, Ravizza R, Gariboldi M, Monti E. Photodynamic effects of porphyrin and chlorin photosensitizers in human colon adenocarcinoma cells. *Bioorg Med Chem* 2004; 12: 4853–4860.
58 Jimenez-Oses G, Garcia JI, Silva AMG, Santos ARN, Tome AC, Neves MGPMS, Cavaleiro JAS. Mechanistic insights on the site selectivity in successive 1,3-dipolar cycloadditions to *meso*-tetraarylporphyrins. *Tetrahedron* 2008; 64: 7937–7943.
59 Rokitskaya TI, Zaitsev AV, Ol'shevskaya VA, Kalinin VN, Moisenovich MM, Agapov II, Antonenko YN. Boronated derivatives of chlorin e_6 and fluoride-containing porphyrins as penetrating anions: a study using bilayer lipid membranes. *Biochemistry (Moscow)* 2012; 77: 975–982.
60 Shtil AA, Ol'shevskaya VA, Zaitsev AV, Petrova AS, Markova AA, Tatarskiy VV., Puchnina SV, Suldin AV, Kalinin VN, Miyoshi N. Boronation: what is next? Polyfluorinated carboranylchlorin, a potent antitumor photoradiosensitizer with advantageous synthesis and pharmacological properties. Paper presented at the Euroboron 5 conference, Moscow-Suzdal, 4–8 September 2016.
61 Spikes JD. Quantum yields and kinetics of the photobleaching of hematoporphyrin, Photofrin II, tetra(4-sulfonatophenyl)-porphine and uroporphyrin. *Photochem Photobiol* 1992; 55: 797–808.
62 Ochsner M. Photophysical and photobiological processes in the photodynamic therapy of tumours. *J Photochem Photobiol: B Biol* 1997; 39: 1–18.
63 Fabre B, Clark JC, Vicente MGH. Synthesis and electrochemistry of carboranylpyrroles. toward the preparation of electrochemically and thermally resistant conjugated polymers. *Macromolecules* 2006; 39: 112–119.
64 Hiramatsu R, Kawabata S, Tanaka H, Sakura Y, Suzuki M, Ono K, Miyatake SI, Kuroiwa T, Hao E, Vicente MGH. Tetrakis(*p*-carboranylthio-tetrafluorophenyl)chlorin (TPFC): application for photodynamic therapy and boron neutron capture therapy. *J Pharma Sci* 2015; 104: 962–970.
65 Sivaev IB, Bregadze VI. Polyhedral boranes for medical applications: current status and perspectives. *Eur J Inorg Chem* 2009; 11: 1433–1450.
66 Barth RF, Vicente MGH, Harling OK, Kiger III WS, Riley KJ, Binns PJ, Wagner FM, Suzuki M, Aihara T, Kato I, Kawabata S. Current status of boron neutron capture therapy of high grade gliomas and recurrent head and neck cancer. *Rad Oncol* 2012; 7: 146–167.
67 Brown SB, Brown EA, Walker I. The present and future role of photodynamic therapy in cancer treatment. *Lancet Oncol* 2004; 5: 497–508.

68 Bhupathiraju NVSDK, Rizvi W, Batteas JD, Drain CM. Fluorinated porphyrinoids as efficient platforms for new photonic materials, sensors, and therapeutics. *Org Biomol Chem* 2016; 14: 389–408.

69 Barth RF, Coderre JA, Vicente MGH, Blue TE. Boron neutron capture therapy of cancer: current status and future prospects. *Clin Cancer Res* 2005; 11: 3987–4002.

70 Josefsen LB, Boyle RW. Photodynamic therapy: novel third-generation photosensitizers one step closer. *Br J Pharmacol* 2008; 154: 1–3.

71 Ol'shevskaya VA, Nikitina RG, Zaitsev AV, Luzgina VN, Kononova EG, Morozova TG, Drozhzhina VV, Ivanov OG, Kaplan MA, Kalinin VN, Shtil AA. Boronated protohaemins: synthesis and in vivo antitumour efficacy. *Org Biomol Chem* 2006; 4: 3815–3821.

72 Ozcelik S, Gul A. Boron-containing tetrapyrroles. *Turk J Chem* 2014; 38: 950–979.

73 O'Connor AE, Gallagher WM, Byrne AT. Porphyrin and nonporphyrin photosensitizers in oncology: preclinical and clinical advances in photodynamic therapy. *Photochem Photobiol* 2009; 85: 1053–1074.

74 Luzgina VN, Ol'shevskaya VA, Sekridova AV, Mironov AF, Kalinin VN, Pashchenko VZ, Gorokhov VV, Tusov VB, Shtil AA. Synthesis of boron-containing derivatives of pyropheophorbide *a* and investigation of their photophysical and biological properties. *Russ J Org Chem* 2007; 43: 1243–1251.

75 Ratajski M, Osterloh J, Gabel D. Boron-containing chlorins and tetraazaporphyrins: synthesis and cell uptake of boronated pyropheophorbide *a* derivatives. *Anti-Cancer Agents Med Chem* 2006; 6: 159–166.

76 Ol'shevskaya VA, Zaitsev AV, Savchenko AN, Kononova EG, Petrovskii PV, Kalinin VN. Synthesis of boronated derivatives of pheophorbide *a*. *Doklady Chem* 2008; 423: 294–298.

77 Golovina GV, Rychkov GN, Ol'shevskaya VA, Zaitsev AV, Kalinin VN, Kuzmin VA, Shtil AA. Differential binding preference of methylpheophorbide *a* and its diboronated derivatives to albumin and low density lipoproteins. *Anti-Cancer Agents Med Chem* 2013; 13: 639–646.

78 Kuchin AV, Mal'shakova MV, Belykh DV, Ol'shevskaya VA, Kalinin VN. Synthesis of boronated derivatives of chlorin e_6 with amide bond. *Doklady Chem* 2009; 425: 80–83.

79 Ol'shevskaya VA, Nikitina RG, Savchenko AN, Malshakova MV, Vinogradov AM, Golovina GV, Belykh DV, Kutchin AV, Kaplan MA, Kalinin VN, Kuzmin VA, Shtil AA. Novel boronated chlorin e_6-based photosensitizers: synthesis, binding to albumin and antitumour efficacy. *Bioorg Med Chem* 2009; 17: 1297–1306.

80 Omarova EO, Nazarov PA, Firsov AM, Strakhovskaya MG, Arkhipova AY, Moisenovich MM, Agapov II, Ol'shevskaya VA, Zaitsev AV, Kalinin VN, Kotova EA, Antonenko YN. Carboranyl-chlorin e_6 as a potent antimicrobial photosensitizer. *PLoS One* 2015; 11: e0141990.

81 Nikitina RG, Kaplan MA, Olshevskaya VA, Rodina JS, Drozhzhina VV, Morozova TG. Photodynamic therapy with boronated chlorin as a photosensitizer. *J Cancer Sci Ther* 2011; 3: 216–219.

82 Asano R, Nagami A, Fukumoto Y, Yazama F, Ito H, Sakata I, Tai A. Synthesis and biological evaluation of new chlorin derivatives as potential photosensitizers for photodynamic therapy. *Bioorg Med Chem* 2013; 21: 2298–2304.

83 Grin MA, Titeev RA, Brittal DI, Chestnova AV, Feofanov AV, Lobanova IA, Sivaev IB, Bregadze VI, Mironov AF. Synthesis of cobalt bis(dicarbollide) conjugates with natural chlorins by the Sonogashira reaction. *Russ Chem Bull* 2010; 59: 219–224.

84 Efremenko AV, Ignatova AA, Borsheva AA, Grin MA, Bregadze VI, Sivaev IB, Mironov AF, Feofanov AV. Cobalt bis(dicarbollide) *versus closo*-dodecaborate in boronated chlorin e_6 conjugates: implications for photodynamic and boron-neutron capture therapy. *Photochem Photobiol Sci* 2012; 11: 645–652.
85 Volovetskiy AB, Shilyagina NY, Dudenkova VV, Pasynkova SO, Ignatova AA, Mironov AF, Grin MA, Bregadze VI, Feofanov AV, Balalaeva IV, Maslennikova AV. Study of the tissue distribution of potential boron neutron capture therapy agents based on conjugates of chlorin e_6 aminoamide derivatives with boron nanoparticles. *Biophysics* 2016; 61: 133–138.
86 Bregadze VI, Semioshkin AA, Las'kova JN, Berzina MY, Lobanova IA, Sivaev IB, Grin MA, Titeev RA, Brittal DI, Ulybina OV, Chestnova AV, Ignatova AA, Feofanov AV, Mironov AF. Novel types of boronated chlorin e_6 conjugates. *Appl Organometal Chem* 2009; 23: 370–374.
87 Olshevskaya VA, Savchenko AN, Golovina GV, Lazarev VV, Kononova EG, Petrovskii PV, Kalinin VN, Shtil AA, Kuz'min VA. New boronated derivatives of purpurin-18: synthesis and intereaction with serum albumin. *Doklady Chem* 2010; 435: 328–333.
88 Semioshkin AA, Sivaev IB, Bregadze VI. Cyclic oxonium derivatives of polyhedral boron hydrides and their synthetic applications. *Dalton Trans* 2008; 977–992.
89 Sivaev IB, Semioshkin AA, Bregadze VI. New approach to incorporation of boron in tumor-seeking molecules. *Appl Radiat Isotop* 2009; 67: S91–S93.
90 Grin MA, Brittal DI, Tsiprovskiy AG, Bregadze VI, Mironov AF. Boron-containing conjugates of natural chlorophylls. *Macroheterocycles* 2010; 3: 222–227.

3.4

Nanostructured Boron Compounds for Boron Neutron Capture Therapy (BNCT) in Cancer Treatment

Shanmin Gao,[1] Yinghuai Zhu,[2,] and Narayan Hosmane[3,*]*

[1] School of Chemistry and Materials Science, Ludong University, Yantai, China
[2] School of Pharmacy, Macau University of Science and Technology, Taipa, Macau, China
[3] Department of Chemistry and Biochemistry, Northern Illinois University, DeKalb, Illinois, USA
* Corresponding authors

3.4.1 Introduction

Nanomaterials have emerged as one of the most fruitful areas in cancer treatment and are considered as a medical boon for the diagnosis, treatment, and prevention of cancer [1]. Also, paralleled by advances in chemistry, biology, pharmacy, nanotechnology, medicine, and imaging, several different systems have been developed in the last decade in which disease diagnosis and therapy are combined (Figure 3.4.1a) [2,3]. This is stimulating the development of a diverse range of nanometer-sized objects that can recognize cancer tissue, enabling visualization of tumors, delivery of anticancer drugs, and/or the destruction of tumors by different therapeutic techniques. By choosing the right size, shape, coating, and charge as well as targeting moiety (Figure 3.4.1b), the fate of the particles in the body can be well predicted [4]. Newer nanomaterials are garnering increasing interest as potential multifunctional therapeutic agents; these drugs are conferred novel properties, by virtue of their size and shape [5].

Emerging inorganic nanomaterials such as mesoporous silica, magnetic and gold nanoparticles, carbon nanotubes, quantum dots, and polymeric nanoparticles have been widely used in biomedical research, with great optimism, for cancer diagnosis and therapy. Such nanoparticles possess unique optical, electrical, magnetic, and/or electrochemical properties. With such properties along with their impressive nanosize, these particles can be targeted to cancer cells, tissues, and ligands efficiently and monitored with extreme precision in real time [7].

Over the past decade, a variety of emerging nanomaterials and technology, including magnetic nanomaterials, quantum dots, surface-enhanced Raman scattering (SERS) technology, carbon nanomaterials, lipopolyplex nanoparticles, nano- and microbubbles, protein-based nanomaterials, mesoporous silica nanoparticles, and neutron capture therapy (NCT), have been developed for tumor imaging and therapy.

Boron-Based Compounds: Potential and Emerging Applications in Medicine, First Edition.
Edited by Evamarie Hey-Hawkins and Clara Viñas Teixidor.
© 2018 John Wiley & Sons Ltd. Published 2018 by John Wiley & Sons Ltd.

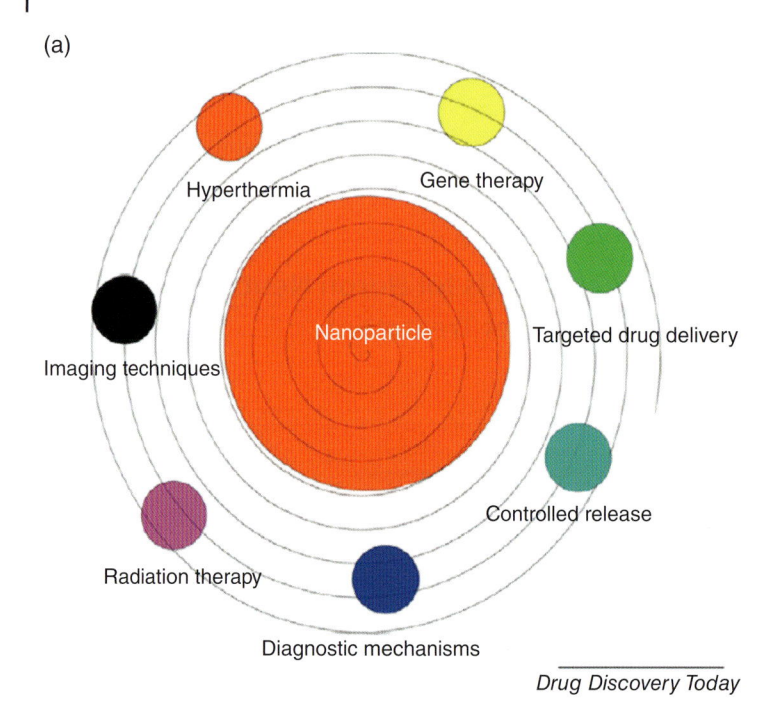

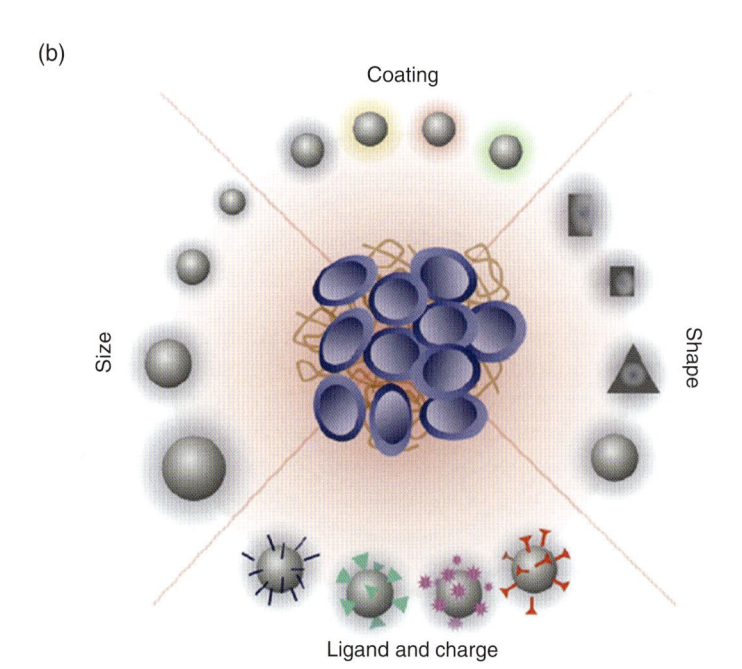

Figure 3.4.1 (a) Different axes of nanoparticle applications as theranostic agents. *Source*: Ref. [6], © 2012 Elsevier Ltd. All rights reserved. (b) The main properties influencing the distribution, elimination, and targeting of particles to tumors. *Source*: Ref. [4], © 2011 Elsevier Inc. All rights reserved.

Such nanomaterials possess unique optical, electrical, magnetic, and/or electrochemical properties. In addition, the enhanced permeability and retention (EPR) is significantly contributing to selective (or preferential) accumulation of nanoparticles in tumor. Such accumulation can be even improved by attachment of specific targeting moieties to the surface of the nanoparticles [8]. Tumor cells have an increased vascular permeability and a decrease in their lymphatic drainage system, which leads to the passive accumulation of macromolecular drugs in these neoplastic cells.

3.4.2 Boron Neutron Capture Therapy (BNCT)

Many people consider NCT to be a new and innovative approach to the treatment of cancer that is still in its formative stage. However, the basic idea behind this approach has been around for 80 years, almost as long as the idea of the existence of the neutron. The basic approach in NCT was outlined in 1936 by Gordon L. Locher [9], when he formulated his binary concept of treating cancer: "In particular, there exist the possibilities of introducing small quantities of strong neutron absorbers into the regions where it is desired." Three isotopes, ^{10}B, ^{155}Gd, and ^{157}Gd, are the ones that have been most studied in applications of NCT [10].

At present, nuclear reactors are the most appropriate source of nuclear beams of sufficient intensity for NCT use, and there are only a limited number of nuclear reactors that produce the high-quality neutron beams that can be used in medical treatment [11]. Accelerator-based neutron sources are based on neutrons produced when a charged particle, usually a proton, strikes a suitable target [12].

3.4.2.1 Principles of BNCT

The first developments of NCT were based on molecular boron compounds such as boronphenylalanine (BPA) [13], carborane derivatives [14], and macrocycles (porphyrins or phthalocyanins) functionalized with organic derivatives of boron [15]. However, the induced cytotoxicity of these NCT agents is limited by their molecular structure, which prevents a sufficient accumulation in cancerous cells. Since their residence time in the blood is expected to be longer than for molecular boron compounds, the use of boron-based dendrimers and nanoparticles has recently been proposed [16].

It is generally accepted that BNCT is a useful binary cancer treatment, in which the delivery agents containing ^{10}B are selectively transported into tumor cells and then irradiated with thermal neutrons of appropriate energy. As detailed in Equation 3.4.1, the ^{10}B nucleus adsorbs a neutron to form an excited ^{11}B nucleus that decays, emitting an α-particle ($^{4}He^{2+}$) and a $^{7}Li^{3+}$ ion with high energy. The linear energy transfer of these heavily charged particles has a range of approximately one cell diameter (5–9 μm) [17], which confines radiation damage to the cell from which they arise, hence minimizing cytotoxic effects on the surrounding tissue.

$$^{10}B + {}^{1}n \rightarrow \left[{}^{11}B \right] \rightarrow {}^{4}He^{2+}(\alpha) + {}^{7}Li^{3+} + 2.31\,\text{MeV} \quad (3.4.1)$$

Advantages of BNCT therapy include: (1) boron-10 has a nuclear capture cross section of 3838 barn, which is more than three orders of magnitude higher than those of other nuclei commonly found in living tissue; (2) unlike other natural radioactive elements used in radiotherapy [18], the ^{10}B employed in BNCT treatment is nonradioactive; and (3) due to its high electrophilicity, boron can be easily incorporated into compounds containing a hydrolytically stable linkage [19].

In theory, BNCT treatment should cause only minimal damage to the boron-free region, provided the ^{10}B atoms are delivered to a targeted region with high selectivity and dosed with a sizeable neutron flux. However, under neutron irradiation, damage caused by the reactions within normal tissue by nitrogen and hydrogen needs to be considered in order to define the dose-limiting toxicity. To be effective in BNCT, the necessary boron concentration has been estimated at 10^9 ^{10}B atoms (natural abundance: 19.9%) per cell, or approximately 35 μg ^{10}B/g of tissue [20]. To date, it remains a big challenge to selectively deliver the ^{10}B agents into a tumor cell so as to reach the required high ^{10}B concentration. Under these conditions, it has been estimated that approximately 85% of the radiation damage arises from the neutron capture reaction. To avoid unnecessary damage to healthy tissue in the path of the neutron beam, the surrounding tissues should contain less than 5 μg ^{10}B/g of tissue.

It is well recognized that glioblastoma multiforme is one of the most malignant brain tumors [13,21]. It may aggressively infiltrate the brain and, thus, is rarely able to be removed completely by a surgical procedure. Up to the present, the lethal glioblastoma multiforme has been practically incurable [13,21]. Unlike other tumors, it was observed that the brain tumors tend not to metastasize to other organs in the body. Therefore, it is reasonable to expect that the patient's life could be significantly increased through successfully eradicating the malignant tumors in the brain. Principally, BNCT is employed to treat brain tumors because the effective technology maximizes the therapeutic effect in the tumor, while minimizing neurotoxicity in adjacent brain parenchyma [22]. Very recent reports demonstrate that BNCT may also be effective in other cancer therapies, such as treating neck tumors [23]. To conduct a successful BNCT treatment, it is crucial to harvest both sufficient neutron flux and boron concentration in tumors. To meet the requirement of boron concentration, researchers have examined both small-molecule-based and macromolecule-based boron agents in recent decades [24–27]. Unfortunately, none of these agents have been examined in clinical treatment owing to their low selectivity and/or toxicity. To date, two small molecules, $Na_2B_{12}H_{11}SH$ (BSH) and boronophenylalanine (BPA; Figure 3.4.2), remain the only clinically used agents. Nevertheless, clinical results from the two compounds are not universally attractive because of their low tumor-to-brain tissue and tumor-to-blood ^{10}B ratios [28–32].

Nanotechnology has been highlighted to be of great interest in various research areas. Currently, advanced nanoscaled materials are well developed with many

p-borono-L-phenylalanine   Sodium mercaptododecaborate

Figure 3.4.2 *p*-Borono-l-phenylalanine (BPA) and sodium mercaptododecaborate (BSH). *Source*: Ref. [19].

pharmaceutical applications, including acting as a drug carrier [33–35]. In BNCT, in contrast to the classical small agents that contain fewer boron atoms, highly boron-enriched nanocomposites allow the selective delivery of the required amounts of boron to tumors. Therefore, various nanoscaled boron-enriched delivery agents have been synthesized and explored as boron carriers [36–40]. Research in this specific area is developing rapidly.

3.4.2.2 Liposome-Based BNCT Agents

Liposomes are small, spherical vesicles composed of membranes of phospholipids. The phospholipids are molecules with a hydrophilic head and a hydrophobic tail. Cell membranes are composed of such molecules arranged in two layers. When these membranes are disrupted, they can reassemble as extremely small spheres, usually as bilayers (liposomes). Liposomes have the potential of delivering large amounts of boron to cancer cells [37]. In addition, modification of the liposome surface by reacting with NH_2-terminated polyethylene glycol (PEGylation) or attachment of antibodies or receptor groups can enhance the delivery of therapeutic molecules [37].

Another, similar vesicle is low-density lipoprotein (LDL), which is a major carrier of cholesterol. Cancer cells avidly absorb LDL as a source of cholesterol for their rapidly dividing cells. The LDL can be isolated and their cholesterol core replaced by hydrophobic carboranes [41]. *In vitro* studies of hamster V-79 cancer cells have found that use of such boronated LDLs resulted in intercellular concentrations of ~240 μg ^{10}B/cell, which is about 10× the amount needed for effective BNCT [41]. The use of drug-laden vesicles, such as liposomes or LDLs, also takes advantage of a general phenomenon of the EPR effect [8]. The macromolecules may also contain groups that are preferentially taken up by cancer cells. For example, the presence of folate ions on the surface of boron-containing liposomes greatly enhances the boron uptake in human KB squamous epithelial cancer cells, which have overexpressed foliate acceptors [42]. Other examples are the reconstituted LDLs, which still retain their ability to bond to LDL-specific sites on the tumor cells. These liposomes and modified LDLs have the ability to deliver massive amounts of ^{10}B to cancer cells and have been termed "supertankers" for boron delivery [43].

Owing to their enormous flexibility and diversity, liposomes have been considered among the most outstanding drug carriers for the selective delivery of multifarious loads to tumors. Liposomes can be conveniently functionalized with antibodies, peptides, and so forth. Therefore, drugs encapsulated in the vehicles can be selectively taken up and highly accumulated in tumor cells [44,45]. Similar to other drugs, BNCT agents can also be encapsulated inside liposomes that circulate throughout the bloodstream, where the drug is then released via diffusion through the liposome or by liposomal degradation. In recent decades, liposomes have been evaluated *in vivo* as BNCT delivery agents [37,39,40]. These new delivery systems could reduce the amount of boron required for successful BNCT treatment and circulate in the blood for longer periods by avoiding the reticulo-endothelial system, which causes phagocytosis. Additionally, magnetic liposomes loaded with both drugs, and ferromagnetic materials have been investigated as BNCT agents [46,47]. Magnetic liposomes have been designed that demonstrate a selectivity for carrying doxorubicin to the tumor cells [48]. Interestingly, liposomes involving fluorescence-labeled *closo*-dodecaborane lipids

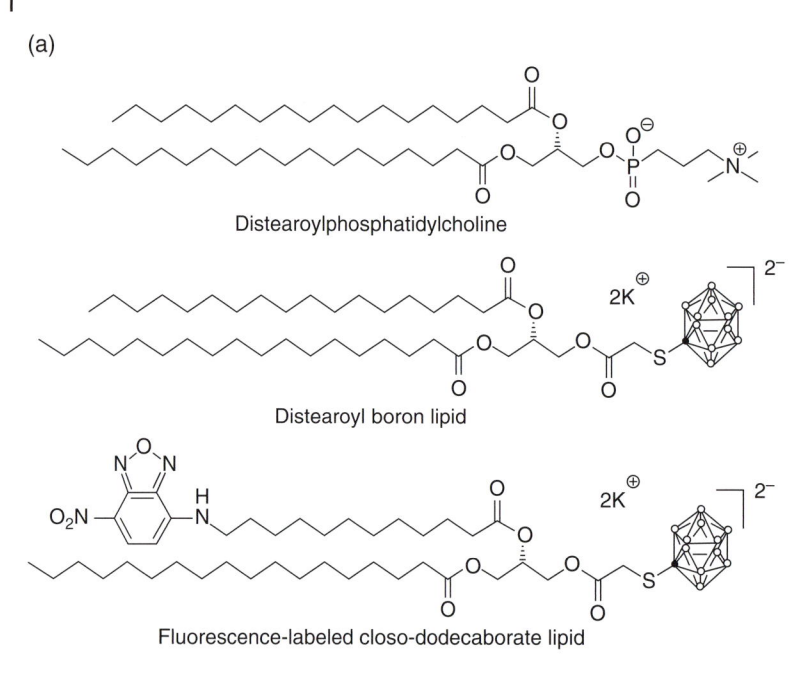

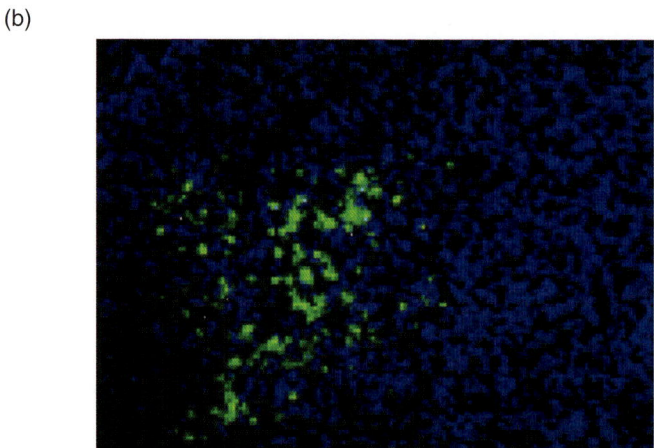

Figure 3.4.3 *closo*-Dodecaborate lipid. (a) Fluorescence-labeled *closo*-dodecaborate lipid and (b) *in vivo* image of the fluorescence-labeled *closo*-dodecaborate lipid-labeled liposomes (green) in the colon of 26 tumor-bearing mice (tumor tissues were stained in blue). *Source*: Reproduced with permission from Ref. [50].

(Figure 3.4.3) have been prepared and then selectively delivered to the tumor tissue, rather than hypoxic regions, in the colons of 26 tumor-bearing mice. It was also demonstrated that dodecaborate and fatty-acid species in fluorescence-labeled *closo*-dodecaborane lipid (FL-SBL), as represented in Figure 3.4.3a, can be readily cleaved under biological conditions, with the boron concentrations in various organs decreasing sufficiently to be undetectable [49]. It is possible to encapsulate both BSH and magnescope inside the distearoyl boron lipid liposomes [50]. The BSH-encapsulating distearoyl boron lipid liposomes displayed improved boron delivery capability and significant *in vivo* antitumor effects after thermal NCT [50].

In conclusion, liposomes have demonstrated their potential as BNCT agents, although as yet no products are commercially available for use in clinical treatment. Issues related to liposome stability, large-scale production, and uptake by the reticuloendothelial system have to be resolved before liposome-based BNCT delivery systems can be meaningfully involved in future study. Basically, liposomes are both physically and chemically unstable; the phospholipid-based components are prone to hydrolysis and/or oxidation. Considering the complicated synthetic processes in which, for example, the sonication method and reduced-pressure evaporation method are employed, it is difficult to scale up liposome production. After systemic administration, liposome-based carriers could be endocytosed by cells of the mononuclear phagocyte systems. Therefore, it is important to develop "smart" liposomes and prolong their blood circulation times.

3.4.2.3 Carbon Nanotubes

Carbon nanotubes (CNTs), which were discovered by Sumio Ijima in 1991 [51], are currently attracting wide attention in both academia and industry because of their unique properties, such as nanoscaled size, cylindrical arrangement of carbon atoms, low mass density, high thermal stability, excellent conductivity, and superb mechanical properties, as well as their potential for applications [52,53]. It is well recognized that there are two major types of CNTs: single-walled CNTs (SWCNTs) and multiwalled CNTs, which differ according to the number of layers of graphene sheet that encapsulate the tubes [54]. The stability and flexibility of CNTs are likely to prolong the circulation time and the bioavailability of these macromolecules, thus enabling highly effective gene and drug therapies.

Our earlier research in this area has led to designing the *nido*-carborane units-appended water-soluble SWCNTs for BNCT application (Figure 3.4.4). For comparison, tumor, blood, lung, liver, and spleen samples were collected and analyzed. The nanocomposites provide a favorable tumor-to-blood ratio of 3.12:1.00 and a boron concentration of 21.5 µg/g of tumor in the 48 h period after administration [55]. Furthermore, it was observed that retention in tumor tissue was higher than in the blood and other tissues. Similar to other CNT-based drug carriers, it is crucial for CNT-based BNCT delivery agents to further improve their water solubility and biocompatibility.

3.4.2.4 Boron and Boron Nitride Nanotubes

Another suggested nanomaterial for use as a BNCT agent is the boron nanotube (BNT). The first successful synthesis of a single-wall BNT was achieved by the reaction of BCl_3 and H_2 over an Mg-MCM-41 catalyst [56]. The nanotubes had diameters of ~3 nm and lengths of ~16 nm. Unfortunately, the materials were quite sensitive to high-energy electron beams, and hence detailed structural characteristics could not be obtained. However, if the nanotubes can be functionalized to make them water-soluble, such structures should prove to be powerful BNCT carriers.

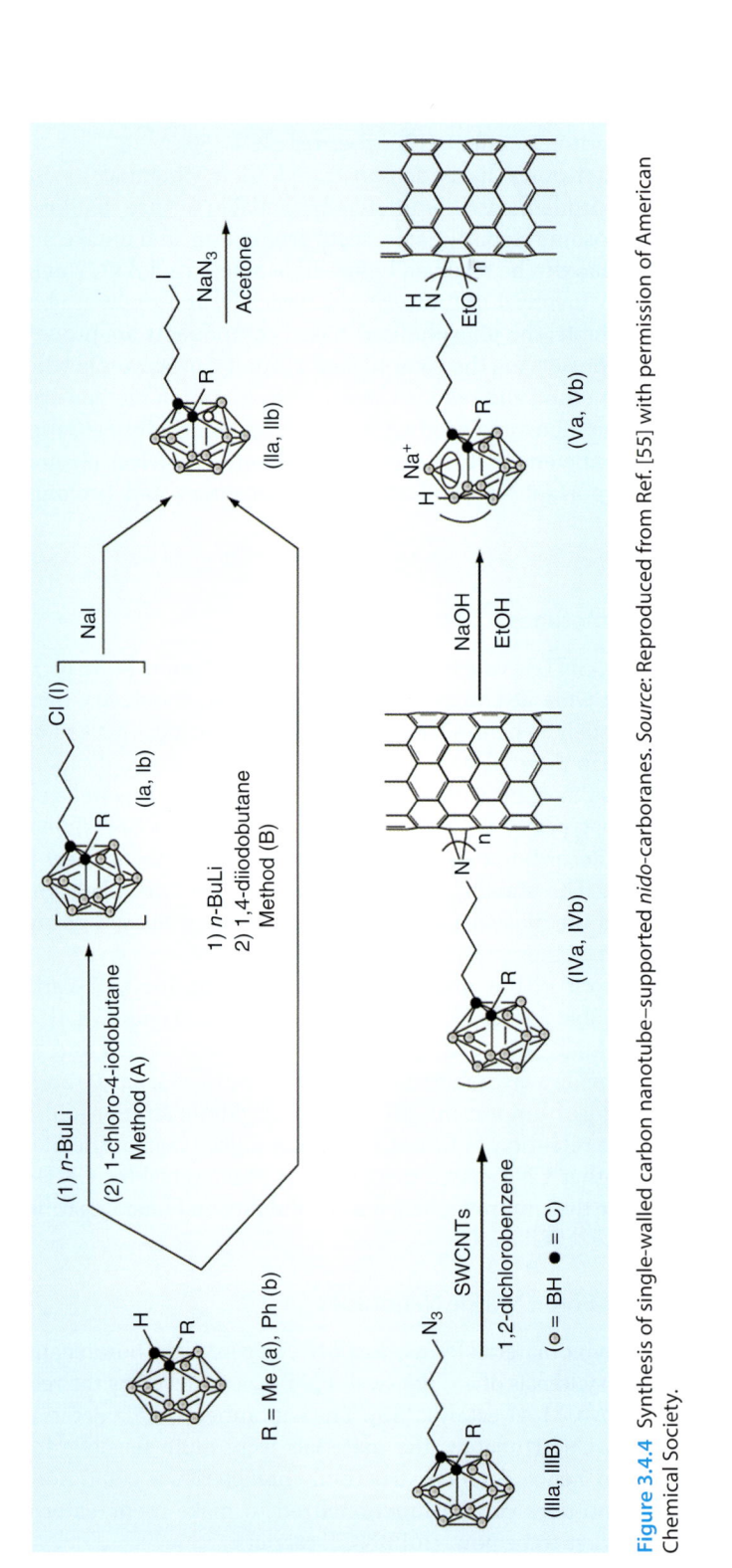

Figure 3.4.4 Synthesis of single-walled carbon nanotube–supported *nido*-carboranes. *Source:* Reproduced from Ref. [55] with permission of American Chemical Society.

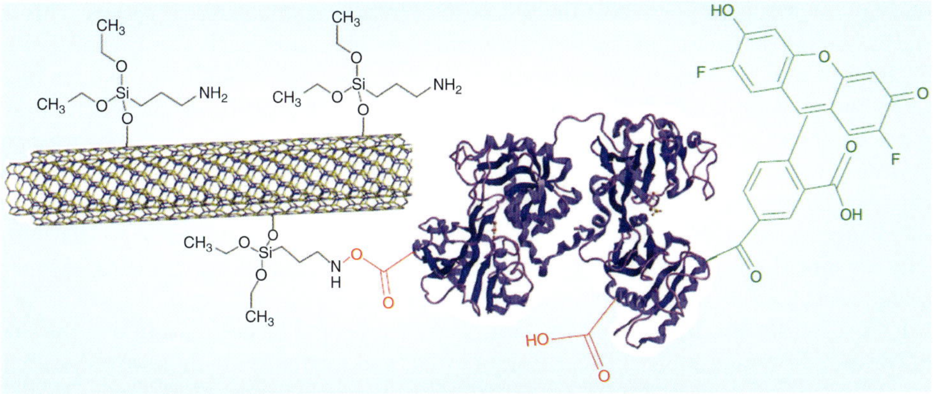

Figure 3.4.5 Transferrin-grafted boron–nitride nanotube. *Source*: Reproduced from Ref. [61] with permission of Elsevier B.V. All rights reserved.

In contrast to BNTs, boron–nitride nanotubes (BNNTs) have been demonstrated to be useful drug delivery agents. Boron–nitride is isoelectronic with carbon; thus, BNNTs are isosteres of CNTs. In comparison to CNTs, BNNTs have been shown to be nontoxic to HEK293 cells [57], and they can be functionalized to promote water solubility. A number of methods of functionalizing BNNTs include interacting them noncovalently with glycodendrimers [57], coating them with polyethyleneimine (PEI) [58] or poly-l-lysine (PLL), or reacting them with substituted quinuclidine bases [59]. The PLL-coated BNNT could be further reacted with folic acid to give the foliate-conjugated nanomaterial F-PLL-BNNT. *In vitro* studies showed that the F-PLL-BNNT bioconjugant is selectively localized in human glioblastoma multiforme T98G cells compared with the healthy human primary fibroblasts as controls [60].

Ciofani and coworkers reported that transferrin can be successfully grafted with BNNTs as represented in Figure 3.4.5. The nanocomposites demonstrated enhanced and targeted cellular uptake of BNNTs on primary human endothelial cells [61]. Considering the transferrin receptor is highly expressed by brain capillaries to mediate the delivery of iron to the brain, the transferrin–BNNT delivery agent is expected to access the brain via the blood–brain barrier (BBB).

Although BNTs have the potential to be ideal BNCT agents, there remain many issues that need to be resolved before BNTs can be applied to drug delivery. The major challenges are in the large-scale fabrication of BNTs and in the development of strategies to make BNTs water-soluble.

3.4.2.5 Magnetic Nanoparticles-Based BNCT Carriers

One of the major problems in any type of cancer chemotherapy is that of directing the drug to the tumor and avoiding healthy tissue. Since all chemotherapeutic drugs are by nature cytotoxic, localization of these drugs in the close vicinity of the tumor could result in the use of lower drug concentrations. This can be done by attaching the drug to a biomolecule that is overused in the malignant cell or to some receptor molecule that is overexpressed in the cancer cells. Another potential way to increase the efficacy of a cancer drug is to physically direct it to the tumor by some external means. This is the basic approach in magnetically targeted therapy. In this approach, the drug of choice is attached to a biocompatible magnetic nanoparticle carrier, usually in the form of a ferrofluid, and is injected into the patient via the circulatory system. When these particles enter the bloodstream, external, high-gradient magnetic fields can be used to concentrate the complex at a specific target site within the body. Once the drug or carrier is correctly concentrated, the drug can be released, via either enzymatic activity or changes in physiological conditions, and be taken up by the tumor cells [62]. The advantage of this methodology is that decreased amounts of cytotoxic drugs would be required, thereby decreasing unwanted side effects.

It has been shown that particles as large as 1–2 μm could be concentrated at the site of intracerebral rat glioma-2 (RG2) tumors [62], and a later study demonstrated that 10–20 nm magnetic particles were even more effective in targeting these tumors in rats [63]. Accordingly, magnetic targeting in humans demonstrated that the infusion of ferrofluids was well tolerated in most patients, and the ferrofluid could be successfully directed to advanced sarcomas without associated organ toxicity. Therefore, application of this technique can be considered appropriate vectors for the use of BNCT treatment [63].

Our recent work has demonstrated that encapsulated magnetic nanocomposites with a high load of carborane cages can be synthesized and their biodistribution patterns in cancer cell lines can be evaluated [64]. In this work, commercially available MNPs of iron oxides, covered by starch, have been enriched with the carborane cages, 1-R-2-butyl-*ortho*-$C_2B_{10}H_{10}$ (R = Me, **3**; Ph, **4**), by catalytic azide–alkyne cycloaddition reactions (Figure 3.4.6) [64]. Boron concentrations in tissue have been examined, and the results are demonstrated in Figure 3.4.7a and 3.4.7b [64]. It was shown (Figure 3.4.7a) that boron concentrations at different time intervals in the tumor are less than 14.7 μg/g of tumor, with a slow elimination after 30 h in the absence of an external magnetic field. However, as Figure 3.4.7b demonstrates, in the presence of an external magnetic field, the boron concentration in the tumor reached a high value of 51.4 μg/g of tumor with tumor-to-normal tissue ratios of approximately 10:1. According to transmission electron microscopic (TEM) images (Figure 3.4.8) [64], the entrapped magnetic nanocomposites aggregated inside the tumor. The exact mechanism of the accumulation of magnetic nanocomposite carriers within tumor cells has not yet been determined. Compared to the results without the external magnetic field, it appears that

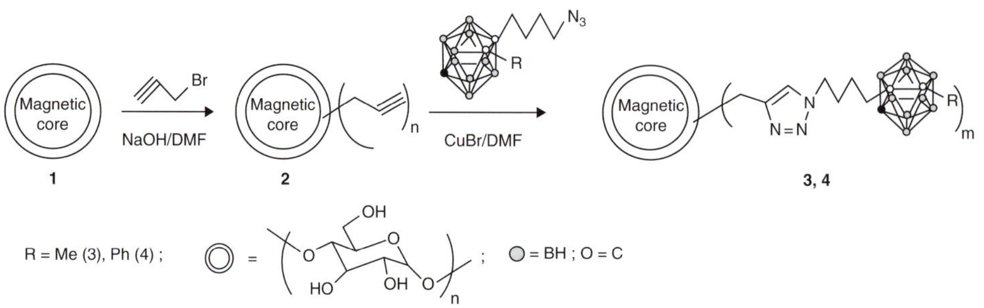

Figure 3.4.6 Synthesis of encapsulated magnetic nanocomposites. *Source*: Reproduced from Ref. [64] with permission from Hindawi Publishing Corporation. https://www.hindawi.com/journals/jnm/2010/409320/. Licensed under CC 3.0.

(a) (b)

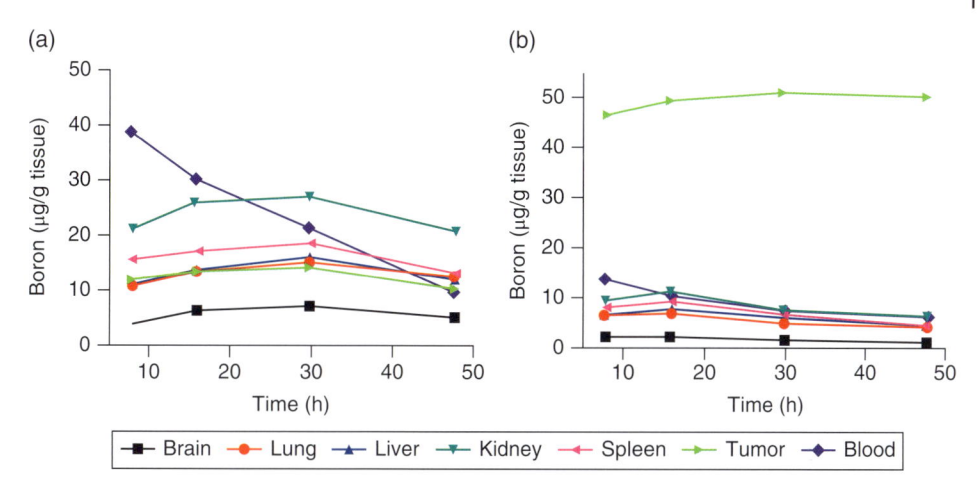

Legend: Brain — Lung — Liver — Kidney — Spleen — Tumor — Blood

Figure 3.4.7 Boron concentration distribution in tissues using compound **3**. (a) Without external magnetic field, and (b) with external magnetic field. *Source*: Reproduced from Ref. [64] with permission from Hindawi Publishing Corporation. https://www.hindawi.com/journals/jnm/2010/409320/. Licensed under CC 3.0.

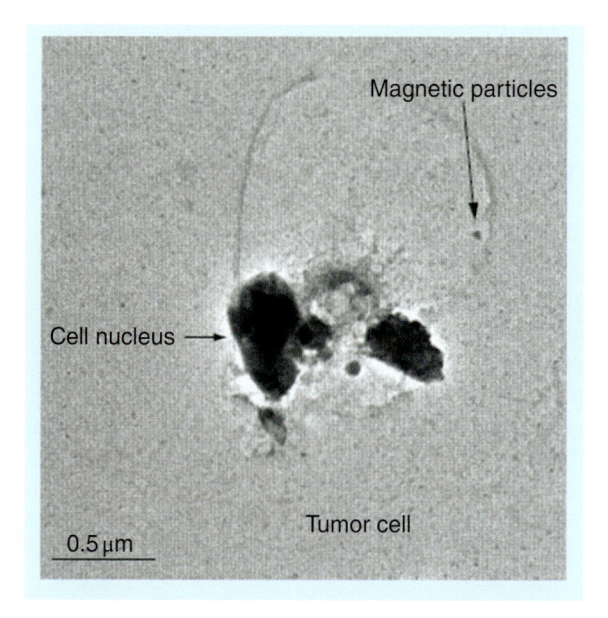

Figure 3.4.8 Transmission electron microscopy (TEM) image representing the magnetic cores of compound **3** within tumor cells. *Source*: Reproduced from Ref. [64] with permission from Hindawi Publishing Corporation. https://www.hindawi.com/journals/jnm/2010/409320/. Licensed under CC 3.0.

introduction of an external magnetic field plays a key role in an enhanced accumulation, with approximately a threefold increase in nanoparticle concentrations entrapped within the tumor. In this regard, it should be pointed out that even aggregated boron nanoparticles should be therapeutically effective for BNCT. The preliminary results provide new hope for a successful NCT and for the useful combination of the drugs with BNCT, magnetic resonance imaging (MRI), or thermotherapy characteristics.

3.4.2.6 Other Boron-Enriched Nanoparticles

Other boron-containing nanoparticles derived from boron carbides [65,66], block-copolymers [67,68], boron powder [69], borosilicates [70], and mercaptocarborane-capped gold nanoparticles [71,72] have also been reported. Commercially available boron carbide has been successfully functionalized with a synthetic dye called lissamine and the transacting transcriptional activator peptide; the resulting nanocomposites (<100 nm) can be translocated into B16 F10 malignant melanoma cells in a high amount of 1 wt% [65]. The particles have been reported to show significant proliferative inhibition for both particle-loaded and unloaded closely neighboring cells after neutron irradiation [65]. The functionalized boron carbide nanoparticles also show positive effects to the *in vivo* growth inhibition of an aggressive solid tumor, B16-OVA melanoma, by neutron capture [66]. However, no information is available regarding *in vivo* selectivity of this type of delivery agent.

Polymer-based boron-containing nanoparticles were prepared by radical copolymerization of acetal-poly(ethyleneglycol)-*block*-poly(lactide)-methacrylate with 4-vinylbenzyl substituted *closo*-carborane (Figure 3.4.9) [67,68]. Compared with particles obtained from self-assembly (non-cross-linked) rather than copolymerization, the particles from copolymerization of 1,2-bis(4-vinylbenzyl)-*closo*-carboranes demonstrated the following advantages [67]: improved stability in the presence of serum proteins without notable leakage in 50 h, extended blood circulation time, and increased tumor accumulation of up to 5.4% injection dose/g. Similar results are reported from the same group for particles produced from self-assembly and copolymerization of mono-4-vinylbenzyl-substituted *closo*-carborane [68].

More straightforward methods of using commercially available elemental boron powder to prepare boron nanoparticles have been reported [69]. After milling and dopamine coating, the boron nanoparticles, with a size of approximately 40 nm, were prepared. The particles did not demonstrate toxicity to murine macrophage cells as claimed [69]. Borosilicate nanoparticles with a size range of between 100 and 200 nm were prepared by milling a xerogel of $2SiO_2$–B_2O_3 [70]. After functionalization with folic acid, the resulting particles demonstrated increased incorporation in the tumor cells and hemocompatibility [70]. Cioran *et al.* prepared 2 nm gold nanoparticles capped with mercaptocarborane ligands. These monolayer-protected clusters (MPCs) qualitatively showed significant toxicity and the ability to penetrate into most cell compartments, with a strong tendency of finally residing inside membranes [71]. Uptake of the particles by a human cancer cell line has been demonstrated, and their intracellular fate shows the preferential location of the particles in vesicles within membranes and in the nucleus.

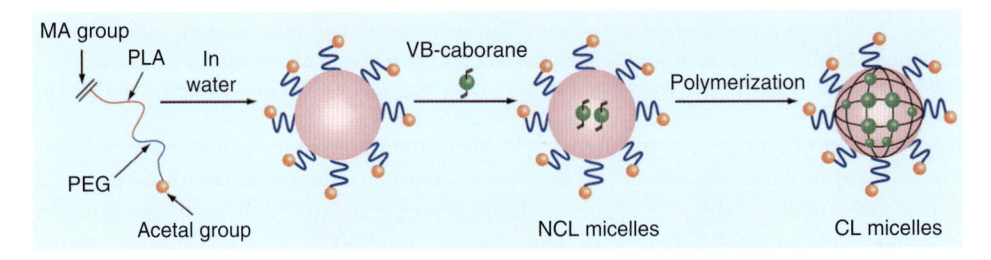

Figure 3.4.9 Preparation of carborane-containing polymer nanoparticles. *Source*: Reproduced from Ref. [67] with permission of Kluwer Academic Publishers.

New electrochemical sensors and diagnostic tools may arise from such nanoparticles. The material should also be attractive as a boron-rich agent for BNCT. Ciani *et al.* obtained new boron carriers built with *ortho*-carborane functionalized gold nanoparticles (GNPs). The interaction between carboranes and the gold surface was assured by one or two SH-groups directly linked to the boron atoms of the $B_{10}C_2$ cage. To improve cell uptake, the hydrophilic character of carborane functionalized GNPs was enhanced by further coverage with an appropriately tailored diblock copolymer (PEO-b-PCL). This polymer also contained pendant carboranes to provide anchoring to the pre-functionalized GNPs. *In vitro* tests, carried out on osteosarcoma cells, showed that the final vectors possessed excellent biocompatibility joint to the capacity of concentrating boron atoms in the target, which is encouraging evidence to pursue applications *in vivo*. Biological trials, performed *in vitro* on rat osteosarcoma cells, showed that these engineered gold nanoparticles had no appreciable toxicity and were able to accumulate boron atoms inside the cells, which is very promising for their application in BNCT [72].

3.4.3 Summary and Outlook

Nanomaterial-based drug systems provide the advantage of being able to penetrate cell membranes through minuscule capillaries in the cell wall of rapidly dividing tumor cells, while at the same time having low cytotoxicity toward normal cells. Nanomaterials have been found to have favorable interaction with the brain blood vessel endothelial cells of mice, and thus they might have the possibility of being transported to other brain tissues, making them potential NCT agents. In recent years, nanomaterial-based BNCT agents were initially employed for untreatable brain cancers, glioblastoma multiforme, but, to date, it has not been conclusively demonstrated that BNCT is superior to standard treatments for glioblastoma multiforme. At best, BNCT treatment might extend the time before regrowth, but it does not provide a true cure. Thus, efforts should also be made to treat cancers other than glioblastoma multiforme. Much effort has been directed toward developing nanomaterial-based BNCT agents in recent decades. Lipidic, polymeric, and magnetic nanoparticles; carbon nanotubes; boron and boron–nitride nanotubes; boron carbides; boron powder; borosilicates; and mercaptocarborane-capped gold nanoparticles have been widely studied. However, among all of the new boron carriers developed in the past 50 years, none have, as yet, made it to clinical trials. It will be necessary to carry out further *in vivo* studies and clinical trails. For boron-enriched magnetic nanocomposites, strategies are required to counter their tendency for embolization, and also their unclear cytotoxicity must be resolved. More advanced forms of BNTs can be anticipated as being of high interest in terms of their synthesis technology and water-solubility, and future applications should be significantly improved.

References

1 Barreto JA, O'Malley W, Kubeil M, Graham B, Stephan H, Spiccia L. Nanomaterials: applications in cancer imaging and therapy. Adv. Mater., 2011, **23**, H18–H40 (2011); Gao S, Fu R, Hosmane N. Nanomaterials for boron and gadolinium neutron capture therapy for cancer treatment. Pure Appl. Chem., **87**, 123–134 (2015).

2 Lammers T, Aime S, Hennink WE, Storm G. http://pubs.acs.org/doi/abs/10.1021/ar200019c-notes-1#notes-1; Kiessling F. Theranostic nanomedicine. Acc. Chem. Res., **44**, 1029–1038 (2011); Hosmane, NS (Editor). Boron Science: New Technologies and Applications. CRC Press: Boca Raton, FL (2011).
3 Jokerst JV, Gambhir SS. Molecular imaging with theranostic nanoparticles. Acc. Chem. Res., **44**, 1050–1060 (2011).
4 Grimm J, Scheinberg DA. Will nanotechnology influence targeted cancer therapy? Semin Radiat Oncol., **21**, 80–87 (2011).
5 Scheinberg DA, Villa CH, Escorcia FE, McDevitt NR. Conscripts of the infinite armada: systemic cancer therapy using nanomaterials. Nat. Rev. Clin. Oncol., **7**, 266–276 (2010).
6 Ahmed N, Fessi H, Elaissari A. Theranostic applications of nanoparticles in cancer. Drug Discov. Today, **17**, 928–934 (2012).
7 Yang F, Jin C, Subedi S, Lee CL, Wang Q, Jiang YJ, Li J, Di Y, Fu DL. Emerging inorganic nanomaterials for pancreatic cancer diagnosis and treatment. Cancer Treat. Rev., **38**, 566–579 (2012).
8 Maeda H, Seymour LW, Miyamoto Y. Conjugates of anticancer agents and polymers: advantages of macromolecular therapeutics in vivo. Bioconjugate Chem., **3**, 351–362 (1992).
9 Locher, GL. Biological effects and therapeutic possibilities of neutrons. Am. J. Roentgenol. Radium Ther., **36**, 1–13 (1936).
10 Zhu YH, Yan KC, Maguire JA, Hosmane NS. Boron-Based Hybrid Nanostructures: Novel Applications of Modern Materials. Hybrid Nanomaterials: Synthesis, Characterization, and Applications. BPS Chauhan (Editor). John Wiley & Sons: New York, 181–195 (2001).
11 Busse PM, Harling OK, Palmer MR, Kiger WR III, Kaplan J, Kaplan I, Chuang CF, Goorley JT, Riley KJ, Newton TH, Cruz GAS, Lu XQ, Zamenhof RG. A critical examination of the results from the Harvard-MIT NCT Program Phase I clinical trial of neutron capture therapy for intracranial disease. J. Neuro-Oncol., **62**, 111–121 (2003).
12 Kreiner AJ, Paolo DH, Burlon AA, Kesque JM, Valda AA, Debray M E, Giboudot Y, Levinas P, Fraiman M, Romeo V, Somacal HR, Minsky DM. Accelerator-based boron neutron capture therapy and the development of a dedicated tandem-electrostatic-quadrupole. AIP Conf. Proc., **947**, 17 (2007), http://scitation.aip.org/content/aip/proceeding/aipcp/10.1063/1.2813801.
13 Barth RF, Coderre JA, Vicente MGH, Blue TE. Boron neutron capture therapy of cancer: current status and future prospects. Clin. Cancer Res., **11**, 3987–4002 (2005).
14 Valliant JF, Guenther KJ, King AS, Morel P, Schaffer P, Sogbein OO, Stephenson KA. The medicinal chemistry of carboranes. Coord. Chem. Rev., **232**, 173–230 (2002).
15 Bregadze VI, Sivaev IB, Gabel D, Wohrle D. Polyhedral boron derivatives of porphyrins and phthalocyanines. J. Porphyrins Phthalocyan., **5**, 767–781 (2001).
16 Parrott MC, Marchington EB, Valliant JF, Adronov A. Synthesis and properties of carborane-functionalized aliphatic polyester dendrimers. J. Am. Chem. Soc., **127**, 12081–12089 (2005).
17 Soloway AH, Tjarks W, Bauman BA, Rong FG, Barth RF, Codogni IM, Wilson JG. The chemistry of neutron capture therapy. Chem. Rev., **98**, 1515–1562 (1998).
18 Wara WM, Bauman GS, Sneed PK, *et al.* Brain, brain stem, and cerebellum. In: Principles and Practice of Radiation Oncology, 3rd ed. CA Perez and LW Brady (Editors). Lippincott-Raven: Philadelphia, 777–828 (1998).

19. Zhu YH, Hosmane NS. Applications and perspectives of boron-enriched nanocomposites in cancer therapy. Future Med. Chem., **5**, 705–714 (2013).
20. Fairchild RG, Bond VP. Current status of ^{10}B-neutron capture therapy: enhancement of tumor dose via beam filtration and dose rate, and the effects of these parameters on minimum boron content: a theoretical evaluation. Int. J. Radiat. Oncol. Biol. Phys., **11**, 831–840 (1985).
21. Laws ER, Shaffrey ME. The inherent invasiveness of cerebral gliomas: implications for clinical management. Int. J. Dev. Neurosci., **17**, 413–420 (1999).
22. Perry A, Schmidt RE. Cancer therapy-associated CNS neuropathology: an update and review of the literature. Acta Neuropathol., **111**, 197–212 (2006).
23. Kouri M, Kankaanranta L, Seppälä T, Tervo L, Rasilainen M, Minn H, Eskola O, Vähätalo J, Paetau A, Savolainen S, Auterinen I, Jääskeläinen J, Joensuu H. Undifferentiated sinonasal carcinoma may respond to single-fraction boron neutron capture therapy. Radiother. Oncol., **72**, 83–85 (2004).
24. Renner MW, Miura M, Easson MW, Vicente MGH. Recent progress in the syntheses and biological evaluation of boronated porphyrins for boron neutron-capture therapy. Anticancer Agents Med. Chem., **6**, 145–157 (2006).
25. Shelly K, Feakes DA, Hawthorne MF, Schmidt PG, Krisch TA, Bauer WF. Model studies directed toward the boron neutron-capture therapy of cancer: boron delivery to murine tumors with liposomes. Proc. Natl. Acad. Sci. USA, **89**, 9039–9043 (1992).
26. Radel PA, Kahl SB. Enantioselective synthesis of l- and d-carboranylalanine. J. Org. Chem., **61**, 4582–4588 (1996).
27. Kabalka GW, Yao ML. The synthesis and use of boronated amino acids for boron neutron capture therapy. Anticancer Agents Med. Chem., **6**, 111–125 (2006).
28. Fukuda H, Hiratsuka J, Honda C, Kobayashi T, Yoshino K, Karashima H, Takahashi J, Abe Y, Kanda K, Ichihashi M, Mishima Y. Boron neutron capture therapy of malignant melanoma using 10B-paraboronophenylalanine with special reference to evaluation of radiation dose and damage to the skin. Radiat. Res., **138**, 435–442 (1994).
29. Coderre JA, Elowitz EH, Chadha M, Bergland R, Capala J, Joel DD, Liu HB, Slatkin DN, Chanana AD. Boron neutron capture therapy for glioblastoma multiforme using p-boronophenylalanine and epithermal neutrons: trial design and early clinical results. J. Neurooncol., **33**, 141–152 (1997).
30. Elowitz EH, Bergland RM, Coderre JA, Joel DD, Chadha M, Chanana AD. Biodistribution of *p*-boronophenylalanine inpatients with glioblastoma multiforme for use in boron neutron capture therapy. Neurosurgery, **42**, 463–469 (1998).
31. Hatanaka H, Nakagawa Y. Clinical results of long-surviving brain tumor patients who underwent boron neutron capture therapy. Int. J. Radiat. Oncol. Biol. Phys., **28**, 1061–1066 (1994).
32. Van Rij CM, Wilhelm AJ, Sauerwein W, van Loenen A. Boron neutron capture therapy for glioblastoma multiforme. Pharm. World Sci., **27**, 92–95 (2005).
33. Jain RK. Normalizing the tumor vasculature with anti-angiogenic therapy: a new paradigm for combination therapy. Nat. Med., **7**, 987–989 (2001).
34. Santini J, Cima M, Langer R. A controlled-release microchip. Nature, **397**, 335–338 (1999).
35. Sartor O, Dineen MK, Perez-Marreno R, Chu FM, Carron GJ, Tyler RC. An eight-month clinical study of LA-2575 30.0 mg: a new 4-month, subcutaneous delivery system for leuprolide acetate in the treatment of prostate cancer. Urology, **62**, 319–323 (2003).

36 Chen W, Mehta SC, Lu DR. Selective boron drug delivery to brain tumors for boron neutron capture therapy. Adv. Drug Delivery Rev., **26**, 231–247 (1997).
37 Wu G, Barth RF, Yang W, Lee RJ, Tjarks W, Backer MV, Backer JM. Boron containing macromolecules and nanovehicles as delivery agents for neutron capture therapy. Anti-Cancer Agents Med. Chem., **6**, 167–184 (2006).
38 Yih TC, Al-Fandi M. Engineered nanoparticles as precise drug delivery systems. J. Cell Biochem., **97**, 1184–1190 (2006).
39 Zhu Y, Koh CY, Maguire JA, Hosmane NS. Recent developments in boron neutron capture therapy driven by nanotechnology. Curr. Chem. Biol., **1**, 141–149 (2007).
40 Hosmane NS, Maguire JA, Zhu Y, Takagaki M. Boron and Gadolinium Neutron Capture Therapy for Cancer Treatment. World Scientific Publishing, Singapore, 52–88 (2012).
41 Laster BH, Kahl SB, Popenoe EA, Pate DW, Fairchild RG. Biological efficacy of boronated low-density lipoprotein for boron neutron capture therapy as measured in cell culture. Can. Res., **51**, 4588–4593 (1991).
42 Pan XQ, Wang HQ, Shukla S, Sekido M, Adams DM, Tjarks W, Barth RF, Lee RJ. Boron-containing folate receptor-targeted liposomes as potential delivery agents for neutron capture therapy. Bioconjugate Chem., **13**, 435–442 (2002).
43 Hawthorne MF. The role of chemistry in the development of boron neutron capture therapy of cancer. Angew. Chem. Int. Ed., **32**, 950–984 (1993).
44 Gabizon A, Price DC, Huberty J, Bresalier RS, Papahadjopoulos D. Effect of liposome composition and other factors on the targeting of liposomes to experimental tumors: biodistribution and imaging studies. Cancer Res., **50**, 6371–6378 (1990).
45 Torchilin P. Recent advances with liposomes as pharmaceutical carriers. Nat. Rev. Drug. Discov., **4**, 145–160 (2005).
46 Ristori S, Oberdisse J, Grillo I, Donati A, Spalla O. Structural characterization of cationic liposomes loaded with sugar-based carboranes. Biophys. J., **88**, 535–547 (2005).
47 Yanagie H, Maruyama K, Takizawa T, Ishida O, Ogura K, Matsumoto T, Sakurai Y, Kobayashi T, Shinohara A, Rant J, Skvarc J, Ilic R, Kuhne G, Chiba M, Furuya Y, Sugiyama H, Hisa T, Ono K, Kobayashi H, Eriguchi M. Application of boron-entrapped stealth liposomes to inhibition of growth of tumor cells in the *in vivo* boron neutron-capture therapy model, Biomed. Pharmacother., **60**, 43–50 (2006).
48 Nobuto H, Sugita T, Kubo T, Shimose S, Yasunage Y, Murakami T, Ochi M. Evaluation of systematic chemotherapy with magnetic liposomal doxorubicin and a dipole external electromagnet. Int. J. Cancer, **109**, 627–635 (2004).
49 Nakamura H, Ueda N, Hyun SB, Ueno M, Tachikawa S. Design and synthesis of fluorescence-labeled *closo*-dodecaborate lipid: its liposome formation and *in vivo* imaging targeting of tumors for boron neutron capture therapy. Org. Biomol. Chem., **10**, 1374–1380 (2012).
50 Koganei H, Ueno M, Tachikawa S, Tasaki L, Ban HS, Suzuki M, Shiraishi K, Kawano K, Yokoyama M, Maitani Y, Ono K, Nakamura H. Development of high boron content liposomes and their promising antitumor effect for neutron capture of cancers. Bioconj. Chem., **24**, 124–132 (2013).
51 Iijma S. Helical microtubules of graphitic carbon. Nature, **354**, 56–58 (1991).
52 De Volder MFL, Tawfick SH, Baughman RH, Hart AJ. Carbon nanotubes: present and future commercial applications. Science, **339**, 535–539 (2013).

53 Cuenca A, Jiang H, Hochwald S, Delano M, Cance W, Grobmyer S. Emerging implications of nanotechnology on cancer diagnostics and therapeutics. Cancer, **107**, 459–466 (2006).
54 Aqel A, Abou El-Nour KMM, Ammar RAA, Al-Warthan A. Carbon nanotubes, science and technology part (I) structure, synthesis and characterization. Arabian J. Chem., **5**, 1–23 (2010).
55 Zhu YH, Ang TP, Carpenter K, Maguire J, Hosmane N, Takagaki M. Substituted carborane-appended water soluble single-wall carbon nanotubes: new approach to boron neutron capture therapy drug delivery. J. Am. Chem. Soc., **127**, 9875–9880 (2005).
56 Ciuparu D, Klie RF, Zhu YM, Pfefferle L. Synthesis of pure boron single-wall nanotubes. J. Phys. Chem. B, **108**, 3967–3969 (2004).
57 Chen X, Wu P, Rousseas M, Okawa D, Gartner Z, Zettl A, Bertozzi CR. Boron nitride nanotubes are noncytotoxic and can be functionalized for interaction with proteins and cells. J. Am. Chem. Soc., **131**, 890–891 (2009).
58 Ciofani G, Raffa V, Menciassi A, Cushieri A. Cytocompatibility, interactions, and uptake of polyethyleneimine-coated boron nitride nanotubes by living cells: confirmation of their potential for biomedical applications. Biotechnol. Bioeng., **101**, 850–858 (2008).
59 Maguer A, Leroy E, Bresson L, Doris E, Loiseau A, Mioskowski C. A versatile strategy for the functionalization of boron nitride nanotubes. J. Mater. Chem., **19**, 1271–1275 (2009).
60 Ciofani G, Raffa V, Menciassi A, Cushieri A. Folate functionalized boron nitride nanotubes and their selective uptake by glioblastoma multiforme cells: implications for their use as boron carriers in clinical boron neutron capture therapy. Nanoscale Res. Lett., **4**, 113–121 (2009).
61 Ciofani G, Del Turco S, Genchi GG, D'Alessandro D, Basta G, Mattoli V. Transferrin-conjugated boron nitride nanotubes: protein grafting, characterization, and interaction with human endothelial cells. Int. J. Pharm., **436**, 444–453 (2012).
62 Alexiou C, Arnold W, Klein RJ, Parak FG, Hulin P, Bergemann C, Erhardt W, Wagenpfeil S, Lubbe AS. Locoregional cancer treatment with magnetic drug targeting. Cancer Res., **60**, 6641–6648 (2000).
63 Sincai M, Ganga D, Ganga M, Argherie D, Bica D. Antitumor effect of magnetite nanoparticles in cat mammary adenocarcinoma. J. Magn. Magn. Mater., **293**, 438–441 (2005).
64 Zhu Y, Lin Y, Zhu YZ, Lu J, Maguire JA, Hosmane NS. Boron drug delivery via encapsulated magnetic nanocomposites: a new approach for BNCT in cancer treatment. J. Nanomater., 2010, 409320 (**2010**).
65 Mortensen MW, Björkdahl O, Sørensen PG, Hansen T, Jensen MR, Gundersen HJG, Bjornholm T. Functionalization and cell uptake of boron carbide nanoparticles: the first step toward T cell-guided boron neutron capture therapy. Bioconj. Chem., **17**, 284–290 (2006).
66 Petersen MS, Petersen CC, Agger R, Sutmuller M, Jensen MR, Sorensen PG, Mortensen MW, Hansen T, Bjornholm T, Gundersen HJ, Huiskamp R, Hokland M. Boron nanoparticles inhibit tumour growth by boron neutron capture therapy in the murine B16-OVA model. Antican. Res., **28**, 571–576 (2008).
67 Sumitani S, Oishi M, Nagasaki Y. Carborane confined nanoparticles for boron neutron capture therapy: improved stability, blood circulation time and tumor accumulation. React. Funct. Polym., **71**, 684–693 (2011).

68 Sumitani S, Oishi M, Yaguchi T, Murotani H, Horiguchi Y, Suzuki M, Ono K, Yanagie H, Nagasaki Y. Pharmacokinetics of core-polymerized, boron-conjugated micelles designed for boron neutron capture therapy for cancer. Biomaterials, **33**, 3568–3577 (2012).
69 Gao Z, Walton NI, Malugin A, Ghandehari H, Zharov I. Preparation of dopamine-modified boron nanoparticles. J. Mater. Chem., **22**, 877–882 (2012).
70 Grandi S, Spinella A, Tomasi C, Bruni G, Fagnoni M, Merli D, Mustarelli P, Guidetti G F, Achilli C, Balduini C. Synthesis and characterization of functionalized borosilicate nanoparticles for boron neutron capture therapy applications. J. Sol-Gel. Sci. Technol., **64**, 358–366 (2012).
71 Cioran AM, Musteti AD, Teixidor F, Krpetic Z, Prior IA, He Q, Kiely CJ, Brust M, Vinas C. Mercaptocarborane-capped gold nanoparticles: electron pools and ion traps with switchable hydrophilicity. J. Am. Chem. Soc., **134**, 212–221 (2012).
72 Ciani L, Bortolussi S, Postuma I, Cansolino L, Ferrari C, Panza L, Altieri S, Ristori S. Rational design of gold nanoparticles functionalized with carboranes for application in boron neutron capture therapy. Int. J. Pharmaceut., **458**, 340–346 (2013).

3.5

New Boronated Compounds for an Imaging-Guided Personalized Neutron Capture Therapy

Nicoletta Protti,[1,2] Annamaria Deagostino,[3] Paolo Boggio,[3] Diego Alberti,[4] and Simonetta Geninatti Crich[4,]*

[1] *Department of Physic, University of Pavia, Pavia, Italy*
[2] *Nuclear Physics National Institute (INFN), Unit of Pavia, Pavia, Italy*
[3] *Department of Chemistry, University of Torino, Torino, Italy*
[4] *Department of Molecular Biotechnology and Health Sciences, University of Torino, Torino, Italy*
* Corresponding author: Department of Molecular Biotechnology and Health Sciences, University of Torino, via Nizza 52, 10126, Torino, Italy. Email: simonetta.geninatti@unito.it

3.5.1 General Introduction on BNCT: Rationale and Application

Boron neutron capture therapy (BNCT) is an experimental form of binary radiotherapy that selectively targets and damages tumor cells, even in the very challenging scenario in which the malignancy is infiltrating into the surrounding normal tissue or is spreading in the whole organ. The physical principle of BNCT is very elegant and pretty old (dated back to 1936, in the publication by Gordon Locher [1]), but the therapy is still facing important critical issues to be solved before being exploited in its whole potential within clinical practice.

The rationale of the treatment relies on the nuclear capture reaction induced by thermal neutrons (energy <0.5 eV) in the less abundant stable isotope of boron, ^{10}B, which produces a couple of highly energetic, short-range secondary particles (see Figure 3.5.1).

The probability of such reaction is estimated by the nuclear physics quantity called microscopic cross section (σ), which for ^{10}B equals 3840 barn when neutrons of energy equal to 25 meV (thermal energy) interact with ^{10}B. The selectivity of BNCT is mainly due to the high linear energy transfer (LET) particles, an α-particle and a recoil nucleus of lithium-7, emitted after ^{10}B capture reaction. In fact, these particles have the right energy (total maximum kinetic energy = 2.79 MeV) to induce lethal effects in the cell loaded by ^{10}B [2], while their short ranges (the maximum path a charged particle can cross in a fixed material) are less than or comparable to the mean cell diameter (3.5 and 8.5 µm in water for lithium and α-particles, respectively). Thus, it allows a cell-level-selective radiation treatment, driven mainly by ^{10}B microscopic distribution before

Boron-Based Compounds: Potential and Emerging Applications in Medicine, First Edition.
Edited by Evamarie Hey-Hawkins and Clara Viñas Teixidor.
© 2018 John Wiley & Sons Ltd. Published 2018 by John Wiley & Sons Ltd.

$$^{10}B + n \rightarrow \begin{cases} \alpha(1.47 \text{ MeV}) + {}^7\text{Li} (0.84 \text{ MeV}) + \gamma(0.48 \text{ MeV}) & 94\% \\ \alpha(1.78 \text{ MeV}) + {}^7\text{Li} (1.01 \text{ MeV}) & 6\% \end{cases}$$

Figure 3.5.1 ^{10}B capture reaction induced by thermal neutrons.

neutron irradiation. Besides, low-energy neutrons taken alone are practically unable to induce serious biological effects in cells and tissues.

Therefore, BNCT is a binary radiotherapy theoretically capable of administering a unique form of tumor-cell-selective treatment, almost impossible to gain by all the other available radiotherapy modalities. Practically, the effectiveness of BNCT is deeply connected to the optimized accumulation of selective ^{10}B-containing agent(s) in malignant cells as well as to the delivery of a suitable thermal neutron flux even at the deepest regions of the mass to be treated. Furthermore, the recent initiation of clinical trials employing hospital-based accelerators will favor new and more numerous BNCT-treated patients. Simultaneously, huge amounts of innovative ^{10}B-vectors are continuously proposed to gain an improved selective targeting of cancer cells [3,4]. The first clinical trials investigating BNCT efficacy against malignant gliomas started in the early 1950s at the Brookhaven Graphite Research Reactor facility [5], using several nonselective ^{10}B vectors (borax, sodium-pentaborate, etc.). Around the beginning of the 1960s, a second BNCT facility at the Massachusetts Institute of Technology started its own clinical trials on brain tumor patients [6]. The median survival and prognostic outcomes of these trials were largely comparable to those obtained at that time using conventional radiotherapy. Unfortunately, acute side effects were reported due to ^{10}B-vector toxicity or irradiation procedures, and in 1961 these trials were stopped.

BNCT clinical application restarted in 1968 in Japan thanks to Professor Hatanaka, who introduced a new ^{10}B-vector, sodium borocaptate Na$_2$B$_{12}$H$_{11}$SH (BSH, Figure 3.5.2b), which selectively targets brain tumor cells by a passive loading mechanism that exploits the damaged blood–brain barrier at the tumor site. BNCT was used as an intraoperative radiotherapy, thus improving the dose distribution and reducing the thermal neutron attenuation in the first slabs of tissues. Professor Hatanaka reported a five-year survival rate of 58% in patients affected by grade 3 and 4 malignant glioma, leading to a worldwide new interest in BNCT [7]. In 1987, the second milestone was delivered by Professor Mishima, who treated superficial malignant melanomas (thus

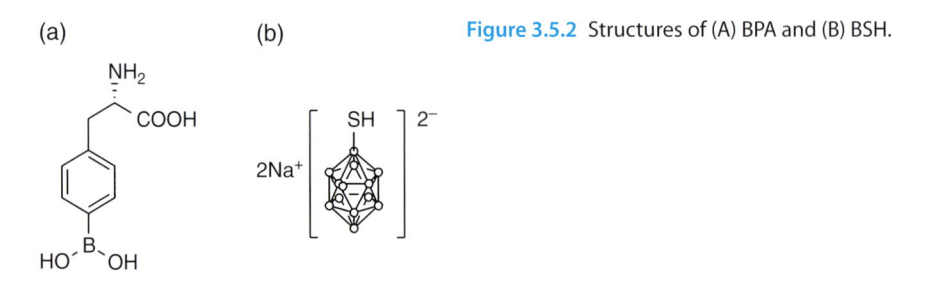

Figure 3.5.2 Structures of (A) BPA and (B) BSH.

starting BNCT application to non-brain tumors) using the other most important drug in BNCT, p-boronophenylalanine (BPA, Figure 3.5.2a) [8].

On the basis of these efforts, several BNCT facilities were opened in the USA, Latin America, Europe, and East Asia during the 1990s, exploiting epithermal neutron beams extracted from research nuclear reactors to treat deep-seated tumors [9]. At the end of this exponential phase, the clinical indications for the therapy went far beyond merely central nervous system malignancies and included skin melanomas, head and neck tumors, pleural mesothelioma, and hepatocellular carcinoma.

Nowadays, very few centers offer BNCT as an alternative cancer treatment, proposing the therapy as phase I/II clinical trials to highly selected patients often affected by high-grade malignancies and generally suffering from very poor prognosis. This clearly implies significant challenges in getting a proved and consolidated demonstration of BNCT as cancer treatment among the whole clinicians' community. It is likely connected to the intrinsic high complexity of the BNCT modality and to the obstacles connected with the construction and managing of nuclear reactors. Although they are presently the only available sources of stable epithermal neutron beams with the required high intensity to perform patient irradiation in reasonable times, the BNCT community is approaching a completely new phase thanks to the effort of several research teams involved in the development of next-generation BNCT neutron sources, based on low-energy particle accelerators [10–15]. Undoubtedly, the availability of accelerator-driven facilities will improve BNCT clinical trial diffusion because: (1) accelerator technology is suitable for installation inside hospitals, and clinicians working in a standard radiotherapy department are skilled in this technology; (2) costs of low-energy accelerator construction and maintenance are modest if compared to other conventional radiotherapy medical devices, and they are greatly lower than those of nuclear reactors; (3) properly designed accelerator-based facilities are affected by very low levels of radioactivity once the beam is turned off, making the installation and subsequent maintenance of the machine much easier than that of research reactors; and (4) by changing the physical reactions from which neutrons are produced, the beam energy spectrum can be easily peaked at intermediate energies (typically, for clinical use, in the epithermal range) perfectly suitable for treating deep-seated tumors and able to significantly spare superficial tissues, thus improving the quality of patient treatment.

Finally, the diffusion of BNCT as a prominent therapy is prevented by two further main reasons: (1) the lack of properly tumor-selective boron carriers and (2) the absence of noninvasive methods for the evaluation of boron concentration in diseased and healthy tissues of patients undergoing neutron irradiation. In fact, BPA and BSH, the commonly investigated clinical agents, do not have sufficient tumor selectivity, and new targeted agents are desperately needed. Furthermore, since the measurement of local boron concentration is crucial to determine the optimal neutron irradiation time and calculate the delivered radiation dose in a personalized manner, new BNCT agents that are detectable with recently developed imaging techniques are also needed. Two problems are that boron distribution varies from patient to patient, and large uncertainties exist in the tumor-to-blood boron concentration ratio. This chapter will describe different strategies that greatly enhance the chance of success with this potentially ideal form of treatment for many types of cancers.

3.5.2 Imaging-Guided NCT: Personalization of the Neutron Irradiation Protocol

The effectiveness of BNCT strongly depends on the amount of energy locally deposited by the ^{10}B neutron capture reaction. This energy is proportional to the product of ^{10}B concentration and thermal neutron flux inside the tumor volume at the irradiation time. Currently, the local real-time measurement of these quantities is a big challenge that is hardly being addressed by the BNCT researchers' community.

Several techniques and methods have been developed and refined to precisely measure ^{10}B concentration in healthy and pathological tissues [16,17]. The main drawback is the lack of a clinically available real-time technique to measure ^{10}B and neutron flux spatial distributions.

Nowadays, the approach used in clinical trials to prescribe a therapeutic dose to the tumor is an indirect method based on pharmacokinetics models, which predict the ^{10}B concentration in the tumor, surrounding tissues, and peripheral blood as a function of time and initial injected ^{10}B doses. Blood samples are taken immediately after starting boron infusion and after regular time intervals up to the end of the treatment. ^{10}B concentration is usually measured by inductively coupled plasma–mass spectrometry (ICP-MS). Combining the blood ICP-MS result with the mathematical model of ^{10}B uptake and wash-out, the current ^{10}B concentration in the tumor is inferred [18]. Once the ^{10}B concentration is known, it is used within BNCT-dedicated treatment planning systems (TPSs) to calculate doses. Limitations and drawbacks of such indirect approaches are quite clear. Several clinical experiences reported that the follow-up did not show the expected outcomes and also that different posttreatment results were observed even after application of the same BNCT protocols. These differences can be explained by a lack of precision in dosimetry assessment that could be improved by a deeper knowledge of boron concentration during irradiation.

In this section, attention will be devoted mainly to the exploitation of high-sensitivity imaging techniques, namely positron emission spectroscopy (PET), single-photon emission computed tomography (SPECT), and magnetic resonance imaging (MRI) and spectroscopy (MRS), to overcome this fundamental limitation. Examples of optical imaging applications also will be described.

3.5.2.1 Positron Emission Tomography

PET represents a significant advance in cancer imaging and has improved the management of patients. In PET imaging, positrons are emitted by a radionuclide, and they interact with an electron in an annihilation process. This process produces two coincident 511 keV photons, which are detected simultaneously in a detector ring [19]. During the last decade, several PET radiopharmaceuticals entered the clinic, and PET became an important tool for the staging of cancer patients and assessing response to therapy. The physical integration of PET and computed tomography (CT) in hybrid PET-CT scanners allows combined anatomical and functional imaging. PET can be used for tumor staging and to predict tumor response to the therapy, detect early recurrence, and monitor the therapeutic treatment. PET scans necessitate the injection of a small amount of biologically significant material like glucose or oxygen that has been labeled with radionuclides such as carbon-11, nitrogen-13, oxygen-15, and fluoride-18. The most commonly used isotope in PET scans is fluorine-18 because of its long half-life of approximately 110 min and the low energy of its emitting positrons, which contributes to high-resolution image acquisition [20]. FDG (fluorine-18 combined with deoxy-glucose) is the most used radiotracer in clinical practice. When a PET tracer, such as fluorine-18, is coupled to a BNCT agent, it yields an interesting route for its localization and quantification before and after the therapeutic treatment. In this context, a fluorine-18 boronophenylalanine analog (^{18}F-FBPA) (Figure 3.5.3) was first proposed for *in vivo* evaluations of BPA biodistribution in brain tumors [21].

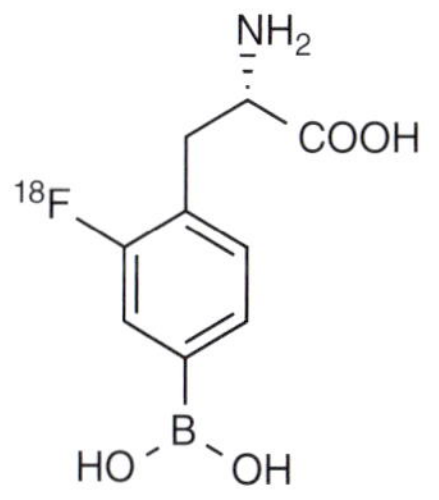

Figure 3.5.3 Fluorine-18 boronophenylalanine analog (^{18}F-FBPA).

In ^{18}F-FBPA, the radiolabeled fluorine atom is embedded into BPA in place of a hydrogen atom, and this may slightly increase the lipophilicity in comparison to the parent unmodified compound. Most of the papers of the last ten years on PET-guided BNCT applications concern the determination of the boron concentration in tumor and normal tissues based on ^{18}F-BPA molecular imaging [22]. One of the most important BNCT application fields regards the treatment of recurrent head and neck malignancies, as it avoids severe impairment of orofacial structures and functions. In this context, a successful application of PET-guided BNCT on a patient bearing a recurrent submandibular gland cancer was reported by Aihara and coworkers [23]. A complete regression in the tumor, and no acute or chronic complications for 1.5 years, was observed. The BPA-accumulating capacity of the tumor by ^{18}F-BPA was determined before BNCT. The tumor–normal tissue boron concentration ratio was 2.9, superior to 2.5 and expected to respond well to BNCT. The same method was then applied four years later on four patients with local recurrence or metastasis to the regional lymph nodes after completion of initial treatments, including surgery, chemotherapy, and radiotherapy [24]. All patients showed at least a tentative partial response, while a marked improvement in quality of life was seen in one patient.

Recently, a comparison of the accumulation of ^{18}F-BPA with that of ^{18}F-fluorodeoxyglucose (^{18}F-FDG) in head and neck cancers has been reported to assess the usefulness of ^{18}F-FDG PET for screening candidates for BNCT [25] ^{18}F-FDG is the most popular PET tracer and is available in many hospitals. Twenty patients with pathologically proven malignant tumors of the head and neck underwent both whole-body ^{18}F-BPA PET–CT and ^{18}F-FDG PET–CT within two weeks of each other. The accumulation of ^{18}F-FDG was significantly correlated with that of ^{18}F-BPA, demonstrating that ^{18}F-FDG PET might be an effective screening method for selecting patients with head and neck cancer for BNCT treatment.

Like head and neck tumors, meningiomas and schwannomas associated with neurofibromatosis-2 (NF2) also are malignancies that cannot be completely controlled by conventional surgery, radiotherapy, and chemotherapy. Clearly, novel treatment modalities, such as BNCT, should be considered in severe NF2, where multiple tumors cause early

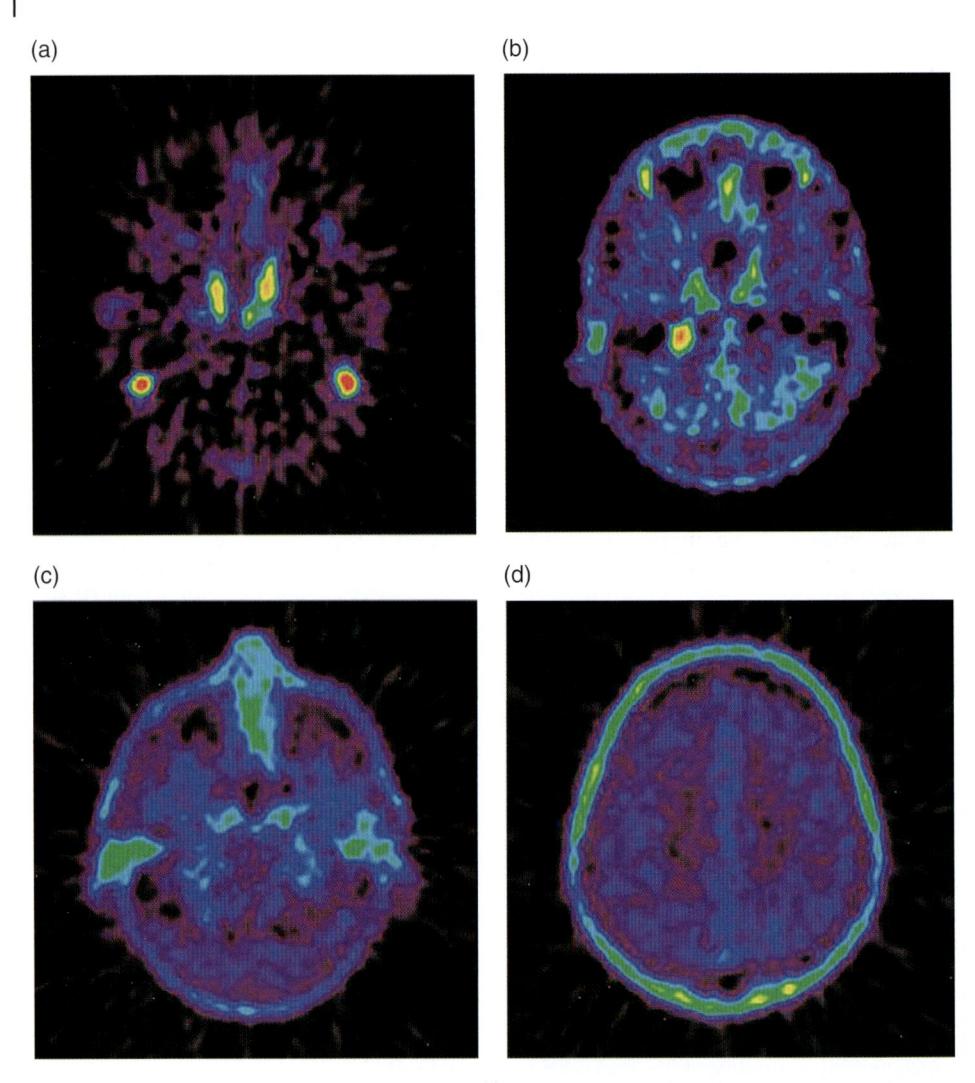

Figure 3.5.4 (a) Cerebral blood volume and (b–d) ^{18}F-BPA PET images of a patient with a sporadic right vestibular schwannoma. In spite of the small volume of the tumor, which measured 16 mm in diameter, the schwannoma is well visualized with ^{18}F-BPA in (b), whereas the blood volume image obtained with inhaled [^{15}O]CO shows only major cranial vessels. In (c), increased tracer uptake is seen in nasal mucosa and the parotid glands, while (d) indicates that the scalp has a higher uptake than the brain. This is in line with skin and mucosal toxicity seen in BNCT, where BPA is found to accumulate in these normal tissues. *Source*: Havu-Auren *et al.* (2007), fig. 3, p. 90. Reproduced with permission of Springer-Verlag.

morbidity and mortality. As a first test, the uptake of ^{18}F-BPA was studied in ten patients with sporadic or NF2-associated meningiomas and schwannomas [26]. All tumors accumulated ^{18}F-BPA and could be visualized. A model including three parameters – K1 (transport), k2 (reverse transport), and k3 (intracellular metabolism) – was applied to better illustrate ^{18}F-FBPA uptake kinetics. Tracer input function and cerebral blood volume were measured (Figure 3.5.4). ^{18}F-BPA uptake in tumor and brain was assessed with a three-compartmental model and graphical analysis. Not necessarily all

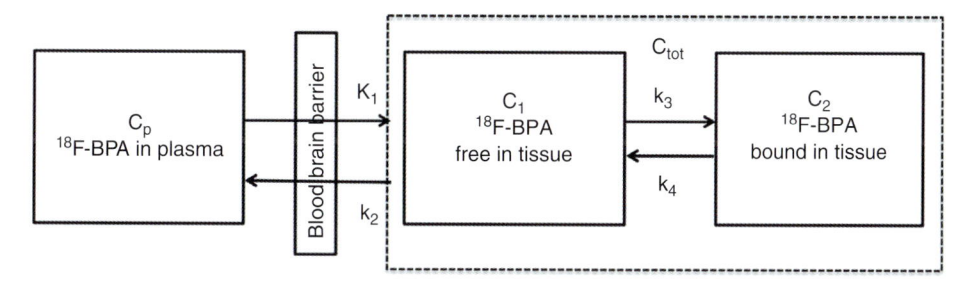

Figure 3.5.5 A 3-compartment model for ^{18}F-BPA uptake. The rate constants K_1, k_2, k_3, and k_4 define the transport between the central compartment Cp (plasma); the tissue compartment C1 represents nonspecifically bound ^{18}F-BPA; and the deeper tissue compartment C2 represents ^{18}F-BPA bound in the tissue (cell). K_1 and k_2 are the rate constants for forward and reverse transport of ^{10}B BPA across the blood–brain barrier, respectively. k_3 and k_4 are the anabolic and the reverse process rate constants. *Source*: Koivunoro et al. (2015), fig. 1, p. 190. Reproduced with permission of Elsevier.

schwannomas and meningiomas continue to metabolize ^{18}F-BPA beyond transport more than the normal brain does. Therefore, the authors outlined the need for PET imaging to individually assess the applicability of BNCT using BPA and suggested that some but not all schwannomas and meningiomas may be suitable for BNCT in spite of similar presentation in morphological imaging (Figure 3.5.4). Finally, they hypothesized that meningiomas and schwannomas can be therapeutically affected by BNCT at lower BPA gradients than those needed for malignant gliomas, suggesting the use of low-dose BNCT, possibly fractionated, as a totally new concept based on the increased vascularity and proliferative capacity of meningioma vessels.

The application of this closed three-compartment model (Figure 3.5.5), based on dynamic ^{18}F-BPA–PET studies to estimate the BPA concentrations in the tumor and the normal brain with time, was extended to a total of 98 patients with glioma treated with BPA-F-mediated BNCT in Finland from 1999 to 2011 [27].

The authors concluded that constant tissue-to-blood boron concentration ratios might not be valid approximations at the time of neutron irradiation. Its validity remains to be verified for 2-h BPA infusions delivering clinically useful BPA doses (290 mg/kg or higher). The same model was then adopted by Menichetti et al. [28].

In alternative, Michiue and coworkers [29] recently proposed a mercapto-*closo*-undecahydrododecaborate fused with a short arginine peptide (BSH-3R) that allows the penetration of glioma cell membranes. To monitor the pharmacokinetic properties of these agents *in vivo*, BSH and BSH-3R were linked to ^{64}Cu-DOTA as a PET probe. BSH-3R appeared to be the ideal boron compound for clinical use during BNCT, and its development is essential for pharmacokinetic imaging.

3.5.2.2 Single-Photon Emission Computed Tomography

To solve the problem of real-time measurement of ^{10}B distribution, a useful tool is the 478 keV photon emitted after ^{10}B capture reaction by the excited ^{7}Li recoil nucleus (see Figure 3.5.1). The possibility to perform a real-time counting of the reaction rate on ^{10}B through the detection of this photon represents an attractive strategy to solve the described challenge of monitoring the therapeutic dose, and it would allow the real-time optimization of the irradiation plan, thus improving the effectiveness of BNCT and the therapy personalization for each patient.

Considering the following relationship:

$$D \sim \int n_B \sigma \Phi \, dV \qquad (3.5.1)$$

where D = the dose; n_B = the density of ^{10}B nuclei; σ = the microscopic cross section of the ^{10}B capture reaction; Φ = the neutron flux; and V = the volume where the dose must be measured. It follows that the counting of 478 keV γ-rays means a direct estimation of the product $n_B \Phi$ and consequently of the dose D. Due to the emission of a single photon, the development of a BNCT-dedicated SPECT is the logic consequence. Presently, a few groups worldwide are investigating the possibility to develop such a SPECT system [30–32], using mainly solid-state detectors.

The main advantage of this technique is that the probe required for the imaging is exactly the same molecule used for therapy, thus avoiding further radioactivity delivered to the patient and the development of a specific tracer.

3.5.2.3 Magnetic Resonance Imaging and Spectroscopy

Nuclear magnetic resonance (NMR) detection of BNCT agents can be pursued by both imaging (MRI) and spectroscopy (MRS) modalities. Nuclei used for these applications are ^1H, ^{19}F, and $^{10}B/^{11}B$.

3.5.2.3.1 ^1H-MRI

^1H-MRI has been proposed to assess the biodistribution of BNCT compounds. Although its sensitivity is lower in comparison to nuclear and optical modalities, the high spatial resolution (<100 µm) of ^1H-MRI can provide detailed morphological and functional information, and the absence of radiation makes it safer than techniques based on the use of radioisotopes. Furthermore, as MRI does not use decaying isotopes, the observation time window is significantly larger and the scan can be repeated without any toxic effect detected. An MRI signal is dependent on the longitudinal (T_1) and transverse (T_2) relaxation times of water protons, and therefore the contrast in an MR image arises mainly from differences in the relaxation times of tissue water protons as a consequence of the interaction with biological macromolecules and membranes. In both clinical and experimental applications, the endogenous contrast can be altered by the use of contrast agents (CAs) that are able to decrease T_1 and T_2 of water protons in the tissues where they distribute. Most of the contrast agents used in clinical settings are polyaminocarboxylate complexes of Gd^{3+}. The ligands are multidentate (seven or eight donor atoms), forming complexes with very high thermodynamic and kinetic stabilities, thus limiting the release of the highly toxic free metal ions. Thus, these agents can be used to carry out an indirect boron quantification upon their linking to the neutron capture compound by measuring the MRI signal intensity enhancement, which is directly proportional to the local concentration of the Gd-containing probe. Interestingly, a given cell can be visualized by MRI when the number of Gd^{3+} complexes is of the order of 10^8–10^9 per cell (i.e., the same concentration of ^{10}B atoms needed to provide an effective BNCT treatment). Different types of dual probes, containing both a paramagnetic ion for MRI detection and ^{10}B atoms for BNCT, have been reported in the last ten years. Unfortunately, the presence of the MRI probe caused a dramatic change in their

Figure 3.5.6 Structure of (A) Gd-DTPA-BPA and (B) AT101.

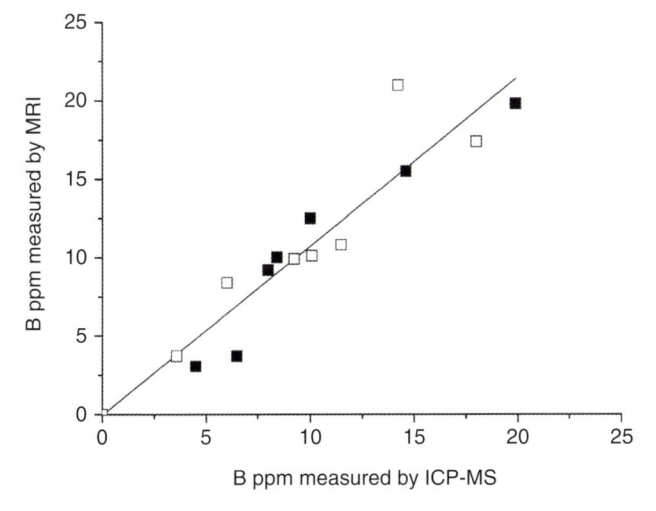

Figure 3.5.7 Correlation between intracellular boron concentrations measured by ICP-MS and MRI. Boron concentrations were measured in B16 murine melanoma cells (■) and HepG2 human hepatocarcinoma cells (□) after 16 h of incubation at different Gd-B-L/LDL particle concentrations. *Source*: Adapted from Geninatti Crich S. et al. (2011), fig. 2, p. 8482.

biodistribution. For example, the intratumor boron concentration of BPA conjugated to the Gd-DTPA complex (Figure 3.5.6a) was significantly reduced with respect to BPA alone, as a consequence of its decreased affinity for BPA receptors [33]. This problem can be overcome by using nanosized delivery agents (i.e., liposomes, micelles, polymers, proteins, etc.) able to simultaneously deliver both BNCT and MRI agents without any alteration of the nanoparticle biodistribution and interaction with specific receptors. In this context, a dual MRI–BNCT probe, containing a carborane linked to a Gd-DOTA monoamide complex for MRI and an aliphatic chain (AT101, Figure 3.5.6b) for the binding to a biological nanocarrier represented by low-density lipoproteins (LDLs), has been recently reported [34].

LDLs were exploited as nanosized carriers for highly proliferating tumor cells that overexpress LDL receptors, namely human hepatoma (HepG2), murine melanoma (B16), and human glioblastoma (U87). Cellular labeling experiments were performed, and the MRI assessment of the amount of boron taken up by tumor cells was validated by ICP-MS measurements (Figure 3.5.7).

BNCT was performed inside the thermal column of the TRIGA Mark II reactor at the University of Pavia on B16 melanoma tumor-bearing mice after administration of the

B/Gd-containing agent. An intratumor boron concentration of more than 30 ppm was measured by MRI 3–4 h after boron administration (Figure 3.5.8). BNCT treatment resulted in a dramatic reduction in tumor growth over that observed in untreated animals [35].

The same adduct, LDL/AT101, was then used for the MRI-guided treatment of lung metastases generated by intravenous (i.v.) injection of a Her2+ breast cancer cell line (i.e., TUBO) in BALB/c mice (Figure 3.5.8) and transgenic EML4-ALK mice that were used as the primary tumor model [36]. Tumor masses of boron-treated mice increased markedly slower than those of the control group. Other synthetic strategies have been accomplished to insert a cholesterol moiety, shorten the preparation by using the hydroboration reaction, or introduce a triazole unit via a Huisgen reaction [37–39].

3.5.2.3.2 MRS Spectroscopy

^{19}F MRI is a useful method to investigate the pharmacokinetics of fluorinated drugs. The ^{19}F nucleus has a 100% abundance, and its NMR sensitivity is 83% of that of ^{1}H (with constant noise). There is a negligible endogenous ^{19}F MRI signal from the body, as the physiological concentration of detectable mobile fluorine is below the detection limit (usually, less than 10^{-3} µmol/g wet tissue weight). This lack of background signal provides ^{19}F MRI with a potentially extremely high contrast-to-noise ratio and specificity, if a fluorinated compound can be introduced as an exogenous contrast agent [40]. *In vivo* experiments using the C6 rat glioma model demonstrated that ^{19}F MRI in combination with ^{1}H MRI can selectively map the biodistribution of BPA labeled with a ^{19}F atom (^{19}F-BPA). Furthermore, correlation between the results obtained by ^{19}F MRI on rat brain and ^{19}F MRS on blood samples showed the maximum ^{19}F-BPA uptake in the C6 glioma model at 2.5 h after infusion, determining the optimal irradiation time. The improved ^{19}F-BPA–fructose complex uptake in C6 tumor-bearing rats after L-DOPA pretreatment was also observed using ^{19}F MRI [41].

3.5.2.3.3 ^{10}B and ^{11}B NMR

Both natural boron isotopes ^{11}B (80% natural abundance) and ^{10}B (20% natural abundance) are detectable by NMR. ^{11}B has spin 3/2, a quadrupole moment of 4.06 fm^2, and a relative gyromagnetic ratio of 0.32, whereas ^{10}B has spin 3, a quadrupole moment of 8.46 fm^2, and a relative gyromagnetic ratio of 0.107. Since ^{10}B is the isotope active for BNCT, the molecules used for the therapy are ^{10}B enriched, but in consideration of the ^{11}B natural abundance in comparison to ^{10}B isotopes, higher sensitivity and better spectral resolution of ^{11}B should be expected. Otherwise, the unique relaxation properties of spin 3 might cause the T_2 of ^{10}B to be longer than ^{11}B T_2 at the same molecular site. For this reason, both boron isotopes have been proposed in MRI–BNCT applications. One recent study reports the study of metabolism of a BPA–fructose complex or BSH solutions in a clinical trial by ^{10}B NMR [42]. Two patients, who suffered from squamous cell carcinoma of head and neck, were infused with ^{10}B-enriched BNCT agents. Urine samples were periodically collected and analyzed by ^{10}B NMR spectroscopy. The results revealed time-dependent metabolic changes of the administered compounds. BPA–fructose dissociated to the constituents BPA and fructose, and the borate group was partly cleaved from BPA. BSH was partly aggregated to a dimer form, BSSB. These observations were previously reported for cultured cells and animal models and were confirmed in Ref. [43] in human cancer patients. Also, ^{10}B and ^{11}B NMR-based

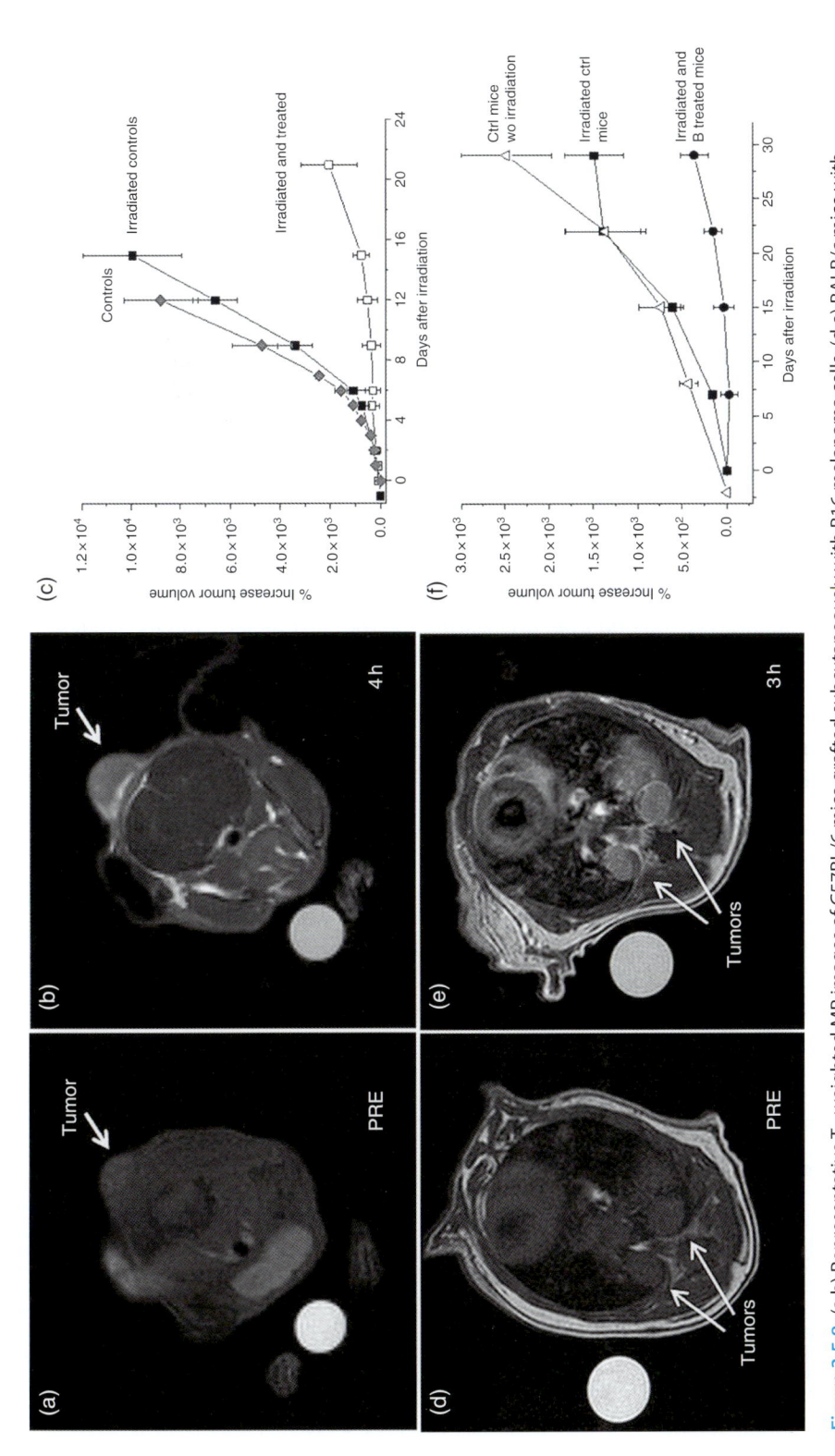

Figure 3.5.8 (a,b) Representative T_1-weighted MR images of C57BL/6 mice grafted subcutaneously with B16 melanoma cells. (d,e) BALB/c mice with pulmonary metastases generated by the injection of breast cancer cells (TUBO). Images were acquired (a,d) before, (b) 4 h after, and (e) 3 h after the administration of AT101/LDL particles. The arrows indicate tumor regions. The graphs show the percentage of tumor volume increase, measured by MRI on (c) B16 tumors and (f) lung metastasis. Error bars indicate the SD. *Source:* (a,b) Adapted from Geninatti Crich *et al.* (2011), fig. 3, p. 8483. (c) Adapted from Geninatti Crich *et al.* (2011), fig. 4, p. 8484. (d,e) Adapted from Alberti *et al.* (2015), fig. 3, p. 746. (f) Adapted from Alberti *et al.* (2015), fig. 5, p. 748.

spectroscopic methods have the advantage of not requiring any boron carrier chemical modification; however, their sensitivity is significantly lower than that of PET, SPECT, and MRI and currently does not allow the quantification of the boron amount accumulated at the tumor site *in vivo* during standard BNCT treatments.

3.5.2.4 Optical Imaging

Optical imaging (OI) uses non-ionizing radiation ranging from ultraviolet to infrared light. In preclinical cancer research, it is primarily exploited for localizing tumors and metastases as well as for monitoring disease progression. One of the fundamental advantages of OI use in biomedical research is the accessibility to interactions between light and tissue and the corresponding photophysical and photochemical processes at the molecular level. Generally, *in vivo* OI is based on the illumination of a target tissue with a light source of a specific wavelength or wavelength range that is able to excite fluorophores. The excitation light has to penetrate through several tissue layers to reach the fluorophores and is, therefore, partially reflected and scattered. Photons are also absorbed by various types of molecules and tissue components [44]. Some examples of fluorescent-labeled potential BNCT agents have been recently described, but few of them have then been used in preclinical or clinical studies. Figure 3.5.9a shows a benzo[b]acridin-12(7H)-one fluorophore bearing a carboranyl moiety, and its biological effectiveness as a BNCT agent in cancer treatment has been reported [45].

The cellular uptake of these novel compounds into U87 human glioblastoma cells was evaluated by boron analysis (ICP-MS) and confocal microscopy. The compound entered the cells and deposited a sufficient amount of boron atoms (2.8×10^{10} ^{10}B atoms per cell) by fulfilling the requirement of low cytotoxicity in the absence of neutron irradiation. Its cellular trafficking was studied by time-lapse confocal fluorescence microscopy in live cells by taking advantage of its fluorescence properties (Figure 3.5.10). Remarkably, carboranylmethylbenzo[b]acridin-12(7H)-one presented considerably high activity in the U87 cells when combined with neutron irradiation.

Another interesting example is represented by the fluorescence-labeled *closo*-dodecaborane lipid (FL-SBL; Figure 3.5.9b) synthesized from (S)-(+)-1,2-isopropylideneglycerol [46] and loaded into highly sensitive DSPC liposomes. No BNCT experiments were accomplished, but a preliminary *in vivo* imaging study of tumor-bearing mice revealed that the FL-SBL-labeled DSPC liposomes were delivered to the tumor tissue

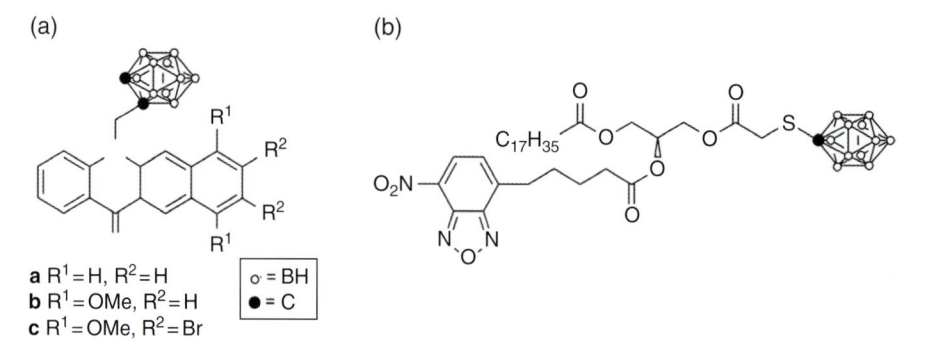

Figure 3.5.9 Structures of (a) benzo[b]acridin-12(7H)-one and (b) FL-SBL.

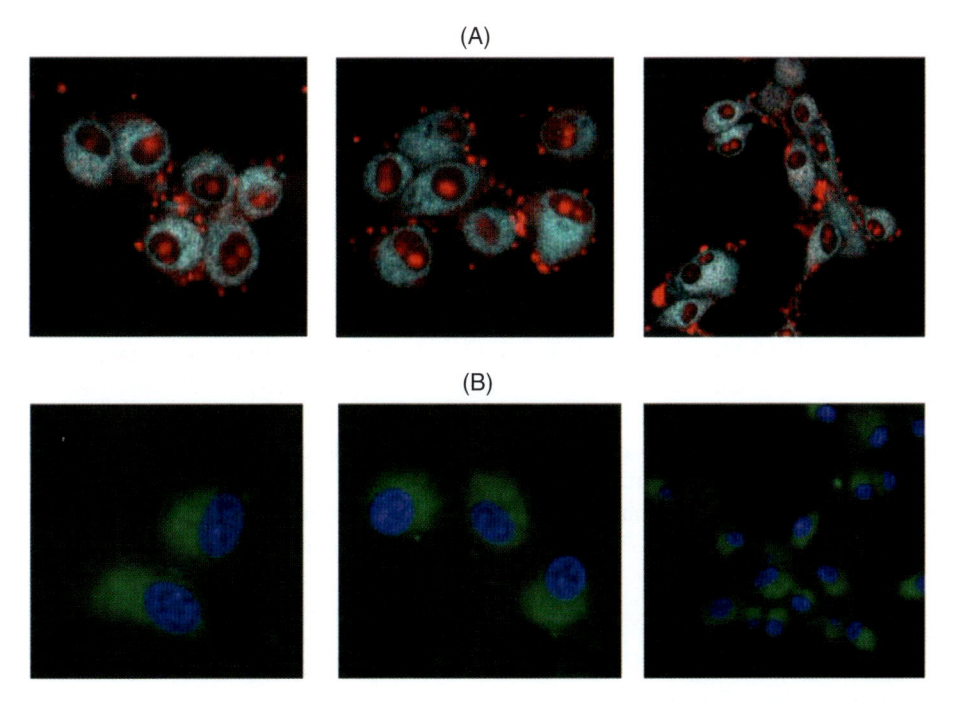

Figure 3.5.10 Fluorescent live-cell imaging of the carboranylmethylbenzo[b]-acridones: (Panel A) Figure 3.5.9a compound **a** and (Panel B) Figure 3.5.9a compound **c** in the U87 cells visualized by time-lapse confocal microscopy. The whole cell or cell nuclei were stained with (Panel A) DHE shown in red or (Panel B) Hoechst 33342 shown in blue. After incubation with the dyes for 10 min, the cells were washed with phosphate-buffered saline (PBS), and the compounds were added at 200 μM. Images in (Panel A) blue or (Panel B) green channels were acquired after a 30-min time period. *Source*: Da Silva *et al.* (2014), fig. 5, p. 5205. Reproduced with permission of The Royal Society of Chemistry.

but not distributed to hypoxic regions. Very recently, Bregadze and coworkers reported the synthesis and *in vitro* studies of boronated Zn(II) phthalocyanine. A multistep reaction sequence starting with cyclotetramerization of 4-(3,5-dimethoxyphenoxy) phthalonitrile in the presence of zinc(II) acetate was exploited; its detailed intracellular distribution and localization were studied using confocal laser scanning. The boronated phthalocyanine was found to accumulate in A549 human lung adenocarcinoma cells. Its maximal cytoplasmic concentration was achieved at an extracellular concentration of $32 \pm 3\,\mu M$ [47].

3.5.2.5 Boron Microdistribution

Knowledge concerning ^{10}B microdistribution in normal and tumor tissues is of critical importance for BNCT's success. Many analytical approaches, such as prompt γ-ray spectroscopy and atomic emission spectroscopy, yield average boron concentrations from gross tissue specimens and, therefore, have limited utility for studying the microdistribution of boron delivery agents. Conversely, the double-focusing magnetic sector-dynamic secondary ion mass spectrometer (SIMS-based ion microscopy) is particularly well suited for boron microlocalization studies [48]. The technique maintains the spatial integrity of the analyte sputtered from the surface of the sample, producing images of isotopic distribution, which can be related to tissue histology with a resolution comparable to that of a high-quality light microscope. The boron analysis by the ion microscope is not affected by elemental speciation (the free or bound chemical form of boron) within the biological matrix. In this context can be placed the interesting approach of Chandra and coworkers [49], where SIMS was used for the assessment of different boron pools in the nucleus and cytoplasm of human GBM (glioblastoma multiforme) cells following exposure to BPA and to the new agent, *cis*-ABCPC (1-amino-3-boronocyclopentanecarboxylic acid). Moreover, α-autoradiography has generally been used to detect ^{10}B intracellular microdistribution. In fact, autoradiography images generated by the individual tracks of the α- and lithium particles provide fundamental information on the spatial distribution and localized dosimetry of boron in tissues. Using this technique, Portu *et al.* [50] explored potential changes in boron microdistribution in a hamster cheek pouch oral cancer model, using neutron capture autoradiography, to evaluate if the distribution of GB-10 (decahydrodecaborate) was altered by prior application of BPA–BNCT in sequential BNCT protocols.

3.5.3 Targeted BNCT: Personalization of *in vivo* Boron-Selective Distribution

The development of new, more selective boron delivery agents is the other fundamental need for the future progress of BNCT [51]. In fact, although the currently clinically used boron delivery agents (BSH and BPA) have shown good therapeutic results, they lack specificity for tumors with respect to healthy cells. This is one of the limiting factors that prevent the diffusion of this therapy in mainstream medicine. Thus, one of the most effective ways to optimize BNCT is to increase the tumor–normal tissue ratio achieved with BPA and BSH. In the past decade, much progress has been made in phenotyping human tumors. Knowledge of cancer at the molecular level has greatly increased, and this has prompted new targeted therapies for cancer. Precise therapeutic protocols exploit changing cell membrane receptor and transporter expression, enzymatic activity, and increased cell metabolism (Figure 3.5.11) [52]. This section will describe different strategies used in the last ten years to improve tumor boron-selective delivery, dividing the different boron carriers into small and nanosized, respectively, and highlighting advantages and disadvantages of their use.

3.5.3.1 Small-Sized Boron Carriers

There are many examples reported in the literature of small boron carriers targeted to specific receptors or transporters expressed on tumor cells. The use of delivery agents with a size <10 nm has the advantage of a more accurate molecular characterization of the carrier and an easier extravasation and perfusion of the tumor mass. However, it is obvious that the smaller size prevents them from carrying huge amounts of boron, thus reducing their efficiency and requiring the exploitation of high-capacity transporters and receptors for their cell internalization.

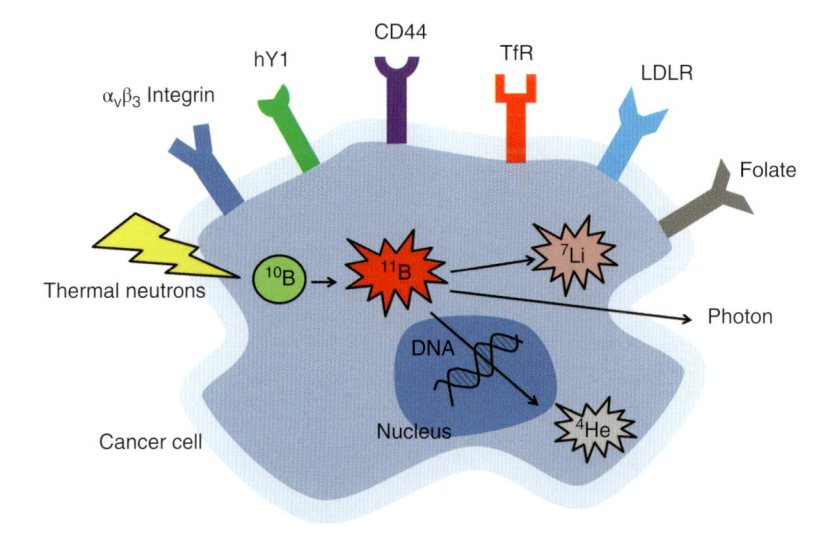

Figure 3.5.11 Schematic representation of tumor cell destruction by BNCT using targeted boron carriers.

Boron carriers (summarized in Table 3.5.1) are functionalized with groups recognizing specifically the target epitope, such as peptides [53,54], nucleosides [55,56], carbohydrates [57], unnatural amino acids [58,59], or other molecules of biological interest [60]. Unfortunately, only a few of them have been used for *in vivo* tumor irradiation [55,56].

The use of peptides is particularly suitable for these applications as a consequence of their high selectivity for transmembrane receptors overexpressed on tumor cells. Due to their specific expression on proliferating endothelial and tumor cells of various origins, integrins $\alpha_v\beta_3$ are one of the most attractive targets for antitumor drug delivery and tumor imaging [61]. Integrins regulate a diverse array of cellular functions related to progression, angiogenesis, and metastasis in the tumor microenvironment. On this basis, Dubey and coworkers [53] developed a promising trifunctional theranostic agent, DC-1 (Figure 3.5.12), comprising a nonpeptidic integrin ligand, a monomethine cyanine dye, and a carborane dendritic wedge for BNCT, allowing a very high boron content (60 boron atoms/molecule).

A scaffold-based L-lysine was used to combine all three modalities in one construct. Another promising approach exploited by Ahrens *et al.* [54] is represented by peptide ligands selectively targeting distinct G protein-coupled receptors that are highly expressed in tumor tissue. Analogs of neuropeptide Y (NPY) bind and activate the human Y1 receptor subtype (hY1 receptor), which is found in 90% of breast cancer tissue and in all breast-cancer-derived metastases. Each peptide carries 30 boron atoms. The boron uptake in human Y1R-expressing cells was more than tenfold higher than the threshold limit value of 10^9 boron atoms per cell, which is necessary for successful BNCT treatments.

An alternative molecular target is thymidine kinase-1 (TK1), a cytosolic deoxynucleoside kinase whose activity is only found in proliferating cells; it is distributed and expressed in a wide variety of malignant tumors [62]. Barth and coworkers [55,56]

Table 3.5.1 Summary of novel small-sized boron delivery systems for BNCT

Boron carrier	Receptor	Tumor model	Boron ppm in tumor	Tumor–organ ratio	BNCT	Ref.
Peptide: RGD-mimetic	αvβ3 integrins	WM115 and MCF7 cells	ND	Tumor/liver = 3 Tumor/lung = 6	No	53
Neuropeptide	hY1 and hY2	HEK293_hY1R_EYFP; HEK293_HA_hY2R_EYFP cells	17.8×10^9 and 11.3×10^9 boron atoms per cell	ND	No	54
Carboranyl nucleoside	Thymidine kinase	F98 glioma-bearing rats	27.6	Tumor/brain = 10.6	Yes	55 56
5-thio-D-glucopyranose	Glucose transporters	SK-Hep1 human hepatocellular carcinoma	1×10^{11} boron atoms per cell	ND	No	57
Unnatural amino acids (AA)	AA transporters	B16 melanoma F98 rat glioma	31.7	Tumor/brain = 15.1	No	58 59

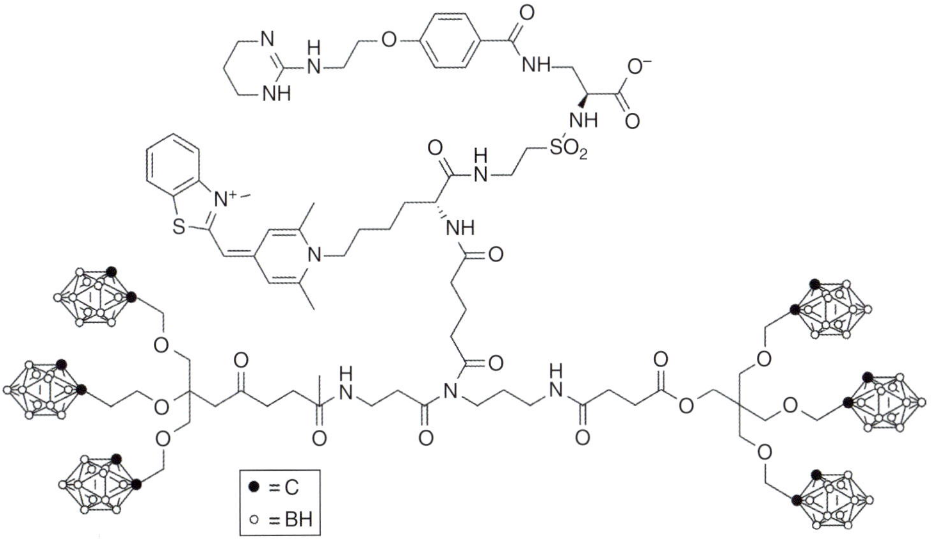

Figure 3.5.12 Schematic representation of the trifunctional agent DC-1.

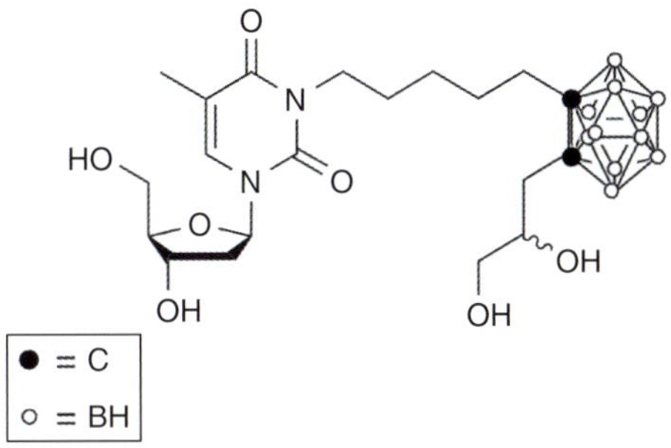

Figure 3.5.13 Schematic representation of 3-[5-{2-(2,3-dihydroxyprop-1-yl)-o-carboran-1-yl} pentan-1-yl] thymidine (N5–2OH).

exploited TK1 substrates, which are selectively trapped in tumor cells after selective phosphorylation, as boron delivery agents for brain tumors (Figure 3.5.13). BNCT has been performed on mice bearing L929 TK1(+) tumors, which had received thymidine targeted compound (by intratumor injection). They showed an average 15-fold reduction in mean tumor volume on day 30 after implantation. In contrast, animals bearing L929 tumors not expressing TK1 showed modest reductions in tumor volumes, which were not significantly different from those of irradiated animals bearing TK1(+) tumors without boron treatment. The reduction in tumor volumes correlated with a marked decrease in the number of proliferating cells. The last example is a new class of boronated unnatural cyclic amino acids [58,59] showing a good selectivity toward tumors in animal and cell culture models, higher than that of currently used agents in clinical BNCT. One of these amino acids, ABCPC, has shown a tumor-to-blood ratio of 8 and a tumor-to–normal brain ratio of nearly 21 in a melanoma-bearing mouse model.

3.5.3.2 Nanosized Boron Carriers

Nanosized carriers, with a diameter ranging from 50 to 200 nm, can be loaded with a large amount of boron atoms to be specifically delivered into tumors, thus avoiding intratumor injection, which has many disadvantages. In fact, the most appropriate Gd administration is i.v., because it allows the accumulation in tumors through active (receptor-mediated endocytosis) or passive (enhanced permeation and retention [EPR]) targeting while avoiding systemic distribution in normal tissues. The literature on the use of nanoparticles to selectively deliver drugs, vaccines, and imaging agents has exponentially increased in recent years, and they appear to be good candidates for the specific delivery of boron agents to tumor cells.

Liposomes are the most-used nanosized boron carriers [63]. They are spherical vesicles composed of a unilamellar lipid bilayer and are able to transport their contents encapsulated in their internal aqueous cavity and/or intercalated in the phospholipid bilayer, thus allowing the transport of both hydrophilic and hydrophobic boron-containing compounds to tumors (Figure 3.5.14).

A nonspecific class of liposomes, *long-time circulating polyethylene glycol (PEG)-coated liposomes*, is able to accumulate in solid tumors as a consequence of microvascular permeability and defective lymphatic drainage. The extent of passive extravasation directly depends on the prolonged residence time of liposomes in the bloodstream. In the absence of liposome active targeting, the boron-containing molecule must first be released in the tumor extracellular matrix in order to allow its diffusion into cells. Alternatively, the active liposomes' targeting to tumor cells is obviously a promising strategy to improve drug delivery to tumors. Table 3.5.2 shows some examples of liposomes functionalized with specific ligands that are able to bind and be internalized through receptors expressed on tumor vasculature or on tumor cells [38,64–67]. In all the formulations herein reported, a significant increase of tumor-to–healthy tissue boron ratio was observed. Alternatively, a widely used

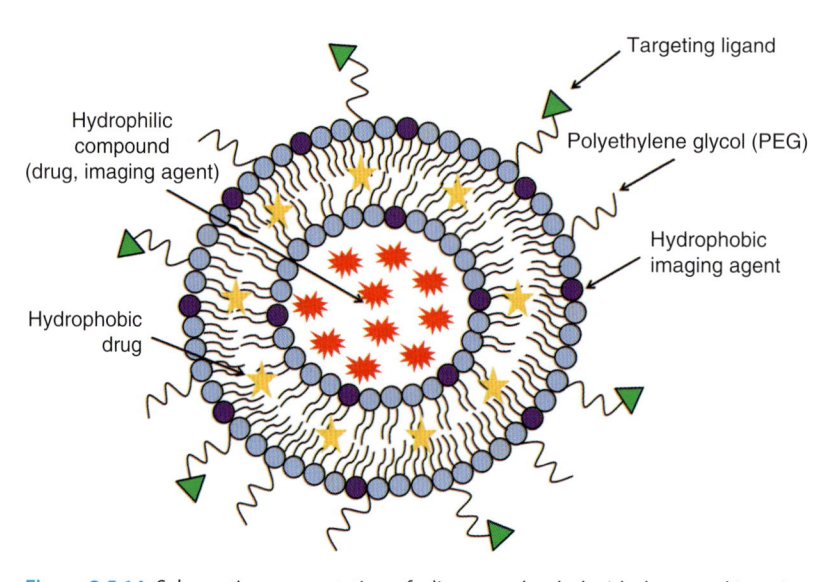

Figure 3.5.14 Schematic representation of a liposome loaded with drugs and imaging agents.

Table 3.5.2 Summary of novel nanosized boron delivery systems for BNCT

Boron carrier	Receptor	Tumor model	Boron ppm in tumor	BNCT	Ref.
Hyaluronan-targeted liposomes	CD44	AB22 mesothelioma-bearing mice	10	Yes	64
Transferrin-targeted liposomes	TfR1	Colon 26 tumor-bearing mice F98 glioma tumor-bearing rats	35/82	Yes/no	65,66
Nontargeted liposomes	Passive targeting	Hamster cheek pouch oral cancer model	67	Yes	67
Folate-targeted liposomes	Folate receptor	Human ovarian cancer cells (IGROV-1)	38	Yes	38
Low-density lipoproteins (LDLs)	LDL receptor	B16 melanoma and lung metastasis	30.5/41	Yes	35,36

natural boron nanocarrier is made up of LDLs that accumulate in tumor cells characterized by an upregulation of LDL receptors together with a good selectivity between tumor and healthy tissues [35,36]. LDLs have been loaded with a Boron–Gd probe to perform imaging-guided BNCT. *In vivo* MR image acquisition showed that the amount of boron taken up in the tumor region was 30.5 and 41 ppm for mouse melanoma and lung metastasis generated by mammary adenocarcinoma, respectively. After neutron irradiation, tumor growth was followed for 30–40 days by MRI. Tumor masses of boron-treated mice increased markedly slower than those of the control group. Finally, a recent paper by Tietze *et al.* [68] describes a promising strategy to deliver boron in tumor tissues based on the use of magnetically directed nanoparticles. This implies the use of i.v.-administered superparamagnetic iron-oxide nanoparticles (SPIONs) coated with boron-containing molecules, which are attracted by an external magnetic field that is directed to the tumor tissue. This leads to an enhanced local enrichment of particles in the area of interest due to bypassing the metabolic barriers, RES uptake, and dilution effects.

3.5.4 Combination of BNCT with Other Conventional and Nonconventional Therapies

In recent years, much attention has been devoted to combinations of different therapeutic modalities, as possible strategies to treat cancer. This relies on the evidence that although a large majority of chemotherapeutic protocols and radiotherapies can considerably reduce tumor masses, they often fail in causing their complete regression as shown by a high number of tumor recurrence cases. Moreover, the time-dependent development of chemoresistance and radioresistance by a minor cell population within the tumor and the nonspecific toxicity toward normal cells are the other major limitations of standard therapies.

3.5.4.1 Chemotherapy

Although the concurrent administration of chemotherapy and radiotherapy is a standard protocol for the treatment of many cancers, few examples about the combination of BNCT with chemotherapy have been reported until now. Carborane-loaded nanoparticles are good candidates to achieve this goal, because the formed hydrophobic core can be efficiently loaded with hydrophobic drugs. Also, owing to the presence of hydridic B–H units, carboranes are involved in particular kinds of interactions, such as dihydrogen bonding [69,70]. As a result, carboranes are able to interact with anticancer drugs to increase loading contents and provide synergistic effect through the bond of B–H···H–X, where X is N, O, or S. In this context, Xiong et al. [71] proposed poly(ethylene glycol)-b-poly(L-lactide-co-2-methyl-2(2-dicarba-closo-dodecarborane) propyloxycarbonylpropyne carbonate) (PLMB) that self-assembled into nanoparticles in water as a consequence of the carborane hydrophobicity. PLMB nanoparticles showed prolonged blood circulation time and enhanced accumulation of boron species at the tumor site. Owing to the dihydrogen bonds between carborane and DOX, DOX was readily encapsulated, and DOX@PLMB nanoparticles exhibited a pH-dependent release behavior *in vitro*. Neutron irradiation showed the highest therapeutic efficacy compared to individual chemotherapy or BNCT *in vivo*.

3.5.4.2 Photodynamic Therapy (PDT)

Other promising results have been obtained by the combination of PDT and BNCT to eliminate unresectable glioblastoma cells that invade adjacent normal brain tissue. PDT, similarly to BNCT, is a binary modality for tumor treatment that relies on the selective accumulation of a sensitizer within tumor tissue, followed by its activation upon irradiation with red light [72–74]. A PDT and BNCT combination using a single drug has several advantages, including increased therapeutic effect due to the targeting of different cellular components and/or mechanisms of tumor cell destruction. Conjugation of porphyrins with boron polyhedra results in compounds comprising the photosensitizing porphyrin moiety and the boron cluster capable of capturing neutrons. This type of compound is promising not only for PDT, but also for BNCT.

3.5.4.3 Standard Radiotherapy

To the authors' knowledge, all the reported studies about the combination of BNCT with X-ray external beam radiotherapy (XRT) are regarding brain tumors, in particular GBM. This is due to the still very poor prognosis of GBM patients despite the continuous improvements in available treatments, primarily based on adjuvant radiotherapy and adjuvant chemotherapy following initial surgery. For GBM patients, postoperative fractionated XRT is often considered. In fact, prolonged overall survival in selected patients has been reported when radiosurgery, stereotactic radiotherapy, proton beam radiation, or other conformal radiotherapies, or the addition of a radiosurgical boost prior to external radiotherapy, have been administered in combination with the primary resection [75,76]. Regardless, fractionated radiotherapy cannot control GBM because of the limitations in dose delivery due to normal brain tolerance. In this context, BNCT may improve local control and survival after the initial treatment of the malignancy thanks to its cell-selective feature.

The first experimental study of subsequent administration of BNCT and a boost of X-ray irradiation was reported by Barth and coworkers [77]. The study was based on the premise that the high-LET radiations associated with BNCT would kill both hypoxic and oxygenated cells, thus reducing the tumor burden and improving the cytotoxic effect of the following X-ray irradiation. The F98 glioma model and MRA 27 human melanoma model of metastatic brain tumor were induced in nude rats. The ^{10}B-carriers were a combination of BSH and BPA or BPA alone, respectively. Two routes of ^{10}B administration were tested, i.v. and intracarotid (i.c.) injection, the latter being more effective and bringing almost 50% higher values of ^{10}B concentrations. Neutron irradiations were performed at the Brookhaven National Laboratory Medical Research Reactor 14 days after tumor cells implantation and 2.5 h after the end of ^{10}B infusion. Then, approximately 7–10 days later, subsets of rats treated with BNCT were irradiated with a total dose of 15 Gy delivered in 5 Gy daily fractions using 6 MV photons from a conventional linear accelerator. The outcomes reported a significant therapeutic gain obtained by the combination of BNCT with the X-ray boost in comparison with control groups of untreated, photon-alone, or neutron-alone irradiated animals. On the base of this preclinical work, the trial conducted at Osaka Medical College [78] was designed, involving a cohort of 21 patients affected by newly diagnosed GBM who underwent either BNCT alone (protocol 1) or BNCT followed by fractionated XRT of 20 to 30 Gy (protocol 2). The study exploited the epithermal neutron beam at the Kyoto University Research Reactor to improve the distribution of thermal neutrons in deep-seated tumors without craniotomy. To deliver ^{10}B in the malignant cells, both BSH and BPA were used to improve ^{10}B accumulation in all the tumor cells thanks to the different accumulation mechanisms. In protocol 2, XRT was started 2 weeks after neutron irradiation (single-dose BNCT), using a 2 Gy daily fraction regimen for a total of 20 to 30 Gy. The goal of this radiotherapy boost was to reduce the chances of local recurrence, by compensating the possible heterogeneous distribution of boron compounds (in particular, in microscopic nodules infiltrating the healthy tissue surrounding the tumor bulk) as well as the limited penetration of low-energy neutrons, especially in the deepest part of the malignancy. The whole cohort of patients treated with BNCT ($n = 21$) had a median survival time (MST) of 15.6 months after diagnosis. This number must be compared with the MST of the control group ($n = 27$), equal to 10.3 months. After careful statistical analysis and comparisons among risk groups, the authors concluded that the BNCT group actually shows a longer survival rate at the cost of limited toxicities (the most notable one being radio-induced brain necrosis). Also, the second study about the combination of BNCT with external beam X-ray irradiation addressed newly diagnosed GBM tumors, and was carried out at the University of Tsukuba and Tokushima University, using the reactor-based neutron beams of Japanese Research Reactor-4 (JRR-4) of the Japanese Atomic Energy Agency (JAEA) [79]. The aim of the study was to investigate the survival benefits, safety, and dose distribution in tumor and normal tissue of GBM patients who underwent either an intraoperative NCT (IO-NCT, protocol 1, $n = 7$) or an external beam NCT (EB-NCT, protocol 2, $n = 8$). The boron vectors were: BSH alone for protocol 1, and the same administration of BSH for protocol 2 in combination with BPA. In both protocols, BNCT was administered in a single fraction, using an epithermal or an epithermal-thermal mixed beam. Due to a deeper location at the time of irradiation of the tumors enrolled in protocol 2, additional conventional photon irradiation (at a total dose of 30 Gy in 15 fractions or 30.6 Gy in 17 fractions) was delivered to high-intensity areas of T_2-weighted MR images to compensate for the differences in dose distributions between the two protocols. The results indicated a median overall survival (OS) and a median time to tumor progression (TTP) of 25.7 and 11.9 months, respectively, considering the whole cohort of patients. The one- and two-year survival rates were 80 and 53.3%, respectively. No serious toxicity related to the boron vector was recorded as well as no serious adverse events. Comparing the two-year survival rate of IO-NCT (42.9%) to that of a highly improved (two boron drugs, post-neutron XRT boost) but far less invasive EB-NCT protocol (62.5%), the authors concluded that their results suggest that postoperative EB-NCT combined with XRT boost is effective for newly diagnosed glioblastoma, with the further advantage of almost null invasiveness.

3.5.5 Conclusions

In this chapter, different approaches to further develop B-containing agents have been described, with the final goal of increasing BNCT competitiveness with respect to other routine tumor treatment protocols. The optimization of neutron irradiation time and the estimation of the delivered radiation dose can be performed by measuring local B concentration in the tumor and in the surrounding tissues by high-resolution imaging techniques in real time just before and during the neutron irradiation. The personalization of neutron therapy obtained by the combination of imaging and therapy and the development of more specifically targeted B carriers open new routes to bridge the gap between the research and the clinical use of this alternative radiotherapy. Finally, the combination of BNCT with other conventional and nonconventional therapies can increase the probability of reducing tumor recurrence.

References

1 Locher GL. Biological effects and therapeutic possibilities of neutrons. Am J Roentgenol Radium Ther. 1936; 36(1):1–13.
2 Soyland C, Hassifjel SP. Survival of human lung epithelial cells following in vitro alpha-particle irradiation with absolute determination of the number of alpha-particle traversals of individual cells. Int J Rad Biol. 2000; 76:1315–1322.
3 Hosmane SN, editor. Boron Science: New Technologies and Applications. Boca Raton, FL: CRC Press, 2012.
4 Gao SM, Hosmane NS. Dendrimer and nanostructure supported carboranes and metallacarboranes: an account. Russ Chem Bull Intl Ed. 2014; 63:788–810.
5 Farr LE, Sweet WH, Robertson JS, Foster CG, Locksley HB, Sutherland DL, et al. Neutron capture therapy with boron in the treatment of glioblastoma multiforme. Am J Roent Ther Nucl Med. 1954; 71:279–293.
6 Asbury AK, Ojeman RG, Nielsen SL, Sweet WH. Neuropathological study of fourteen cases of malignant brain tumor treated by boron-10 slow neutron capture radiation. J Neuropathol Exp Neurol. 1972; 31(2):278–303.
7 Hatanaka H. Clinical results of boron neutron capture therapy. Basic Life Sci. 1990; 54(15):15–21.

8. Mishima Y, Ichihashi M, Hatta S, Honda C, Yamamura K, Nakagawa T, et al. First human clinical trial of melanoma neutron capture. Diagnosis and therapy. Strahelenther Onkol. 1989; 165(2–3):251–254.
9. Sauerwein WA, Wittig A, Moss R, Nakagawa Y, editors. Neutron Capture Therapy: Principles and Applications. Berlin: Springer-Verlag, 2012.
10. Tanaka H, Y Sakurai, M Suzuki, S Masunaga, T Mitsumoto, K Fujita, et al. Experimental verification of beam characteristics for cyclotron-based epithermal neutron source (C-BENS). Appl Radiat Isot. 2011; 69(12):1642–1645.
11. Cartelli D, Capoulat ME, Bergueiro J, Gagetti L, Suàrez Anzorena M, del Grosso MF, et al. Present status of accelerator-based BNCT: focus on developments in Argentina. Appl Radiat Isot. 2015; 106(12):18–21.
12. Green S. Development in accelerator based boron neutron capture therapy. Radiat Phys Chem. 1998; 51(4–6):561–569.
13. Kumada H, Kurihara T, Yoshioka M, Kobayashi H, Matsumoto H, Sugano T, et al. Development of beryllium-based neutron target system with three-layer structure for accelerator-based neutron source for boron neutron capture therapy. Appl Radiat Isot. 2015; 106(12):78–83.
14. Sorokin I, Taskaev S. A new concept of a vacuum insulation tandem accelerator. Appl Radiat Isot. 2015; 106(12):101–103.
15. Pisent A, Colautti P, Esposito J, De Nardo L, Conte V, Agosteo D, et al. Progress on the accelerator based SPES-BNCT project at INFN Legnaro. J Phys Conf Ser. 2006; 41:391–399.
16. Wittig A, Michel J, Moss RL, Stecher-Rasmussen F, Arlinghaus HF, Bendel P, et al. Boron analysis and boron imaging in biological materials for boron neutron capture therapy (BNCT). Crit Rev Oncol Hematol. 2008; 68:66–90.
17. Salt C, Lennox AJ, Takagaki M, Maguire JA, Hosmane NS. Boron and gadolinium neutron capture therapy. Russ Chem Bull Intl Ed. 2004; 53:1871–1888.
18. Savolainen S, Kortesniemi M, Timonen M, Reijonen V, Kuusela L, Uusi-Simola J, et al. Boron neutron capture therapy (BNCT) in Finland: technological and physical prospects after 20 years of experience. Phys Med. 2013; 29:233–248.
19. Adam MJ, Wilbur DS. Radiohalogens for imaging and therapy. Chem Soc Rev. 2005;34(2):153–163.
20. Fowler JS, Wolf AP. Working against time: rapid radiotracer synthesis and imaging the human brain. Acc Chem Res. 1997; 30(4):181–188.
21. Ishiwata K, Ido T, Mejia AA, Ichihashi M, Mishima Y. Synthesis and radiation dosimetry of 4-boron-2-(18F)fluoro-D,L-phenylalanine: a target compound for PET and boron neutron capture therapy. Appl Radiat Isot. 1991; 42(4):325–328.
22. Evangelista L, Jory G, Martini D, Sotti G. Boron neutron capture therapy and 18F-labelled borophenylalanine positron emission tomography: a critical and clinical overview of the literature. Appl Radiat Isot. 2013; 74:91–101.
23. Aihara T, Hiratsuka J, Morita N, Uno M, Sakurai Y, Maruhashi A, et al. First clinical case of boron neutron capture therapy for head and neck malignancies using F-18-BPA PET. Head Neck. 2006; 28:850–855.
24. Ariyoshi Y, Miyatake SI, Kumura Y, Shimahara T, Kawabata S, Nagata K, et al. Boron neuron capture therapy using epithermal neutrons for recurrent cancer in the oral cavity and cervical lymph node metastasis. Oncol. Rep. 2007; 18:861–866.
25. Tani H, Kurihara H, Hiroi K, Honda N, Kono Y, Arai Y, et al. Correlation of (18)F-BPA and (18)F-FDG uptake in head and neck cancers. Radiother Oncol. 2014; 113:193–197.

26. Havu-Auren K, Kiiski J, Lehtio K, Eskola O, Kulvik M, Vuorinen V, et al. Uptake of 4-borono-2- F-18 fluoro-L-phenylalanine in sporadic and neurofibromatosis 2-related schwannoma and meningioma studied with PET. Eur J Nucl Med Mol I. 2007; 34:87–94.
27. Koivunoro H, Hippelainen E, Auterinen I, Kankaanranta L, Kulvik M, Laakso J, et al. Biokinetic analysis of tissue boron (10B) concentrations of glioma patients treated with BNCT in Finland. Appl Radiat Isot. 2015; 106:189–194.
28. Menichetti L, Cionini L, Sauerwein WA, Altieri S, Solin O, Minn H, et al. Positron emission tomography and (18F)BPA: a perspective application to assess tumor extraction of boron in BNCT. Appl Radiat Isot. 2009; 67:S351–S354.
29. Iguchi Y, Michiue H, Kitamatsu M, Hayashi Y, Takenaka F, Nishiki T, et al. Tumor-specific delivery of BSH-3R for boron neutron capture therapy and positron emission tomography imaging in a mouse brain tumor model. Biomaterials. 2015; 56:10–17.
30. Hale B, Katabuchi T, Hayashizaki N, Terada K, Igashira M, Kobayashi T. Feasibility study of SPECT system for online dosimetry imaging in boron neutron capture therapy. Appl Radiat Isot. 2014; 88:167–170.
31. Manabe M, Sato F, Murata I. Basic detection property of an arrya-type CdTe detector for BNCT-SPECT. Measurement and analysis of anti-coincidence events. Appl Radiat Isot. 2015; doi:10.1016/j.apradiso.2015.11.003.
32. Winkler A, Koivunoro H, Reijonen V, Auterinen I, Savolainen S. Prompt gamma and neutron detection in BNCT utilizing a CdTe detector. Appl Radiat Isot. 2015; 106:139–144.
33. Takahashi K, Nakamura H, Furumoto S, Yamamoto K, Fukuda H, Matsumura A, et al. Synthesis and in vivo biodistribution of BPA–Gd–DTPA complex as a potential MRI contrast carrier for neutron capture therapy. Bioorg Med Chem. 2005; 13(3):735–743.
34. Aime S, Barge A, Crivello A, Deagostino A, Gobetto R, Nervi C, et al. Synthesis of Gd(III)-C-palmitamidomethyl-C'-DOTAMA-C-6-o-carborane: a new dual agent for innovative MRI/BNCT applications. Org Biomol Chem. 2008; 6:4460–4466.
35. Geninatti Crich S, Alberti D, Szabo I, Deagostino A, Toppino A, Barge A, et al. MRI-guided neutron capture therapy by use of a dual gadolinium/boron agent targeted at tumour cells through upregulated low-density lipoprotein transporters. Chem Eur J. 2011; 17:8479–8486.
36. Alberti D, Protti N, Toppino A, Deagostino A, Lanzardo S, Bortolussi S, et al. A theranostic approach based on the use of a dual boron/Gd agent to improve the efficacy of boron neutron capture therapy in the lung cancer treatment. Nanomedicine. 2015; 11(3):741–750.
37. Toppino A, Bova ME, Geninatti-Crich S, Alberti D, Diana E, Barge A, et al. A carborane-derivative "click" reaction under heterogeneous conditions for the synthesis of a promising lipophilic MRI/GdBNCT agent. Chem Eur J. 2013; 19:720–727.
38. Alberti D, Toppino A, Geninatti-Crich S, Meraldi C, Prandi C, Protti N, et al. Synthesis of a carborane-containing cholesterol derivative and evaluation as a potential dual agent for MRI/BNCT applications. Org Biomol Chem. 2014; 12:2457–2467.
39. Boggio P, Toppino A, Geninatti-Crich S, Alberti D, Marabello D, Medana C, et al. The hydroboration reaction as a key for a straightforward synthesis of new MRI-NCT agents. Org Biomol Chem. 2015; 13:3288–3297.
40. Ruiz-Cabello J, Barnett BP, Bottomley PA, Bulte JW. Fluorine (19F) MRS and MRI in biomedicine. NMR Biomed. 2011; 24(2):114–129.

41 Fasano F, Campanella R, Migneco LM, Pastore FS, Maraviglia B. In vivo F-19 MRI and F-19 MRS of F-19-labelled borophenylalanine-fructose complex on a C6 rat glioma model to optimize boron neutron capture therapy (BNCT). Phys. Med. Biol. 2008; 53(23):6979–6989.

42 Bendel P, Wittig A, Basilico F, Mauri PL, Sauerwein W. Metabolism of borono-phenylalanine–fructose complex (BPA–fr) and borocaptate sodium (BSH) in cancer patients—results from EORTC trial 11001. J Pharm Biomed Anal. 2010; 51(1):284–287.

43 Elhanati G, Salomon Y, Bendel P. Significant differences in the retention of borocaptate monomer (BSH) and dimer (BSSB) in malignant cells. Cancer Lett. 2001; 172:127–132.

44 Etrych T, Lucas H, Janouskova O, Chytil P, Mueller T, Mader K. Fluorescence optical imaging in anticancer drug delivery. J Control Release. 2016; 226:168–181.

45 Da Silva AF, Seixas RS, Silva AM, Coimbra J, Fernandes AC, Santos JP, *et al.* Synthesis, characterization and biological evaluation of carboranylmethylbenzo[b]acridones as novel agents for boron neutron capture therapy. Org Biomol Chem. 2014; 12(28):5201–5211.

46 Nakamura H, Ueda N, Ban HS, Ueno M, Tachikawa S. Design and synthesis of fluorescence-labeled closo-dodecaborate lipid: its liposome formation and in vivo imaging targeting of tumors for boron neutron capture therapy. Org Biomol Chem. 2012; 10(7):1374–1380.

47 Birsöz B, Efremenko AV, Ignatova AA, Gül A, Feofanov AV, Sivaev IB, *et al.* New highly-boronated Zn(II)-phthalocyanine: synthesis and in vitro study. Biochem Biophys J Neutron Ther Cancer Treat. 2013; 1:8–14.

48 Smith DR, Chandra S, Barth RF, Yang W, Joel DD, Coderre JA. Quantitative imaging and microlocalization of boron-10 in brain tumors and infiltrating tumor cells by SIMS ion microscopy: relevance to neutron capture therapy. Cancer Res. 2001;61(22):8179–8187.

49 Chandra S, Ahmad T, Barth RF, Kabalka GW. Quantitative evaluation of boron neutron capture therapy (BNCT) drugs for boron delivery and retention at subcellular scale resolution in human glioblastoma cells with imaging secondary ion mass spectrometry (SIMS). J Microsc. 2014; 254(3): 146–156.

50 Portu A, Molinari AJ, Thorp SI, Pozzi EC, Curotto P, Schwint AE, *et al.* Neutron autoradiography to study boron compound microdistribution in an oral cancer model. Int J Radiat Biol. 2015; 91(4):329–335.

51 Hosmane NS. Boron and Gadolinium Neutron Capture Therapy for cancer treatment. Singapore: World Scientific Publishing, 2012.

52 Luderer MJ, de la Puente P, Azab AK. Advancements in tumor targeting strategies for boron neutron capture therapy. Pharm Res. 2015; 32(9):2824–2836.

53 Dubey R, Kushal S, Mollard A, Vojtovich L, Oh P, Levin MD, *et al.* Tumor targeting, trifunctional dendritic wedge. Bioconjug Chem. 2015; 26(1):78–89.

54 Ahrens VM, Frank R, Boehnke S, Schütz CL, Hampel G, Iffland DS, *et al.* Receptor-mediated uptake of boron-rich neuropeptide y analogues for boron neutron capture therapy. ChemMedChem. 2015; 10(1):164–172.

55 Barth RF, Yang W, Wu G, Swindall M, Byun Y, Narayanasamy S, *et al.* Thymidine kinase 1 as a molecular target for boron neutron capture therapy of brain tumors. Proc Natl Acad Sci USA. 2008; 105(45):17493–17497.

56 Barth RF, Yang W, Nakkula RJ, Byun Y, Tjarks W, Wu LC, et al. Evaluation of TK1 targeting carboranyl thymidine analogs as potential delivery agents for neutron capture therapy of brain tumors. Appl Radiat Isot. 2015; 106:251–255.

57 Šnajdr I, Janoušek Z, Takagaki M, Císařová I, Hosmane NS, Kotora M. Alpha (α-) and beta (β-)carboranyl-C-deoxyribosides: syntheses, structures and biological evaluation. Eur J Med Chem. 2014; 83:389–397.

58 Chandra S, Barth RF, Haider SA, Yang W, Huo T, Shaikh AL, et al. Biodistribution and subcellular localization of an unnatural boron-containing amino acid (Cis-ABCPC) by imaging secondary ion mass spectrometry for neutron capture therapy of melanomas and gliomas. PLoS One. 2013; 8(9):e75377.

59 Kabalka GW, Yao ML, Marepally SR, Chandra S. Biological evaluation of boronated unnatural amino acids as new boron carriers. Appl Radiat Isot. 2009; 67(7–8 Suppl.):S374–S379.

60 El-Zaria ME, Ban HS, Nakamura H. Boron-containing protoporphyrin IX derivatives and their modification for boron neutron capture therapy: synthesis, characterization, and comparative in vitro toxicity evaluation. Chemistry. 2010; 16(5):1543–1552.

61 Danhier F, Le Breton A, Préat V. RGD-based strategies to target alpha(v) beta(3) integrin in cancer therapy and diagnosis. Mol Pharm. 2012; 9(11):2961–2673.

62 Arner ES, Eriksson S. Mammalian deoxyribonucleoside kinases. Pharmacol Therapeut. 1995; 67:155–186.

63 Nakamura H. Liposomal boron delivery for neutron capture therapy. Methods Enzymol. 2009; 465:179–208.

64 Sasai M, Nakamura H, Sougawa N, Sakurai Y, Suzuki M, Lee CM. Novel hyaluronan formulation enhances the efficacy of boron neutron capture therapy for murine mesothelioma. Anticancer Res. 2016; 36(3):907–911.

65 Maruyama K, Ishida O, Kasaoka S, Takizawa T, Utoguchi N, Shinohara A, et al. Intracellular targeting of sodium mercaptoundecahydrododecaborate (BSH) to solid tumors by transferrin-PEG liposomes, for boron neutron-capture therapy (BNCT). J Control Release. 2004; 98(2):195–207.

66 Miyata S, Kawabata S, Hiramatsu R, Doi A, Ikeda N, Yamashita T, et al. Computed tomography imaging of transferrin targeting liposomes encapsulating both boron and iodine contrast agents by convection-enhanced delivery to F98 rat glioma for boron neutron capture therapy. Neurosurgery. 2011; 68(5):1380–1387.

67 Heber EM, Kueffer PJ, Lee MW Jr, Hawthorne MF, Garabalino MA, Molinari AJ, et al. Boron delivery with liposomes for boron neutron capture therapy (BNCT): biodistribution studies in an experimental model of oral cancer demonstrating therapeutic potential. Radiat Environ Biophys. 2012; 51(2):195–204.

68 Tietze R, Unterweger H, Dürr S, Lyer S, Canella L, Kudejova P, et al. Boron containing magnetic nanoparticles for neutron capture therapy – an innovative approach for specifically targeting tumors. Appl Radiat Isot. 2015; 106:151–155.

69 Grimes RN, editor. Carboranes, 2nd ed. London: Elsevier Academic, 2011.

70 Gassin PM, Girard L, Martin-Gassin G, Brusselle D, Jonchère A, Diat O, et al. Surface activity and molecular organization of metallacarboranes at the air-water interface revealed by nonlinear optics. Langmuir. 2015; 31(8):2297–2303.

71 Xiong H, Zhou D, Qi Y, Zhang Z, Xie Z, Chen X, et al. Doxorubicin-loaded carborane-conjugated polymeric nanoparticles as delivery system for combination cancer therapy. Biomacromolecules. 2015; 16(12):3980–3988.

72 Hiramatsu R, Kawabata S, Tanaka H, Sakurai Y, Suzuki M, Ono K, *et al.* Tetrakis(p-carboranylthio-tetrafluorophenyl)chlorin (TPFC): application for photodynamic therapy and boron neutron capture therapy. J Pharm Sci. 2015; 104(3):962–970.

73 Asano R, Nagami A, Fukumoto Y, Miura K, Yazama F, Ito H, *et al.* Synthesis and biological evaluation of new BSH-conjugated chlorin derivatives as agents for both photodynamic therapy and boron neutron capture therapy of cancer. J Photochem Photobiol B. 2014; 140–149.

74 Efremenko AV, Ignatova AA, Borsheva AA, Grin MA, Bregadze VI, Sivaev IB, *et al.* Cobalt bis(dicarbollide) versus closo-dodecaborate in boronated chlorin e(6) conjugates: implications for photodynamic and boron-neutron capture therapy. Photochem Photobiol Sci. 2012; 11(4):645–652.

75 Amelio D, Lorentini S, Schwarz M, Amichetti M. Intensity-modulated radiation therapy in newly diagnosed glioblastoma: a systematic review on clinical and technical issues. Radiother Oncol. 2010; 97(3):361–369.

76 Mizumoto M, Tsuboi K, Igaki H, Yamamoto T, Takano S, Oshiro Y, *et al.* Phase I/II trial of hyperfractionated concomitant boost proton radiotherapy for supratentorial glioblastoma multiforme. Int J Radiat Oncol Biol Phys. 2010; 77(1):98–105.

77 Barth RF, Grecula JC, Yang W, Rotaru JH, Nawrocky M, Gupta N, *et al.* Combination of boron neutron capture therapy and external beam radiotherapy for brain tumors. Int J Radiot Oncol Biol Phys. 2004; 58(1):267–277.

78 Kawabata S, Miyatake S, Kuroiwa T, Yokoyama K, Doi A, Iida K, *et al.* Boron neutron capture therapy for newly diagnosed glioblastoma. J Radiat Res. 2009; 50:51–60.

79 Yamamoto T, Nakai K, Kageji T, Kumada H, Endo K, Matsuda M, *et al.* Boron neutron capture therapy for newly diagnosed glioblastoma. Radiother Oncol. 2009; 91:80–84.

3.6

Optimizing the Therapeutic Efficacy of Boron Neutron Capture Therapy (BNCT) for Different Pathologies: Research in Animal Models Employing Different Boron Compounds and Administration Strategies

Amanda E. Schwint,[1,2] Andrea Monti Hughes,[1,2] Marcela A. Garabalino,[1] Emiliano C.C. Pozzi,[1,3] Elisa M. Heber,[1] and Veronica A. Trivillin[1,2]

[1] *Department of Radiobiology, National Atomic Energy Commission, San Martín, Buenos Aires Province, Argentina*
[2] *National Research Council (CONICET), Buenos Aires, Argentina*
[3] *Department of Research and Production Reactors, Ezeiza, Buenos Aires Province, Argentina*

3.6.1 BNCT Radiobiology

Boron neutron capture therapy (BNCT) is a binary treatment modality for cancer that involves the selective accumulation of boron carriers in tumors, followed by irradiation with a thermal or epithermal neutron beam. The high linear energy transfer (LET) α-particles and recoiling ^{7}Li nuclei resulting from the capture of a thermal neutron by a [^{10}B] nucleus have a high relative biological effectiveness (RBE). Their short range in tissue (6–10 μm) would limit the damage largely to cells containing [^{10}B]. In this way, BNCT would potentially target neoplastic tissue selectively, sparing healthy tissue [1–3]. Although there are a number of nuclides that have a high propensity for absorbing low-energy or thermal neutrons (high neutron capture cross-sections), [^{10}B] is the most attractive because it is non-radioactive and readily available, comprising approximately 19.9% of naturally occurring boron, and the particles emitted by the capture reaction are largely high LET with combined path lengths of approximately one cell diameter. Also, the well-understood chemistry of boron allows it to be readily incorporated into a wide variety of different chemical structures (e.g., [4]).

Since BNCT involves biochemical rather than geometric targeting, relying on the preferential incorporation of boron carriers to tumor cells, it is ideally suited to treat undetectable micrometastases [5], infiltrating malignant cells and foci of malignant transformation in field-cancerized tissue, with minimum damage to healthy tissues in the treatment volume [6]. The planning target volume (PTV) can be enlarged to include potential undetectable target cells in the organ to be treated, relying on the preferential incorporation of boron to transformed cells.

BNCT is a mixed-field irradiation. The radiation doses delivered to tumor and normal tissue during BNCT originate in the energy deposition from ionizing radiation with different LET characteristics. Although the neutron capture cross-sections for the

Boron-Based Compounds: Potential and Emerging Applications in Medicine, First Edition.
Edited by Evamarie Hey-Hawkins and Clara Viñas Teixidor.
© 2018 John Wiley & Sons Ltd. Published 2018 by John Wiley & Sons Ltd.

elements in normal tissue are several orders of magnitude lower than for [^{10}B], two of these, hydrogen and nitrogen, are present in such high concentrations that their neutron capture contributes significantly to the total absorbed dose. In addition to the α and ^{7}Li high-LET products that give rise to the tumor-specific boron dose component, a nonspecific background dose results from: (1) low-LET γ-rays in the neutron beam, (2) low-LET γ-rays resulting from the capture of thermal neutrons by hydrogen atoms [^{1}H(n,γ)^{2}H], (3) high-LET protons produced by the scattering of fast neutrons when a hardened epithermal neutron beam spectrum is employed, and (4) high-LET protons resulting from the capture of thermal neutrons by nitrogen atoms [^{14}N(n,p)^{14}C]. The biologically effective dose will depend on the RBE for the different dose components that contribute to the total irradiation dose. RBE depends on the LET of radiation. High-LET, densely ionizing particles induce direct damage to DNA (i.e., mainly double-strand breaks that are largely irreparable and lethal for the cell). Direct damage is not influenced by the oxygenation level of the tissue or its proliferative status. Conversely, low-LET radiation induces mainly indirect damage via the action of free radicals on DNA. This effect is largely reparable and causes sublethal damage to the cell. Indirect damage is enhanced by the presence of oxygen and is greater in proliferating cells with unwinding, exposed DNA strands. In the particular case of the boron dose component, RBE is termed the *compound biological effectiveness* (CBE) factor and depends on the RBE of α-particles and ^{7}Li ions and the microdistribution of [^{10}B] in a particular tissue [2,3]. BNCT protocols should ideally maximize the tumor-selective boron radiation component and minimize the nonselective background dose to improve the therapeutic ratio [2]. Maximizing and optimizing the delivery of boron to tumor are the most effective ways to do this.

3.6.2 An Ideal Boron Compound

An ideal boron compound will be nontoxic at therapeutic dose levels, will accumulate selectively in tumor cells, will target all tumor cells homogeneously, and will deliver [^{10}B] to tumor efficiently. Preferential accumulation of boron in tumor contributes to the therapeutic advantage of BNCT for tumor versus normal tissue. The importance of homogeneous targeting of boron to tumor lies in the fact that the tumor cells poorly loaded with boron will be less responsive or altogether refractory to treatment and will lead to therapeutic failure (e.g., [7]). Absolute boron content in tumor tissue must be high enough (10^9 atoms [^{10}B]/cell) to allow sufficient [^{10}B](n,α)^{7}Li capture reactions to occur for the effect to be lethal. Also, at a given tumor–healthy tissue boron concentration ratio, high absolute [^{10}B] concentrations are an asset because they allow for shorter irradiation times and the concomitant reduction in background dose that affects tumor and healthy tissue alike [2]. An ideal boron carrier will clear rapidly from blood and normal tissue but will persist in tumor long enough to allow for neutron irradiation. Finally, the microdistribution of the ideal [^{10}B] carrier will place the [^{10}B] atoms close to a therapeutically useful target such as DNA. The short ranges of α and lithium particles make the microdistribution of the boron relative to the subcellular target of critical radiobiological significance [8–10]. Since the [^{10}B] isotope only has a natural abundance of 19.9%, BNCT agents must be enriched with [^{10}B] during synthetic preparation to be maximally effective [11].

Developing an ideal boron compound that fulfills all these requirements is a huge challenge. Within this context, the international community has devoted (and continues to do so) much effort and resources to search for the "ideal" boron compound that would potentially replace the three "imperfect" compounds currently authorized for use in humans: boronophenylalanine (BPA), sodium borocaptate (BSH), and decahydrodecaborate (GB-10). Although these compounds delivered as single agents in the traditional way (as a single intravenous injection or infusion prior to a single neutron irradiation) have shown therapeutic potential for different pathologies [12–18], there is clearly opportunity and need for improvement. To date, no other boron compounds have reached the stage for evaluation in a clinical biodistribution study. Once a new [^{10}B] carrier is identified as promising from cell culture studies, it still faces many hurdles, beginning with biodistribution studies in appropriate experimental tumor models and *in vivo* evaluation of toxicity, followed by experimental *in vivo* radiobiological studies. If the results warrant evaluation in a clinical scenario, clinical biodistribution studies will follow. They are extremely costly, are of no direct benefit to participants, and must comply with the stringent requirements of regulatory agencies [19]. Only if this sequence of studies affords encouraging data will the corresponding authorizations be sought to enter the boron compound in a clinical trial. Typically, in the United States for example, of 10,000 general medicinal compounds that are developed, only five enter clinical trials and only one is finally approved by the US Food and Drug Administration (FDA) for treatment. The process "from bench to bedside" typically takes ten years and costs over US$1000 million [20].

Optimizing the delivery of [^{10}B] compounds currently authorized for use in humans is an excellent short- and medium-term strategy. It will help to bridge the gap between research and clinical application. The knowledge gained will also be applicable to potentially better [^{10}B] compounds if and when they are developed.

3.6.3 Clinical Trials, Clinical Investigations, and Translational Research

Clinical studies of BNCT for glioblastoma multiforme, melanoma, head and neck tumors, and liver metastases have been performed or are underway in the United States, Japan, Europe, Argentina, and Taiwan (e.g., [16,21–24]), employing nuclear reactors as the neutron source. More recently, clinical investigations have been performed in Japan on BNCT for recurrent hepatic and gastrointestinal cancer [25], locally recurrent lung cancer [26], and extramammary Paget's disease [27]. Ongoing clinical trials in Japan include BNCT for lung cancer. To date, the clinical results have shown a therapeutic advantage, with identified opportunities for improvement. The more recent development of accelerators (e.g., [28,29]), to be used as the neutron source for BNCT, paves the way for more widespread clinical trials for different tumors.

The need to optimize BNCT for different pathologies requires translational research in adequate experimental models. In particular, experimental models are employed to test novel boron compounds and explore different strategies to optimize the delivery of boron compounds approved for use in humans. The first stage in the *in vivo* evaluation of the therapeutic potential of a boron carrier is to perform boron biodistribution studies in an appropriate experimental model. Boron biodistribution studies are essential to design and plan useful BNCT preclinical and, ultimately, clinical research protocols. In particular, they identify potentially useful boron compounds and administration protocols and enable the choice of the optimum time post administration for the boron carrier to perform neutron irradiation, seeking to maximize tumor boron levels while minimizing boron concentration in healthy tissue and blood. To date, there is no clinically practical, online, noninvasive way to evaluate boron concentration during irradiation for BNCT. Thus, dose calculations are based on boron content values in blood, tumor, and normal tissue obtained from biodistribution studies performed beforehand [8]. At most, blood samples can be taken just before irradiation to measure blood boron concentration and infer tissue boron concentration, assuming the tumor–blood boron concentration ratios established in previously performed biodistribution studies hold true [21]. In the specific case of experimental models, dose calculations are based on the mean values obtained from biodistribution studies in separate sets of animals (e.g., [30]). In this sense, it is important to bear in mind that large intratumor, intertumor, and intersubject variations in gross boron content values have been reported (e.g., [31,32]). These variations must be accounted for in dose calculation and dose prescription, to avoid exceeding the radiotolerance of the healthy tissues within the treatment volume and/or underdosing the tumor. We face the challenge of performing a reliable determination of the time of maximal boron accumulation in tumor in a patient to establish the optimum timeframe and exposure time for neutron irradiation for that particular patient. This challenge could be potentially overcome by positron emission tomography (PET)-guided BNCT using a dual-modality agent that would allow for monitoring of the real-time boron accumulation within the patient's tumor. One example of a dual-modality BNCT agent is 4-borono-2-^{18}F-fluoro-phenylalanine (^{18}F-BPA) [11].

Although boron biodistribution studies are essential and orientative, we cannot stress enough the importance of actual radiobiological studies in appropriate experimental models to determine the efficacy of boron carriers and administration protocols. The combined biological effects of a heterogeneous distribution of [^{10}B] and a mixed-field irradiation with high- and low-LET components can only be investigated in studies involving irradiation [33,34]. As exemplified in Section 3.6.4, a boron carrier that might be ruled out based solely on biodistribution studies might prove successful in actual BNCT studies [10].

3.6.4 Boron Carriers

In the 1950s and early 1960s, boric acid and its derivatives were used in clinical trials for malignant gliomas [35,36]. However, the outcome of these trials was disappointing because these chemical compounds failed to target tumor selectively and exhibited poor tumor retention [37], high blood boron concentration resulted in excessive damage to normal brain vasculature and the scalp [4], and the penetration of thermal neutron beams was insufficient to treat deep-seated brain tumors [2]. Since the late 1950s, efforts began to synthesize tumor-selective boronated agents. At that time, the focus was on boronated agents that would accumulate in brain tumors surrounded by a disrupted blood–brain barrier (BBB) and not cross the intact BBB into the normal brain.

BPA ($C_9H_{12}{}^{10}BNO_4$), a low-molecular-weight boron carrier, was synthesized in the late 1950s. Taking advantage of its structural similarity with melanin precursors, a BNCT research program for melanoma was initiated with BPA with encouraging results [38]. BPA was first used clinically to treat patients with cutaneous malignant melanomas in 1988 [39,40]. Later studies showed that BPA is in fact taken up by active transport, making use of the L-amino-acid cell transporter system [41]. Multiple experimental and clinical studies went on to show that BPA targets different types of tumors selectively, including brain tumors (e.g., [2,16,30,42]), conceivably due to the increased metabolic activity of tumor cells and upregulation of L-type amino acid transport expression at the cell membrane [43]. In this way, BPA would target tumor selectively on a "cell-by-cell" basis. The downside to this advantage for BNCT is that BPA targets different tumor cell populations with varying metabolic activity relatively heterogeneously. Subpopulations of quiescent cancer cells may have reduced BPA uptake [44]. Tumor cells insufficiently loaded with boron will fail to respond to BNCT (e.g., [45,46]). Conversely, the initial notion that BPA could not be used for BNCT of brain tumors because it traverses the intact brain barrier was later challenged by the fact that this ability (coupled to preferential targeting of tumor cells) may well be essential for effective BNCT of highly infiltrative brain tumors such as glioblastoma multiforme [2]. The BPA-fructose complex (designed to improve BPA solubility) for BNCT of patients was first used in 1994 to treat glioblastoma [47]. Since then, BPA-fructose has become the more commonly used clinical boron delivery agent for both intra- and extracranial tumors, with some very good results [34].

BSH ($Na_2{}^{10}B_{12}H_{11}SH$), also a low-molecular-weight boron carrier, was first used in the 1960s for the treatment of high-grade gliomas with BNCT [48]. BSH does not cross the intact BBB in the normal brain but accumulates in brain tumors due to the fact that the blood vessels in intracranial tumors lack a properly functioning BBB. The relatively low tumor-to-blood boron concentration ratios in patients with malignant brain tumors [49] suggest that passive diffusion is the primary mode of accumulation of BSH in tumors. BSH has also been reported to be carried in the blood as a disulfide complex with serum albumin [50]. Some studies suggest that BSH would accumulate more in mostly necrotic areas [51]. Handling and storage procedures for BSH should avoid the formation of oxidation products that might be toxic. BSH was initially proposed to treat brain tumors as a stand-alone boron agent [52] because its capacity to target tumors selectively would rely on a disrupted BBB around tumor [53]. However, it has also been used to treat recurrent head and neck tumors in combination with BPA, with good results [54,55]. The advantages of the combined administration of [^{10}B] compounds with different properties and uptake mechanisms are discussed further here.

The simple sodium salt of $closo$-$B_{10}H_{10}{}^{2-}$ (GB-10) is a largely diffusive low-molecular-weight agent that, like BSH, does not traverse the intact BBB (e.g., [31]). Due to this property, GB-10 was initially proposed as a boron agent for the treatment of brain tumors with BNCT and for BNCT-enhanced fast neutron therapy (BNCT/FNT) of extracranial tumors such as non-small-cell lung cancer [56]. Although GB-10 is approved for its use in humans, it has only been used clinically in two patients within the context of a clinical trial for brain tumors in 1961 [57] and in a pharmacokinetic trial based on the measurement of boron concentration in blood samples only, performed in volunteer subjects with glioblastoma multiforme or non-small-cell lung cancer [58]. GB-10 is readily soluble, easy to handle, and nontoxic at high concentrations.

As described further here, more recently our group has explored the value of GB-10 as a boron carrier for BNCT, both administered as a stand-alone agent and administered jointly with BPA, in experimental tumor models that do not involve the BBB.

While all three boron compounds – BPA, BSH, and GB-10 – have been proved to be nontoxic at therapeutic doses and are approved for use in humans, only BPA and BSH have been used clinically to a significant extent. Currently, BPA and BSH are the only two boron compounds used in clinical trials. The results of these clinical trials indicate that BNCT mediated by these boron carriers is safe and can play an important role in cancer therapy in certain cases. Although clinical trials with BPA and BSH have shown encouraging results [34], there is undoubtedly room for improvement. BPA and BSH are far from "ideal" boron compounds, mainly due to their modest tumor selectivity.

The development of new, more selective boron delivery agents has been, and still is, at the center of the efforts of the BNCT international community. While more penetrating reactor epithermal neutron beams have been developed and optimized, and although accelerator technology for BNCT is up and running, the boron compounds approved for use in humans fall short of fulfilling the requirements of an ideal boron compound. Over the past 30 years, several classes of boron-containing compounds have been designed and synthesized in attempts to fulfill these requirements. Recent efforts have involved incorporating boron delivery agents into tumor-targeting molecules such as peptides, proteins, antibodies, nucleosides, sugars, porphyrins, liposomes, and nanoparticles. The low-molecular-weight boron delivery agents explored include boronated natural and unnatural amino acids [59] and boron-containing linear and cyclic peptides with high tissue-penetrating properties [15]. Boron-containing purines, pyrimidines, thymidines, nucleosides, and nucleotides have also been investigated (e.g., [60]). Porphyrin derivatives containing boron have been explored due to their low toxicity and natural affinity for tumors [61]. Boron-containing chlorins, bacteriochlorins, and phthalocyanines are promising dual agents for both BNCT and photodynamic therapy of tumors [15]. Other boron-containing DNA-binding molecules, including alkylating agents, DNA intercalators, minor-groove binders and polyamines [62], and boron-containing sugars [63], have also been studied. Among the high-molecular-weight boron delivery agents, monoclonal antibodies for overexpressed receptors on tumor cells such as vascular endothelial growth factor (VEGFR), somatostatin receptors, and the epidermal growth factor receptor (EGFR) have been extensively studied by Barth and coworkers (e.g., [64,65]).

One of the more promising routes would seem to be nanoscale drug delivery systems using liposomes and nanoparticles. Liposomes are efficient drug delivery vehicles that are able to deliver large quantities of a wide range of encapsulated agents selectively to tumor tissue [66]. Tumor blood vessels resulting from angiogenesis and vasculogenesis are structurally and functionally abnormal [67]. These aberrant blood vessels allow small liposomes (<100 nm) to pass through, allowing for selective tumor targeting. Furthermore, the lymphatic drainage of solid tumors is generally deficient. The substances that diffuse into tumor tissue persist in the interstitial space [68] for prolonged periods due to the known enhanced permeability and retention (EPR) effect [69]. Because liposomes are appropriately sized, they may take advantage of the EPR effect, and the incorporated agent(s) need not necessarily exhibit tumor affinity. Additionally, the fact that the serum half-life of an encapsulated drug is longer than that of the free drug makes it possible to use a lower dose. Because the liposome preserves the structural integrity of the drug, toxicity is often reduced [70]. Small unilamellar liposomes in particular are potentially useful boron delivery vehicles for BNCT and have been extensively studied by Hawthorne and coworkers [70–74] and other groups [75–82]. They can encapsulate aqueous solutions of sodium salts of polyhedral borane anions and/or incorporate lipophilic boron-containing moieties embedded within the bilayer membrane [83]. Additional classes of boron-containing nanoparticles have been investigated as boron carriers for BNCT [84,85].

As stated above, despite their potential value, none of these compounds have got beyond the laboratory bench. Admittedly, the procedures to bring a new drug into clinical practice are extremely costly and lengthy. Improvement in the delivery of boron compounds approved for their use in humans and therapeutic strategies devised to optimize BNCT using these same compounds would be the best way to move forward in the short and medium term [10,34,86]. Concomitant international efforts to develop and identify novel boron carriers will also benefit from the knowledge derived from these studies.

3.6.5 Optimizing Boron Targeting of Tumors by Employing Boron Carriers Approved for Use in Humans

Different groups have explored different ways in which to improve the delivery and distribution of BPA and BSH in experimental models and human subjects with varying success. One example involves the combined administration of BPA and BSH [87]. The combined administration of [^{10}B] compounds with different properties and uptake mechanisms contributes to more homogeneous tumor targeting and helps to overcome the potential toxicity of higher doses of each of the compounds given alone. Combinations of agents may be superior to any single agent [32,87,88]. Additional techniques used to improve boron accumulation in brain tumors are BBB disruption by a hyperosmotic agent such as mannitol, intracarotid administration of the boron compound, and convection-enhanced delivery [11,89]. Slow infusions of BPA would improve boron targeting of infiltrating tumor cells in the brain [19]. Pre-loading with an amino acid analog such as L-3,4-dihydroxyphenylalanine (L-DOPA), which is structurally similar to BPA and enters the cell through the L-type amino acid transport system, would improve the subsequent accumulation of BPA via an antiport (exchange) mechanism [90]. Mild temperature hyperthermia would improve delivery by increasing blood flow [91].

3.6.6 BNCT Studies in the Hamster Cheek Pouch Oral Cancer Model

3.6.6.1 The Hamster Cheek Pouch Oral Cancer Model

Efforts by our group have shown the pivotal importance of *in vivo* studies in appropriate experimental models to contribute to the knowledge of BNCT radiobiology and the optimization of BNCT for different pathologies. Much of our work has been performed in the hamster cheek pouch oral cancer model. The relatively poor overall five-year survival rate for malignancies of the oral cavity [92] and the fact that radical surgery often results in large tissue defect [93] pose the need for more effective and selective therapies. Studies in appropriate experimental models are pivotal to progress in this field. The hamster cheek pouch model of oral cancer was proposed by our group to explore, for the first time, the feasibility of applying BNCT to the treatment of head and neck cancer [30,94]. Our first experimental studies preceded the first clinical trial of BNCT for head and neck malignancies [54] and provided evidence, for the first time, of the efficacy of BNCT to treat oral cancer at an experimental level.

The hamster cheek pouch oral cancer model is the most widely accepted model of oral cancer [95–97]. The hamster cheek pouch anatomically resembles a pocket that is easily accessible to local tumor induction by topical application of the carcinogen dimethyl-1,2-benzanthracene (DMBA) 0.5% in mineral oil 2–3 times a week for three months (e.g., [12,46]). The pouch can be readily everted for local treatment such as irradiation and macroscopic follow-up (Figure 3.6.1). Carcinogenesis protocols induce precancerous changes and squamous cell carcinomas that closely resemble human lesions [98]. This model poses an advantage in that tumors are induced by a process that mimics the process of malignant transformation, giving rise to tumors surrounded by precancerous tissue. Precancerous tissue is not available in animal models based on tumor growth from transformed cells implanted in healthy tissue. The possibility to study precancerous tissue is essential in head and neck cancer studies. Multiple primary tumors that arise from field-cancerized tissue are a known phenomenon in head and neck cancer [98–101]. Once the primary tumors have been treated, second primary tumors often develop from field-cancerized tissue, contributing to local-regional recurrence and therapeutic failure [102]. The possibility of studying precancerous tissue is also clinically relevant given its role as a potentially dose-limiting tissue. Precancerous tissue is more radiosensitive than normal tissue [103,104] and can develop dose-limiting mucositis. Oral mucositis refers to inflammatory, erosive/ulcerative oral mucosal lesions and necrosis [88,105]. In a clinical scenario, oral mucositis limits the dose that can be administered with BNCT to head and neck and brain tumors [16,23] and is a frequent dose-limiting side effect during conventional radiotherapy for advanced head and neck tumors, affecting approximately 80% of the patients [106,107]. Also, oral mucositis could enhance tumorigenesis (e.g., [88,108]). Despite its incidence and clinical relevance, no effective way to prevent or treat mucositis is currently available [107,109].

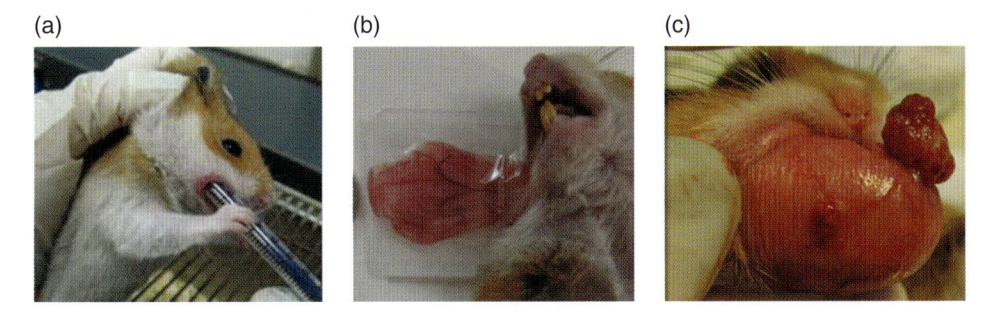

Figure 3.6.1 (a) Topical application of the carcinogen DMBA 0.5% in mineral oil in the hamster cheek pouch. The pouch can be readily everted for local treatment, such as irradiation (b) and macroscopic follow-up (c).

Table 3.6.1 Summary of experimental conditions and outcomes corresponding to the protocols as indicated

Protocol or tumor response	BPA–BNCT 1	BPA–BNCT 2	GB10–BNCT 3	(GB10+BPA)–BNCT 4	Seq–BNCT 5	BPA-BNCT Th+ 6	BPA-BNCT Th− 7	Th+Seq-BNCT 8	GB10-BNCT No EP 9	GB10-BNCT EP 10	MAC-TAC-BNCT Single: 4 w 11	MAC-TAC-BNCT Double: 16 w 12
Nuclear reactor	RA-6	RA-3	RA-6	RA-6	RA-3	RA-3	RA-3	RA-3	RA-3	RA-3	RA-3	RA-3
Tumor dose (Gy)	5.2	5.6	7	8.5	10	4	4.3	10	2.4	4	21	21 × 2
Precancerous tissue dose (Gy)	3.9	3.7	7	4.2	9	3	2.7	9	2.7	2.3	5	5 × 2
Complete response	78%	46%	41%	66%	76%	56%	43%	87%	6%	46%	33%	38%
Partial response	13%	27%	30%	27%	19%	28%	24%	13%	42%	46%	37%	50%
Overall response	91%	73%	70%	93%	95%	84%	67%	100%	48%	92%	71%	88%
Mucositis in precancerous tissue	80% ≥G3	100% G3/G4	0% ≥G3	0% ≥G3	35% G3/G4 22% G4	0% ≥G3	80% G3/G4	38% G3 0% G4	0% ≥G3	11% G3	0% ≥G3	0% ≥G3

Note: In all cases, the data correspond to 28 days post treatment, except in the case of MAC–TAC–BNCT, for which follow-up data correspond to 4 weeks post treatment in the case of a single treatment (column 11) and 16 weeks in the case of a double treatment (column 12).

3.6.6.2 BNCT Mediated by BPA

Our initial studies validated the hamster cheek pouch model for BNCT studies, and showed that BPA was taken up selectively by tumor tissue and, to a lesser degree, by precancerous tissue. Mean boron concentration was 37 ± 18 ppm for tumor tissue and 20 ± 6 ppm for precancerous tissue at 3.5 h post intraperitoneal (i.p.) or intravenous (i.v.) administration of BPA (15.5 mg [^{10}B]/kg), and the mean boron concentration ratios were 2.4:1 for tumor–normal pouch and 3.2:1 for tumor–blood [30]. These absolute and relative boron concentration values would be therapeutically useful. They comply with the guidelines for potential therapeutic efficacy we previously established based on international data and our own studies (e.g., [8]). For a boron compound administration protocol to be potentially useful and worthy of evaluation in *in vivo* radiobiological studies, the boron compound(s) should not be toxic at the administered dose, absolute boron concentration in tumor should be higher than 20 ppm, and boron concentration ratios of tumor–normal tissue and tumor–blood should be equal to or higher than 1.

We then went on to report the first evidence of the usefulness of BNCT for the treatment of head and neck cancer in an experimental model. We assessed the response of hamster cheek pouch tumors, precancerous tissue, and normal pouch tissue to BNCT mediated by BPA (BPA-BNCT) using the previously assessed BPA administration protocol [30] and neutron irradiation with the hyperthermal or mixed beam (thermal + epithermal) of the RA-6 Nuclear Reactor at the Bariloche Atomic Center [110]. BPA–BNCT at 5.2 Gy absorbed dose to tumor induced complete tumor remission by 15 days post treatment in 78% of the tumors and partial remission in an additional 13% of tumors. No toxicity in normal tissue and moderate/severe, albeit reversible, mucositis in precancerous tissue were observed [94]. The data are summarized in Table 3.6.1 (column 1), and a representative example is shown in Figure 3.6.2. The influence of beam spectrum on outcome is exemplified by performing a similar experiment at the thermal facility of the RA-3 Nuclear Reactor. As summarized in Table 3.6.1, tumor response at RA-3 (column 2) [111] was lower than at RA-6 (column 1) [94], at similar dose levels.

3.6.6.3 BNCT Mediated by GB-10 or by GB-10 + BPA

Within the context of optimizing the use of boron compounds approved for use in humans to yield an improved therapeutic advantage, we performed a biodistribution study with the boron compound GB-10 in the hamster cheek pouch oral cancer model [31]. As described in this chapter, GB-10 is a largely diffusive agent that has been proposed for BNCT of brain tumors. It has also been proposed for BNCT-enhanced fast neutron therapy (BNCT/FNT). Capture of the thermal neutron component within a predominantly fast neutron beam could selectively enhance cytotoxicity in deeply located tumors such as lung cancer. In this way, this treatment could be applicable to a wider scope of patients [112]. Because of the steep nature of the dose–response curve for fast neutrons, a selective increase in dose to tumor with high-LET radiation has the potential for a substantial increase in the probability of tumor control [113].

Although *a priori* we did not expect GB-10 to be a selective stand-alone boron delivery agent in tumors other than brain tumors, our initial rationale was to evaluate its biodistribution to explore its potential value as one component of a combined

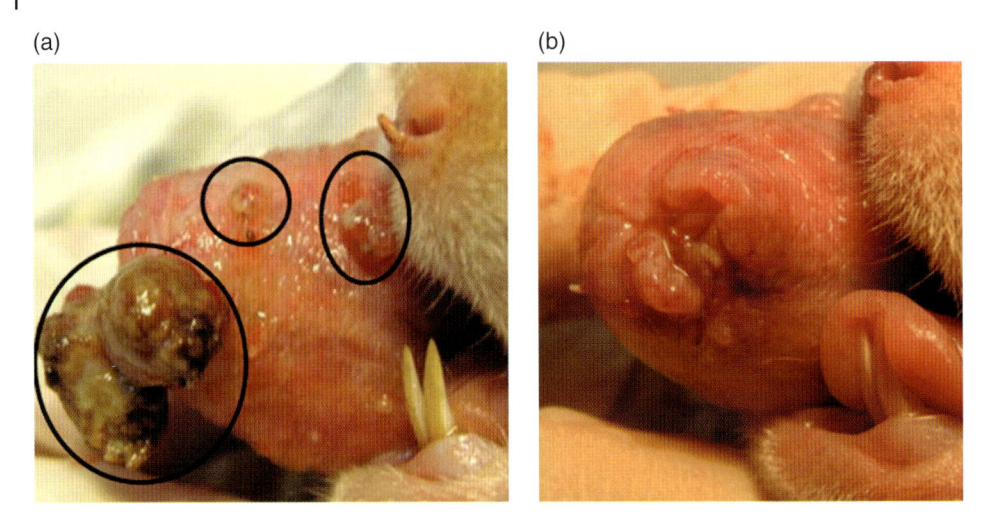

Figure 3.6.2 An example of BNCT therapeutic effect on tumors: (a) Cancerized cheek pouch bearing tumors before BPA–BNCT at the RA-6 Nuclear Reactor. (b) Cancerized cheek pouch of the same animal 28 days after BNCT, exhibiting tumor remission with slight mucositis in precancerous tissue.

administration protocol (GB-10 and BPA administered jointly), aimed at targeting different cell populations of a heterogeneous tumor. GB-10 alone (50 mg [^{10}B]/kg administered intravenously), as expected, failed to target oral tumors selectively. However, it achieved therapeutically useful boron concentration levels in tumor at 3 h post intravenous administration (i.e., 32 ± 21 ppm). At the same timepoint, the boron concentration value was 34 ± 17 ppm in precancerous tissue, 22 ± 7 ppm in normal pouch tissue, and 32 ± 6 ppm in blood [31]. We concluded that despite the fact that GB-10 uptake is not selective in tumors other than brain tumors, homogeneous deposition of GB-10 [32] in different tumor populations and high absolute tumor boron concentration values may render this compound useful for BNCT treatment of head and neck tumors. The consistently low brain and spinal cord values would contribute to reducing toxicity in these organs when they are included in the planning target volume.

We then went on to explore biodistribution values for different combinations of BPA and GB-10 administered jointly. The rationale for the joint administration of these compounds would be to profit from the selective, albeit heterogeneous [30,32], uptake of BPA by tumor tissue and the homogeneous distribution of GB-10 in different tumor areas. The best results were obtained with the protocol that involved intravenous administration of GB-10 as a bolus injection at a dose of 34.5 mg [^{10}B]/kg coupled to BPA administered at a total dose of 31 mg [^{10}B]/kg as fractionated i.p. injections over a 3-h period to simulate an infusion. Three hours post administration of GB-10 and 1.5 h after the last i.p. injection of BPA, the selected times for neutron irradiation based on pharmacokinetic data [31], boron concentration was 63 ± 21 ppm in tumor, 41 ± 14 ppm in precancerous tissue, 38 ± 18 ppm in normal pouch tissue, and 30 ± 14 ppm in blood. The combined (GB-10 + BPA) administration protocol delivered boron to tumor 3.3-fold more homogeneously than BPA alone, and the GB-10 protocol targeted tumor 1.8-fold more homogeneously than the BPA protocol. However, the degree of homogeneity in the deposition of boron in precancerous and normal tissue was similar for all three administration protocols, suggesting that the gross boron distribution in these relatively homogeneous tissues was relatively homogeneous for BPA, GB-10, and GB-10 + BPA [32].

As described here, we reported the first evidence of success of BNCT for the treatment of oral cancer in the hamster cheek pouch model using BPA as the boron carrier. Although no damage to normal tissue was observed, dose-limiting reversible mucositis in precancerous tissue was an unwanted side effect associated with therapeutic success. In addition, complete remissions in the larger tumors were not observed, conceivably due to heterogeneous tumor targeting by BPA [30,32,94]. Within this context, we performed a low-dose *in vivo* BNCT study in the hamster cheek pouch oral cancer model at the hyperthermal RA-6 neutron beam to explore the therapeutic potential of GB-10 combined with BPA.

Low-dose BNCT mediated by GB-10 alone at 5.3 Gy absorbed dose to tumor produced significant tumor control without radiotoxic effects in normal tissue and precancerous tissue around tumor. This was a surprising finding that did not seem to correlate with the classical notion that tumor lethality is associated with unacceptable normal tissue radiotoxicity if boron is not delivered selectively to tumor. Low-dose BNCT at 4.3 Gy absorbed dose to tumor using GB-10 and BPA administered jointly also induced significant tumor control with no radiotoxic effects on normal tissue and precancerous tissue [114]. As described, the combined administration of GB-10 and BPA achieved high boron concentration values in tumor, thus allowing for a reduction in irradiation time for the same tumor dose, with the concomitant reduction in the background dose that affects tumor and normal tissues similarly. The reduction of dose to normal and precancerous tissue relative to tumor and the absence of radiotoxic effects suggested the possibility of significantly escalating dose to tumor.

A follow-on study was performed at RA-6 at higher dose levels. Overall tumor response expressed as partial remission + complete remission 28 days post treatment was 70% for GB-10–BNCT at 8 Gy absorbed dose to tumor, 93% for (GB-10 + BPA)–BNCT at 8.5 Gy absorbed dose to tumor [12], and 91% for BPA–BNCT at 5.2 Gy absorbed dose to tumor (dose-limiting mucositis in precancerous tissue precluded dose escalation with BPA–BNCT) [94]. (GB-10 + BPA)–BNCT and GB-10–BNCT were able to achieve complete remission in the larger tumors (>100 mm^3), while BPA–BNCT failed to induce complete remission in this tumor volume range. This would conceivably be due to more homogeneous tumor boron targeting by GB-10 and to the benefits afforded by the combined administration of GB-10 and BPA [12]. Only transient, slight mucositis was observed in precancerous tissue in the case of (GB-10 + BPA)–BNCT and GB-10–BNCT protocols. As described in this chapter, in the case of the BPA–BNCT protocol (at a lower absorbed tumor dose), mucositis in precancerous tissue was moderate/severe. No mucositis was observed in normal tissue for any of the three protocols. The data are summarized in Table 3.6.1 (columns 1, 3, and 4).

The most surprising finding was that despite the fact that GB-10 does not target hamster cheek pouch tumors selectively, GB-10–BNCT induced 70% tumor response with virtually no precancerous or normal tissue radiotoxicity. Light microscopy analysis showed that GB-10–BNCT selectively damages aberrant tumor blood vessels, sparing precancerous and normal tissue blood vessels. The radiosensitivity of aberrant tumor blood vessels would be greater than that of precancerous and normal tissue blood vessels. In terms of biological considerations, the tumor blood vessel structure and function are altered. Tumor vessels are tortuous and dilated, and their walls exhibit fenestrae, vesicles, and transcellular holes; widened interendothelial junctions; and a discontinuous or absent basement membrane [115]. These features would render tumor vessels more sensitive to radiation-induced damage than precancerous and normal tissue blood vessels, thus providing a selective mechanism of tissue damage for a chemically nonselective boron compound. These findings suggested a new paradigm in BNCT radiobiology. In the case of GB-10–BNCT, selective tumor lethality would result from selective blood vessel damage (and ensuing tissue damage) rather than from the selective uptake of the boron compound in tumor. In the case of (GB-10 + BPA)–BNCT, in addition to achieving more homogeneous tumor boron targeting by the combined administration of GB-10 and BPA [32], this protocol would optimize therapeutic efficacy by combining the vascular targeting mechanisms of GB-10–BNCT [12] and the cellular targeting mechanisms of BPA–BNCT [30,94].

These findings stress the pivotal importance of actual radiobiological studies to assess the therapeutic efficacy of BNCT mediated by a boron compound under study. Based on biodistribution studies alone, GB-10 would have been ruled out as a stand-alone boron agent for BNCT of head and neck tumors. It took an actual *in vivo* radiobiological study to reveal its potential value as a boron carrier for BNCT of head and neck tumors.

3.6.6.4 Sequential BNCT

As described, we demonstrated that BPA–BNCT, (GB-10 + BPA)–BNCT, and GB-10–BNCT induced good tumor response (expressed as complete remission + partial remission) in the hamster cheek pouch oral cancer model employing the RA-6 hyperthermal neutron beam [12,94,110] and the RA-3 thermal neutron facility [111,116]. Attempting to improve therapeutic efficacy, in particular complete tumor remission, at no extra cost in terms of toxicity in dose-limiting precancerous tissue, we devised and explored a novel approach to BNCT termed *Sequential BNCT* (Seq-BNCT). Irradiations were performed at the RA-3 thermal facility employing a lithium-6 carbonate shielding to protect the body of the animal while the cheek pouch is everted out of the enclosure [111]. Seq-BNCT involves the sequential application of BPA–BNCT (BPA at a dose of 15.5 mg [^{10}B]/kg), followed by GB-10–BNCT (GB-10 at a dose of 50 mg [^{10}B]/kg) with an interval of 24 (Seq-24h-BNCT) or 48h (Seq-48h-BNCT) between the two applications. Thus, Seq-BNCT is a new way of combining the contributions of BPA and GB-10. The Sequential modality was devised based on notions of BNCT radiobiology contributed by studies by our group and others (e.g., [2,12,114]). Seq-BNCT involves the use of two boron agents with different properties and complementary mechanisms of action, conceivably contributing to a more homogeneous, therapeutically successful targeting of heterogeneous tumor cell populations. The brief interval between applications would favor targeting with GB-10 in the second application. It is known that interstitial fluid pressure (IFP) is elevated in most human and experimental tumors [117] mainly as the result of unregulated angiogenesis, and is partly responsible for the distribution of blood-borne therapeutic agents [118]. A decrease in IFP has been reported to occur shortly after irradiation [119] and improves the uptake of therapeutic agents [120]. In addition, the induction of void space by cancer cell death would also enhance intratumoral delivery [121]. Seq-BNCT would profit from this effect in terms of improving GB-10 tumor targeting 24 or 48 h after BPA–BNCT. Thus, GB-10 would have a better chance of targeting a tumor when it is administered as part of the Sequential protocol than when it is administered with BPA in the joint GB-10 + BPA protocol. The 24 or 48 h interval between applications is short enough to preclude tumor cell repopulation [122] and could favor targeting of the tumor cells that were refractory to the first application. Regarding dose-limiting mucositis in precancerous tissue, Seq-BNCT would favor a reduction in the severity of mucositis or, at least, would not exacerbate mucositis compared to a single application at the same total dose employing (GB-10 + BPA)–BNCT. Because mucositis is a multistage process initiated by mucosal injury and associated with an increased production of inflammatory cytokines that cause direct mucosal damage and initiate positive feedback loops [123], the 24 or 48 h interval between BNCT applications might conceivably allow the inflammatory process to partially subside before the second dose is delivered, precluding the exacerbation of mucositis.

A comparison of tumor control and mucositis in dose-limiting precancerous tissue was performed for Seq-BNCT (Seq-24 h-BNCT and Seq-48 h-BNCT) versus (GB-10 + BPA)–BNCT. The single application of BNCT was for the same total tumor absorbed dose (9.9 ± 2 Gy) as Seq-BNCT (BPA–BNCT at 3.9 ± 1.8 Gy to tumor, followed by GB-10–BNCT at 6.0 ± 3.4 Gy to tumor) [124]. Here, we must point out that the neutron irradiations in this case were performed at the RA-3 thermal nuclear facility and cannot be compared with our earlier studies at the hyperthermal neutron beam at RA-6 due to the differences in neutron spectrum and γ components that influence biological response [3,110,116].

At 28 days post treatment, Seq-24 h-BNCT and Seq-48 h-BNCT induced, respectively, overall tumor responses (partial remission + complete remission) of 95 and 91%, while overall tumor response for (GB-10 + BPA)–BNCT was significantly lower at 75%. Complete remission was higher for Seq-24 h-BNCT and Seq-48 h-BNCT (76 and 68%, respectively) than for (GB-10 + BPA)–BNCT (50%). No statistically significant differences were observed between Sequential protocols. The data for Seq-24 h-BNCT are summarized in Table 3.6.1 (column 5). The incidence of complete remissions was higher in small (<10 mm^3) tumors than in medium (10–100 mm^3) and large (>100 mm^3) tumors for both Seq-BNCT protocols and (GB-10 + BPA)–BNCT, a characteristic feature in this model and in cancer therapy generally. However, this effect was exacerbated in the case of the Sequential protocols. This finding could be attributed to the fact that IFP increases with the size of tumor, conceivably impairing the distribution of boron compounds more in the larger tumors. The comparative therapeutic benefit of Seq-BNCT might be less robust for the larger tumors. Fluence matched beam-only groups (no boron compound administration) were studied for each of the protocols to assess the effect of background dose. Overall tumor response for the beam-only protocols ranged from 3 to 14%, with no complete remissions. These values were within the same range as the spontaneous partial remissions observed in the cancerized, nontreated group (16%).

The severity of mucositis was evaluated semiquantitatively according to an oral mucositis scale [6] based on macroscopic features adapted for the carcinogen-treated hamster cheek pouch from the World Health Organization (WHO) classification for oral mucositis in human subjects [125] and the six-point grading system for normal hamster cheek pouches of Ref. [126]. The Sequential protocols and (GB-10 + BPA)–BNCT induced reversible mucositis in the dose-limiting precancerous tissue that peaked at 14 days post treatment and had resolved by 21–28 days post treatment. The incidence of Grade 3/4 mucositis (moderate/severe) at 14 days post treatment was 35% for the Sequential protocols taken together (no statistically significant differences were observed between the two Sequential protocols) compared to 60% for the (GB-10 + BPA)–BNCT protocol. No toxicity was observed in normal tissue with any of the protocols. In terms of toxicity to precancerous tissue, there were no adverse effects associated with the improved tumor response induced by the Sequential protocols.

Within the context of recent and ongoing BNCT trials for head and neck malignancies that showed encouraging tumor control associated with dose-limiting mucositis [16,23], the search for novel BNCT strategies that improve tumor control at no extra cost in terms of mucositis is particularly relevant. Sequential BNCT would be a clinically promising BNCT modality for head and neck cancer that uses boron carriers approved for their use in humans and enhances tumor response at no extra cost in terms of toxicity in dose-limiting precancerous tissue [124].

3.6.6.5 Tumor Blood Vessel Normalization to Improve Boron Targeting for BNCT

As described in this chapter, it is well known that tumor blood vessels are characteristically hyperpermeable, leading to an elevation in IFP. In addition, proliferating tumor cells exert compressive forces on blood vessels [127]. The abnormal structure and function of tumor blood vessels impair blood flow and effective convective fluid transport, leading to defective distribution of blood-borne therapeutic agents [67]. Tumor blood vessel normalization by tailored administration of anti-angiogenic agents that downregulate vascular endothelial growth factor (VEGF), overexpressed in the majority of solid tumors, would lead to less leaky, less dilated, and less tortuous vessels; decreased IFP; increased tumor oxygenation; and improved penetration of drugs in tumors [67]. Within this context, reversible tumor blood vessel normalization prior to boron compound administration would conceivably improve boron targeting and BNCT therapeutic efficacy. Given that the anti-angiogenic monoclonal antibodies employed to induce blood vessel normalization in human subjects, rats, and mice (e.g., [67]) cannot be used in hamsters due to lack of cross-antigenicity, we developed a technique to transiently normalize aberrant tumor blood vessels in the hamster cheek pouch oral cancer model by employing thalidomide as an anti-angiogenic drug [45]. We assessed, for the first time, the effect of tumor blood vessel normalization prior to the administration of BPA on the therapeutic efficacy and potential toxicity of BNCT mediated by BPA in the hamster cheek pouch oral cancer model at the RA-3 Nuclear Reactor [46].

Blood vessel normalization was induced by two doses of thalidomide in tumor-bearing hamsters on two consecutive days. All studies in the thalidomide-treated animals were performed 48 h after the first dose of thalidomide, previously established as the window of normalization [45]. It is known from chemotherapy studies that no therapeutic benefit is obtained when the drug is administered outside the time window of normalization in each case [128]. One group of animals was treated with BPA–BNCT after treatment with thalidomide (Th + BPA–BNCT) and compared with a group treated with BPA–BNCT alone (Th – BPA–BNCT), both at an absorbed tumor dose of 4 ± 1 Gy. Overall tumor response (complete remission + partial remission) at 28 days post treatment was significantly higher for Th + BPA–BNCT than for Th – BPA–BNCT (i.e., 84% vs. 67%). The data are summarized in Table 3.6.1 (columns 6 and 7). Complete tumor response for small tumors (<10 mm^3) was significantly higher in thalidomide-pretreated animals than in animals treated with BPA–BNCT alone (73% vs. 49%). Enhanced tumor control could be attributed to transient normalization of aberrant blood vessels and temporary restoration of adequate blood flow, enough to deliver BPA at therapeutically useful levels to otherwise inaccessible tumor areas. It has been shown that the short-term administration of an anti-angiogenic agent can prune inefficient vascular sprouts [129], yielding more efficient flow paths.

Pretreatment with thalidomide did not induce statistically significant changes in overall tumor control induced by beam only, confirming that the primary therapeutic benefit of blood vessel normalization would be associated with improved tumor boron targeting.

Our working hypothesis – that transient restoration of vascular function in tumors would improve drug (BPA) delivery – led us to expect an increase in absolute boron concentration in tumors of animals treated with thalidomide prior to the administration of BPA. However, gross boron measurements by inductively coupled plasma mass spectrometry (ICP-MS) failed to reveal statistically significant differences in gross boron content between Th + and Th – animals (20 ± 8 ppm and 24 ± 6 ppm, respectively). Although this finding seemed initially counterintuitive, it is known that vascular hyperpermeability and leakiness in aberrant tumor blood vessels lead to interstitial hypertension and temporally and spatially heterogeneous blood flow. In this way, drugs will accumulate in areas that already have a sufficient amount, while inaccessible regions will continue to be inaccessible [130]. In this sense, values of gross boron concentration tell us little about the adequacy and therapeutic usefulness of boron distribution, although, admittedly, gross boron values are used for dosimetric calculations. In fact, the primary aim of blood vessel normalization is not to increase total drug uptake. Instead, it seeks to distribute drugs effectively to a larger proportion of tumor cells by fixing the flawed delivery system [67]. Within this context, it is not surprising that tumor gross boron values did not exhibit statistically significant differences induced by blood vessel normalization. α-Particle spectrometry and neutron autoradiography studies in hamster cheek pouch tumors of Th + and Th – animals showed that pretreatment with thalidomide did not increase the absolute boron content in oral tumors but improved BPA boron-targeting homogeneity [131]. BPA distribution was heterogeneous, with preferential accumulation in parenchyma, while pretreatment with thalidomide induced a more homogeneous distribution of the boron compound [9]. These findings suggest that the improvement in tumor response elicited by aberrant blood vessel normalization prior to administration of BPA could be ascribed to an improvement in BPA distribution in tumors [46].

Regarding potential radiotoxicity, pretreatment with thalidomide reduced the incidence of Grade 3/4 mucositis in dose-limiting precancerous tissue from 80% in the Th – BPA–BNCT group to 0% in the Th + BPA–BNCT group. Given that cytokines are involved in the pathogenesis of oral mucositis [132], this protective effect can be ascribed to the cytokine inhibitor activity of thalidomide [133]. The fact that thalidomide was approved by the FDA in 1997 to combat a series of conditions (e.g., [134]) fits in well with our efforts to optimize boron targeting with compounds approved for their use in humans. However, blood vessel normalization techniques in general (not amenable to be used in hamsters but approved for their use in humans) merit investigation as a way of optimizing the therapeutic advantage of BNCT in patients.

3.6.6.6 Tumor Blood Vessel Normalization + Seq-BNCT

As described in this chapter, we demonstrated that Seq-BNCT (BPA–BNCT followed by GB-10–BNCT 24 or 48 h later) improves the tumor control achieved with (GB-10 + BPA)–BNCT without increasing radiotoxicity in precancerous tissue, and that blood vessel normalization prior to BPA administration for BNCT improves tumor control. We decided to combine both modalities in an attempt to improve efficacy in the hamster cheek pouch oral cancer model without enhancing mucositis in dose-limiting precancerous tissue. The working hypothesis was that by combining both treatment modalities, we would favor homogeneous distribution of the boron compounds in tumor and improve tumor response without additional radiotoxicity in precancerous tissue. One group of hamsters was treated with thalidomide as previously described [45,46]. In the window of normalization, the animals were treated with Seq-BNCT (i.e., BPA–BNCT followed by GB-10–BNCT with a 48 h interval at a total prescribed dose to tumor of 9.9 Gy at RA-3) [124]. Although overall tumor response was already remarkably high for Seq-BNCT (91%), pretreatment with thalidomide to normalize blood vessels increased overall tumor response at 28 days post irradiation to 100% for Th + Seq-BNCT. Likewise, pretreatment with thalidomide significantly increased complete tumor remission from 68% for Seq-BNCT to 87% for Th + Seq-BNCT [9]. The combined administration of blood vessel normalization and Seq-BNCT further reduced mucositis in precancerous tissue (22% Grade 4 mucositis in Seq-BNCT vs. 0% Grade 4 mucositis in Th + Seq-BNCT). The data are summarized in Table 3.6.1 (columns 5 and 8). This effect on mucositis would be the result of the combined protective effects of thalidomide and Seq-BNCT as described here. The combined application of blood vessel normalization and Seq-BNCT would profit from the benefits of both modalities.

No changes were detected in tumor BPA gross boron concentration as measured by ICP-MS after blood vessel normalization [46] or in tumor GB-10 gross boron content after blood vessel normalization followed by BPA–BNCT (the first step in Seq-BNCT) [9]. However, an advantage of the synergism between both modalities would be the increase in GB-10 boron targeting homogeneity induced by pretreatment for aberrant blood vessel normalization followed by BPA–BNCT, as revealed by our neutron autoradiography studies [9]. This effect would be coupled to the previously described increase in BPA boron homogeneity following blood vessel normalization [46]. These findings would further stress the direct association between tumor response and the degree of boron targeting homogeneity.

3.6.6.7 Electroporation + BNCT

As described above, a critical aspect of the therapeutic efficacy of BNCT is the biodistribution and microdistribution of [^{10}B] in tumor and in the dose-limiting normal and precancerous tissues in the target volume. Electroporation (or electropermeabilization) involves the localized application of pulsed electric fields and can act as a nonspecific system to administer therapeutic agents and improve their delivery. In particular, electrochemotherapy, devised to improve tumor delivery of chemotherapeutic agents, was approved by the European Community to treat cutaneous and subcutaneous tumors [135,136]. Within this context, electroporation was proposed as a way of improving tumor boron targeting for BNCT. For example, electroporation has been used *in vitro* to improve BSH uptake in SCCVII tumor cells [137].

Continuing our efforts to improve boron targeting with boron compounds approved for their use in humans, we evaluated if electroporation could improve the targeting of [^{10}B] in BNCT mediated by the boron compound GB-10 administered intravenously, thus increasing tumor response *in vivo* in the hamster cheek pouch oral cancer model. The GB-10–BNCT protocol was particularly interesting to work on because it induces only mild mucositis in precancerous tissue with moderate tumor control. The idea of improving tumor control of a protocol that is virtually "cost-free" in terms of toxicity was particularly compelling. BNCT mediated by GB-10 (50 mg [^{10}B]/kg), at an absorbed dose of 4 Gy to tumor at RA-3, induced 48% overall tumor control (complete remission + partial remission) at 28 days post treatment with only mild mucositis in precancerous tissue. When electroporation was performed individually in each pouch tumor available for study 10 mins after administration of GB-10, and irradiation was performed 2 h 50 mins later at the same neutron fluence as for GB-10–BNCT, tumor control increased significantly to 92% at virtually no additional cost in terms of precancerous tissue mucositis. The data are summarized in Table 3.6.1 (columns 9 and 10). Furthermore, electroporation significantly improved complete remission of small tumors (<10 mm^3) from 7 to 65%. Seeking to understand the mechanisms involved in this robust improvement in tumor response, we performed boron biodistribution studies. The results of these studies revealed that electroporation increased GB-10 tumor gross boron concentration from 10 ± 2 ppm to 20 ± 10 ppm. While this increase is remarkable in and of itself, ongoing neutron autoradiography studies at our lab would suggest that changes in the parenchyma/stroma microdistribution of GB-10 in tumor would be at the root of electroporation-enhanced tumor control. The fact that no changes in gross boron content or boron microdistribution were observed in precancerous tissue would explain how the enhancement in tumor control induced by electroporation would bear no cost in terms of toxicity in dose-limiting precancerous tissue [138].

3.6.6.8 Assessing Novel Boron Compounds

Different groups worldwide have devoted (and continue to do so) considerable efforts and resources to the design and synthesis of boron compounds for BNCT. These compounds are typically tested first *in vitro* in terms of uptake and toxicity. While *in vitro* studies are contributory, it is mandatory to perform *in vivo* biodistribution studies in appropriate animal models followed by actual radiobiological studies. Biodistribution studies are orientative but are far from conclusive regarding actual biological effect, as exemplified above.

While a large proportion of our efforts have been devoted to optimizing the delivery of boron compounds authorized for use in humans to facilitate extrapolation to a clinical scenario, we have also assessed the value of novel boron compounds in collaboration with the international groups that synthesize them. Perhaps the most eloquent example of these studies is our assessment of boronated liposomes for BNCT in the hamster cheek pouch oral cancer model.

As described in this chapter, liposomes are efficient drug delivery vehicles that can selectively deliver large quantities of a wide variety of [^{10}B] agents to tumor tissue. Small liposomes (<100 nm) pass through aberrant tumor vessels [67] and passively accumulate by the EPR effect [139]. Hawthorne and coworkers recently demonstrated the therapeutic efficacy of a boron-rich liposomal system, MAC–TAC liposomes, incorporating K[$nido$-7-CH$_3$(CH$_2$)$_{15}$-7,8-C$_2$B$_9$H$_{11}$] (MAC) in the lipid bilayer and encapsulating a concentrated aqueous solution of the hydrophilic species Na$_3$[ae-B$_{20}$H$_{17}$NH$_3$] (TAC) in the aqueous core for treating BALB/c mice bearing EMT6 mammary adenocarcinomas [140]. With the aim of exploring the feasibility, safety, and efficacy of BNCT mediated by MAC–TAC liposomes to treat head and neck cancer, our first step was to perform a biodistribution study with MAC–TAC liposomes in the hamster cheek pouch oral cancer model. Absolute tumor boron values for MAC–TAC liposomes administered intravenously at 18 mg [^{10}B]/kg peaked to 67 ± 16 ppm at 48 h post administration with very favorable ratios of tumor boron concentration relative to precancerous and normal tissue (approximately 13:1 and 6:1, respectively) [83]. In vivo BNCT studies were performed at RA-3, prescribing 5 Gy absorbed dose to precancerous tissue (which resulted in an absorbed dose to tumor of 21 Gy). In these studies, dose was prescribed to precancerous tissue as the "organ at risk" in the treatment volume, based on previously reported data on mucositis in dose-limiting precancerous tissue (e.g., [12,94]). MAC–TAC–BNCT yielded an overall tumor response of 70% after 28 days. In contrast, the beam-only protocol (no boron compound administration) showed that the background dose induced an overall tumor response of only 28%. A second application of BNCT with re-administration of liposomes at an interval of 4, 6, or 8 weeks resulted in sustained tumor response rates of 70–88%, of which the complete remission ranged from 37 to 52%. The data are summarized in Table 3.6.1 (columns 11 and 12). Because of the good therapeutic outcome, it was possible to extend the follow-up period of the BNCT treated groups to 16 weeks after the first treatment.

Our working hypothesis to test the double application of BNCT was that the interval between treatments would allow the inflammatory process associated with mucositis to partially subside before the second application is delivered, thus avoiding unacceptable levels of mucositis that might be observed in precancerous tissue exposed to two full BNCT applications. In addition, it is known that inflammation associated with mucositis can induce tumor promotion, activating precancerous lesions [108,141]. Within this context, the fact that BNCT can be applied twice without enhancing mucositis is an asset not only in terms of sustaining tumor control and preventing toxicity but also in terms of inhibiting the development of novel tumors that would model the development of second primary tumors in field-cancerized oral mucosa, a frequent cause of therapeutic failure. In terms of therapeutic efficacy, lengthening overall treatment time in conventional (i.e., low-LET) radiotherapy is known to reduce toxicity at the expense of reducing tumor control probability [142]. However, in the case of BNCT that involves a mixed radiation field of high- and low-LET radiation, a repeat treatment would allow for boron retargeting of cell populations originated in cells that were refractory to the first treatment and subsequently proliferated (e.g., [88]). The intervals between treatments would minimize tumor cell repopulation.

A remarkable finding was that only mild mucositis was observed in dose-limiting precancerous tissue associated with a robust and sustained 70–88% tumor response [143].

3.6.7 BNCT Studies in a Model of Oral Precancer in the Hamster Cheek Pouch for Long-Term Follow-up

While many of our studies had proved the efficacy of BNCT to treat existing tumors, we still faced the challenge to achieve and demonstrate a long-term inhibitory effect on the development of tumors from precancerous tissue without exceeding the radiotolerance of this tissue. The clinical relevance of addressing this issue lies in the fact that second primary tumor locoregional recurrences that arise in field-cancerized tissue are a frequent cause of therapeutic failure [102]. In head and neck cancer, there is a risk of approximately 20% for second primary tumors [144]. Furthermore, the constraints imposed on therapeutic protocols by the dose-limiting nature of precancerous tissue must be assessed. Within this context, the hamster cheek pouch oral cancer model poses a unique advantage in that it allows for the study of tumors and field-cancerized tissue, unlike tumor models based on the growth of implanted tumor cells in healthy tissue. However, the aggressiveness of this model as employed in short-term tumor control studies (e.g., [12,46,94,124]) precludes long-term follow-up. Repeated application of DMBA induces hepatotoxicity [145] and causes animal decline. Given the need for long-term follow-up to better evaluate the effect of BNCT on precancerous tissue in terms of the development of secondary primary tumors from precancerous tissue, we developed a less aggressive model of precancerous tissue in the hamster cheek pouch that allows for long-term studies and mimics the kinetics of the phenomenon of field cancerization in humans more closely [146]. The hamster cheek pouch oral precancer model is induced similarly to the oral cancer model, but topical application of the carcinogen DMBA, 0.5% in mineral oil, is performed twice a week for six weeks to complete a total of 12 applications, compared to 36 applications to induce the oral cancer model for short-term tumor control studies. This less aggressive carcinogenesis protocol guarantees tumor development in ≥80% of the animals by three months after the end of the induction protocol but allows for eight-month follow-up.

Using this model, we first performed boron biodistribution studies employing GB-10 (50 mg [^{10}B]/kg), BPA (15.5 mg [^{10}B]/kg), and the combined administration of GB-10 (34.5 [^{10}B]/kg) and BPA (31 [^{10}B]/kg), and revealed that there were no statistically significant differences in boron concentration with the oral cancer model used for short-term tumor control studies that have been described in this chapter [103]. The first *in vivo* studies with this model were performed at RA-6 and were designed to assess the long-term potential inhibitory effect of BNCT on the development of new tumors (not present at the time of irradiation) and evaluate mucositis in precancerous tissue. We prescribed the following absorbed doses to precancerous tissue: 7.2 ± 1.8 Gy for GB-10–BNCT, 4.3 ± 1.8 Gy for BPA–BNCT, and 4.4 ± 1.5 Gy for (GB-10 + BPA)–BNCT. Once more, we were evaluating protocols that employed boron carriers approved for their use in humans and attempting to profit from the benefits of combining boron compounds as outlined in this chapter. All three protocols induced a statistically significant reduction in tumor development from precancerous tissue, reaching a maximum inhibition of 77–100%. The inhibitory effect of BPA–BNCT and (GB-10 + BPA)–BNCT persisted at 51% at the end of the follow-up period of eight months, whereas for GB-10–BNCT it faded after two months. Beam only (background dose) elicited a significant but transient (two months) reduction in tumor development [103]. While GB-10–BNCT was effective to control tumors in the oral cancer model [12], its inhibitory effect on tumor development from precancerous tissue in the oral precancer model was poor. The overall more conserved structure and function of precancerous tissue blood vessels compared to aberrant tumor blood vessels [147] would reduce biological efficacy of GB-10–BNCT in precancerous tissue compared to its known efficacy in tumor based on a selective effect on aberrant tumor blood vessels [12]. At eight months post treatment with BPA–BNCT or (GB-10 + BPA)–BNCT, the precancerous pouches that did not develop tumors had regained the macroscopic and histological appearance of normal (non-cancerized) pouches. The inhibitory effect on precancerous tissue would be due to cellular and/or vascular targeting of foci of precancerous tissue at a high risk of malignant transformation [148], the effect on the microenvironment of these foci [149], or both. In addition, vascular targeting by GB-10–BNCT in the (GB-10 + BPA)–BNCT protocol might affect the process of angiogenesis, pivotal to malignant transformation.

While a therapeutic effect was observed in precancerous tissue associated with early, slight/moderate reversible mucositis, normal tissue did not exhibit any effect whatsoever at the same neutron fluence. This differential effect cannot be attributed to differences in gross boron concentration between precancerous and normal tissue as revealed by biodistribution studies [103]. However, preferential microlocalization (undetectable in terms of gross boron concentration measurements) in foci of precancerous tissue at a higher risk of malignant transformation could contribute to this differential response. Our observations would suggest that the difference in response between normal tissue and precancerous tissue, at the same neutron fluence and the same gross boron concentration, would be largely due to their differences in CBE for GB-10, BPA, and GB-10 + BPA.

This study provided evidence, for the first time, that BNCT induces a long-term partial inhibitory effect on tumor development from precancerous tissue with no normal tissue radiotoxicity and without exceeding precancerous tissue tolerance. Furthermore, we showed that BNCT is capable of reverting at least the histological hallmarks of premalignancy. Thus, the BNCT protocols that were previously proved effective to control established tumors would also inhibit locoregional recurrences caused by the development of tumors in precancerous tissue, suggesting a novel application of BNCT [103].

We then went on to explore the long-term effect of a double application of BNCT (full-dose re-irradiation) mediated by BPA or GB-10 + BPA with a six-week interval between applications. A double application or re-treatment must not be confused with a fractionated treatment. In the case of a double application, dose is prescribed not to exceed the maximum tolerated dose to the "organ at risk" (or dose-limiting tissue) for each application. In the case of a fractionated treatment, dose is prescribed not to exceed the maximum tolerated dose to the organ at risk (or dose-limiting tissue) for the fractionated treatment as a whole. Following this line of thought, if standard radiotherapy is applied prescribing dose to the maximum tolerated dose to the organ at risk (or dose-limiting tissue), a re-treatment in the case of tumor recurrence would not be possible.

In a clinical scenario, a double application of BNCT (full-dose re-treatment) would potentially serve several purposes: to improve tumor control, treat a local recurrence, inhibit the development of second primary tumors from field-cancerized tissue, and/or treat second primary tumors that developed after the first BNCT treatment. The potential benefits of a double application of BNCT must be weighed against the cost in terms of mucositis in dose-limiting precancerous tissue given that oral mucositis is a dose-limiting consideration in BNCT of brain tumors (e.g., [42,150]) and head and neck cancer [16,94].

Of these clinically relevant aspects, we employed the hamster model of precancer to study the potential inhibitory effect of double BNCT on the development of new tumors (not present at the time of irradiation) from precancerous tissue and potentially dose-limiting mucositis in this tissue [104]. We performed double applications of BPA–BNCT and (GB-10 + BPA)–BNCT at 4 ± 1 Gy absorbed dose to precancerous tissue at RA-3 with a six-week interval. Here, we must point out that this study was performed at RA-3, while the single BNCT study [103] described above was performed at RA-6. Differences in irradiation components between reactors preclude a direct comparison between these studies.

We followed these groups for eight months, comparing them to a control cancerized, time-matched group left untreated. An early 100% maximum inhibition was observed for both double BNCT protocols versus control two weeks post treatment. The inhibitory effect of double BNCT on tumor development from precancerous tissue against control persisted for eight months for double BPA–BNCT and for three months for double (GB-10 + BPA)–BNCT. Double beam only (background dose) exerted a transient inhibitory effect for two months. The cancerized pouches in the double BNCT groups that had not developed tumors after eight months post double BNCT were macroscopically and histologically similar to the pouches that had never been cancerized, and differed from the cancerized pouches after eight months. The precancerous tissue treated with double BPA–BNCT or double (GB-10 + BPA)–BNCT only exhibited mild reversible mucositis that peaked one week after the first and second applications and resolved to Grade 0/1 by three weeks after each application. While none of the animals in the double (GB-10 + BPA)–BNCT group reached Grade 2 mucositis (slight), 33% and 25% of the animals in the double BPA–BNCT groups reached Grade 2 mucositis after the first and second applications, respectively. Grade 3 mucositis is considered an acceptable degree of toxicity in clinical trials [16]. An extremely valuable finding was that the second application of BNCT did not exacerbate the mucositis observed after the first treatment. It must be noted that re-treatment was performed after mucositis induced by the first application had resolved to Grade 0/1.

In the conditions of this study, we noted a slight therapeutic advantage for double BPA–BNCT versus double (GB-10 + BPA)–BNCT, associated with a slightly higher (but still low) degree of mucositis in precancerous tissue. However, the data for both protocols are encouraging and worthy of assessment in a clinical scenario [104]. Shortening the interval between applications from six to two weeks in an attempt to minimize repopulation in the target tissue (precancerous tissue) did not exacerbate mucositis but failed to improve therapeutic effect [88]. Our working hypothesis was that the shortest interval that did not result in severe mucositis would be the most therapeutically effective option because it would avoid repopulation as much as possible. However, our findings did not confirm our hypothesis. It is known that inflammation-induced tumor promotion can lead to the activation of precancerous lesions [108,141,151], and that chronic inflammation has been described as one of the hallmarks of cancer, acting on any stage of tumorigenesis (e.g., [152]). Within this context, the fact that the second application was applied earlier, before the moderate mucositis that developed after the first application had resolved, could promote tumor development. Hence, a pivotal aspect of double applications of BNCT is the interval between applications.

The best therapeutic effect was afforded by a double application of (GB-10 + BPA)–BNCT at an absorbed dose to precancerous tissue of 5 Gy in each of two applications administered four weeks apart. Inhibition of tumor development versus control was 100% up to two months post treatment and persisted at 63% eight months post treatment. Mucositis was slight in the dose-limiting field-cancerized tissue in 67% of the cases. Our studies suggest that the interval must not be shorter than 3–4 weeks, the minimum time period necessary to allow mucositis to resolve after the first application. Given that mucositis is dose-limiting, the use of radioprotectors such as histamine (a compound also approved for use in humans) coupled to BNCT was explored with good results in this model, showing a reduction in BNCT-induced mucositis in precancerous tissue without jeopardizing therapeutic efficacy [6].

The salient finding of these studies [88,104] was that retreatment with BPA–BNCT and (GB-10 + BPA)–BNCT is possible and therapeutically useful at no additional cost in terms of radiotoxicity. Thus, a double application of BNCT would be a treatment option as a pre-established protocol or scheduled as a single application followed by re-treatment in the case of recurrence.

3.6.8 BNCT Studies in a Model of Liver Metastases in BDIX Rats

Patients with multifocal, nonresectable, bilobar liver metastases from colorectal cancer who do not respond to chemotherapy can only be offered palliative treatment. This lack of therapeutic option is particularly disappointing considering that in most cases, the primary tumor in the colon can be excised and liver is the only site of metastatic spread [153]. Normal liver radiosensitivity jeopardizes the possibility of effectively treating liver metastases without exceeding normal liver radiotolerance (e.g., [154]). Furthermore, liver poorly tolerates the large treatment volumes involved in conventional radiotherapy using photons or electrons [155]. Recent interest in BNCT for the treatment of liver metastases has risen. Because this technique is based on biological targeting rather than geometric conformation, it would be adequate to treat undetectable micrometastases, a major challenge in oncological therapy [156]. *Ex-situ* BNCT mediated by BPA (systemic administration of the boron carrier, removal and neutron irradiation of the isolated organ, followed by autograft) controlled metastatic liver nodules in two patients [13]. Some biodistribution studies have been performed in liver tumor experimental models (e.g., [157–160]) and in liver metastases patients [161–163]. Some attempts have also been made to perform BNCT studies in experimental models employing endpoints indirectly related to liver control [153,155]. Studies in an experimental liver metastases model contribute to the understanding of BNCT radiobiology for this pathology and to the optimization and design of therapeutically safe and useful BNCT protocols.

We adapted a liver metastases model from Refs. [158,164]. Subcapsular inoculations of syngeneic colon cancer cells (DHD/K12/TRb) were performed by laparotomy in BDIX rats to induce the development of subcapsular tumor nodules that simulate metastases but are more amenable to follow-up [165]. Two weeks post inoculation,

100% of the animals developed localized, measurable, vascularized tumor nodules, with no peritoneal or pulmonary dissemination [8]. This model was used for biodistribution and *in vivo* BNCT studies.

Our first step was to perform boron biodistribution studies in this model. We evaluated a total of 11 administration protocols for BPA and GB-10 alone or combined at different dose levels and employing different administration routes, in keeping with the radiobiological notions contributed by our studies in the hamster cheek pouch oral cancer model. We assessed gross boron content in blood, tumor tissue, and a wide variety of potentially dose-limiting healthy tissues. This study revealed that six of the protocols assessed were potentially therapeutic, according to the established guidelines described here (no manifest toxicity, absolute boron concentration in tumor >20 ppm, and boron concentration ratios of tumor–normal tissue and tumor–blood ≥1): the single-administration protocols BPA (31.0 mg [^{10}B]/kg) i.p. administration, BPA (46.5 mg [^{10}B]/kg) i.p., and BPA (46.5 mg [^{10}B]/kg) i.p. + i.v.; and the combined boron compound administration protocols BPA (15.5 mg [^{10}B]/kg) i.v. + GB-10 (50 mg [^{10}B]/kg) i.v., BPA (31.0 mg [^{10}B]/kg) i.p. + GB-10 (34.5 mg [^{10}B]/kg) i.v., and BPA (46.5 mg [^{10}B]/kg) i.p. + GB-10 (20 mg [^{10}B]/kg) i.v. [8].

We then went on to assess the therapeutic efficacy and potential toxicity of BNCT in the liver metastases model in BDIX rats by employing one of the protocols that proved potentially therapeutic in our biodistribution studies (BPA at 46.5 mg [^{10}B]/kg i.p. + i.v.). Irradiations were performed at the RA-3 thermal facility employing a lithium-6 carbonate shield to protect the body of the animal while exposing the liver area through a collimated aperture. Due to the high boron content of the kidney at the time of irradiation (3 h post administration of BPA) [8], acrylic tabs were used to artificially distance the kidneys from the collimator aperture during irradiation and thus minimize kidney radiotoxicity. Given the presence of boron in the intraperitoneal liquid as a result of i.p. administration, the peritoneal cavity was flushed with warm saline solution prior to irradiation to remove residual BPA and minimize intestinal toxicity. The total absorbed dose administered with BPA–BNCT was 13 ± 3 Gy in tumor and 9 ± 2 Gy in healthy liver. We demonstrated an unequivocal response at 3 weeks post BNCT, with tumor remission in 100% of the animals compared to 0% remission in animals treated with beam only (background dose) or left untreated (but submitted to the same manipulation). Three weeks post BNCT, the tumor surface area posttreatment/pretreatment ratio was 0.46 ± 0.20 for BPA–BNCT, 2.7 ± 1.8 for beam only, and 4.5 ± 3.1 for the untreated group. The pretreatment tumor nodule mass of 48 ± 19 mg fell significantly to 19 ± 16 mg for BPA–BNCT but rose significantly to 140 ± 106 mg for beam only and to 346 ± 302 mg for the untreated group. A representative example is shown in Figure 3.6.3. No clinical, macroscopic, or histological liver toxicity was observed [5].

In a follow-on study [166], at 5 weeks post treatment at similar dose levels, we showed that the tumor surface posttreatment/pretreatment ratio was 0.45 ± 0.20 for BPA–BNCT, 7.8 ± 4.1 for beam only, and 12.2 ± 6.6 for the untreated group, and that the tumor nodule weight was 7.3 ± 5.9 mg for BPA–BNCT, 960 ± 620 mg for beam only, and 750 ± 480 mg for the untreated group. At this dose level, BPA–BNCT achieved a 99% reduction in tumor mass compared to the untreated group, with no associated liver toxicity. Histological grading correlated with the macroscopic endpoints evaluated.

A dose–response analysis was performed to establish threshold doses for an effective treatment [166]. In view of potential differences between prescribed and administered

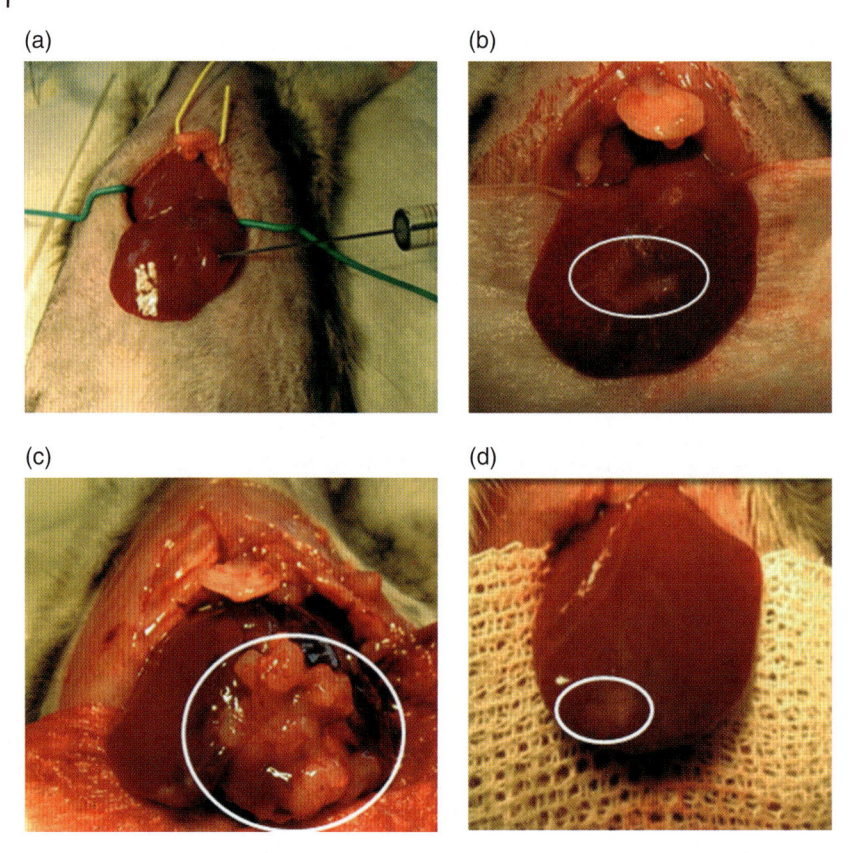

Figure 3.6.3 (a) Subcapsular inoculation of syngeneic colon cancer cells (DHD/K12/TRb) in the liver of BDIX rats. (b) Tumor nodule, 2 weeks post inoculation (pretreatment). (c) Untreated tumor (time-matched with BNCT-treated tumor). (d) Tumor, 3 weeks post BNCT.

doses, retrospective dose assessment was performed in each animal. Our working hypothesis was that these differences would be due to the well-known spread in boron concentration values (e.g., [8]), lack of reproducibility in intraperitoneal drug administration, and variations in tumor neutron flux resulting mainly from differences in positioning the animal within the shielding device. Tumor neutron flux was measured *in situ*, and pre-irradiation blood boron concentration was evaluated in each animal for retrospective dose estimation. Potential threshold doses for some degree of tumor response and significant tumor control were established at 6.1 Gy (1.8 Gy boron dose) and 9.2 Gy (5.6 Gy boron dose) absorbed dose, respectively.

We can conclude that BPA–BNCT is therapeutically effective to treat liver metastases with no liver toxicity within the study period. BNCT would offer two potential advantages over external photon radiotherapy. First, BNCT can treat both visible and undetectable liver tumors, whereas conformal radiotherapy can only treat visible liver tumors that are delineated by the physician in the treatment plan. Second, BNCT can treat multiple liver tumors without exceeding normal liver tolerance. In contrast, when 3D conformal radiotherapy is applied to the treatment of more than three liver nodules, the risk of liver failure is a significant concern [167].

3.6.9 BNCT Studies in a Model of Diffuse Lung Metastases in BDIX Rats

Metastatic lung disease is still a leading cause of death. Surgery, radio, and chemotherapy have failed to improve survival satisfactorily, and the overall prognosis for patients is poor. In conventional external radiotherapy, it is difficult to deliver therapeutic doses to malignant cells without causing radiation pneumonitis in the healthy lung [168]. Within this context, the search for more selective and less toxic treatment strategies is warranted, particularly in view of the marked radiosensitivity of the healthy lung [169]. BNCT has been proposed for the treatment of diffuse, nonresectable tumors in the lung. BNCT can offer a dose gradient between tumor and normal cells if $[^{10}B]$ atoms accumulate preferentially in tumor cells. Furthermore, due to the fact that BNCT is based on biological rather than geometric targeting, it would be well suited to treat diffuse micrometastases. In addition, with BNCT it is unnecessary to adjust for breathing motions [170].

The effects of BNCT mediated by BPA on the normal lung of Fischer 344 rats have been assessed to establish RBE and CBE factors [171]. In addition, functional and histological changes in the normal lung of Fischer 344 rats after BNCT mediated by BPA were assessed [172]. BPA biodistribution studies were performed in an experimental model of lung metastases of colon carcinoma DHD/K12/TRb cells in syngeneic BDIX rats using neutron autoradiography and α-spectrometry to assess boron concentration and boron microdistribution, respectively [170,173]. In the same model, Bakeine *et al.* [169] established the feasibility of using BNCT mediated by BPA in an *in vitro/in vivo* system. Suzuki *et al.* [168] studied the effect of BNCT mediated by BPA in an experimental model of ectopic tumors, implanted in the thoracic cavity of mice to mimic pleural mesothelioma.

Suzuki *et al.* [174] went on to treat two patients with diffuse or multiple pleural tumors. The same group then treated a patient with recurrent lung cancer in the previously irradiated chest wall with two fractions of BNCT [26]. All of these studies have afforded encouraging results but undoubtedly leave room for improvement.

Within this context, based on the work of Bortolussi *et al.* [170], we tailored the model of colon carcinoma diffuse lung metastases in BDIX rats to perform biodistribution studies with five different administration protocols employing the boron compounds BPA and GB-10, administered alone or in combination [175]. DHD/K12/TRb colon carcinoma cells were injected in the jugular vein of syngeneic BDIX rats. Three weeks was the time post-injection selected for biodistribution studies, because it guaranteed reproducible development of abundant vascularized lung metastases while preserving enough healthy lung tissue for evaluation. Based on these biodistribution studies, we identified two administration protocols worthy of evaluation in *in vivo* BNCT studies: BPA (46.5 mg $[^{10}B]$/kg) i.p. and the combined administration of BPA (31 mg $[^{10}B]$/kg) i.p. + GB-10 (34.5 mg $[^{10}B]$/kg) i.v. The boron concentration values for the BPA protocol, 3 h post administration, were 23 ± 7 ppm for metastases, 12 ± 7 ppm for lung, and 14 ± 2 ppm for blood, and the boron concentration values for the BPA + GB-10 protocol were 33 ± 9 ppm for metastases, 28 ± 6 ppm for lung, and 32 ± 6 ppm for blood.

To date, only BPA has been explored as a boron carrier for BNCT of diffuse lung metastases. We went on to perform *in vivo* BNCT studies in the diffuse lung metastases model at RA-3 employing the BPA and combined BPA + GB-10 administration protocols. We designed and built a lithium-6 carbonate shielding device to protect the body of the rat from thermal neutrons while allowing the delivery of a therapeutically useful and uniform dose to the lungs [176]. The effect of two dose levels (minimum absorbed dose to tumor 4 Gy and 8 Gy) was assessed two weeks after treatment, employing the endpoints of: percentage surface of lung lobes occupied by metastases at a macroscopic level, percentage area occupied by metastases in histological sections of lung lobes, and percentage lung weight/body weight. All three endpoints revealed that both BNCT protocols at both dose levels induced a halt in tumor growth. The beam-only (background dose) protocol failed to induce any effect in comparison with untreated animals with lung metastases. Survival studies showed that BNCT extended the life of the animals by 35% compared to untreated animals bearing lung metastases. No ostensible clinical, macroscopic, or histological toxicity was observed in the lung of BNCT-treated animals [177], suggesting that it would be possible to escalate the dose to optimize efficacy. At these dose levels, it was not possible to detect differences in efficacy between the BPA–BNCT protocol and the (GB-10 + BPA)–BNCT protocol.

3.6.10 BNCT Studies in a Model of Arthritis in Rabbits

Rheumatoid arthritis (RA) is an autoimmune disease that is characterized by the accumulation and proliferation of inflammatory cells in the synovial joint lining, resulting in the formation of pannus tissue that invades and destroys adjacent articular cartilage, ligaments, and bone [178]. In this way, the normal synovial lining is replaced by a highly vascularized mass of inflammatory tissue. The first line of treatment for RA consists of reducing synovial inflammation with drugs. However, several joints are often refractive to this type of treatment. Surgical removal of excess inflamed tissue is considered the most effective treatment for persistent synovitis in RA patients. Although both open and arthroscopic synovectomy allow for 80% removal of the inflamed synovial membrane and grant alleviation of the symptoms for 2–5 years, the synovial membrane is replenished within six months after surgery and inflammation recurs [179]. A less invasive option is radiation synovectomy using β-emitting radionuclides injected directly into the joint space. Success rates of up to 80% have been reported for 2–5 years. However, this procedure is not widely used due to potential healthy tissue irradiation caused by leakage of the β-emitter away from the joint [180]. Pathological synovium is the target in RA. Because this tissue and local malignancies share common features (e.g., [181–183]), BNCT has been proposed as an alternative approach to synovectomy [74,180] and termed *boron neutron capture synovectomy* (BNCS). The application of BNCS to the treatment of arthritis has been investigated in the antigen-induced arthritis (AIA) rabbit stifle joint 72 h post treatment employing potassium dodecahydrodecaborate as the boron carrier [179]. Van Lent *et al.* [184] demonstrated selective depletion of macrophages in the synovial lining of normal murine knee joints five days post treatment with BNCS mediated by boronated liposomes. BNCS would avoid the problems associated with the leakage of β-emitters from the joint while still profiting from the advantages of radiation synovectomy relative to surgery: increased potential to destroy all the inflamed tissue, reduced risk of blood clots and infection, the need for only local anesthesia, and the fact that the treatment is carried out on an outpatient basis with no associated pain or discomfort and would require no rehabilitation [180,185].

Three different boron carriers administered directly in the joint have been explored experimentally for BNCS, namely $K_2B_{12}H_{12}$ [180] and two boronated liposomes [74,184]. Each of these compounds has advantages and disadvantages for BNCS. However, importantly, none of these boron carriers have been approved for their use in humans. An ideal boron compound for BNCS should be nontoxic and biochemically stable. Residence time in the synovium should be enough to allow for irradiation to be completed. Boron uptake in target tissue (pathological synovium) and healthy tissues (e.g. cartilage, muscle, tendon, and skin) should be such that a therapeutic radiation dose can be delivered to the target tissue without exceeding the radiotolerance of the healthy tissues.

We performed biodistribution studies for the first time with boron compounds authorized for their use in humans (GB-10 and BPA) in a model of AIA in New Zealand rabbits. The diameter of a human finger is approximately the same as the diameter of the arthritic rabbit knee. In this sense, the rabbit experimental model is better suited for BNCS studies than the murine models. The animals were submitted to induction of AIA in the knee joints in keeping with a modification of a standard protocol [186]. Briefly, AIA was induced by two successive intradermal immunizations, 15 days apart, with ovoalbumin emulsion, 1:1 in complete Freund's adjuvant. Ten days later, two successive injections, 10 days apart, were performed in the knee articulation of both hind legs of each rabbit. Approximately 50–60 days after the first immunization, the onset of the disease was verified by clinical examination (swelling, and pain or tenderness on palpation) and magnetic resonance imaging (MRI) studies as described in Refs. [187,188]. Histopathological studies revealed that the clinical and MRI signs of disease were associated with the presence of inflammatory infiltrate, synovial hyperplasia, fibrosis, and/or pannus.

We tested intra-articular (i.a.) administration protocols employing BPA (0.7 mg [^{10}B]) i.a. or GB-10 (5 mg [^{10}B] or 50 mg [^{10}B]) i.a. in a time range of 13–85 min post administration, and i.v. (systemic) administration protocols employing BPA (15.5 mg [^{10}B]/kg) i.v., GB-10 (50 mg [^{10}B]/kg) i.v., or the combined administration of BPA (15.5 mg [^{10}B]/kg) + GB-10 (50 [[^{10}B]]/kg) 3 h post administration [185]. The i.a. administration protocols at <40 min post administration for BPA and GB-10 and the i.v. administration protocols for GB-10 and GB-10 + BPA at 3 h post administration exhibited therapeutically useful boron concentrations (>20 ppm) in pathological synovium.

Having evidenced that both i.a. and systemic administration protocols exhibited potentially useful boron biodistribution values, we decided to focus on the i.a. administration protocols. Fifteen minutes post administration, the boron concentration values for the BPA i.a. protocol were 159 ± 65 ppm for synovium, 128 ppm for cartilage, 2 ± 1 ppm for muscle, 3 ± 1 ppm for skin, and 0.5 ± 0.03 ppm for blood. At the same time-point, the boron concentration values for the GB-10 i.a. protocol were 378 ± 256 ppm for synovium, 206 ppm for cartilage, 7 ± 3 ppm for muscle, 157 ± 62 ppm for skin, and 3.6 ± 0.5 ppm for blood. The low blood boron values for the i.a. protocols revealed scarce release of the boron carrier from the articulation into the bloodstream. With the i.a. protocols, boron uptake is maximized in the target volume, reducing the shielding requirements for neutron irradiation. An additional advantage is that the majority of articular joints are located far from the body's radiation-sensitive organs. The fact that therapeutically useful boron concentrations are retained in the synovium up to 40 min post injection guarantees that the residence time of boron in the joint would be sufficient for the irradiation times (approximately 10 min) needed to reach therapeutically useful doses. The administration of the boron compounds to a site immediately adjacent to the diseased tissue results in extremely high uptake levels. Synovium boron concentration achieved with the i.a. protocols was 5–10 times the "ideal" target tissue boron concentration of 30 ppm established for the treatment of target tissue with BNCT (e.g., [2,12]). As previously mentioned, high absolute boron concentration in target tissue allows for shorter irradiation times to reach the desired dose in target tissue. A reduction in irradiation time reduces background dose and maximizes the boron component of the dose. Admittedly, the spread in boron concentration values was large. This is an issue of concern in BNCT in general because it precludes adequate treatment planning, imposes constraints on the analysis of biological responses as a function of dose, and can cause underdosing or overdosing (e.g., [3]). Within this context, dose prescription to healthy tissues must be conservative to ensure that their radiotolerance is not exceeded. In the case of the i.a. administration, the spread in values is particularly high, perhaps due to potential variations in the local injection procedure. Nevertheless, the advantages of an i.a. administration protocol would conceivably outweigh the disadvantages of a larger spread in boron concentration values. If an *in vivo*, noninvasive, online method to measure boron concentration in tissues was available in the future, the characteristic spread in boron concentration values would be less of a concern.

Since an important objective of a synovectomy procedure is to spare articular cartilage from permanent damage, cartilage was considered as the dose-limiting tissue in dosimetric analysis. The biodistribution data posed a considerable concern in that synovium boron-targeting selectivity versus cartilage was marginal. Based on the biodistribution data alone, an actual *in vivo* BNCS study might have been dismissed on the basis of potential damage to cartilage. It must be stressed that, whereas in the treatment of a malignant disease moderate/severe side effects might be acceptable, in the case of a disease that is not life-threatening, the safety requirements are more stringent. Extensive research has shown that articular cartilage is one of the least sensitive structures to radiation damage [189]. Detailed dosimetric calculations suggested that it would be possible to deliver therapeutically useful doses to synovium without significant damage to cartilage despite the very slight differences in boron concentration between synovium and cartilage [185].

Within this context, we performed low-dose BNCS studies mediated by BPA or GB-10 administered i.a. in AIA rabbits [190], in keeping with the administration protocols selected from the biodistribution studies described above [185]. Fifteen minutes post administration of the boron compounds, irradiations were performed with the thermal beam of the RA-1 Reactor (Buenos Aires, Argentina) at a thermal neutron flux of approximately 1.6×10^8 n/cm^{2}sec to the target area (knee joint). The geometric setup involves no body shielding, and we relied on boron retention in the joint to exert a selective effect in the pathological articulation. The animals were placed face-up on a device designed ad hoc to transport the rabbit to the irradiation position, with their hind legs toward the core of the Reactor. Irradiations lasted 10 min, resulting in an approximate neutron fluence of 2.9×10^{11} n/cm^2 to deliver

2.4 Gy or 3.9 Gy to synovium for BPA and GB-10, respectively. Untreated AIA animals and healthy animals were used as controls.

Throughout the follow-up period of two months, the rabbits did not exhibit any clinical signs of toxicity. A follow-up accomplished two months after BNCS showed that the hind leg knee joints of all the rabbits treated with BPA–BNCS or GB-10–BNCS were no longer swollen or painful on palpation. Serum levels of the pro-inflammatory cytokines tumor necrosis factor-α and interferon-γ decreased post BNCS in 67 and 83% of the rabbits, respectively. However, the difference between mean pre- and posttreatment values did not reach statistical significance, conceivably because BNCS is a local treatment and a robust systemic effect might take longer to achieve. A follow-up accomplished after two months showed that the MRI images of the AIA knee joints treated with BNCS were similar to those of control healthy joints in 100% of the cases (i.e., with no areas of necrosis or peri-articular effusion) but markedly different from untreated AIA joints that exhibited hydroarthrosis in the joint space, alterations in subchondral bone, and alterations in the peri-articular soft tissue. The histological analysis of the synovial membranes obtained postmortem, two months after treatment, revealed that in 70–100% of the fields corresponding to cases of AIA joints treated with BNCS, the histological features were similar to those of healthy joints (i.e., no synovial hyperplasia, scarce or no lymphoplasmocytic infiltrate, and no alterations in vascularization) but very different from those of untreated AIA joints, which exhibited synovial hyperplasia, angiogenesis, edema, and abundant inflammatory infiltrate, as previously described in Sanchez-Pernaute et al. [186].

The follow-up in this study was enough to show reversal of clinical symptoms, MRI and histological features of AIA, with no evidence of toxicity. Both BPA–BNCS and GB-10–BNCS, even at these very low dose levels, were therapeutically effective with no ostensible differences between the protocols. While BPA–BNCT would target malignant tissue on a cell-by-cell basis, GB-10–BNCT would mainly target aberrant blood vessels while preserving normal blood vessels in healthy tissue [12]. It is well known that neovascularization of the rheumatoid synovium is essential to perpetuate an angiogenic disease such as RA [191,192]. Within this context, GB-10 as a boron carrier for BNCS, alone or in combination with BPA, would be particularly suited to treat RA. It would be particularly contributory to explore the therapeutic efficacy of the combined i.a. administration of GB-10 + BPA.

This study [190] suggests that considerably lower doses to target tissue than anticipated from radiation synovectomy studies would be necessary to achieve a therapeutic effect with BNCS mediated by i.a. administration of BPA or GB-10. This is an asset and minimizes (or altogether prevents) associated toxicity to healthy tissues, such as cartilage, and would allow for re-treatment if necessary.

3.6.11 Preclinical BNCT Studies in Cats and Dogs with Head and Neck Cancer with no Treatment Option

Preclinical trials in animals with spontaneous tumors and clinical trials in humans are performed on the patients that are most difficult to treat. A clinical trial is designed to explore the feasibility, safety, and potential toxicity of a new treatment modality and, if possible, to monitor therapeutic efficacy, improve the clinical condition of the patient, and prolong survival. The patients included in a clinical trial are those that have no treatment option and/or have been refractory to standard treatments. Hence, the chances of showing efficacy are smaller than when a treatment modality is used as a first line of therapy. This issue must be considered when we compare the outcome of a standard treatment with that of an experimental treatment.

An experimental treatment that can be applied safely to patients that have already been treated with standard therapy will be advantageous. BNCT offers a dose gradient between tumor (or target tissue in general) and dose-limiting tissues in terms of selective boron targeting, and higher RBE and CBE values for target tissue than for healthy tissue. This feature is an asset in terms of minimizing the radiation dose that is delivered to healthy tissues that have already been exposed to standard radiotherapy.

Earlier and ongoing BPA–BNCT preclinical trials, in terminal cats and dogs with head and neck cancer that do not have a treatment option, seek to contribute clinically representative data to the knowledge of BNCT for head and neck cancer. Our data to date on boron biodistribution studies and BNCT studies in three cats treated with low-dose BPA–BNCT at RA-1 [193], three cats treated at a higher dose level of BPA–BNCT at RA-6 [194], and one dog treated at RA-6 reveal the efficacy of BPA–BNCT to improve the clinical condition of the animals and partially control tumors with only slight associated toxicity in healthy tissues. Very importantly, these studies showed that an animal patient whose tumor recurred locally several months after BNCT can be re-treated with a second full dose of BNCT with excellent results in terms of tumor control with only mild associated toxicity. Full-dose re-treatment with standard radiotherapy would not be an option due to toxicity constraints. However, our translational work in the hamster cheek pouch oral cancer and oral precancer models (described in this chapter) encouraged us to assess retreatment with BNCT, leading us to demonstrate its feasibility, safety, and efficacy.

3.6.12 Future Perspectives

The recent initiation of BNCT clinical trials employing hospital-based accelerators rather than nuclear reactors as the neutron source will allow for new and more numerous trials for different pathologies. *In vivo* translational research in adequate experimental animal models will continue to contribute to the optimization of BNCT for different pathologies and the design of safe and effective treatment protocols. Research into novel, potentially "closer to ideal" boron carriers and studies performed to optimize boron targeting with boron carriers approved for use in humans are complementary strategies to optimize BNCT. The combined application of BNCT and other treatment modalities is a promising approach.

References

1 Locher, L 1936. Biological effects and therapeutic possibilities of neutrons. *Am. J. Roentgenol.*, vol. 36, pp. 1–13.

2. Coderre, JA & Morris, GM 1999. The radiation biology of boron neutron capture therapy. *Radiat. Res.*, vol. 151, pp. 1–18.
3. Hopewell, JW, Morris, GM, Schwint, A, & Coderre, JA 2011. The radiobiological principles of boron neutron capture therapy: a critical review. *Appl. Radiat. Isot.*, vol. 69, pp. 1756–1759.
4. Barth, RF, Coderre, JA, Vicente, MGH, & Blue, TE 2005. Boron neutron capture therapy of cancer: current status and future prospects. *Clin. Cancer Res.*, vol. 11, pp. 3987–4002.
5. Pozzi, ECC, Cardoso, JE, Colombo, LL, Thorp, S, Monti Hughes, A, Molinari, AJ, Garabalino, MA, Heber, EM, Miller, M, Itoiz, ME, Aromando, RF, Nigg, DW, Quintana, J, Trivillin, VA, & Schwint, AE 2012. Boron neutron capture therapy (BNCT) for liver metastasis: therapeutic efficacy in an experimental model. *Radiat. Environ. Biophys.*, vol. 51, pp. 331–339.
6. Monti Hughes, A, Pozzi, ECC, Thorp, SI, Curotto, P, Medina, VA, Martinel Lamas, DJ, Rivera, ES, Garabalino, MA, Farías, RO, Gonzalez, SJ, Heber, EM, Itoiz, ME, Aromando, RF, Nigg, DW, Trivillin, VA, & Schwint, AE 2015a. Histamine reduces boron neutron capture therapy-induced mucositis in an oral precancer model. *Oral Dis.*, vol. 21, pp. 770–777.
7. Chanana, AD, Capala, J, Chadha, M, Coderre, JA, Diaz, AZ, Elowitz, EH, Iwai, J, Joel, DD, Liu, HB, Ma, R, Pendzick, N, Peress, NS, Shady, MS, Slatkin, DN, Tyson, GW, & Wielopolski, L 1999. Boron neutron capture therapy for glioblastoma multiforme: interim results from the phase I/II dose-escalation studies. *Neurosurgery*, vol. 44, no. 6, pp. 1182–1192; discussion 1192–1193.
8. Garabalino, MA, Monti Hughes, A, Molinari, AJ, Heber, EM, Pozzi, ECC, Cardoso, JE, Colombo, LL, Nievas, S, Nigg, DW, Aromando, RF, Itoiz, ME, Trivillin, VA, & Schwint, AE 2011. Boron neutron capture therapy (BNCT) for the treatment of liver metastases: biodistribution studies of boron compounds in an experimental model. *Radiat. Environ. Biophys.*, vol. 50, pp. 199–207.
9. Molinari, AJ, Thorp, SI, Portu, AM, Saint Martin, G, Pozzi, EC, Heber, EM, Bortolussi, S, Itoiz, ME, Aromando, RF, Monti Hughes, A, Garabalino, MA, Altieri, S, Trivillin, VA, & Schwint, AE 2015. Assessing advantages of sequential boron neutron capture therapy (BNCT) in an oral cancer model with normalized blood vessels. *Acta Oncol.*, vol. 54, no. 1, pp. 99–106.
10. Schwint, AE & Trivillin, VT 2015. 'Close-to-ideal' tumor boron targeting for boron neutron capture therapy is possible with 'less-than-ideal' boron carriers approved for use in humans. *Ther. Deliv.*, vol. 6, no. 3, pp. 269–272.
11. Luderer, MJ, de la Puente, P, & Azab, AK 2015. Advancements in tumor targeting strategies for boron neutron capture therapy. *Pharm. Res.*, vol. 32, no. 9, pp. 2824–2836.
12. Trivillin, VA, Heber, EM, Nigg, DW, Itoiz, ME, Calzetta, O, Blaumann, H, Longhino, J, & Schwint, AE 2006. Therapeutic success of boron neutron capture therapy (BNCT) mediated by a chemically non-selective boron agent in an experimental model of oral cancer: a new paradigm in BNCT radiobiology. *Radiat. Res.*, vol. 166, pp. 387–396.
13. Zonta, A, Prati, U, Roveda, L, Ferrari, C, Zonta, S, Clerici, AM, Zonta, C, Pinelli, T, Fossati, F, Altieri, S, Bortolussi, S, Bruschi, P, Nano, R, Barni, S, Chiari, P, & Mazzini, G 2006. Clinical lessons from the first applications of BNCT on unresectable liver metastases. *J. Phys. Conference Series*, vol. 41, pp. 484–495.

14. Matsumura, A, Yamamoto, T, Tsurubuchi, T, Matsuda, M, Shirakawa, M, Nakai, K, Endo, K, Tokuue, K, & Tsuboi, K 2009. Current practices and future directions of therapeutic strategy in glioblastoma: survival benefit and indication of BNCT. *Appl. Radiat. Isot.*, vol. 67, no. 7–8 Suppl, pp. 12–14.
15. Barth, RF, Vicente, MG, Harling, OK, Kiger, WS 3rd, Riley, KJ, Binns, PJ, Wagner, FM, Suzuki, M, Aihara, T, Kato, I, & Kawabata, S 2012. Current status of boron neutron capture therapy of high grade gliomas and recurrent head and neck cancer. *Radiat. Oncol.*, vol. 7, pp. 146.
16. Kankaanranta, L, Seppälä, T, Koivunoro, H, Saarilahti, K, Atula, T, Collan, J, Salli, E, Kortesniemi, M, Uusi-Simola, J, Välimäki, P, Mäkitie, A, Seppänen, M, Minn, H, Revitzer, H, Kouri, M, Kotiluoto, P, Seren, T, Auterinen, I, Savolainen, S, & Joensuu, H 2012. Boron neutron capture therapy in the treatment of locally recurred head-and-neck cancer: final analysis of a phase I/II trial. *Int. J. Radiat. Oncol. Biol. Phys.*, vol. 82, no. 1, pp. 67–75.
17. Suzuki, M, Kato, I, Aihara, T, Hiratsuka, J, Yoshimura, K, Niimi, M, Kimura, Y, Ariyoshi, Y, Haginomori, S, Sakurai, Y, Kinashi, Y, Masunaga, S, Fukushima, M, Ono, K, & Maruhashi, A 2014. Boron neutron capture therapy outcomes for advanced or recurrent head and neck cancer. *J. Radiat. Res.*, vol. 55, pp. 146–153.
18. Barth, RF 2015. From the laboratory to the clinic: how translational studies in animals have led to clinical advances in boron neutron capture therapy. *Appl. Radiat. Isot.*, vol. 106, pp. 22–28.
19. Barth, RF 2009. Boron neutron capture therapy at the crossroads: challenges and opportunities. *Appl. Radiat. Isot.*, vol. 67, no. 7–8 Suppl, pp. S3–S6.
20. Paul, SM, Mytelka, DS, Dunwiddie, CT, Persinger, CC, Munos, BH, Lindborg, SR & Schacht, AL 2010. How to improve R&D productivity: the pharmaceutical industry's grand challenge. *Nat. Rev. Drug Discov.*, vol. 9, pp. 203–214.
21. González, SJ, Bonomi, MR, Santa Cruz, GA, Blaumann, HR, Calzetta Larrieu, OA, Menéndez, P, Jiménez Rebagliati, R, Longhino, J, Feld, DB, Dagrosa, MA, Argerich, C, Castiglia, SG, Batistoni, DA, Liberman, SJ, & Roth, BM 2004. First BNCT treatment of a skin melanoma in Argentina: dosimetric analysis and clinical outcome. *Appl. Radiat. Isot.*, vol. 61, no. 5, pp. 1101–1105.
22. Miyatake, S, Kawabata, S, Hiramatsu, R, Furuse, M, Kuroiwa, T, & Suzuki, M 2014. Boron neutron capture therapy with bevacizumab may prolong the survival of recurrent malignant glioma patients: four cases. *Radiat. Oncol*, vol. 9, pp. 6.
23. Wang, LW, Chen, YW, Ho, CY, Hsueh Liu, YW, Chou, FI, Liu, YH, Liu, HM, Peir, JJ, Jiang, SH, Chang, CW, Liu, CS, Wang, SJ, Chu, PY, & Yen, SH 2014. Fractionated BNCT for locally recurrent head and neck cancer: experience from a phase I/II clinical trial at Tsing Hua Open-Pool Reactor. *Appl. Radiat. Isot.*, vol. 88, pp. 23–27.
24. Yanagie, H, Higashi, S, Seguchi, K, Ikushima, I, Fujihara, M, Nonaka, Y, Oyama, K, Maruyama, S, Hatae, R, Suzuki, M, Masunaga, S, Kinashi, T, Sakurai, Y, Tanaka, H, Kondo, N, Narabayashi, M, Kajiyama, T, Maruhashi, A, Ono, K, Nakajima, J, Ono, M, Takahashi, H, & Eriguchi, M 2014. Pilot clinical study of boron neutron capture therapy for recurrent hepatic cancer involving the intra-arterial injection of a (10) BSH-containing WOW emulsion. *Appl. Radiat. Isot.*, vol. 88, pp. 32–37.
25. Yanagie, H, Higashi, S, Seguchi, K, Ikushima, I, Oyama, K, Nonaka, Y, Maruyama, S, Hatae, R, Sairennji, T, Takahashi, S, Suzuki, M, Masunaga, S, Kinashi, T, Sakurai, Y, Tanaka, H, Maruhashi, A, Ono, K, Nakajima, J, Ono, M, Takahashi, H, & Eriguchi, M 2012. Pilot clinical study of boron neutron capture therapy for recurrent hepatic cancer

and gastric cancer. Oral communication at the 15th International Congress on Neutron Capture Therapy, Tsukuba, Japan, September 10–14.

26 Suzuki, M, Suzuki, O, Sakurai, Y, Tanaka, H, Kondo, N, Kinashi, Y, Masunaga, SI, Maruhashi, A, & Ono, O 2012. Reirradiation for locally recurrent lung cancer in the chest wall with boron neutron capture therapy (BNCT). *Int. Cancer Conf. J.*, vol. 1, no. 4, pp. 235–238.

27 Sasaoka, S 2012. The first clinical trial of Boron Neutron Capture Therapy using ^{10}B-para-boronophenylalanine for treating extra-mammary Paget's disease. poster communication presented at the 15th International Congress on Neutron Capture Therapy, Tsukuba, Japan, September 10–14.

28 Kreiner, AJ, Castell, W, Di Paolo, H, Baldo, M, Bergueiro, J, Burlon, AA, Cartelli, D, Vento, VT, Kesque, JM, Erhardt, J, Ilardo, JC, Valda, AA, Debray, ME, Somacal, HR, Sandin, JC, Igarzabal, M, Huck H, Estrada, L, Repetto, M, Obligado, M, Padulo, J, Minsky, DM, Herrera, M, Gonzalez, SJ, & Capoulat, ME 2011. Development of a tandem-electrostatic-quadrupole facility for accelerator-based boron neutron capture therapy. *Appl. Radiat. Isot.*, vol. 69, no. 12, pp. 1672–1675.

29 Yoshioka, M, Kurihara, T, Kobayashi, H, Matsumoto, H, Matsumoto, N, Kumada, H, Matsumura, A, Sakurai, H, Tanaka, S, Sugano, T, Hashirano, T, Nakashima, H, Nakamura, T, Kiyanagi, Y, Hiraga, F, Ohba, T, & Okazaki, K 2014. Construction of Accelerator-based BNCT facility at Ibaraki Neutron Medical Research Center. *Proceedings of 16th International Congress on Neutron Capture Therapy*, Helsinki, Finland, pp. 66–67.

30 Kreimann, EL, Itoiz, ME, Dagrosa, A, Garavaglia, R, Farías, S, Batistoni, D, & Schwint, AE 2001a. The hamster cheek pouch as a model of oral cancer for boron neutron capture therapy studies: selective delivery of boron by boronophenylalanine. *Cancer Res.*, vol. 61, no. 24, pp. 8775–8781.

31 Heber, E, Trivillin, VA, Nigg, D, Kreimann, EL, Itoiz, ME, Rebagliati, RJ, Batistoni, D, & Schwint, AE 2004. Biodistribution of GB-10 (Na$_2$^{10}B$_{10}$H$_{10}$) compound for boron neutron capture therapy (BNCT) in an experimental model of oral cancer in the hamster cheek pouch. *Arch. Oral Biol.*, vol. 49, no. 4, pp. 313–324.

32 Heber, EM, Trivillin, VA, Nigg, DW, Itoiz, ME, Gonzalez, BN, Rebagliati, RJ, Batistoni, D, Kreimann, EL, & Schwint, AE 2006. Homogeneous boron targeting of heterogeneous tumors for boron neutron capture therapy (BNCT): chemical analyses in the hamster cheek pouch oral cancer model. *Arch. Oral Biol.*, vol. 51, no. 10, pp. 922–929.

33 Wittig, A, Collette, L, Moss, R, & Sauerwein, WA 2009. Early clinical trial concept for boron neutron capture therapy: a critical assessment of the EORTC trial 11001. *Appl. Radiat. Isot.*, vol. 67, no. 7–8, pp. S59–S62.

34 Moss, RL 2014. Critical review, with an optimistic outlook, on Boron Neutron Capture Therapy (BNCT). *Appl. Radiat. Isot.*, vol. 88, pp. 2–11.

35 Farr, LE, Sweet, WH, Robertson, JS, Foster, CG, Locksley, HB, Sutherland, DL, Mendelsohn, ML, & Stickley, EE 1954. Neutron capture therapy with boron in the treatment of glioblastoma multiforme. *Am. J. Roentgenol. Radium. Ther. Nucl. Med.*, vol. 71, pp. 279–293.

36 Locksley, HB, & Farr, LE 1955. The tolerance of large doses of sodium borate intravenously by patients receiving neutron capture therapy. *J. Pharmacol. Exp. Ther.*, vol. 114, no. 4, pp. 484–489.

37 Kabalka, GW, & Yao, ML 2006. The synthesis and use of boronated amino acids for boron neutron capture therapy. *Anticancer Agents Med. Chem.*, vol. 6, no. 2, pp. 111–125.

38. Mishima, Y, Ichihashi, M, Nakanishi, T, Tsui, M, Ueda, M, Nakagawa, T, & Suzuki, T 1983. Cure of Malignant melanoma by single thermal neutron capture treatment using melanoma seeking compounds: ^{10}B/melanogenesis interaction to *in vitro/in vivo* radiobiological analysis to preclinical studies. In RG Fairchild & G Brownell (eds), *Proceedings of the First International Symposium on Neutron Capture Therapy*, Brookhaven National Laboratory, Upton, NY, pp. 355–364.
39. Mishima, Y, Honda, C, Ichibashi, M, Obara, H, Hiratsuka, J, Fukuda, H, Karashima, H, Kand, KTK, & Yoshino, K 1989a. Treatment of malignant melanoma by single neutron capture therapy with melanoma-seeking ^{10}B-compound. *Lancet*, vol. 1, pp. 388–389.
40. Mishima, Y, Ichihashi, M, Tsuji, M, Hatta, S, Ueda, M, Honda, C, & Suzuki, T 1989b. Treatment of malignant melanoma by selective thermal neutron capture therapy using melanoma-seeking compound. *J. Invest. Dermatol.*, vol. 92, pp. 321S–325S.
41. Wittig, A, Sauerwein, WA, & Coderre, JA 2000. Mechanisms of transport of p-borono-phenylalanine through the cell membrane *in vitro*. *Radiat. Res.*, vol. 153, no. 2, pp. 173–180.
42. Kankaanranta, L, Seppälä, T, Koivunoro, H, Välimäki, P, Beule, A, Collan, J, Kortesniemi, M, Uusi-Simola, J, Kotiluoto, P, Auterinen, I, Serèn, T, Paetau, A, Saarilahti, K, Savolainen, S, & Joensuu, H 2011. L-boronophenylalanine-mediated boron neutron capture therapy for malignant glioma progressing after external beam radiation therapy: a Phase I study. *Int. J. Radiat. Oncol. Biol. Phys.*, vol. 80, no. 2, pp. 369–376.
43. Wongthai, P, Hagiwara, K, Miyoshi, Y, Wiriyasermkul, P, Wei, L, Ohgaki, R, Kato, I, Hamase, K, Nagamori, S, & Kanai, Y 2015. Boronophenylalanine, a boron delivery agent for boron neutron capture therapy, is transported by ATB0,+, LAT1 and LAT2. *Cancer Sci.*, vol. 106, no. 3, pp. 279–286.
44. Kawabata, S, Miyatake, S, Nonoguchi, N, Hiramatsu, R, Iida, K, Miyata, S, Yokoyama, K, Doi, A, Kuroda, Y, Kuroiwa, T, Michiue, H, Kumada, H, Kirihata, M, Imahori, Y, Maruhashi, A, Sakurai, Y, Suzuki, M, Masunaga, S, & Ono, K 2009. Survival benefit from boron neutron capture therapy for the newly diagnosed glioblastoma patients. *Appl. Radiat. Isot.*, vol. 67, no. 7–8 Suppl, pp. S15–S18.
45. Molinari, AJ, Aromando, RF, Itoiz, ME, Garabalino, MA, Monti Hughes, A, Heber, EM, Pozzi, EC, Nigg, DW, Trivillin, VA, & Schwint, AE 2012a. Blood vessel normalization in the hamster oral cancer model for experimental cancer therapy studies. *Anticancer Res.*, vol. 32, no. 7, pp. 2703–2709.
46. Molinari, AJ, Pozzi, EC, Monti Hughes, A, Heber, EM, Garabalino, MA, Thorp, SI, Miller, M, Itoiz, ME, Aromando, RF, Nigg, DW, Trivillin, VA, & Schwint, AE 2012b. Tumor blood vessel "normalization" improves the therapeutic efficacy of boron neutron capture therapy (BNCT) in experimental oral cancer. *Radiat. Res.*, vol. 177, no. 1, pp. 59–68.
47. Diaz, AZ 2003. Assessment of the results from the Phase I/II boron neutron capture therapy trials at the Brookhaven National Laboratory from a clinician's point of view. *J. Neurooncol.*, vol. 62, pp. 101–109.
48. Hatanaka, H 1986. Clinical experience of boron-neutron capture therapy for gliomas: a comparison with conventional chemo-immuno-radiotherapy. in H Hatanaka (eds), *Boron-Neutron Capture Therapy for Tumors*, Nishimura Co. Ltd., Niigata, pp. 349–379.
49. Gabel, D, Preusse, D, Haritz, D, Grochulla, F, Haselsberger, K, Fankhauser, H, Ceberg, C, Peters, HD, & Klotz, U 1997. Pharmacokinetics of $Na_2B_{12}H_{11}SH$ (BSH) in patients

with malignant brain tumours as prerequisite for a phase I clinical trial of boron neutron capture. *Acta Neurochir. (Wien)*, vol. 139, no. 7, pp. 606–611; discussion 611–612.

50 Bauer, WE, Bradshaw, KN, & Richards, TL 1992. Interaction between boron containing compounds and serum albumin observed by nuclear magnetic resonance. in BJ Allen & BV Harrington (eds), *Progress in Neutron Capture Therapy for Cancer*, Plenum Press, New York, pp. 339–343.

51 Patel, H, & Sedgwick, EM 2000. BPA & BSH accumulation in experimental tumors. *Proceedings of the Ninth International Symposium on Neutron Capture Therapy for Cancer*, Osaka, Japan, pp. 59–60.

52 Nakagawa, Y, Kageji, T, Mizobuchi, Y, Kumada, H, & Nakagawa, Y 2009. Clinical results of BNCT for malignant brain tumors in children. *Appl. Radiat. Isot.*, vol. 67, no. 7–8 Suppl, pp. S27–S30.

53 Ono, K, Kinashi, Y, Suzuki, M, Takagaki, M & Masunaga, SI 2000. The combined effect of electroporation and borocaptate in boron neutron capture therapy for murine solid tumors. *Jpn. J. Cancer Res.*, vol. 91, no. 8, pp. 853–858.

54 Kato, I, Ono, K, Sakurai, Y, Ohmae, M, Maruhashi, A, Imahori, Y, Kirihata, M, Nakazawa, M, & Yura, Y 2004. Effectiveness of BNCT for recurrent head and neck malignancies. *Appl. Radiat. Isot.*, vol. 61, no. 5, pp. 1069–1073.

55 Kato, I, Fujita, Y, Maruhashi, A, Kumada, H, Ohmae, M, Kirihata, M, Imahori, Y, Suzuki, M, Sakrai, Y, Sumi, T, Iwai, S, Nakazawa, M, Murata, I, Miyamaru, H, & Ono, K 2009. Effectiveness of boron neutron capture therapy for recurrent head and neck malignancies. *Appl. Radiat. Isot.*, vol. 67, no. 7–8 Suppl, pp. S37–S42.

56 Stelzer, KJ, Gavin, PR, Risler, R, Kippenes, H, Hawthorne, MF, Nigg, DW, & Laramore, GE 2001. Boron neutron capture-enhanced fast neutron therapy (BNC/FNT) for non-small cell lung cancer in canine patient. in MF Hawthorne (eds), *Frontiers in neutron capture therapy*, New York: Kluwer Academic/Plenum Publishers, pp.735–739.

57 Sweet, WH, Soloway, AH, & Brownell, GL 1963. Boron-slow neutron capture therapy of gliomas. *Acta Radiol.*, vol. 1, pp. 114–121.

58 Diaz, A, Stelzer, K, Laramore, G, & Wiersema, R 2002. Pharmacology studies of $Na_2{}^{10}B_{10}H_{10}$ (GB-10) in human tumor patients. in MW Sauerwein, R Moss & A Wittig (eds), *Research and Development in Neutron Capture Therapy*, Bologna: Monduzzi Editore, International Proceedings Division, pp. 993–999.

59 Kabalka, GW, Yao, ML, Marepally, SR, & Chandra, S 2009. Biological evaluation of boronated unnatural amino acids as new boron carriers. *Appl. Radiat. Isot.*, vol. 67, pp. S374–S379.

60 Barth, RF, Yang, W, Wu, G, Swindall, M, Byun, Y, Narayanasamy, S, Tjarks, W, Tordoff, K, Moeschberger, ML, Eriksson, S, Binns, PJ, & Riley, KJ 2008. Thymidine kinase 1 as a molecular target for boron neutron capture therapy of brain tumors. *Proc. Natl. Acad. Sci. USA*, vol. 105, pp. 17493–17497.

61 Renner, MW, Miura, M, Easson, MW, & Vicente, MG 2006. Recent progress in the syntheses and biological evaluation of boronated porphyrins for boron neutron-capture therapy. *Anticancer Agents Med. Chem.*, vol. 6, pp. 145–157.

62 Crossley, EL, Ziolkowski, EJ, Coderre, JA, & Rendina, LM 2007. Boronated DNA binding compounds as potential agents for boron neutron capture therapy. *Mini Rev. Med. Chem.*, vol. 7, pp. 303–313.

63 Sibrian-Vazquez, M, & Vicente, MGH 2011. Boron tumor-delivery for BNCT: recent developments and perspectives. in NS Hosmane (ed), *Boron Science: New Technologies & Applications*, CRC Press, pp. 203–232.
64 Wu, G, Yang, W, Barth, RF, Kawabata, S, Swindall, M, Bandyopadhyaya, AK, Tjarks, W, Khorsandi, B, Blue, TE, Ferketich, AK, Yang, M, Christoforidis, GA, Sferra, TJ, Binns, PJ, Riley, KJ, Ciesielski, MJ, & Fenstermaker, RA 2007. Molecular targeting and treatment of an epidermal growth factor receptor positive glioma using boronated cetuximab. *Clin. Cancer Res.*, vol. 13, pp. 1260–1268.
65 Yang, W, Barth, RF, Wu, G, Kawabata, S, Sferra, TJ, Bandyopadhyaya, AK, Tjarks, W, Ferketich, AK, Moeschberger, ML, Binns, PJ, Riley, KJ, Coderre, JA, Ciesielski, MJ, Fenstermaker, RA, & Wikstrand, CJ 2006. Molecular targeting and treatment of EGFRvIII-positive gliomas using boronated monoclonal antibody L8A4. *Clin. Cancer Res.*, vol. 12, pp. 3792–3802.
66 Feakes, DA 2011. Design and development of polyhedral borane anions for liposomal delivery. in NS Hosmane (ed), *Boron Science: New Technologies and Applications*, CRC Press, vol. 12, pp. 277–292.
67 Jain, RK 2005. Normalization of tumor vasculature: an emerging concept in antiangiogenic therapy. *Science*, vol. 307, pp. 58–62.
68 Kharaishvili, G, Simkova, D, Bouchalova, K, Gachechiladze, M, Narsia, N, & Bouchal, J 2014. The role of cancer-associated fibroblasts, solid stress and other microenvironmental factors in tumor progression and therapy resistance. *Cancer Cell Int.*, vol. 14, pp. 41.
69 Matsumura, Y, & Maeda, H 1986. A new concept for macromolecular therapeutics in cancer chemotherapy: mechanism of tumoritropic accumulation of proteins and the antitumor agent smancs. *Cancer Res.*, vol. 46, no. 12 Pt 1, pp. 6387–6392.
70 Li, T, Hamdi, J, & Hawthorne, MF 2006. Unilamellar liposomes with enhanced boron content. *Bioconjug. Chem.*, vol. 17, no. 1, pp. 15–20.
71 Shelly, K, Feakes, DA, Hawthorne, MF, Schmidt, PG, Krisch, TA, & Bauer, WF 1992. Model studies directed toward the boron neutron-capture therapy of cancer: boron delivery to murine tumors with liposomes. *Proc. Natl. Acad. Sci. USA*, vol. 89, no. 19, pp. 9039–9043.
72 Feakes, DA, Shelly, K, Knobler, CB, & Hawthorne, MF 1994. $Na_3[B_{20}H_{17}NH_3]$: synthesis and liposomal delivery to murine tumors., *Proc. Natl. Acad. Sci. USA*, vol. 91, no. 8, pp. 3029–3033.
73 Feakes, DA, Shelly, K, & Hawthorne, ME 1995. Selective boron delivery to murine tumors by lipophilic species incorporated in the membranes of unilamellar liposomes. *Proc. Natl. Acad. Sci. USA*, vol. 92, pp. 1367–1370.
74 Watson-Clark, RA, Banquerigo, ML, Shelly, K, Hawthorne, MF, & Brahn E 1998. Model studies directed toward the application of boron neutron capture therapy to rheumatoid arthritis: boron delivery by liposomes in rat collagen-induced arthritis. *Proc. Natl. Acad. Sci. USA*, vol. 95, no. 5, pp. 2531–2534.
75 Pan, XQ, Wang, H, Shukla, S, Sekido, M, Adams, DM, Tjarks, W, Barth, RF, & Lee, RJ 2002. Boron-containing folate receptor-targeted liposomes as potential delivery agents for neutroncapture therapy. *Bioconjug. Chem.*, vol. 13, no. 3, pp. 435–442.
76 Carlsson, J, Kullberg, EB, Capala, J, Sjöberg, S, Edwards, K, & Gedda, L 2003. Ligand liposomes and boron neutron capture therapy. *J. Neurooncol.*, vol. 62, no. 1–2, pp. 47–59.

77 Masunaga, S, Kasaoka, S, Maruyama, K, Nigg, D, Sakurai, Y, Nagata, K, Suzuki, M, Kinashi, Y, Maruhashi, A, & Ono, K 2006. The potential of transferrin-pendant-type polyethyleneglycol liposomes encapsulating decahydrodecaborate-(10)B (GB-10) as (10)B-carriers for boron neutron capture therapy. *Int. J. Radiat. Oncol. Biol. Phys.*, vol. 66, no. 5, pp. 1515–1522.

78 Miyajima, Y, Nakamura, H, Kuwata, Y, Lee, JD, Masunaga, S, Ono, K, & Maruyama, K 2006. Transferrin-loaded nido-carborane liposomes: tumor-targeting boron delivery system for neutron capture therapy. *Bioconjug. Chem.*, vol. 17, no. 5, pp. 1314–1320.

79 Altieri, S, Barth, RF, Bortolussi, S, Roveda, L, & Zonta, A 2009. Thirteenth International Congress on Neutron Capture Therapy. *Appl. Radiat. Isot.*, vol. 67, no. 7–8 Suppl., pp. S1–S2.

80 Nakamura, H 2009. Liposomal boron delivery for neutron capture therapy. *Methods Enzymol.*, vol. 465, pp. 179–208.

81 Shirakawa, M, Yamamato, T, Nakai, K, Aburai, K, Kawatobi, S, Tsurubuchi, T, Yamamoto, Y, Yokoyama, Y, Okuno, H, & Matsumura, A 2009. Synthesis and evaluation of a novel liposome containing BPA-peptide conjugate for BNCT. *Appl. Radiat. Isot.*, vol. 67, no. 7–8 Suppl, pp. S88–S90.

82 Ueno, M, Ban, HS, Nakai, K, Inomata, R, Kaneda, Y, Matsumura, A, & Nakamura, H 2010. Dodecaborate lipid liposomes as new vehicles for boron delivery system of neutron capture therapy. *Bioorg. Med. Chem.*, vol. 18, no.9, pp. 3059–3065.

83 Heber, EM, Kueffer, PJ, Lee, MW Jr, Hawthorne, MF, Garabalino, MA, Molinari, AJ, Nigg, DW, Bauer, W, Monti Hughes, A, Pozzi, EC, Trivillin, VA, & Schwint, AE 2012. Boron delivery with liposomes for boron neutron capture therapy (BNCT): biodistribution studies in an experimental model of oral cancer demonstrating therapeutic potential. *Radiat. Environ. Biophys.*, vol. 51, no. 2, pp. 195–204.

84 Yinghuai, Z, Cheng, Yan, K, & Maguire, JA 2007. Recent developments in boron neutron capture therapy driven by nanotechnology. In HS Hosmane (eds), *Boron Science: New Technologies and Applications*, CRC Press, vol.1, pp. 147–163.

85 Nagasaki, Y 2012. Design of boron-containing nanoparticle for high performance BNCT. Oral communication presented at the 15th International Congress on Neutron Capture Therapy, Tsukuba, Japan, September 10–14.

86 Cruickshank, GS, Ngoga, D, Detta, A, Green, S, James, ND, Wojnecki, C, Doran, J, Hardie, J, Chester, M, Graham, N, Ghani, Z, Halbert, G, Elliot, M, Ford, S, Braithwaite, R, Sheehan, TM, Vickerman, J, Lockyer, N, Steinfeldt, H, Croswell, G, Chopra, A, Sugar, R, & Boddy, A 2009. A cancer research UK pharmacokinetic study of BPA-mannitol in patients with high grade glioma to optimise uptake parameters for clinical trials of BNCT. *Appl. Radiat. Isot.*, vol. 67, no. 7–8 Suppl., pp. S31–S33.

87 Ono, K, Masunaga, S, Suzuki, M, Kinashi, Y, Takagaki, M, Akaboshi, M 1999. The combined effect of boronophenylalanine and borocaptate in boron neutron capture therapy for SCCVII tumors in mice. *Int. J. Radiat. Oncol. Biol. Phys*, vol. 43, no. 2, pp. 431–436.

88 Monti Hughes, A, Pozzi, EC, Thorp, S, Garabalino, MA, Farías, RO, González, SJ, Heber, EM, Itoiz, ME, Aromando, RF, Molinari, AJ, Miller, M, Nigg, DW, Curotto, P, Trivillin, VA, & Schwint, AE 2013. Boron neutron capture therapy for oral precancer: proof of principle in an experimental animal model. *Oral Dis.*, vol. 19, no. 8, pp. 789–795.

89 Barth, RF, Yang, W, Rotaru, JH, Moeschberger, ML, Boesel, CP, Soloway, AH, Joel, DD, Nawrocky, MM, Ono, K, & Goodman, JH 2000. Boron neutron capture therapy of brain

tumors: enhanced survival and cure following blood-brain barrier disruption and intracarotid injection of sodium borocaptate and boronophenylalanine. *Int. J. Radiat. Oncol. Biol. Phys.*, vol. 47, pp. 209–218.

90 Capuani, S, Gili, T, Bozzali, M, Russo, S, Porcari, P, Cametti, C, D.Amore, E, Colasanti, M, Venturini, G, Mariaviglia, B, Lazzarino, G, & Pastore, FS 2008. L-DOPA preloading increases the uptake of borophenylalanine in C6 glioma rat model: a new strategy to improve BNCT efficacy. *Int. J. Radiat. Oncol. Biol. Phys.*, vol. 72, no. 2. Pp. 562–567.

91 Masunaga, SI, Sakurai, Y, Tano, K, Tanaka, H, Suzuki, M, Kondo, N, Narabayashi, M, Watanabe, T, Nakagawa, Y, Maruhashi, A, & Ono, K 2014. Effect of bevacizumab combined with boron neutron capture therapy on local tumor response and lung metastasis. *Exp. Ther. Med.*, vol. 8, pp. 291–301.

92 Mehrotra, R, Ibrahim, R, Eckardt, A, Driemel, O, & Singh, M 2011. Novel strategies in head and neck cancer. *Curr. Cancer Drug Targets*, vol. 11, pp. 465–478.

93 Khuri, FR, Lippman, SM, Spitz, MR, Lotan, R, & Hong, WK 1997. Molecular epidemiology and retinoid chemoprevention of head and neck cancer. *J. Natl. Cancer Inst.*, vol. 89, pp. 199–211.

94 Kreimann, EL, Itoiz, ME, Longhino, J, Blaumann, H, Calzetta, O, & Schwint, AE 2001b. Boron neutron capture therapy for the treatment of oral cancer in the hamster cheek pouch model. *Cancer Res.*, vol. 61, no. 24, pp. 8638–8642.

95 Salley, JJ 1954. Experimental carcinogenesis in the cheek pouch of the Syrian hamster. *J. Dent. Res.*, vol. 33, no. 2, pp. 253–262.

96 Vairaktaris, E, Spyridonidou, S, Papakosta, V, Vylliotis, A, Lazaris, A, Perrea, D, Yapijakis, C, & Patsouris, E 2008. The hamster model of sequential oral oncogenesis. *Oral Oncol.*, vol. 44, no. 4, pp. 315–324.

97 Nagini, S 2009. Of humans and hamsters: the hamster buccal pouch carcinogenesis model as a paradigm for oral oncogenesis and chemoprevention. *Anticancer Agents Med. Chem.*, vol. 9, no. 8, pp. 843–852.

98 Monti-Hughes, A, Aromando, RF, Pérez, MA, Schwint, AE, & Itoiz, ME 2015b. The hamster cheek pouch model for field cancerization studies. *Periodontol. 2000*, no. 67, pp. 292–311.

99 Slaughter, D, Southwick, H, & Smejkal, W 1953. "Field cancerization" in oral stratified squamous epithelium: clinical implications of multicentric origin. *Cancer*, vol. 6, pp. 963–8.

100 Dhooge, IJ, De Vos, M, & Van Cauwenberge, PB 1998. Multiple primary malignant tumors in patients with head and neck cancer: results of a prospective study and future perspectives., *Laryngoscope*, vol. 108, no.2, pp. 250–256.

101 Jaiswal, G, Jaiswal, S, Kumar, R, & Sharma, A 2013. Field cancerization: concept and clinical implications in head and neck squamous cell carcinoma. *J. Exp. Ther. Oncol.*, vol. 10, pp. 209–214.

102 Smith, BD, & Haffty, BG 1999. Molecular markers as prognostic factors for local recurrence and radioresistance in head and neck squamous cell carcinoma. *Radiat. Oncol. Investig.*, vol. 7, no. 3, pp. 125–144.

103 Monti Hughes, A, Heber, EM, Pozzi, E, Nigg, DW, Calzetta, O, Blaumann, H, Longhino, J, Nievas, SI, Aromando, RF, Itoiz, ME, Trivillin, VA, & Schwint, AE 2009. Boron neutron capture therapy (BNCT) inhibits tumor development from precancerous tissue: an experimental study that supports a potential new application of BNCT. *Appl. Radiat. Isot.*, vol. 67, no. 7–8 Suppl, pp. S313–S317.

104 Monti Hughes, A, Pozzi, EC, Heber, EM, Thorp, S, Miller, M, Itoiz, ME, Aromando, RF, Molinari, AJ, Garabalino, MA, Nigg, DW, Trivillin, VA, & Schwint, AE 2011. Boron Neutron Capture Therapy (BNCT) in an oral precancer model: therapeutic benefits and potential toxicity of a double application of BNCT with a six-week interval., *Oral Oncol.* vol. 47, no. 11, pp. 1017–1022.

105 Lalla, RV, Choquette, LE, Curley, KJ, Dowsett, RJ, Feinn, RS, Hegde, UP, Pilbeam, CC, Salner, AL, Sonis, ST, & Peterson, DE 2014. Randomized double-blind placebo-controlled trial of celecoxib for oral mucositis in patients receiving radiation therapy for head and neck cancer. *Oral Oncol.*, vol. 50, no. 11, pp. 1098–1103.

106 Sonis, ST 2004. A biological approach to mucositis. *J. Support Oncol.*, vol. 2, pp. 21–32, discussion 35–6.

107 Jensen, SB, & Peterson, DE 2014. Oral mucosal injury caused by cancer therapies: current management and new frontiers in research. *J. Oral Pathol. Med.*, vol. 43, pp. 81–90.

108 Pérez, MA, Raimondi, AR, & Itoiz, ME 2005. An experimental model to demonstrate the carcinogenic action of oral chronic traumatic ulcer. *J. Oral Pathol. Med.*, vol. 34, pp. 17–22.

109 Sonis, ST 2009. Mucositis: the impact, *biology and therapeutic opportunities of oral Mucositis. Oral Oncol.*, vol. 45, pp. 1015–1020.

110 Calzetta, O, Blaumann, H, & Longhino, J 2002. RA-6 reactor mixed beam design and performance for NCT trials. Presented in at the 10th International Congress on Neutron Capture Therapy, Essen, Germany, September 8–13.

111 Pozzi, E, Nigg, DW, Miller, M, Thorp, SI, Heber, EM, Zarza, L, Estryk, G, Monti Hughes, A, Molinari, AJ, Garabalino, M, Itoiz, ME, Aromando, RF, Quintana, J, Trivillin, VA, & Schwint, AE 2009. Dosimetry and radiobiology at the new RA-3 reactor boron neutron capture therapy (BNCT) facility: application to the treatment of experimental oral cancer. *Appl. Radiat. Isot.*, vol. 67, no. 7–8 Suppl, pp. S309–S312.

112 Laramore, GE, Wootton, P, Livesey, JC, Wilbur, DS, Risler, R, Phillips, M, Jacky, J, Buchholz, TA, Griffin, TW, & Brossard, S 1994. Boron neutron capture therapy: a mechanism for achieving a concomitant tumor boost in fast neutron radiotherapy. *Int. J. Radiat. Oncol. Biol. Phys.*, vol. 28, pp. 1135–1142.

113 Stelzer, KJ, Lindsley, KL, Cho, PS, Laramore, GE, & Griffin TW 1997. Fast neutron radiotherapy: the University of Washington experience and potential use of concomitant boost with boron neutron capture. *Radiat. Prot. Dosimetry*, vol. 70, pp. 471–475.

114 Trivillin, VA, Heber, EM, Itoiz, ME, Nigg, D, Calzetta, O, Blaumann, H, Longhino, J, & Schwint, AE 2004. Radiobiology of BNCT mediated by GB-10 and GB-10 + BPA in experimental oral cancer. *Appl. Radiat. Isot.*, vol. 61, no. 5, pp. 939–945.

115 Carmeliet, P, & Jain, RK 2000. Angiogenesis in cancer and other diseases. *Nature*, vol. 407, no. 6801, pp. 249–257.

116 Miller, M, Quintana, J, Ojeda, J, Langan, S, Thorp, S, Pozzi, E, Sztejnberg, M, Estryk, G, Nosal, R, Saire, E, Agrazar, H, & Graiño, F 2009. New irradiation facility for biomedical applications at the RA-3 reactor thermal column. *Appl. Radiat. Isot.*, vol. 67, no. 7–8 Suppl, pp. S226–S229.

117 Jain, RK 1987. Transport of molecules across tumor vasculature. *Cancer Metastasis Rev.*, vol. 6, pp. 559–593.

118 Yeo, SG, Kim, JS, Cho, MJ, Kim, KH, & Kim, JS 2009. Interstitial fluid pressure as a prognostic factor in cervical cancer following radiation therapy. *Clin. Cancer Res.*, vol. 15, no. 19, pp. 6201–6207.

119 Znati, CA, Rosenstein, M, Boucher, Y, Epperly, MW, Bloomer, WD, & Jain, RK 1996. Effect of radiation on interstitial fluid pressure and oxygenation in a human tumor xenograft. *Cancer Res.*, vol. 56, no. 5, pp. 964–968.

120 Buchegger, F, Rojas, A, Delaloye, AB, Vogel, CA, Mirimanoff, RO, Coucke, P, Sun, LQ, Raimondi, S, Denekamp, J, Pèlgrin, A, Delaloye, B, & Mach, JP 1995. Combined radioimmunotherapy and radiotherapy of human colon carcinoma grafted in nude mice. *Cancer Res.*, vol. 55, no. 1, pp. 83–89.

121 Nagano, S, Perentes, JY, Jain, RK, & Boucher, Y 2008. Cancer cell death enhances the penetration and efficacy of oncolytic herpes simplex virus in tumors. *Cancer Res.*, vol. 68, no. 10, pp. 3795–3802.

122 Hopewell, JW, Nyman, J, & Turesson, I 2003. Time factor for acute tissue reactions following fractionated irradiation: a balance between repopulation and enhanced radiosensitivity. *Int. J. Radiat. Biol.*, vol. 79, no. 7, pp. 513–524.

123 Mais, K 2006. Mucositis from radiotherapy to the head and neck: an overview. *Nursing*, vol. 1, pp. 18–20.

124 Molinari, AJ, Pozzi, EC, Monti Hughes, A, Heber, EM, Garabalino, MA, Thorp, SI, Miller, M, Itoiz, ME, Aromando, RF, Nigg, DW, Quintana, J, Santa Cruz, GA, Trivillin, VA, & Schwint, AE 2011. "Sequential" boron neutron capture therapy (BNCT): a novel approach to BNCT for the treatment of oral cancer in the hamster cheek pouch model. *Radiat. Res.*, vol. 175, no. 4, pp. 463–472.

125 López Castaño, F, Oñate-Sánchez, RE, Roldán-Chicano, R, & Cabrerizo-Merino, MC 2005. Measurement of secondary mucositis to oncohematologic treatment by means of different scale. *Med. Oral Patol. Oral Cir. Bucal.*, vol. 10, no. 5, pp. 412–421.

126 Sonis, ST, Peterson, RL, Edwards, LJ, Lucey, CA, Wang, L, Mason, L, Login, G, Ymamkawa, M, Moses, G, Bouchard, P, Hayes, LL, Bedrosian, C, & Dorner, AJ 2000. Defining mechanisms of action of interleukin-11 on the progression of radiation induced oral mucositis in hamsters. *Oral Oncol.*, vol. 36, no. 4, pp. 373–381.

127 Fukumura, D, & Jain, RK 2007. Tumor microvasculature and microenvironment: targets for antiangiogenesis and normalization. *Microvasc. Res.*, vol. 74, pp. 72–84.

128 Segers, J, Di Fazio, V, Ansiaux, R, Martinive, P, Feron, O, Wallemacq, P, & Gallez, B 2006. Potentiation of cyclophosphamide chemotherapy using the anti-angiogenic drug thalidomide: importance of optimal scheduling to exploit the 'normalization' window of the tumor vasculature. *Cancer Lett.*, vol. 244, no. 1, pp. 129–135.

129 Tong, RT, Boucher, Y, Kozin, SV, Winkler, F, Hicklin, DJ, & Jain, RK 2004. Vascular normalization by vascular endothelial growth factor receptor 2 blockade induces a pressure gradient across the vasculature and improves drug penetration in tumors. *Cancer Res.*, vol. 64, pp. 3731–3736.

130 Jain, RK 1999. Understanding barriers to drug delivery: high resolution in vivo imaging is key. *Clin. Cancer Res.*, vol. 5, pp. 1605–1606.

131 Bortolussi, S, Altieri, S, Protti, N, Stella, S, Ballarini F, Bruschi P, et al., 2010. ^{10}B measurement by alpha spectrometry and ^{10}B imaging by neutron autoradiography as a contribution to the understanding of BNCT radiobiology in oral cancer and liver metastases animal models. *Proceedings of the 14th. International Congress on Neutron Capture Therapy*, Buenos Aires, Argentina, pp. 59–62.

132 Sonis, ST 1998. Mucositis as a biological process: a new hypothesis for the development of chemotherapy-induced stomatotoxicity. *Oral Oncol.*, vol. 34, pp. 39–43.

133 Lima, V, Brito, GAC, Cunha, FQ, Reboucas, CG, Falcao, BAA, Augusto, RF, Souza, ML, Leitão, BT, & Ribeiro, RA 2005. Effects of the tumor necrosis factor-alpha inhibitors pentoxifylline and thalidomide in short-term experimental oral mucositis in hamsters. *Eur. J. Oral Sci.*, vol. 113, pp. 210–217.

134 Teo, S, Morgan, M, Stirling, D, & Thomas, S 2000. Assessment of the in vitro and in vivo genotoxicity of Thalidomid (thalidomide). *Teratog. Carcinog. Mutagen.*, vol. 20, pp. 301–311.

135 Mir, LM 2006. Bases and rationale of the electrochemotherapy. *EJC Suppl.*, vol. 4, pp. 38–44.

136 Olaiz, N, Signori, E, Maglietti, F, Soba, A, Suárez, C, Turjanski, P, Michinski, S, & Marshall, G 2014. Tissue damage modeling in gene electrotransfer: the role of pH. *Bioelectrochemistry*, vol. 100, pp. 105–111.

137 Ono, K, Kinashi, Y, Masunaga, S, Suzuki, M, & Takagaki, M 1998. Electroporation increases the effect of borocaptate (10B-BSH) in neutron capture therapy. *Int. J. Radiat. Oncol. Biol. Phys.*, vol. 42, no. 4, pp. 823–826.

138 Garabalino, MA, Olaiz, N, Pozzi, ECC, Thorp, S, Curotto, P, Itoiz, ME, Aromando, R, Portu, A, Saint Martin, G, Monti Hughes, A, Trivillin, V.A, Marshall, G, & Schwint, AE 2015. La electroporación aumenta el control tumoral inducido por la terapia por captura neutrónica en boro (BNCT) mediada por GB-10 en el modelo de cáncer bucal en hámster. *Oral communication presented at the XLII Reunión Anual de la Asociación Argentina de Tecnología Nuclear (AATN)*, Ciudad Autónoma de Buenos Aires, Argentina.

139 Fang, J, Nakamura, H, & Maeda, H 2011. The EPR effect: unique features of tumor blood vessels for drug delivery, factors involved, and limitations and augmentation of the effect. *Adv. Drug Deliv. Rev.*, vol. 63, no. 3, pp. 136–151.

140 Kueffer, PJ, Maitz, CA, Khan, AA, Schuster, SA, Shlyakhtina, NI, Jalisatgi, SS, Brockman, JD, Nigg, DW, & Hawthorne, MF 2013. Boron neutron capture therapy demonstrated in mice bearing EMT6 tumors following selective delivery of boron by rationally designed liposomes. *Proc. Natl. Acad. Sci. USA*, vol. 110, no. 16, pp. 6512–6517.

141 Grivennikov, SI, Greten, FR, & Karin, M 2010. Immunity, *inflammation, and cancer*. *Cell*, vol. 140, pp. 883–899.

142 Dörr, W, Schlichting, S, Bray, MA, Flockhart, IR, & Hopewell, JW 2005. Effects of dexpanthenol with or without Aloe vera extract on radiation-induced oral mucositis: preclinical studies. *Int. J. Radiat. Biol.*, vol. 81, pp. 243–250.

143 Heber, EM, Hawthorne, MF, Kueffer, PJ, Garabalino, MA, Thorp, SI, Pozzi, EC, Monti Hughes, A, Maitz, CA, Jalisatgi, SS, Nigg, DW, Curotto, P, Trivillin, VA, & Schwint AE 2014. Therapeutic efficacy of boron neutron capture therapy mediated by boron-rich liposomes for oral cancer in the hamster cheek pouch model. *Proc. Natl. Acad. Sci. USA*, vol. 111, no. 45, pp. 16077–16081.

144 Hoebers, F, Heemsbergen, W, Moor, S, Lopez, M, Klop, M, Tesselaar, M, & Rasch, C 2011. Reirradiation for head-and-neck cancer: delicate balance between effectiveness and toxicity. *Int. J. Radiat. Oncol. Biol. Phys.*, vol. 81, pp. e111–e118.

145 Letchoumy, PV, Chandra Mohan, KV, Kumaraguruparan, R, Hara, Y, & Nagini, S 2006. Black tea polyphenols protect against 7,12-dimethylbenz[a]anthracene-induced hamster buccal pouch carcinogenesis. *Oncol. Res.*, vol. 16, pp. 167–178.

146 Heber, EM, Monti Hughes, A, Pozzi, EC, Itoiz, ME, Aromando, RF, Molinari, AJ, Garabalino, MA, Nigg, DW, Trivillin, VA, & Schwint, AE 2010. Development of a model of tissue with potentially malignant disorders (PMD) in the hamster cheek pouch to explore the long-term potential therapeutic and/or toxic effects of different therapeutic modalities', *Arch. Oral Biol.*, vol. 55, no. 1, pp. 46–51.

147 Lurie, AG, Tatematsu, M, Nakatsuka, T, Rippey, RM, & Ito, N 1983. Anatomical and functional vascular changes in hamster cheek pouch during carcinogenesis induced by 7, 12-dimethylbenz(a)anthracene. *Cancer Res.*, vol. 43, no. 12 Pt 1, pp. 5986–5994.

148 Heber, EM, Aromando, RF, Trivillin, VA, Itoiz, ME, Nigg, DW, Kreimann, EL, & Schwint, AE 2007. Therapeutic effect of boron neutron capture therapy (BNCT) on field cancerized tissue: inhibition of DNA synthesis and lag in the development of second primary tumors in precancerous tissue around treated tumors in DMBA-induced carcinogenesis in the hamster cheek pouch oral cancer model. *Arch. Oral Biol.*, vol. 52, no. 3, pp. 273–279.

149 Laconi, E, Doratiotto, S, & Vineis, P 2008. The microenvironments of multistage carcinogenesis. *Semin. Cancer Biol.*, vol. 18, no. 5, pp. 322–329.

150 Coderre, JA, Morris, GM, Kalef-Ezra, J, Micca, PL, Ma, R, Youngs, K, & Gordon, CR 1999. The effects of boron neutron capture irradiation on oral mucosa: evaluation using a rat tongue model. *Radiat. Res.*, vol. 152, no. 2, pp. 113–118.

151 Lewis, CE, & Pollard, JW 2006. Distinct role of macrophages in different tumor microenvironments. *Cancer Res.*, vol. 66, pp. 605–612.

152 Multhoff, G, & Radons, J 2012. Radiation, inflammation, and immune responses in cancer. *Front. Oncol.*, vol. 2, pp. 58.

153 Nano, R, Barni, S, Chiari, P, Pinelli, T, Fossati, F, Altieri, S, Zonta, C, Prati, U, Roveda, L, & Zonta, A 2004. Efficacy of boron neutron capture therapy on liver metastases of colon adenocarcinoma: optical and ultrastructural study in the rat. *Oncol. Rep.*, vol. 11, pp. 149–154.

154 Jirtle, RL, Michalopoulos, G, Strom, SC, Deluca, PM, & Gould, MN 1984. The survival of parenchymal hepatocytes irradiated with low and high LET radiation. *Br. J. Cancer.*, vol. 49, no. 6, pp. 197–201.

155 Suzuki, M, Masunaga, S, Kinashi, Y, Takagaki, M, Sakurai, Y, Kobayashi, T, & Ono, K 2000. The effects of boron neutron capture therapy on liver tumors and normal hepatocytes in mice. *Jpn. J. Cancer Res.*, vol. 91, pp. 1058–1064.

156 Cardoso, JE, Trivillin, VA, Heber, EM, Nigg, DW, Calzetta, O, Blaumann, H, Longhino, J, Itoiz, ME, Bumaschny, E, Pozzi, E, & Schwint, AE 2007. Effect of boron neutron capture therapy (BNCT) on normal liver regeneration: towards a novel therapy for liver metastases. *Int. J. Radiat. Biol.*, vol. 83, no. 10, pp. 699–706.

157 Pinelli, T, Altieri, S, Fossati, F, Zonta, A, Ferrari, C, Prati, U, Roveda, L, Ngnitejeu, Tata, S, Barni, S, Chiari, P, Nano, R, & Ferguson, DM 2001. Operative modalities and effects of BNCT on liver metastases of colon adenocarcinoma: a microscopical and ultrastructural study in the rat. In MF Hawthorne (eds.), *Frontiers in neutron capture therapy*, New York: Kluwer Academic/Plenum Publishers, pp. 1427–1440.

158 Roveda, L, Prati, U, Bakeine, J, Trotta, F, Marotta, P, Valsecchi, P, Zonta, A, Nano, R, Facoetti, A, Chiari, P, Barni, S, Pinelli, T, Altieri, S, Braghieri, A, Bruschi, P, Fossati, F, & Pedroni, P 2004. How to study boron biodistribution in liver metastases from colorectal cancer. *J. Chemother.*, vol. 16, pp. 15–18.

159 Suzuki, M, Masunaga, S, Kinashi, Y, Nagata, K, Sakurai, Y, Nakamatsu, K, Nishimura, Y, Maruhashi, A, & Ono, K 2004. Intra-arterial administration of sodium borocaptate (BSH)/lipiodol emulsion delivers B-10 to liver tumors highly selectively for boron neutron capture therapy: experimental studies in the rat liver model. *Int. J. Radiat. Oncol. Biol. Phys.*, vol. 59, pp. 260–266.

160 Liao, AH, Chou, FI, Kuo, YC, Chen, HW, Kai, JJ, Chang, CW, Chen, FD, & Hwang JJ 2010. Biodistribution of phenylboric acid derivative entrapped lipiodol and 4-borono-2-18 F-fluoro-L-phenylalaninefructose in GP7 TB liver tumor bearing rats for BNCT. *Appl. Radiat. Isot.*, vol. 68, no. 3, pp. 422–426.

161 Altieri, S, Braghieri, A, Bortolussi, S, Bruschi, P, Fossati, F, Pedroni, P, Pinelli, T, Zonta, A, Ferrari, C, Prati, U, Roveda, L, Barni, S, Chiari, P, & Nano, R 2004. Neutron radiography of human liver metastases after BPA infusion. Proceedings of the 11th *International Congress on Neutron Capture Therapy,* Boston, USA, October 11–15.

162 Wittig, A, Malago, M, Collette, L, Huiskamp, R, Bührmann, S, Nievaart, V, Kaiser, GM, Jöckel, KH, Schmid, KW, Ortmann, U, & Sauerwein, W 2008. Uptake of two 10B-compounds in liver metastases of colorectal adenocarcinoma for extracorporeal irradiation with boron neutron capture therapy. *Int. J. Cancer*, vol. 122, pp. 1164–1171.

163 Cardoso, J, Nievas, S, Pereira, M, Schwint, A, Trivillin, V, Pozzi, E, Heber, E, Monti Hughes, A, Sanchez, P, Bumaschny, E, Itoiz, M, & Liberman, S 2009. Boron biodistribution study in colorectal liver metastases patients in Argentina. *Appl. Radiat. Isot.*, vol. 67, pp. S76–S79.

164 Caruso, M, Panis, Y, Gagandeep, S, Houssin, D, Salzmann, J, & Klatzmann, D 1993. Regression of established macroscopic liver metastases after in situ transduction of suicide gene. *Proc. Natl. Acad. Sci. USA*, vol. 90, pp. 7024–7028.

165 De Jong, GM, Aarts, F, Hendriks, T, Boerman, OC, & Bleichrodt, RP 2009. Animal models for liver metastases of colorectal cancer: research review of preclinical studies in rodents. *J. Surg. Res.*, vol. 154, pp. 167–176.

166 Pozzi, EC, Trivillin, VA, Colombo, LL, Monti Hughes, A, Thorp, SI, Cardoso, JE, Garabalino, MA, Molinari, AJ, Heber, EM, Curotto, P, Miller, M, Itoiz, ME, Aromando, RF, Nigg, DW, & Schwint, AE 2013. Boron neutron capture therapy (BNCT) for liver metastasis in an experimental model: dose–response at five-week follow-up based on retrospective dose assessment in individual rats. *Radiat. Environ. Biophys.*, vol. 52, no. 4, pp. 481–491.

167 Suzuki, M, Sakurai, Y, Hagiwara, S, Masunaga, S, Kinashi, Y, Nagata, K, Maruhashi, A, Kudo, M, & Ono, K 2007a. First attempt of boron neutron capture therapy (BNCT) for hepatocellular carcinoma. *Jpn. J. Clin. Oncol.*, vol. 37, no. 5, pp. 376–381.

168 Suzuki, M, Sakurai, Y, Masunaga, S, Kinashi, Y, Nagata, K, Maruhashi, A, & Ono, K 2007b. A preliminary experimental study of boron neutron capture therapy for malignant tumors spreading in thoracic cavity. *Jpn. J. Clin. Oncol.*, vol. 37, no. 4, pp. 245–249.

169 Bakeine, GJ, Di Salvo, M, Bortolussi, S, Stella, S, Bruschi, P, Bertolotti, A, Nano, R, Clerici, A, Ferrari, C, Zonta, C, Marchetti, A, & Altieri, S 2009. Feasibility study on the utilization of boron neutron capture therapy (BNCT) in a rat model of diffuse lung metastases. *Appl. Radiat. Isot.*, vol. 67, pp. S332–S335.

170 Bortolussi, S, Bakeine, JG, Ballarini, F, Bruschi, P, Gadan, MA, Protti, N, Stella, S, Clerici, A, Ferrari, C, Cansolino, L, Zonta, C, Zonta, A, Nano, R, & Altieri, S 2011. Boron uptake measurements in a rat model for boron neutron capture therapy of lung tumours. *Appl. Radiat. Isot.*, vol. 69, no. 2, pp. 394–398.

171 Kiger III, WS, Lu, XQ, Harling, OK, Riley, KJ, Binns, PJ, Kaplan, J, Patel, H, Zamenhof, RG, Shibata, Y, Kaplan, ID, Busse, PM, & Palmer, MR 2004. Preliminary treatment planning and dosimetry for a clinical trial of neutron capture therapy using fission converter epithermal neutron beam. *Appl. Radiat. Isot.*, vol. 61, pp. 1075–1081.

172 Kiger, JL, Kiger 3rd, WS, Riley, KJ, Binns, PJ, Patel, H, Hopewell, JW, Harling, OK, Busse, PM, & Coderre, JA 2008. Functional and histological changes in rat lung after boron neutron capture therapy. *Radiat. Res.*, vol. 170, no. 1, pp. 60–69.

173 Altieri, S, Bortolussi, S, Bruschi, P, Fossati, F, Vittor, K, Nano, R, Facoetti, A, Chiari, P, Bakeine, J, Clerici, A, Ferrari, C, & Salvucci, O 2006. Boron absorption imaging in rat lung colon adenocarcinoma metastases. *J. Instrum.*, vol. 41, pp. 123–126.

174 Suzuki, M, Endo, K, Satoh, H, Sakurai, Y, Kumada, H, Kimura, H, Masunaga, S, Kinashi, Y, Nagata, K, Maruhashi, A, & Ono, K 2008. A novel concept of treatment of diffuse or multiple pleural tumors by boron neutron capture therapy (BNCT). *Radiother. Oncol.*, vol. 88, no. 2, pp. 192–195.

175 Trivillin, VA, Garabalino, MA, Colombo, LL, González, SJ, Farías, RO, Monti Hughes, A, Pozzi, EC, Bortolussi, S, Altieri, S, Itoiz, ME, Aromando, RF, Nigg, DW, & Schwint, AE 2014a. Biodistribution of the boron carriers boronophenylalanine (BPA) and/or decahydrodecaborate (GB-10) for Boron Neutron Capture Therapy (BNCT) in an experimental model of lung metastases. *Appl. Radiat. Isot.*, vol. 88, pp. 94–98.

176 Razetti, A, Farías, RO, Thorp, SI, Trivillin, VA, Pozzi, EC, Curotto, P, Schwint, AE, & González, SJ 2014. Design, construction and application of a neutron shield for the treatment of diffuse lung metastases in rats using BNCT. *Appl. Radiat. Isot.*, vol. 88, pp. 50–54.

177 Trivillin, VA, Serrano, A, Garabalino, MA, Colombo, LL, Pozzi, ECC, Monti Hughes, A, Curotto, P, Thorp, S, Farías, RO, González, SJ, Bortolussi, S, Altieri, S, Itoiz, ME, Aromando, RF, Nigg, DW, & Schwint, AE 2015. BNCT for the treatment of tumors in lung: therapeutic efficacy and survival in an experimental model of colon carcinoma lung metastases. *Proceedings of the 15th International Congress of Radiation Research*, Kyoto, Japan, May 25–29.

178 Firestein, GS 2003. Evolving concepts of rheumatoid arthritis, *Nature*, vol. 423, pp. 356–61.

179 Shortkroff, S, Binello, E, Zhu, X, Thornhill, TS, Shefer, RE, Jones, AG, & Yanch, JC 2004. Dose response of the AIA rabbit stifle joint to boron neutron capture synovectomy. *Nucl. Med. Biol.*, vol. 31, pp. 663–670.

180 Yanch, JC, Shortkroff, S, Shefer, RE, Johnson, S, Binello, E, Gierga, D, Jones, AG, Young, G, Vivieros, C, Davison, A, & Sledge, C 1999. Boron neutron capture synovectomy: treatment of rheumatoid arthritis based on the $^{10}B(n,\alpha)^7Li$ nuclear reaction. *Med. Phys.*, vol. 26, no. 3, pp. 364–375.

181 Sweeney, SE, & Firestein, GS 2004. Rheumatoid arthritis: regulation of synovial inflammation. *Int. J. Biochem. Cell Biol.*, vol. 36, pp. 372–378.

182 Szekanecz, Z, & Koch, AE 2009. Angiogenesis and its targeting in rheumatoid arthritis. *Vascul. Pharmacol.*, vol. 51, pp. 1–7.

183 Mor, A, Abramson, SB, & Pillinger, MH 2005. The fibroblast-like synovial cell in rheumatoid arthritis: a key player in inflammation and joint destruction. *Clin. Immunol.*, vol. 115, pp. 118–128.

184 Van Lent, PLEM, Krijger, GC, Hofkens, W, Nievaart, VA, Sloetjes, AW, Moss, RL, Koning, GA, & Van Der Berg, WB 2009. Selectively induced death of macrophages in the synovial lining of murine knee joints using ^{10}B-liposomes and neutron capture synovectomy. *Int. J. Radiat. Biol.*, vol. 85, pp. 860–871.

185 Trivillin, VA, Abramson, DB, Bumaguin, GE, Bruno, LJ, Garabalino, MA, Monti Hughes, A, Heber, EM, Feldman, S, & Schwint, AE 2014b. Boron neutron capture synovectomy (BNCS) as a potential therapy for rheumatoid arthritis: boron biodistribution study in a model of antigen-induced arthritis in rabbits. *Radiat. Environ. Biophys.*, vol. 53, pp. 635–643.

186 Sanchez-Pernaute, O, Lopez-Armada, MJ, Calvo, E, Diez-Ortego, I, Largo, R, Egido, J, & Herrero-Beaumont, G 2003. Fibrin generated in the synovial fluid activates intimal cells from their apical surface: a sequential morphological study in antigen-induced arthritis. *Rheumatology*, vol. 42, pp. 19–25.

187 Mortarino, P, Goy, D, Palena, A, Abramson, D, Toledo, J, Zapata, M, Sarrió, L, Fracalossi, NM, Jamin, A, Cointry, G, & Feldman, S 2009. Tratamiento oral con hidrolizados enzimáticos de colágeno (HEC) ejerce efectos beneficiosos sobre la evolución de la artritis a nivel experimental y clínico, disminuyendo el fenómeno articular. Correlato con disminución de anticuerpos antiproteína citrulinada. *Medicina*, vol. 68, no. II, pp. 117.

188 Toledo, J, Abranson, D, Acosta, N, Mortarino, P, Goy, D, Gonzalez, EM, García Tentella, B, Palena, Alfonso, A, Fracalossi, M, Sarrió, L, Zingoni, N, Cointry, GR, & Feldman, S 2009. Terapias emergentes en artritis. *Medicina*, vol. 69, no. 1, pp. 263.

189 Takahashi, S, Sugimoto, M, Kotoura, Y, Oka, M, Sasai, K, Abe, M, & Yamamuro, T 1992. Long-lasting tolerance of articular cartilage after experimental intraoperative radiation in rabbits. *Clin. Orthop. Relat. Res.*, vol. 275, pp. 300–305.

190 Trivillin, VA, Bruno, LJ, Gatti, DA, Stur, M, Garabalino, MA, Monti Hughes, A, Castillo, J, Scolari, H, Schwint, AE, & Feldman, S 2016. Boron neutron capture synovectomy (BNCS) as a potential therapy for rheumatoid arthritis: radiobiological studies at RA-1 Nuclear Reactor in a model of antigen-induced arthritis in rabbits. *Radiat. Environ. Biophys.*, 55(4):467–475.

191 Szekanecz, Z, & Koch, AE 2001. Update on synovitis. *Rheumatoid Arthritis*, vol. 3:53–63.

192 Taylor, PC, & Feldmann, M 2004. Rheumatoid arthritis: pathogenic mechanisms and therapeutic targets. *Drug Discov. Today*, vol. 1, no. 3, pp. 289–295.

193 Rao, M, Trivillin, VA, Heber, EM, Cantarelli, MA, Itoiz, ME, Nigg, DW, Rebagliati, RJ, Batistoni, D, & Schwint, AE 2004. BNCT of 3 cases of spontaneous head and neck cancer in feline patients. *Appl. Radiat. Isot.*, vol. 61, pp. 947–952.

194 Trivillin, VA, Heber, EM, Rao, M, Cantarelli, MA, Itoiz, ME, Nigg, DW, Calzetta, O, Blaumann, H, Longhino, J, & Schwint, AE 2008. Boron neutron capture therapy (BNCT) for the treatment of spontaneous nasal planum squamous cell carcinoma in felines. *Radiat. Environ. Biophys.*, vol. 47, no. 1, pp. 147–155.

Index

Note: Page numbers referring to figures are in *italics* and those referring to tables are in **bold**

a

cis-ABCPC (1-amino-3-boronocyclopentanecarboxylic acid)   402
aberrant (tumor) blood vessels   427, 430, 431, 436
accelerator(s)   390–391, 409
activity
  biological   63, 72, 74, 78, 81–83, 93
  cytotoxic   79, 82–84, 90–92, 94
adamantane   4, 5, 36, 113, 120
adenosine derivatives   28
adenosine receptors   28
aggregation   309
allometric scaling   244
all-trans-retinoic acid   11
alpha autoradiography   233, 402
α–particles   389, 402, 416
Alzheimer's disease   7
1α,25-dihydroxyvitamine D_3 13
amphiphilicity   320
androgen antagonist   7, 9
androgen receptor   7
anti-androgen withdrawal syndrome   11
anticancer activity   23
anticancer agent   7, 72, 91
anticancer profile   90
anti-HCMV agents   24
anti-HCMV drugs   24
anti-infectious disease drugs   23
anti-prostate cancer   9, 10
antiviral activity   23, 25
antiviral drugs   23
AR-ER dual ligand   9
arthritis   442
asborin   63, 70, 91–93
  nido-   91
atomic emission spectroscopy   401
azide   146, 150, 152

b

background dose   417, 427, 429, 434, 435, 437, 439, 442, 444
bacteria   165
barn   389
beam
  EB–NCT (external beam NCT)   409
  epithermal neutron   391, 409
benzo[b]acridin-12(7H)-one   400
benzophenone   39
benzoxaboroles   178, 179
B–H bond
  hydridic   75
bicalutamide   9, 10
bicyclooctane   4, 5
binding
  affinity   6, 73, 76, 82, 94, 95
  pockets   61, 80
biocompatibility   327
biodistribution   73, 77, 78, 83, 84, 88, 93, 324

Boron-Based Compounds: Potential and Emerging Applications in Medicine, First Edition.
Edited by Evamarie Hey-Hawkins and Clara Viñas Teixidor.
© 2018 John Wiley & Sons Ltd. Published 2018 by John Wiley & Sons Ltd.

biologically effective dose   417
bio-membranes (biological membranes)   159, 160, 170
bio-orthogonality *see* strategy, bio-orthogonal chemistry
blood–brain barrier (BBB)   61, 72, 74–76, 80, 299, 390, *395*
blood vessel damage, selective   428
BNCT *see* boron neutron capture therapy (BNCT)
BNCT, clinical investigations   418
BNCT, clinical studies/trials   390–392, 418–421, 423, 437, 445, 446
BNCT, double application   434, 436–438
BNCT, preclinical studies/trials   445, 446
BNCT radiobiology   416, 422, 428, 438
BNCT, therapeutic efficacy of   428, 432
BNCT, translational research   418, 446
BNNT (boron-nitride nanotube)   379
BODIPY-containing phenylboronic acids   185–188
BODIPYs   318
boric acid   176, 177, 419
boronated porphyrins   344, 347, 349–351
boron-based COSAN   169
boron biodistribution studies   418, 419, 433, 435, 439, 446
boron carriers   416, 419, 421, 422, 430, 435, 443, 446
boron carrier/s, selective accumulation in tumor   416
boron cluster(s)   21–25, 27, 28, 30, 72, 235, 236, 244–254, 258, 259
  decapped   70
  11-vertex   65
boron compound, ideal   417, 418, 421, 443
boron compound, nonselective   428
boron compounds, approved for use in humans   418, 421, 422, 425, 430, 433, 435, 446
boron compounds, combined administration   420, 422, 427, 428, 432, 435, 441, 443
boron-containing nucleosides   20
boron content, gross   419, 431–433, 439
boron, homogeneous targeting   417, 427, 428, 431, 432
boron, microdistribution/microlocalization   417, 432, 433, 436, 441

boron neutron capture synovectomy (BNCS)   442–445
boron neutron capture therapy (BNCT)   4, 60, 72, 84, 87, 88, 95, 167, 169, 232, 259, 271–274, 277, 279, 282–286, 343, 344, 347, 349, 350, 352, 353, 355, 356–358, 360–362, 371, 373–375, 377, 379–381, 383, 389–398, *394*, 400–405, **404**, **407**, 408, 409, 416
  agent(s)   61, 63, 72, 83, 85, 391, 393, 396, 398, 400
  borax   390
  boron microdistribution   401–402
  imaging guided   407
BPA (L-*para*-boronophenylalanine)   234, 235, 254, 373, 374, *390*, 391, 393–395, 397, 398, 402, 409, 416, 418, 420
BSH (disodium mercapto-undecahydro-*closo*-dodecaborate)   60, 84, 89, 234–236, 251, 254, 255, 360, 361, 374, 376, 377, 390, 391, 395, 398, 402, 409, 418, 420

C

cage rearrangement   64, 67–69, 71, 77
calixarene   112
cancer   61, 63, 87, 88, 90, 91
carbamate   145, 146, 150, 152
carbohydrate
  radiolabelled   85
carbonate   145, 146, 150
carbonic anhydrase   116
carboplatin   152
carborane-containing nucleosides   23
carborane(s)   397, 403, 408
  3-amino-*o*-carborane   347, 348, 356, 358, 364
  1-azidomethyl-*o*-carborane   347, 349
  1-carba-*closo*-monocarbon carborane   344
  carborane epoxides   345
  cesium 1-hydroxymethyl-*closo*-monocarbadodecaborate   357, 362, 364
  cesium 1-mercapto-1-carba-*closo*-dodecaborate   352
closo-   6, 60, 62–64, 66, 69–72, 74–76, 80, 82, 85, 87, 88, 93
  cluster   70, 84, 85, 93, 94

carborane(s) (cont'd)
  1-formyl-*o*-carborane 345
  1-hydroxyethyl-*o*-carborane 357
  9-hydroxymethyl-carboranes 356, 357
  1-hydroxymethyl-*o*-carborane 356, 357, 362
  3-hydroxy-*o*-carborane 348
  9-isocyanato-carboranes 347
  1-isocyanato-*o*-carborane 347
  1-mercapto-carboranes 352–354
  9-mercapto-carboranes 347, 348, 352
  meta- or *m*- 4, 5, 45, 46, 51, 60, 61, 67, 69
  nido- 25, 60–62, 66–69, 71, 72, 77, 85, 88, 90, 91, 93
  nido-carborane(−1) see carborane(s), *nido*-
  nido-$[C_2B_9H_9R_2]^{2-}$ see dicarbollide
  nido-$[C_2B_9H_{10}R_2]^{-}$ see dicarba-*nido*-dodecahydroundecaborate
  ortho- or *o*- 4, 5, 36, 43, 49, 53, 60, 61, 67, 69, 72, 115
  para- or *p*- 4, 5, 25, 60, 61, 71
  1-trifluoromethanesulfonylmethyl-1-carba-*closo*-dodecaborate cesium 344, 358
  1-trifluoromethane sulfonylmethyl-*o*-carborane 344, 358
carboranylphenol 14
3-carboranyl thymidine analogs 271
carcinogen dimethyl-1,2-benzanthracene (DMBA) 423, 435
carcinogenesis 7, 423, 435
cats and dogs 445, 446
Cb^{2-} see dicarbollide
$[C_2B_9H_{11}]^{2-}$ see dicarbollide
cell death 349, 350, 352, 353, 355, 358, 359
cell membrane
  transport 61
cellular targeting (mechanisms) 428
chaotropic 111, 113
chaperone activity 40, 47, 48
charge density 110, 119
chemical probes 39
chemoresistance 407
chemotherapeutic nucleosides 30
chemotherapy 393, 408

chlorambucil 152
chlorin 314, 344, 355, 356, 358, 359, 361, 362
  bacteriochlorin 355, 356
  fluorinated carboranylchlorin 352, 354, 355
  methylpheophorbide *a* 357, 358
  purpurin 18
  pyropheophorbide *a* 356
  water-soluble boronated chlorins 351, 360
chronic lymphocytic leukemia 26
click chemistry 25, 146, 150, 152
click reaction 39
closomer 141–158
CNS (central nervous system) 73–75, 76, 78
CNT (carbon nanotubes) 377–379
cobaltabis(dicarbollide) see COSAN
cobaltacarborane 310
cobaltacarborane, full-sandwich see COSAN
$[3,3'\text{-}Co(1,2\text{-}C_2B_9H_{11})_2]^{-}$ see COSAN
collimator 240–242
colorimetric sensors 197
combretastatin A-4 (CA-4) 53
compound biological effectiveness (CBE) 417, 436, 441, 446
computed tomography (CT) 392, 393
confocal microscopy 400–401
conjugate
  cholesterol 84
  peptide 78, 85
conjugation 146, 150
COSAN [cobaltabis(dicarbollide)] 63, 65, 72, 78, 79, 81, 83, 84, 109, 117, 119, 132, 159–169, 170, 305, 361, 362, 364
Cp^{-} see cyclopentadienyl
cross section (σ) 389, 396
cryo-transmission electron microscopy (cryoTEM) 118, 161, 163
3-carboranyl thymidine analogues (3CTAs) 271–286
cucurbituril 112, 115
4'-cyanostilbene-4-boronic acid 182, 183
cyclodextrin 112, 114

cyclooxygenase (COX)   91, 117
cyclopentadienyl   64, 71, 73–75, 79, 83
cyclotriveratrylene   112
cytomegalovirus   24
cytotoxicity *see* activity, cytotoxic

d

dansyl   142, 146
deboronation   61, 67, 69–71, 85, 87, 91, 93
decahydrodecaborate   418, 420
decapping *see* deboronation
decarborane coupling   39, 45, 50
delivery vehicle   78, 84, 89
dendrimer   141, 145, 146, 306
depression   7
D-glucose   175, 176, 189–195
diagnostics   72, 141, 148, 152
diboronic acid sensors   189–195
dicarba-*closo*-dodecaborane *see also* carborane(s), *closo*-
dicarba-*nido*-dodecahydroundecaborate   61, 66, 67, 72
dicarbollide   61, 66, 68, 71, 72, 78, 79, 84
  bis-   63, 79, 83
  cluster   74
  complex   69, 74
  ligand   74–76, 82, 83, 95
dictyostelium   164–166, 169
dictyostelium discoideum   164
diffuse lung metastases in BDIX rats   441
dihydrogen bond   111, 117, 128
4'-dimethylaminostilbene-4-boronic acid   182, 183
1,2-dioleoyl-*sn* glycero-3-phosphocholine (DOPC)   161, 162
dioxane ring opening   27
dipalmitoylphosphatidylcholine (DPPC)   118, 119
1,2-diphytanoyl-*sn*-glycero-3-phosphocholine (DphPC)   161
direct damage   417
dispersion   129
dispersion interactions   117
DLC (fluorescent delocalized lipophilic cation)   87, 89–91

DMPC (dimyristoylphophatidylcholine)   89, 118
DNA   305
docking   127
docking program   120
dodecaborate   113, 115
dodecahydro-*closo*-dodecaborate   141, 143
dodecahydroxy-*closo*-dodecaborate   141–143, 144
dose calculations   419
dose distribution   390, 409–410
dose-limiting precancerous tissue   429, 430, 432–434, 437
dosimetry   392, 402
drug
  anticancer   87
  design   21, 23, 72, 74, 78, 82, 91, 95, 109
DSBL (distearoyl boron lipid)   84
DSPC (distearoylphosphatidylcholine)   84
dye displacement   113

e

EDA (energy decomposition analysis)   65, 66
EINS (electrophile-induced nucleophilic substitution)   63
electron energy loss spectroscopy
electroporation   432, 433
enhanced permeability and retention effect (EPR)   233, 406, 421, 434
enzyme   61, 95
  inhibition   79, 80
enzyme inhibitor   116
epolactaene *tert*-butyl ester (ETB)   48
ER-binding affinity   13
estradiol   3, 5
estrogen agonist   5
estrogen antagonist   7
estrogen receptor   3, 5
experimental models   418, 419, 422, 423, 438

f

FDG (fluorine-18 combined with deoxy-glucose)   393

ferracarborane   81
ferrabis(dicarbollide) (FESAN)
   165, 166
field-cancerized tissue/oral mucosa   416,
   423, 434–436
FL-SBL (fluorescence-labeled boron
   lipid)   84
fluorescence carbohydrate sensors   179
fluorescent   141, 146, 152, 308
fluorescent materials   205
fluorinated carboranylporphyrins   350,
   352, 353, 355
folic acid   326
force field   120
fragment-based drug design   120
FRET sensors   196

g
Gd-DTPA-BPA   397
glioblastoma   298
gliomas   307
glucose   312
guanidine   70, 71, 74, 80, 81

h
hamster cheek pouch oral cancer model
   422, 423, 425, 427, 428, 430,
   432–435, 439
HCMV   24
head group   119
heat shock protein-60 (HSP60)   40,
   47, 49
HEK293 cells   162–164, 165, 169
HeLa cells   38, 52, 164, 165
HIF1 inhibitor(s)   36, 47, 55
high boron content   78, 84
high-LET (linear energy transfer)
   298, 416
HIV-1 protease   132
HIV protease inhibitor   109
hospital based accelerators   446
host-guest chemistry   115
human carbonic anhydrase   132
hybrid
   organic-inorganic   65, 72, 78, 83
hydride   62, 63
hydrogen bonding   3, 6, 111, 113, 127

hydroneutral   121
hydrophilic   121
hydrophobic *see also* hydrophobicity
hydrophobic constant   13
hydrophobic effect   113
hydrophobic interaction   3, 6
hydrophobicity   15, 61, 63, 75, 80, 83, 85,
   93, 94
hydroxycarborane   112
hydroxyflutamide   9
hydroxylation   141, 143, 148, 150, 152
hypoxia-inducible factor (HIF)
   35, 49, 51
hypoxia response element (HRE)   35, 38,
   46, 52, 55

i
IC_{50} (half maximal inhibitory
   concentration)   82–86, 90–92, 95
^{124}I-COSAN   167, 168
^{125}I-COSAN   167, 168
I_2-COSAN [3,3'-Co(8-I-1,2-$C_2B_9H_{10}$)$_2$]$^-$
   165, 166, 168
ICT sensors   180
imaging   308
imaging agent(s) *see* MRI contrast
   agent
implicit solvent   131
indicator displacement   115
indirect damage   417
indoborin   94
indomethacin   94, 95, 120
inductively coupled plasma–atomic
   emission spectroscopy   233
inductively coupled plasma–mass
   spectrometry (ICP–MS)   161, 162,
   392, 397
inertness
   chemical   87
   kinetic   73
inflammatory processes   30
inhibition   316
inhibitor   80, 83, 85
   AKR   63
   COX   91, 95
   5HT1A subtype   74, 76
   isoform-specific   79

NOS   79, 80, 82
  PSMA   86
injection   315
integrin   150
intracellular   52, 159, 166, 170, 274, 282, 284, 305, 314, 324, 362, 382, 401–402
IO–NCT (intraoperative NCT)   409
ionic bonding   3
I-PEG-COSAN   167
^{124}I-PEG-COSAN   167, 168
^{125}I-PEG-COSAN   168
isolobal   64
isomerization *see* cage rearrangement
isoquinoline sulfonamide inhibitor   116
isotopic exchange   243, 246, 247

l

laloxifene   7, 8
laser post-ionization secondary neutral mass spectrometry   233
LDLs (low-density lipoproteins)   397–399, **407**
LET (linear energy transfer) particles   389, 409
ligand
  Cb^{2-} 64, 65
  Cp^- 64–66
  metal–   65, 66, 69, 71, 72
  nitrosyl   78
linear energy transfer   232, 233
lipid bilayer   118
lipids   119, 120, 300, 321, 324, 360, 375
lipophilic *see* lipophilicity
lipophilicity   75, 76, 80, 85, 87, 88, 299
liposome(s)   84, 89, 90, 118, 160, 162, 163, 322, 397, 400, 406, **407**, 421, 422, 433, 434, 442, 443
lithium   389, 402
lithium-6 carbonate shield   439
  ^{7}Li nuclei   416
liver metastases in BDIX rats, model   438
logP *see* octanol-water distribution coefficient
low-density lipoprotein   307
lysine   152
L-lysine, scaffold based   403

m

magnetic resonance   144, 146
magnetic resonance imaging (MRI)   73, 234, 243, 396–400, 407
  contrast agents (CAs)   146, 150, 396, 406
  signal intensity enhancement   396
magnetic resonance spectroscopy (MRS)   396, 398
$^{99m}Tc^I$ (technetium-99m, technetium(I)-99m)   68, 69, 71–75, 77, 79
$^{99m}Tc^I$-dicarbollide complex   69, 75, 77
magnetic resonance spectroscopic imaging (MRSI)   234
manassantins   49
m-chlorophenyl chloroformate   146
medicinal chemistry   13, 20–21, 23, 38, 30, 60, 78, 109, 120, 126, 175, 236, 271
melanoma   352, 354, 355, 359, 360
membrane-like structures   159
metallabisdicarbollides   109
metallacarborane(s)   20, 60, 62, 64–66, 68, 69, 72, 79, 80, 82, 95, 96
  isomers   71
  radiolabelled   70, 76
metallocene   64, 65, 79
microwave   42, 45, 50, 54, 68, 68, 71, 74, 93
  heating   68, 71, 74, 93
MIP (Molecular Imaging Probes)   72–74, 75
  target-specific   76
molecular imaging   205, 238, 258
monoclonal antibodies   305
monodisperse   142, 145, 152
monolayer nanovesicles   159
MRI *see* magnetic resonance imaging (MRI)
MTT assay   38
multifunctional   142, 148
mutated receptor   10

n

nanomaterials   371, 373, 383
nanomolecular   155
nanoparticles   324, 371, 373, 379, 381–383, 397, 402, 406–407, 408, 421, 422
nanosized carriers *see* nanoparticles
nanostructures   148
nanovesicles   159, 161, 170

nanowires 325
naphthylboronic acids 184, 185
necrosis 350, 352, 359
neutron autoradiography 431–433, 441
neutron capture radiography 233
neutron capture therapy (NCT) 109, 118
nicotinamide phosphoryltransferase 120
nido-carborane(−1) *see* carborane(s), *nido*
nido-[$C_2B_9H_9R_2$]$^{2-}$ *see* dicarbollide
nido-[$C_2B_9H_{10}R_2$]$^-$ *see* dicarba-*nido*-dodecahydroundecaborate
NMR (nuclear magnetic resonance) 396, 398
^{11}B 67, 68, 91
^{11}B,^{11}B 2D 67
noncovalent immobilization 115
NOS (nitric oxide synthases) 79–81
inhibition 82
NPY (neuropeptide Y) 403–**404**
nuclear imaging 235, 238, 239, 243, 244, 254, 256, 259
nuclear receptor 3
nuclear receptor ligand 16
nucleophile 62, 63, 67, 69, 70, 78
nucleoside(s) 25, 27, 304, 403
nucleoside analogs 20, 23, 24, 26
nucleoside-boron cluster conjugates 21
nucleoside derivatives 27
nucleoside drugs 20

o

octanol-water distribution coefficient 14, 120
OI (optical imaging) 400
oral mucositis 423, 429, 431, 437
ortho-carborane(s) *see* carborane(s), *ortho*
osteoporosis 7
oxonium-derivative 25

p

para-carborane *see* carborane(s), *para*
paramagnetic 205
partition coefficient (log*D*) 75
payload 141, 148, 155
PC (phophatidylcholine) 89
PDT *see* photodynamic therapy (PDT)

PEG (polyethylene glycol) 150, 311, 406
peptide(s) 302, 395, 403–**404**
permeability 313
PET (positron emission tomography) 73, 167–169, 235, 238, 240–242, 243, 247, 249, 252–256, 257, 392–395, 400
PET sensors 186
pharmacokinetic 243, 244, 253, 254, 258, 259
pharmacophore 3, 61, 66, 72, 80, 82, 83, 91, 93–95, 121
phenylboronic acid 177, 178
1-phenyl-*o*-carborane 113
π 13
photodynamic therapy (PDT) 343, 344, 349, 350, 352, 355–358, 360, 361, 408
photosensitizers 309, 343
phthalocyanine 314
piperazine 68, 75, 76
polyanionic 317
polyarginine 302
polyethylene glycol *see* PEG
polyhedral 67
polyhedral borane 141, 143
porphyrins 300, 408
  5-(4-aminophenyl)-10,15,20-triphenyl-porphyrin 344
  5,10,15,20-tetrakis(4-aminophenyl) porphyrin 344
  5,10,15,20-tetrakis(pentafluorophenyl) porphyrin 350
  5,10,15,20-tetraphenylporphyrin 344
  5,10,15-tris(4-nitrophenyl)-20-phenylporphyrin 344
positron emission tomography *see* PET (positron emission tomography)
positron range 248
precancerous tissue, tumors surrounded by 423
prolymphocytic leukemia 26
prompt γ-ray spectroscopy 234, 401
proof-of-principle 82, 83
prostate cancer 148, 150
prosthetic group 66, 73, 74, 84, 85
proton beam radiation 408

PSMA (prostate-specific membrane antigen)   85
purinergic receptor   30

q
QM/MM   117
QSAR   13–15

r
radioactive decay   239–241, 253
radioastatination   245
radiobiological studies, *in vivo*   418, 419, 425, 428, 433
radiobromination   250
radiofluorination   248, 249
radiohalogenation   245
radio-imaging   72, 83, 95
radio-imaging agents   78
  target-specific   73, 74, 79
radioiodination   **246, 247, 249, 250**
radiolabeling   243–245, 247, 248, 251–253, 254, 256, 258, 259
radiometallation   245, 251
radiometals   **245, 251, 252**
radionuclide   73, 84, 85, 239, 240, 243, 245, 247, 251, 253, 258, 259
radiopharmaceuticals   68, 74, 85, 205
radioprotectors   438
radioresistance   407
radiosurgery   408
radiotherapeutic agents   72, 79, 83
radiotherapy   72, 79, 83, 393, 408–409
RAMAN   162–164
RAMAN spectroscopy   162
ratios, boron concentration   419, 420, 425, 439
reactive oxygen species (ROS)   29, 343, 344, 359
receptor   63, 73, 76, 85
  adrenergic   74–76
  biological   74
  5-HT1A, antagonist   75
  serotonin   74, 76
receptor-mediated endocytosis   406
reconstruction   242, 243
regression analysis   15

Re^I (rhenium, rhenium(I))   68, 69, 71–75, 77–79
relationship
  structure-activity   80, 95
relative biological effectiveness (RBE)/RBE   416, 417, 441, 446
replacement
  bio-isosteric   74, 78
reporter assay   38, 47
RES (reticuloendothelial system)   407
retinoic acid receptor   11
retinoic acid X receptor   11
retreatment   438, 446
reverse transcription polymerase chain reaction (RT-PCR)   38, 53, 55
rhenacarborane   69, 76, 78, 79
ruthenacarborane   82, 83
RXR antagonist   12

s
salting-in   113
sandwich
  full-   65, 81, 82
  half-   65, 72, 79, 82, 83, 95, 96
  mixed-   65, 66, 71, 72, 79, 81–83, 95, 96
sarcoma   355, 358–360
scintillation crystal   240, 241
selective estrogen receptor modulator   7
sensitivity   234, 239
sequential BNCT   428, 430
SERM   7
short range secondary particles   389
σ-hole bonding 128
SIMS (secondary ion mass spectrometry) based ion microscopy   401–402
size exclusion   150, 152
sodium borocaptate (BSH) *see* BSH
sodium pentaborate   390
Sonogashira coupling   45, 53
spatial resolution   234, 242
specific activity   244, 246
SPECT (single-photon emission computed tomography)   73, 235, 238, 240–242, 243, 253, 258, 395–396, 400
spermidine   312
spermine   312

SPIONs (superparamagnetic iron-oxide nanoparticles) 407
spontaneous tumors 445
squamous cell carcinomas 423
stability 327
strategy
  bio-orthogonal chemistry 87
  pre-targeting 84, 85
  target-vector recognition 84, 85, 92
substitution
  electrophilic 62
  mono 61
  oxidative 62
  pattern 80, 82, 96
  selective 61
sulforhodamine-B 152
superchaotropic 111, 113
supramolecular 109, 115
Suzuki–Miyaura-type diboron coupling 36
syngeneic colon cancer cells/DHD/K12/TRb 438, 441

t

T1 (longitudinal relaxation time) 396–399
T2 (transverse relaxation time) 396, 398, 410
tamoxifen 7, 8
target
  biological 72–74, 78, 83, 91, 94, 95
  -specific, recognition 83
target identification 40
technetacarborane 74, 78
technetium(I)-^{99m}Tc-dicarbollide complex see ^{99m}TcI-dicarbollide complex
TEM (transmission electron microscopy) 380, 381
testosterone 9
thalidomide 430–432
therapeutic 141, 145, 148
thermal neutrons 389–390, *403*, 409
thymidine 304
thymidine kinase 274, 275, 280, 281
thymidine-kinase-1 (TK1) 274–277, 279–285

transcriptional activity 38, 47, 49, 51
tricarbollide 71, 79, 82
tumor 73, 79, 83–85, 87, 88
  breast 398, 403, **404**
  colon **407**
  glioma 395, 397, 398, 400–401, 402, **404**, **407**, 408–410
  hamster cheek pouch oral 402, **407**
  head and neck 393, 398
  hepatocellular carcinoma 391, 397, **404**
  lung 398, 399, 401, **407**
  melanoma(s) 397, *399*, **404**, 405, 407
  meningiomas 393–395
  ovarian **407**
  pleural mesothelioma **407**
  schwannomas 393–395
  submandibular gland cancer 393
tumor blood vessel normalization 430
tumor lethality, selective 428
tumor-targeting 311

u

Ullmann-type reaction 51
uptake
  ^{10}B 88, 90
  ^{10}B–^{11}B 94

v

vascular targeting (mechanisms) 428, 436
vector 70, 76, 78, 84, 85
  targeting 73, 75, 78
  tumor-targeting 60
vertex differentiation 148, 152
vesicles 159–163, 166, 169, 170
vitamin D receptor ligand 12

w

weakly coordinating 110
western blot 40
window of normalization 430, 432

x

X-ray 408–409

z

Zn(II) phthalocyanine 401